D0845960

SECOND EDITION

Physical Geography Today

A PORTRAIT OF A PLANET

Physical Geography Today

A PORTRAIT OF A PLANET

SECOND EDITION

Robert A. Muller
Louisiana State University

Theodore M. Oberlander
University of California, Berkeley

CRM

RANDOM HOUSE

Second Edition
98765432

Library of Congress Cataloging in Publication Data

Muller, Robert A.
Physical geography today.

Principle contributor of the 1st (1974) ed. R.J. Kolenkow.
Bibliography: p.
Includes index
I Physical geography. I. Oberlander, Theodore, joint author. II. Kolenkow, Robert J. Physical geography today. III. Title.
GB55.K64 1978 910'.02 77-28213

ISBN: 0-394-32088-3
Cover art: *The Bridge of San Luis Rey, the Andes* by Frederick E. Church, oil on canvas; 31 x 48 inches. Courtesy of Kennedy Galleries, Inc.

Typography: Chestnut House
Manufactured in the United States of America

Preface

There are many different ways to perceive the world—artists, poets, physicists, architects, economists, and historians all have their own unique perspectives. Geographers, too, have their particular view of the earth—one that focuses on the arrangements of natural and man-made features at the earth's surface, and on the interactions among the processes that create and modify these features. A single academic quarter or semester does not provide an adequate opportunity to study both the natural and man-made features of the earth, so this book centers its attention upon physical geography, which is more than a composite of physical sciences such as meteorology, climatology, biology, and geology. Our view is that physical geography is both a description of what exists at the surface of the earth and an explanation of how and why physical processes have acted to produce these varying phenomena. The scope of physical geography extends from the atmosphere through the hydrosphere and biosphere to the upper portion of the lithosphere, but this text concentrates upon the surface layer where land, sea, and air meet and interact, and where life occurs. The organizing theme is that the processes that influence surface phenomena involve both energy and materials. Energy continuously cascades through different physical systems, ranging in scale from the global atmosphere to a single convection cell, or from the global biosphere to a city park or a grove of trees. Materials such as oxygen, carbon, and water flow through these systems in never-ending cycles that involve inputs, outputs, storages, and quasi-equilibrium states. The various physical systems constantly interact and respond to one another, exchanging energy and materials, each gain in one system constituting a loss to another.

We begin with a discussion of the earth's evolution, placing the planet within a framework of time and space. The second chapter introduces energy and moisture inputs and transformations within the earth-atmosphere system, and the cycling of energy, moisture, and other materials. Emphasis is placed on the significance of energy and moisture interactions to most environmental subsystems. These two chapters serve as a foundation for the main body of the text, which begins with Chapter 3.

The ensuing chapters trace the cascades of energy and moisture through the atmosphere to the earth's surface, and explain the resultant atmospheric and oceanic circulations that distribute energy and moisture over the planet. Following this, the local water budget and the relationships of water budget components to environmental responses such as vegetation, drainage, and soils are introduced, with the water budget theme being picked up in several subsequent chapters. We next progress to global patterns of climate, including a discussion of the uses and limitations of climatic classification systems. The interrelationships of climate and vegetation are explored, followed by an analysis of the energetics of plant productivity. The following chapter, dealing with soils, once more emphasizes climatic interactions, but it also serves as a bridge linking the atmospheric and biotic subsystems to interactions and responses of the land surface in terms of geomorphological processes and resulting landforms.

The final six chapters explore the processes that give rise to the earth's highly variable surface configuration. Following a conceptual overview stressing the energetics of erosion and deposition, we deal with the processes by which running water redistributes surface materials and creates erosional and depositional landforms, the nature of lithospheric materials and the role of the earth's internal energy in creating geologic structures that influence the character of landscapes, the effect of differential climatic inputs, the unique consequences of glaciation, and the effects of ocean waves and currents along the edge of the land.

Our aim has not been to introduce and define every term used by physical geographers, nor to touch on every phenomenon that enters into the realm of physical geography. Rather than a skimming of the whole, we have sought to develop an understanding of the most important facts and relationships by extended treatment of those topics that we feel are of paramount importance. Some of these have not previously been explored in beginning textbooks in physical geography. Short case studies have been scattered through the text to highlight human responses to the seemingly changeless, yet so variable, environment.

The text has been structured to fit approximately within an academic quarter or semester. Together with a selection of additional readings, it can also serve for a two-semester sequence. Each chapter has been developed so that students can understand its essential concepts from reading and study of the text, illustrations, and extended figure captions. Hence, some chapters can be assigned for reading and study only, without loss of continuity and development of basic concepts.

The appendixes provide discussions of some of the basic tools of the geographer, including maps and images resulting from remote sensing techniques. These materials can be adapted to the needs of particular courses. The glossary, which clarifies technical terms, also serves as a reference and index. The illustrations are correlated with the text, and, with their extended captions, are integral to the presentation of essential concepts. A large number of maps and diagrams have been especially prepared for this text. The metric system with English equivalents is used to express numerical data in the text and graphs.

The first edition of *Physical Geography Today* was put together by the collective effort of a team of physical geographers and physical scientists. This second edition represents a complete reworking of the text by two of the original participants. All chapters have been revised and many have been completely rewritten, and the level of the presentation has been elevated throughout. The new edition reflects more of the personal perspective of the present authors, assisted by in-depth reviews by many scholars, who are acknowledged in the following section. The only significant reordering of topics in the present edition is near the middle of the book, where climatic classification is now followed immediately by regional climates and associated vegetation distributions. The treatment of soils now follows that of ecological energetics and is transitional to the discussions of landforms in the later chapters. The four-color impact and, hopefully, the readability of the first edition have been preserved, with new material introduced to keep pace with current concepts and interests in physical geography and the other environmental disciplines. Much has been learned about the earth during the mid-1970s, but much more needs to be known and assimilated if the earth's productivity is to be maintained or even increased in the face of increasing human demands on a finite planet.

Acknowledgments

The second edition has been reorganized and rewritten by the authors from the contributions to the first edition by the original team of scholars. In addition to the authors of the second edition, the original group included Reid A. Bryson, University of Wisconsin at Madison; Douglas B. Carter, now at the University of Illinois, Chicago Circle; R. Keith Julian; Robert J. Kolenkow; Robert P. Sharp, California Institute of Technology; and M. Gordon Wolman, The Johns Hopkins University.

The authors wish to acknowledge the assistance of the following reviewers, who contributed detailed commentaries on the outline and individual chapter drafts of the second edition:

John J. Alford, Western Illinois University; John T. Andrews, University of Colorado; A. John Arnfield, Ohio State University; Harry P. Bailey, University of California, Riverside; Everette Bannister, Michigan State University; Roger G. Barry, University of Colorado; Robert B. Batchelder, Boston University; Kenneth L. Bowden, Northern Illinois University; Brian T. Bunting, McMaster University; Stanley A. Changnon, Jr., Illinois State Water Survey, Atmospheric Sciences Section; Donald R. Coates, State University of New York at Binghamton; Arnold Court, California State University, Northridge; Frank F. Cunningham, Simon Fraser University; Joe R. Eagleman, University of Kansas; H. F. Garner, Rutgers University; Bruce G. Gladfelter, University of Illinois, Chicago Circle; Jay R. Harman, Michigan State University; William Imperatore, Appalachian State University; Laurence S. Kalkstein, University of Delaware; Martin Kellman, York University; Arleigh H. Laycock, University of Alberta; M. John Loeffler, University of Colorado; David S. McArthur, San Diego State University; S. B. McCann, McMaster University; John R. Mather, University of Delaware; David H. Miller, University of Wisconsin, Milwaukee; Robert Nunley, University of Kansas; E. D. Ongley, Queen's University; Robert W. Pease, University of California, Riverside; Norbert P. Psuty, Rutgers University; Peter J. Robinson, University of North Carolina at Chapel Hill; Robert V. Ruhe, Indiana University; David M. Sharpe, Southern Illinois University, Carbondale; M. L. Shelton, University of California, Davis; Olav Slaymaker, University of British Columbia; Kenard E. Smith, Virginia Polytechnic Institute; Thomas R. Vale, University of Wisconsin, Madison; Jack R. Villmow, Northern Illinois University; Hartmut Walter, University of California, Los Angeles; Wayne M. Wendland, University of Illinois, Urbana-Champaign.

We wish to especially thank Robert B. Batchelder, John R. Mather, and David S. McArthur for their advice and overall critical reviews of much of the manuscript. Of course, the authors alone are responsible for the integrity of the text.

We worked very closely with three people at Random House, and we wish to thank them for their enthusiasm and unstinting support. Charles E. Stewart, former senior editor of the College Department, encouraged us to undertake the second edition and coordinated the overall organization and reviews of the text. Suzanne Thibodeau, project editor, sharpened both our writing and our thinking with her meticulous attention to consistency and structure, and helped us, week after week, to stay closer to our schedules. Elaine Rosenberg, project editor, attended with good humor to the mass of details vital to the production process. We also wish to express our appreciation to our wives, Jeanne and Lucille, for their direct help, encouragement, and patience, and finally to the students who over the years unknowingly helped us to organize and reorganize our thoughts about physical geography.

Overview

Contents

Photographic "Illumination" by © Glen Heller, 1973. (Courtesy, Images Gallery, New York)

ONE

In the Beginning

The earth has been the scene of continuous change since its formation about 5 billion years ago. And it continues to change today because of the dynamic interplay of energy with air, water, and land. Physical geography is the study of the processes that have shaped the surface of the earth.

THERE IS SUCH a reassuring familiarity in the natural features of our planet—clouds and trees, hills and seacoasts—that it is easy to take them for granted. But the earth's surface, unlike that of the moon, Mercury, or Mars, is the scene of ceaseless change. Some of the changes are easily observed—the sun warming the earth, clear skies coming after a rainstorm, plant growth slowing as winter approaches. Less obvious changes are the gradual eroding of hillsides by water and the wearing away of seaside cliffs by pounding waves. Change is the essence of nature on planet earth.

The thin layer of land, water, and air on the surface of the earth is man's natural home. It is also the dynamic arena in which atmosphere, rock, soil, water, and vegetation interact under the driving force of energy from the sun and from the earth's internal heat. Physical geography is the study of the human environment at the surface of the earth and the forces that shape that environment. It is both a description of *what* is at the earth's surface and an explanation of *how* and *why* physical processes act to shape the earth.

The discipline embraces a wide range of natural phenomena, from the circulation of the upper atmosphere to the chemical processes that help form soils. Its primary focus is on interactions—the interaction of vegetation with climate, the interaction of the sun's radiant energy with the atmosphere, the interaction of water with rock and soil. Physical geography touches on several disciplines—geology, meteorology, climatology, biology, pedology, oceanography (Figure 1.1)—but it is more than a synthesis of other sciences. Because it emphasizes the physical processes and principles of interactions, it provides a perspective to the study of the earth's surface that is impossible with any one field alone.

Physical geography is also concerned with the interactions between man and the earth. We depend on the earth for food and materials, and in turn, our agriculture and industry cause changes on the earth. The scale and tempo of human activities are steadily increasing in response to expanding needs for food, water, and energy, and as a result, our impact on the environment is becoming ever greater. Some of these activities have produced unexpected and occasionally undesirable changes. A knowledge of the interrelatedness of physical processes can enlarge our understanding of the effects on the earth of human activities.

The earth is a dynamic planet, and its present state represents the results of continuing changes over long periods of time. This chapter places the earth in the time scale of geologic and evolutionary processes. Although most of this book discusses the forces that shape our existing environment, a look at the earth's history should make it clear that the present is part of a continuum of ongoing change.

THE CHANGING EARTH AND THE TIME SCALE OF CHANGE

Physical, chemical, and biological processes are steadily transforming the earth's surface. Some of the transformations are visible on the human time scale of a few years or decades: new volcanoes form, beaches come and go, shallow ponds become filled with vegetation. Other processes work so slowly that their consequences become apparent only over immense spans of time.

The mountain ranges along the western coast of the United States have been lifted up from sea level during the last 1 or 2 million years and continue to rise today. Even the familiar patterns of the continents and oceans change with time. Recent research indicates that all of the earth's landmasses were grouped together in one or two supercontinents until about 250 million years ago. These gigantic landmasses then gradually fragmented into separate continents that have since been moving imperceptibly across the earth's surface, driven, it is generally believed, by convection currents in the semi-molten material beneath the earth's surface. The Atlantic Ocean is a relatively new feature and is expanding as North and South America gradually move away from Europe and Africa.

The time scale of geologic change goes back about 5 billion years, to the time when the earth was formed. About 3 billion years ago, the first life forms appeared, marking the start of biologic evolution. During the greater part of its history, life remained on a relatively simple level. About a billion years ago, the only life forms on earth were

Figure 1.1 (opposite) To describe the surface of the earth, physical geography draws upon the specialized knowledge of many disciplines. These include *geodesy,* the study of the shape of the earth; *hydrology,* the study of water's role on the earth; *pedology,* the study of soils; and *geomorphology,* the study of the earth's landforms. The unique contribution of physical geography is its concern with the interactions among different parts of the earth—how the earth's relation to the sun influences climate, for example, and how climate in turn affects the development of systems such as vegetation and soils.

Climatology
Meteorology
Hydrology
Pedology
Plant
Geography
Ecology
Astronomy
Geology
Geodesy
Geomorphology
Cartography
Physical Geography
Physical
Oceanography

simple plants in the sea. More complex forms of life evolved in response to the stresses of particular environments. Life, in turn, acted on the environment, modifying it and introducing new processes that redirected the course of the earth's development.

Although it is true that large portions of the earth's history still remain conjectural because only a few remnants of early events are available for study, nevertheless, we are not completely in the dark regarding these periods. Assuming that processes of change in the past operated according to the same fundamental principles which hold true today, scientists can often reconstruct events about which they have little direct knowledge. The great changes that have occurred in the earth's life forms and landforms through the course of time have left a variety of traces, including sediment layers, many kinds of rocks, fossilized plants and animals, radioactive decay products, and ancient magnetism in rocks. With careful observation, experimentation, and scientific reasoning, it has been possible to use such clues to reconstruct a plausible account of some of the major events in the history of the earth.

FORMATION OF THE UNIVERSE

Peoples in many lands have developed creation stories which assert that the universe did not always exist but was created at a definite point in time. Interestingly, present scientific knowledge points to a similar conclusion. There was once a time when all the matter and energy in the universe was compressed into a single nucleus, or "cosmic egg." Pressure within the nucleus was so great that individual atoms—and hence elements—could not exist. Energy soon heated this primordial matter to a temperature of billions of degrees, and the nucleus exploded into a fireball. As a result of this "big bang," the nucleus began expanding. The primordial matter separated into

Figure 1.2 Our own galaxy, the Milky Way. To gain some idea of the scale of the universe, bear in mind that the Milky Way is a disk of stars about 100,000 light years in diameter (one light year equals about 9.6×10^{12} kilometers, or 6×10^{12} miles). The total number of galaxies observable in space is estimated to be in the billions.

atomic particles, and the extreme heat caused these to fuse into elements. However, as expansion continued, the temperature decreased considerably, and the initial period of element formation came to an end.

For many millions of years after the big bang, there were no galaxies or stars. Vast clouds of atoms swirling through space were the only remnants of the cataclysm. Gradually, the force of gravitation caused the atoms to amass into denser clouds, from which the galaxies would evolve (Figure 1.2). These centers of aggregation tended to increase in size as they swept up more atoms from space. Atoms gained speed as the gravitational pull of a massive cloud drew them in, and when atoms collided with one another, their energy was converted to heat. The largest clouds reached temperatures high enough to initiate self-sustaining nuclear reactions. These clouds began to glow as stars; our sun was among them.

If the big-bang theory is correct, the glowing fireball would have given off vast amounts of energy in the form of radiation. The degraded remnants of this radiation should still be present everywhere in the universe. The fact that such radiation has recently been detected gives powerful support to the big-bang hypothesis. There is other evidence as well. Among the most convincing is the fact that, according to observations of light that the earth receives from distant galaxies, it appears that the universe is still expanding today, with stars and other bodies rushing away from each other at great speeds.

By extrapolating the measured rate of the universe's expansion back in time, scientists have estimated that the big bang occurred about 20 billion years ago, or 20×10^9 years ago (meaning that 10 is multiplied by itself 9 times before being multiplied by 20—a way of abbreviating large numbers to facilitate calculations). It is difficult to imagine the passage of 20×10^9 years. To put the time span in perspective, it is helpful to reduce it in scale by means of an analogy. Imagine that the time period from the big bang to the present corresponds to one 24-hour day, and that the beginning of the universe occurred at midnight 24 hours ago. On such a scale, each second of the 24-hour day corresponds to about 200,000 years of actual time. Studies of our sun's energy-generating processes indicate that its active life probably began about 6 billion, or 6×10^9, years ago. According to our analogy, this would have been about 17 hours after the formation of the universe.

Formation of the Solar System

The only planets that we are able to observe in the universe are the nine within our own solar system. It seems likely that many other stars have planetary systems as well, but since planets give off comparatively little radiation and shine only by reflected light, they are outside the reach of our instruments. Therefore whatever theories are proposed to explain the formation of planetary systems have to be based solely on the one example available.

Numerous models have been devised to account for the origin of the planets in our solar system, but none agrees fully with all the known facts. The American chemist Harold Urey, a Nobel Prize winner, has formulated one of these models. In its broad outlines, it is subscribed to by most contemporary scientists. Urey presents a picture in which a massive swirling cloud of matter condensed at its center to form the sun. The outer portions of the cloud then broke up into eddies, which contracted into swarms of bodies called *planetesimals.* The planetesimals were smaller than the present planets, but some may have been as large as the moon. Later, the swarms of planetesimals came together under gravitational attraction and formed planets. Urey has suggested that the moon might be a planetesimal

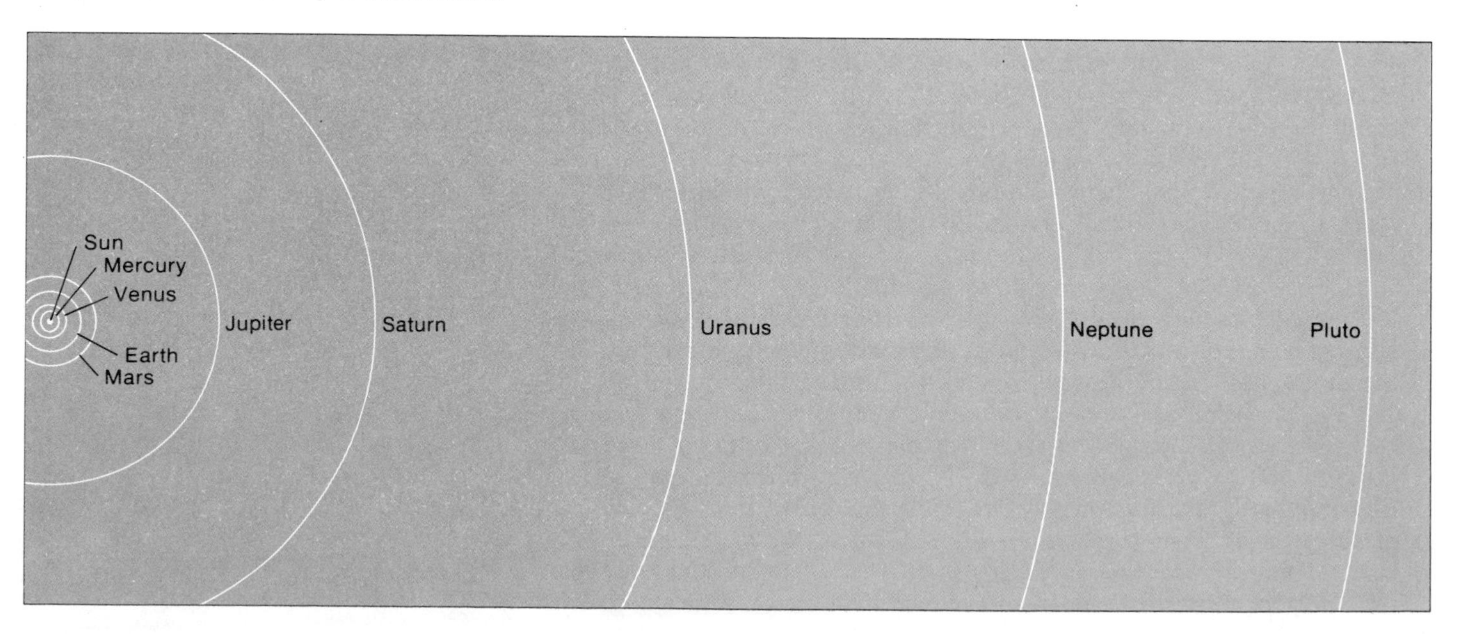

Figure 1.3 This diagram shows the distances of the planets from the sun in correct scale. The planets themselves are not shown because their size would be disproportionate to the scale of their orbits. Between the orbits of Mars and Jupiter are swarms of asteroids, or minor planets, which may be fragments of earlier planets. The chunks of matter that form the asteroid belt range from dust particles to one lump that is as large as the British Isles.

The orbits of the planets lie nearly in the same plane, and every planet revolves about the sun in the same direction, which is consistent with the hypothesis that planetary bodies originally condensed from a whirling disk of gas. The inner planets, Mercury, Venus, Earth, and Mars, together with Pluto, are denser than the other planets because they are too small to have retained much of the hydrogen that was present when the solar system was formed.

trapped in the gravitational pull of the earth. Evidence from radioactive dating, which is discussed later in this chapter, places the time of the earth's formation at about 4.6×10^9 years ago, which would be about 6 P.M., or three-quarters through our hypothetical day.

The original cloud of matter that formed the solar system had a swirling, or rotational, motion that was imparted to the planets. Two kinds of rotational motion occur in the solar system: the planets revolve about the sun in approximately circular paths, and they rotate on their own axes. The fact that every planet revolves about the sun in the same direction and that the orbits of the planets lie in nearly the same plane is consistent with the theory that planetary bodies originally condensed from a whirling disk of gas (Figure 1.3).

The rotation imparted to the earth at its formation is responsible for our cycle of night and day. It also has important consequences for the distribution of incoming solar energy over the earth's surface and for the development of weather systems. As a matter of chance, the earth's rotational axis was set at a considerable angle to its orbital motion. How this tilt influences solar energy distribution to the earth and how it determines the seasons are discussed in Chapter 3.

Structure of the Earth

It is believed that earth materials accumulated at comparatively low temperatures and that the earth was never fully molten. In Urey's model, the final slow build-up of the earth from colliding planetesimals would not have heated the whole planet significantly. The earth's interior slowly became hotter after its formation because of energy released from radioactive elements. *Radioactivity* arises from instability in the nucleus of an atom, where most of the atom's mass is concentrated. The instability causes the nucleus to change form spontaneously, releasing energy. When the interior of the earth became hot, materials flowed and diffused from place to place. Lighter materials rose toward the earth's surface and heavier substances sank toward the center. The earth differentiated into three principal concentric shells, or zones: the *crust*, the *mantle*, and the *core* (see Figure 1.4, pp. 10–11). These shells differ from one another in size, chemical make-up, and density.

The outermost shell, or crust, varies in thickness from about 10 to 70 kilometers (km), or 6 to 40 miles. It is composed of the solid rock material that forms the continents and ocean floors. The continents are formed of rock types known collectively as *sial*, which consists primarily of compounds of silicon and aluminum. Beneath the continents and ocean floors lies denser rock called *sima*, composed mainly of compounds of silicon and magnesium. The sial and sima together form the crust.

The second shell, the mantle, contains most of the earth's mass and extends to a depth of some 2,900 km (1,800 miles). The mantle is composed of high-density rock material, derived mostly from iron and magnesium. Although temperatures in the mantle are high enough to melt this material, pressures are so great at these depths that most of the mantle is essentially rigid.

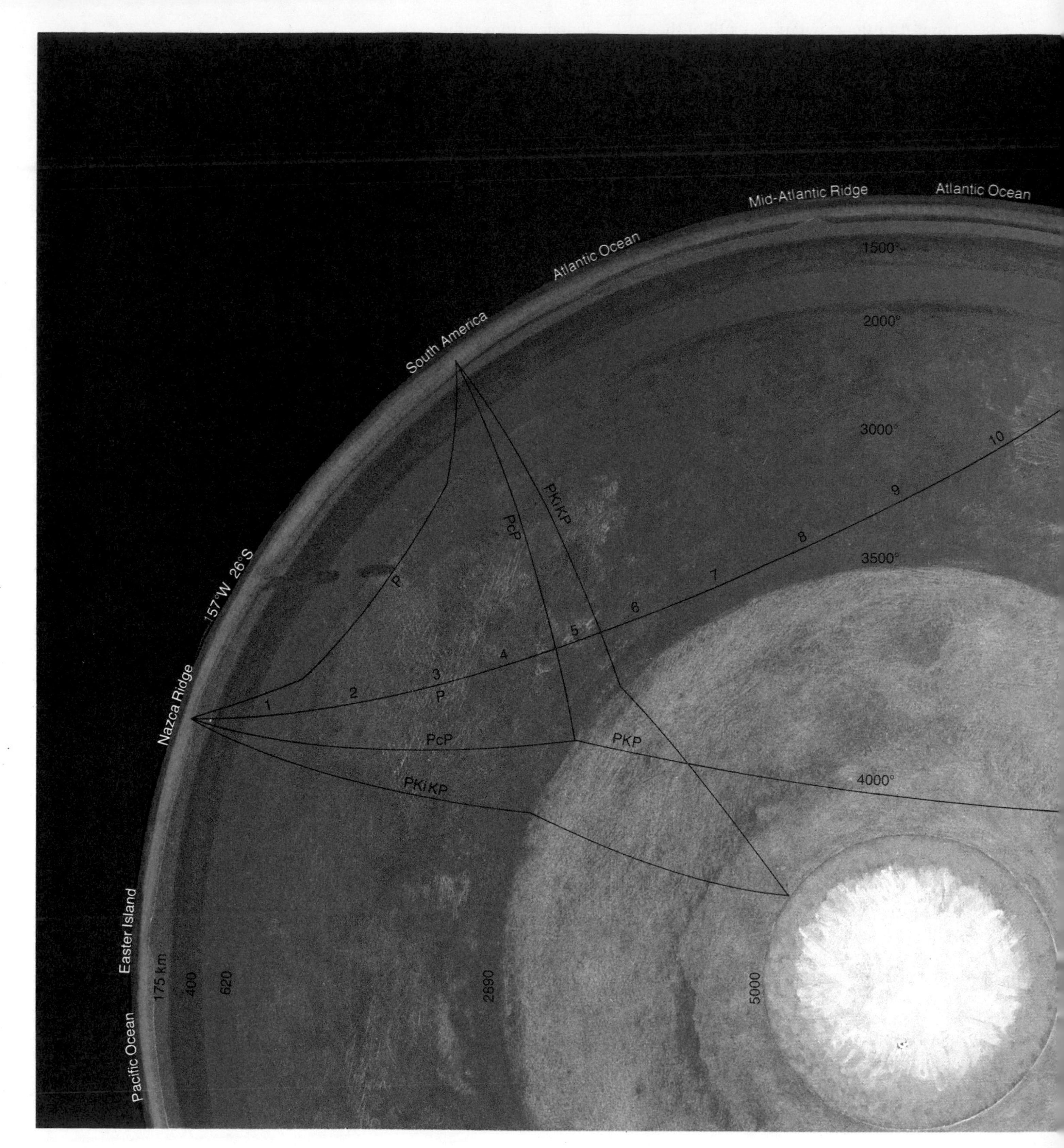
Mid-Atlantic Ridge
Atlantic Ocean
Atlantic Ocean
South America
157°W 26°S
Nazca Ridge
Easter Island
Pacific Ocean
175 km
400
620
2890
5000
1500°
2000°
3000°
3500°
4000°
P
PcP
PKiKP
PKP
1
2
3
4
5
6
7
8
9
10

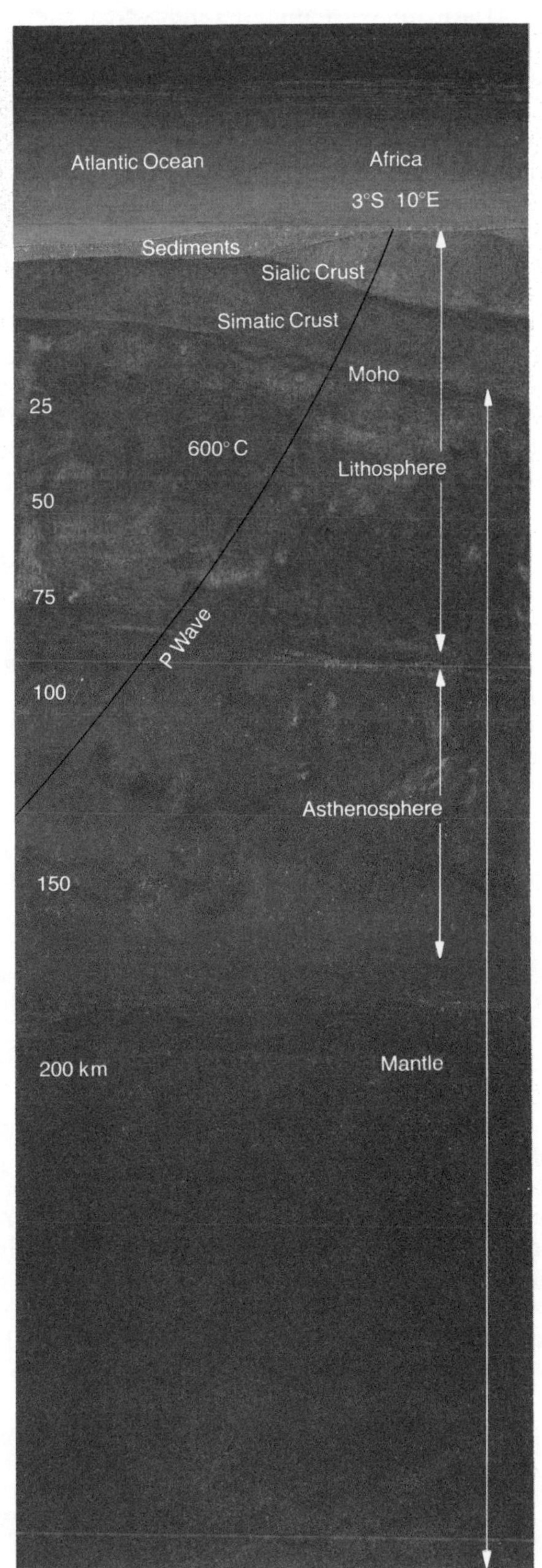

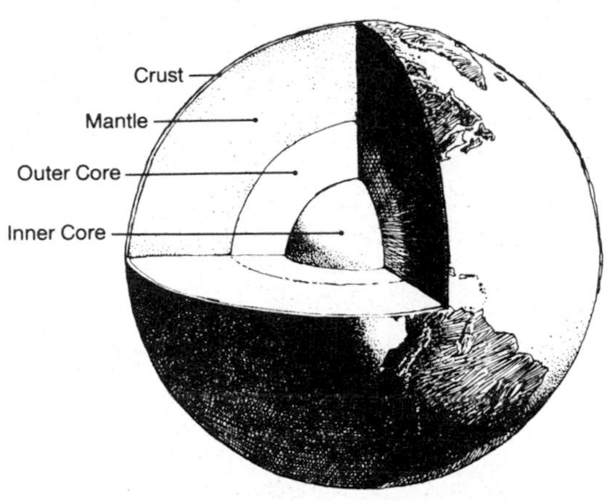

Figure 1.4 (far left) The earth in cross section. The earth may be visualized as a series of concentric shells having different properties. The *crust* varies in thickness from about 10 km (6 miles) under the oceans to 40 km or more (25 miles) under the continents. The second shell, or *mantle,* extends to a depth of about 2,900 km (1,800 miles), at which point the *outer core* begins. At 5,000 km (3,000 miles) below the surface is the *inner core,* believed to be an alloy of iron and nickel (that possibly includes some silicon or sulfur).

(near left) The boundary between the crust and mantle is the *Mohorovičić discontinuity* (Moho). The chemical composition of the mantle varies, probably differing beneath ocean basins and continents. The outer 200 km (125 miles) of the earth may be separated into two classes of material: the zone of low strength and low seismic velocity is called the *asthenosphere;* the more rigid layers that overlie it, which include both the crust and the upper mantle, are called the *lithosphere. Lithospheric plates* may be visualized as rigid slabs that are free to move on the low-strength asthenosphere. It is important to note that lithospheric plates include both crust and mantle and that a single plate may include both oceanic and continental crust. These concepts are discussed further in Chapter 11.

Our knowledge of the inner constitution of the earth has come primarily from a study of the waves generated by earthquakes. These waves, called *seismic waves,* travel through the earth at speeds that vary according to the properties of the material through which they pass. The cross section shows seismic waves of various types, labeled *P,* that have been generated by an undersea earthquake on the Nazca Ridge southwest of the Peruvian coast. The numbers on the path of the *P* wave that travels to a point 3°S 10°E on the African coast represent the time elapsed in minutes. *PcP* denotes waves that are reflected back to the earth's surface at the boundary between the mantle and the outer core. *PKiKP* waves are reflected back from the inner core. *PKP* represents waves that pass through the mantle and the core. The study of seismic data is our most powerful tool for understanding the earth's interior composition and has led to the conclusion that the inner core seems to be solid whereas the outer core acts like a liquid.

The mantle surrounds the core, which has a radius of about 3,400 km (2,100 miles). This part of the earth has such a high density that it is presumed to be composed of iron and nickel compounds that sank slowly downward, replacing the lighter elements. The core itself is divided into two parts: an outer core, which is believed to be molten, and an inner core, which seems to be solid.

Flows of molten rock from erupting volcanoes reveal that temperatures within the interior of the earth exceed 1200°Celsius (C) or 2200°Fahrenheit (F). The disintegration of radioactive elements generates heat that keeps the outermost fringe of the mantle hot enough to be plastic and to be capable of internal motion. This gives the crust the freedom of movement important in geologic processes. The amount of heat that radioactivity generates in the earth decreases with time as radioactive elements disintegrate into stable forms; the amount of heat supplied by radioactivity was 15 times greater in the early days of the earth than it is today.

EVOLUTION OF THE EARTH'S ATMOSPHERE AND LANDFORMS

There is strong evidence that any atmosphere the earth possessed at the time of its formation was not retained long. One of the chief indications of this is the fact that neon is practically nonexistent on the earth but is relatively abundant on the sun. The difference is puzzling because the earth and the sun are thought to have been formed from the same cloud of atoms, which would make their compositions similar initially. Neon is an inert gas—that is, it forms no chemical compounds—so it would be found primarily in the atmosphere. The relative lack of neon in the earth's atmosphere indicates that most of the original atmosphere was lost. Urey has suggested that when atoms from the original swirling cloud condensed to form solid planetesimals, enough heat was generated to drive away most of the atmospheric gases even before the planets took shape.

The Early Atmosphere

The earth lost its original atmosphere, but a new atmosphere began to form as gases released from molten rock escaped to the earth's surface from the interior. The composition of this second atmosphere was probably similar to the gases present-day volcanoes release: water vapor, carbon dioxide, sulfurous gases, and nitrogen. Eventually, much of the water vapor condensed and fell to the ground as rain. The early atmosphere probably contained mostly carbon dioxide, with some water vapor and nitrogen and only traces of oxygen gas, which would make it similar to the present atmosphere of Venus.

The Early Landforms

The earth's surface 3 or 4×10^9 years ago must have had a rather forbidding appearance, somewhat like the volcanic landscape in Figure 1.5. Volcanoes erupted, spreading molten rock over the land. The

Figure 1.5 The appearance of the earth's surface before the advent of life on the land may have been similar to this raw volcanic wasteland on the slopes of Haleakala Crater on the island of Maui, Hawaiian Islands.

ground was completely barren, and the earth was devoid of life. Water vapor released from the earth's interior collected in the atmosphere and then fell as rain, forming lakes and seas. The presence of moisture made possible chemical reactions that weakened the structure of solid rocks at the earth's surface, so that large rock fragments and mineral grains littered the ground. There was no organic plant matter to help form true soils, and as water from rainstorms flowed down hillsides, it carried rock debris and dissolved minerals down to the seas. Windstorms swept sand and dust across the landscape. As wind and water cleared debris from the ground, fresh surfaces of rock were exposed and the weathering process continued.

The early landscape probably had the steep slopes, exposed rock faces, and piles of debris characteristic of present-day deserts or volcanic regions, but the complete lack of vegetation and soil and the abundance of rainfall and flowing water gave the early landscape an appearance that does not exist anywhere on the earth today.

If wind and water had been the only agents at work on the landscape, the land areas of the earth would have been worn down to low, featureless plains in a few tens of millions of years. But the earth's

Figure 1.6 (left) These rocks represent sediments deposited originally in horizontal beds by water. The upper layers are inferred to be more recent than lower layers because of the relative position of each bed.
(right) These rocks in England have been deformed slowly by pressures associated with the earth's internal energy. The buildings at the top of the bluff gives an idea of the scale of deformation.

internal energy also modified the land surfaces. Mountain chains formed when internal energy lifted up the earth's crust more rapidly than external forces could wear it down—processes and interactions that are still going on today (see Figure 1.6).

APPEARANCE AND DEVELOPMENT OF LIFE

About 3×10^9 years ago—with only 3½ hours left to go on our 24-hour clock—the history of the earth took a unique turn. Life appeared. The origin of life has not yet been fully explained in terms of physical laws. However, in 1953 Stanley Miller, working with Harold Urey at the University of Chicago, carried out an experiment that indicates how life might possibly have arisen from inanimate matter. Miller took mixtures of simple gases such as ammonia, methane, hydrogen, and water vapor and allowed them to react together for several days under the stimulation of an electric discharge. He discovered that the reacting gases produced complex molecules. Among the reaction products, he was able to identify amino acids, the building blocks of the proteins which are found in every living cell. Miller's work showed that given sufficient energy to break chemical bonds, relatively simple chemical substances can recombine to form the complex molecules required in life processes.

Conditions 3×10^9 years ago could have been roughly analogous to the conditions in Miller's experiments. The earth's early atmosphere contained a variety of gases, and the energy to disrupt the molecules could have come from lightning and from intense, energetic ultraviolet light from the sun. Rain washed many of the reaction products out of the atmosphere into the oceans, and in time the oceans became an "organic broth" rich with synthesized complex molecules. If such molecules appeared today nearly anywhere on the earth's surface, they would be quickly consumed by bacteria and other microorganisms. But on the early earth, no living things existed which could digest these molecules and break them down into simpler chemical compounds, and so they simply accumulated. Over the millennia, the "broth" grew thicker.

Although these organic molecules were the building blocks of life and collisions brought them together into various combinations, these groupings were not actually alive. In order to qualify as living, an aggregation of molecules must be able to do two things: it must be able to absorb energy from its surroundings, and it must be able to reproduce. Eventually, however, a combination of molecules appeared which was capable of these two functions. For energy, the earliest organisms probably depended on the chemical reactions that resulted from the breakdown of the complex molecules present in ocean water. Some present-day organisms exist in much the same way. Yeast cells, for example, break down carbohydrates into simpler substances and live off the energy that is released.

The Fossil Record

The origin of life on the earth remains a matter of speculation because the first life forms left no known traces. Most of our knowledge of the history of life on earth comes from the *fossil record.* Wind and flowing water carried fine rock debris from the early landforms and deposited it in layers of sediment on the land and on the ocean floor, covering over the remains of plants and animals. As more and more debris accumulated, the underlying layers were compacted and consolidated into rock. The rock preserved the shape of the organisms. Eventually, water infiltrating the rock dissolved the organic remains. More resistant compounds simultaneously took their place, however, and maintained their forms for long ages (Figure 1.7, p. 16). The many layers of sedimentary rock exposed at the Grand Canyon of the Colorado River are examples of successive deposits of fossil-bearing sediment. It can be inferred that in undisturbed sedimentary rock formations, the earliest fossils are found in the deepest layers. Layers of sedimentary rock therefore contain a partial history of life on the earth.

The fossil record is not a complete catalog of all the life forms that have appeared on the earth, however. Because soft tissues are rarely preserved, most fossils only record the shape of such hard remains as bones, shells, and woody stems. Furthermore, since the process of

Figure 1.7 This fossil is exceptional because the soft tissues of plants and animals are rarely preserved. This squidlike organism has been so well preserved in rock that even its ink and ink sac are identifiable.

sedimentation is variable, there may be long periods in which no deposition at all occurs in a particular location. And the later wearing away of sedimentary rock may destroy large portions of the local fossil record.

Organic Evolution

The earliest living things of which we have direct evidence are simple plants which flourished about 3×10^9 years ago. This early form of vegetation closely resembled the blue-green algae which can be seen today floating on the surface of ponds in the summer. In sharp contrast to the first life forms, which could take in nourishment only by ingesting organic molecules present in sea water, these early plants were specialized organisms capable of producing food energy using only carbon dioxide, water, and the sun's radiant energy. The difference between these two types of organisms must necessarily raise the question: how did complex plants arise from earlier, simpler forms of life? Also, what was the origin of higher forms of animal life, such as dinosaurs and human beings?

In 1859 the British naturalists Charles Darwin and Alfred Russel Wallace independently set forth the idea of *organic evolution,* which provides insight into the process of biological change. According to Darwin and later researchers, there is inherent variability in the mechanism of heredity, as well as accidental alterations in the controls of heredity. This means that an organism does not produce offspring which are perfect copies of itself; rather, offspring generally differ in certain ways both from their parents and from other members of their species. In a given environment, the individuals that are most efficient at such things as gathering food and escaping predators are most likely to survive, reproduce, and pass along some of their characteristics to their offspring. If a species is successful, and if the environment remains relatively unchanged, individuals with a hereditary make-up like that of most other members of their species will be favored, and the species itself will undergo little further biological evolution. According to the fossil record, opossums, for example, have changed little in the past 60 million years.

If, on the other hand, the physical environment changes—perhaps because of a change in climate—individual organisms with a hereditary make-up different from that of most other members of the species may have a slight advantage in competing for survival, a principle that Darwin called *natural selection*. The history of the peppered moth in Britain provides a modern example of natural selection. Prior to 1850, before industry was widespread, most peppered moths were light in color with dark speckles. They blended with the tree lichens on which they rested and became inconspicuous to predator birds. A small proportion of the moths were of a dark variety. After the rise of industrialization, air pollution killed many of the lichens and darkened the trees with soot, so that in industrial areas the dark moths were less conspicuous to birds. By 1900 almost all peppered moths around industrial centers were dark (see Figure 1.8, p. 18).

The light and dark peppered moths still belong to a single species. That is, the two varieties are still able to interbreed and produce viable offspring. However, if environmental conditions continue to change, the descendants of the favored members of a species may undergo still further natural selection. Over a sufficient period of time, the biological changes that have accumulated may result in the development of a new species. Darwin and Wallace proposed that it is this process which is responsible for producing the many different species of plants and animals on the earth today, all of which can be traced back ultimately to the first primitive organisms.

INTERACTION OF LIFE WITH THE ENVIRONMENT

With the evolution of new forms of life, new forces became available to interact with and shape the environment. The appearance of green plants 3×10^9 years ago—at 8:30 P.M. on the 24-hour clock—set new processes in motion that literally changed the face of the earth. The first plants were confined to the shallow layers of the sea, where water shielded them from the sun's damaging ultraviolet light. The atmosphere was predominantly carbon dioxide, nitrogen, and water vapor, and the green plants used carbon dioxide dissolved in the water to produce carbohydrates by *photosynthesis*, a process that involves capturing energy from sunlight.

In the course of photosynthesis, water molecules are split and oxygen is released. Plant cells use some of the oxygen in their life processes and give off the remainder. Because oxygen is chemically

Figure 1.8 (left) The photograph shows the light and dark forms of the peppered moth at rest on a lichen-covered tree located in a nonindustrial area in England. The light form, which is far less visible to predator birds than the dark form, is the dominant type in the region.

(right) The photograph shows the light and dark forms at rest on a tree that is free of lichens and darkened by industrial pollution. In this region the dark form is the dominant type because it is less visible and has a greater chance of survival than the light form.

reactive, most of the oxygen the early plants released probably did not remain in the atmosphere long; it combined chemically with iron-bearing minerals in the earth's crust. Eventually, when the chemical reactions of minerals with oxygen were complete, oxygen gas began to accumulate in the atmosphere. Today, the earth's atmosphere is 20 percent oxygen, most of it released by green plants through geologic time.

Beginning about 600 million years ago—at 11:15 P.M. on the 24-hour clock—the fossil record testifies to the sudden appearance of complex, shelled, multicellular marine animals. The particular evolutionary steps that spanned the gap between these animals and the earlier plant life from which they developed remain unknown. Major evolutionary adaptations seem to have occurred comparatively rapidly—perhaps within few million years or so, which is too short a time interval to leave a clear record on the fossil calendar. The conditions that must have led to the appearance of animal life, however, are well understood. Part of the oxygen that was produced by early aquatic plants dissolved in the sea water. The availability of oxygen for respiration and of plants for food opened up new possibilities for the development of higher forms of life. When organisms capable of exploiting these new conditions finally appeared, natural selection caused them to multiply very rapidly.

Animals and plants helped to change the environment in other ways. Aquatic plants incorporated carbon from carbon dioxide into the carbohydrates formed during photosynthesis. The early marine animals that fed on these plants incorporated carbon and other materials into their hard shells. When the animals died, their shells became part of sedimentary mineral deposits. As more and more carbon was removed from the ocean in this way, it was replaced by atmospheric carbon dioxide. As a result, the amount of carbon dioxide in the atmosphere was greatly diminished. Today, carbon dioxide constitutes only a small fraction of the earth's atmosphere; there are only about 300 molecules of carbon dioxide for every million molecules of air.

The presence of oxygen in the earth's atmosphere led to the formation of *ozone*. Ozone, which has a molecular structure consisting of three oxygen atoms as opposed to oxygen's two, forms a layer high in the present atmosphere that absorbs much of the cell-destroying ultraviolet radiation given off by the sun. The early atmosphere contained little or no oxygen. Therefore, no ozone layer was present to prevent ultraviolet light from reaching the earth's surface, which was uninhabitable as a consequence. However, once the lethal radiation was blocked, the land became accessible to life. About 400 million years ago—only 30 minutes before midnight on the 24-hour clock—forests of giant seed ferns spread across the land. At about the same time, amphibious animals appeared and rapidly multiplied.

When plants moved onto the land, new processes began to operate, and the landscape changed considerably. The chemical attack by plant acids and the mechanical pressure of growing roots accelerated the fragmentation of rocks. The fine rock particles mixed with organic matter to form soils, and a cover of soil and vegetation softened the harsh slopes of barren rock. Protected by plant cover, the land was able for the first time to resist the erosive action of rain and flowing water. As a result, the rate of sediment production decreased significantly.

Living organisms were also responsible for the formation of mineral deposits. Vigorous plant growth in swampy areas led to the accumulation of thick layers of plant remains, which compacted first to peat and later to coal. Coal formation, which began about 350 million years ago, fixed an enormous amount of atmospheric carbon in mineral form, contributing to the decrease in the carbon dioxide content of the atmosphere.

Present-day deposits of limestone, such as the White Cliffs of Dover in England, are largely formed from the remains of shelled

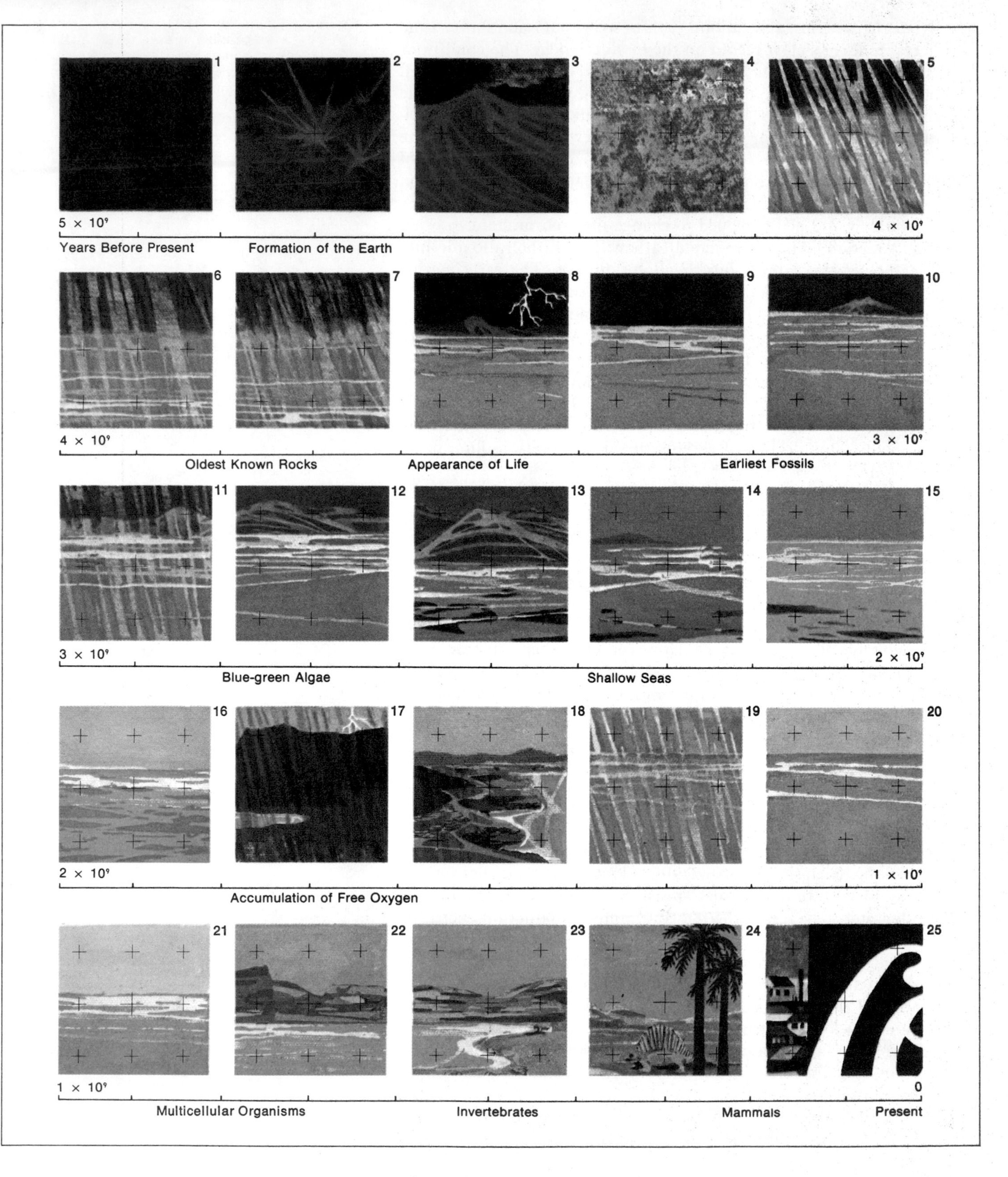
1
2
3
4
5
5 × 10⁹
4 × 10⁹
Years Before Present
Formation of the Earth
6
7
8
9
10
4 × 10⁹
3 × 10⁹
Oldest Known Rocks
Appearance of Life
Earliest Fossils
11
12
13
14
15
3 × 10⁹
2 × 10⁹
Blue-green Algae
Shallow Seas
16
17
18
19
20
2 × 10⁹
1 × 10⁹
Accumulation of Free Oxygen
21
22
23
24
25
1 × 10⁹
0
Multicellular Organisms
Invertebrates
Mammals
Present

marine organisms. When the animals died, their shells, composed of calcium carbonate, were deposited on the ocean floor, and after compaction they became limestone. Subsequent uplifts of the earth's crust raised many of the limestone beds above sea level. Most of the limestone deposits on the earth today were laid down 70 to 100 million years ago, about 8 minutes before midnight on the 24-hour clock.

Throughout the earth's history, interactions between living organisms and the physical environment have altered the environment greatly. But the environment too has been a powerful agent of change. Changing climates and ecological conditions have affected the evolutionary development of plants and animals, often in profound and dramatic ways. About 130 million years ago, for example, there was a flourishing of giant reptiles, descendants of the earlier amphibians (see Figure 1.10, pp. 22–23). These animals dominated the land until about 50 million years ago—just 4 minutes before midnight on the 24-hour clock—at which time they became extinct, and modern mammals appeared.

Man is a relative newcomer on the earth. The earliest known manlike remains are from about 2.6 million years ago, 11 seconds before midnight. But humans are not intruders on an alien earth. No less than other organisms, we have been molded by processes as natural as those that built the mountains, and our interaction with the physical environment influenced our development. Like all animals we are adapted to breathing the gases currently present in the atmosphere, and our eyes respond only to wavelengths of solar radiation that easily penetrate the atmosphere to reach the surface below. Our interactions with the environment will continue to affect our development as a species.

Figure 1.9 (opposite) Time-lapse history of the earth. If it were possible to have set up a camera that could record an image from the same location every 200 million years, then these twenty-five frames would represent some of the visible changes that might have occurred during the earth's 5-billion-year development.
1. Nothing can yet be discerned as gravitational forces begin to consolidate the diffuse material that will form the earth.
2. After the initial accretion of the earth, meteorites are impacting on the planet, forming craters and adding new material.
3. The impacts have led to heating of the outer layers of the earth, and there is extensive flow of molten material.
4. Gases locked in the earth's interior escape to form an atmosphere. The early atmosphere is composed largely of carbon, carbon dioxide, nitrogen, and hydrogen.
5–7. Precipitation from the early atmosphere helped form the primordial seas.
8. Primitive life forms appear. They are nourished by organic molecules in the oceans.
9–10. The fossil record begins. Early forms of plant life are preserved in sedimentary rock.
11–12. Blue-green algae flourish and molecular oxygen slowly begins to accumulate in the atmosphere.
13–20. The oxygen level continues to increase; carbon dioxide decreases. The ozone layer begins to form, preparing the way for life on the land.
21–22. Multicellular, shelled marine organisms appear.
23. Invertebrates evolve; their hard parts are preserved in the rock record.
24. Land plants and animals begin their explosive development.
25. Humans appear and civilization begins its rise.

DATING THE EARTH

Before 1800 people generally believed that the earth's landscape had been formed by a series of violent catastrophes that had lasted only a few thousand years. Toward the end of the eighteenth century, James Hutton, an enlightened Scottish naturalist, took a fresh look at the mountains and streams of his native land and interpreted what he saw as the products of erosion and mountain uplift. He realized that given enough time, these slow processes could have produced the landscapes of the British Isles. He concluded that the eras of geologic time must be much longer than had been thought. Similarly, Charles Darwin concluded that evolution required time spans on the order of hundreds of millions of years.

To check the ideas of Hutton and Darwin, scientists in the nineteenth century attempted to develop an accurate way to measure geologic time. The fossil record gave the correct succession of the geologic eras, but it did not indicate their actual ages. So the rate of sediment deposition was measured to calculate how long it would have taken to build up the known thicknesses of sedimentary rock layers. This method failed because the rate of sedimentation has varied with time and because the record contains many gaps.

The accurate natural timekeeper that scientists sought was found

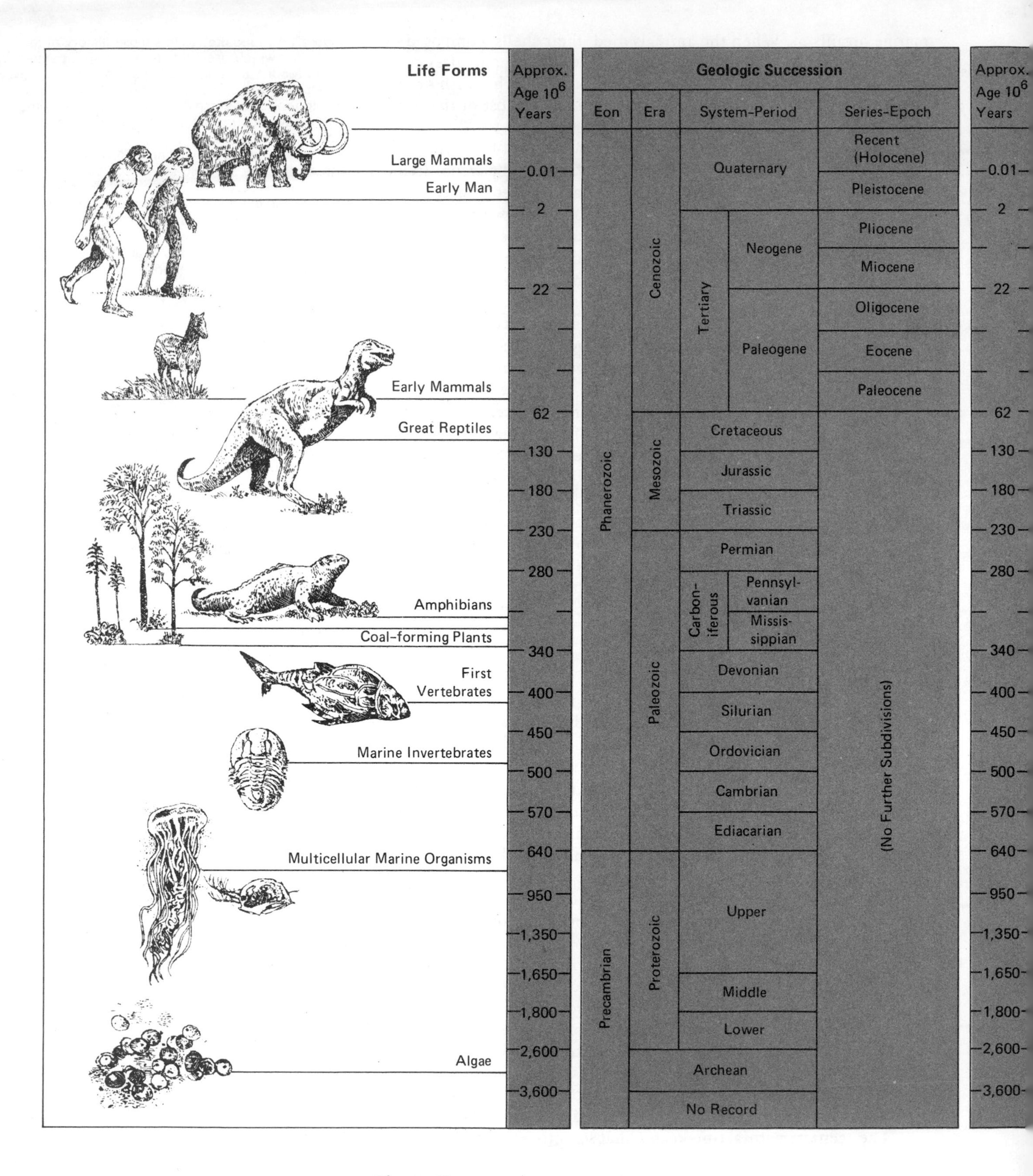
Life Forms
Approx. Age 10^6 Years
Large Mammals
Early Man
Early Mammals
Great Reptiles
Amphibians
Coal-forming Plants
First Vertebrates
Marine Invertebrates
Multicellular Marine Organisms
Algae
0.01
2
22
62
130
180
230
280
340
400
450
500
570
640
950
1,350
1,650
1,800
2,600
3,600
Geologic Succession
Eon
Era
System-Period
Series-Epoch
Phanerozoic
Precambrian
Cenozoic
Mesozoic
Paleozoic
Proterozoic
Quaternary
Tertiary
Neogene
Paleogene
Cretaceous
Jurassic
Triassic
Permian
Carbon-iferous
Pennsyl-vanian
Missis-sippian
Devonian
Silurian
Ordovician
Cambrian
Ediacarian
Upper
Middle
Lower
Archean
No Record
Recent (Holocene)
Pleistocene
Pliocene
Miocene
Oligocene
Eocene
Paleocene
(No Further Subdivisions)

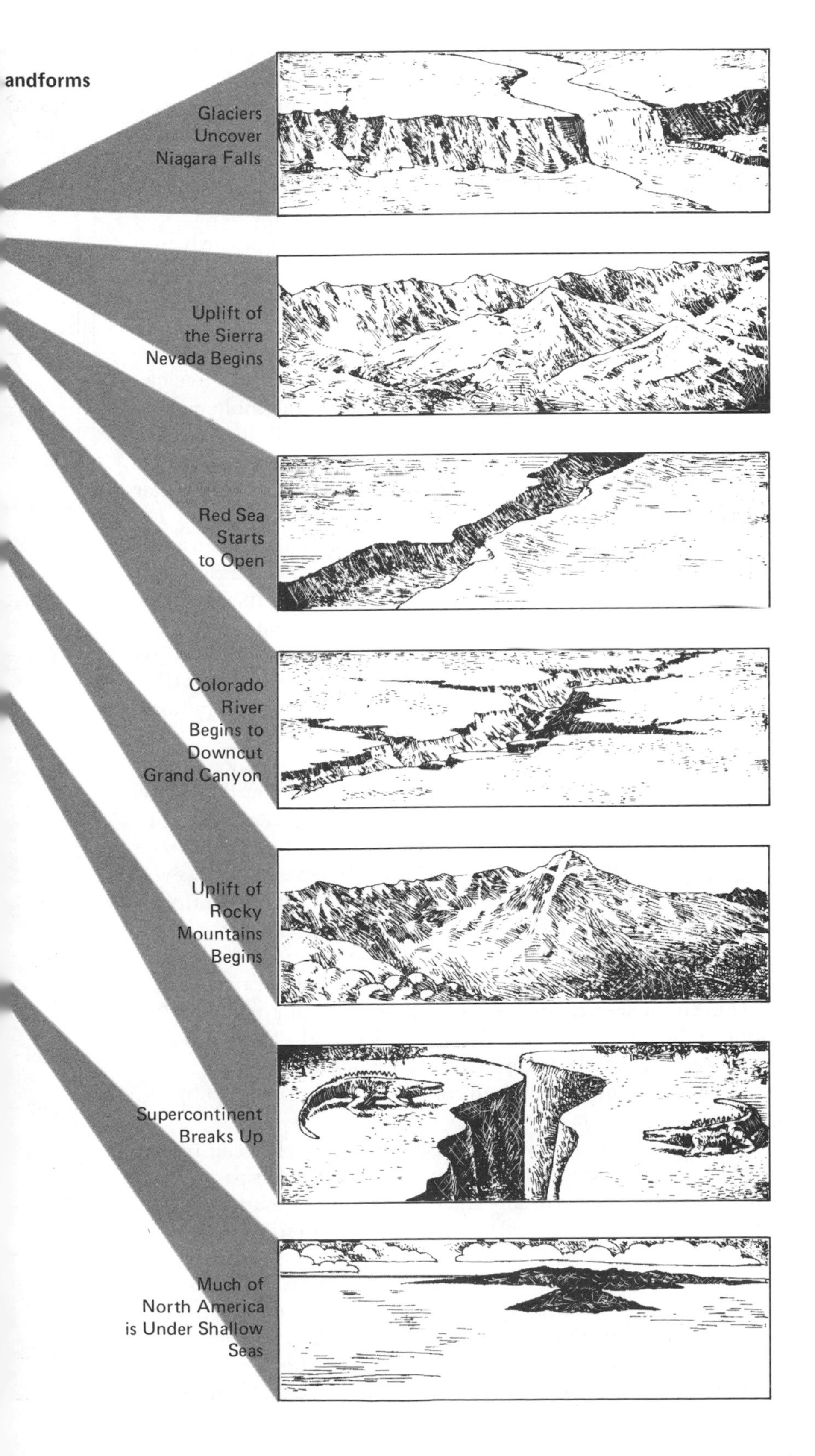

Figure 1.10 The geologic time scale. On the right of the chart, selected events in the earth's geologic history are illustrated. To the left are shown selected life forms and the approximate times when they first appeared or were dominant. A general trend from less complex to more complex life forms reflects the processes of organic evolution. Note that the time scales on the figure are not evenly spaced but emphasize the most recent 600 million years. (After *Adventures in Earth History,* edited by Preston Cloud, W. H. Freeman Company. Copyright © 1970.)

early in the twentieth century when it was discovered that the passage of time could be measured using the decay rates of *radioactive isotopes.* An isotope is one of a number of different forms of a single chemical element, each having a differently structured atom with different degrees of stability and different masses. Isotopes can be designated by the name of the element and a whole number representing the mass of each of its atoms, such as uranium 238 and uranium 234. A radioactive isotope is one that decays spontaneously into another element at a certain predictable rate. The rate of decay is characterized by the *half-life,* which is the time required for half of an initial amount of radioactive atoms to decay. The half-life is unaffected by changes in temperature, pressure, or other external factors.

Suppose a rock contained an initial amount of a radioactive element such as uranium 238 when it solidified from the molten state. Uranium 238 decays with a half-life of 4.5×10^9 years through a chain of steps to the stable end product lead 206. Each uranium atom that decays eventually results in the formation of one atom of lead. Later measurement of the relative numbers of uranium 238 and lead 206 atoms in the rock allows the elapsed time to be computed, because the rate of decay is known. For example, if the rock contains equal amounts of uranium 238 and lead 206, the elapsed time since the rock solidified must have been one half-life, or 4.5×10^9 years.

If the rock initially contained lead 206, the uranium 238/lead 206 method will not be accurate. It is therefore necessary to use more than one radioactive element to establish a date. The most useful radioactive clocks for geologic time are listed in Table 1.1.

Table 1.1. Half-Lives of Radioactive Elements

Radioactive Clock	Half-Life (years)
Rubidium 87/Strontium 87	5.0×10^{10}
Uranium 238/Lead 206	4.5×10^9
Potassium 40/Argon 40	1.3×10^9
Uranium 235/Lead 207	0.7×10^9
Carbon 14	5730

Source: Gray, Dwight E. (ed.) 1972. *American Institute of Physics Handbook.* 3rd ed. New York: McGraw-Hill.

Using radioactive dating methods, it is possible to make a fairly accurate determination of the age of the earth. Among the oldest rocks on earth are some found in Greenland. Their age, according to measurements of their radioactivity, is about 3.8×10^9 years. It can be assumed, then, that the earth is at least 3.8×10^9 years old. However, since erosion may have destroyed all traces of rocks formed prior to those in Greenland, it is possible that the earth is considerably older. According to currently accepted models of the creation of the solar system, meteorites were formed at the same time as the earth. Radioactive dating of meteorites should therefore be an accurate method of determining the earth's true age. This method gives the earth an age of 4.6×10^9 years. A direct estimate of the earth's age from measurements of the total uranium and lead content of the crust also points to an age of 4.6×10^9 years.

Carbon 14 is another radioactive isotope which has been important for dating geologic events, particularly those of the past 40,000 years or so. Carbon is a useful substance for age determination since it is a constituent of virtually all plant and animal matter. Carbon 14—a radioactive form of carbon which is continuously formed in the earth's atmosphere by the action of cosmic rays on nitrogen 14—makes up a definite proportion of all the carbon ingested by plants and animals. When an organism dies, it ceases to take in carbon. The amount of carbon 14 in its cells steadily declines due to radioactive decay, but the amount of ordinary carbon remains constant. Thus, measuring the ratio of radioactive carbon to ordinary carbon allows the age of the sample to be inferred. Carbon 14 dating has provided highly accurate data about relatively recent events in the history of the earth. For example, by dating wood from trees that were killed by the advance of glacial ice, it has been possible to fix the date of the ice's advance through a particular area.

A number of other dating methods are useful, particularly for measuring relatively short time spans of thousands of years. Most of the methods have been used to determine climatic conditions in the recent geologic past. One method uses the variation in the annual growth rings of trees to estimate average weather conditions. Trees more than 4,000 years old have been found among the bristlecone pines of California and Nevada.

Another dating method uses *varves,* which are sediment layers deposited in lakes by streams issuing from melting glaciers. Varves have characteristic annual layers because coarse sediment is deposited directly from meltwaters during the summer while fine sediment settles out later in the year when the waters are still. In Scandinavia, varve analysis has been extended more than 10,000 years into the past, and correlation of varve patterns from different localities has shed light on the course of glacial recession in northern Europe.

THE GEOLOGIC TIME SCALE

The geologic ages can be distinguished from one another and put into the correct time sequence by examining the fossil record in stratified rocks. Using radioactivity and other dating techniques, scientists have been able to assign approximate but relatively firm dates to this sequence. The result is a coherent history of the earth which is not only useful as an organizing tool but is also valuable because it facilitates an understanding of the evolutionary interactions between materials and processes through time, without which neither the present composition of the atmosphere nor the features of the earth's surface can be adequately explained.

The largest divisions in this scheme of geologic succession (see Figure 1.10, pp. 22–23) are known as *eons.* The Precambrian eon, which spans the first 85 percent of the earth's history, corresponds to the period before higher forms of life evolved, when the fossil record is absent or indistinct. The Phanerozoic eon takes in everything from the close of the Precambrian to the present. The eons are divided into *eras.* The end of each era has been marked by episodes of massive mountain-building activity, called *revolutions* or *orogenies* by geologists. In North America, for example, the Appalachian Mountain structure is associated with the close of the Paleozoic era, and the Rocky Mountain structure with the Mesozoic era. Eras are further subdivided into *periods* and *epochs,* and rocks from these units of time are known as *systems* and *series* respectively. Note that some of the ages are named after the regions where their identifying fossils were first studied; the Jurassic, for instance, is named after the Jura Mountains of western Europe. In addition to the mountain-uplift episodes that marked divisions between eras, there have been other landscape-altering events controlled largely by climate, such as the erosion of rocks by flowing water and the onset or retreat of glaciation. For example, the Quaternary period was marked by much colder climates and successive ice sheets which extended southward to near the present-day Ohio and Missouri river valleys, significantly altering the landscape.

As we have seen, the earth is an active, restless planet. During its long history, it has been transformed many times and in many different ways. None of these changes would have been possible had there not been abundant sources of energy available and efficient systems for transporting and exchanging this energy. In the next chapter we will examine some of the ways the earth receives and utilizes its energy.

SUMMARY

The earth has been the scene of continuous change since its formation about 5 billion years ago. The earth today is far different from the earth of billions of years ago. The advent of life, possibly by chance molecular combinations, introduced new processes of change. The actions of photosynthesizing plants significantly modified the early atmosphere by giving off oxygen and thereby providing the means for animal respiration. Radiant energy from the sun powers most processes on the earth's surface, and change on the surface continues today because of interactions among solar energy, water, rock materials, soil, vegetation, animals, and man.

Volcanic activity is dramatic evidence of internal sources of energy from within or below the earth's crust, but internal energy is also responsible for the almost imperceptible growth of mountain systems and the lateral movement of the continents and their associated plates. The landforms of the earth's surface are shaped by continuous interactions between internal energy sources and the more obvious "external" sources, such as gravity, precipitation, running water, moving ice, wind, and waves.

Investigations using the fossil record and radioactive dating have identified many, but by no means all, of the events of the earth's history. Most change is slow enough not to be apparent within a single human lifetime, but from the perspective of the entire span of history, major changes of mountain systems and life forms apparently occurred during relatively short periods of time. The geologic time scale has been organized by scientists so that these major mountain-building revolutions are placed at the close of long eras and periods when crustal disturbances were much less significant.

Physical geography is the study of these interactions at the surface of the earth. In this book we will emphasize energy exchanges, processes, and environmental responses at or near the earth's surface. The next chapter focuses on energy and interactions within a systems framework, and the continuous recycling of earth materials.

Review Questions

1. What is the central focus of physical geography and how might it differ from other specialized environmental disciplines?
2. What are some of the reasons why scientists believe that the universe had a definite beginning in time?
3. Organize the major events of earth history into a hypothetical 24-hour day.
4. Compare the earth's early atmosphere and landforms with those of the present day.
5. In what ways was the earth's early physical environment transformed by the appearance and evolution of living organisms?
6. What role has water played in the changes the earth has undergone in the last 4.6×10^9 years?
7. What are the various techniques used to date the major events of earth history?

Further Reading

Abell, George O. *Exploration of the Universe.* 3rd ed. New York: Holt, Rinehart and Winston, 1975. (738 pp.) This introductory text often used in astronomy courses is especially useful for recent ideas about the formation of the universe and solar system.

Dott, Robert H., Jr., and **Roger L. Batten.** *Evolution of the Earth.* New York: McGraw-Hill Book Co., 1971. (649 pp.) This introductory text traces the evolution of crustal features, flora, and fauna from a geological perspective.

Eicher, Donald L. *Geologic Time.* Englewood Cliffs, N.J.: Prentice-Hall, Inc., 1968. (150 pp.) This brief monograph, part of the Foundations of Earth Science series, surveys each of the methods used in dating crustal materials and features.

Whipple, Fred L. *Earth, Moon, and Planets.* 3rd ed. Cambridge, Mass.: Harvard University Press, 1968. (297 pp.) One of the Harvard University paperbacks on astronomy, this text offers a nonmathematical approach emphasizing recent advances. Nevertheless, it needs to be updated to account for current investigations of the planets.

The Starry Night by Vincent Van Gogh, 1889. (Collection, the Museum of Modern Art, New York; acquired through the Lillie P. Bliss bequest)

Everything on the earth—its hills and valleys, its trees and rivers, its farms and cities—is unified and driven by energy. Larger swirls push the atmospheric processes; smaller eddies shape the land; the ceaseless flow of energy continually rearranges the smallest particles of matter. Energy from the sun and from within the earth flows through the earth's systems in dynamic processes and patterns.

TWO
The Dynamic Planet

WITHOUT ENERGY, the earth would be lifeless and unchanging, because energy is the basis for all movement and all change. It is the muscle that makes things go, and no process can operate without the expenditure of energy. But energy alone is insufficient to promote change; there also must be something for the energy to act on. The moon receives radiant energy from the sun, but little change occurs on the moon because it possesses no atmosphere, water, or life to use energy in processes of change.

Physical geography is concerned with the ways energy works on the thin skin of air, water, rock, soil, and plants that covers the earth's surface. How different systems such as vegetation, the atmosphere, landforms, and the oceans respond to energy and how they interact by transferring energy to one another are major themes of this book. To set the stage for later chapters, this chapter discusses energy—where it comes from, the forms it takes—and it points out the general ways systems respond to energy. The properties of water are also described, because many systems exchange energy by means of moisture. In later chapters we will discuss components of the earth's principal systems—from green plants to ocean waves—and we will examine how energy and moisture interact within particular systems to produce change.

Processes of change on the earth are powered by two sources of energy: the radiant energy from the sun and the internal heat energy of the earth. The sun's radiant energy provides the power to drive most of the geographic processes important on the human time scale of days or years, such as the movement of moisture through the atmosphere by weather systems and the growth of vegetation. With the help of energy from the sun, green plants manufacture carbohydrates from water and carbon dioxide. Heat from the sun's rays lifts water molecules from the ocean's surface into the atmosphere, where they may form clouds and rainstorms. Solar energy warms the air and land, and the heated air circulates in wind and weather patterns. Energy from the sun also supplies the motive force to transport water from place to place on the earth's surface, an action with significant consequences for climate, vegetation, soil, and erosion.

The sudden shock of an earthquake or the appearance of a new volcanic island emerging from the ocean in an interval of a few months is a spectacular reminder that the earth's internal energy is also at work producing change. Most of the changes induced by internal energy, such as the raising of mountains or the widening of ocean basins, produce important effects only over long spans of geologic time. Erosion processes also tend to be slow, because rock offers great resistance to the action of wind and water. Many of the earth's landforms are shaped both by the earth's internal energy and by external energy from the sun.

Before proceeding to a discussion of energy and its effect on the environment, we turn first to the basic questions of what energy is and what forms of energy are most important in terms of changes at the surface of the earth.

ENERGY AND WORK

Energy is the capacity for doing work or producing change. It can assume many different forms and act in a great variety of ways. The sun, for example, transmits radiant energy to the earth, and one of its forms is visible light. A teakettle filled with boiling water contains heat energy. When a wristwatch is wound, its mainspring acquires elastic energy. A tree uses chemical energy to grow. A baseball flying through the air has energy of motion, or kinetic energy. A loose rock on the side of a cliff contains gravitational, or potential, energy. The amount of energy available is also a crucial factor in determining interactions and responses in the physical environment. For example, the availability of solar energy is basic to the climate of each region of the earth.

All processes in the physical world may be viewed in terms of energy and work. When the sun's energy causes a molecule of water to leave the surface of the ocean and enter the atmosphere, work is done to break the molecule away from the water and lift it through the air. As it rises, it gains potential energy. Later, when the molecule falls to earth as part of a raindrop, its potential energy is converted to kinetic energy. If the raindrop happens to fall upon a lump of soil, part of this kinetic energy will be used to do the work of breaking off bits of the soil. For some of the many forms of energy and the ways in which energy interacts with the systems of the earth see Figure 2.1, pp. 32–33.

Forms of Energy

All forms of energy are capable of performing work, but not all of them are equally important to physical geography. Electrical energy, for example, performs a great variety of work in homes and factories, but its importance in the natural environment is limited. The following is a review of the forms of energy which are of greatest significance to the processes and interactions taking place at the surface of the earth.

Radiant energy from the sun is responsible for heating the atmosphere and the earth's surface. Solar radiation, which contains more than visible light, travels through space at a speed of about 300,000 km (186,000 miles) per second, reaching our planet in about 9⅓ minutes. Solar radiation falls on the surface of the earth with varying intensity, and this has important consequences for the environment. It is the geographical distribution of solar radiation by seasons which powers the weather and climate systems of the earth. Chapter 3 will deal in greater detail with the principles of solar radiation and the exchanges of radiant energy at the earth's surface and in the atmosphere.

Heat energy is the energy associated with the random motion of atoms and molecules in a substance. In general, the energy of the atoms and molecules can be measured by temperature; the hotter an object is, the more vigorously its atoms jostle about and the greater its heat energy. We need to make a distinction between heat energy and temperature, however. A cup of hot coffee, 50°C (122°F), has a higher temperature than a bathtub of warm water, 35°C (95°F), but more energy is stored in the bathtub because of its larger volume than in the coffee cup. We also need to make a further distinction between *sensible heat,* which can be felt or measured directly with a thermometer, and *latent heat,* which represents heat energy temporarily stored in water vapor. We will see later in this chapter that latent heat is converted to sensible heat when water vapor condenses and becomes water droplets or ice crystals.

Gravitational energy—a form of energy associated with altitude—is constantly encountered in everyday life. It refers to the energy an object possesses by virtue of its elevation. Gravitational energy is often called *potential energy* because it represents the potential to do work. A parcel of air aloft in the atmosphere, for example, has the potential to subside to near the earth's surface unless it is acted upon by an upward-directed force. For objects near the surface of the earth, gravitational energy can be considered proportional to altitude and mass. A large rock near the top of a mountain has more potential energy and will make a bigger splash in a lake below than a smaller rock located part way down, both because of its greater altitude and its greater mass.

Energy of motion, or *kinetic energy,* is associated with any moving object—the higher the speed, the greater the energy it possesses. For a given speed, however, kinetic energy is proportional to the mass of the object or substance. For example, a volume of air and an equal

6
8
9
7
4
10
13
12
11
16
17
15
14

Figure 2.1 This painting interprets some of the forms of energy important in physical geography, and it illustrates ways in which energy interacts with systems on the earth.

Processes of change on the earth are powered by two sources of energy: the radiant energy from the sun (1) and the internal heat energy of the earth (14). Radiant energy drives the motion of the atmosphere (2) by causing temperature differences on the earth's surface. The wind in turn transfers energy to the ocean, producing ocean currents and waves (3). Ocean currents and the moving atmosphere help distribute energy from the hot equatorial regions of the earth to the cool polar regions. The energy of waves actively shapes coastlines (4) by eroding rocks and by transporting sand to and from beaches.

On the seas and on the land, energy from the sun frees water molecules to enter the atmosphere (5). The evaporated water is transported by the moving atmosphere and returns to the earth's surface as precipitation (6). Some of the water falling on the land runs into streams (7). As a stream flows toward lower altitudes, its gravitational, or potential, energy is used in the work of altering the landscape. Flowing water is a powerful agent of change because of its power to erode the land and to transport and deposit sediment. In cold climates or at high altitudes, precipitation may take the form of snow (8). Accumulated snow produces glaciers (9), which are thick sheets of moving ice that scour the landscape.

Vegetation (10) absorbs solar radiant energy and transforms it to stored chemical energy by the process of photosynthesis. Some of the stored energy is passed on to animals (11), which ultimately depend on green plants as their source of food energy. Vegetation, moisture, and rock materials interact to form soils (12). The energy in part of the vegetation becomes stored as fossil fuels, which constitute the main power source for modern industrial society (13).

The energy released in radioactivity (15) contributes significantly to the internal heat energy of the earth. The temperature of the earth increases with increased depth below the surface, and the geothermal energy obtained by tapping this source of heat may become an important source of power. The plastic, partly molten rock underlying the earth's crust allows heat currents to move and deform the crust. On the local scale, deformations of the crust can result in breaks or faults (16). On the global scale, motion of the crust results in the drifting of continents, the opening of ocean basins, and the uplift of mountains (17).

At a boundary where an oceanic crustal plate converges upon a continental plate (18), the oceanic plate may be deflected downward into the deeper layers of the earth. Stresses cause deformation of the plates; deep fissures allow molten rock to come to the surface in active volcanoes (19).

volume of ocean water may be moving at the same speed, but the volume of water will have greater kinetic energy because of its greater mass. This can be seen from the spectacular erosion and destruction occasionally wrought along coastlines by massive ocean waves during storms.

Chemical energy is a form of energy stored in the bonds which join the molecules of a chemical compound. When substances react chemically, chemical energy is either released, stored, or converted to another form. Green plants, for example, utilize solar radiation, carbon dioxide from the air, and water from the soil to store chemical energy in their tissues in the complex process of photosynthesis.

Energy Transformations

You feel warm when you stand in the sunlight because radiant energy from the sun is being converted to heat energy in your skin. When you run, you use stored chemical energy to produce energy of motion. When you go down a hill on a bicycle without pedaling, you gain speed as your altitude decreases because your gravitational energy is being converted to energy of motion. The principal reason that energy takes so many different forms is that it cannot be destroyed; it can only be converted from one form to another.

Just as energy cannot be destroyed, neither can it be created. No matter how many energy transformations take place, the total amount of energy in the universe remains constant. The electrical energy generated by a coal-burning power plant, for example, is gained only at the expense of the chemical energy in the expended fuel. No way has ever been found to do useful work except by utilizing some form of energy, but the gain of one kind of energy is always accompanied by the loss of another kind.

Although energy is never created or destroyed, it is frequently degraded into forms that cannot be used efficiently. Imagine throwing a snowball at a wall. Your muscular energy gives the snowball energy of motion, but when the snowball hits the wall, it comes to a stop and loses this energy. Where does the energy go? Before the snowball strikes the wall, each of its atoms has the same average speed as that of the snowball. After it hits the wall, the atoms are jostled about. Their motion is transformed into a random, vibrating motion, with each atom moving in a different direction. This random motion of atoms represents heat energy, and it has been gained at the expense of the snowball's energy of motion.

In the case of the snowball, kinetic energy is converted to sensible heat, which warms the snow stuck to the wall, the wall itself, and the nearby air. The generation of heat energy accompanies *all* natural processes involving the transfer of energy. When various kinds of energy, such as chemical, kinetic, or solar radiant energy, are converted to other forms, some of the energy being transferred is inevitably converted into heat. But because heat energy is associated with the random motion of atoms, it is impossible to transform it into other forms with perfect efficiency. Some of the heat is dissipated and serves only to warm the surrounding area. The fact that energy transfers always involve the conversion of some energy into dissipated heat means that useful energy can never be transferred with perfect efficiency.

Man's Use of Energy

Energy must frequently go through a long chain of transformations before it is finally converted into a form which man can use to accomplish work. The chain begins with the sun's energy falling on green plants. Part of the energy of sunlight is stored as chemical energy through the process of photosynthesis. If a person eats a plant, he obtains chemical energy from it. Or a plant may decompose and eventually become a fossil fuel. Fossil fuels—coal, oil, and gas—represent radiant energy from the sun stored as chemical energy; they were produced from green plants and microscopic marine animals that lived tens to hundreds of millions of years ago. The heat from burning a fossil fuel or the wood of a plant can be used to drive a steam engine or an electric generator, yielding electrical energy. The electricity can then be used to light a light bulb, to turn an electric motor, or to heat a stove.

Fossil fuels constitute the prime energy source for industrial societies. But it is a source with definite limitations. It took many tens of millions of years for the earth's coal deposits to accumulate, but in the past 100 years a significant fraction of the known fossil fuel deposits has been consumed to meet the energy needs of industrialization. At the present rate of withdrawal, the known coal reserves will be depleted in less than 800 years. World oil production is estimated to peak around 1990, declining thereafter to insignificant rates by about 2050.

Not all sources of useful energy are the products of long chains of energy transformations. Nuclear energy, for example, is energy that has been stored in the structure of atoms since the "big bang" 20×10^9 years ago when the elements were formed. Nuclear energy is released when one kind of atomic nucleus spontaneously decays and changes into another kind that has less internal energy. Although a promising but controversial substitute for fossil fuels, nuclear energy could not satisfy our energy needs indefinitely because the supply of nuclear fuel is limited. Ultimately, other energy sources will have to be found. One possibility is to utilize radiant energy directly from the sun. Much effort will probably be expended during the next few decades toward exploiting this abundant and "clean" energy source. But because nuclear energy also powers the sun, the supply of solar energy is limited. In a few billion years, when the supply of suitable atoms is exhausted, the sun will no longer radiate energy.

Units of Energy

Before going on to consider the role of energy in the environment, we must briefly discuss the ways in which energy is measured. Since energy is the capacity to do work, measurements of energy are invariably expressed in terms of the amount of work which a particular quantity of energy will accomplish. The measurement system for energy employs a number of different units—ergs, joules, calories, foot-pounds, watts, and British thermal units. The unit of energy most frequently encountered in physical geography is the *gram calorie*, often called simply the *calorie*. The calorie is defined as the amount of energy which, if converted entirely to heat energy, would raise the temperature of 1 gram of water from 14.5°C to 15.5°C. Although the term "calorie" is often used, the energy content of food is actually measured in kilogram (kg) calories. One kilogram calorie equals 1,000 gram calories. A 100-watt light bulb consumes energy at the rate of approximately 25 calories per second. When the sun's rays are perpendicular to a hypothetical surface at the top of the earth's atmosphere, the sun delivers approximately 2 calories of radiant energy per square centimeter (cm^2) per minute. (There are approximately 6.5 cm^2 per sq. in. of area.) Recently, some scientists have begun measuring radiant energy in watts per square meter, but we will continue to use calories in this text.

WATER: AN ENERGY CONVERTER

Water plays a fundamental role on the earth's surface because it is one of the principal means by which systems exchange energy. Chemically, water is a compound of two of the most common elements in the universe, hydrogen and oxygen, in the ratio of two atoms of hydrogen for every one of oxygen (H_2O). The unique properties of water enable it to convert energy from one form to another and to store and redistribute energy from one place to another. For example, when water flows downhill, its gravitational energy is converted to kinetic energy, some of which is expended in the work of loosening and transporting rock debris (see Chapter 12). Glaciers, which are essentially rivers of frozen water, alter the landscape in much the same fashion (see Chapter 15). Strong winds transfer their energy to the ocean surface in the form of waves. These may travel thousands of miles, transporting kinetic energy from oceanic storms to faraway coasts, where the energy is expended in the work of shoreline modification (see Chapter 16).

Water also fulfills an important function in the atmospheric and oceanic circulation systems which continuously transport energy from warmer to colder regions of the earth. But in order to understand how these functions are carried out, we must first investigate the behavior of water in its three basic forms: as a liquid, as water vapor, and as ice.

Changes of Phase

The three physical forms that a substance can take—solid, liquid, and gas—are called *phases*. Water is the only common substance that occurs naturally in all three phases. In order for any substance to change phase, a certain amount of heat energy must either be absorbed or released by the molecules of that substance. Energy is absorbed when a substance changes from a solid to a liquid and from a

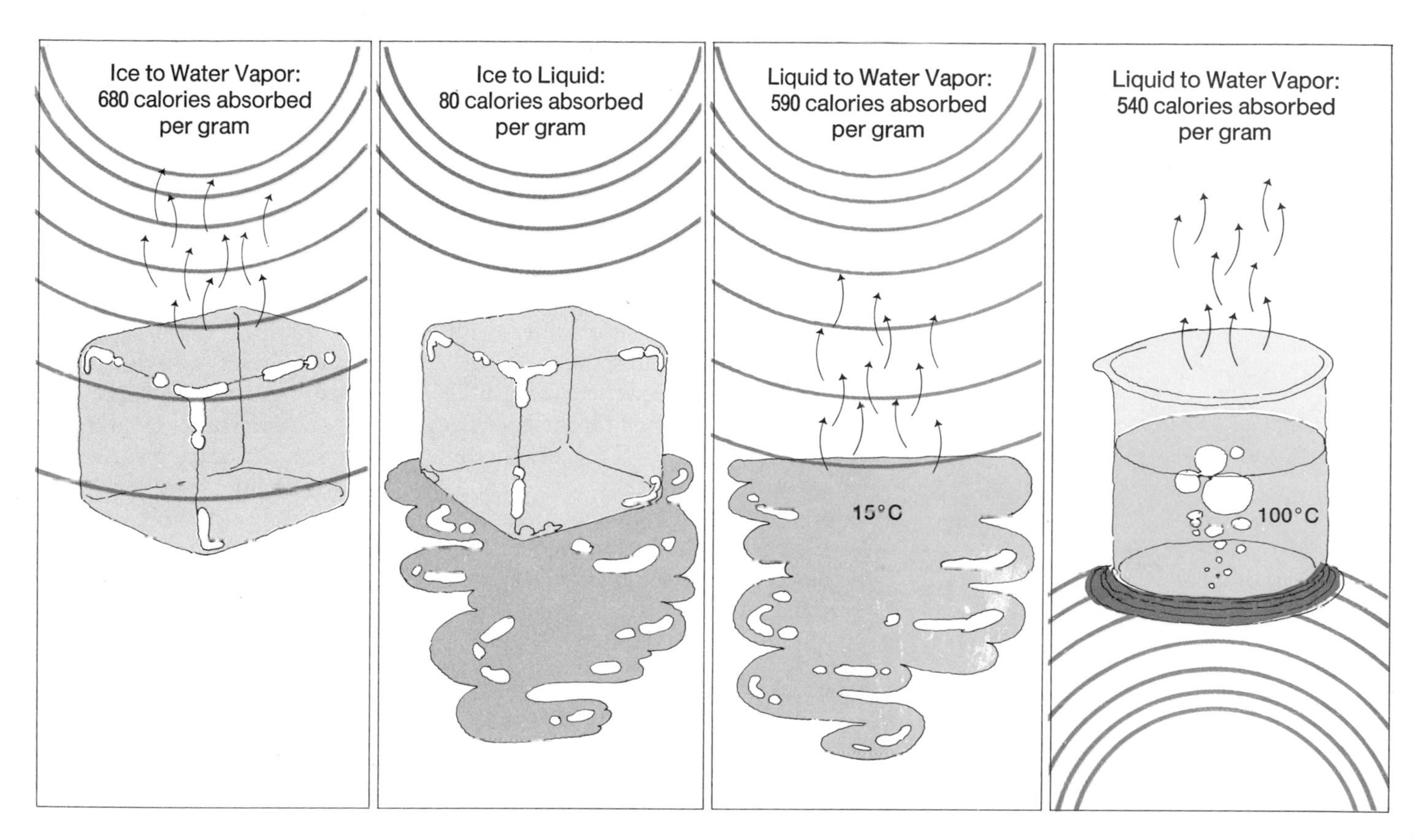

Figure 2.2 Significant energy changes occur when water changes from one physical state to another. Energy must be added to change water from a physical state in which the water molecules are tightly bound to a state in which they are more loosely bound, even if the temperature of the water remains the same. To change 1 gram of water from solid ice directly into gaseous water vapor requires adding approximately 680 calories of energy, called the latent heat of sublimation. The change from solid to liquid requires 80 calories per gram, called the latent heat of fusion. The energy required to change 1 gram of liquid water to gaseous vapor, the latent heat of vaporization, varies slightly with the temperature of the water. For water at 15°C the latent heat of vaporization is 590 calories per gram, and for water at 100°C it is a 540 calories per gram.

Energy must be released to change water from a state in which the molecules are loosely bound to a state in which they are more tightly bound. The amount of energy that must be released is numerically equal to the latent heats described above; for example, 590 calories must be removed from 1 gram of water vapour at 15°C to form 1 gram of liquid water at the same temperature.

liquid to a gas. Energy is released when a substance changes in the opposite direction. This energy is known as *latent heat*—latent in the sense that the molecules which are changing phase do not actually change in temperature.

Under normal conditions, the melting point of ice and the freezing point of liquid water are both 0°C (32°F) (Figure 2.2). However, ice at its melting point will not become liquid unless enough energy is added to break the molecules away from the strong mutual attraction they have for one another in the solid state. To melt 1 gram of ice at 0°C requires adding 80 calories of energy. Conversely, to freeze 1 gram of liquid water at 0°C requires removing 80 calories of energy from the water. The energy that is absorbed or released in these changes is called the *latent heat of fusion*.

Energy is also required to change liquid water to vapor at the same temperature. This energy is called the *latent heat of vaporization,* and it varies slightly with the temperature of the water. For water at an average outdoor temperature of 15°C (59°F), the latent heat of vaporization is 590 calories per gram; for water at the boiling point of 100°C (212°F), it is 540 calories. The molecules that escape the liquid as the water evaporates carry with them absorbed latent heat, and this energy can be stored in the water vapor for long periods of time and circulated over great distances.

When molecules of water vapor in the air condense to form droplets at 15°C, 590 calories of energy are released per gram of water.

This energy is called the *latent heat of condensation* and is numerically the same as the latent heat of vaporization. It is the principal source of energy in thunderstorms and hurricanes.

To change 1 gram of solid ice at 0°C directly into gaseous water vapor at the same temperature requires the absorption of approximately 680 calories of energy. For the reverse process of water vapor to ice, the same amount of energy must be released. The energy required in both these changes is called the *latent heat of sublimation*, and it represents the latent heats of fusion and vaporization combined.

The latent heat of water is much greater than the latent heat of other common substances at or near the earth's surface, and this has important consequences for weather and climatic conditions. Because the energy required for the latent heat of vaporization comes from the local environment, the evaporation of water at the earth's surface results in significant local cooling. In moist climatic regions with water available in lakes, streams, and the soil, much of the incoming solar radiation during the daytime is utilized for evaporation. In dry climatic regions, in contrast, most of the incoming solar radiation must be converted to sensible heat, and local temperatures consequently tend to be higher.

Properties of Water

In addition to its ability to change phase at the temperatures ordinarily encountered on the earth, water possesses other properties of importance in physical geography. One of water's outstanding qualities is its capacity to act as a reservoir of heat energy. The amount of heat energy required to raise the temperature of 1 gram of a substance at normal atmospheric pressure by 1°C is known as the *specific heat* of the substance. The specific heat of liquid water is nearly equal to 1 calorie per gram of water per degree Celsius over the temperature range from freezing to boiling. The specific heat of water is about 5 times greater than the specific heat of soil or air; that is, the amount of heat that will raise the temperature of a body of water by 1°C (1.8°F) will raise the temperature of an equal weight of soil by about 5°C (9°F). Stated another way, water must absorb about 5 times more heat energy than soil or air in order to rise in temperature the same amount.

Comparing the specific heat of water, soil, and air, however, does not take into account their relative densities. Air is much less dense than water; thus under normal conditions, 1 gram of air will occupy a much greater volume than 1 gram of water. Therefore, in physical geography, it is often more useful to compare the thermal properties of substances in terms of a unit of volume, such as a cubic centimeter (cm^3), rather than a unit of mass, such as a gram. The term used to express the ability of a substance to absorb heat in relation to its volume is *heat capacity*. Comparing the heat capacity of water to that of other common substances on the earth's surface can tell us more about the role that water plays in the environment. Because of the very low density of air relative to water, the heat capacity of water is many times greater than that of air, and the heat capacity of water is still 2 to 3 times greater than that of soil. Its high heat capacity means that water can store large amounts of energy without excessive changes in temperature, but absorption of small amounts of energy by air results in substantial temperature increases. This is one of the reasons why temperatures fluctuate much less in the oceans than on land.

Water has other remarkable properties which affect its interaction with other components of the environment. Nearly a universal solvent, water dissolves almost anything to some extent. Water molecules produce strong electric forces that hold substances dissolved in the water in solution, but the water molecules do not react chemically with the dissolved substance. Because water can carry materials in solution without affecting them chemically, it is an ideal fluid for transporting substances through the circulatory systems of plants and animals.

Water's ability to act as a solvent plays a significant part in *weathering*, which is the breakdown and alteration of rock materials at or near the earth's surface. Limestone can be dissolved away completely by water, and areas with extensive limestone beds are noted for the caves and subterranean channels which subsurface water has created. In terms of weathering, however, water's physical effect on surface rocks is as important as its chemical action. For example, the pressure water exerts when it freezes in a crack and expands can be great enough to split a rock. Such frost weathering is a major cause of rock disintegration in climates where the temperature range allows repeated freezing and thawing.

THE ATMOSPHERE: AN ENERGY TRANSPORTER

The earth's atmosphere is central in physical geography because it is the medium that transports energy in the form of warm air and water vapor over the earth's surface. Because the atmosphere is a mixture of gases, familiarity with the behavior of gases provides a foundation for understanding the detailed discussions of atmospheric phenomena and atmospheric circulation that appear in Chapters 4 and 5.

Unlike the molecules in liquids or in solids, the molecules that constitute a gas are not strongly bound to one another. They are constantly moving, and they frequently collide with one another and with neighboring surfaces. When a gas molecule collides with a surface, it exerts a push on the surface. The total force molecular collisions exert on a unit area of surface at any given time is called the *pressure* of the gas.

The pressure the earth's atmosphere exerts decreases rapidly with height; at an altitude of 5 or 6 km (3 or 3.6 miles), the atmospheric pressure is only half the pressure at sea level (see Figure 2.3). The molecules at sea level must exert enough of an upward push or force to sustain the weight of the entire atmosphere, whereas at an altitude of 5 or 6 km, the molecules must sustain the weight of only half as much air. Standard atmospheric pressure at sea level is specified as 760 millimeters (mm), or 29.9 in., of mercury, which is the height of the mercury column that can be supported by standard atmospheric pres-

Figure 2.3 The pressure exerted by air in the atmosphere decreases rapidly with increased altitude. As the graph indicates, pressure at sea level is approximately 1,000 millibars, but at an altitude of 5 km (3 miles) it has fallen to about 540 millibars, about half the sea level value. Note that the pressure on the graph is not evenly spaced in pressure units but is instead marked off in successive powers of ten. Such a *logarithmic scale* is useful when graphing a quantity that varies over a wide range; for instance, the logarithmic scale on this graph indicates a pressure range from less than 10 millibars to 1,000 millibars. The pressure at sea level can be thought of as the force per unit area exerted by air molecules at sea level to support the weight of a column of the earth's atmosphere. The graph shows the standard value of atmospheric pressure at each altitude; actual atmospheric pressure at a particular location on a particular day can be a few percent higher or lower than the standard value. If the air is denser than usual, the pressure is higher than normal, and if the air is less dense, the pressure is lower than normal.

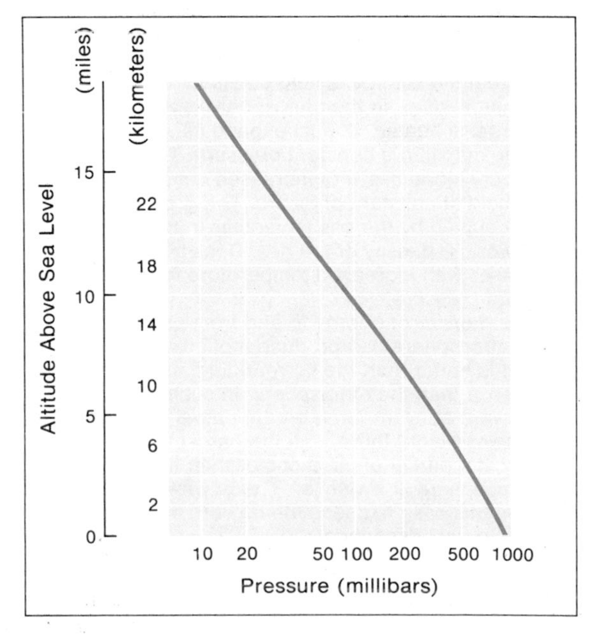

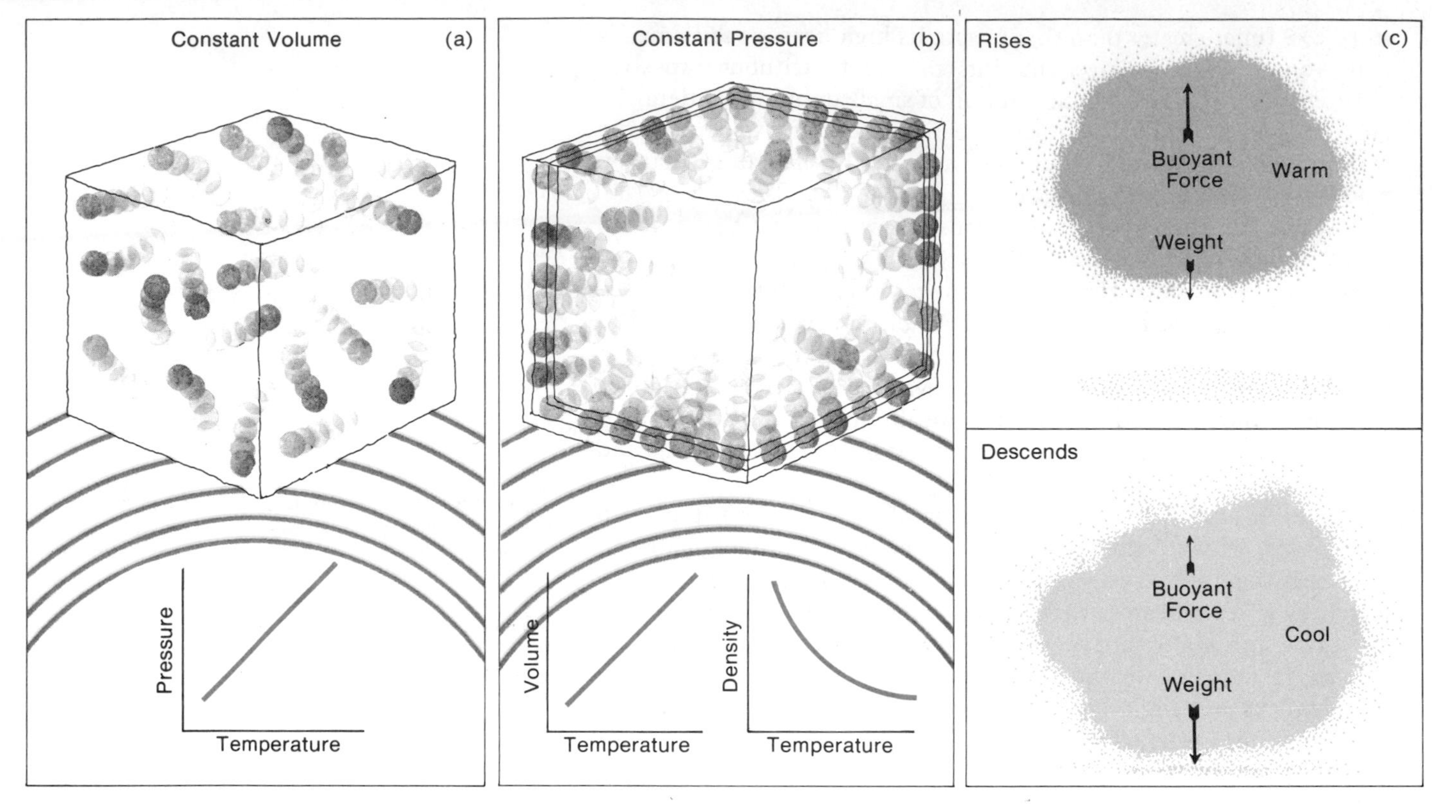

Figure 2.4 (a) The pressure, volume, and temperature of a fixed amount of gas are interdependent; a change in one of the three quantities is always accompanied by a change in one or both of the others. Suppose, for example, that a gas is confined in a sealed box so that its volume is constant. Changing the temperature of the gas changes the pressure the gas exerts on the walls of the box; the gas molecules move more rapidly at high temperatures and therefore collide more forcefully and more frequently with the walls. For gases under normal conditions, pressure increases with increased temperature if the volume is held constant.
(b) A parcel of air in the atmosphere is unconfined. If the parcel is heated, it must expand its volume in order to maintain a constant pressure. The volume of the parcel increases with increased temperature for a gas held at constant pressure. The greater the volume occupied by the gas molecules in the parcel, the lower the density of the gas. Density therefore decreases with increased temperature for a gas at constant pressure.
(c) The transfer of energy to and from parcels of air in the atmosphere can produce motion. A parcel of air that is hotter than the surrounding atmosphere is less dense than the atmosphere; in such a situation, the upward buoyant force on the parcel exceeds the downward weight force, and the parcel rises. Conversely, if a parcel of air is cooler than the surrounding atmosphere, it is denser. The downward weight force in this case exceeds the upward buoyant force, and the parcel descends.

sure. In meteorology, pressure is usually quoted in *millibars* because it allows for easier calculations. Standard atmospheric pressure is equivalent to approximately 1,000 millibars.

The pressure, volume, and temperature of a fixed amount of gas are interdependent; a change in one of the three quantities is always accompanied by a change in one or both of the others (see Figure 2.4). Suppose, for example, that a fixed amount of gas is confined in a sealed box so that its volume is constant. If the temperature of the gas is increased, the pressure is also increased because the gas molecules will move more rapidly and therefore collide more frequently and more forcefully with the walls of the box.

But if the gas is not sealed in a container, a rise in temperature will result in a corresponding increase in volume, while pressure will tend to remain constant. For example, a parcel of air in the atmosphere is unconfined. If a parcel of air just above the earth's surface is heated, its volume expands in order to maintain its internal pressure. Because the air in the heated parcel has fewer molecules in a given volume than the surrounding cooler air, the parcel is less dense than the surrounding air. In this situation, the upward buoyant force on the parcel exceeds the downward weight force, and the parcel begins to rise like a hot air balloon. Conversely, if a parcel of air is cooler than the surrounding atmosphere, it is denser; the downward weight force exceeds the upward buoyant force, and the parcel descends. Because of differences in density and buoyancy, sensible heat is redistributed vertically through the lower atmosphere.

This vertical redistribution of heat energy causes variations in pressure and sets in motion a horizontal circulation of the air commonly known as wind. Winds act to transfer energy from one region of the earth to another. For example, they can transfer warm air from the equatorial regions, which receive a greater amount of solar radiation, to polar regions, which receive much less. Winds can also redistribute latent heat present in water vapor, transporting it thousands of kilometers from regions where evaporation has occurred to regions where the vapor condenses back to water or ice. The circulation of the atmosphere and its significance in energy transfer are evaluated in Chapter 5.

THE EARTH'S SYSTEMS

The earth's atmosphere, with its ability to circulate and redistribute energy, can be seen as a system. So can the waters of the earth (the *hydrosphere*), with their ability to convert energy from one form to another. A system is any collection of interacting objects. The earth as a whole can be considered a single system, but it contains so many phenomena interacting in such a variety of ways that understanding them all at one time becomes extremely complex. We can simplify the task considerably by subdividing the world into smaller systems, such as climate, vegetation, soils, and landforms.

Viewing the world in terms of systems is a useful way to organize knowledge because systems, whatever their nature, have many basic features in common (see Figure 2.5). Nearly all systems at or near the earth's surface are *open systems.* In an open system, energy and material *inputs* enter the system from outside its boundaries and are transformed in interactions within the system to new forms. These either remain in storage for a time or leave the system as *outputs* of material and energy to become part of other systems. In a *closed system,* on the other hand, every element is affected only by other elements within the system; a wristwatch powered by a small chemical battery is an example. Although closed systems are not common at the earth's surface, the earth as a whole has elements of both closed and open systems. For example, the earth receives energy in the form of solar radiation and it reflects energy back into space. However, with the exception of meteorites, almost no material enters or leaves the earth's boundaries.

Examining any one system in terms of its inputs and outputs will reveal the high degree of interdependence of all systems. Consider the commonplace occurrence of waves breaking on a beach. As the waves break on the beach, their impact results in the moving of sand particles; sand is shunted along the beach because of the oblique direction of the waves' approach; sand moves into the beach at one end and out of the beach at the other end; the beach changes daily in form as sand arrivals exceed or fall below sand losses. The inputs to the system consist of the sand that forms the beach and the kinetic energy which drives the waves. The output is the day-to-day change in the form of the beach.

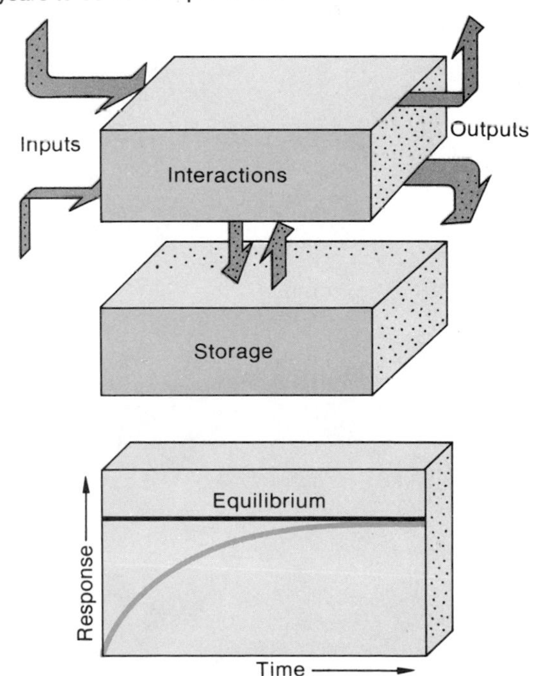

Figure 2.5 A system is any collection of interacting objects. Physical geography is concerned with a wide variety of systems, including landforms, soils, vegetation, and the atmosphere. Thinking about the world in terms of systems is useful because systems have many features in common, some of which are illustrated in this schematic diagram. The inputs to a system represent energy and material received from outside the system's boundaries. The inputs are transformed by system interactions to new forms that are either placed in storage for a time or transferred to other systems as outputs of material and energy. Consider a field of corn as an example of a system: the inputs include radiant energy from the sun, carbon dioxide from the atmosphere, and moisture and nutrients from the soil. Part of the input is stored as chemical energy in the process of photosynthesis. Outputs include the water vapor returned to the atmosphere by plant transpiration, the free oxygen released in photosynthesis, and the chemical energy that is eventually used as food.

The lower diagram illustrates schematically one of the ways that systems respond in time to an input. In this mode of systems behavior, the output does not reach full strength immediately upon application of an input. Time is required for a steady condition to be achieved. Some systems, such as the atmosphere, respond rapidly to new inputs; other systems, such as soils, may require hundreds or thousands of years to come to equilibrium.

But by tracing these inputs and outputs farther, we can see that the beach system is really part of a much larger picture. For example, the kinetic energy of the waves was originally imparted by wind stress on the surface of the water many kilometers away. The wind resulted from differences in the receipt of solar energy in the atmosphere, causing pressure variations and thus circulation of the air. Solar radiation itself is an output of atomic reactions taking place within the sun. The material input of the beach system—sand—is an output of a complex erosional system on the land. Finally, the output of the beach system—the changing form of the sand deposits—has important consequences for the living organisms that exist in that environment.

Ultimately, all systems on the earth are linked by their interactions. The output from one system represents the input to another system. Rainfall, for example, is an output of the atmospheric system of water delivery as well as an input to vegetation and landform systems on the earth's surface.

The Equilibrium Concept

Variations in the inputs to a system generally produce short-term changes, yet the long-term behavior of the system often remains constant. Many systems tend toward such an *equilibrium* state in which continued inputs of energy and material produce essentially no change in one or more outputs. Consider a stream flowing toward the sea. When excess rain causes flooding, the stream may rapidly deepen its bed by erosion. But when the flow returns to normal, the stream channel returns to its original shape and size by depositing sediment to fill the eroded portions of the bed. The average form of the channel remains much the same over long time periods, although there are short-term fluctuations.

The time required for a system to reestablish equilibrium after undergoing a disturbance is known as its *relaxation time.* This can vary widely from system to system. The stream bed, for example, may resume its former shape within hours or days after being scoured by a flood. But when a volcano erupts and covers the ground with mineral ash, a much longer relaxation time is necessary. First, the physical character and chemical composition of the ash gradually change in response to atmospheric inputs of energy and moisture. Later, when plants begin to grow and organic matter becomes mixed with the ash, another input becomes active and the composition changes further. Over time the ash is transformed to a true soil in equilibrium with its surroundings. However, this usually requires tens to hundreds of years.

Some systems, after undergoing a disturbance, do not return to their original state but assume a new form. In such cases, relaxation time designates the amount of time required for a system to change from one state of equilibrium to another in response to changed inputs. If a river is dammed, the amount of sand and silt it transports to the sea is often significantly reduced. A beach on the coast near the river's mouth then receives less sand to replace the sand carried away by ocean waves. The configuration and size of the beach will change for a few years or decades after the dam is built until the beach disappears or a new equilibrium is achieved.

Some systems respond to inputs of energy or material by storing some of the energy or material. Rocks that are under tension or compression deep in the earth store energy much as a compressed spring stores energy. An earthquake occurs when the stored energy is suddenly released.

Many natural systems are vulnerable to stresses or inputs that result from the increased tempo of human industrial and agricultural activities. By understanding the principles and interactions of a system, we gain insight into where the system is most sensitive to change and what the effect of altered inputs will be.

CYCLES OF MATERIALS

In terms of energy, the earth as a whole is an open system. There is a continuous input of solar energy arriving at the top of the earth's atmosphere and, simultaneously, an equal quantity of energy radiated back into space. We know that the earth's energy input and output must be approximately equal because we can observe no significant long-term temperature change on the earth's surface. However, between the time that energy from the sun arrives at our planet and the time it returns into space, it takes part in a great variety of complex energy exchanges and serves as the driving force behind numerous systems on the earth's surface.

The materials that make up these systems, on the other hand, are finite and unvarying. With the exception of meteorite falls, the earth has acquired no new material since it formed 4.6×10^9 years ago. Nor has it given up material, except for the loss of a few gas molecules to space. In terms of material, then, the earth may be considered a closed system.

Since materials on the earth are limited, they must be continually recycled and reused in order for the earth's systems to continue functioning. The passage of materials through cycles makes renewal and change possible without a continual supply of new materials. The rate at which materials are recycled, however, varies greatly from system to system. For example, ocean water can be cycled into the atmosphere as water vapor, then back to the ocean again as precipitation, all in the space of a few hours. However, if that same water vapor happens to fall as snow in a polar region and become part of a glacier, it may remain in that frozen state for thousands of years. Some of the earth's more lengthy cycles involve other materials. For example, there are massive underground beds extending westward from New York to Ohio and Michigan in which oceanic salts have been "in storage" for several hundred million years. Eventually, erosion of the North American continent will recycle these salts back to the oceans. The earth's crust itself is subject to recycling processes. In Chapter 11 we will see how the earth's crust is ever so slowly reabsorbed into the mantle, and how the mantle yields up new crustal materials equally slowly.

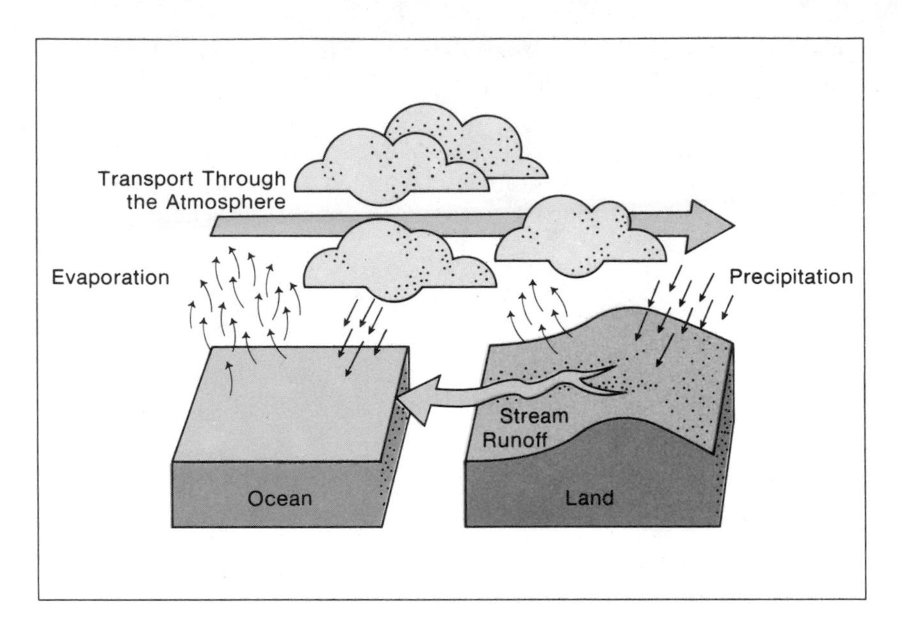

Figure 2.6 The hydrologic cycle is a key element in physical geography because water has important interactions with systems such as vegetation and landforms. The processes involved in the cycling of water include the evaporation of water from the oceans, the evaporation and transpiration from the continents, the transport of water vapor through the atmosphere, and the return of water from the atmosphere to the surface as precipitation. The annual evaporation loss from the oceans exceeds the amount they gain from precipitation, but the deficit is made up, on the average, by streams flowing from the continents into the oceans.

The Hydrologic Cycle

Rivers and streams carry water endlessly from the continents to the oceans. If there were no "return flow" back to the continents, there would eventually be no water on the lands. The series of events by which the earth's supply of fresh water is renewed is known as the *hydrologic cycle* and is illustrated in Figure 2.6. Ocean water evaporates into the atmosphere, and the water vapor is transported by atmospheric circulation over the continents, where it condenses into cloud droplets. The water falls on the land as precipitation, and some of it drains into streams and rivers to return once more to the ocean. The cycle—ocean to atmosphere to continent and back again to ocean—is completed.

This description is highly simplified, and in actuality there are many subcycles within the larger cycle. For example, not all evaporated ocean water falls as precipitation on the continents; some of it falls directly on the oceans themselves. Nor does all evaporation take place at ocean surfaces; a considerable amount of water enters the atmosphere from the land and through the leaves of plants in a process called *evapotranspiration*. But regardless of these variations, the basic operation of the cycle remains constant—to continually replenish water in the atmosphere and at the earth's surface. Water used in agricultural irrigation, in industry, in the home, is not really lost, but eventually returns to the oceans after passing through subsystems of the cycle. The hydrologic cycle is of great significance to physical geography, and aspects of it will be investigated further in later chapters of this book.

The earth's supply of water is located in several sectors of the environment. An undetermined amount of water is chemically bound in minerals within the earth. From time to time, volcanic activity releases some of this water to the surface in the form of vapor. The salt water in the ocean basins constitutes about 97 percent of the free water on the earth. The next largest store of water is in glaciers, particularly the polar ice sheets. Most of the remaining water is stored in porous rock not far below the surface of the ground, forming the groundwater that is tapped by wells. Lakes, ponds, and rivers contain only a fraction of 1 percent of the total free water on the earth. Relatively little water is stored in the atmosphere at any given moment. Indeed, if all the water vapor in the atmosphere were to fall as rain to the earth's surface, it would amount to a layer of water only about 2.5 centimeters (cm), or 1 inch, in depth.

The ability of systems to store materials affects the operation of cycles. On the average, water that evaporates into the atmosphere from the land or from the oceans remains stored in the atmosphere for a week or two, which is enough time for the water to be transported long distances. Water that falls as snow on the Antarctic icecap, however, may be incorporated into glacial ice for thousands or millions of years before it returns to the oceans.

The Carbon Cycle

The carbon cycle, like the hydrologic cycle, is of vital importance to life forms on the earth's surface. It is also quite complex, comprising a number of subcycles which vary considerably in their rate of turnover. In terms of living organisms, the most significant aspect of the carbon cycle involves carbon dioxide (CO_2), a gas which makes up about .03 percent of the atmosphere by volume. Green plants utilize atmospheric CO_2 along with water and solar radiation to manufacture carbohydrates in the process of photosynthesis. The reserves of CO_2 in the atmosphere are not depleted, however, because animals and plants are continually expelling CO_2 as a waste product of *respiration*, the process by which nutrients are oxidized to produce energy. The role of plants in the environment is discussed in more detail in Chapter 9.

Atmospheric CO_2 represents only a small fraction of the carbon present in the environment. A great quantity of CO_2 is stored in solution in the oceans, while deposits of fossil fuels, particularly coal, oil, and natural gas, account for another sizable percentage. Much additional CO_2 has been released into the atmosphere since fossil fuels became the primary energy source during the Industrial Revolution beginning in the nineteenth century. Some scientists have become concerned about this added discharge of CO_2 into the atmosphere because CO_2 is a strong absorber of thermal radiation directed upward from the earth's surface. The absorbed radiation contributes to raising the temperature of the lower atmosphere (see Chapter 3). Recent estimates suggest about a 0.3°C global warming of the atmosphere due to a 10 percent increase of CO_2.

Concern about possible atmospheric warming has prompted studies of the CO_2 levels in the atmosphere. The results of one such study, which was conducted over a period of 15 years, are shown in Figure 2.7, p. 46. A testing site near the summit of Mauna Loa on the island of Hawaii was selected because it is distant from any substantial local source of CO_2. The figure shows a long-term increase of CO_2. The increase from 1850, when massive industrialization was beginning,

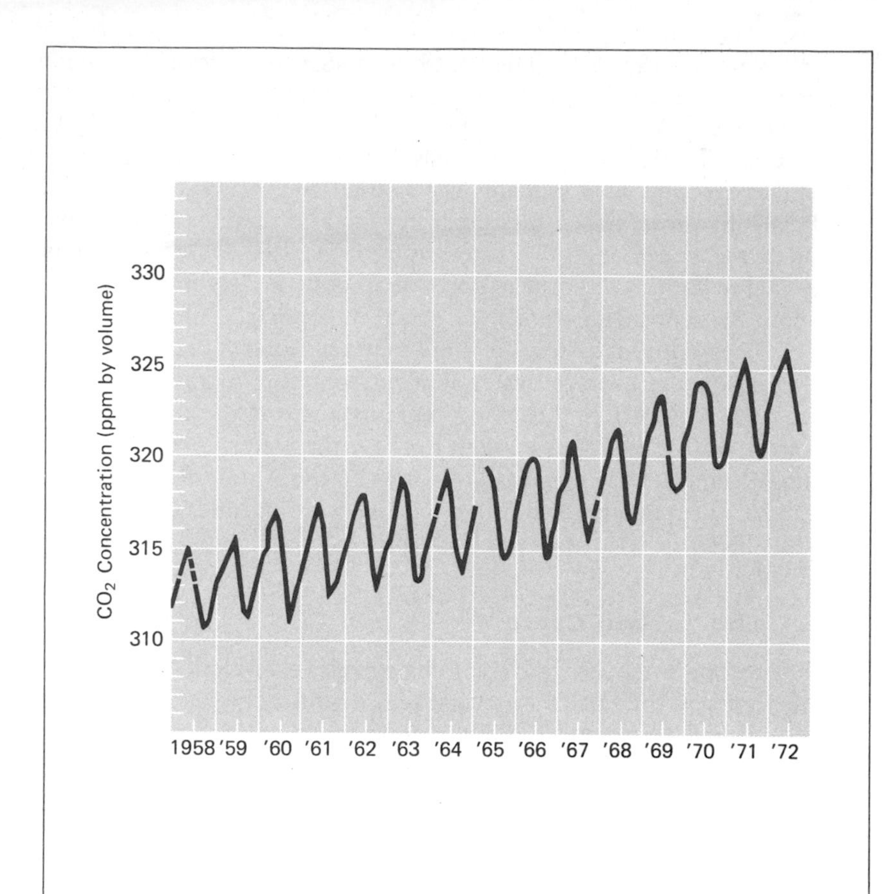

Figure 2.7 Mean monthly concentrations of CO_2 at Mauna Loa, Hawaii, at a site distant from any major local source of CO_2. The cyclical pattern is due primarily to the seasonal increases and decreases of vegetation in the northern hemisphere, but the long-term trend upward is obvious.

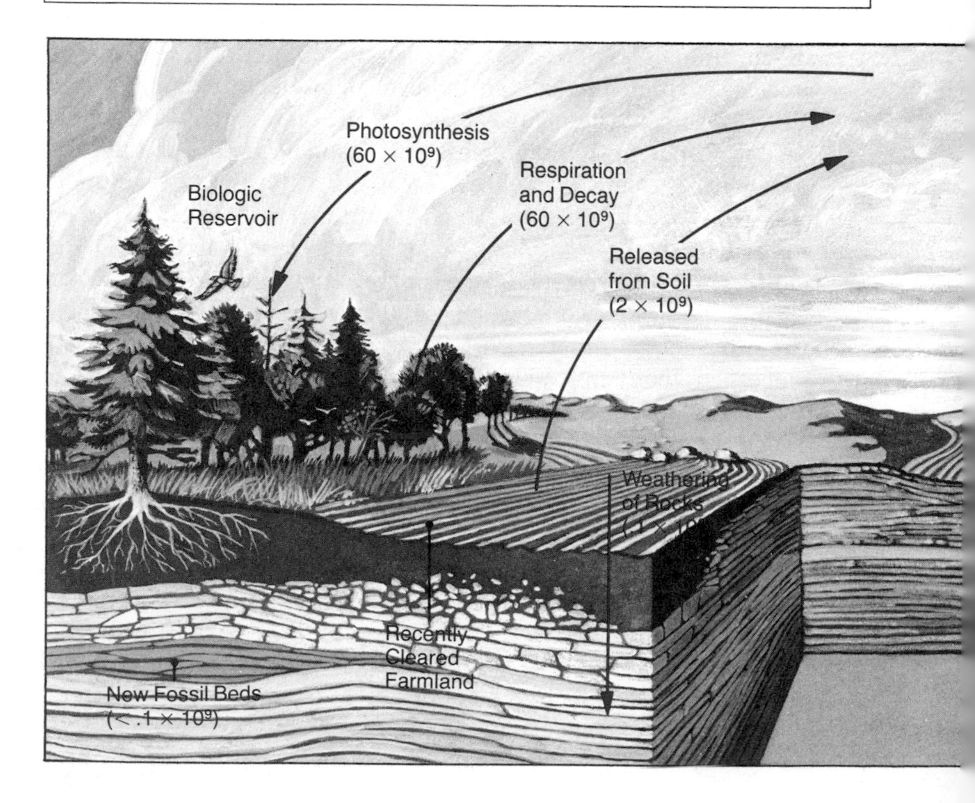

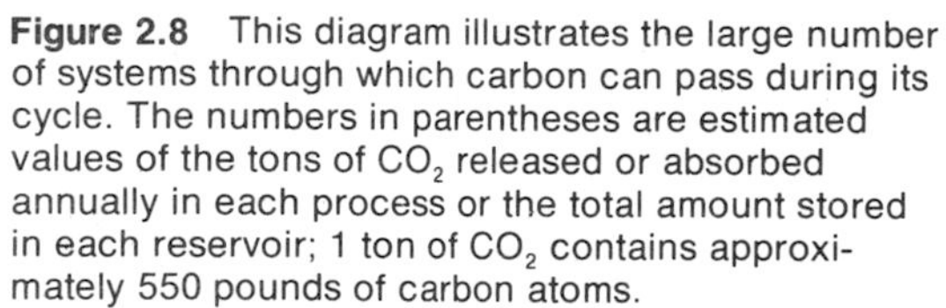

Figure 2.8 This diagram illustrates the large number of systems through which carbon can pass during its cycle. The numbers in parentheses are estimated values of the tons of CO_2 released or absorbed annually in each process or the total amount stored in each reservoir; 1 ton of CO_2 contains approximately 550 pounds of carbon atoms.

Transfer of CO_2 to and from the atmosphere is an essential part of the carbon cycle. CO_2 enters the atmosphere primarily from the respiration and decay of organisms, from the burning of fossil fuels, and from the CO_2 dissolved in the oceans. Photosynthesizing green plants and the oceans absorb CO_2 from the atmosphere. Some of the carbon absorbed by plants is locked into long-term storage as fossil fuels, and some of the carbon absorbed by the ocean is stored in sediment beds of marine shells.

The transfer of CO_2 to and from organisms is believed to be nearly in balance over the year. The CO_2 content of the atmosphere, however, is increasing by several percent each decade. This increase is the result of man's industrial activities; it would be even greater if it were not for the CO_2 taken up by the oceans. (After "Carbon Dioxide and Climate" by Gilbert N. Plass, copyright © 1959 by Scientific American, Inc. All rights reserved.)

until 1974 has been estimated at 13 percent; this excess is projected to rise to about 22 percent by 1990. Nevertheless, the increase is only about half of the estimated human input into the atmosphere. Where is the remaining CO_2?

Most of the "missing" CO_2 is believed to be stored in solution in the oceans. This storage capability of the oceans and perhaps of vegetation has reduced the thermal atmospheric impact of the burning of fossil fuels, but a continued increase in CO_2 levels has the potential for creating significant global warming in the future.

BUDGETS OF ENERGY, WATER, AND OTHER MATERIALS

Inputs, outputs, storage, and balance constitute parts of the *budget* of a system, a concept that has proved useful in physical geography for analyzing the distribution of energy or materials in complex systems. As the name suggests, the budget concept employs a procedure analogous to a financial accounting where income and savings are balanced against expenses.

As a simple illustration of the budget concept, take the example of an ordinary outdoor swimming pool. The pool gains water from the city water supply and from rainfall. These constitute its input. Assuming that it does not have a drainage outlet, the pool loses water only by evaporation into the atmosphere and by the splashing of water onto the surrounding deck by swimmers. These constitute its output. According to the budget method, all of the water must be accounted

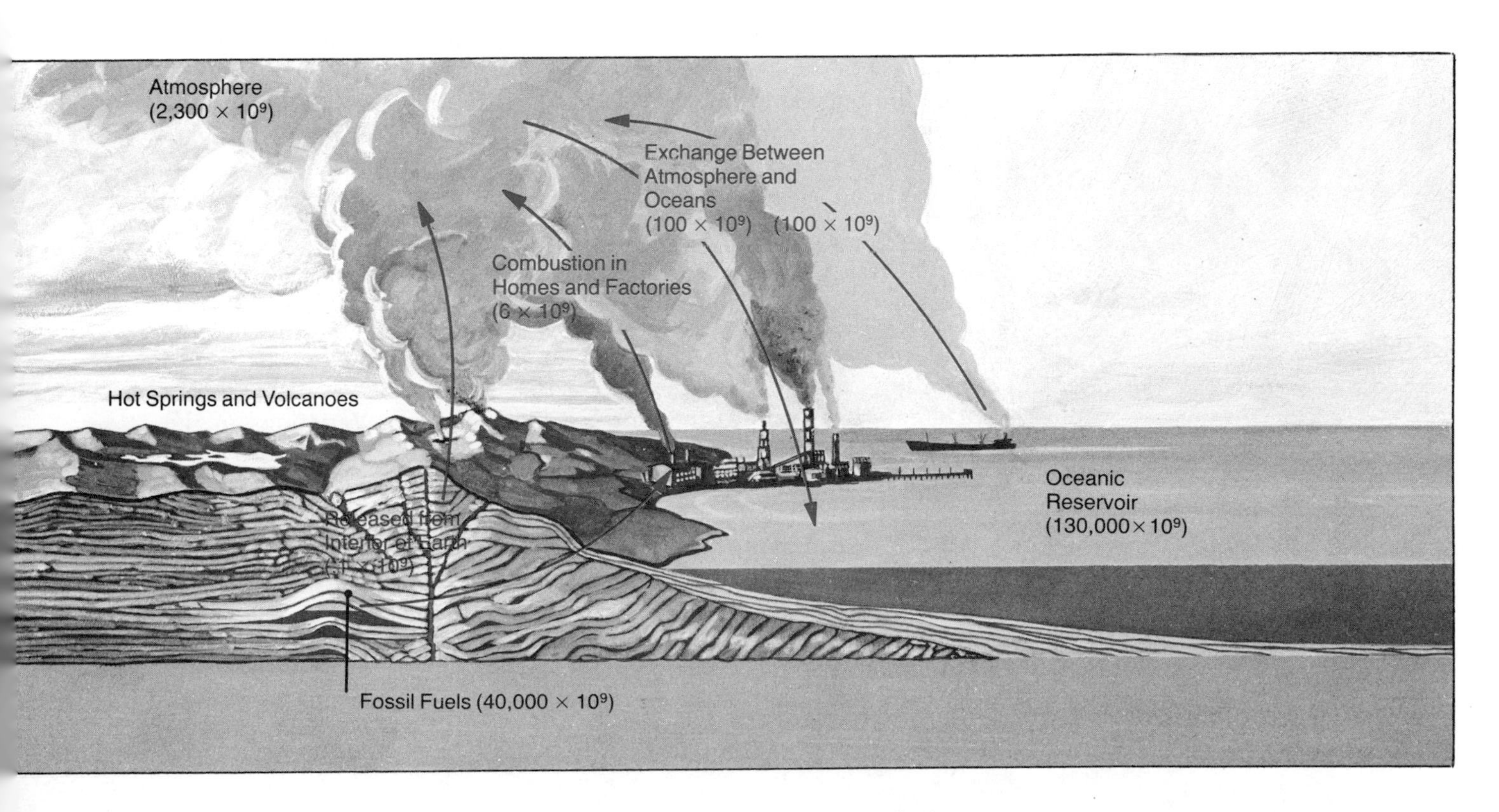

for. Differences between measured inputs and water-level changes represent unmeasured water losses through evaporation and splashing. However, for the budget to be a useful quantitative tool, a detailed knowledge of all interactions in the system is necessary. If the bottom of the pool is cracked, for example, the leaks need to be included as output in the budget. Similarly, rainfall that drains into the pool from the surrounding deck area must be figured as part of the pool's input.

The budget concept is particularly well adapted for evaluating water used by vegetation, the effects of droughts, irrigation needs, and runoff of flood waters into stream channels. The water budget can be calculated on both a local and a global level and is integrally linked with the concept of the hydrologic cycle. A more detailed discussion of these ideas is presented in Chapter 6.

Budgets of other materials can be equally interesting and useful. The carbon budget, discussed earlier in terms of the carbon cycle, indicates that CO_2 released into the atmosphere by the burning of fossil fuels is in part absorbed by the world's oceans. Similarly, budgets of river and beach sands have proven to be useful for analyses of navigational and recreational resource problems.

Budgets can also be drawn up to compare the input and output of energy in a system. The outdoor swimming pool, for example, absorbs incoming solar radiation, which it stores as heat energy. This can be budgeted against radiation losses to the atmosphere and the loss of latent heat through evaporation. Energy input and output will be reflected in the changing temperature of the pool water and can be calculated on both daily and annual time scales. As in the case of the water budget, all interactions of the system must be accounted for. For example, there may be some small exchange of energy through the walls of the pool to the surrounding soil. The most important application of the energy budget in physical geography is in calculating the sun's input of radiant energy to the earth and its availability to environmental systems. Examples of global and local energy budgets are examined in Chapter 3.

THE CLASSIFICATION OF PHYSICAL SYSTEMS

Although materials such as water are cycled through systems and can be used over and over again, the atmospheric delivery of water is not uniform over the surface of the earth. Energy too is available in different quantities in different geographical areas. Variations in the amounts of energy and moisture delivered to different regions cause systems in each region to follow various lines of development, which leads to a regional differentiation of climate, vegetation, soils, and landforms on the earth's surface. The earth exhibits such a variety of landscapes and climates that it would be impossible to study them in detail without some sort of classification system.

When the earth is considered as a whole, similarities as well as differences among regions become apparent. The southwestern deserts in the United States, for example, are similar in some ways to the deserts of central and southwestern Asia. One of the tasks of physical geography is to identify similarities and differences among regions and the natural systems at work within them. Systems that are subjected to similar inputs make similar responses; thus the problems encountered, the lessons learned, and the solutions developed in one area may be applied to others of a similar type. This can help us avoid the costly mistakes that have already led to environmental degradation in many parts of our planetary home. But before we can do this, we must devise classification systems that differentiate the important characteristics of the earth's natural systems.

In later chapters we will see examples of several classification schemes. Because classification schemes represent a human attempt to simplify nature, they cannot possibly take into account all the subtle variables in the physical environment. Any attempt to draw a boundary between one region and another neglects the continuous grading of properties that are actually observed in the world. Southern California, for example, receives only a few inches of rain each year and is "dry"; northern California receives many inches of rain each year and is "moist." Where is the dividing line between California's "dry" and "moist" regions? There is no clear boundary between one region and another, and classification schemes can never embody the infinite degree of variation actually present on the earth's surface. Nevertheless, they form a useful framework for thinking about the world.

SUMMARY

Physical geography is especially concerned with the processes which modify the surface of the earth. This work can be accomplished only by the expenditure of energy, which can take many forms. Energy forms which are of interest in physical geography include solar radiation, heat energy, gravitational or potential energy, kinetic energy, chemical energy, and the latent heat energy associated with the changes of phase of water. Radiant energy arrives from the sun continuously, and physical geographers study the various ways it is received at the earth's surface and converted to other energy forms.

Water is of fundamental importance in our planetary environment. By changing from a liquid to a gas, it serves to convert energy to a latent form, the latent heat of vaporization, which can then be transported vertically and horizontally in the atmosphere to be released elsewhere as the latent heat of condensation. Water also serves to moderate temperature changes at the earth's surface and in the atmosphere for both daily and annual time scales. Water is the source of most nutrients for plant and animal life. It also transmits work-producing kinetic energy in the forms of running water, slow-moving glacial ice, and waves.

In order to compensate for local energy gains and losses, the atmosphere transports energy—in the forms of sensible heat and latent heat present in water vapor—from place to place. Energy income, storage, and outgo, as well as the atmospheric and oceanic circulations, can be interpreted within a systems framework. The earth and atmosphere together have elements of both open and closed systems, receiving solar radiation as income and returning radiational energy to space, but exchanging virtually no matter with space. At or near the earth's surface, however, energy cascades through open systems and subsystems, in which there are large exchanges of materials as well. The outputs of one system usually provide the inputs of many other systems.

Materials such as water and carbon are continuously cycled and recycled through systems, but the time during which materials may remain in storage varies spectacularly. Although the sum total of matter remains essentially constant, materials may be combined and recombined into forms which are sometimes more and sometimes less valuable to human activity. The income, outgo, and storage of energy and materials can be evaluated within a budget framework. Varying amounts of energy and moisture are delivered to the regions of the earth, and varying amounts of materials are available for human use. The complex mosaic of distributional patterns necessitates regional classifications of climate, vegetation, soils, and landforms.

Review Questions

1. Describe the various forms of energy which are significant to physical geography.
2. What part does heat play in energy exchanges?
3. How are energy flows at the earth's surface and in the atmosphere related to changes of phase of water?
4. The heat capacity of water is greater than that of any other common substance. How is this fact related to climatic variability?
5. Explain what happens when a parcel of unconfined air is heated; when it is cooled.
6. How is the increase in atmospheric carbon dioxide due to the consumption of fossil fuels believed to be related to long-term climatic variability?
7. How might the river systems and oceans be altered by changes in the amounts of snow and ice stored in glaciers and ice sheets on the continents?
8. Discuss examples of inputs, outputs, and interactions of some local systems. Can they be applied to local and regional planning?
9. How might it be possible to evaluate the climate of a region in terms of energy and moisture availability?
10. On the basis of the heat capacity of water, how might the climate of coastal regions be different from that of the continental interior?
11. Why is it becoming increasingly vital for man to understand the budgets, cycles, and storage times of the various earth materials?

Further Reading

Chorley, Richard J., ed. *Water, Earth, and Man.* London: Methuen & Co., 1969. (588 pp.) This is a most useful collection of 38 selections, mostly by British authors, dealing with various aspects of the hydrologic cycle. The selections are at introductory and intermediate levels and complement a number of chapters in this text.

______, and **Peter Haggett,** eds. *Physical and Informational Models in Geography.* London: Methuen & Co., 1969. (315 pp.) This paperback includes Parts I, II, and V of the more comprehensive *Models in Geography,* first published in 1967. The focus is much more theoretical than in *Water, Earth, and Man.*

______, and **Barbara A. Kennedy.** *Physical Geography: A Systems Approach.* London: Prentice-Hall International, Inc., 1971. (370 pp.) This pioneering text ranges from the introductory to the advanced, with particular emphasis on geomorphology within systems-theory and statistical-model frameworks. There is some duplication of the two preceding works.

Davis, John C. *Statistics and Data Analysis in Geology.* New York: John Wiley & Sons, Inc., 1973. (550 pp.) This specialized but introductory text is included because the data processing techniques and statistical models are also appropriate for physical geography. The text is especially useful for those who wish to learn about a technique or skill on their own.

Weyl, Peter K. *Oceanography: An Introduction to the Marine Environment.* New York: John Wiley & Sons, Inc., 1970. (535 pp.) This book complements *Physical Geography Today* very well. The chapters on oceanic salts and geochemical cycles are especially pertinent to Chapter 2 of this text.

Case Study:

Understanding Nature

The world around us is a wonderfully complex mosaic, the patterns of which have aroused curiosity perhaps from the very beginnings of humankind. What is the sky, the earth, the ocean? Why do rivers flow as they do, and why does the land take such extraordinary configurations as it extends to the horizon? And how does each piece of the mosaic fit with all the others in the ever-changing natural world?

In our attempts to sort through the earth's myriad phenomena, we tend to construct *models* in our mind to help us understand what things and events are and how they are interconnected. With models we find a way of perceiving order in nature, of gaining deeper knowledge of why and how things happen as they do. Consider the model for "tree." One of us might think of a pine tree; another, an elm or a centuries-old redwood. Even though these are different trees, they nevertheless have enough similarities to be understood by all of us as belonging to the same category, and we share an understanding of that model.

Now consider the ecologist, whose model may be a little more complex: he views a "tree" as a marvelous microenvironment, where different creatures occupy different niches from roots to treetop, where intricate energy exchanges take place between the inhabitants and the surrounding environment. His model explains much more because he has probed a little deeper into a seemingly simple part of the world and discovered its larger dimensions.

All fields of inquiry claim many models that are useful in explaining the world on many levels, from many vantage points. For example, there are innumerable ways to look at and understand the nature of one of the earth's most important and most common substances, water.

(below left) *Wherever rainfall runs off the land, it erodes its own drainage system. This map, which is an example of a* graphic model, *shows the land area drained by the Mississippi River system (see also Chapter 12). Graphs, photographs, and diagrams are types of graphic models used to aid our conceptualization of the physical world.*

(below right) *Drainage patterns may also be simulated by computers, as this example of a* mathematical model *shows. This model is used to study the effects of various types of earth surface on the development of stream patterns and to predict changes that will occur in underground supplies of water as water is added or withdrawn. Computer models can also be used to simulate changes in variables such as precipitation or runoff to see the effect on a total system. In this illustration, the curved lines represent contours of elevation and the blue lines show the development of a drainage pattern on homogeneous material (see Chapter 12). The water flows down the slope in the direction of the arrow. Computer models are used in the oceanic and atmospheric systems as well and in other branches of science where complex systems with many variables are involved.*

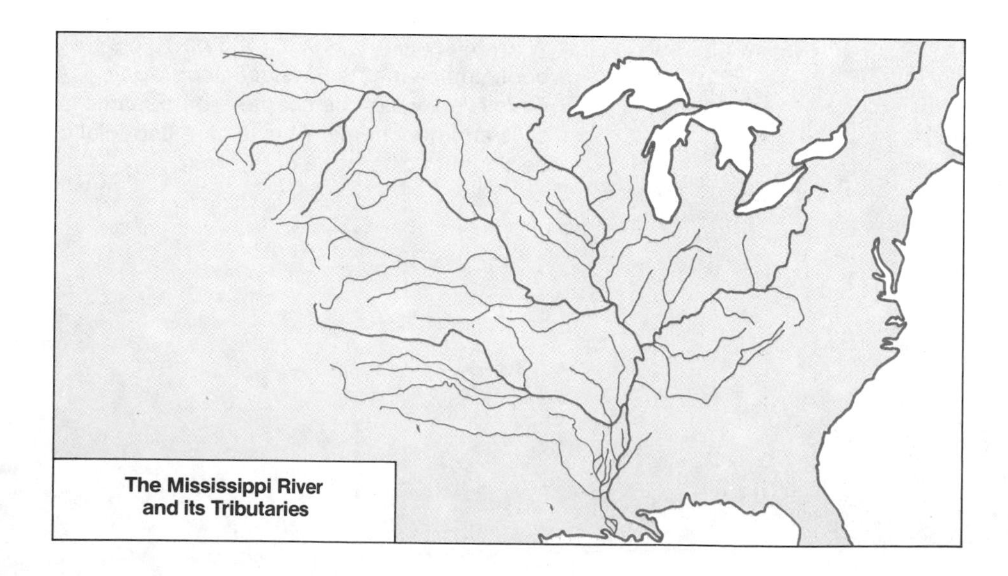

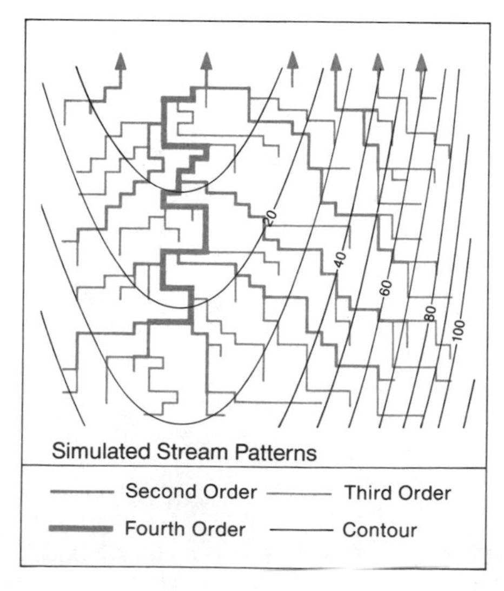

Water covers three-fourths of the earth's surface; it is also beneath the surface, permeating the soil and rock, and in the atmosphere, circulating in the form of vapor.

Different types of models describe water's different properties. A chemist uses a molecular model to explain water's fundamental makeup of two hydrogen atoms and one oxygen atom. A physical scientist is interested in water's unique ability to form three states of matter. A physical geographer, however, is concerned with water as a transporter of heat energy through the atmosphere and as an agent of change on the earth's surface. So models have been developed that describe how water flows across the earth in streams, rivers, and glaciers, how it circulates in the atmosphere and oceans, how it affects climate, vegetation, and soil formation, and how it shapes the landscape.

The concepts of "cycle" and "system" and "budget" are important components of the models in physical geography. The hydrologic cycle mentioned earlier and described more fully in later chapters is a model to explain how all the earth's original supply of water is continually recycled. The same water is transported time and time again from the land and oceans into the air, and then back to earth again. The local water budget is a model to describe the availability of water for various processes and purposes in the environment. Water shapes the earth's surface features, and several of the models useful for analyzing flowing water are illustrated here.

Keep in mind that the models described in this book are the physical geographer's perceptions of the world—a world we can come to know much better through educated eyes.

An actual replica to scale of the Mississippi River system is an example of a physical model. *It gives engineers answers to real life emergencies that arise in flood conditions. During the flood of 1973 (see Case Study following Chapter 12), this model was a valuable addition to computer models in calculating the amount of water-rise that would accompany the opening of a spillway. By using automatic instruments to make graphic recordings of water levels at critical points, inputs and outputs can be tested on the model, and the effects of proposed engineering projects can be predicted. Built and operated by the United States Corps of Engineers, the accurate miniature model of the river basin is laid out on 220 acres of land near Clinton, Mississippi. The portion shown in this photograph is the Atchafalaya River basin in southern Louisiana.*

Number 8 by Mark Rothko, 1952. (From the collection of Mr. and Mrs. Burton Tremaine, Meriden, Connecticut)

THREE

Energy and Temperature

All life processes are driven by complex exchanges of visible and invisible forms of radiant energy from the sun. The heat of the tropics and the cold of the polar zones are consequences of the earth-sun relation in space and of energy transfers on the earth.

RADIANT ENERGY FROM THE SUN is the most important source of power for the processes that shape the surface of the earth. Radiant energy lifts water from the oceans to the atmosphere, drives the passing clouds, keeps the earth at a comfortable temperature, and supplies energy to green plants. If the earth continually received solar energy and did not dissipate any energy, the oceans would boil and the land would be scorched. But the average temperature of the earth remains much the same from year to year, because, on the average, the earth returns as much energy to space as it receives from the sun.

The amount of energy the sun delivers to a particular location, however, depends on the position of the sun in the sky. When the sun is low in the sky, as at sunset or during a northern winter, its rays have less heating power. The atmosphere also influences the amount of energy received from the sun. The summer sun at midday brings sunbathers out to the beach, but if a passing cloud blocks the sun's rays, a chill sweeps over.

Each place on earth therefore does not receive an equal amount of solar radiation. Over the year, tropical regions receive a greater amount of energy than they radiate back into space; polar regions, on the other hand, return more energy to space than they receive from the sun. We know, of course, that the tropical regions are not progressively heating up nor the polar regions cooling off. This is because atmospheric and oceanic circulations redistribute the energy which the earth receives, bringing excess energy in the form of warm air and water from the tropics to the poles and transporting cool air and water toward the equator.

Since energy cannot be destroyed, all the solar radiation reaching the earth can be accounted for. In this chapter we will examine the interactions that influence the earth's gain and loss of radiant energy.

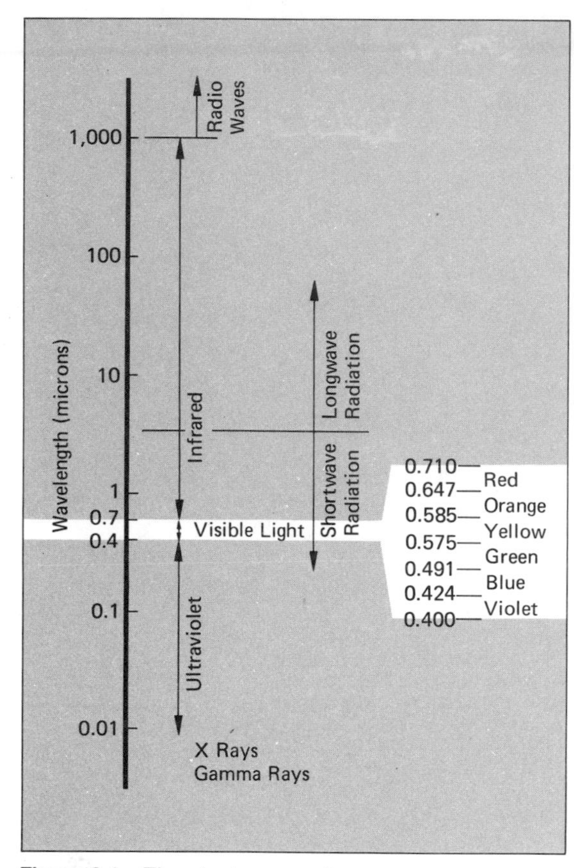

Figure 3.1 The electromagnetic spectrum is conventionally divided on the basis of wavelength. The divisions most important in physical geography are infrared radiation, visible light, and ultraviolet radiation. Infrared radiation consists of wavelengths intermediate between the wavelengths of radio waves and those of red light. Visible light consists of wavelengths between approximately 0.7 and 0.4 micron, and ultraviolet light corresponds to wavelengths shorter than 0.4 micron. At extremely short wavelengths, the ultraviolet portion of the spectrum merges into the x-ray and gamma ray divisions.

If radiation has a wavelength longer than 4 microns, it is called longwave radiation; if the wavelength is shorter than 4 microns, it is called shortwave radiation. The distinction is useful in physical geography because solar radiant energy input to the earth is mostly at wavelengths shorter than 4 microns, and the energy radiated from the earth is at wavelengths longer than 4 microns (see Figure 3.2).

THE ELECTROMAGNETIC SPECTRUM AND SOLAR RADIATION

Visible light is not the only kind of radiant energy which the sun emits; it is just the one form human eyes are capable of detecting. The principal characteristic that all forms of solar radiant energy have in common is that they travel in waves. Energy which is transferred in the form of waves is known as *electromagnetic radiation.*

Electromagnetic radiation can be classified according to wavelength—the distance between the crest of one wave and that of the next. The unit of measurement that is generally used to specify wavelength is the *micron,* which is equal to one millionth of a meter. Figure 3.1 displays the various types of radiation according to wavelength, an arrangement known as the *electromagnetic spectrum.* The figure shows that visible light from the sun possesses wavelengths ranging from 0.4 to 0.7 of a micron. We see the various wavelengths of visible light as different colors. These colors appear as a rainbow when solar radiation in the visible portion of the spectrum is refracted, or bent, and then reflected by water droplets in the atmosphere. Although we cannot sense them with our eyes, there are important classes of solar radiation with wavelengths longer and shorter than those of visible light. Ultraviolet radiation, X-rays, gamma rays, and cosmic rays are characterized by wavelengths shorter than those of visible light, and infrared radiation and radio waves have longer wavelengths than visible light.

Every object, whether warm or cold, emits electromagnetic radiation because of the jostling of its molecules. The kind of radiation an object emits, however, depends on the temperature of the radiating surface. One of the basic laws of radiation states that the wavelength of maximum emission decreases as the temperature of the radiating surface increases. In other words, the hotter the object, the shorter the wavelength of the radiation which it emits. For example, when you turn on the heating element of an electric stove, the coil remains dull black while it is warming up. During this time it is emitting infrared radiation, which you can feel as heat but which you cannot observe visually. As it grows hotter, the wavelength of maximum emission decreases, shifting over into the visible portion of the spectrum; the coil glows dull red, and then bright red. If the coil could be heated still further without melting it, it would begin to glow yellow-white like the sun.

But on the surface of the earth or in the atmosphere, such temperatures occur only under man-made conditions. Normally, environmental temperatures range between 233° and 313°Kelvin (K), or −40° and 104°F. (See Appendix I for an explanation of the Kelvin temperature scale.) Objects at such temperatures emit electromagnetic radiation with a wavelength of maximum emission of 4 microns or more. The surface of the sun, on the other hand, has a temperature of approximately 6,000°K (10,800°F). At this temperature, most of its radiation is at wavelengths of less than 4 microns. Hence, we can distinguish between solar radiation and radiation emitted by terrestrial sources as shortwave and longwave radiation respectively.

Another law of radiation which is significant to the study of physical geography states that an object's total emission over the entire electromagnetic spectrum is directly related to the temperature of the object in degrees K, raised to the fourth power. In other words, a very hot object, such as the sun, not only has a shorter wavelength of maximum emission than a cooler object, such as the earth, but it also emits radiation with far greater intensity. We shall see the importance of these facts when we explore the earth-sun relationships which control the distribution of incoming solar radiation over the globe.

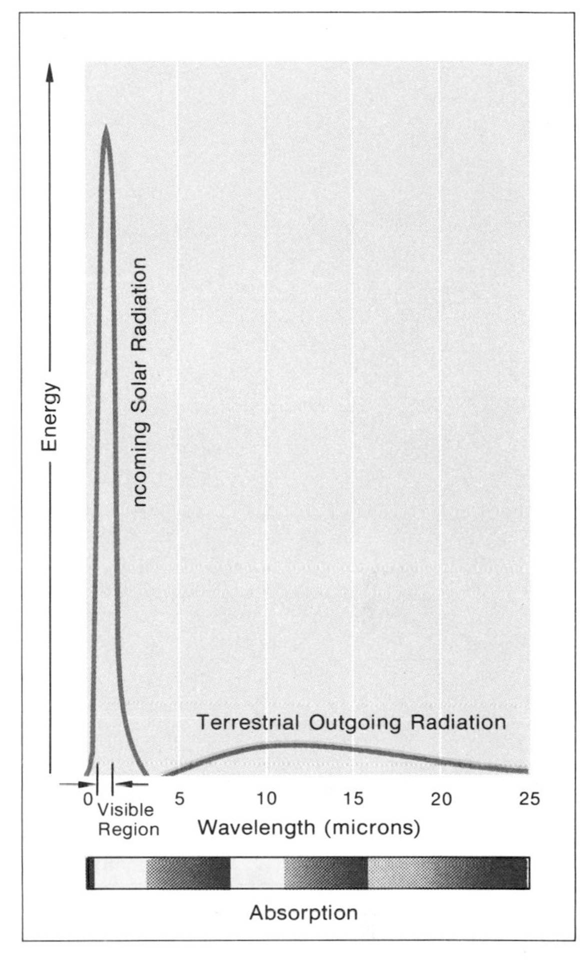

Figure 3.2 The solar radiant energy received by the earth is distributed over a band of wavelengths. Because of the high surface temperature of the sun, 6,000° K(10,800°F), energy from the sun is emitted primarily at wavelengths shorter than 4 microns, with much of the energy concentrated in the visible region of the spectrum. In contrast, the longwave radiation emitted by the earth is confined to wavelengths longer than 4 microns and has a broad peak at about 10 microns, because of the earth's comparatively low average surface temperature of 285°K (54°F).

The lower portion of the figure indicates the degree to which atmospheric gases, primarily carbon dioxide and water vapor, absorb electromagnetic energy near the earth's surface. Wavelength bands of strong absorption are shown in red, and bands of relative transparency are indicated by yellow. The lower atmosphere is relatively opaque to longwave radiation, so that much of the radiant energy emitted by the earth's surface is absorbed there. The atmosphere is relatively transparent to electromagnetic radiation in the band from 8.5 to 11 microns, and radiant energy in this band can escape to space if the sky is clear. (After Dobson, 1963)

SOLAR ENERGY INPUT TO THE EARTH

The sun emits radiant energy at the same average rate in all directions. Because the earth is relatively small, it intercepts only a small fraction of the total energy the sun emits. If we divide the total radiant energy from the sun into a billion equal units, the earth intercepts only ½ of 1 unit. But this small fraction is still a huge quantity of energy, amounting to 2.6×10^{18} calories per minute. The solar radiant energy intercepted by the earth in 1 minute is about equal to the total electric energy generated on the earth in 1 year.

Not all of this radiant energy reaches the surface of the earth because the earth's atmosphere modifies the solar radiation which passes through it. For example, a considerable portion of the ultraviolet radiation falling on the earth is filtered out by the ozone layer in the upper atmosphere. Even before this process takes place, however, two factors operate to control the distributional patterns of incoming solar radiation at the top of the atmosphere—the number of hours of daylight and the altitude at which the sun appears above the horizon. Both of these factors are determined by the earth-sun relationship.

The Earth-Sun Relationship

Like every planet, the earth follows a fixed path, or *orbit*, around the sun. The orbit is elliptical, so that the distance of the earth from the sun varies slightly through the year, causing the radiant energy input to vary as well. However, since the orbit is nearly circular, the earth-sun distance never varies more than 1.8 percent from the average of 149.5×10^6 km (92.9×10^6 miles), and the range of fluctuation in solar energy received is not very great either, departing no more than 7 percent from the average. The earth intercepts the greatest amount of solar radiation in early January, which is when its orbit brings it closest to the sun, and least in July, when it is farthest from the sun.

In addition to annual variations in energy input caused by the earth's elliptical orbit, there is a daily cycle caused by the rotation of the planet on its axis. The earth completes approximately 365¼ rotations for each revolution about the sun. Because the earth is essentially spherical in shape, the sun illuminates only half of the globe at any time. The boundary line between light and dark halves is the *circle of illumination* (see Figure 3.3).

The Earth's Tilted Axis

As the earth rotates on its axis, each part of its surface becomes light, then dark, and then light again. If the earth's axis of rotation were perpendicular to the plane of its orbit, every place on the earth would have 12 hours of daylight and 12 hours of darkness each day of the year, and the seasonal variation from long summer days to brief winter days would not occur. But the earth's axis is not perpendicular to the orbital plane; it tilts at an angle of 23½° from the perpendicular, so the duration of daylight at a given location varies over the year.

Figure 3.3 illustrates that as the earth journeys around the sun, the orientation of the earth's axis remains essentially fixed. Both the seasons and the variation in daylight hours from equator to poles are direct consequences of this phenomenon. In addition to having a direct bearing on the length of daylight, the fixed tilt of the earth's axis affects the elevation of the sun above the horizon. About December 21 or 22, for example, the sun appears to be directly overhead at noon at latitude 23½°S. This parallel is called the *tropic of Capricorn*, and the particular instant when the sun is directly overhead at this parallel is known as the *winter solstice*. The sun never appears directly overhead at any latitude south of the tropic of Capricorn. In the northern latitudes at winter solstice, the sun is low in the southern sky, and from any location south of the tropic of Capricorn, the sun always appears toward the north. The European explorers who navigated south of the tropic of Capricorn for the first time were surprised to see the sun on the right at midday as they sailed toward the west.

As Figure 3.3 shows, the proportion of each parallel of latitude that lies in the illuminated part of the globe varies as one travels from north to south. The fraction of the parallel that is illuminated indicates the relative length of day and night at that latitude. For example, consider the situation at winter solstice, the shortest day of the year in northern latitudes. North of the Arctic Circle (latitude 66½°N), the parallels lie completely in darkness and the sun is not visible at all. Areas situated within the Arctic Circle, such as Greenland and northern Scandinavia, experience 24 hours of darkness on this date. South of this latitude, an increasingly greater proportion of each parallel lies in sunlight, which means that, as one moves southward, the days become longer and longer. Night continues to be longer than day, however, until one arrives at the equator (0° latitude) (see Table 3.1). Exactly half the equator lies in the illuminated zone and half in the zone of darkness, so that day and night are of equal duration. South of

Table 3.1. Length of Daylight During Winter Solstice*

Latitude	Daylight
90°N	0
80°N	0
70°N	0
60°N	5 hr 52 min
50°N	8 hr 4 min
40°N	9 hr 20 min
30°N	10 hr 12 min
20°N	10 hr 55 min
10°N	11 hr 32 min
0°	12 hr 7 min
10°S	12 hr 28 min
20°S	13 hr 5 min
30°S	13 hr 48 min
40°S	14 hr 40 min
50°S	15 hr 56 min
60°S	18 hr 8 min
70°S	24 hr
80°S	24 hr
90°S	24 hr

* Not counting twilight. Daylight includes time when at least the upper edge of the sun's disk is above the horizon.
Source: Adapted from List, Robert J. (ed.) 1951. *Smithsonian Meteorological Tables.* 6th rev. ed. Washington, D.C.: Smithsonian Institution Press.

the equator, proportionately more of each parallel lies in the illuminated zone, and therefore day is longer than night. Finally, south of the Antarctic Circle (latitude 66½°S), the sun is always above the horizon at winter solstice, and there is no night at all.

As the earth traverses its orbit, the relation of its axis to the sun changes. Figure 3.3 shows that about June 21, six months after winter solstice, the sun is overhead at noon at latitude 23½°N, the parallel known as the *tropic of Cancer.* The instant the sun is overhead is called the *summer solstice.* At that time of year, the extreme north receives constant illumination, while latitudes south of 66½°S are in constant darkness.

At two intermediate times—about March 21 and September 21—the sun is overhead at noon at the equator. The instant the sun passes over the equator is called the *vernal equinox* (March) or the *autumnal equinox* (September). Only at these times of the year does the circle of illumination pass through the poles and coincide with meridians of longitude (see Figure 3.3). At the time of the equinoxes, day and night are of equal duration everywhere on earth.

The altitude of the sun above the horizon determines the intensity of the solar beam reaching the earth. The position of the sun in the sky at any time and at any location can be calculated by taking into account the position of the earth in its orbit, the tilt of its axis, and its spherical shape. The results of such calculations can be charted for each latitude. The two solar position charts shown in Figure 3.4, p. 60, give the sun's position in terms of its *altitude* and *azimuth,* a system for specifying location in the sky analogous to the system of latitude and longitude on the earth.

Suppose you stand facing north with your arm outstretched, pointing north, and then turn your body clockwise and raise your arm until you are pointing at the sun. The angle through which you turn your body is the sun's azimuth, and the angle through which you raise your arm is the sun's altitude. Azimuth is specified as an angle between 0° and 360°, starting from north and going clockwise. Altitude is specified as an angle between 0° and 90°, starting from the level horizon. Thus an azimuth of 90° and an altitude of 45° mean that the sun is due east and is halfway between the horizon and the point directly overhead.

One of the things you can read from a solar altitude and azimuth chart is the maximum height of the sun above the horizon. Using the charts in Figure 3.4, you can verify that at summer solstice the maximum altitude of the sun above the horizon is about 83° in New Orleans, Louisiana, and about 68° in Minneapolis, Minnesota. At winter solstice, the maximum altitude of the sun is about 36° in New Orleans and only about 21° in Minneapolis.

How does the sun's altitude affect the intensity of solar radiation falling on the earth's surface? Imagine holding a square of cardboard in a stream of sunlight. If you hold the card perpendicular to the sun's rays, the card intercepts the maximum possible amount of energy, as shown in the bottom left section of Figure 3.3. If you turn the card edge-on toward the rays, the card receives little or no energy. At intermediate angles, intermediate amounts of energy strike the card.

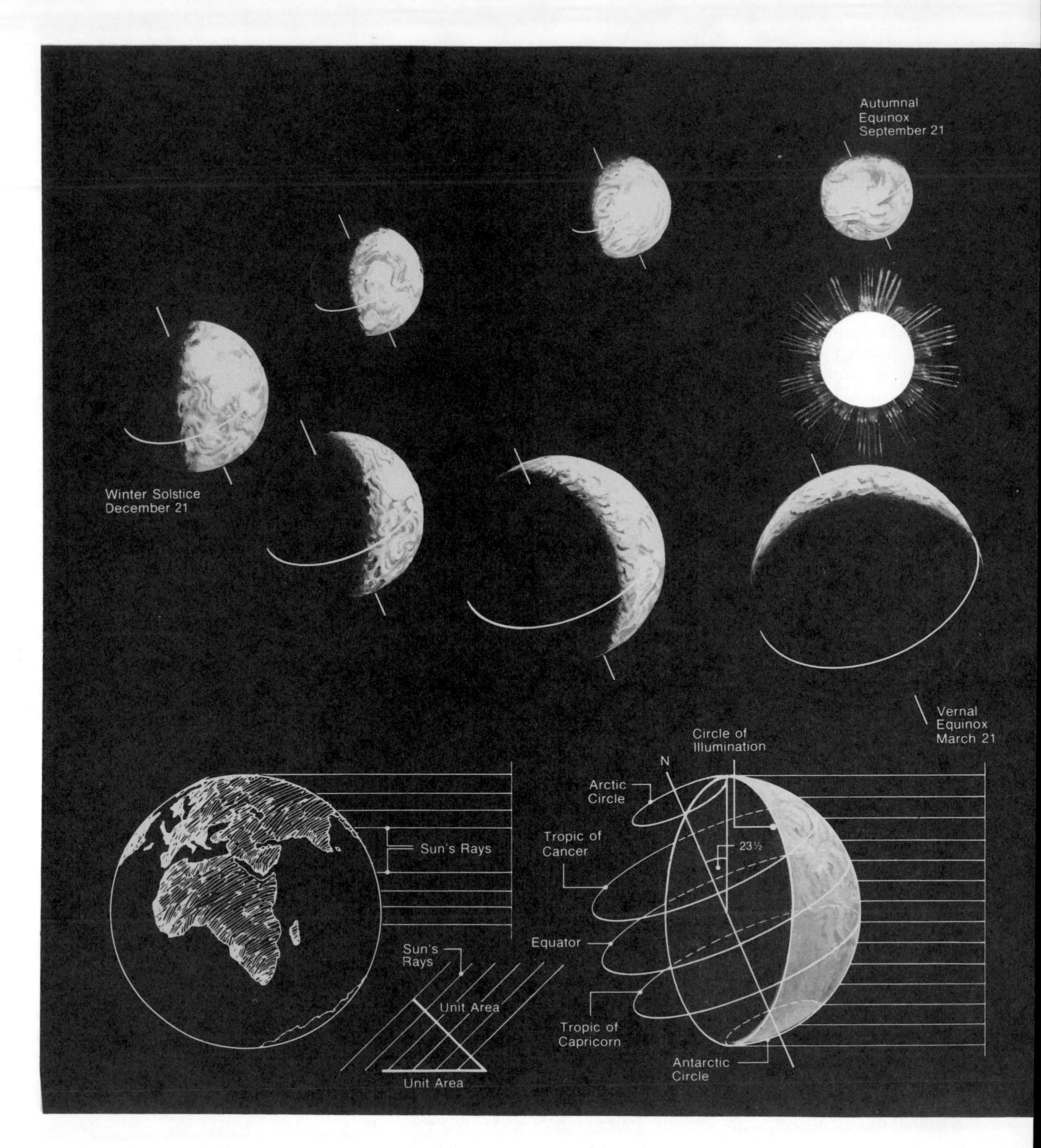
Autumnal
Equinox
September 21
Winter Solstice
December 21
Vernal
Equinox
March 21
Circle of
Illumination
N
Arctic
Circle
Tropic of
Cancer
23½
Sun's Rays
Equator
Sun's
Rays
Unit Area
Unit Area
Tropic of
Capricorn
Antarctic
Circle

Figure 3.3 (top) The amount of solar energy that reaches a given location at the top of the atmosphere depends on the distance between the earth and the sun and on the orientation of the earth. The top half of the diagram shows the earth at twelve different times during the year. The maximum distance of the earth from the sun is 152×10^6 km (94.4×10^6 miles) and occurs early in July. The minimum distance is 147×10^6 km (91.4×10^6 miles) and occurs early in January. Because solar energy input to the earth varies inversely as the square of the earth-sun distance, the amount of solar radiation the earth receives is 1.07 times greater in early January than in July. The earth's axis makes an angle of $23\frac{1}{2}°$ with a perpendicular to the plane of the earth's orbit. As the earth circles the sun, the orientation of the axis remains the same. At the time of the winter solstice, the sun is overhead on the tropic of Capricorn in the southern hemisphere, and at the time of the summer solstice, it is overhead on the tropic of Cancer in the northern hemisphere. At the time of the vernal and autumnal equinoxes, the sun is overhead at the equator.

(bottom left) The amount of solar radiant energy that a unit area on the earth's surface intercepts depends on the angle between the sun's rays and the plane of the area. As the sketch shows, a unit area intercepts the greatest amount of radiant energy when the area is perpendicular to the rays.

(bottom center) The solar radiant energy input to a location on the earth during a given 24-hour period depends partly on the duration of daylight. At any single moment, one-half of the earth is illuminated by the sun. The circle of illumination is the boundary between the light and dark regions of the earth. Because of the tilt of the earth's axis, the duration of daylight is in general different at different locations. The diagram illustrates the situation at winter solstice.

(bottom right) The surface of the earth has in principle been divided into two sets of intersecting grid lines to enable locations on the earth to be specified accurately and without ambiguity. The *parallels of latitude* are circles parallel to the plane of the equator. A particular parallel of latitude is specified by its angular distance north or south of the equator, with the equator designated as latitude 0°. The north pole has latitude 90°N, for example.

The second set of grid lines, the *meridians of longitude,* are circles drawn with the earth's axis as a diameter. One meridian, chosen to be the *prime meridian,* is designated as 0° longitude. In the United States, and in many other countries, the prime meridian is taken to be the meridian on which the astronomical observatory in Greenwich, England, is situated. Any other meridian is specified by its angular distance east or west of the prime meridian, ranging from 0° to 180°.

For precision, a degree can be divided into 60 minutes of angular measure, and a minute can be divided into 60 seconds. The latitude and longitude of Washington, D.C., for example, can be written 38 degrees 54 minutes North, 77 degrees 2 minutes West, or in abbreviated form as 38°54′N, 77°02′W.

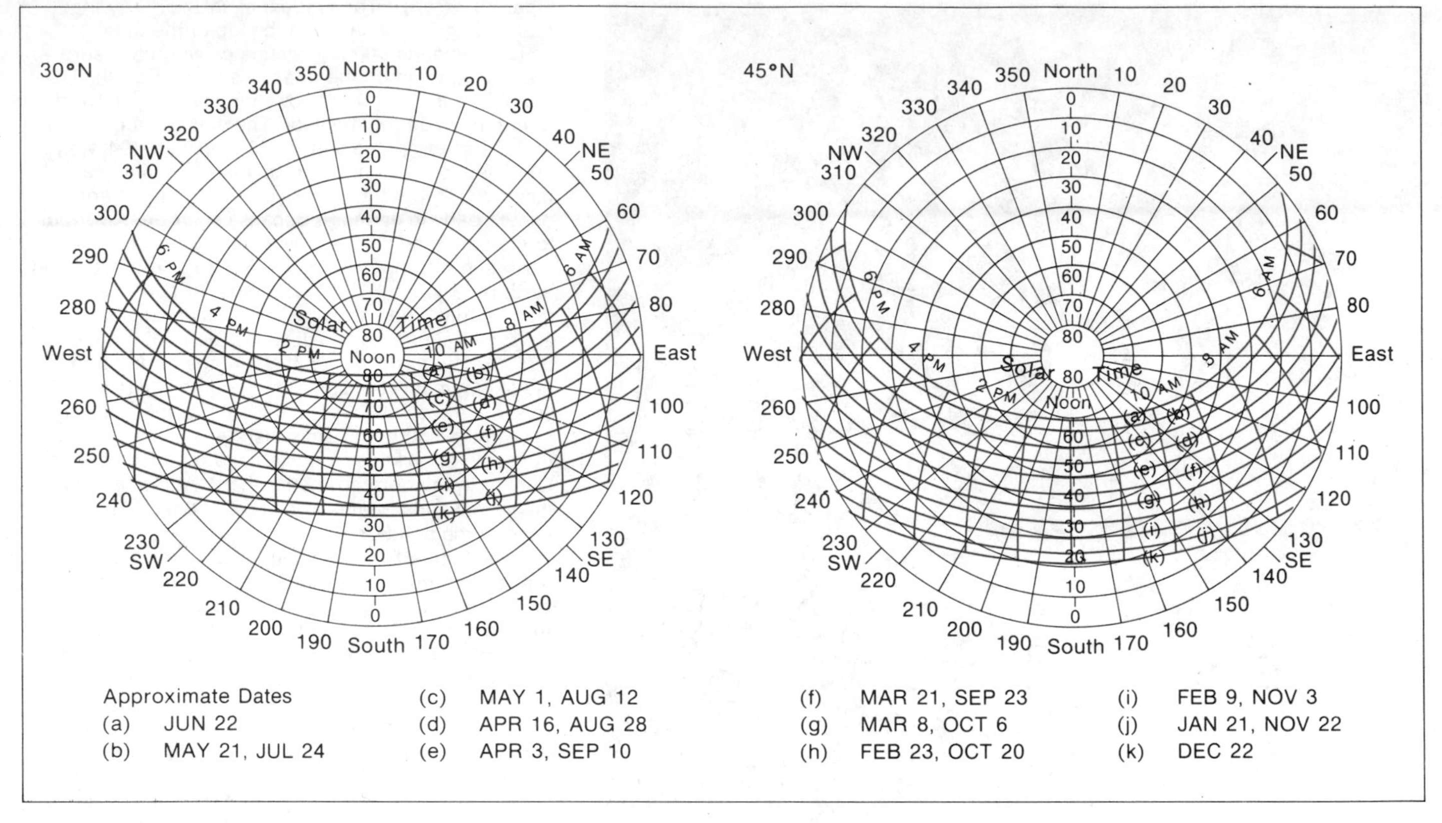

Figure 3.4 The solar altitude and azimuth chart on the left is for latitude 30°N, the latitude of New Orleans, Louisiana, and the chart on the right is for latitude 45°N, the latitude of Minneapolis, Minnesota. The charts represent the dome of the sky as it would appear to an observer standing at the center of the circle. The numbers around the rim of the circle stand for azimuth, or degrees from north. The vertical line of numbers gives the altitude of the sun in degrees, starting with 0° at the horizon. To find the sun's position in the sky at a given time for these two latitudes, first select the red grid line corresponding to the date, and then find the point of intersection with the red line representing the time of day. Intermediate dates and times can be estimated by interpolation. The sun's altitude can be read from the scale of concentric circles, and the sun's azimuth can be read from the scale of radial lines. On March 21 at latitude 30°N, for example, the sun's altitude at 11:00 A.M. is 57° above the horizon, and its azimuth is approximately 150° from true north.

The duration of daylight can be estimated from solar position charts by finding the times on a given day when the sun's altitude is greater than 0°. On December 22 at latitude 45°N, the sun has an altitude greater than 0° between 7:45 A.M. and 4:15 P.M. Because these charts show the geometric position of the center of the sun, the duration of sunlight is a few minutes longer than these times suggest. (After List, 1951)

A farmer concerned about his wheat crop wants to know how much energy falls on his particular acres, not how much total energy the earth receives. The farmer's field is analogous to the card. When the sun is low in the sky, less energy is delivered to a given area than when the sun is high. The relative amount of energy falling on a particular area depends on the angle at which the incoming rays strike the area. Sunbathers recognize this principle by propping themselves in lounge chairs to be more perpendicular to the sun's rays. The strawberry fields in Switzerland and the vineyards in West Germany are planted on steep valley slopes for the same reason. When the sun is 60° above the horizon, a patch of ground on a steep 30° slope facing the sun receives nearly 15 percent more energy input than a horizontal patch of the same area.

Calculating Daily Solar Radiation at the Top of the Atmosphere

Radiant energy from the sun strikes a unit area at the top of the earth's atmosphere at a rate known as the *solar constant*. Assuming that the unit area is perpendicular to the sun's rays and that the earth and the sun are separated by their average distance, the accepted value of the solar constant is 1.94 calories per sq cm per minute. Scientists, however, generally express the solar constant in a simpler form by employing a measurement known as the *langley*. One langley equals 1 calorie per sq cm. The solar constant can therefore be expressed as 1.94

langleys per minute. This value is an estimate, but new research is expected to refine it further.

Consider now a unit area of 1 sq cm at the top of the atmosphere. The amount of radiant energy striking the area depends on the value of the solar constant, the duration of daylight, the sun's altitude, and the true distance between the earth and the sun. The total energy striking the area in one day can be found in principle by calculating the position of the sun for every minute of the day and then using the angle appropriate to each time to find the amount of energy delivered to the area in each minute. The total energy delivered in a day is the sum of the energies received during each minute.

Figure 3.5 shows the total daily solar radiation input to a horizontal square centimeter at the top of the atmosphere for any latitude throughout the year. The number on each curve represents energy input in langleys per 24 hours. For example, the energy input at latitude 40°N on November 1 is approximately 480 langleys in 24 hours. The shaded areas on the chart represent times of continuous darkness.

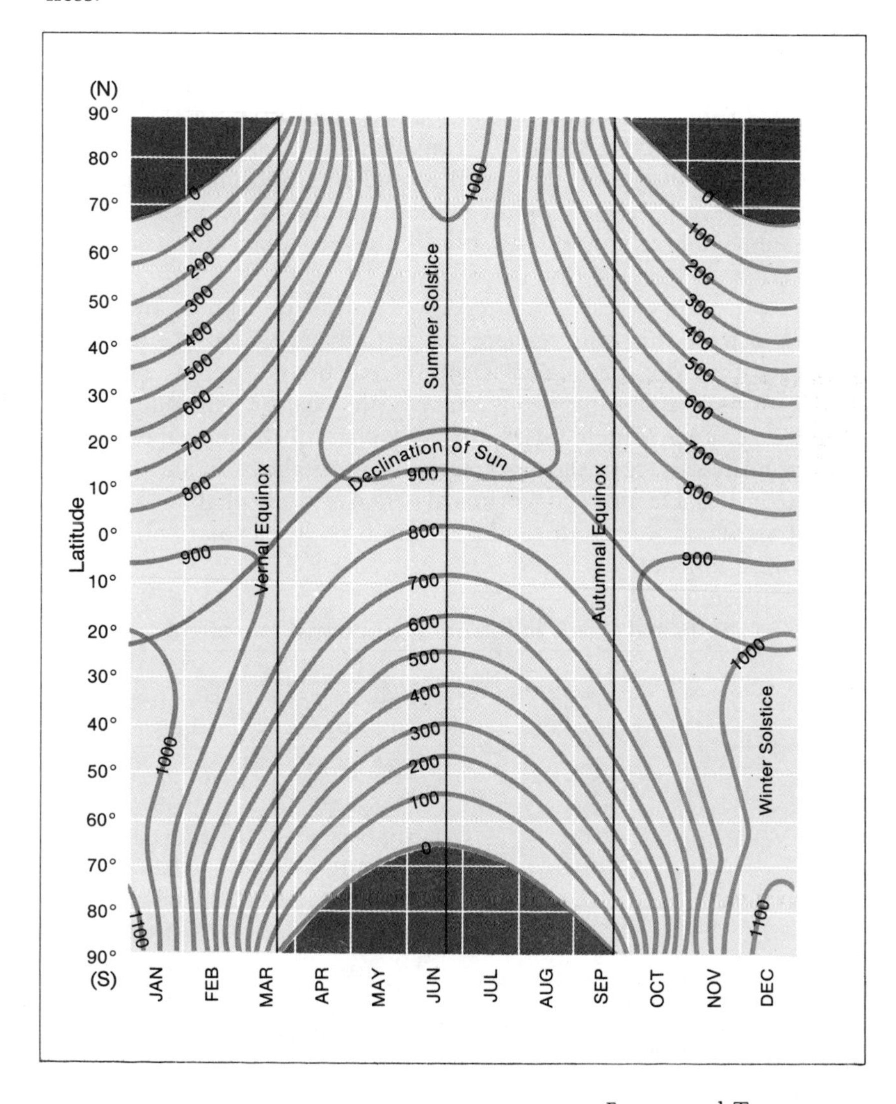

Figure 3.5. This graph shows solar radiant energy input to the top of the atmosphere. The horizontal scale is marked off in months of the year, from January at the left end of the scale through December at the right. The vertical scale lists latitude north and south of the equator. The curved contour lines give the solar radiant energy input to the top of the atmosphere in units of langleys per 24 hours, based on a value of 1.94 langleys per minute for the solar constant. On the equator at summer solstice, for example, the radiant energy input is approximately 800 langleys in 24 hours. The shaded areas of the diagram poleward of latitude 66½° represent times of perpetual darkness when there is no solar radiant energy input.

The energy input into the northern and southern hemispheres is not perfectly symmetrical. At summer solstice in the northern hemisphere (June 21), locations at latitude 15°N receive about 900 langleys per 24 hours, but at winter solstice in the southern hemisphere (December 21), locations at latitude 15°S receive about 970 langleys per 24 hours. The reason is that the earth is nearer the sun in January than in July, and the energy input to the earth is 7 percent higher in January than in July.

The solar declination, the orange curve in the graph, represents the latitude where the solar altitude at noon is 90°; the solar declination ranges between 23½°N and 23½°S, with the declination at 0°, the equator, on the equinoxes.

The amount of energy input varies more at the poles throughout the year than at the equator, where the input is relatively uniform from day to day. Locations at the equator never receive as much as 900 langleys in 24 hours, but the energy input near the south pole sometimes exceeds 1,100 langleys in 24 hours. The reason for this is that at winter solstice the south pole receives continuous sunlight the whole day, while at the equator daylight never much exceeds 12 hours. Averaged over a full year, however, locations at the equator receive nearly two and one-half times as much energy as areas near the south pole.

Interaction of Solar Radiation With the Atmosphere

Before solar radiation can reach the surface of the earth, it must pass through the atmosphere. The earth's atmosphere is a mixture of gases, particles, and clouds that responds selectively to electromagnetic radiation at various wavelengths. The atmosphere therefore plays a central role in determining the earth's response to solar radiation.

The chemical makeup of the atmosphere is given in Table 3.2. We have already seen in Chapter 1 how events early in the earth's history caused the introduction of nitrogen, oxygen, and carbon dioxide into the atmosphere. Neon, helium, and krypton are inert, chemically unreactive gases, and the small amounts present today are probably remnants from the earth's original atmosphere. Argon, another inert gas, but one which is more abundant, seems to have been produced largely by the decay of radioactive potassium in the earth's crust.

The nitrogen, oxygen, and inert gases in the atmosphere are well mixed and have the same relative proportions at 25 km (15 miles) altitude as they do at sea level. Other constituents, such as carbon dioxide, water vapor, and ozone, vary with location, altitude, and time. Water vapor, which enters the atmosphere through evapotranspiration at the surface of the earth, rarely appears above an altitude of 10 km (6 miles). Ozone occurs primarily at an altitude of about 25 km (15 miles), where it is produced by reactions between solar radiation and oxygen molecules.

Table 3.2. Principal Gases in the Earth's Lower Atmosphere

Gas	Molecular Formula	Number of Molecules of Gas per Million Molecules of Air	Proportion (percent)
Nitrogen	N_2	7.809×10^5	78.09
Oxygen	O_2	2.095×10^5	20.95
Water Vapor	H_2O	variable	variable
Argon	Ar	9.3×10^3	0.93
Carbon Dioxide	CO_2	330.0	0.03
Neon	Ne	18.0	1.8×10^{-3}
Helium	He	5.0	5.0×10^{-4}
Krypton	Kr	1.0	1.0×10^{-4}

Source: Adapted from List, Robert J. (ed.) 1951. *Smithsonian Meteorological Tables.* 6th rev. ed. Washington, D.C.: Smithsonian Institution Press.

The atmosphere is conventionally divided into "spheres," primarily according to variations of temperature with altitude. In the *troposphere*—the portion of the atmosphere nearest the earth's surface—temperature decreases as altitude increases. On the average, there is a decrease in temperature of 6.5°C for every km (3.6°F for every 1,000 ft) increase in height. Temperature in the *stratosphere,* on the other hand, increases with altitude, and the point of changeover between the two spheres is known as the *tropopause.* Figure 3.6 shows further temperature/altitude relationships at higher elevations in the meso- and thermospheres.

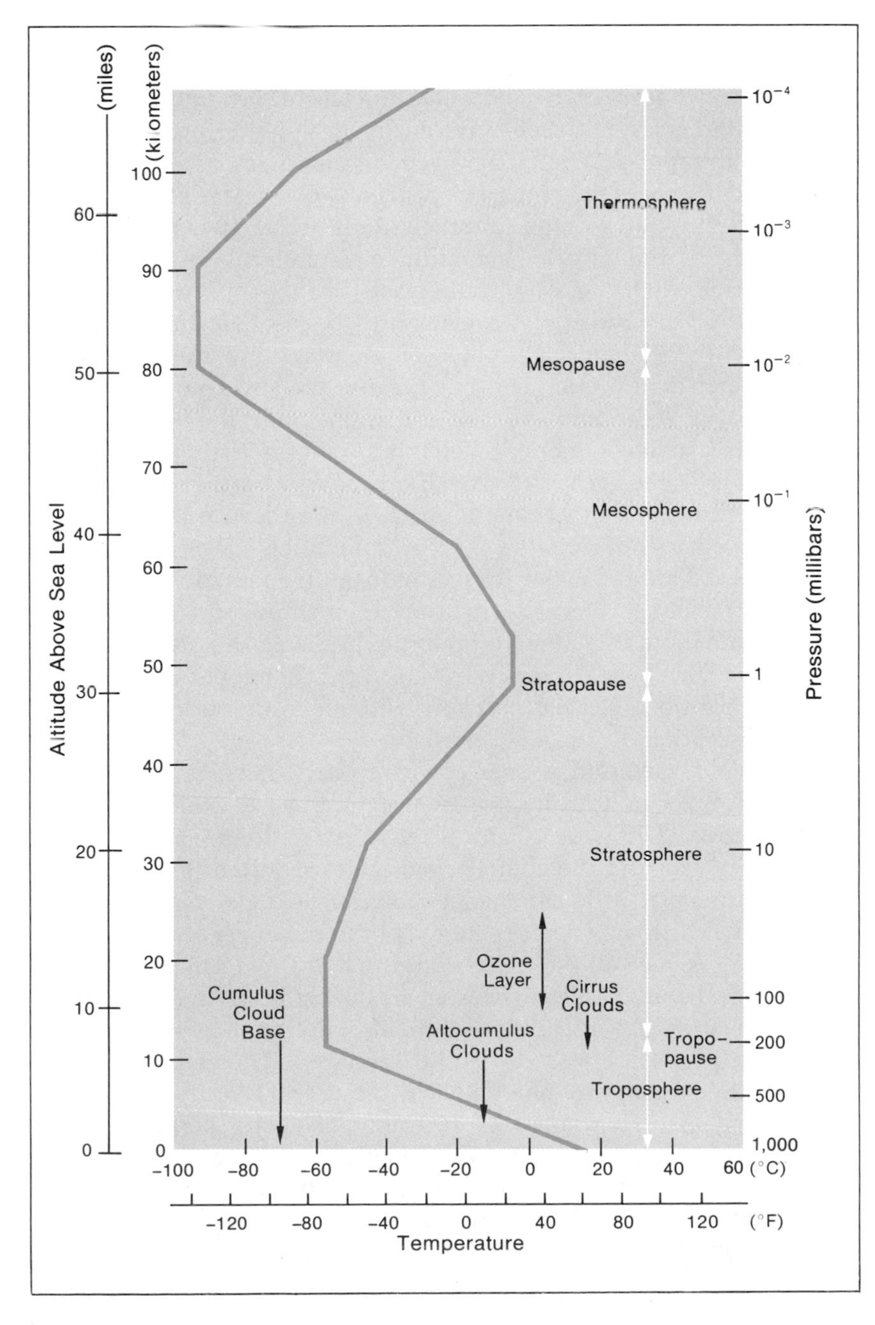

Figure 3.6 The atmosphere is conventionally divided into layers, primarily according to the variation of temperature with altitude. The *troposphere,* which is the lowest layer of the atmosphere, decreases in temperature with increased altitude at the average rate of 6.5°C per km (3.6°F per 1,000 ft). The troposphere is warmest at the earth's surface, on the average, because it is heated primarily from below by the transfer of heat energy from the surface. Air in the troposphere contains water vapor, and most clouds and weather phenomena are confined to this layer. Temperature increases with increased altitude in the *stratosphere,* largely because atmospheric gases such as ozone absorb a portion of the radiant energy incident from the sun. Air in the stratosphere is nearly devoid of water vapor, so clouds seldom form there.

The temperature curve in the diagram is made up of straight line segments and corresponds to the values assumed for the 1962 United States standard atmosphere. A standard atmosphere is meant to represent the average state of the atmosphere, but the actual temperatures and pressures on a given day may differ from the standard values. (After The United States Air Force, 1965)

Electromagnetic radiation from the sun interacts with gas molecules in the atmosphere by *absorption* and by *scattering*. In absorption, which is usually a selective process dependent on wavelength, a molecule takes up radiant energy and converts the energy to a different form. In scattering, which can occur for any wavelength, gas molecules divert beams of solar radiation into other directions. Because of scattering, the earth's surface receives light from the sky in addition to light coming directly from the sun. The moon, in contrast, has no atmosphere to scatter sunlight, which is why shadows on the moon are so harsh. Gas molecules in the earth's atmosphere scatter blue light much more effectively than red light, so most of the light scattered from the sky is blue.

All wavelengths of electromagnetic radiation are scattered to some extent by dust particles and water droplets in the atmosphere as well as by the gases, but each wavelength of radiation behaves differently in absorption. When longwave radiation is absorbed by a gas, for example, it causes the molecules of the gas to vibrate. The vibrating molecules collide with one another and heat energy is produced. Energy transfer from longwave radiation to gas molecules is selective, however. Each kind of gas molecule reacts strongly only with longwave radiation at particular wavelengths. Water vapor and carbon dioxide are the principal absorbers of longwave radiation (wavelengths greater than 4 microns). But because the sun's surface temperature is so high, most of the radiation from the sun is at wavelengths less than 4 microns, which is too short for any molecules in the atmosphere to act as effective absorbers. As we shall see later in this chapter, nearly all the radiation that is absorbed by the earth's atmosphere comes not from the sun but from much cooler terrestrial surfaces, which are the major sources of longwave radiation.

The wavelengths of visible light are even shorter than those of longwave radiation, therefore no molecule in the earth's atmosphere can absorb visible light effectively. Visible light therefore travels long distances through a clear atmosphere without losing much energy through absorption.

Ultraviolet radiation is energetic and often transfers sufficient energy to a molecule to rearrange its structure or even to break the molecule apart. Ultraviolet light is harmful to living organisms, because it can disrupt the complex molecules required for life processes. Fortunately, little ultraviolet radiation actually reaches the surface of the earth; most of it is filtered out in the upper atmosphere. In the outer layers of the earth's atmosphere, molecular oxygen strongly absorbs ultraviolet light with wavelengths shorter than 0.2 micron. The ozone layer strongly absorbs ultraviolet light with slightly longer wavelengths (between 0.2 and 0.3 micron). The absorption of energy in the upper atmosphere and in the ozone layer heats these layers to temperatures comparable to ground level temperatures.

With the exception of ultraviolet radiation, then, most of the incoming solar radiation is able to pass through a clear atmosphere and reach the earth's surface. However, water vapor and carbon dioxide molecules absorb most of the longwave radiation emitted by ter-

Figure 3.7 (opposite) This diagram traces 100 units of shortwave solar radiation incident at the top of the earth's atmosphere and its interactions with the atmosphere and the ground. The numbers represent average global values. The interactions of incoming shortwave radiation are shown on the left, and the interactions of outgoing longwave radiation are shown on the right. All of the indicated interactions occur at a given location during daylight hours, but at night, when there is no solar radiant energy input, only the longwave interactions occur.

The earth as a whole is neither gaining nor losing radiant energy; for every 100 units of solar radiant energy received, an average of 33 + 67 = 100 units are returned to space. Similarly, energy is in balance at the earth's surface, on the average. The surface receives 45 + 98 = 143 units of energy and loses 113 + 30 = 143 units. Can you use the diagram to show that the average gain and loss of energy by the atmosphere are also in balance?

Note that the 98 units of longwave radiant energy returned to the ground by the atmosphere are important components in the energy balance at the ground. If it were not for the energy contributed by the atmosphere, the ground would cool to the temperature where the radiated energy was in balance with the absorbed energy. In the absence of an atmosphere, the average temperature of the earth would be approximately −20°C (−4°F). (After Riehl, 1972)

restrial sources, and this absorption has profound effects on the energy exchanges between the earth's surface and the atmosphere.

THE ENERGY BALANCE OF THE EARTH-ATMOSPHERE SYSTEM

We have seen how solar radiation strikes different areas on the earth's surface in varying amounts and intensities and how the atmosphere interacts with incoming solar radiation. Because the earth is neither heating up nor cooling off, the amount of energy the earth receives from the sun must be equal to the amount it radiates back into space.

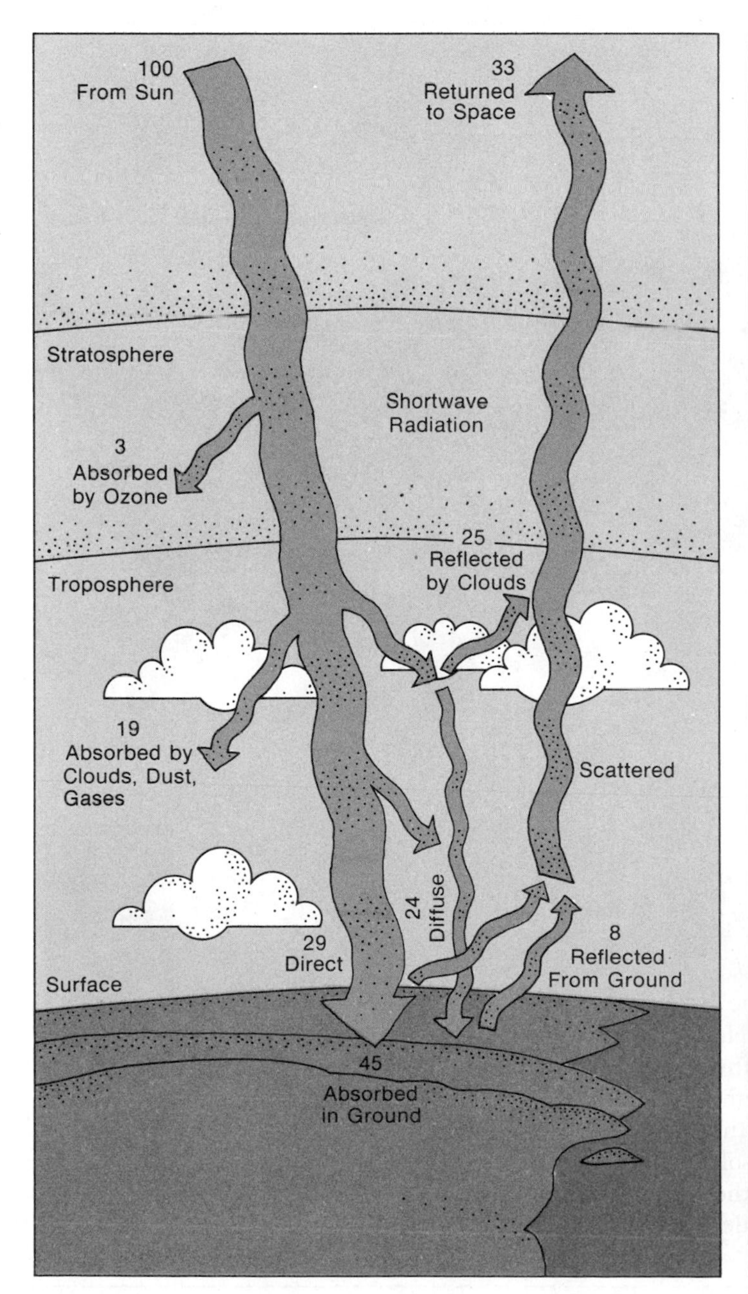

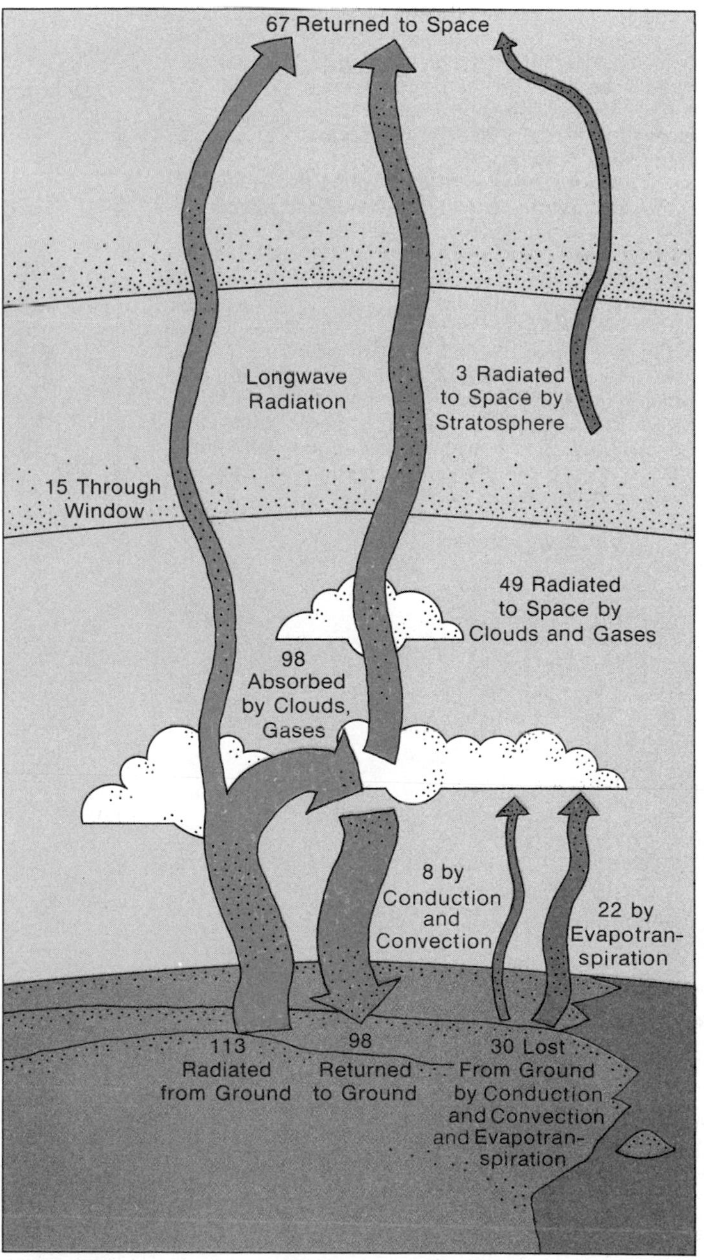

Figure 3.8 The average solar radiation received at the ground in the United States during January and July is shown in units of langleys per day. (After *The National Atlas of the United States of America,* 1970)

The average solar radiation received globally at the ground-sea surface during December and June is shown in units of langleys per day. (After Löf, Duffie, and Smith, 1966)

During December and January, when the sun is overhead in the southern hemisphere, the solar radiation received in the northern hemisphere decreases rapidly with increased latitude. Southern Florida, for example, receives more than 300 langleys per day during January on the average, but northern Minnesota receives only 150 langleys per day. Solar radiation input is small at high latitudes during the winter because of the short duration of daylight and because of the low altitude of the sun in the sky. The solar radiation input to a given location is larger when skies are clear than when skies are cloudy; for this reason the arid southwestern United States receives more solar radiation than cloudy locations at the same latitude.

During the summer months (June to September in the northern hemisphere, December to March in the southern hemisphere), the solar radiation input exhibits little variation with latitude. Locations in northern Canada receive the same input as locations in Texas. The increased duration of daylight with increased latitude compensates for the lower altitude of the sun toward the pole. The variation in solar radiation input between different locations during the summer is caused primarily by differences in the degree of cloudiness of the atmosphere. Where the isolines (lines connecting points of equal value) are dashed, the data are missing or incomplete. Data for Greenland are unavailable for June.

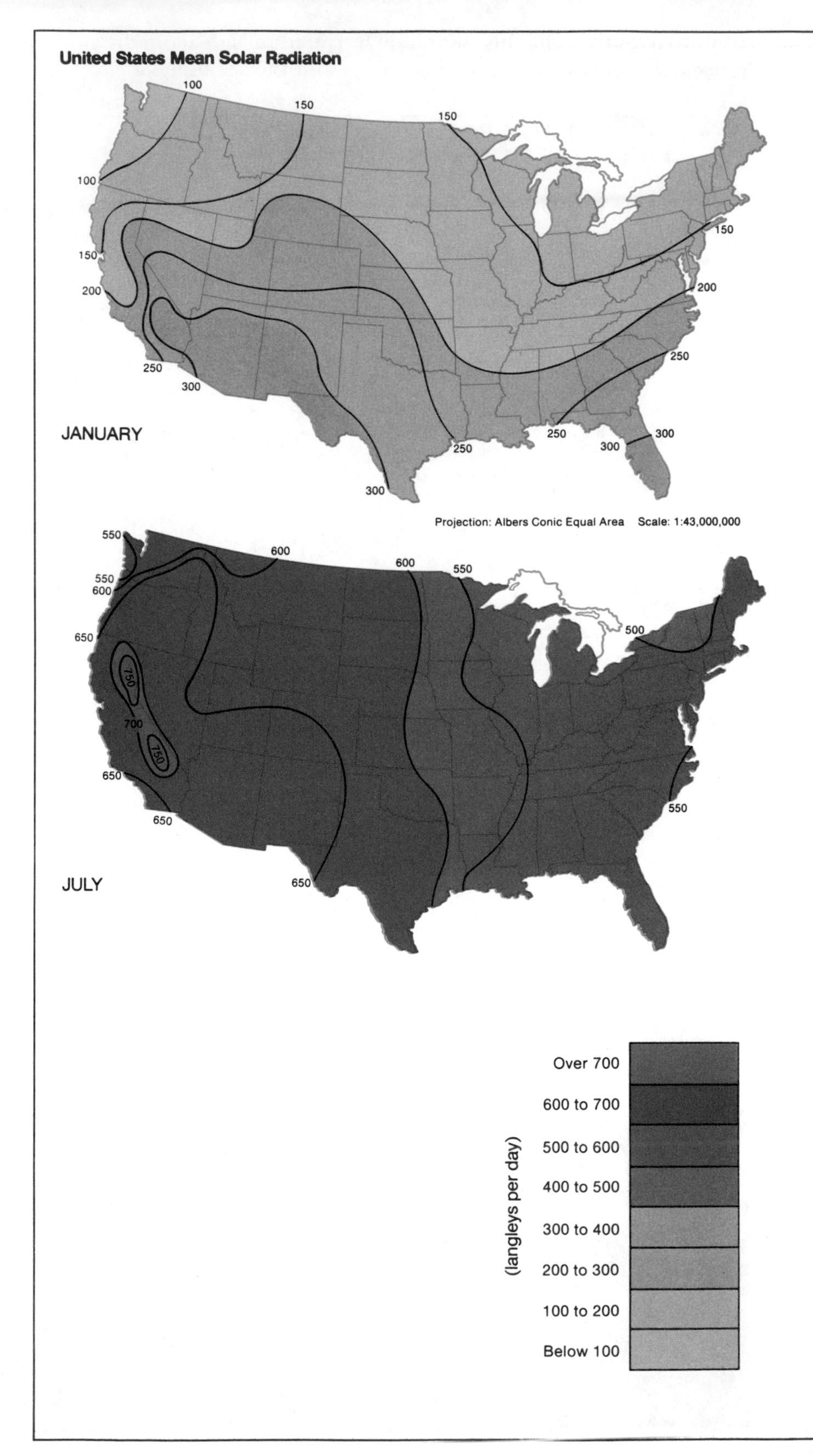

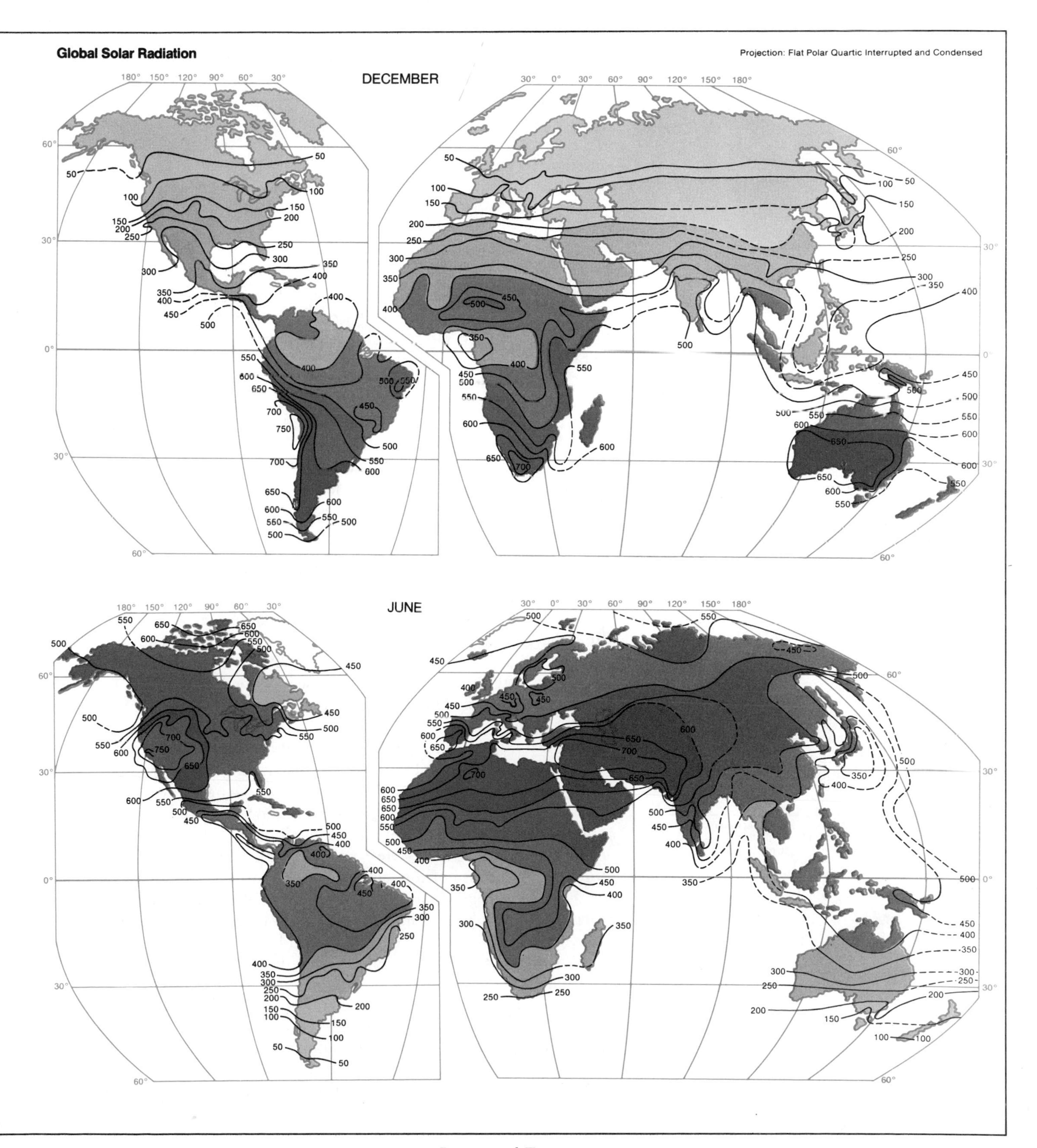
Global Solar Radiation
Projection: Flat Polar Quartic Interrupted and Condensed
DECEMBER
JUNE

This section examines the ways in which the earth loses energy and thereby maintains an energy balance. It is important to keep in mind that while there is considerable seasonal and geographical variation in energy exchanges within the earth-atmosphere system, we will be considering a grand average for the entire globe over a year. For this purpose, we will assume, as in Figure 3.7, p. 65, that all the solar radiation intercepted by the earth and atmosphere over a year is equivalent to 100 units of energy.

Reflection and Albedo

Air travelers flying at high altitudes may have noticed the dazzling brightness of the clouds below. Thick, puffy summer clouds are particularly efficient at reflecting shortwave radiation, throwing back as much as 90 percent of the solar energy that falls on their upper surfaces. The fraction of shortwave radiation an object reflects is called its *albedo*. With the exception of totally black objects, every object reflects radiation to some extent. The albedo of a shiny mirror is nearly 100 percent, although most surfaces in the environment have albedos considerably less than this (see Table 3.3). It is particularly important to take the earth's albedo into account in analyzing the energy balance because in the process of reflection no radiation is absorbed; it is simply redirected. Moreover, in contrast to scattering, which redirects radiation in all directions, reflection generally occurs in a single direction. The radiation reflected from a cloud usually returns to space, although it may be reflected downward again by another cloud higher up. Of the solar radiation that reaches the upper atmosphere, about 33 percent is returned to space through reflection. Atmospheric gases and dust absorb about 22 percent. The remaining 45 percent is absorbed by the ground (see Figure 3.7).

Table 3.3. Albedo of Various Surfaces

Surface	Albedo (percent)
Fresh Snow	80–95
Dense Stratus Clouds	55–80
Ocean (Sun Near Horizon)	40
Ocean (Sun Halfway up Sky)	5
Bare Dark Soil	5–15
Bare Sandy Soil	25–45
Desert	25–30
Dry Steppe	20–30
Meadow	15–25
Tundra	15–20
Green Deciduous Forest	15–20
Green Fields of Crops	10–25
Coniferous Forest	10–15

Sources: List, Robert J. (ed.) 1951. *Smithsonian Meterological Tables.* 6th rev. ed. Washington, D.C.: Smithsonian Institution Press. Budyko, M. I. 1974. *Climate and Life.* D. H. Miller, tr. New York: Academic Press, pp. 54–55.

Clouds have a significant effect on solar energy input to the earth's surface. Figure 3.8, pp. 66–67, shows the average solar radiation input to a horizontal surface on the ground for the United States and for the globe in winter and in summer. The amount of energy input to the top of the atmosphere depends only on day length and solar altitude. The irregularities in the pattern of energy reaching the ground are caused primarily by differences in cloudiness. The southwestern deserts in the United States are farther north than Florida but receive more energy because of the lack of clouds. China and the deserts of North Africa are both at latitude 25°N, but the African deserts receive more than one and one-half times as much solar energy input over the course of a year because of extensive cloudiness over China.

A portion of the radiant energy reaching the ground is immediately reflected without transfer of energy. The albedo of the ground can vary from 90 percent for a fresh snowcover to a few percent for an ocean. Furthermore, the albedo at a given location varies from season to season as snow comes and goes and as bare fields become covered with crops (see Table 3.3). The average annual distribution of solar radiation in the earth-atmosphere system by latitudes is shown in Figure 3.9.

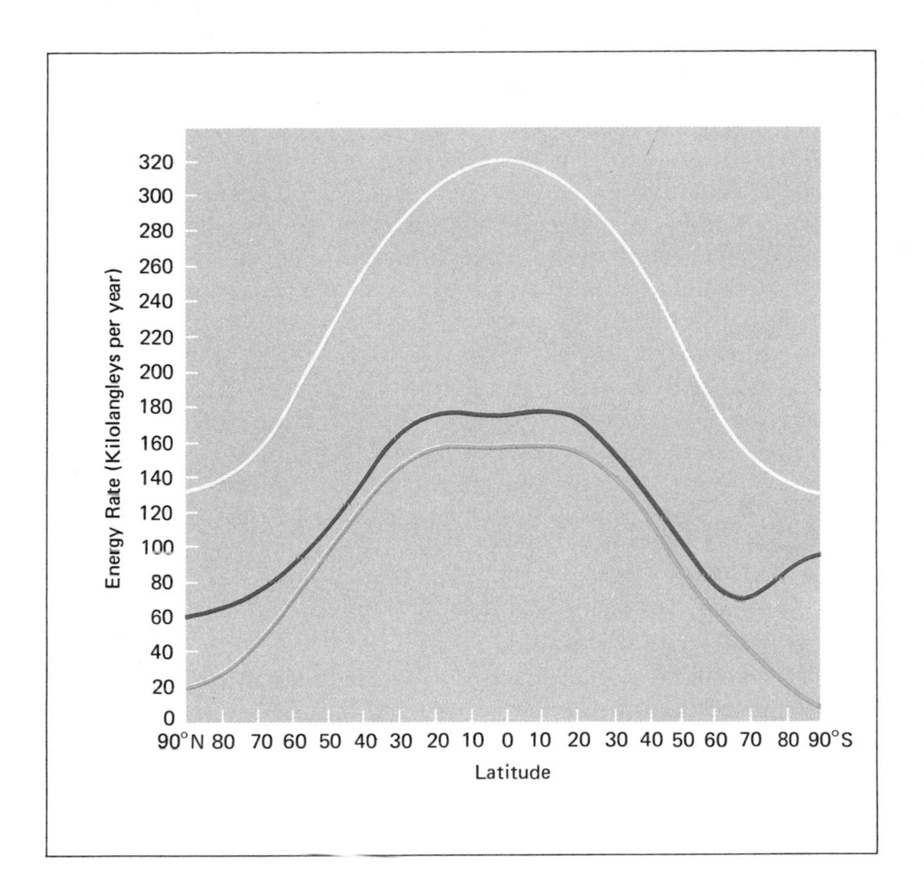

Figure 3.9 This graph shows the average annual distribution of solar radiation in the earth-atmosphere system by latitudes. The yellow curve represents incoming solar radiation at the top of the atmosphere, the red curve the incoming solar radiation just above the earth's surface, and the green curve the net solar radiation gain of the surface. In polar regions the combined effects of low solar altitudes, clouds, snow cover, and icecaps are evident. (After Sellers, 1965)

Energy Absorption and Atmospheric Exchanges

Because it absorbs radiation from the sun, the earth's surface is warm—15°C (59°F) on the average—and it emits longwave radiation continuously. If the earth had no atmosphere, all of the emitted longwave radiation would escape into space, and the unlit half of the earth's surface would rapidly become cold after sunset. This does not occur, however, because the atmosphere acts like a greenhouse, helping the earth's surface to retain much of its heat.

Because water vapor and carbon dioxide molecules are the only molecules in the lower atmosphere capable of absorbing longwave radiation, these minor constituents of the atmosphere play a major part in the earth's utilization of radiant energy. The absorbed radiant energy heats the lower atmosphere to a temperature comparable to the temperature of the earth's surface. The atmosphere also radiates longwave radiation; some of this radiation is directed upward and is lost in space, but most of it returns to the surface of the earth. Although the earth's surface loses energy by longwave radiation, it regains an average of nearly 80 percent of the lost energy by reradiation from the lower atmosphere. On the average, therefore, the longwave radiation exchanges between the surface and the lower atmosphere result in a net radiative loss for the surface and a gain for the atmosphere.

The atmosphere's greenhouse effect varies according to the amount of cloudiness present. Clouds contribute to the heating of the atmosphere by strongly absorbing longwave radiation emitted from the earth's surface. They contribute to the heating of the earth by returning much of the longwave radiation to the surface. Like the atmospheric gases, clouds emit longwave radiation, much of which is received by the ground. The earth's surface cools more rapidly on a clear night than it does when there is a thick cloud cover.

None of the gas molecules in the atmosphere is an efficient absorber of radiation in the wavelength range from 8.5 to 11 microns; thus the atmosphere is transparent to this band of wavelengths. Radiation emitted in this band escapes to space through a sort of *atmospheric window* when the sky is free of clouds.

The balance of incoming and outgoing energy in the earth-atmosphere system is illustrated in Figure 3.7. The left-hand diagram traces the average attenuation of solar radiation (100 units) as it passes downward from the top of the atmosphere to the earth's surface. Except for the absorption of ultraviolet radiation by the ozone layer, the atmosphere does not strongly absorb the incoming solar radiation. However, because a considerable amount of shortwave radiation is returned to space through scattering and reflection, only 45 units are actually absorbed by the ground.

The right-hand diagram traces the course of longwave radiation as it is emitted from the earth's surface and returned to space. Note that the 45 units of shortwave radiation which were absorbed by the ground are now reradiated as 113 units of longwave radiation. This increase is due to the greenhouse effect of the atmosphere. Clouds and gases in the troposphere absorb 98 units of longwave radiation from the ground and reradiate them downward. These 98 units plus the 45 units of solar radiation originally absorbed by the ground add up to 143 units. Thirty of these units are transferred to the atmosphere through processes which will be discussed in the following section, leaving 113 units to be radiated as longwave radiation from the surface.

As the left-hand diagram shows, the troposphere absorbs 19 units of incoming solar energy; these 19 units plus the 30 units gained from the ground make 49 units of energy that the atmosphere radiates as longwave radiation into space. Add the 15 units that escape through the atmospheric window and the 3 units that are radiated to space by the stratosphere, and the total longwave loss to space equals 67 units. Recall that of the 100 units of solar radiation entering the earth-atmosphere system, 33 were returned to space by scattering and reflection. This left a total of 67 units of incoming solar energy. The gain of 67 units is matched by the longwave loss of 67 units. The earth-atmosphere system is in balance.

Conduction, Convection, and Evapotranspiration

When two objects at different temperatures are in contact, heat flows by *conduction* from the warmer to the colder object. This principle has important consequences for the transfer of heat energy at the surface of the earth. The earth's surface absorbs solar radiation and converts it to heat energy; as a result, the ground becomes warm. When the top layer of the ground is warmer than the deeper layers, as at midday, heat energy flows by conduction deeper into the ground. Late in the day or at night, when radiation losses have cooled the top layer, heat energy flows from the warm, deeper layers back toward the surface.

Conduction also takes place between the ground and the atmosphere. Conductional heating of the air next to the ground helps to dissipate the net radiation gain of the surface and to compensate the net radiation loss of the troposphere. As shown in Figure 3.7, an average of 8 units of energy is transferred in this way from the earth to the atmosphere. *Convection*—the vertical movement of parcels of warm air—increases the efficiency of this process by exchanging the warmer surface air with cooler air aloft. At times, however, the air becomes warmer than the ground with which it is in contact. At such times, conduction causes heat to flow from the air downward to the surface.

Another energy exchange taking place between the earth's surface and the troposphere involves the change of water and ice to water vapor. As we saw in Chapter 2, about 590 calories are required to vaporize 1 gram of water at the average temperature of the earth's surface. There are several processes by which the vaporization of water is brought about. The roots of plants absorb water from the soil and pass water vapor out through openings in their leaves in a process called *transpiration*. Evaporation from water surfaces and directly from the soil also produces water vapor. Since transpiration and evaporation are difficult to measure independently, they are considered jointly as *evapotranspiration*. Figure 3.7 shows that, on the average, 22 units of energy are transferred by evapotranspiration from the earth's surface to the troposphere. This energy is stored in the water vapor in the form of latent heat. It becomes sensible heat, capable of warming the surrounding air, only when the water vapor condenses back into water droplets elsewhere in the atmosphere.

Between them, evapotranspiration and convectional heating of the air serve to dissipate the remainder of the energy not reflected or radiated back by the earth's surface. Photosynthesis, which is discussed in detail in Chapter 9, utilizes a minuscule proportion of the incoming solar radiation, an amount so small that it is not necessary to take it into account in an analysis of the energy balance of the earth-atmosphere system.

Latitudinal Differences in Solar Income and Radiational Losses

The energy balance of the earth-atmosphere system shown in Figure 3.7 represents an annual average for all areas of the globe. If we examine particular regions of the earth, however, we discover that energy exchanges are not in balance and that they vary considerably according to latitude. As shown in Figure 3.10, equatorial and tropical regions receive more solar energy than they return to space through reflection and longwave radiation. Polar regions, on the other hand, lose more energy through reflection and longwave radiation than they receive.

Figure 3.10 The graph shows average annual global values of absorbed solar radiant energy and emitted longwave radiation at each latitude in the northern hemisphere for the earth's surface and the atmosphere together. Between the equator and latitude 38°N, absorption exceeds emission, and there is a net input of energy to the earth. Poleward of 38°N, emission exceeds absorption, and there is a net loss of energy. The temperature in the lower latitudes therefore rises until the rate of heat flow toward the poles is sufficient to carry away the excess energy. (After Hare, 1966)

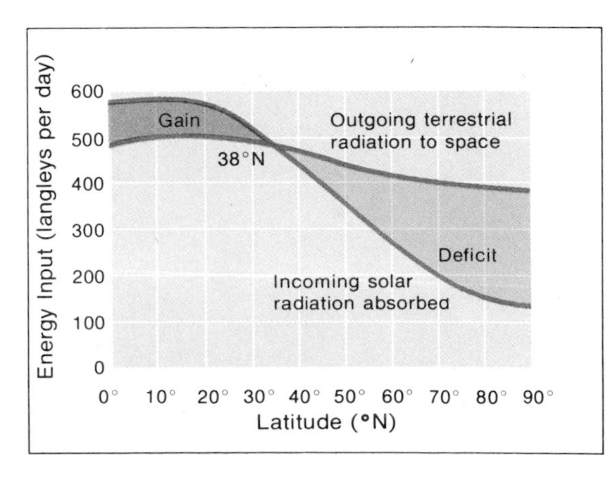

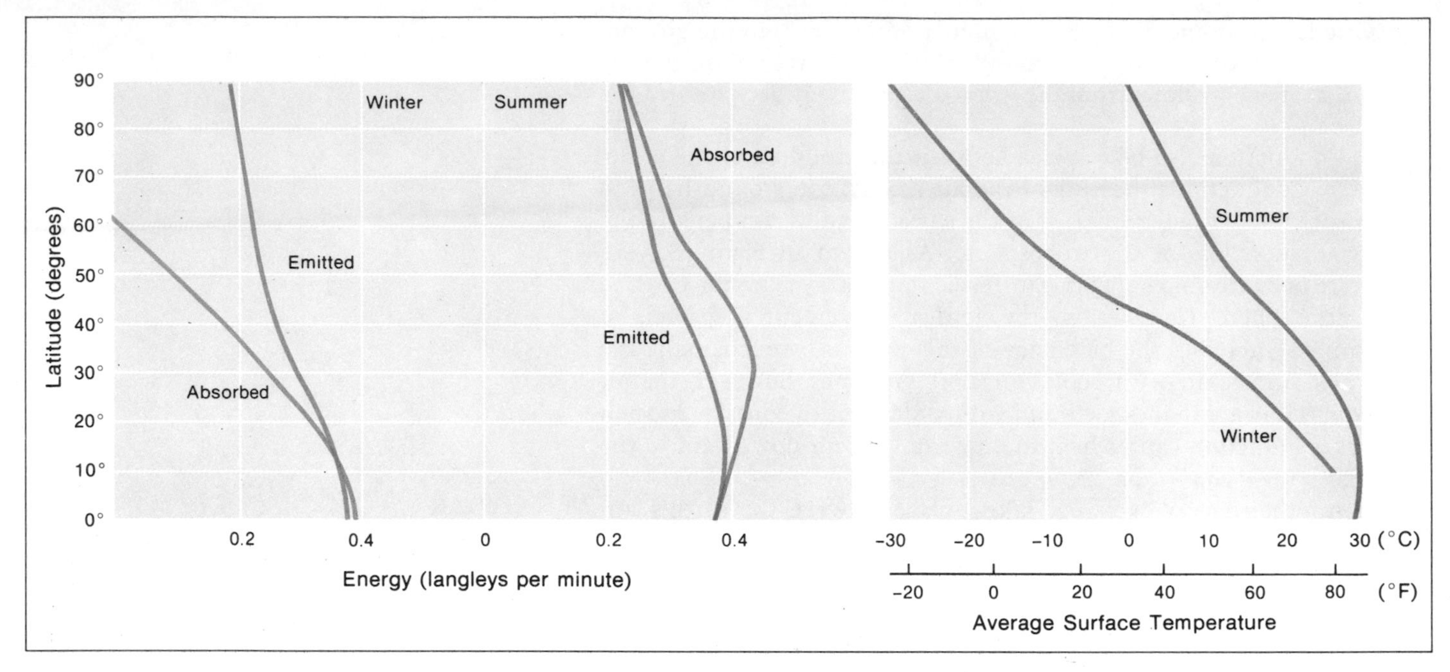

Figure 3.11 Energy exchange and temperature in the northern hemisphere. The diagram on the left shows the average rates at which the earth and the atmosphere absorb solar radiant energy and emit longwave radiation to space. During the winter, absorption exceeds emission only at latitudes near the equator, but during the summer the rate at which energy is absorbed exceeds the rate of emission at every latitude. Much of the excess absorbed energy warms the oceans and is stored there.

The diagram on the right shows the average surface temperature of the atmosphere at a height of approximately 2 meters above the surface. The temperature difference between low and high latitudes is greater during the winter than during the summer. The temperature differential helps to power atmospheric motion and a poleward flow of energy from the tropics. (After Riehl, 1972)

These regional gains and losses of energy are compensated by atmospheric and oceanic circulations which transport warm air and water (and latent heat in the form of atmospheric water vapor) toward the poles, and colder air and water toward the equator. There is a strong seasonality to the poleward transport of energy. Figure 3.11 diagrams average energy exchanges and temperatures in the northern hemisphere. In winter, only the zone extending 20° north of the equator registers a gain of energy, but during summer, the entire northern hemisphere gains more radiational energy than is lost to space. Atmospheric circulation, therefore, tends to be more vigorous during winter than summer because of the greater temperature differences between low and high latitudes.

ENERGY BUDGETS AT THE EARTH'S SURFACE

The energy exchanges between the sun and space and the earth and atmosphere have been measured with increasing accuracy in recent years by an array of meteorological and geophysical satellites. The energy exchanges that occur at the earth's surface can be measured by more routine techniques. Our knowledge of local energy budgets is based chiefly on measurements of radiation exchanges, atmospheric circulation, temperature, moisture, and evapotranspiration which have been collected at various experimental stations. By interpreting these data in the light of the principles of energy exchange discussed in this chapter, scientists are able to gain a better understanding of how the energy balance at a particular geographical location is modified by such factors as the degree of cloudiness, the nature of the vegetation cover, the presence of large bodies of water, and even the activities of man.

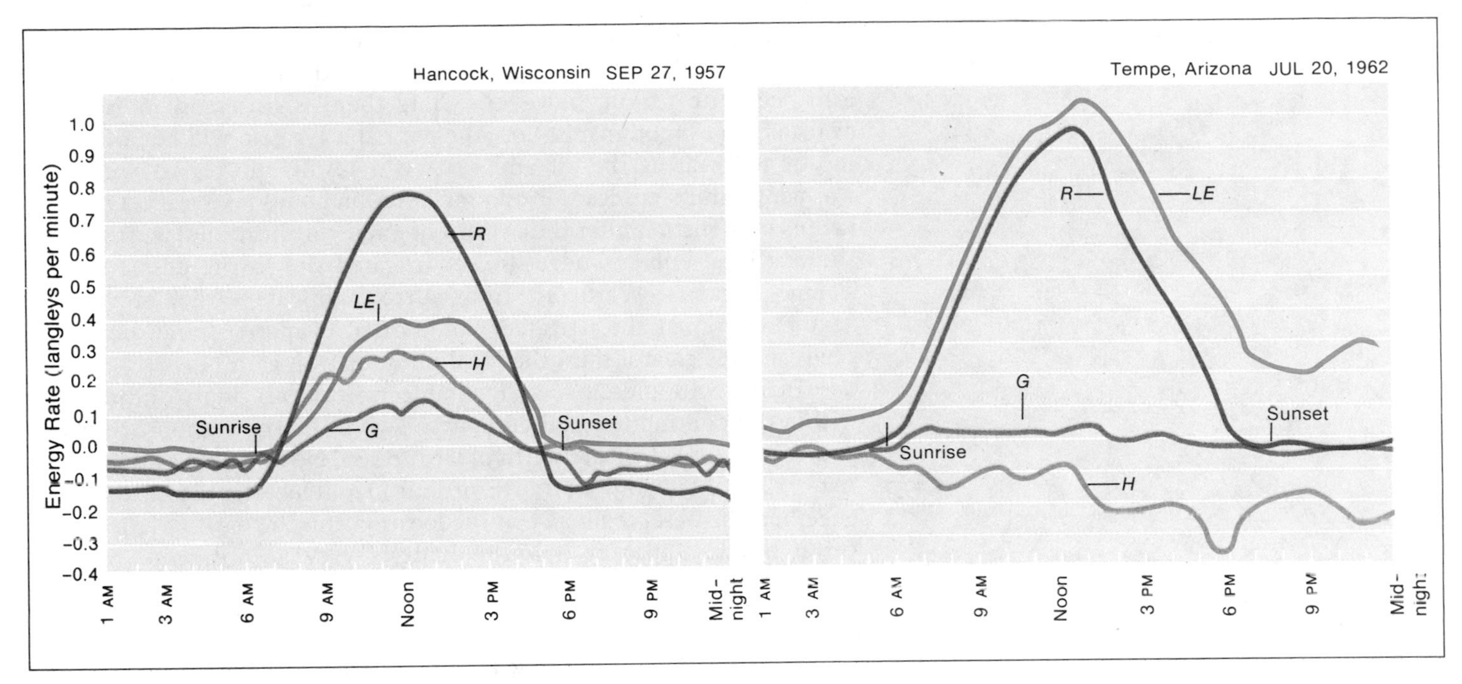

Figure 3.12 The local energy budget and its principal components during a 24-hour period are shown for Hancock, Wisconsin, a moist midlatitude location, and for Tempe, Arizona, an irrigated desert location. The radiation balance *R* is the difference between the rate at which solar radiant energy is absorbed at the ground and the net loss of longwave radiation in the exchange between the surface and the atmosphere. During most of the day, the radiation balance is positive, which indicates a net flow of radiant energy toward the surface of the ground. During the night *R* is negative because the net flow of radiant energy is away from the surface.

Energy cannot accumulate at the surface, and the excess or deficiency of radiant energy flow to the surface can be accounted for as the sum of three components: *LE*, the energy removed by the evaporation and transpiration of water; *H*, the energy flowing from the ground to the air by conduction and convection; and *G*, the energy that flows into the soil from the surface of the ground.

The *LE*, *H*, and *G* curves have been inverted to facilitate comparison with the net radiation. At noon at Hancock, for example, *LE*, *H*, and *G* represent energy losses from the surface, and their sum is equal to the net radiation gain. At Hancock and Tempe at midday, *LE* is negative because of evaporation and plant transpiration. The small positive value of *LE* before sunrise at Hancock indicates the condensation of water vapor on the ground as dew. At Hancock, *H* is negative at midday, which indicates that heat from the ground is warming the air. At Tempe, rapid evaporation from the irrigated plot keeps the ground cooler than the air so that heat flows from the air over the dry landscape to the irrigated ground, making *H* positive. (After Sellers, 1965)

Local Energy Budgets

On an annual world-wide average, energy input to the earth's surface is equal to energy loss from the surface. But how energy is used at any particular location depends on the prevailing environmental conditions. The diagrams in Figure 3.12 represent a schematic analysis of the chief factors involved in local energy budgets. The net radiation is partitioned three ways: the energy flow from the surface to the atmosphere because of evapotranspiration; the energy flow from the surface to the air by conduction and convection; and the energy flow between the surface and layers deeper in the soil. A very small portion of incoming solar energy not shown in the figure is utilized directly for photosynthesis.

During a sunny day at Hancock, Wisconsin, for example, more radiation, both shortwave and longwave, is received than is reflected or radiated back into space (see Figure 3.12). The rate of incoming solar energy rises rapidly, producing a net radiative gain which begins shortly after sunrise, peaks near solar noon, and returns to zero a few minutes before sunset. Thus, during the day, the radiation budget at Hancock can be said to be positive.

At night, however, more energy is lost through longwave radiation than is received, and the radiation budget at Hancock is then negative. Energy flows from the warmer atmosphere and from warmer depths of the ground toward the cooling surface. Thus the lower atmosphere is heated by the surface during the day, when the surface radiation budget is positive, and cooled by the surface at night, when the budget is negative. On the average, heat flows from the surface into the atmosphere, heating the atmosphere from below; it is this vertical transfer of energy that provides one of the driving forces of the weather.

The Hancock example is representative of energy exchanges during fair weather in humid climatic regions of the middle latitudes. In a dry climatic region, however, where there is little soil moisture available for evapotranspiration, the net radiative gain will be utilized mostly for heating the air and soil. As a result, the region will be warmer than if significant evapotranspiration could take place. Other factors may further alter the pattern of energy utilization. Figure 3.12 shows a local energy budget for an irrigated site in the desert near Tempe, Arizona. Warm air from surrounding dry areas provides "extra" energy at the irrigated plot, so that evapotranspiration proceeds at rates greater than the local net radiation gain.

Figure 3.13 presents energy budgets for four more locations within the continental United States. Two of these locations, Yuma and West Palm Beach, are south of latitude 35°N, while the other two, Astoria and Madison, are north of latitude 40°N. Significant differences may be observed between the locations on the basis of latitude, yet it is also apparent that the budgets vary according to other environmental factors as well.

Differential Heating and Cooling of Land and Water

The land surfaces and water bodies in a region respond differently to energy input. This is chiefly due to the fact that the heat capacity of

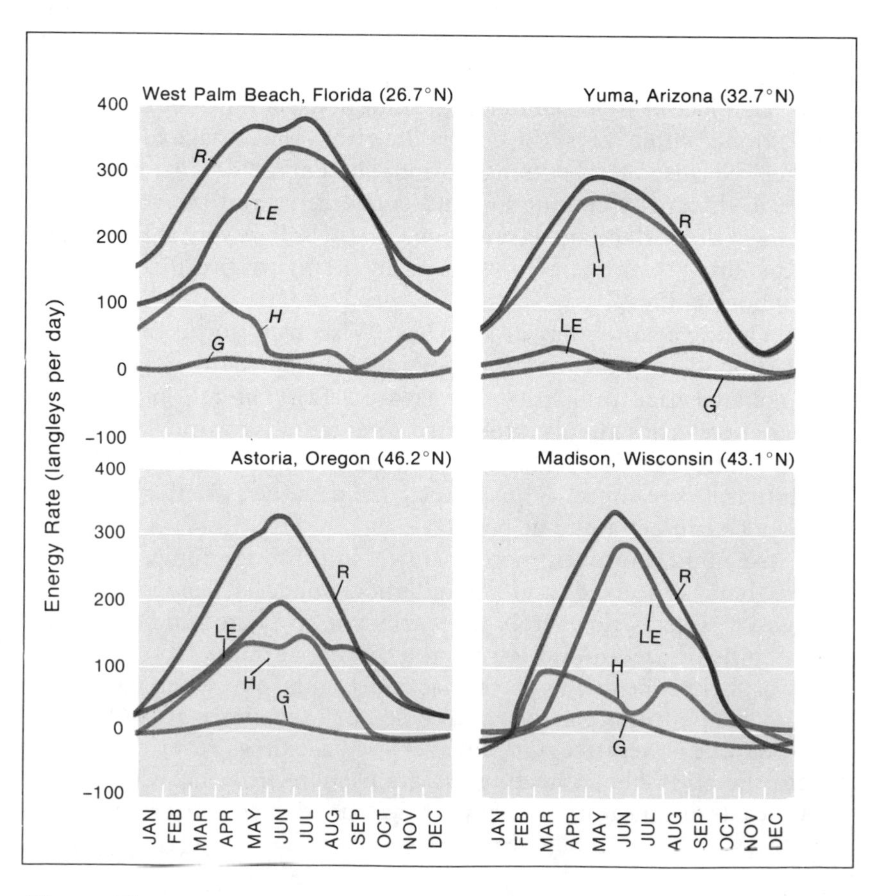

Figure 3.13 The graphs show the annual regime of the local energy budgets at four places in the United States. Madison is representative of humid midlatitude regions with low solar radiation income during winter, West Palm Beach of humid subtropical regions with moderate solar radiation income during winter, Astoria of humid and cloudy maritime climates with much winter cloudiness, and Yuma of hot desert regions with little cloudiness.

The values of *R* are similar at all four locations during summer, but *R* is much smaller at Astoria and Madison during winter because of fewer hours of daylight and low solar altitudes; during winter in Madison, *R* is slightly negative.

LE is usually the largest component of *R* during the growing season. *LE* is zero at Madison during winter, when low temperatures inhibit evapotranspiration, and there is no net radiational gain. *LE* is low throughout the year at Yuma because of the absence of water for evapotranspiration, and most of the net radiational gain goes into heating the air. In each location, energy tends to be conducted from the surface into the soil during spring and back to the surface during fall. (After Sellers, 1965)

land and water vary greatly. Heat capacity can be defined as the amount of heat necessary to raise 1 cu cm of a substance by 1°C relative to water. The heat capacity of ocean water is about 1, but the heat capacity of most land surfaces is only about 0.2 to 0.4. Thus, for a given energy input, a cubic centimeter of continent surface will heat up three to five times more than an equal volume of water.

On a clear summer day at noon, the surface of soil exposed to the sun can easily reach temperatures of 40°C (104°F) or more. Dry beach sand is often too hot for bare feet in the summer. Conduction carries some heat from the hot surface to cool deeper layers, but the process is relatively slow because the air spaces between the grains of soil act as insulation. As a consequence, the heat is concentrated in the topmost soil layers. Normally, most of this heat energy flows into the atmosphere or is utilized in evapotranspiration.

At night, the soil cools rapidly because of longwave radiation. If the sky is not cloudy, the amount of longwave radiation returned to the ground is relatively small, and the soil surface temperature can drop to low values. Cold soil cools the atmosphere at immediate ground level. If there is no wind to mix the air, the air near the ground may be several degrees cooler than the air a few meters higher. Profiles of typical air and ground temperatures during a fair winter day in southern New Jersey are shown in Figure 3.14. Even during the day,

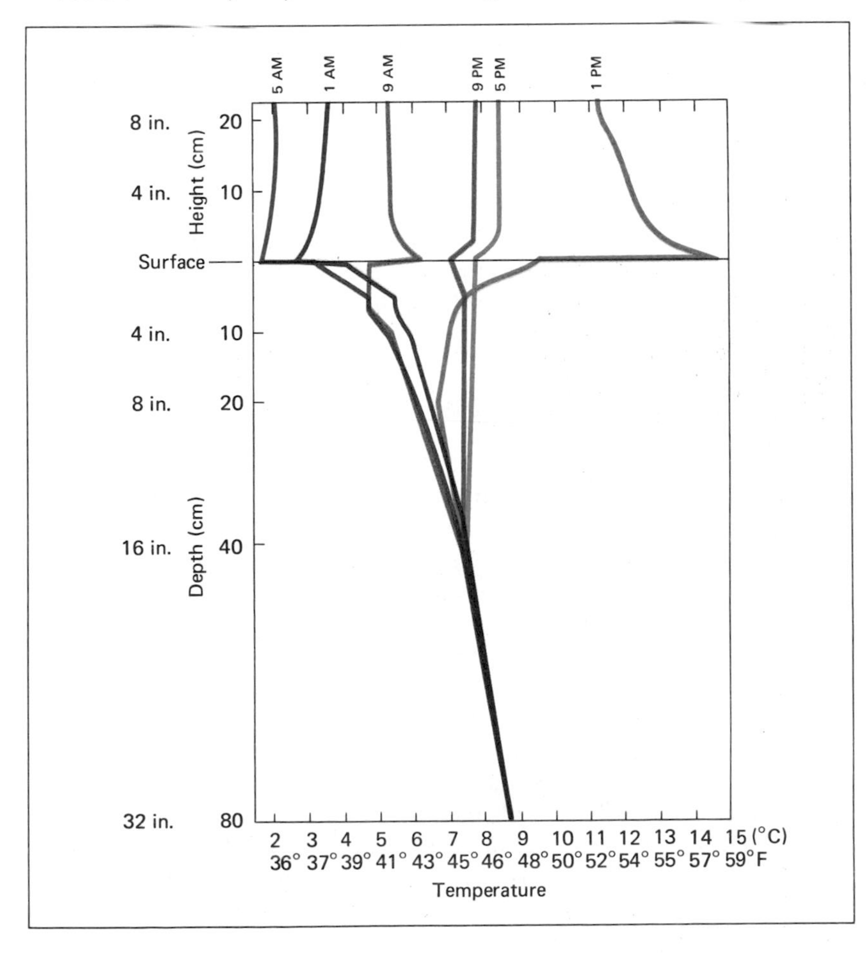

Figure 3.14 This figure illustrates steep temperature gradients immediately above and below the ground surface during fair days with little wind. The data were taken at an experimental site near Seabrook, New Jersey, during December 5, 1956. Between 9:00 A.M. and 1:00 P.M., the warmest temperatures were measured within the first 5 cm above the surface, but during late afternoon, overnight, and early morning, when the local radiation budget was negative, the coldest temperatures were immediately above the surface, which was losing heat by longwave radiational cooling. (After Mather, 1974)

Figure 3.15 The temperature records at Baton Rouge, Louisiana, for October 19 to 23, 1951, illustrate the daily variation in air temperature as the ground heats and cools. The temperature rises in the morning, when the energy budget becomes positive, and reaches a peak in the early afternoon. The radiation gain declines in the afternoon and becomes negative at night, causing the temperature to drop.

On October 19 to 21, skies were clear, but on October 22 and 23, skies were cloudy. On a cloudy day, the daily variation in temperature is less extreme than on a clear day, because clouds reduce solar energy input during the day. Clouds also reduce surface heat loss caused by longwave radiation, so that temperatures do not fall as low on a cloudy night as on a clear one.

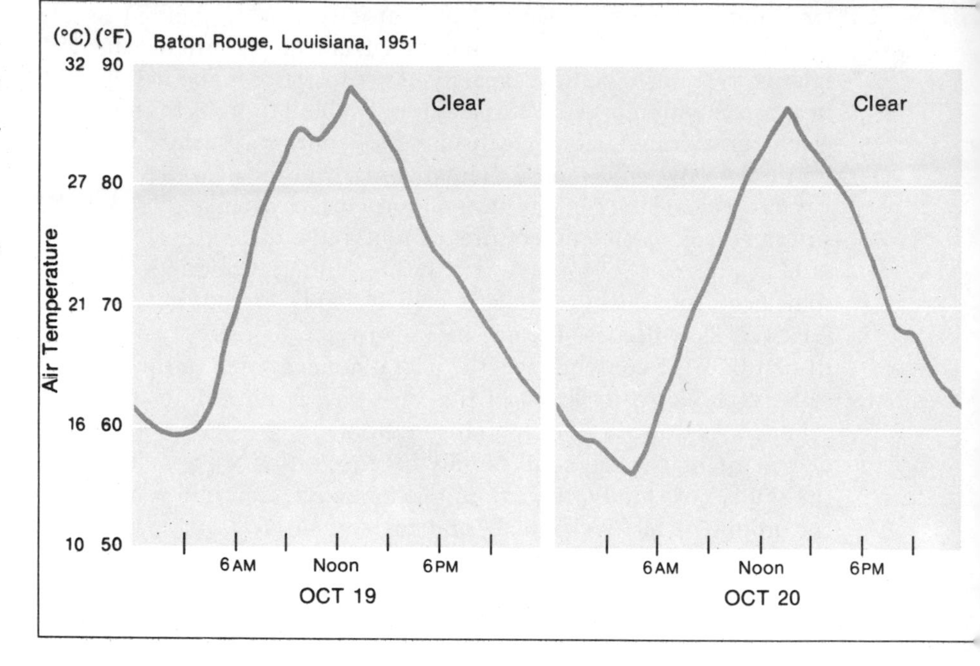

Figure 3.16 (opposite) The average temperature distributions during January and July are shown for three cities in different climatic regions over the four-year period 1935 through 1939. Each point on a temperature distribution tells the number of hours during the given month that the specified temperature was recorded; in San Diego in July, for example, a temperature of 20°C (68°F) was recorded in 57 separate hours.

A narrow temperature distribution, which San Diego, a coastal city, shows, means that the temperature remains comparatively constant. At Elko in July, in contrast, the temperature distribution is broad, reflecting the hot days and cool nights of the cloudless desert. The distributions are moderately broad for Cleveland, a midcontinent city with a moist climate. (After Court, 1951)

the soil cools rapidly if shielded from the sun. The daytime temperature of the soil under a cover of tall plants may be as much as 10°C (18°F) lower than the temperature at the top of the plants.

Surface temperatures of oceans and large water bodies change much less through the year than land surface temperatures. In addition to having a much greater heat capacity than land, water also distributes heat energy much more effectively. Solar radiation can penetrate water to a depth of tens of meters, spreading a given quantity of energy to a larger volume. Ocean waves and currents further distribute heat energy by mixing. As a consequence, local energy budgets are modified substantially by the nature and availability of water at the surface.

Temperature Regimes

The physical principles underlying the energy budget may be applied on a local scale to help explain temperature variations at a given location. Figure 3.15 shows temperature records for a period of several days at Baton Rouge, Louisiana. The air temperature near the ground tends to follow the temperature of the ground. In clear weather, the air and ground temperatures rise markedly during the day and fall at night. The daily rise in temperature begins an hour or so after sunrise, when the radiant energy input finally surpasses losses. The air and ground continue to warm up while the radiation budget is positive. Cooling begins in the late afternoon when radiation losses exceed gains. The graph also shows that during cloudy weather, the temperature changes very little. Clouds reduce the amount of incoming solar radiation during the day and increase the amount of longwave radiation downward from the atmosphere, which smooths out extreme variations in temperature.

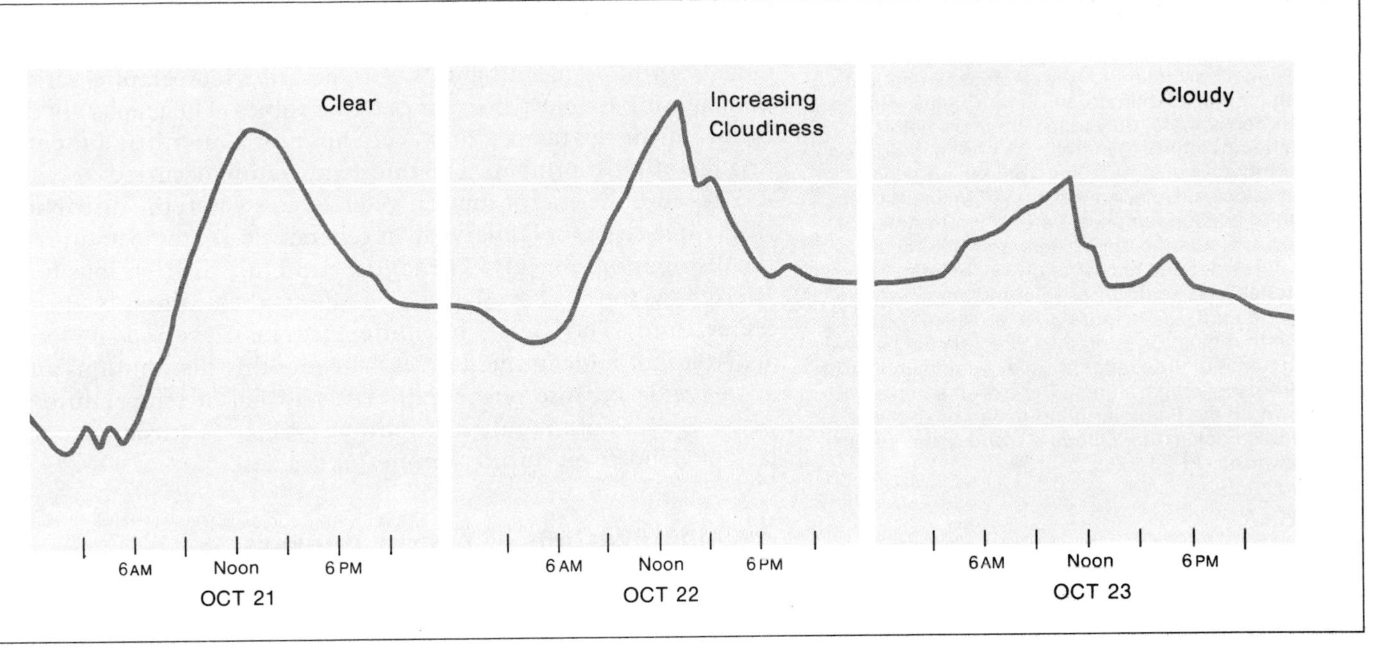

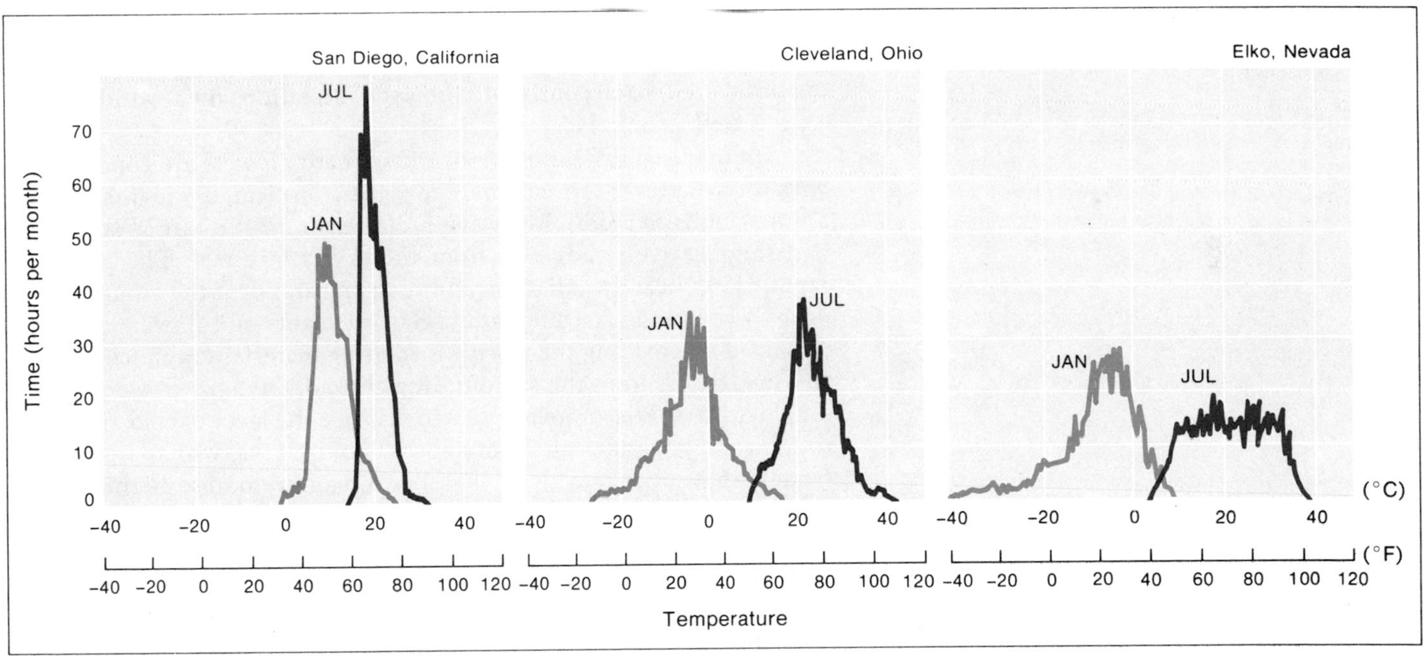

Proximity to a large body of water also tends to smooth out variations in temperature, because the temperature of a large body of water changes little from day to day. Energy transfer between the water and the atmosphere helps to maintain a uniform temperature at coastal cities compared to cities in the heart of a continent.

Monthly temperature records also may be interpreted according to the physical principles that govern the energy budget. Figure 3.16 shows temperature records for January and July at San Diego, California, a coastal city; Cleveland, Ohio, a midcontinent city with frequent

Figure 3.17 (opposite) Global temperature distributions are shown for January and July. Temperatures have been reduced to approximate sea level values by applying a correction for the decrease in temperature with increased altitude. In winter (January in the northern hemisphere, July in the southern hemisphere), temperatures generally depend markedly on latitude and decrease regularly from the equator toward the poles. Note, however, the continental and maritime effects on temperature over North America and the North Atlantic during January. In summer temperatures depend less strongly on latitude. These trends reflect the behavior of solar radiant energy input to the earth (see Figure 3.18, pp. 80–81).

Note that blue tones on this map indicate below-freezing temperatures, and "warmer" colors indicate above-freezing temperatures. The extreme poles are not shown on the flat polar quartic map projections used in this book. (After *Goode's World Atlas*, 1970, and Trewartha, 1968)

cloud cover; and Elko, Nevada, an interior city in the desert. The figure shows the *distribution* of temperature for the two months, which is a more meaningful way to describe temperature variation than quoting simple average or extreme values. The graphs are made by measuring the temperature every hour for a month and then plotting the number of hours a certain temperature occurred.

Because the energy input is reduced in winter, the distributions in all three cities for January fall at generally lower temperatures than the distributions for July. The January and July distributions for San Diego show the most overlap, which reflects the uniformity of coastal temperatures. The January-July differences for Cleveland and for Elko are typical of midcontinent cities. The monthly distributions are relatively wide because of the daily rise and fall of temperature. The extreme width of the July distribution for Elko is attributable to the lack of cloud cover during summer in the desert.

Modifications of Energy Budgets

The complex interaction of the earth-atmosphere system with radiant energy depends on a wide variety of factors. The composition of the atmosphere, the cloudiness of the sky, the albedo of the ground—all influence the utilization of radiant energy. A change in any of these factors can alter the energy balance. Man's activities, both purposeful and accidental, have contributed to such alterations on a world-wide scale as well as on a local one.

On the local scale, gardeners and orchard growers are concerned with ways to prevent frost damage to plants. On clear, dry nights, there are no clouds to absorb longwave radiation from the earth's surface, and so longwave reradiation from the sky is minimal. The ground rapidly loses heat by radiation, and temperatures near the ground can drop to below the freezing point. Garden plants and flowers can be protected by covering them with insulating material to reduce radiative energy loss from the surrounding ground. Orchard growers have employed large fans or helicopters to displace the layer of cold air next to the soil by warmer air from above. Smudge pots, which are small heaters that produce a dense blanket of smoke to absorb and then reradiate longwave radiation back to the ground, are generally less effective protection. Spraying vineyards with water throughout the night can protect them from frost. As the water freezes, 80 calories of heat energy are released per gram of water. Even though ice forms on the plants, the temperature is held to 0°C (32°F), which most plants can endure.

Large cities can produce their own distinctive environments, partly by modifying the radiation balance. Cities can be as much as 7° or 8°C (12° or 14°F) warmer than the surrounding countryside, because the towering city buildings receive considerable radiant energy even when the sun is low in the sky, and the canyonlike streets tend to catch reflected radiation. The dense concrete and bricks of cities have a lower heat capacity than soil, and they also cool less rapidly. Energy released from fuel burning in homes, factories, and automobiles augments the radiant heat input in cities. If the air over a

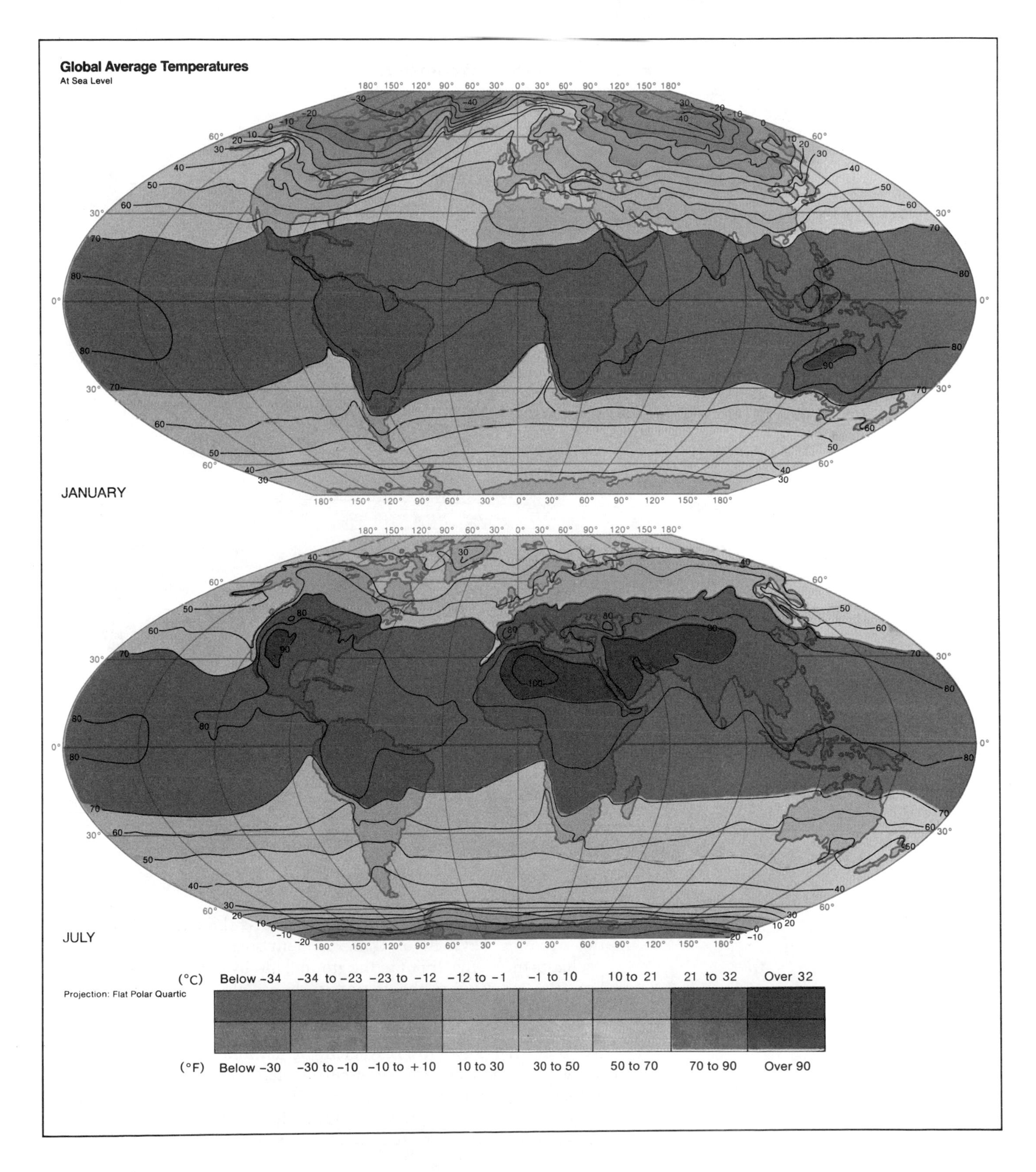

Global Average Temperatures
At Sea Level
JANUARY
JULY
(°C) Below -34 -34 to -23 -23 to -12 -12 to -1 -1 to 10 10 to 21 21 to 32 Over 32
(°F) Below -30 -30 to -10 -10 to +10 10 to 30 30 to 50 50 to 70 70 to 90 Over 90
Projection: Flat Polar Quartic

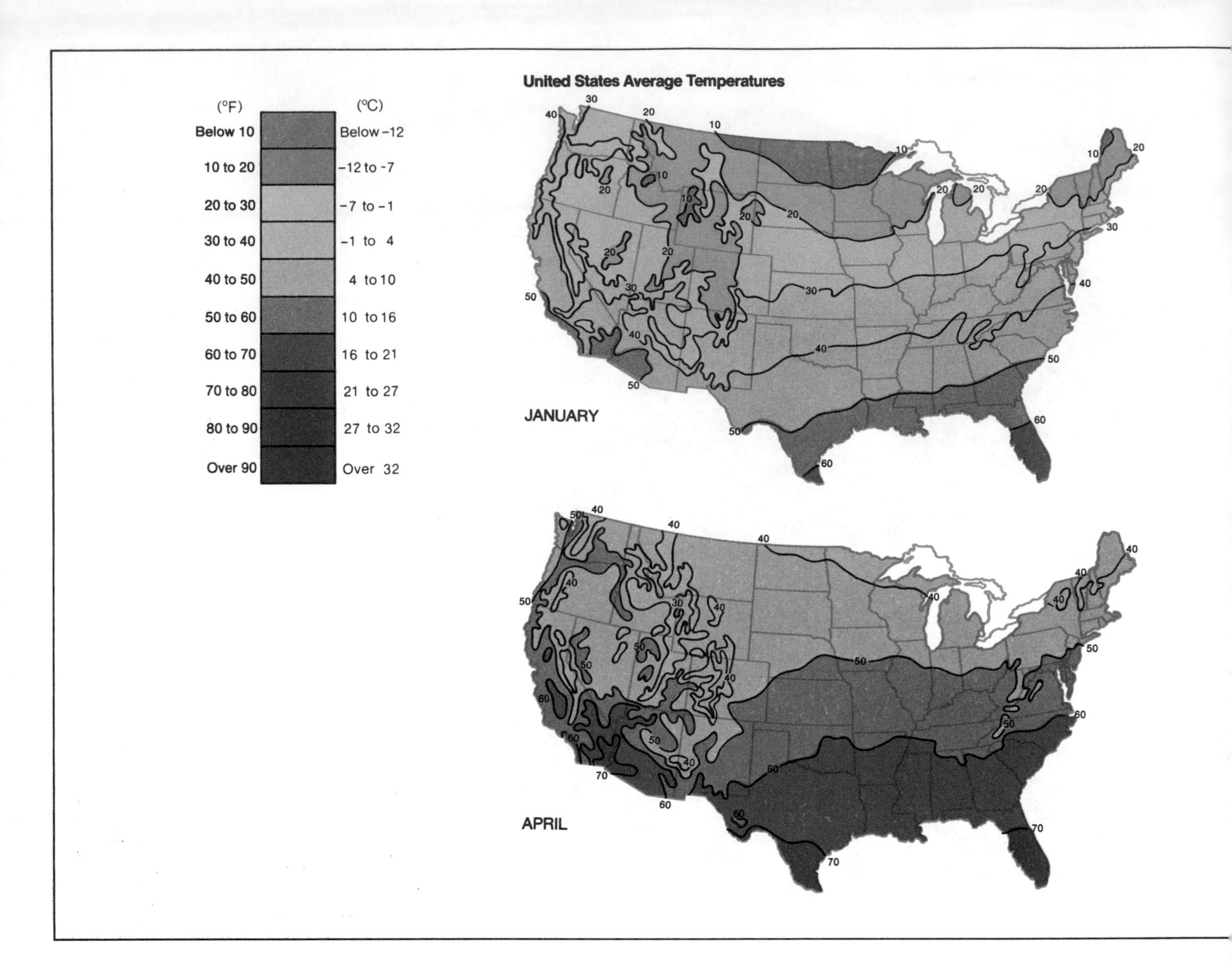

city is dusty or smoky, the radiant energy balance is affected. The opacity of the air, called the *turbidity*, affects the energy balance in two opposing ways: by blocking incident solar radiation, turbidity decreases the energy input to the ground, but by increasing the return longwave radiation to the ground, turbidity also decreases the heat loss.

Human modification of the energy balance did not begin with the Industrial Revolution. Ancient agricultural practices, such as overgrazing and excessive forest cutting, have altered the albedo and evapotranspiration rates over perhaps 20 percent of the land surface. Removal of plant cover has allowed the wind to blow soil particles into the air, which has increased the turbidity of the atmosphere.

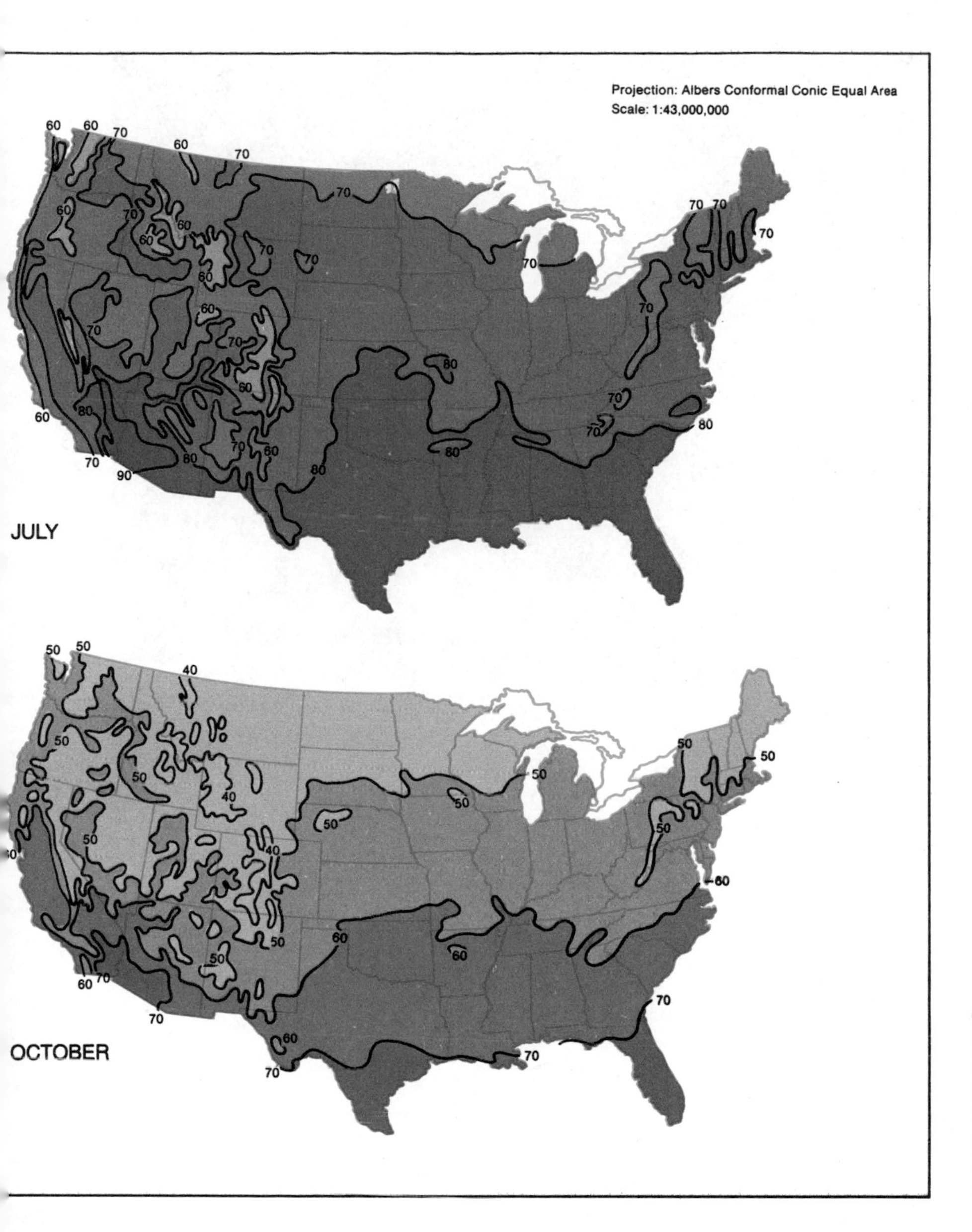

Figure 3.18 United States temperature distributions (in °F) are shown for selected months. In January, temperatures decrease regularly with increased latitude. In July, the variation with latitude is less marked because the input of solar radiant energy is relatively independent of latitude during the summer (see Figure 3.8, pp. 66–67). (After *The National Atlas of the United States of America*, 1970)

The industrial age has intensified man's impact. Jet aircraft deposit exhaust products and make condensation trails high in the atmosphere, which results in greater cloudiness and greater reflection of shortwave radiation. Industrial activities have changed the scattering and absorption of radiation in the earth's atmosphere because they have increased the turbidity and the carbon dioxide content of the atmosphere. The construction of artificial lakes and of roads (roads cover nearly 1 percent of the surface of the United States) has altered albedos and evapotranspiration patterns.

Data from satellite surveys of key parameters, such as albedo, land use and plant cover, and degree of cloudiness over the entire surface of the earth, are required to assess the effects of these changes.

Figure 3.19 The different colors in this *thermogram* of downtown New York City correspond to different temperatures, with red indicating the warmest temperature. The generation of heat within cities and the comparatively high heat capacities of building materials turn cities into *heat islands* that are several degrees warmer than the surrounding countryside. Washington, D.C., for example, is frost-free for a month longer than neighboring rural areas because of man's modifications of the local climate.

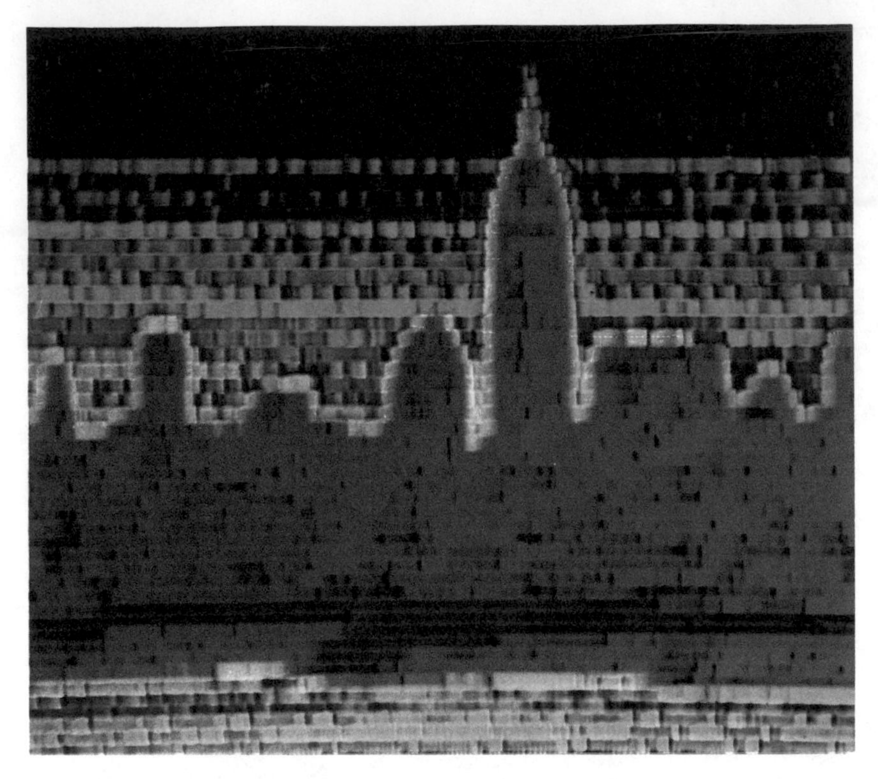

The launching of the Earth Resources Technology Satellite, ERTS-1, in the summer of 1972 was an important first step in this program. From its orbit between the poles, ERTS-1 surveys bit by bit nearly the entire earth rotating beneath. The significance of such surveys is that every kind of weather together with every kind of land or water surface make unique energy-balance combinations, while the energy balance in turn affects the formation of weather systems.

SUMMARY

The sun provides the energy for all life processes on the earth. Solar energy warms the earth's surface and drives the circulation of the atmosphere and oceans, it powers the hydrologic cycle and the movement of weather systems, and it provides energy for photosynthesis. The processes described in this chapter underlie the weather and climate systems developed in later chapters.

The sun transmits energy in the form of electromagnetic radiation, which has a spectrum of wavelengths; radiation from the sun consists of ultraviolet, visible, and infrared (mostly shortwave) radiation. The intensity of the sun's radiant energy input at the top of the atmosphere is expressed by the solar constant.

The sun's location in the sky at any time can be described by altitude and azimuth angles similar to the grid system of parallels of latitude and meridians of longitude on the earth. Because of the tilt of the earth's axis of rotation, the amount of daylight possible at any particular location depends on the time of year and the latitude, and

Further Reading

Bennett, Iven. "Monthly Maps of Mean Daily Insolation for the United States." *Solar Energy,* Vol. IX, No. 3 (1965): 145–152. This article provides a careful analysis of solar radiation data for the United States.

Budyko, M. I. *Climate and Life.* English edition edited by David H. Miller. New York: Academic Press, 1974. Chapters 1 through 3 focus on Budyko's classic analysis of global energy-budget components. This text also offers some overall perspectives on the extensive Soviet research into climate and its consequences.

Geiger, Rudolf. *The Climate Near the Ground.* Translation of the 4th German edition. Cambridge, Mass.: Harvard University Press, 1965. This long-term classic stresses the

the amount of solar radiation delivered to the top of the atmosphere above any particular place over a day depends, in turn, on the number of daylight hours and the sun's altitude above the horizon.

The earth receives solar energy, and it radiates energy back to the atmosphere and space in the form of longwave radiation. Water vapor and carbon dioxide molecules in the lower atmosphere absorb much of the longwave radiation emitted by the earth, heating the atmosphere from below. Longwave radiation reemitted by the atmosphere is a substantial energy input to the earth's surface, and it raises average surface temperatures significantly.

The earth's surface, therefore, gains energy from the sun and from longwave radiation from the atmosphere, and it loses energy by longwave radiation to the atmosphere and space, by conduction and convection of heat into the atmosphere, by evapotranspiration, and by conduction of heat into the ground. The daily variation of air temperature near the ground can be interpreted in terms of the local energy budget, which can be modified by clouds, proximity to large water bodies, and various human activities. The next chapter explains how the sun's energy drives precipitation processes in the atmosphere.

Review Questions

1. Describe the electromagnetic spectrum. What is the relation of temperature to wavelength?
2. What factors determine the amount of solar radiation reaching the top of the earth's atmosphere?
3. Describe the variation of day length from pole to pole at the winter solstice.
4. Describe how the atmosphere intercepts and interacts with incoming solar radiation.
5. Discuss the various ways the surface of the earth loses energy.
6. What is the role of the atmosphere in heating the earth?
7. What local environmental conditions modify a region's energy budget?
8. In energy-budget terms, why does the maximum daily air temperature tend to occur between 2:00 and 6:00 P.M.?
9. Compare the utilization of net radiation in humid and in dry climates.
10. Explain the effects of landmasses and oceans on the seasonal regimes of air temperatures.
11. What are various ways in which the energy balance of the earth-atmosphere system might be modified by human action?

results of field studies, on each of the continents, with particular emphasis on radiation and temperature. Geiger has sometimes been honored as the "father of microclimatology."

Mather, John R. *Climatology: Fundamentals and Applications.* New York: McGraw-Hill Book Co., 1974. Chapter 2 includes basic discussions of radiation and temperature instrumentation and data as well as their applications and limitations.

Miller, David H. "A Survey Course: The Energy and Mass Budget at the Surface of the Earth." Publ. No. 7, Comm. on College Geog., Assoc. of American Geog., 1968. This very useful monograph is organized into study units which proceed from energy exchange processes through local energy budgets to regional synthesis. Emphasis is placed on professional papers from almost every region of the globe.

Neiburger, Morris, James Edinger, and **William Bonner.** *Understanding Our Atmospheric Environment.* San Francisco: W. H. Freeman & Co., 1973. Radiation exchanges in the atmosphere are well described in Chapter 3.

Reifsnyder, William E., and **Howard W. Lull.** "Radiant Energy in Relation to Forests." *Tech. Bull. No. 1344,* U.S. Dept. of Agriculture, 1965. This is an excellent handbook for measurement of radiation exchanges, especially those in and over forests, and analysis and application of data.

Sellers, William D. *Physical Climatology.* Chicago: The University of Chicago Press, 1965. This work focuses almost entirely on fundamental analyses, in both descriptive and mathematical forms, of the radiation and energy budgets at the earth's surface.

Trewartha, Glenn T. *An Introduction to Climate.* 4th ed. New York: McGraw-Hill Book Co., 1968. Regional focus on geographical distributions of solar radiation and temperature is found in Chapters 1, 2, and 7–12.

Seascape Study with Rain Clouds by John Constable, *c.* 1824–1828. (Royal Academy of Arts, London)

Water enters the atmosphere through evaporation, is transported in clouds, and falls again to the earth as rain, snow, sleet, or hail. Condensation and precipitation are the results of a number of complex physical interactions that involve energy and moisture.

FOUR

Moisture and Precipitation Processes

DURING THE AFTERNOON OF June 9, 1972, rain clouds gathered over the Black Hills of South Dakota; rain poured steadily and dumped more than 30 cm (12 in.) of water over several areas in a 6-hour period. Mountain streams became swollen as they coursed down the hillsides; by the time they reached the inhabited valleys below, they had become rampaging torrents, overflowing their banks and sweeping cars and even houses into the turbulent floodwaters. Automobiles and buildings blocked the spillway of a dam just upstream from Rapid City, causing water levels behind the dam to rise an additional 3 meters (10 ft). Late in the evening, the dam burst, and a flood wave causing enormous destruction and death swept suddenly through the city.

A similar incident occurred on the night of July 31, 1976, when extremely heavy rains caused floodwaters to sweep through the narrow canyon of Thompson Creek in the foothills of the Rocky Mountains north of Denver, Colorado. Within a few brief hours, the raging waters destroyed nearly all traces of human habitation.

Figure 4.1 Man depends on the moisture delivered by the atmosphere to sustain his food crops. Because precipitation occurs only when certain conditions are present, the amount of moisture received by a given region may vary markedly from year to year. The coming of rain, or the lack of it, is a central concern of farmers each year.

(right) People in India rejoice in the coming of the summer monsoon rains, which supply the country with much of its annual precipitation.

(opposite) This farm in Oklahoma was abandoned in the 1930s when a succession of drought years made farming impossible. The dry soil was transported by the wind, forming drifts of soil and turning the region into a dust bowl for a period of several years.

What happens in the atmosphere to cause a deluge that results in a disastrous flood on the earth's surface? The water that fell on the Black Hills gathered as molecules of water vapor over the Gulf of Mexico, thousands of kilometers south of South Dakota. Moving masses of air carried the water vapor northward across the Great Plains. Along the way, some of the molecules of water vapor coalesced to form droplets of water a few microns in diameter, which were too small to fall to the ground as rain. Over the Black Hills, the droplets coalesced further to form raindrops. The return of water to the earth's surface in the form of rain completed the chain in which a phenomenal number of molecules—about 10^{22} separate molecules of water vapor—joined to form each raindrop.

The transfer of water from the earth's surface to the atmosphere and back again to the surface constitutes the atmospheric part of the hydrologic cycle. The movement of water through the atmosphere occurs by means of three separate operations: evaporation, condensation, and precipitation. *Evaporation* involves the conversion of liquid water to vapor. *Condensation* involves the conversion of water vapor to water droplets, in the form of either fog, clouds in the atmosphere, or dew at the earth's surface. *Precipitation* refers to the coalescence of water droplets into raindrops, snowflakes, or hailstones that are large enough to fall to the ground. Recall from Chapter 2 that evaporation represents a withdrawal of sensible heat at the earth's surface and its conversion into latent heat, and that in condensation the latent heat

stored in the water vapor is released as sensible heat. These changes of phase that water undergoes in the atmosphere are a significant part of the energy balance of the earth-atmosphere system.

Although about half the earth is covered by clouds at any time, only a small fraction of the clouds possesses the necessary qualifications for generating rain. Precipitation is the result of several complex physical interactions. The conditions that lead to it are so narrowly determined that it should be considered an unusual phenomenon rather than a commonplace event. This chapter describes how the atmosphere is supplied with water vapor and the circumstances that lead to condensation, cloud formation, and precipitation.

TRANSFER OF WATER TO THE ATMOSPHERE

Water is transferred from the earth's surface to the atmosphere by the process of evaporation. When water evaporates, individual water molecules leave the liquid and enter the atmosphere as water vapor, a dry gas. In a liquid, molecules are bound together by attractive forces, but the molecules are also in motion and continually collide with one another. The higher the temperature of the liquid, the more energy the molecules have and the more forcefully they collide. When especially forceful collisions occur near the surface of the liquid, some of the molecules gain enough energy to break away and enter the air. Because the escaping molecules carry energy with them, evaporation tends to be a cooling process. This is why you feel a chill when you step out of a swimming pool even though the air may be warmer than the water. The energy which is carried by the molecules is the latent heat of vaporization discussed in Chapter 2. On the earth's surface, the radiant energy from the sun ultimately supplies the energy required to evaporate water. To evaporate a layer of water 1 cm thick from a pan or a puddle at 15°C (59°F), the average temperature of the earth's surface, requires the expenditure of 590 calories for every sq cm of surface area.

Evaporation from Water Surfaces

If a jar partly filled with water is left open in a room, the water in the jar will eventually evaporate completely. Water molecules that leave the liquid are moving rapidly, and they disperse as water vapor to all parts of the room; only a few molecules return to the jar. Since the rate at which water molecules leave the liquid greatly exceeds the rate at which they return, the water gradually evaporates. Heat from the room provides the energy for the latent heat of vaporization.

Water molecules continue to evaporate even if the jar is tightly sealed. However, since the water vapor cannot disperse widely, the air above the liquid in the jar becomes mixed with increasingly more water molecules. From time to time, they strike the surface of the water and some become bound to the liquid again. Eventually a condition of equilibrium is reached in which the number of water molecures leaving the liquid is balanced by the number of molecules returning. In this state, the air in the jar is said to be *saturated;* it contains as much water vapor as it can hold at its particular temperature.

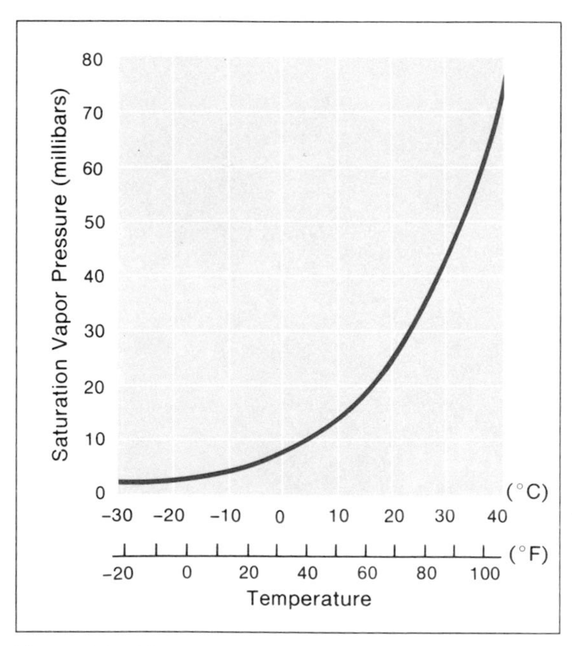

Figure 4.2 The curve shows that the saturation vapor pressure of air increases rapidly with increased temperature. The data extend to temperatures lower than 0°C (32°F), the normal freezing point of water, because small droplets of water can remain liquid at temperatures as low as −40°C (−40°F); this is known as "supercooled" water. (After List, 1971)

Like any other gas, water vapor exerts pressure. The amount of pressure that water molecules exert in the air is called the *vapor pressure.* Vapor pressure is at its maximum when the air is saturated. The maximum vapor pressure at a given temperature is known as the *saturation vapor pressure.* The saturation vapor pressure depends on the temperature of the air, and it increases rapidly as the temperature rises. Figure 4.2 shows that for temperatures found at or near the earth's surface, the saturation vapor pressure nearly doubles for each 10°C increase in temperature.

This relationship between temperature (or energy availability) and saturation vapor pressure becomes clear when we recall that temperature exerts a direct influence on the rate of evaporation from water surfaces. When the temperature of water is increased, the water molecules move more energetically, and a larger proportion of them gains enough energy to leave the liquid. The rate of evaporation increases rapidly with increasing temperature. Because more water molecules are released into warm air and because those molecules move faster and exert a greater force, the saturation vapor pressure of warm air is much greater than that of cold air.

Except in fog or cloud conditions, however, the atmosphere is not usually saturated, and the vapor pressure of the air is less than the maximum possible for its temperature. This situation is important in determining the rate at which evaporation takes place. While the rate of evaporation from water surfaces depends to a great extent on the temperature of the water and air, another important factor is the degree to which the air directly above the surface is saturated with water vapor. In a closed jar, there is no variation in vapor pressure between the air immediately over the water surface and that above. Under normal conditions, however, there is usually a *vapor pressure gradient* from the water through the air above. A gradient is simply the degree to which a quantity varies with distance. Typically, the air over a particular water surface shows a decrease in vapor pressure with increasing altitude. How much evaporation takes place depends on how "steep" the gradient is, or, to put it another way, how rapidly the vapor pressure decreases from the water surface through the air above.

The vapor pressure gradient is chiefly determined by the degree of turbulence and mixing in the atmosphere. For example, when the air is calm, a layer of nearly saturated air forms immediately above the water. Because the difference in vapor pressure within this air layer is small and the gradient is therefore low, or "weak," the rate of evaporation decreases; molecules from the water enter the air nearly as often as they return to the water. During windy weather, however, many water molecules are dispersed through a much deeper layer of the atmosphere, making the vapor pressure gradient steeper, and evaporation rates approach maximum values possible for the existing water temperature and vapor pressure gradient. To maintain the evaporating process, energy for the latent heat of vaporization must come from warmer waters immediately below the surface, from warm air above the surface, or, most typically, from absorbed solar radiation. Evaporation depends, therefore, on energy to maintain the temperature of the water surface, as well as on a vapor pressure gradient into the overlying air.

There are few, if any, routine measurements of evaporation rates from the world's oceans or large lakes; thus the global distribution of evaporation has to be evaluated indirectly by means of energy-budget estimates. In general, average annual evaporation rates are related to latitude and the corresponding patterns of solar radiation gains and longwave radiation losses, as shown in Figure 3.10, p. 71. Mean annual evaporation, therefore, ranges from more than 100 cm (40 in.) over equatorial and subtropical oceans to less than 10 cm (4 in.) over polar oceans.

Evaporation rates vary regionally as well as according to latitude. These regional differences are mainly associated with such factors as relative cloudiness, water temperature, and water vapor content of the atmosphere. The highest annual evaporation rates occur over subtropical oceans, which are characterized by generally fair weather, relatively dry air aloft, and warm waters. Evaporation tends to be somewhat less near the equator, where there is more atmospheric water vapor and therefore a more gradual vapor pressure gradient. A very steep pressure gradient occurs in winter over the Eastern seaboard of the United States as warm subtropical waters of the Gulf Stream persistently move poleward under colder and relatively dry air from the nearby land. The combination of warm water and dry air creates very high evaporation rates throughout this area.

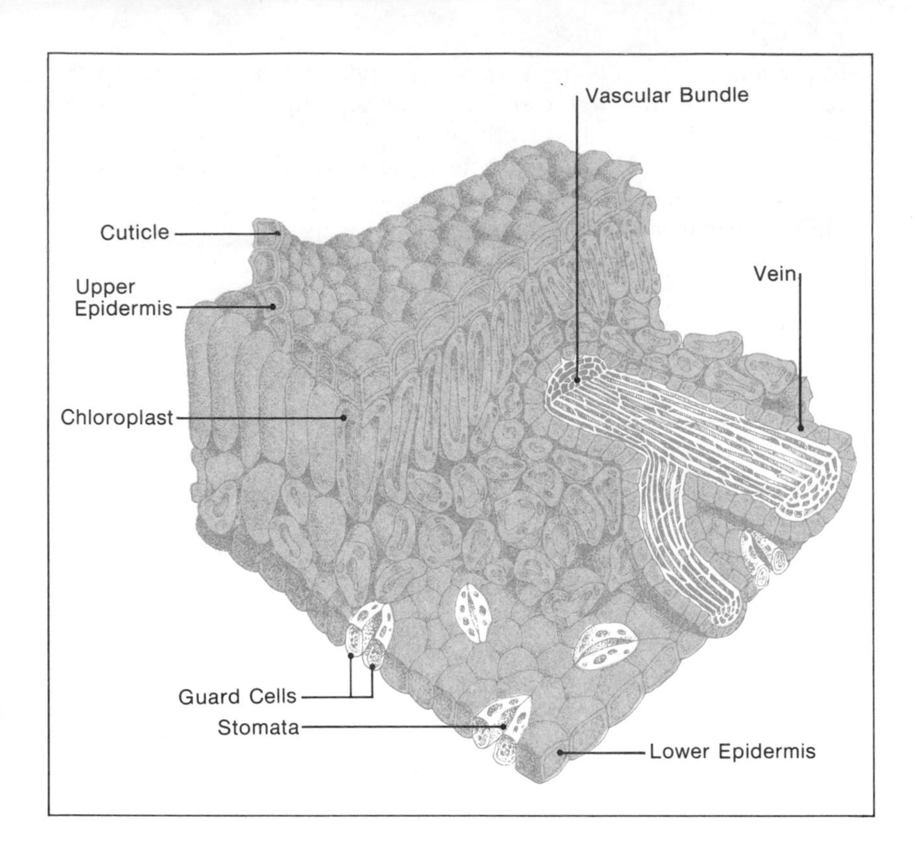

Figure 4.3 The leaves of green plants possess openings known as *stomata* in the bottom layer of protective epidermis. When the leaf is exposed to light, photosynthesis occurs in the chloroplasts, and the stomata are open to allow the entry of carbon dioxide and the exit of oxygen and water vapor. Soil moisture absorbed by the roots is transported to the leaves through the vascular bundles in the veins. The transpiration of water vapor is an essential process in green plants, and most of the water vapor entering the atmosphere in heavily vegetated regions is from plant transpiration rather than from evaporation.

The guard cells around the stomata close the openings if the plant lacks moisture and begins to wilt. In plants adapted to hot, dry climates, the guard cells keep the stomata closed during the hottest part of the day, when water loss by transpiration would be greatest. In some plant species adapted to dry climates, the covering, or cuticle, is particularly thick and waxy to prevent loss of moisture through cell walls.

Evapotranspiration from Land Surfaces

Over lakes and oceans, evaporation occurs directly from the surface of the water. On land where plants are growing, a small part of the water vapor entering the atmosphere comes from direct evaporation of water in the soil, but most is released through openings, called *stomata,* in plant leaves (see Figure 4.3). Plants absorb soil water through their roots. The water and dissolved nutrients from the soil are then transported up the stem of the plant, and most of the water is eventually transpired through the stomata as vapor. The flow of water through a plant is an essential part of its life processes. During the day, when a plant is photosynthesizing, the stomata are open to allow the entry of carbon dioxide; simultaneously, water vapor escapes by transpiration. In the dark, the stomata are closed and there is little transpiration. The energy requirement for transpiration is the same as that for evaporation: 590 calories are required to transpire 1 gram of water at ordinary temperatures.

For the purposes of physical geography, it is useful to evaluate water loss from land surfaces as the combination of transpiration from vegetation and evaporation from soil. These two processes are known collectively as evapotranspiration. In areas with dense vegetation cover, water loss to the atmosphere by transpiration is, on the average, at least two or three times greater than water loss by evaporation. This proportion, however, may vary considerably, affecting the rate of evapotranspiration as a whole. For example, when water is readily available in the soil, evapotranspiration rates are dependent on energy availability—especially solar radiation—and on vapor pressure gradients. Therefore, evapotranspiration rates tend to be high during the day and low at night, and higher during summer than winter. However, during periods between rainfalls when soil moisture becomes depleted, evapotranspiration rates decrease. Because of the many variables involved, it is much more difficult to evalute annual global patterns of evapotranspiration from the continents than to estimate evaporation from the oceans. Evapotranspiration and its effect on the local water budget are treated in more detail in Chapter 6.

MOISTURE IN THE ATMOSPHERE

The air always contains water vapor. At a given temperature, air may contain any amount of water vapor from almost none up to the maximum value set by saturation. As Figure 4.4 shows, air at 10°C (50°F), for example, can contain up to 10 grams of water vapor per cu meter. Ten grams, which is the weight of two American 5-cent pieces, may not seem like much water, but rain clouds that contain 10 grams of water vapor per cu meter have the potential to produce rainfall several centimeters deep on the ground below. The water vapor content of the atmosphere is referred to as *humidity.* Humidity can be described by a number of different measures, and each is useful for certain specific situations.

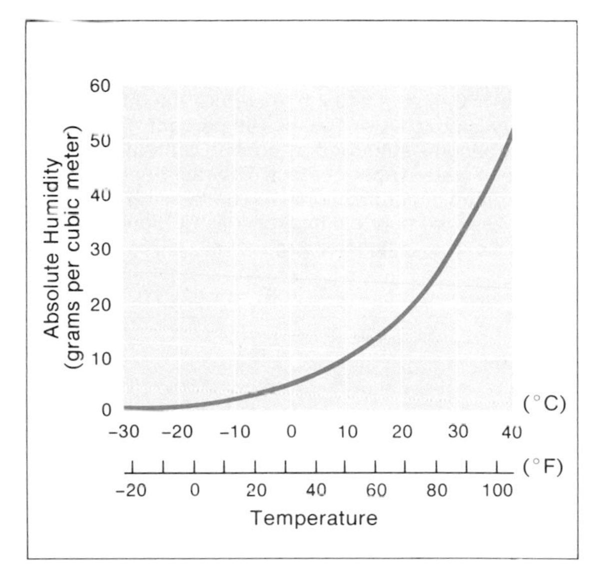

Figure 4.4 The curve shows the maximum amount of water vapor that can be contained in a cubic meter of air at a given temperature. If the air is not saturated, the absolute humidity lies below the curve. The absolute humidity of saturated air rises rapidly with increased temperature, so that warm air is able to contain more water vapor per unit volume than cool air can contain. The shape of the absolute humidity curve is similar to the saturation vapor pressure curve shown in Figure 4.2, p. 88. (After List, 1971)

Measures of Humidity

There are several ways to specify the amount of water vapor present in a quantity of air. The most direct measure of the air's moisture content is the *absolute humidity,* which is the weight of water vapor present in a given volume of air. Absolute humidity is usually expressed in terms of grams of water vapor per cu meter of air. (One cu meter is approximately equal to 35 cu ft.) A disadvantage of the absolute humidity measure is that if the volume of the air changes, as when warming or cooling air expands or contracts, the value of the absolute humidity also changes even though there is no change in water vapor content. Use of the absolute humidity, therefore, tends to be restricted to experiments in physics and engineering where air volume is kept constant.

Relative humidity is the ratio of the amount of water vapor present in a quantity of air compared to the amount that could be held by the same air if it were saturated. To say it another way, relative humidity is the ratio of the *actual vapor pressure* of water vapor in the air to the *saturation vapor pressure.* Usually expressed as a percentage, the relative humidity ratio does not by itself indicate the actual amount of water vapor in the air. For example, hot air at 40°C (104°F) and 50 percent relative humidity contains about 25 grams of moisture per cu meter, but cool air at 20°C (68°F) and 80 percent relative humidity contains only about 14 grams of moisture per cu meter. The cool air in this example has a lower water vapor content but a higher relative humidity than the hot air because the cooler air is closer to saturation. As Figure 4.5, p. 92, shows, air can become saturated by

Figure 4.5 Changes in temperature can affect relative humidity even when the actual moisture content of the air remains unchanged. Consider a hypothetical sample of air taken at 10:00 A.M. with a temperature of 15°C and a vapor pressure of 10 millibars (mb). This sample is represented by point *A*. Saturation vapor pressure for 15°C is about 17mb (point *B*); so the relative humidity is approximately $^{10}/_{17} \times 100$, or 59 percent. Point *C* represents a sample taken the same day at 3:00 P.M. The water vapor content of the air has remained unchanged, but the temperature has risen to 25°C. Saturation vapor pressure for 25°C is about 32 mb (point *D*); so the relative humidity is now $^{10}/_{32} \times 100$, or 31 percent. Relative humidity has decreased as the air has grown warmer. Point *E* represents a sample taken the following morning at 5:00 A.M. Water vapor content is still the same, but the temperature is now 7°C. Saturation vapor pressure for 7°C is about 10 mb; therefore the relative humidity is now $^{10}/_{10} \times 100$, or 100 percent. The sample has become saturated by cooling without any change in water vapor content. The rapid decrease of saturation vapor pressure with decreasing temperature has very important implications for condensation processes. (After Neiburger et al., 1973)

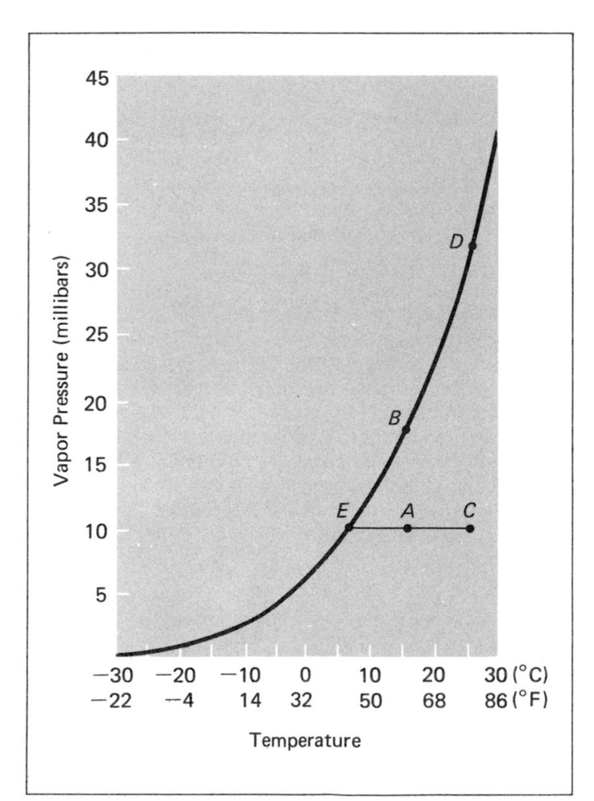

cooling without any change in actual water vapor content; this fact has important consequences in condensation processes.

Because air at different temperatures has different capacities for holding water vapor, a disadvantage of the relative humidity measure is that whenever the temperature changes, the relative humidity also changes. Even if there is little change in the amount of atmospheric water vapor during the course of a day, the relative humidity nevertheless changes markedly (see Figure 4.6).

Two other measures that are sometimes used to characterize the water vapor content of air are the specific humidity and the mixing ratio. The *specific humidity* is the number of grams of water vapor per kg of air. The *mixing ratio* is the ratio of the weight of the water vapor to the weight of the air minus its water vapor. Both of these measures are useful because their numerical value changes only when the amount of water vapor increases or decreases. The specific humidity and mixing ratio are often used in studies of air mass properties and movements; air masses are discussed in Chapter 5.

Distribution of Water Vapor in the Atmosphere

At any given time, only a small fraction of the earth's total water supply is stored in the atmosphere. If all the atmospheric water vapor fell to the earth as rain, it would form a layer of water only 2.5 cm (1 in.) in depth. This is because there is a constant turnover of atmospheric water vapor: water that is gained by evapotranspiration replaces water that has been simultaneously lost by precipitation. Moreover, this cycling process is subject to considerable variation in terms of the amount of water stored and its movement through the atmosphere.

On a fair summer day when the net radiation is positive and the ground is warm and moist, evapotranspiration proceeds rapidly. Atmospheric turbulence, or convection, causes water vapor to disperse quickly as it is carried upward, so that the water vapor content of the air is greater near the ground than a few hundred meters aloft.

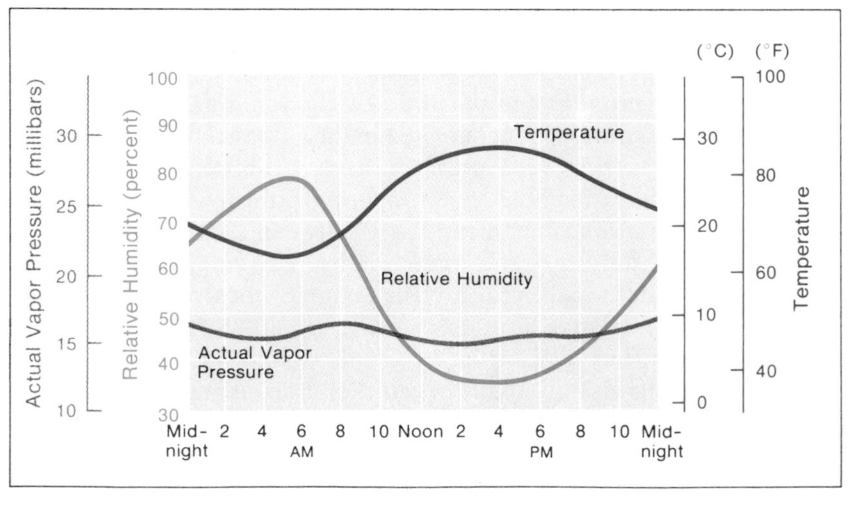

Figure 4.6 During fair weather conditions, the vapor pressure of the air near the ground remains comparatively constant through the day. However, the relative humidity changes markedly through the day as the air temperature changes, because the saturation vapor pressure of air is strongly dependent upon temperature.

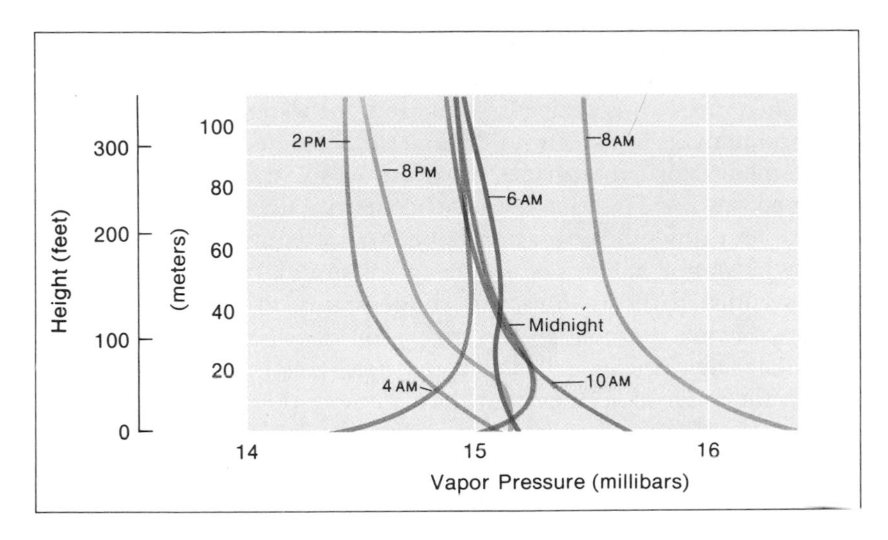

Figure 4.7 Each curve on this graph shows how the vapor pressure in the atmosphere near the ground varies with height at a particular time on a clear summer day. At 8:00 A.M. the vapor pressure decreases rapidly with increased height, indicating a flow of vapor from the surface into the atmosphere. The vapor pressure decreases from 8:00 A.M. to 2:00 P.M. because the air becomes heated during the morning and early afternoon, causing the onset of convection and a consequently more efficient removal of moist air from near the surface. Much of the water vapor remains in the lowest 50 meters of the atmosphere, forming a humid blanket. Water vapor continues to flow upward from the surface during the day while plants are transpiring. By 8:00 P.M. the vapor pressure is nearly constant through the lowest 15 meters of height, indicating that little water vapor is leaving the surface. Later at night, the vapor pressure immediately above the surface increases with increased height, indicating a flow of water vapor from the atmosphere toward the ground. By 6:00 A.M., after sunrise, the flow is once again from the surface to the atmosphere. (After Geiger, 1965)

Another way of describing this situation is to say that the vapor pressure gradient is steep (see Figure 4.7). At night, however, when longwave radiation carries energy from the ground, the air directly above the surface may become cooler than the air higher up. The colder the air becomes, the less water vapor it is able to hold. If the lowest layer of air is chilled to the saturation point, water vapor leaves the saturated air and condenses on the ground as dew or frost or near the surface as fog. In this situation, the vapor pressure gradient undergoes an inversion; the atmospheric water vapor content is slightly lower near the surface than a few tens of meters aloft. The formation of dew, frost, and fog is discussed again later in this chapter.

The water vapor content of the atmosphere varies horizontally as well as vertically. Because temperature affects the amount of water the air can hold, there are marked regional differences in the distribution of water vapor. Average storage over the United States ranges from about 5 cm (2 in.) near the Gulf of Mexico coastline during summer to less than 0.8 cm (0.3 in.) near the Canadian border during winter. This pattern of water vapor distribution is far from stationary, however. Occasional daily rainfalls greater than 15 cm (6 in.) indicate that converging atmospheric circulations are capable of quickly transporting large quantities of water vapor horizontally to a given region.

CONDENSATION AND PRECIPITATION PROCESSES

Condensation is the process by which water in the atmosphere changes phase from water vapor to tiny liquid droplets or ice crystals. When condensation occurs at the earth's surface, it produces dew or frost. When water vapor condenses higher in the atmosphere, the result is a mass of water droplets and ice crystals known as a cloud. Fog is simply a cloud near the ground. The tiny water droplets or ice crystals that result when condensation takes place are too small and light to fall to the ground. Atmospheric turbulence keeps them aloft. Under certain conditions, however, thousands of these microscopic droplets or crystals combine to form raindrops or snowflakes which are then large enough to fall to the ground as precipitation.

When moist air is cooled to saturation, the water vapor in the air no longer remains in stable equilibrium; instead, it tends to condense. The condensation process does not occur easily. If moist air free of dust is cooled, condensation does not occur until the vapor pressure is three or four times the saturation value. Air containing more than its normal saturation amount of water vapor is *supersaturated.* If the air contains dust particles, however, condensation of vapor to water droplets occurs as soon as the air becomes even slightly supersaturated. The particles, called *condensation nuclei,* act as collection centers for water molecules, and they promote the growth of water droplets.

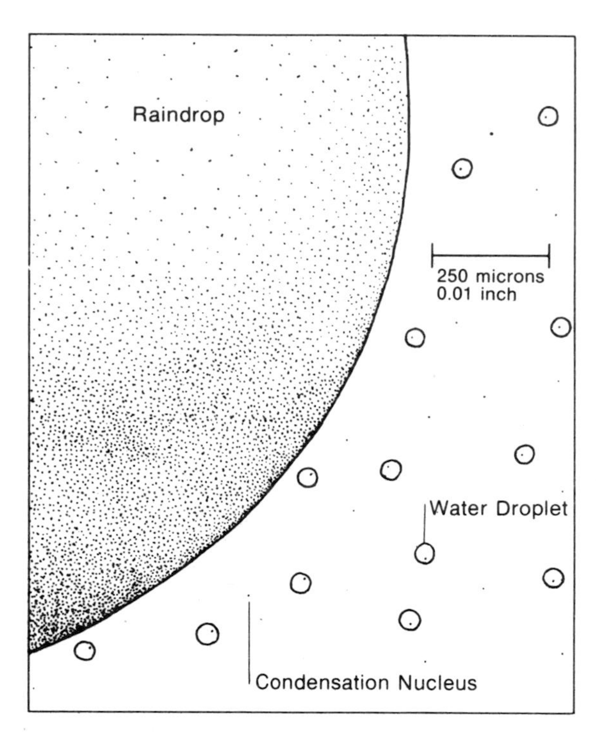

Figure 4.8 This drawing shows to scale the relative sizes of condensation nuclei, water droplets, and a raindrop. Most condensation nuclei are less than 1 micron in radius; water droplets are usually less than 20 microns in radius, and raindrops have radii of 1,000 microns or greater. Each water droplet forms in saturated air by condensation of water vapor on a condensation nucleus, and other processes cause millions of water droplets to coalesce and form a raindrop.

Figure 4.8 shows that condensation nuclei are microscopically small. Very few exceed a radius of 1 micron (1 micron = 10^{-4} cm), and most have radii smaller than 0.1 micron. Giant condensation nuclei—those with radii greater than 1 micron—number only about 1 per cu cm. Principal sources of condensation nuclei in the atmosphere are vapors and smokes from industry and forest fires, salt crystals from sea spray, clay particles from blown soil, and chemicals (particularly sulfates) formed by reactions of trace materials in the air. Polluted air over land contains many thousand such particles per cu cm of air.

Fine particles can remain in the atmosphere for long periods of time; when the volcano Krakatoa near Java erupted in 1883, the dust content of the global upper atmosphere was greater than normal for more than a year afterward. Although the force of gravity pulls small particles in the air toward the earth, the air flow, or drag force of the air, around the particles retards their fall. The drag force increases with the speed of the particle. A falling particle soon reaches a steady maximum speed, called the *settling velocity.* The smaller the particle, the slower the settling velocity. The settling velocity of a 1-micron particle is only a few microns per second. Small particles, such as those in tobacco smoke, remain effectively "suspended" in air because random air currents carry such small particles upward faster than they can fall. Rain tends to wash particles out of the air, but natural processes and industrial activities renew the supply.

Air in the atmosphere is normally dusty enough for condensation to occur. When a parcel of air is cooled to saturation, water vapor readily condenses to droplets, forming a cloud. The droplets in the cloud grow to a radius of 10 or 20 microns before condensation of additional water molecules on the droplets ceases and the droplets stop growing. A typical cloud contains a million such droplets per cu meter, and the average water content of a cloud is about 1 gram of water per cu meter of air.

The tiny droplets that make up clouds can never fall to the ground as rain. A water droplet with a radius of 10 microns has a settling velocity of a fraction of a centimeter per second and would evaporate in the drier air below the cloud after falling only a few centimeters. Under normal conditions, droplets do not grow larger than 10 or 20 microns in radius, nor can they combine to form larger drops. As theoretical and experimental studies have shown, droplets smaller than 20 microns in radius cannot collide and coalesce. When such droplets approach one another, they are forced apart by the air flow between them and cannot merge. Thus most clouds remain as tiny droplets and never form rain.

A water drop must have a radius of at least 100 microns in order to fall from a low cloud to the ground. Thousands of 10-micron droplets must combine to produce one 100-micron drop, and millions of droplets are required to form the average raindrop, which has a radius of 0.1 cm (1,000 microns). The mechanisms that cause water droplets to coalesce to drops large enough to fall from clouds have been studied intensively since the 1940s. Scientists have isolated two separate processes responsible for precipitation. The models which explain these processes are known as the coalescence model and the Bergeron ice crystal model.

The Coalescence Model of Rainfall

As we have seen, intervening air currents prevent average-sized droplets from merging with one another and forming larger droplets. However, if a droplet is oversized—larger than 20 or 30 microns in radius—it may overcome air resistance and grow by coalescing with smaller droplets. The formation of raindrops by coalescence takes place only under certain conditions.

First, droplets grow by coalescence in clouds that are too warm to allow the formation of ice crystals; such clouds are most frequently found in the tropics. Second, these warm clouds must be of a certain size. An embryo raindrop grows by colliding with and sweeping up smaller droplets as it falls through the air. The larger the drop becomes, the faster it falls. Rising currents of air carry the drops upward faster than they can fall out of the cloud. When the growing drop becomes so large that air currents can no longer support it, it falls from the cloud as rain. A droplet requires about ½ hour to grow to raindrop size by coalescence, and rain clouds must be about 1 km (0.6 mile) thick for the growing drops to remain in the cloud long enough to become raindrops. Thinner clouds limit the growth of drops but may produce *drizzle*, a form of precipitation that consists of drops smaller than average raindrops.

Finally, the generation of rain by the mechanism of coalescence depends on the occurrence of oversized droplets, which form only on giant condensation nuclei. Salt particles 1 micron or more in radius make particularly effective condensation nuclei for oversized droplets. One reason that salt particles work so well as condensation nuclei is the *hygroscopic* nature of salt—its natural tendency to take up water vapor. Even far inland, salt particles blown in from the ocean can play an important role in precipitation. Nor is it necessary for the salt particles, or other giant condensation nuclei, to be present in heavy concentration. Because about a million droplets are needed to form an average raindrop, the generation of rain requires the presence of only one oversized droplet for every million small droplets. The average cloud contains about a million small droplets per cu meter; therefore, in order for raindrops to form, only one giant condensation nucleus is needed for each cu meter of cloud.

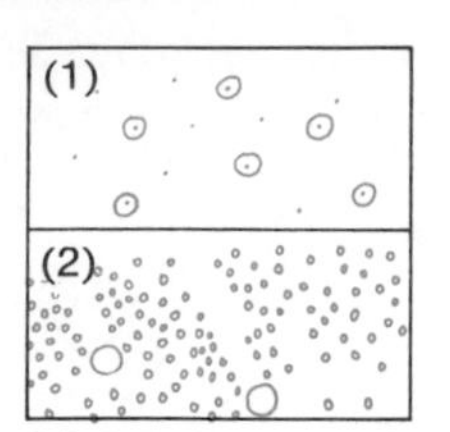

(1) Condensation of Vapor Onto Nuclei and Growth of Droplets From Vapor

(2) Growth of Raindrops by Coalescence With Droplets

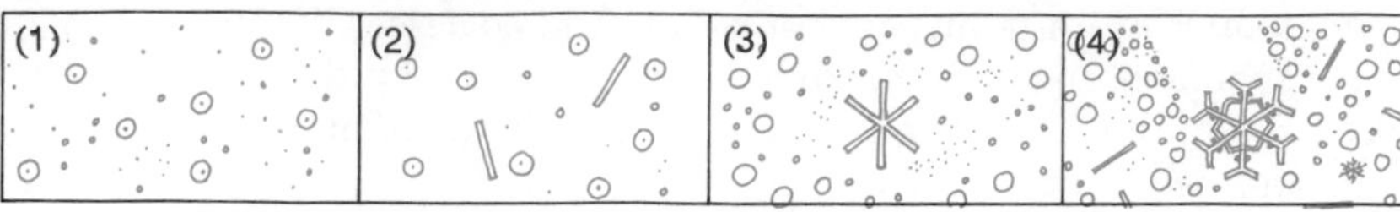

(1) Condensation of Cool Vapor Onto Nuclei and Growth of Supercooled Droplets

(2) Freezing of Supercooled Droplets Onto Ice-Forming Nuclei

(3) Growth of Ice Crystals at the Expense of Water Droplets

(4) Further Growth of Ice Crystals by Coalescence With Ice Crystals and Supercooled Droplets

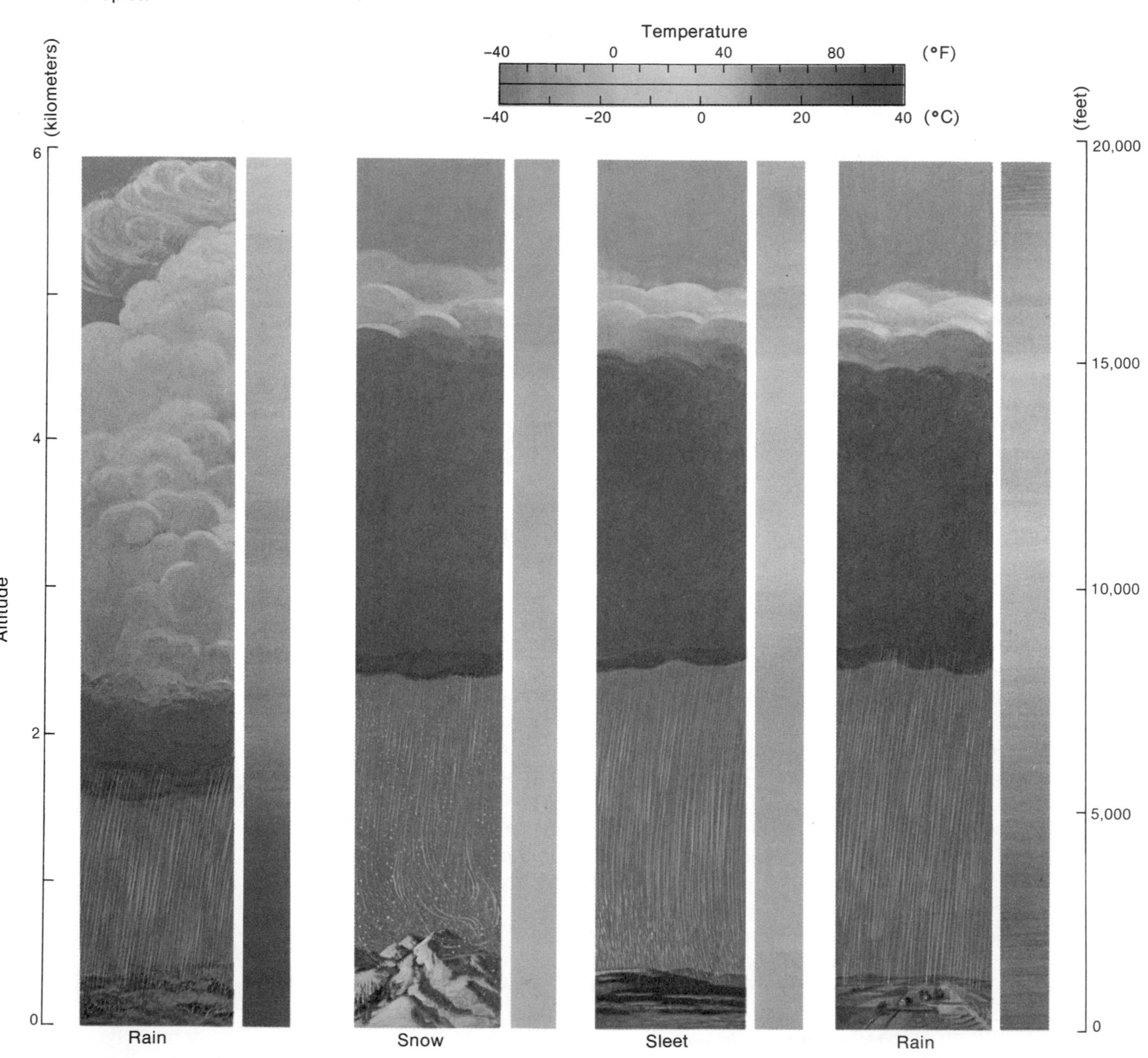

The Bergeron Ice Crystal Model of Rainfall

Coalescence of oversized water droplets accounts for precipitation from clouds that are uniformly warmer than the freezing point of water. But many rain clouds, such as those in the middle latitudes, extend upward to altitudes where the temperature is well below freezing. In 1933 Tor Bergeron, a Swedish meteorologist, proposed a mechanism of droplet growth to explain rainfall from cold clouds.

Water in a relatively large quantity freezes at 0°C (32°F). But water dispersed as fine droplets can remain liquid at temperatures as low as −40°C (−40°F). Water that remains liquid at a temperature below the normal freezing point is called *supercooled*. Because of the supercooling, water droplets in a cloud can remain liquid even when the upper part of the cloud is in air colder than 0°C. Certain particles in the air, known as *ice-forming nuclei*, promote the freezing of supercooled droplets. Once ice crystals have formed on nuclei, they can continue to grow by adding water molecules from the vapor. Water molecules are more strongly bound to ice than to supercooled water droplets; thus ice crystals can grow at the same time that droplets are losing molecules by evaporation. Ice particles can also grow by colliding with supercooled droplets, which freeze onto the ice. As the large ice crystals fall through the lower, warmer parts of the cloud, they may melt and grow larger by coalescing with water droplets. Finally, they fall from the cloud as rain. Ordinary summer raindrops in the middle latitudes often begin life as ice and snow in the upper parts of towering clouds and then melt as they fall. In winter, they fall as snow if the air is not warm enough to melt the ice crystals.

Although effective ice-forming nuclei are relatively rare in nature, only a hundred or so per cu meter are needed to trigger snow from a thick cloud. The formation of ice crystals usually occurs only in parts of the cloud supercooled to −15°C (5°F) or lower. In addition, the size of the cloud and the speed of the air currents must be favorable to the growth of large snowflakes if precipitation is to reach the ground. The numerous factors involved and their interplay are not yet fully understood.

Formation of Hail

Hailstones are spherical balls of ice that fall from tall, vigorous clouds, such as summer thunderheads. They range from minuscule grains to balls the size of a grapefruit. When cut open, hailstones show an onionlike structure of clear and opaque shells of ice. The layering implies that a hailstone is cycled through a cloud many times during its growth (see Figure 4.10, p. 98).

According to the accepted model of hailstone growth, powerful air currents in a thunderstorm cloud carry small ice particles upward through the cloud's central column. If the particles rise through a region that contains a large number of supercooled droplets, a wet layer may form. The slow subsequent freezing produces a layer of clear ice. If the ice particles then pass through a region with only a few small supercooled droplets, the collisions result in instant freezing,

Figure 4.9 (opposite) The principal models that successfully explain precipitation from clouds are the coalescence model and the Bergeron ice crystal model; the main steps of each process are outlined in this figure.

The coalescence model of precipitation is dominant in the tropics, where many clouds are too warm to contain supercooled water droplets and ice crystals. The Bergeron process dominates at higher latitudes, where the upper portions of many clouds are considerably colder than the normal freezing point of water, even in summer.

The type of precipitation that falls from a cloud depends on the mechanism involved and on the variations of temperature in the atmosphere. Precipitation is formed as snow in the Bergeron process, and if the atmosphere beneath the cloud is below freezing, the precipitation reaches the ground as snow. If the snow falls into warm air, however, the precipitation arrives at the ground in the form of rain. The coalescence process cannot produce snow, but either process can produce sleet or ice pellets if raindrops fall through a layer of cold air near the ground and freeze.

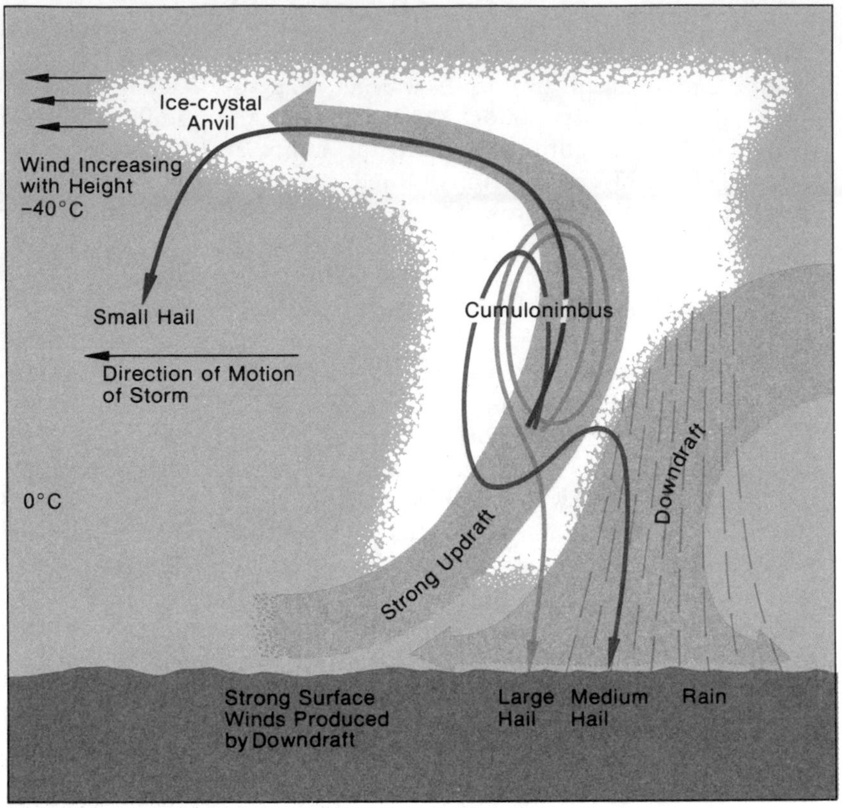

Figure 4.10 This cross section of an intense thunderstorm illustrates the cyclic processes involved in the formation of hailstones. Powerful currents of air in the central updraft carry falling raindrops and ice crystals upward again. Because of the forward motion of the storm, ice particles may be swept up in the updraft a number of times. On each passage through the cloud, the particle gains a new layer of ice, thus becoming a hailstone. Clouds with strong updrafts are able to support hailstones through many cycles until they become too heavy for the air currents to lift and fall to earth. Particles are also blown away from the central column by strong winds aloft, giving the cloud its characteristic anvil shape. (After Flohn, 1969)

producing eventually an opaque or cloudy layer of ice with pockets of trapped air.

Near the top of the cloud, ice particles emerge from the central column of uprushing air and fall downward in slower air currents in the outer parts of the cloud. As they near the base of the cloud, however, they can be swept back into the central column by the cloud's advance. With each cycle through the cloud, the particles gain new layers of ice. After a number of cycles, they grow too large to be supported by the air currents and fall to the ground as hail. Large hail is associated with violent, turbulent thunderstorm clouds because vertical air currents with speeds of 20 to 30 meters per second (45 to 65 miles per hour) are required to carry large hailstones upward.

COOLING OF THE ATMOSPHERE

In order for precipitation to occur, the water vapor content of at least one layer of the atmosphere must be very close to capacity. There are three ways in which this state of saturation can be reached: by adding water vapor to the atmosphere, by cooling the atmosphere (see Figure 4.5), or by a combination of added moisture and cooling. The following sections focus on situations in which cooling, saturation, condensation, cloud formation, and precipitation frequently occur. We begin with the most simple case of nighttime radiational cooling of the air at the ground.

Figure 4.11 Dew forms on plants when the surrounding air cools to the dew point.

Cooling of Surface Air: Dew and Fog

The temperature at which the relative humidity becomes 100 percent is called the *dew point*. If the air cools to the dew point, it cannot remain in stable equilibrium with the amount of moisture it contains; some of the water vapor must condense into liquid. A glass of ice water rapidly becomes covered with condensed moisture in humid summer weather because the moist air near the glass is chilled to the dew point, and it deposits its moisture on the glass.

The same process accounts for the formation of dew. Soil and plants cool rapidly during the night by longwave radiation. As they cool, the air near them also decreases in temperature. If the air cools to the dew point, its water vapor will condense, forming droplets of moisture on the ground and vegetation (see Figure 4.11).

The condensation of moisture at the earth's surface depends primarily on two factors: water vapor content of the air and temperature. Figure 3.12, p. 73, in Chapter 3 shows that during a fair day at Hancock, Wisconsin, the surface experiences a net radiational loss from just before sunset to just after sunrise. The surface continues to cool during this entire period when the net radiation is negative. If the surface air is moist, dew would probably form in the early evening. However, if the air is very dry, condensation might not begin until around sunrise. If condensation occurs when the dew point is lower than the freezing point of water, the water vapor begins to condense as ice crystals, and frost appears on every exposed surface across the

Figure 4.12 (right) This photograph shows San Francisco Bay and Oakland, California, shrouded in fog, with Mount Diablo in the distance to the northeast. Low-lying advection fog frequently occurs in the coastal region around San Francisco as moist air moving eastward over the Pacific Ocean cools to its dew point.

Figure 4.13 (below) This map shows the average annual number of days of fog in the conterminous United States. Fog occurs most frequently along the Pacific Coast and along the coast of New England, where moist air is cooled by cold ocean waters. Fog also occurs frequently in the Appalachian Mountains of the east central United States. It seldom occurs in the warm, dry air of the western deserts. (After Court and Gerston, 1966)

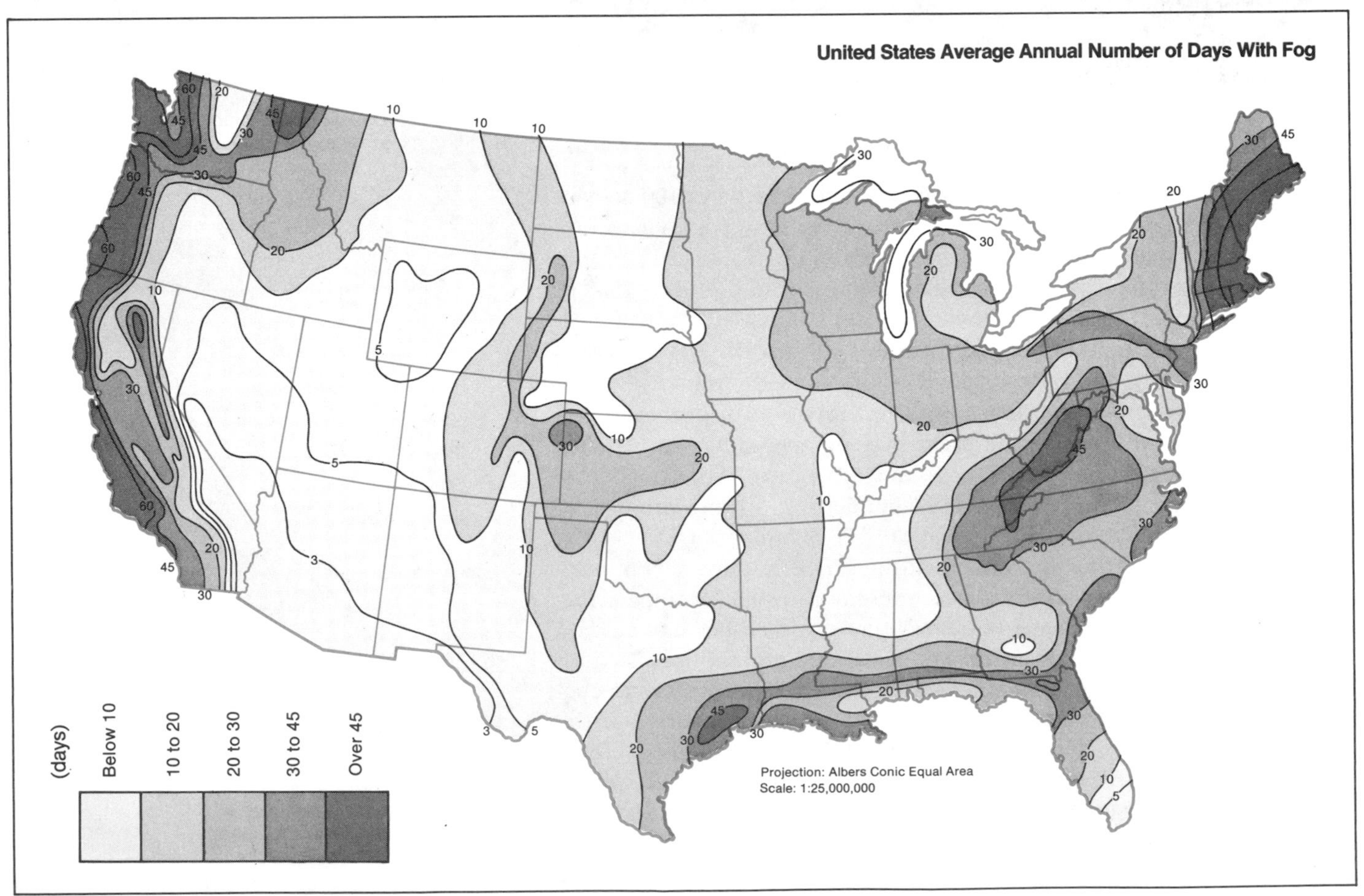

landscape. Condensation of water vapor to liquid makes available approximately 590 calories of heat for every gram of water vapor that condenses. The heat energy made available when dew forms tends to keep the soil from cooling as much as it otherwise might.

A fog is actually a cloud at or near the earth's surface. It occurs when moist air near ground level is cooled to its dew point. There are several ways such cooling can occur: depending on the mechanism involved, fogs can be classified as radiation fogs, advection fogs, or orographic fogs.

Radiation fogs occur at night, when the ground loses heat through longwave radiation. The air near the ground is cooled by contact with the cool ground. If the air happens to be moist, the lowest 10 meters or more may be chilled to the dew point, resulting in a dense fog. Clear, still nights facilitate the formation of radiation fogs because the cool air remains stationary long enough for condensation to take place. If it is windy, the cool air near the ground mixes with warmer upper layers of air, and the dew point may not be reached.

Advection refers to the horizontal movement of air across the earth's surface. When warm, moist air passes over a cool surface, such as snow or cool ocean water, the air may be cooled to its dew point and form advection fog. Large concentrations of cold ocean water located regularly off the Pacific coast and occasionally off the New England coast frequently cause the formation of advection fogs, which may then be swept inland. The map in Figure 4.13 shows that fog is most common adjacent to coastlines.

When advection fog occurs over snow, the result is usually a marked increase in the rate at which snowmelt takes place. Normally, snow melts slowly because it has a high albedo and reflects much of the incoming solar radiation. However, the advection of warm, moist air over snow-covered areas may drastically alter this situation. For example, when air from the Gulf of Mexico, which is high in water vapor, travels over snowfields in Pennsylvania, the warm air is cooled to saturation, and condensation takes place on the snow surface. Every gram of water vapor that condenses makes enough latent heat available to melt more than 7 grams of snow; thus rapid melting of snow can occur even on a cool, overcast day by this process. Condensation melt can supplement the melting that occurs as cool rain falls on snow, and it can produce widespread winter snowmelt floods in areas where the air is often humid in winter, such as near the Great Lakes and in New England.

Orographic fogs are frequently associated with mountain areas, such as the Appalachians (see Figure 4.13). When warm, moist air is forced up mountain slopes, it is often chilled to the dew point, resulting in the formation of fog and cloud. The mechanism of orographic cooling will be discussed in more detail later in this chapter.

Cooling of surface air, either by radiation or advection, is rarely associated with the direct production of precipitation. Cooling and condensation are restricted to a shallow layer of the atmosphere near the surface, and the resulting fogs and shallow clouds are simply not thick enough for the Bergeron ice crystal process or coalescence to be effective. Instead, the air may be filled with small droplets which sometimes fall as drizzle. Precipitation can occur only when large parcels of air cool by being forced to higher elevations. The next sections discuss the various ways air masses can be made to rise.

Adiabatic Cooling

In Chapter 3 we saw that the net radiation is positive during a fair summer day from shortly after sunrise to shortly before sunset. Some of this energy gain goes into heating of the ground surface, and largely because albedo varies from place to place, some spots in the landscape heat up much more than others. Figure 4.14 shows how heated air immediately above local "hot spots" becomes organized into rising bubbles of warm air. Both soaring birds and glider pilots utilize these rising parcels of air, often called *thermals,* to gain altitude. Circling within the updraft, they are carried upward until the air cools and loses its tendency to rise.

This drop in temperature of rising air is caused primarily by a process known as *adiabatic cooling.* The term "adiabatic" refers to the

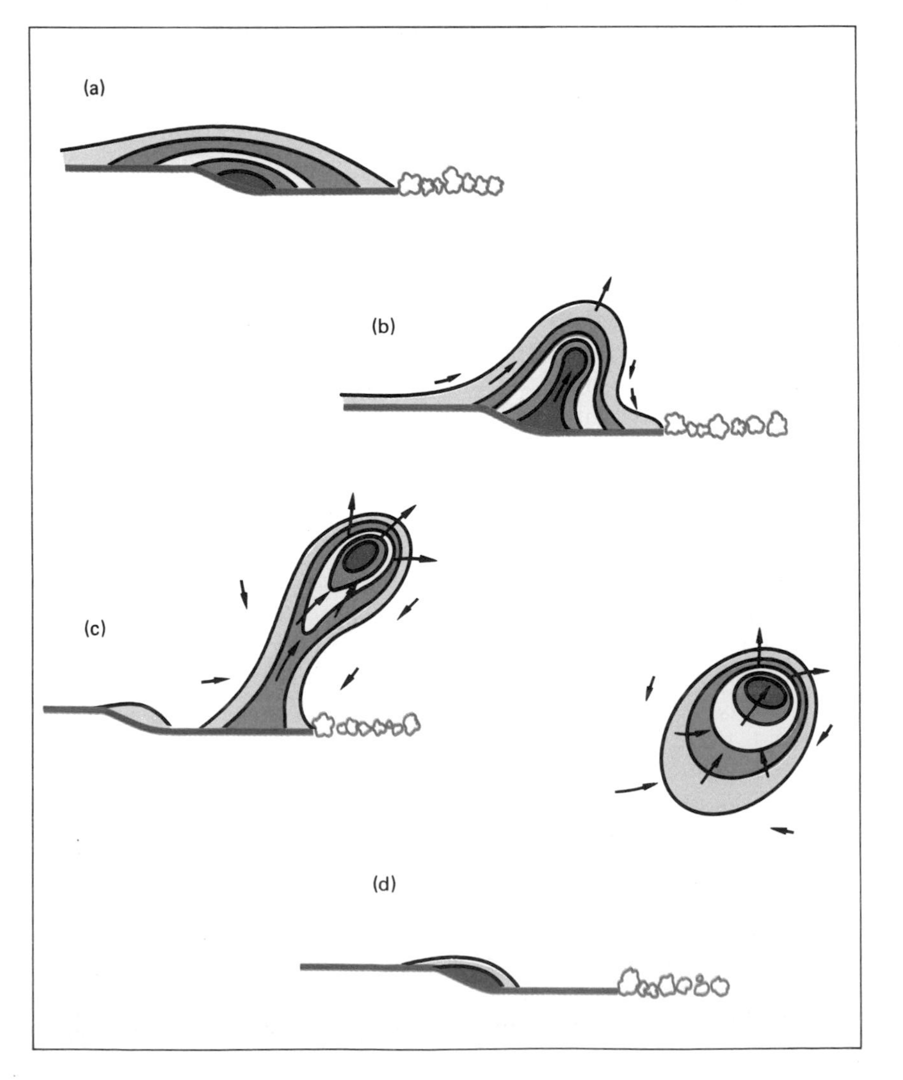

Figure 4.14 A thermal is a rising bubble of air emanating from a local "hot spot" on the earth's surface where the temperature exceeds that of the surrounding area. This differential heating is associated with variations in albedo, orientation of slopes to the sun, and the availability of water for evapotranspiration. The rising thermal initially exchanges little heat with the surrounding air. Each concentric circle, or isotherm, represents an increase of 0.1°C; thus the outer circle is 0.1°C warmer than the general atmosphere, the next circle is 0.2°C warmer, and so on. The highest temperature occurs at the center of the thermal. (After H. Riehl, 1972)

fact that the process occurs without the parcel's receiving or losing energy. Instead, the cooling that takes place depends entirely on the relationship between altitude and atmospheric pressure. As a parcel of warm air rises, the air surrounding it becomes increasingly less dense. In order to make its internal pressure equal to the pressure of the atmosphere, the higher-pressure air in the parcel pushes outward against the air which surrounds it. Because the molecules of the expanding air become more widely diffused, they no longer strike each other as often, and therefore the sensible temperature of the air in the parcel decreases.

The process can also be reversed. When a parcel of cool air descends, it becomes compressed. As a result, the molecules in the parcel collide more frequently and the temperature increases. This is called *adiabatic heating*.

The rate of adiabatic cooling of air that is not saturated and where condensation is not occurring is known as the *dry adiabatic rate*. Its value is 10°C for every km of increasing altitude (5.6°F per 1,000 ft), as illustrated in Figure 4.15. A descending parcel of cool air, subject to adiabatic heating, increases in temperature at the same rate.

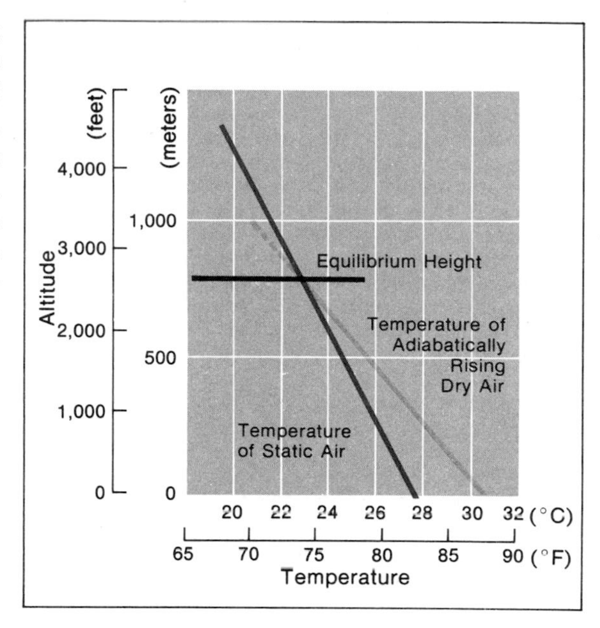

Figure 4.15 When a parcel of dry air rises, it expands in volume in order to equal the pressure of the surrounding atmosphere. The expansion causes the temperature of the parcel to fall. The figure compares the temperature of a typical parcel of rising dry air with the temperature of the air in the surrounding static atmosphere. The temperature of the parcel and the temperature of the static air both decrease with increased altitude, but in general the temperature of the parcel falls more rapidly, as indicated in the figure. The parcel experiences a net upward force, causing it to continue rising, as long as the temperature of the parcel exceeds the temperature of the static air. The equilibrium height of the parcel is near the height where the temperatures are equal.

A rising parcel of air cools at the dry adiabatic rate only until it reaches its dew point and becomes saturated. After this, it continues to cool but at a much reduced rate called the *moist adiabatic rate*. The change in the rate of cooling occurs because some of the water vapor in the parcel condenses to droplets, and the latent heat of condensation is made available to warm the air in the parcel. The added heat increases the buoyancy of the parcel and enables it to rise to greater altitudes. Not all the vapor in the parcel condenses at once. Condensation continues as the parcel rises, and at any given altitude the parcel contains just enough water vapor to maintain saturation. Eventually, the parcel reaches a point of equilibrium where its temperature matches that of the surrounding static air, and here its upward movement ceases. This mechanism can produce clouds that tower to great heights with potential for high-intensity precipitation. The moist adiabatic rate varies with the moisture content of the parcel. In warm, moist air, it may be 5°C per km (2.8°F per 1,000 ft) or even less (see Figure 4.16, p. 104). At very low temperatures, however, it approaches the dry adiabatic rate of 10°C per km.

Thermal Convection

On a fair day, cumulus clouds begin to form at an altitude where warm, moist air reaches saturation and condensation begins to occur. Warm air continues to rise until the equilibrium point is reached and the temperature of the rising parcel of air equals that of the surrounding atmosphere. The equilibrium point marks the greatest height to which the cloud can grow (see Figure 4.16). The vertical movement of air responsible for cloud formation is known as *thermal convection*, and it may vary significantly according to local weather conditions.

Because the atmosphere is heated mostly by longwave radiation from the ground, its temperature normally decreases with increasing altitude. A grand global average for this decrease under all weather

Figure 4.16 (a) A rising thermal or parcel cools at the dry adiabatic rate until condensation begins. The base of the cloud at about 0.7 km marks the altitude at which the thermal cools to the dew point and condensation begins. The parcel then cools at the moist adiabatic rate, which is less than the dry rate. The release of the latent heat of condensation enables the parcel to rise to much higher altitudes before it reaches equilibrium with the surrounding static air. Condensation continues as the parcel rises, and at any given altitude the parcel contains just enough water vapor to keep the air in it saturated.

(b) This shallow inversion, with warmer air above cooler air at the surface, is characteristic of radiational cooling at night. If a parcel were forced to rise, it would come quickly into equilibrium with the surrounding static air. Inversions are associated with atmospheric stability.

(c) With daytime heating, the inversion is much higher aloft, permitting thermals to reach much higher altitudes.

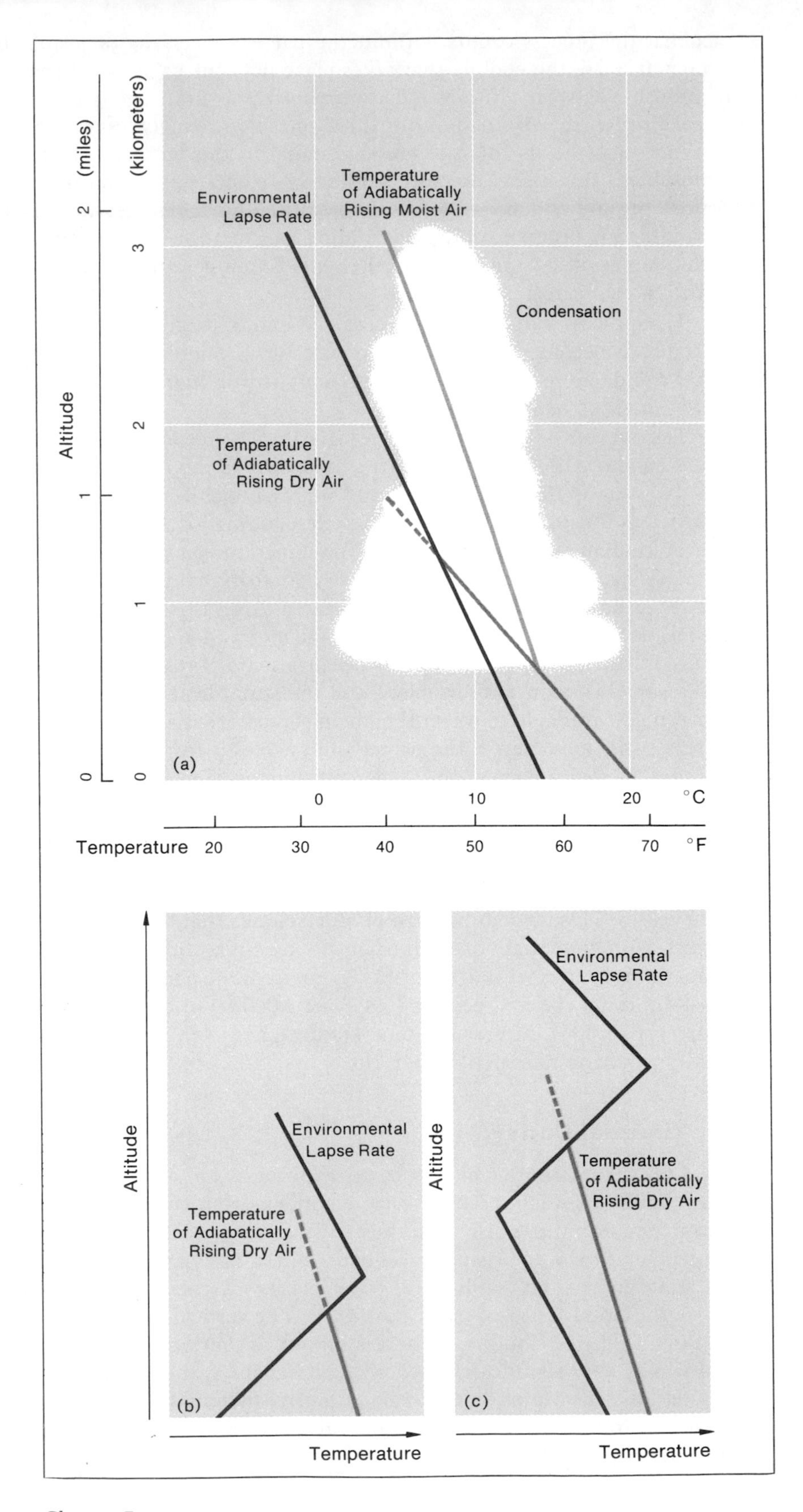

conditions is about 6.5°C per km (3.6°F per 1,000 ft) and is known as the *average lapse rate*. Locally, however, the rate of change of temperature with altitude may deviate considerably from the average. The vertical distribution of temperature on a local basis is known as the *environmental lapse rate*, and it can have important effects on thermal convection and cloud formation.

For example, Figure 4.16b shows warmer air aloft above cooler air at the surface. This temperature inversion is typical of nighttime radiational cooling during fair weather. If a parcel of cool, dry air near the surface were forced to rise, it would cool at the dry adiabatic rate and soon become cooler and denser than the warmer air in the inversion layer above. Being cooler than the surrounding air, the parcel would tend to return downward rather than rise through the inversion. An inversion inhibits upward vertical motion in the atmosphere so that there is little or no chance that thick, vertical clouds capable of producing precipitation will develop. An inversion also tends to act as a lid below which pollutants may accumulate. If the air below an inversion becomes saturated, fog and low stratus clouds usually begin to form (see Figure 4.17f, p. 106). Meteorologists refer to a situation in which temperature inversion inhibits the vertical movement of air as *stability*.

After sunrise, solar radiation generally warms the ground, and the lower atmosphere heats up by thermal convection, thus eliminating the overnight inversion at the surface. A parcel of air can then rise to a much higher altitude in the atmosphere (see Figure 4.16c). If the parcel is cooled to saturation and its water vapor condenses, a fair weather cumulus cloud is formed (see Figure 4.17a). When the environmental lapse rate decreases very rapidly with height, rising parcels will tend to remain warmer and less dense than the surrounding air, and they can progress much farther aloft. The rise of these parcels to great altitudes permits the development of the towering cumulonimbus clouds that are associated with thunderstorms (see Figures 4.16a and 4.17d). Meteorologists refer to the local conditions leading to the development of such clouds as *instability*.

During summer, when rapid heating of the ground tends to eliminate overnight temperature inversions and create conditions of instability, the development of rain-producing cumulus-type clouds follows a definite pattern. Early in the day, the sky is usually clear, but later, scattered wisps of cloud formed by condensation of the water vapor in rising updrafts of moist air begin to appear at an altitude of about 1 km. The wisps increase in size and eventually become large cumulus clouds. The base of the clouds marks the altitude at which the rising moist air has cooled to its dew point.

Cumulus clouds mark the upper section of columns of rising air. A downflow of air between the clouds replaces the air that rises from the earth's surface. For this reason, cumulus clouds never cover the sky completely, and precipitation takes the form of scattered showers or thundershowers which may break out in early afternoon and persist into early evening. Although an individual shower may yield more than 2.5 cm (1 in.) of rain locally, only rarely does this summer instability result in more widespread rains.

a
b
c
d

Figure 4.17 (a) Fair weather cumulus clouds over Baton Rouge, Louisiana. Each cloud forms in a rising thermal with flat and even bases at the altitude where the rising thermals are cooled to the dewpoint and saturation. These clouds tend to be short-lived, and they usually evaporate into the surrounding drier air within an hour or so.

(b) An altocumulus, or "mackerel sky," cloud over Boulder, Colorado, takes the form of a layer of patchy cloud puffs. It forms at moderate altitudes and often heralds the approach of a warm front and accompanying precipitation.

(c) Cirrus clouds are feathery wisps of ice crystals that form at high altitudes. They are frequently the first indication in the sky of the approach of a mid-latitude cyclone system from the west or south.

(d) Cumulonimbus clouds or "thunderheads" rise to great heights and are sometimes associated with heavy rain, flash floods, hail, and tornadoes. This view shows massive thunderstorms organized in an orographic situation with warm, moist mT being forced up and over the Rocky Mountain front at Boulder, Colorado.

(e) The leading edge of a squall line passing over Baton Rouge. The squall line was just ahead of a vigorous cold front, and the angry black clouds are indicative of severe turbulence and damaging winds.

(f) Shallow radiation fog near sunrise in autumn in southeastern Wisconsin.

(g) "Man-made" cumulus over petrochemical plants in the Los Angeles basin. These small clouds are associated with heat and water vapor added to the lower atmosphere as by-products of industrial processes.

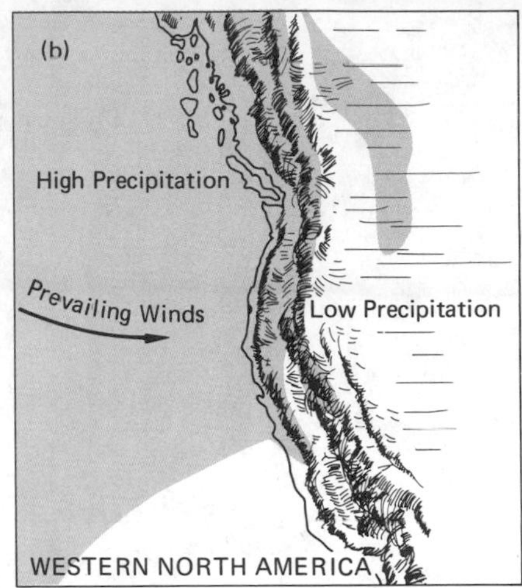

Figure 4.18 (a) Mountains can exert a strong local influence on precipitation because they can force moisture-laden air to rise to cooler high altitudes where condensation can occur.

(b) The mountain ranges along the Pacific coast of the United States intercept moisture transported from the Pacific Ocean by prevailing westerly winds, so that coastal ranges are moist and inland regions are dry.

Orographic Uplift

Mountains can exert a strong local influence on precipitation. When horizontally moving, moisture-laden air is forced to rise to higher altitudes over a mountain barrier, the air cools and condensation may occur. This process is known as *orographic uplift.*

Clouds, fog, and copious precipitation are characteristic of most mountainous regions, much to the disappointment of vacationers. In the western United States, the atmospheric circulation tends to be from west to east; thus the western, or windward, slopes tend to be cloudy and wet. However, when the air descends the eastern slopes, increasing pressure causes it to heat up at the dry adiabatic rate. The higher temperatures evaporate water droplets and the clouds disappear. As a result, the eastern slopes and "rainshadow" valleys tend to be sunny and dry (see Figure 4.18). Not all mountain areas are subject to this pattern of orographic uplift. Along the Appalachian Mountain system in the eastern United States, for example, moist air advances at various times from both east and west, and the ranges are wet on both sides.

A spectacular example of how a mountain can cause large rainfall differences over a short horizontal distance is found on the island of Kauai in the Hawaiian chain. The average rainfall on the windward side of Mount Waialeale is 1,170 cm (460 in.) a year, and it is only 81 cm (32 in.) a year on the lee side. The exceptionally heavy rainfall is produced by orographic cloud that is maintained by steady moisture-laden winds from the northeast.

Frontal Cooling and Convergence

Even in the absence of mountains, uplift of large masses of air with characteristic widespread clouds and precipitation patterns can occur. Except in equatorial and tropical latitudes, atmospheric circulation often brings large masses of warm and cold air into close proximity. Largely because of density differences, the warm and cold air masses tend not to mix, but instead to remain separated along an interface. At this interface there is a rapid transition from the temperature and humidity characteristics of one air mass to those of the other. The interfaces are known as *fronts*, and they mark the place where the warmer, less dense air mass either rides up and over the cooler air, or else is pushed steeply aloft by the colder mass. As it rises, the warm air cools to saturation, clouds form, and precipitation may occur.

Idealized cross sections of fronts are shown in Figure 4.19. Where cold air is advancing horizontally into the former domain of warmer air, the front is designated as cold, and where warm air is advancing horizontally into the former domain of colder air, the front is designated as warm. Figure 4.19a illustrates a typical warm front. The advancing warm air slides over the mass of retreating cold air at the surface. The warm air rises only about 1 km for a horizontal distance of about 100 km; thus the ascent is gentle, unlike flow over a mountain barrier. Cirrus clouds are in the forefront, followed by altostratus and nimbostratus. Widespread but relatively gentle precipitation is usually associated with a warm front. Figure 4.19b illustrates typical cloud distribution along a cold front, where advancing cold air is pushing warm air aloft. The slope of a cold front is steeper than that of a warm front. The warm air rises about 1 km for a horizontal distance of 50 km. As a result of this more abrupt uplift, precipitation is usually more intense and more localized than with a warm front.

Surface air can also rise and cool by *convergence*. In areas of low atmospheric pressure, designated as "lows" on weather maps, surface air flows in an inward-converging spiral toward the center. As air spirals into the low-pressure area, it is forced to rise to allow for additional incoming air, and this frequently results in condensation, clouds, and, ultimately, precipitation. Weather associated with fronts and atmospheric pressure systems and the global precipitation patterns that result from them are discussed in Chapter 5.

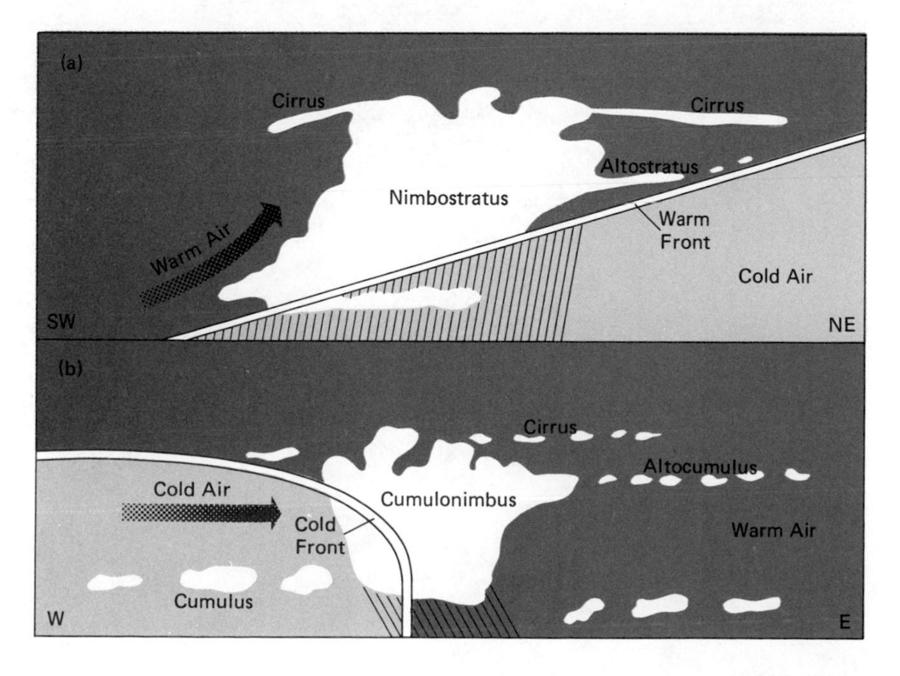

Figure 4.19 (a) When a large mass of warm air moves into a region occupied by colder air, the less dense warm air flows up and over the surface cold air. This atmospheric cross section shows that the warm front slopes gently upward through the lower atmosphere. Characteristic clouds and precipitation from the warm air tend to be widespread, with precipitation usually at light to moderate intensities.

(b) When a cold air mass moves into an area of warmer air, the warm air is pushed upward more sharply. As a result, the slope of the cold front is steeper, and the precipitation associated with it tends to be more intense and localized. (After Byers, 1959)

Figure 4.20 (right) A Buffalo, New York resident attempts to shovel his car out of the snow.
(below) Trees flooded by the creation of the Isabella Reservoir in southern California reappear in drought.

WEATHER MODIFICATION: CLOUD SEEDING AND SOME INADVERTENT CONSEQUENCES OF HUMAN ACTIVITY

Most regions of the earth occasionally experience departures from their customary weather patterns. An area may become wetter, drier, warmer, or colder than usual and remain that way for a period of weeks, months, or sometimes even years. The winter of 1976-1977, for example, was extremely dry from the northern Great Plains westward across the Rocky Mountains to the Pacific Coast. Farmers in western Iowa, ski resort operators in Colorado, and water-resource managers in northern California undoubtedly wished that precipitation could be brought under effective human control. At the same time, weary snow-removal crews in Buffalo, New York, also wished that some control could be exerted over the unusually persistent snow squalls which swept off Lake Erie, eventually bringing normal everyday activities to a standstill during mid-February (see Figure 4.20).

Prolonged periods of less than normal precipitation can have far-reaching social and economic effects. For example, unusually dry weather in the Soviet Union in 1972 forced the Russians to buy large quantities of American grain, and this resulted in serious economic consequences within the United States. Similarly, the effects of the hardships and costs of the winter of 1976-1977 may well persist for several years.

To a certain extent, scientists have learned to exert control over the weather. The key to modern rainmaking is the realization that many clouds possess all the conditions necessary to produce precipitation except a means to trigger the growth of water droplets into raindrops. In warm clouds, only a few oversized droplets per cu meter are required to initiate coalescence into raindrops. In supercooled clouds, a hundred or so ice crystals or ice-forming nuclei per cu meter are enough to trigger the growth of ice crystals. However, in terms of the practical management of precipitation, rainmakers have met with only moderate and controversial success.

In 1946, Vincent Schaefer of the General Electric Research Laboratories discovered that he could stimulate the Bergeron process and change droplets to ice crystals by dropping a piece of dry ice (solid carbon dioxide at a temperature below −40°C) into a chamber containing supercooled water droplets. Field tests by Schaefer and Irving Langmuir showed that a few pounds of crushed dry ice dropped from an airplane onto the top of a supercooled layer cloud could cause light precipitation from the base of the cloud (see Figure 4.21, p. 112). In numerous experiments since then, the summits of cumulus clouds have been seeded with dry ice. Precipitation is usually produced if the cloud is thick enough, if its top is cold, and if the cloud is long-lived—in other words, if the cloud already has the conditions necessary for producing rain.

The tiny crystals in the smoke from burning silver iodide also make excellent ice-forming nuclei. Supercooled droplets will form ice crystals on silver iodide particles at temperatures as high as −4°C (24.8°F); no other known material is as effective. One feature of silver

Figure 4.21 (left) This photograph shows the first conclusive field test of cloud seeding, carried out in 1946 by Irving Langmuir and Vincent Schaefer. The supercooled stratus clouds were seeded by dry ice pellets dropped from an airplane flying around an oval track. In less than 1 hour after seeding, the seeded area cleared because of the induced precipitation.

(right) In this early attempt at rainmaking, electricity from a hand-cranked generator was pumped into clouds in the mistaken belief that the electricity of a lightning flash was somehow responsible for rain.

iodide seeding is that the required number of nuclei can be obtained by burning only a few ounces of silver iodide in burners mounted on aircraft. Using this method, commercial rainmaking operators are able to seed updrafts and the most promising cloud systems directly.

The effectiveness of weather modification operations is difficult to assess. Experiments in seeding individual large cumulus-type clouds in southern Florida have produced dramatic changes in cloud structures as well as large but very local increases in rainfall. Other experiments with cumulus-type cloud systems suggest that precipitation over large regions was not substantially increased by cloud seeding. In fact, in some instances, cloud seeding seemed to cause a decrease in precipitation. In the dry and mountainous West, however, favorable results seem to depend upon careful selection of weather situations where supercooled clouds are forced over mountain barriers; the data indicate that winter-spring snowfall in the mountains can be increased from 10 to 30 percent. Whether increased precipitation in the mountains means that there is a decrease in precipitation in the downwind areas farther east is not fully understood, however.

Rainmaking is not the only aspect of weather modification. Some forms of weather—severe rainstorms, fog, hailstorms, and hurricanes, for example—can be costly to human life and property. Understanding of precipitation processes has been applied in an effort to disrupt conditions that lead to unwanted weather.

Because airports and highways become dangerous or even unusable in heavy fog, economic methods of fog dispersal are being developed. If the fog droplets are supercooled, seeding with dry ice can dissipate the fog as snow crystals, and some airports routinely make use of cold fog dissipation techniques. Fogs in most areas are warm, however, and no inexpensive way of clearing a warm fog has yet been found. Heating the air to a temperature above the dew point will clear up a warm fog, but the energy input required makes the method too costly for general application.

Hailstorms cause so much damage to crops that efforts to disrupt the formation of hail in turbulent clouds are usually worthwhile. Seeding clouds with silver iodide seems either to inhibit further growth of hailstones or to cause the cloud to produce only rain. Because hailstorms tend to be concentrated in certain parts of the country, such as the western Great Plains (see Figure 4.22), it has been feasible to set up hail suppression facilities in particularly susceptible areas. In the Soviet Union, artillery shells and rockets are used to shoot silver iodide directly into clouds, and the government reports a considerable reduction in crop damage caused by hail.

Some efforts have been directed toward modifying the behavior of hurricanes, but the naturally erratic movement and growth of hurricanes make it difficult to determine whether a seeding experiment has reduced the force of storm winds. Hurricane Debbie was seeded by Project Stormfury planes on August 18 and 20, 1969. Maximum winds decreased 31 percent on the first day, increased on the second

Figure 4.22 The map shows the average number of days per year with hail at locations in the conterminous United States, based on weather records from 1899 to 1938. Because hail is damaging to crops and poultry, hailstorms do hundreds of millions of dollars of damage to United States agriculture each year. The frequency of hailstorms is greatest in southeastern Wyoming and northeastern Colorado. Known as Hail Alley, the region is the meeting place where moisture from the Gulf of Mexico, rising air from the hot prairies, and strong winds from the Rocky Mountains spawn intense storms. Some farmers in Hail Alley average only one hail-free crop every five years; experiments are in progress to seed thunderclouds with silver iodide to reduce the chance of hail formation. (After *Yearbook of Agriculture*, 1941)

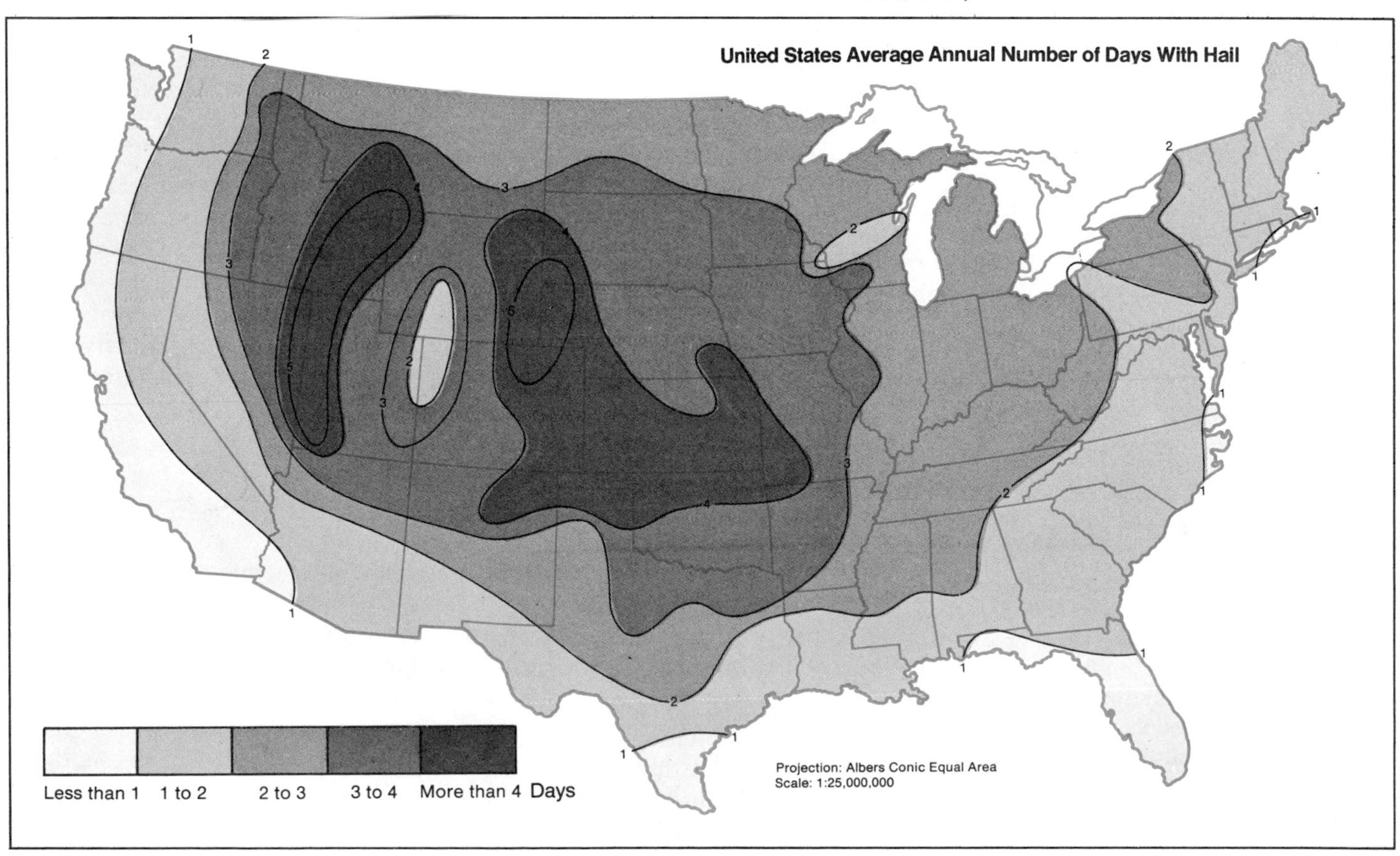

day, and decreased 15 percent again after seeding on August 20. Much more careful research is needed, however, before seeding of hurricanes can become routine.

Although we may gain greater control over the weather in certain instances, there is increasing apprehension about the power of industrialized society to modify the weather inadvertently. An analysis of the long-term precipitation record at La Porte, Indiana, for example, has revealed a pattern of greater precipitation there than in adjacent areas. It has been suggested that the greater precipitation at La Porte is related to the growth of heavy industry in the Chicago metropolitan region. The heavier concentration of particles in the atmosphere—the product of industrial waste—may provide a more abundant supply of condensation nuclei and thus encourage precipitation. It has also been suggested that the decrease in precipitation at La Porte in more recent years is due to the introduction of pollution control devices. These interpretations have been questioned; nevertheless, the controversy has encouraged detailed regional analyses of weather and climate in the vicinity of large industrial cities.

In addition to increased precipitation, other signs of inadvertent weather modification have appeared with unsettling frequency in the vicinity of urban centers in North America and Europe. Most of these effects tend to be greater in winter than in summer—the opposite of the case with precipitation. For example, the volume of atmospheric contaminants is many times greater over the cities, visibility is lower, solar radiation is less, and fog frequencies are higher. In urban areas, temperatures are higher, relative humidities are lower, wind speeds are lower except around tall buildings, thunderstorms are more frequent, and total precipitation is probably 10 to 15 percent greater. These local and regional effects of urbanization do not appear to affect directly the patterns of global circulation at this time.

However, other aspects of inadvertent weather and climate modification may affect the energy balance of the earth-atmosphere system and the resultant global atmospheric circulation. Examples include the addition of carbon dioxide to the atmosphere by the increasing consumption of fossil fuels, the increase of atmospheric particulate matter because of agricultural and industrial activities, and the increase of fluorocarbons and their potential reactions with ozone in the stratosphere. Each of these consequences of human activity needs to be monitored and studied carefully.

SUMMARY

Moisture enters the atmosphere by evaporation from open water surfaces, such as lakes and oceans, and by evapotranspiration from continents. Condensation is the process whereby water vapor returns to a liquid state, and precipitation refers to the accumulation of water or ice into particles big enough to fall to earth. In the local environment, evapotranspiration is a cooling process; latent heat is stored in the water vapor and released again as sensible heat elsewhere in the atmosphere when the water vapor condenses as droplets or ice crystals.

Further Reading

Barry, Roger G., and **Richard J. Chorley.** *Atmosphere, Weather, and Climate.* New York: Holt, Rinehart and Winston, 1970. This basic weather and climate text is written largely from the perspective of the British Isles. Concepts are illustrated frequently by examples from the professional literature. Chapter 2 includes a particularly useful development of evaporation.

The capacity of the atmosphere to hold water vapor decreases with decreasing temperature, and condensation normally takes place when the atmosphere is cooled to the dew point. Condensation begins on microscopic atmospheric nuclei. Precipitation occurs only when a small number of giant droplets grows to raindrop size by sweeping up many smaller droplets, or, in cold clouds, when a few ice crystals grow by accumulating nearby supercooled water droplets. Condensation, clouds, and sometimes precipitation can be associated with five cooling situations: surface radiational cooling, which results in dew or fog; adiabatic cooling, which produces cumulus clouds on fair summer days; orographic cooling, where air is forced over mountain barriers; frontal cooling, when warm, moist air rises over cooler surface air; and convergence, when the inward-spiraling air of a low-pressure system ascends aloft. Artificial nuclei can be introduced into clouds to promote the growth of ice crystals or raindrops, but potential for water-resource management remains controversial.

The atmosphere transports energy, in the forms of latent and sensible heat, and water vapor from place to place over the earth's surface. The general circulation of the atmosphere, which is discussed in Chapter 5, is important in physical geography as the moisture delivery system for the various regions of the earth. The ways by which precipitation is allocated to evapotranspiration, storage, and runoff at the earth's surface are discussed in Chapter 6.

Review Questions

1. Compare the factors affecting the transfer of water to the atmosphere over oceans and continents.
2. What is the significance of the relationship between temperature and the atmosphere's capacity for holding water vapor?
3. What are the mechanisms by which water is returned to the atmosphere from the earth's surface?
4. What variables determine the vapor pressure gradient? What is its significance in terms of evaporation rates?
5. Compare the various measures of atmospheric humidity.
6. Describe the diurnal regime of vapor pressure and relative humidity during fair weather.
7. Discuss the processes by which water vapor is transformed into drops or ice crystals large enough to fall from clouds to the ground as precipitation.
8. Discuss five situations in which atmospheric cooling might eventually result in clouds and precipitation.
9. Why is there very little chance of significant precipitation during radiation fogs?
10. Discuss the diurnal regime of stability and instability near the earth's surface during fair weather.
11. What are the physical bases for atmospheric weather modification?
12. Why is it rather unlikely that there will ever be full agreement about plans for weather modification projects?

Battan, Louis J. *Harvesting the Clouds: Advances in Weather Modification.* Garden City, N.Y.: Anchor Books, Doubleday & Co., 1969. This is an especially readable account of techniques and objectives of weather modification.

Changnon, Stanley A., Jr. "The La Porte Anomaly—Fact or Fiction?" *Bull. Amer. Meteor. Soc.,* Vol. 49 (1968): 4–11. This analysis of an unusual 40-year precipitation record southeast of Chicago has proved somewhat provocative.

———, **Floyd A. Huff,** and **Richard G. Semonin.** "METROMEX: An Investigation of Inadvertent Weather Modification." *Bull. Amer. Meteor. Soc.,* Vol. 52 (1971): 958–967. This is an early summary of research, which evolved from the La Porte study, on the effects of the St. Louis metropolitan region on precipitation processes and patterns.

Holzman, B. G., and **H. C. S. Thom.** "The La Porte Precipitation Anomaly." *Bull. Amer. Meteor. Soc.,* Vol. 51 (1970): 335–337. The authors present a contrasting view of the original 1968 analysis of Changnon.

National Research Council. *Weather & Climate Modification: Problems and Progress.* Washington, D.C.: National Academy of Sciences, 1973. This is an up-to-date and somewhat technical review of research in both planned and inadvertent weather and climate modification.

Neiburger, Morris, James Edinger, and **William Bonner.** *Understanding Our Atmospheric Environment.* San Francisco: W. H. Freeman & Co., 1973. This is an excellent source for basic meteorological concepts in a nonmathematical framework. Chapters 8, 9, and 15 are especially pertinent.

Riehl, Herbert. *Introduction to the Atmosphere.* 2nd ed. New York: McGraw-Hill Book Co., 1972. Recent concepts of atmospheric dynamics are emphasized in this nonmathematical basic work. The almost unique focus on weather and climate variation should be especially useful for students of resource management.

Scorer, Richard, and **Harry Wexler.** *Cloud Studies in Colour.* Oxford: Pergamon Press, 1967. A most interesting set of 122 color plates, mostly from the British Isles, is well coordinated with descriptions and analyses of associated weather events.

Snow Storm—Steam Boat off a Harbour's Mouth Making Signals in Shallow Water and Going by the Land by William Turner. (The Tate Gallery, London)

A furious storm at sea emphasizes the complex interactions of the atmospheric and oceanic systems. Winds drive the surface ocean currents, and the ocean temperatures influence the circulation patterns of the atmosphere.

FIVE

Atmospheric and Oceanic Circulations

THE OUTSTANDING FEATURE OF weather in a midlatitude area like the United States is that it is always changing. A sunny morning in Boston, Massachusetts, for example, may give way to a gray and rainy afternoon. A sultry summer day in Minnesota may see the build-up of the ominous clouds that bring severe thunderstorms and hail.

The extreme variability of weather patterns in North America is not typical of all regions of the earth. Near the equator, for example, the day-to-day weather is comparatively monotonous in its regularity. Weather there usually does not exhibit the successive fair and stormy, warmer and colder periods, each lasting a few days, that sweep across the United States. Yet even in the tropics, dramatic variations in weather do occur. Hurricanes and typhoons disrupt the landscape with battering winds and great quantities of precipitation. Rain-bearing monsoons cause weather in some areas to shift regularly between extremes of aridity and moistness.

But whatever the geographical area—the midlatitudes, the tropics, or the poles—changes in weather are not isolated events. They are linked to larger systems which ultimately encompass the entire globe. For example, a passing summer shower that covers an area of a few square kilometers and lasts only a few minutes is a local event on the scale of the earth's atmosphere. On a larger scale and controlling these local phenomcna are the weather systems, or *secondary circulations*, that move through the middle latitudes and tropics. Such features range up to a few thousand kilometers in size and endure for days before they die. On even larger scales of space and time are the planetary patterns of the *general circulation*, which are comparable in size to the earth and which are persistent features of the atmosphere, though changing somewhat from season to season.

Atmospheric and oceanic circulations are critical factors in determining the earth's weather, both on a local and a global scale. The motions of the atmosphere and oceans are due to solar energy gains in equatorial and tropical regions and energy losses to space in subpolar and polar latitudes; the circulations transport energy in the forms of latent and sensible heat from lower to higher latitudes. The atmospheric circulation also delivers water, derived primarily by evaporation from the oceans, to the continents, but the distribution is far from uniform. Irregularities in the distribution of water and energy not only determine weather variations within local areas but regulate the entire mosaic of global climates as well. The responses of vegetation, soils, and landform systems to the inputs of energy and moisture will be the focus of later chapters. This chapter investigates the systematic behavior of general and secondary circulations and examines their effect on local weather conditions.

FORCES THAT DRIVE THE ATMOSPHERE

The atmosphere is in ceaseless motion, whether churning with a furious storm or puffing an occasional summer breeze. The motion of any parcel of air is determined by the forces acting on it. Although the details of atmospheric motion are complex, some general features can be understood by looking at the forces involved.

A force may be thought of as a push or a pull, analogous to the pushes and pulls we exert with our muscles. According to the fundamental laws of motion, which the English physicist Isaac Newton developed in the seventeenth century, a moving object's speed or direction of motion cannot change unless a force is made to act on the object. Furthermore, horizontal motion is influenced only by forces pushing or pulling in a horizontal direction, and vertical motion is influenced only by forces acting in a vertical direction. The force of gravity pulls every parcel of air downward toward the earth, for example, but gravity has no direct effect on the horizontal motion of air across the earth's surface. The principal horizontal forces that act on a parcel of air in the atmosphere are the pressure, gradient force, the Coriolis force, and friction.

The Pressure Gradient Force

Pressure is force per unit area. If the pressure on one side of a parcel of air is greater than the pressure on the other side, the parcel will be pushed from the higher pressure to the lower pressure region (see Figure 5.1a). The greater the difference in pressure on the two sides of

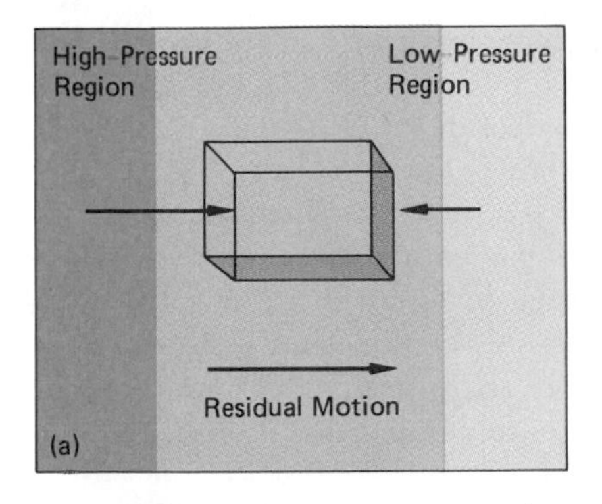

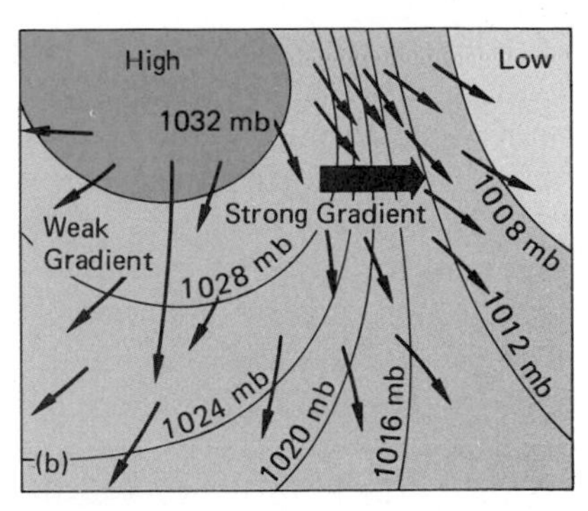

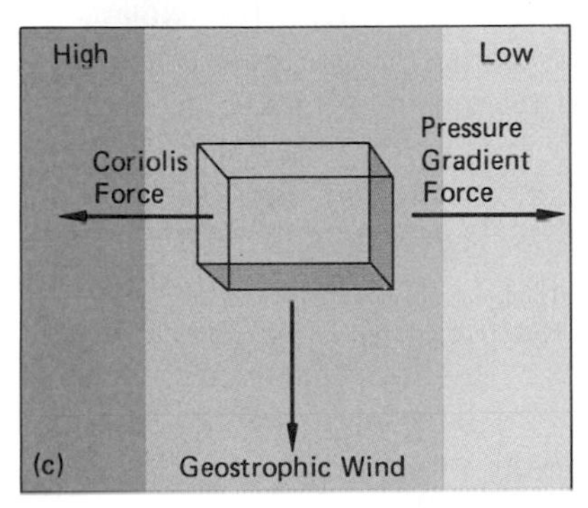

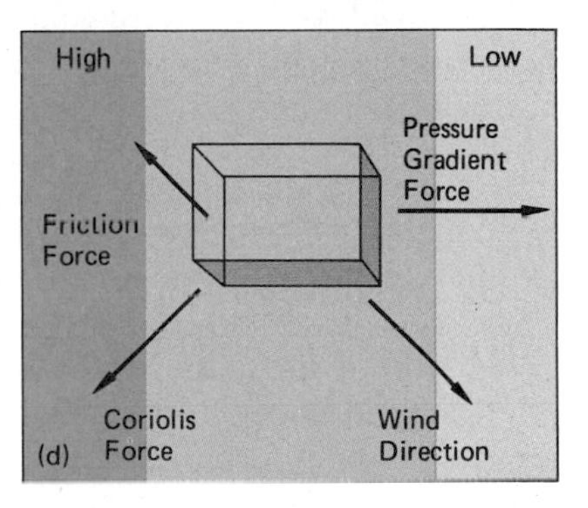

Figure 5.1 (a) The pressure gradient force acts in a direction perpendicular to isobars and pushes a parcel of air away from high pressure toward low pressure. If no other forces were present, the parcel would move toward the low-pressure region.

(b) Winds near the surface of the earth are partially determined by isobar patterns and spacing. Close spacing of isobars is associated with strong horizontal pressure gradients and strong winds.

(c) A parcel of air aloft moving horizontally on the rotating earth experiences a Coriolis force and a pressure gradient force. The Coriolis force acts at right angles to the direction of motion of the parcel. This diagram shows how the Coriolis force can equal the pressure gradient force so that the parcel can move parallel to the isobars. Air motion essentially parallel to isobars is known as a geostrophic wind; the speed of a geostrophic wind is highest where the pressure decreases most rapidly with distance. This diagram is drawn for the northern hemisphere. How would it appear for the southern hemisphere?

(d) Friction acts on air parcels moving near the ground. The diagram shows that when the retarding effect of friction is taken into account, the forces on a parcel of air can be in balance only if the parcel has a component of motion away from high pressure and toward low pressure.

the parcel, the greater the net push and, therefore, the stronger the wind. This "push," caused by the horizontal difference in pressure across a surface, is called the *pressure gradient force.* The direction of the pressure gradient force is always from higher to lower pressure.

On a weather map, meteorologists represent the pressure over a given area with lines called *isobars.* The isobars connect locations on the map which are equal in pressure (see Figure 5.1b). The pressures given may not correspond to the actual values: all readings are corrected to conform to sea level values to compensate for decreases in pressure due to altitude. The pressure gradient force is at right angles to the isobars, from higher to lower pressure, and is strongest where the pressure decreases most rapidly with distance, that is, where the isobar spacing is smallest.

The Coriolis Force

If it were not for the rotation of the earth, winds would simply follow the pressure gradient. But the fact that the earth rotates complicates the motion of the atmosphere. To understand the principle involved, consider two children throwing a baseball on a merry-go-round that is rotating counterclockwise. One child is near the center and the other is near the rim. If the child near the center throws the ball straight toward the child at the rim, the ball will pass to the left of the catcher because the merry-go-round continues to rotate during the ball's flight. From the point of view of the children, the ball moves as though a sidewise force were acting to deflect it, although in actuality the ball moves in a straight line. The same apparent deflection occurs on the rotating earth and is known as the *Coriolis force,* after Gaspard

Figure 5.2 This figure illustrates how the earth's rotation and grid of parallels and meridians are related to the deflection of moving air to the right in the northern hemisphere. Air streams originating at 40°N, 90°W and moving from south to north and from west to east appear nine hours later as southwesterly and northwesterly flows. (After Riehl, 1972)

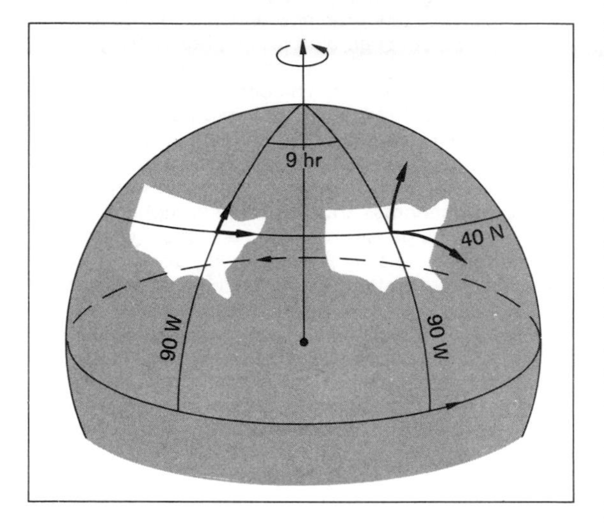

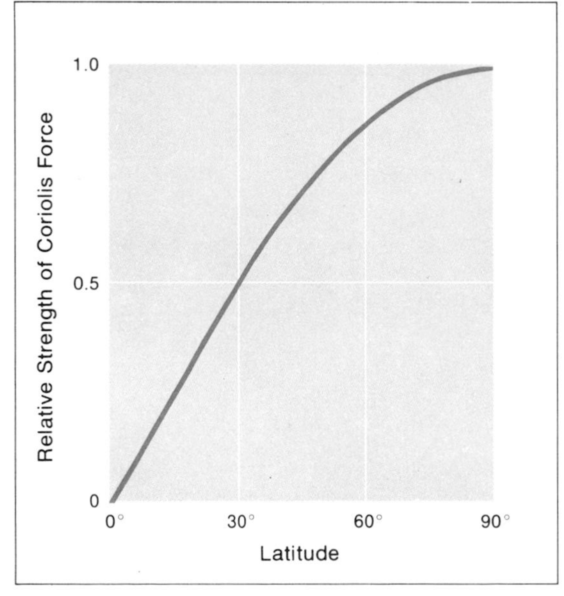

Figure 5.3 For an object such as an air mass moving horizontally on the earth's surface, the strength of the Coriolis deflecting force varies with latitude, as the graph shows.

Coriolis, the nineteenth-century French engineer who introduced the concept. Even though no real force is operating, the concept is convenient for describing the observed deflection of all objects moving freely over the earth's surface, such as air or water.

In the case of moving air, the earth is a rotating system somewhat analogous to the merry-go-round, and a parcel of air moving horizontally in any compass direction is turned or deflected. The Coriolis force acts at right angles to the direction of the flow. In the northern hemisphere, the air is deflected to the right, and in the southern hemisphere, to the left. It is important to remember that deflection to the right or left must always be thought of in terms of looking *downwind* in the direction toward which the air is moving. A full explanation of the Coriolis force requires mathematical formulation, but Figure 5.2 shows that the deflection is related to the rotation of the earth and the earth's grid of parallels and meridians. There is no Coriolis force at the equator and, hence, no deflection. Away from the equator, the Coriolis deflection becomes significant in subtropical latitudes (see Figure 5.3), and it increases to a maximum at the poles. The Coriolis force is also proportional to the speed of the object; thus at any latitude the deflection is greater with stronger winds.

Under ideal conditions where the Coriolis force is equal to the pressure gradient force, a parcel of air would eventually flow at right angles to the pressure gradient and parallel to the isobars. Meteorologists refer to this air movement along isobars as the *geostrophic wind*, and the circulation of upper level air, above the frictional effects of the earth's surface, tends to approach this ideal. Technically, the geostrophic wind is limited to situations with straight isobars, but the airflow aloft remains approximately parallel to curved isobars also. In the middle and higher latitudes of the northern hemisphere, air flows in a clockwise direction around high-pressure areas and in a counterclockwise direction around low-pressure areas (see Figure 5.4). Using these facts, the Dutch meteorologist Buys Ballot in 1857 worked out a law describing the relationship between pressure systems and wind direction aloft. According to Ballot's law, in the northern hemisphere, looking downwind, low pressure will be toward your left and high pressure toward your right. In the southern hemisphere, low pressure will be to your right, high pressure to your left.

The geostrophic wind is especially important to meteorologists concerned with forecasting weather conditions. Upper air measurements of atmospheric pressure, wind speed, and wind direction are taken routinely twice a day at more than 100 weather stations across the continental United States. From these measurements, meteorologists construct weather maps which present reliable estimates for conditions at various levels of the atmosphere throughout the area. These projections of pressure and wind patterns are the basis for predicting weather conditions as much as 72 hours hence.

The Force of Friction

If the energy sources that drive the atmosphere ceased, friction would cause all atmospheric motion to halt in a week or two. The *force of*

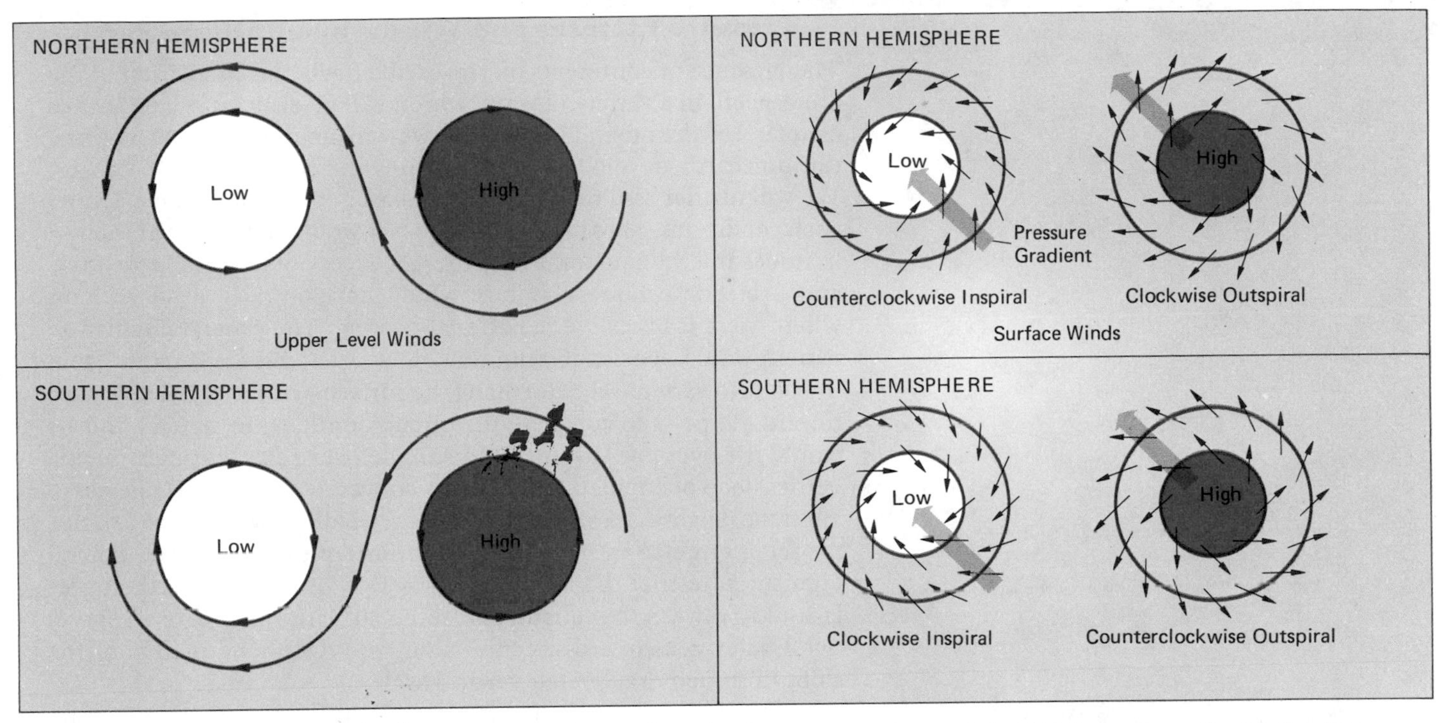

Figure 5.4 This figure shows the relationships between both upper and surface winds for both northern and southern hemispheres. The Coriolis deflection is to the right in the northern hemisphere and to the left in the southern hemisphere. In the northern hemisphere, the resultant geostrophic wind flows parallel to the isobars, with low pressure to the left, high pressure to the right. The surface flow converges toward low pressure and diverges away from high pressure due to frictional effects. Buys Ballot's law should be related to both northern and southern hemisphere diagrams.

friction acts to retard any object moving on the earth or in the atmosphere. For a parcel of air to maintain its movement, it must overcome this force, and as it does so, the direction of its flow is counteracted.

The effect of friction is greater near the earth's surface than it is aloft. Since the Coriolis force is proportional to the speed of an object, near the ground where surface winds are slowed by frictional forces, the Coriolis effect is thereby reduced. Instead of flowing parallel to the isobars, the surface air flows at small angles across the isobars from higher pressure toward lower pressure (see Figure 5.1d). Therefore, surface air converges into an area of low pressure in an inward spiral and diverges away from an area of high pressure in an outward spiral. In the northern hemisphere, the inspiral is counterclockwise, the outspiral clockwise. In the southern hemisphere, the reverse is true (see Figure 5.4). To find the relationship between pressure systems and surface winds, Ballot's law further states: in the northern hemisphere, stand with your back to the wind and rotate 45° to the right; low pressure will be on the left, high pressure on the right. In the southern hemisphere, rotate 45° to the left, and low pressure will be on the right, high pressure on the left.

GENERAL CIRCULATION OF THE ATMOSPHERE

The pressure gradient force, the Coriolis force, and the force of friction are the basic factors which set the atmosphere in motion. But other, large-scale mechanisms are also involved in determining patterns of atmospheric circulation. The following sections describe a generalized model of the global circulation of the atmosphere. This model represents a sort of grand average for one year; the actual circulation at any one moment may be, of course, much different.

Pressure Patterns and Winds on a Uniform Earth

The presence of continents on the earth affects the circulation of the atmosphere in a variety of ways, which will be dealt with later in this chapter. For the moment, however, we will assume the existence of a uniform earth, devoid of surface features and covered only by oceans. We will further assume that the Coriolis effect is not in force. Under such conditions, a broad low-pressure belt would develop over the low latitudes and the equator, where there is excessive radiational heating, and a broad high-pressure belt would develop over polar regions, where there is excessive radiational cooling. The general circulation pattern would consist of wind flow down the pressure gradient from the poles toward the equator along the surface, with a return flow aloft toward the poles to complete the circulation loop. In general, the air would rise over low latitudes and subside over polar latitudes. Such a vertical loop of circulation is called a *convective cell* or a *Hadley cell,* after the English meteorologist George Hadley, who proposed this model of circulation in 1735. Modern interpretations of the general circulation restrict the Hadley cells in each hemisphere to the lower latitudes between the equator and about 30° latitude. Figure 5.5 shows two Hadley cells in cross section along the right-hand margin of the globe, in much exaggerated vertical scale.

If we add rotation to our model of a uniform earth, the Coriolis force comes into play, which in turn alters circulation patterns. In the rotating model, excessive radiational heating still produces a belt of low pressure along the equator. The heated air rises and flows out and aloft toward the poles. Once away from the equator, however, the poleward flowing upper air from the tropics cools by longwave radiation to space and then descends at about 30° latitude in each hemisphere, where dynamic subtropical high-pressure centers are formed. Much of this descending air flows back down the pressure gradient toward the equator and is deflected by the Coriolis force toward the right in the northern hemisphere and toward the left in the southern hemisphere (see Figure 5.5). In both hemispheres, the deflected flow is turned toward the west. Remember that the wind is designated by the direction from which it comes; thus this flow is northeasterly in the northern hemisphere and southeasterly in the southern hemisphere.

The surface pressure and wind systems have traditional names which come from the era of sailing vessels during the seventeenth through the nineteenth centuries. The subtropical highs at 30° latitude —belts of descending warm, dry air, fair weather, and weak surface winds—are often called the *horse latitudes.* Legend has it that when ships were becalmed in these latitudes, the horses were the first to go overboard when supplies of food and fresh water dwindled. Between the horse latitudes and the equator lies an area characterized by northeasterly and southeasterly surface flows known as *trade winds.* The trades are among the most steady and persistent winds on the globe, and they provided sea traders with reliable westward routes over the oceans. The trades converge near the equator into a low-pressure zone of generally light, variable winds and calms known traditionally as the *doldrums,* but referred to by most meteorologists as the intertropical convergence zone (ITC).

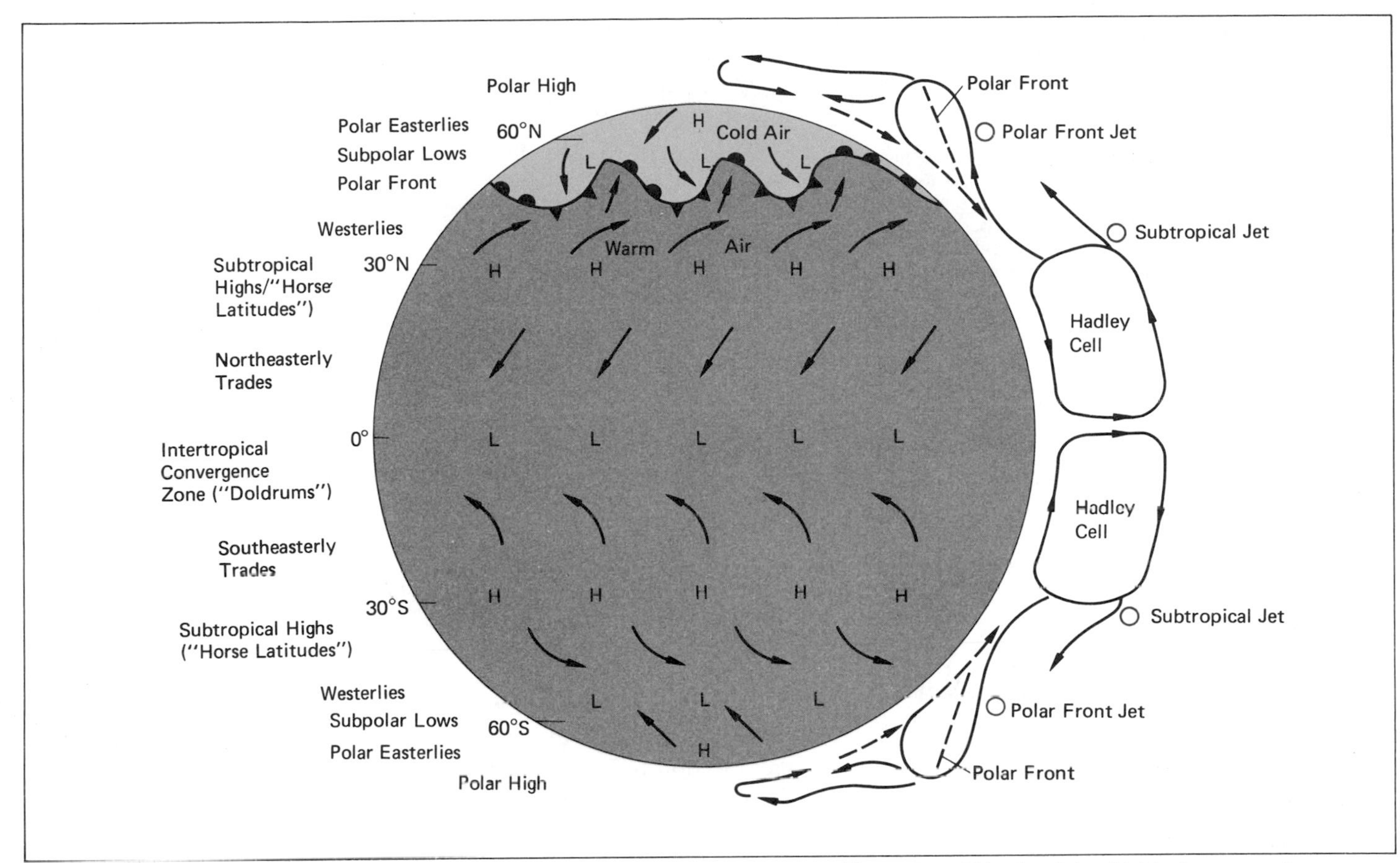

Figure 5.5 This sketch is a highly diagrammatic representation of the general circulation over a homogeneous earth with a water surface. Semipermanent pressure and wind belts are shown on the surface of the sphere, and a much enlarged vertical cross-section is shown on the right-hand hemisphere only. The text contains a description of the pressure and wind systems, and the text and figure should be analyzed together carefully. The polar front is shown only for the northern hemisphere, and it is discussed in more detail in later section of this chapter.

Some of the warm subsiding air of the subtropical highs flows poleward near the surface (see cross section in Figure 5.5). This flow is deflected toward the east to become the southwesterlies of the northern hemisphere and the northwesterlies of the southern hemisphere. Sometimes these winds are simply called *westerlies*. In sailing days, these wind belts provided opportunities for passage eastward.

The surface air within the westerlies is normally of subtropical origin, and when this warm "tropical" air meets colder "polar" air moving toward the equator, the lighter tropical air tends to flow up and over the polar air. The *polar front*, or boundary between the two air masses of different thermal properties, slopes upward toward higher latitudes (see cross section in Figure 5.5). As it reaches cooler regions nearer the poles, the rising warm air is chilled by radiational and adiabatic cooling. As the air drops in temperature, it subsides, forming high-pressure areas known as the *polar highs*. Surface air flows out from the polar highs in both hemispheres and is deflected westward to become the *polar easterlies*. The boundary of the polar easterlies is the polar front, where there is a low-pressure zone, the *subpolar lows*, marked by persistent convergence of polar and tropical air at the surface. From time to time, particularly during winter, the polar front bulges toward the equator, allowing polar air to penetrate to subtropical latitudes. These bulges, known as *polar outbreaks*, are indicated by the broken arrows in the cross section in Figure 5.5.

Figure 5.6 The maps show winter and summer wind patterns for the northern hemisphere in the upper atmosphere near an altitude of 5.5 km (3.4 miles), where the atmospheric pressure is approximately one-half the pressure at sea level. Wind directions and speeds in the upper atmosphere are usually determined by tracking freely drifting radio-equipped balloons that transmit meteorological data to the ground. In winter the upper air winds are strong and form a well-defined pattern of circulation with several undulations. The shaded region indicates the average location of fast-moving jet streams. Many winter storms in the northern latitudes appear to be generated with the help of jet streams. The flow of upper air winds tends to be weaker during summer than during winter. Dashed lines indicate particularly weak flows. (After Riehl, 1972)

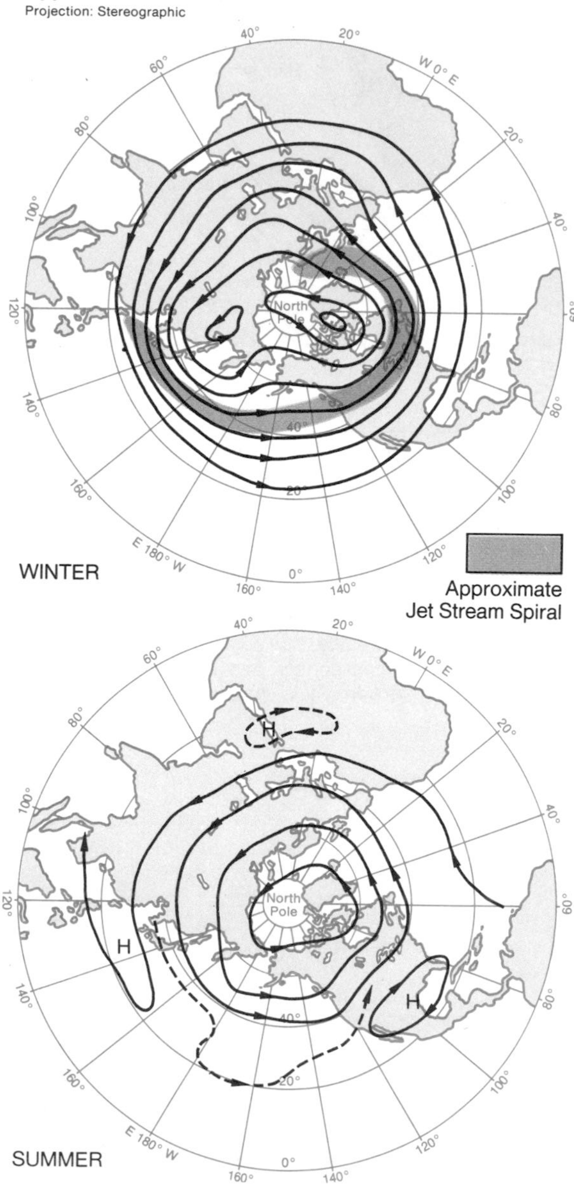

Surface pressure and wind patterns on a uniform, rotating earth can be summarized as follows. There are three broad belts of low pressure and convergence—the subpolar latitudes of each hemisphere and the intertropical convergence zone astride the equator. There are also four belts of high pressure and divergence—the polar and subtropical highs in each hemisphere. In general, low-pressure belts are associated with clouds and precipitation, and high-pressure belts with fair weather. These belts tend to migrate seasonally with shifts in temperature zones, moving poleward in summer and toward the equator in winter. Heat storage in oceans, however, tends to delay the extreme poleward migration by one or two months; thus the belts in the northern hemisphere reach their highest latitudes during July or August.

The Upper Atmosphere and Jet Streams

The circulation pattern of the upper atmosphere is much simpler than that near the surface. Poleward of the subtropical highs, the upper atmosphere flows from west to east in a vast circumpolar vortex. Since friction here is at a minimum, the Coriolis and pressure gradient forces are in balance, and airflow approximates geostrophic wind, with low pressure to the left and high pressure to the right in the northern hemisphere (see Figure 5.6). On the average, strongest flows are concentrated in relatively narrow ribbons called *jet streams.* The cross section in Figure 5.5 shows two jet streams in each hemisphere: the subtropical jets, which are associated with the subtropical highs, and the polar front jets, which are normally above the polar fronts.

Jet streams were first observed in the upper troposphere with the development in the 1940s of aircraft that could fly at altitudes higher than 10 km (6 miles). Jet streams occur at altitudes of between 10 and 15 km and are thousands of kilometers long, hundreds of kilometers wide, and several kilometers thick. The wind speed along the central core of a jet stream may occasionally exceed 300 km per hour (200 miles per hour), fading out to much slower speeds near the edges.

Jet streams gain much of their energy from the temperature difference between cold polar and warm tropical air. The average flow is from west to east, and jet streams generally travel in a wavelike pattern, as shown in Figure 5.7. There are typically three to six waves for each complete pattern around the earth. Usually, the waves remain within certain geographical limits, but at times they undergo increasingly severe oscillations, distributing cold polar air over lower latitudes and warm tropical air over higher latitudes. When the waves of air are displaced for several weeks or longer, the weather of a region under a displaced wave becomes warmer or colder and wetter or drier than normal. Areas of low pressure, known as *troughs,* extend from higher latitudes toward lower latitudes. Colder air flows toward the tropics on the western margins of the troughs, and warmer air flows poleward on the eastern margins. The entire pattern helps to maintain a net poleward flow of energy from equatorial and tropical regions. Eventually, the oscillations may become so extreme that the waves of displaced air are cut off from the main stream, forming cells of rotat-

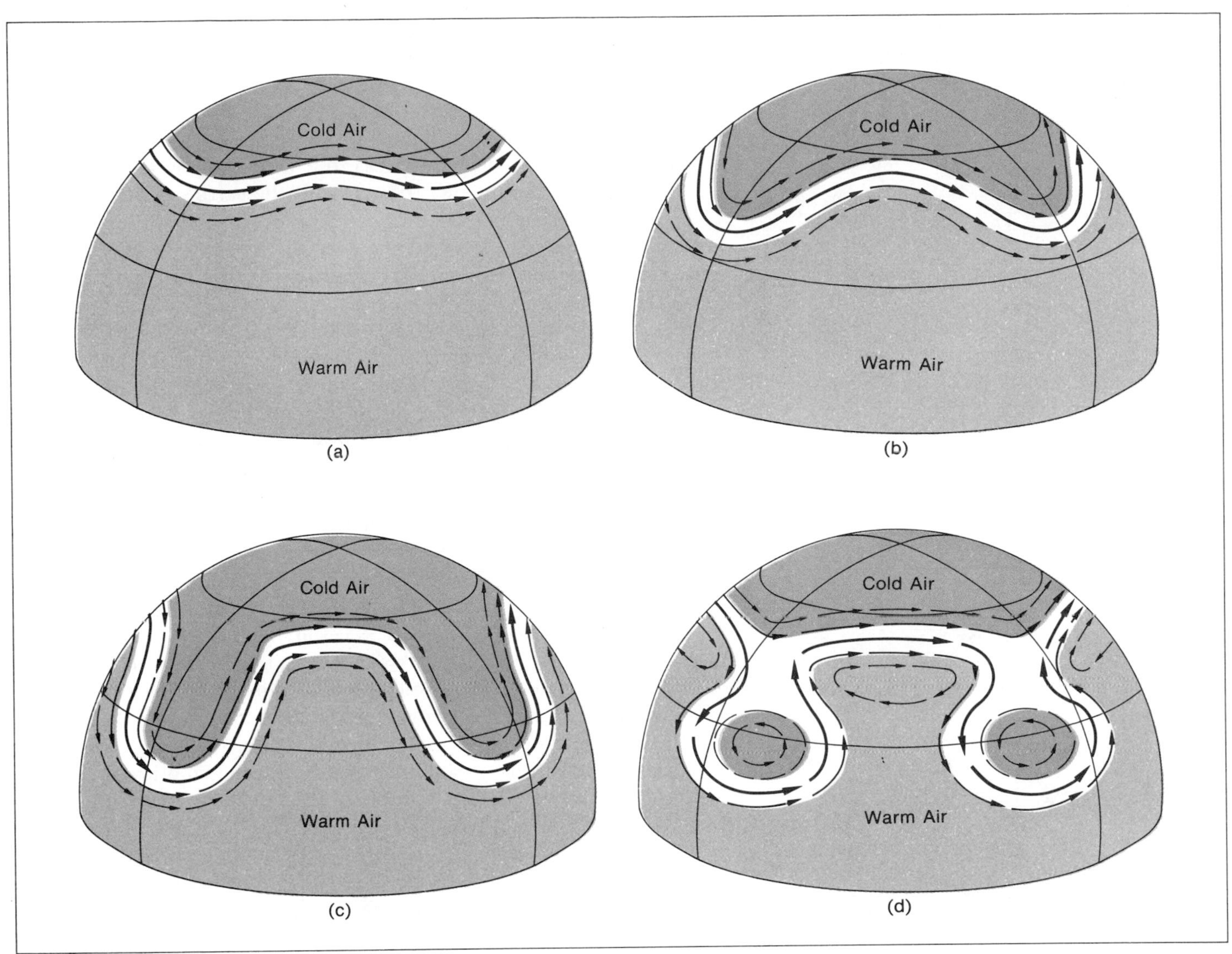

Figure 5.7 Circumpolar jet streams circle the earth at an altitude of 10 to 15 km (6 to 9 miles) and gain much of their energy from the temperature difference between warm tropical and cold polar air. (a) The jet streams appear in conjunction with large, slow-moving undulations, or waves, in the atmosphere. There are typically three to six waves in a complete pattern around the earth. The waves may undergo increasingly severe oscillations (b) (c), until cells of rotating warm and cold air are formed (d). The cells then die away, and the pattern once more resembles that shown in (a); the entire cycle is completed in four to eight weeks.

This mechanism results in the movement of warm air toward the poles and cold air toward the tropics and helps to maintain the poleward flow of energy from equatorial regions. Jet streams and upper air waves also appear to generate weather systems, but the details of the process are not fully understood. (After "The Jet Stream" by Jerome Namias. Copyright © 1952 by Scientific American, Inc. All rights reserved.)

ing warm and cold air. The cells eventually die away, and the pattern once more resembles that shown in Figure 5.7a. A cycle of oscillations in the jet stream is normally completed in one to two months.

Observed Pressure and Wind Systems

The actual average patterns of pressure and wind at the earth's surface are more complicated than the simple belts associated with the uniform earth model. The presence of continents considerably influences the atmospheric circulation. Rock and soil have much smaller specific heats than water; thus the continents become warmer than the oceans in summer and much cooler in winter, especially when snow covered. The higher mountain ranges act as partial barriers to circulation regimes and even disturb the flow of the upper atmosphere.

Especially in the northern hemisphere, the observed pattern of pressure distribution consists of separate cells, or centers, of high and low pressure rather than belts. Nevertheless, these cells tend to be distributed along bands of latitude that are analogous to the pressure belts that would form on a uniform earth. The average direction of surface winds is closely related to the pressure distribution. Winds tend to flow outward from high-pressure regions and inward toward low-pressure regions. In many areas, the winds follow a pattern of geostrophic flow modified by friction with the earth's surface.

The global maps in Figure 5.8, pp. 128–129, show the average direction of surface winds and the average distribution of pressure, adjusted to sea level, for January and July. In the southern hemisphere, there is almost continuous ocean between latitudes 45°S and 65°S. Therefore, the observed circulation tends to be similar to the model for the homogeneous surface, with subpolar lows around the margins of Antarctica and a polar high over the lofty ice sheet that caps the south pole.

In the northern hemisphere, the presence of large landmasses produces patterns of pressure distribution and wind circulation that are more cellular than zonal. In general, high-pressure areas tend to migrate north during summer and south during winter. In winter, for example, when the northern regions of the continents become very cold while the oceans remain relatively warm, the polar high is displaced somewhat in the direction of the equator, forming two centers over the continents—the massive Siberian high over north-central Asia and the much smaller Yukon high over northwestern Canada. Two subpolar lows, the Icelandic and the Aleutian, are situated over the northern hemisphere oceans at similar latitudes. During summer, when the large continents become warmer than the oceans, the subtropical highs over the oceans are better developed, and low-pressure centers evolve over the southwestern United States and southern Asia.

The seasonal shifts in pressure distribution generate seasonal shifts in wind patterns. The inset in Figure 5.8 shows how the mean position of the intertropical convergence zone, which is the meeting place of the trade winds, shifts between January and July. Note that there is relatively little change in the position of the ITC over the Atlantic and eastern Pacific oceans, but that over land areas, especially those regions bordering the Indian Ocean, there is considerable seasonal migration. Associated with the extreme shift in the position of the ITC over eastern and southern Asia is a change in wind direction between winter and summer of approximately 180°. This change in the direction of surface winds has important consequences for the climate in these areas.

A wind system that reverses direction seasonally is known as a *monsoon*. During winter, the Asian continent is cold, and an almost steady flow of dry air from the north brings cool, clear weather conditions to southern and eastern Asia. This airflow, lasting several months, brings the northwest monsoon to China and Japan and the somewhat warmer northeast monsoon to India and southern Pakistan. During summer, when the wind direction reverses, warm, humid air sweeps over the continent from the Indian Ocean and the southwestern Pacific. As the southeast monsoon, the moisture-laden air brings great quantities of precipitation to China and Japan, and as the southwest monsoon, it delivers several months of heavy rainfall to India and Pakistan. There is also a hint of a monsoonlike flow over the Mississippi River valley, with frequent outbreaks of polar air, called "northers," in winter, and humid, sultry air from the Gulf of Mexico during summer.

The global patterns of pressure and wind set in motion secondary patterns, which in turn affect local weather conditions and the distribution of precipitation. In the following sections, we will examine these secondary circulations in the middle latitudes and tropics.

MIDLATITUDE SECONDARY CIRCULATIONS

Weather in the midlatitudes is dominated by the interaction of large air masses possessing very different characteristics. In the northern hemisphere, especially during winter, warm, moist air flowing poleward out of the subtropical highs meets cool, dry air flowing toward the equator from the Siberian and Yukon highs. Where the air masses meet, a front is formed, bringing about characteristic patterns of pressure, temperature, humidity, and precipitation. In addition, airflows frequently become organized into vast, spiraling eddies called cyclones and anticyclones. These may be 1,000 km (600 miles) or more in diameter, and they tend to migrate at varying speeds from west to east. The very changeable weather of the middle latitudes is associated with these migrating systems.

Air Masses

An *air mass* is a large, nearly uniform body of air that moves as a unit, usually in association with secondary high- or low-pressure systems. Air masses are significant to weather forecasting because they may move long distances and still retain the temperature and moisture characteristics of the region where they were first formed. Air masses with polar or tropical characteristics can bring cool, dry polar air or warm, humid tropical air into midlatitude regions.

Air masses are classified into four general categories, depending on whether they originate in polar (cold) or in tropical (warm) regions and over land (continental and dry) or over oceans (maritime and moist).

Extremely cold winter weather in the United States is associated with air masses that form over the snow-covered plains of northern Canada. During the long subarctic winter, the sun remains below the horizon much of the time, and frequent clear skies promote longwave radiational cooling of the snowcover to very low temperatures. The air tends to settle, and if the high pressure persists for days, a bitter cold inversion develops from the surface upward. After a few days, a surface air mass is created that has nearly homogeneous thermal and moisture properties and may extend horizontally for more than 1,000 km (600 miles). This surface air, with low temperatures and humidity, is designated by meteorologists as continental polar air (cP).

Eventually, the cP air breaks out of the source region and moves

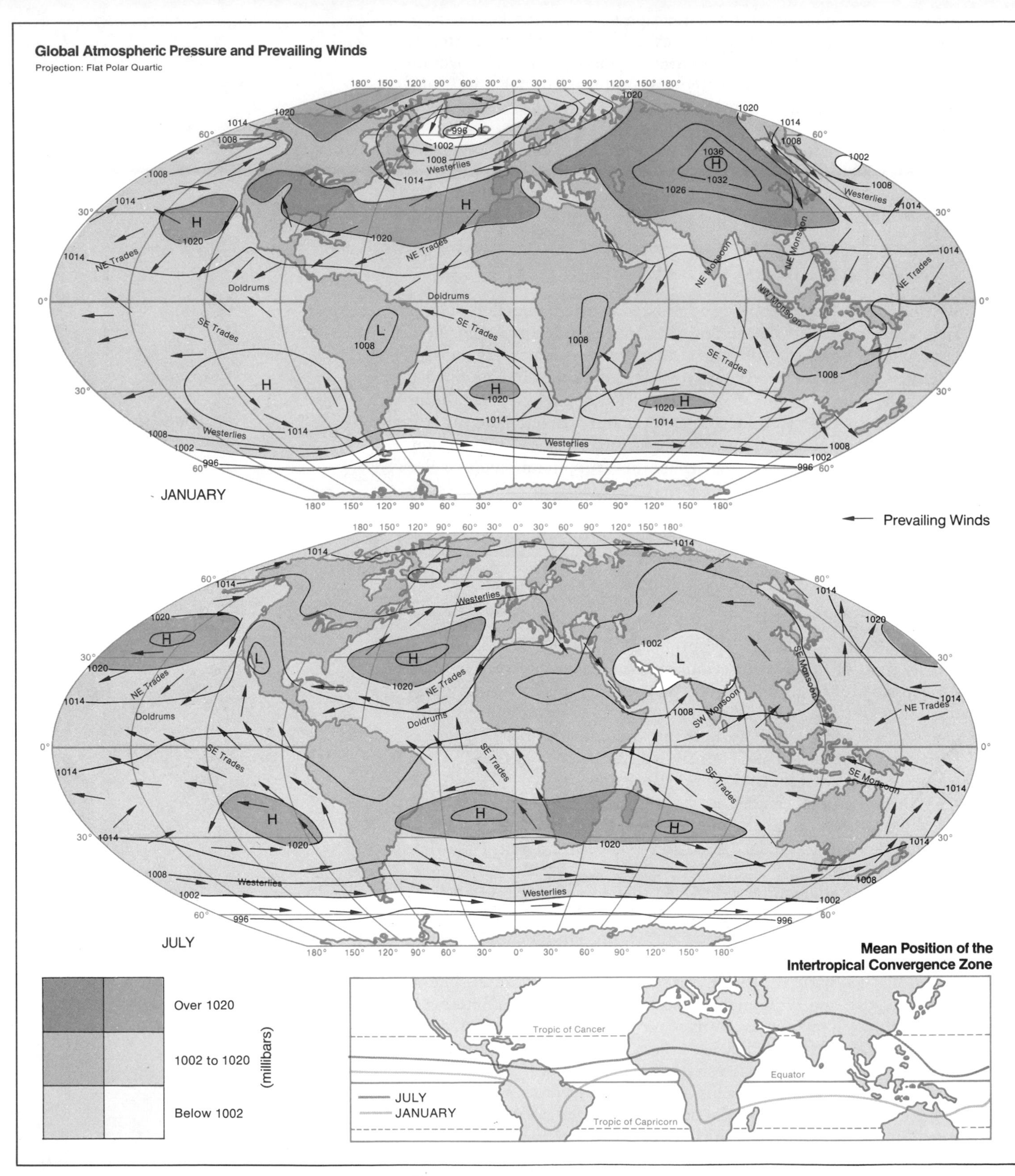
Global Atmospheric Pressure and Prevailing Winds
Projection: Flat Polar Quartic
JANUARY
JULY
Prevailing Winds
NE Trades
SE Trades
Doldrums
Westerlies
NE Monsoon
NW Monsoon
SW Monsoon
SE Monsoon
Over 1020
1002 to 1020
Below 1002
(millibars)
Mean Position of the
Intertropical Convergence Zone
Tropic of Cancer
Equator
Tropic of Capricorn
JULY
JANUARY

Atmospheric Pressure in Polar Regions

Projection: Stereographic

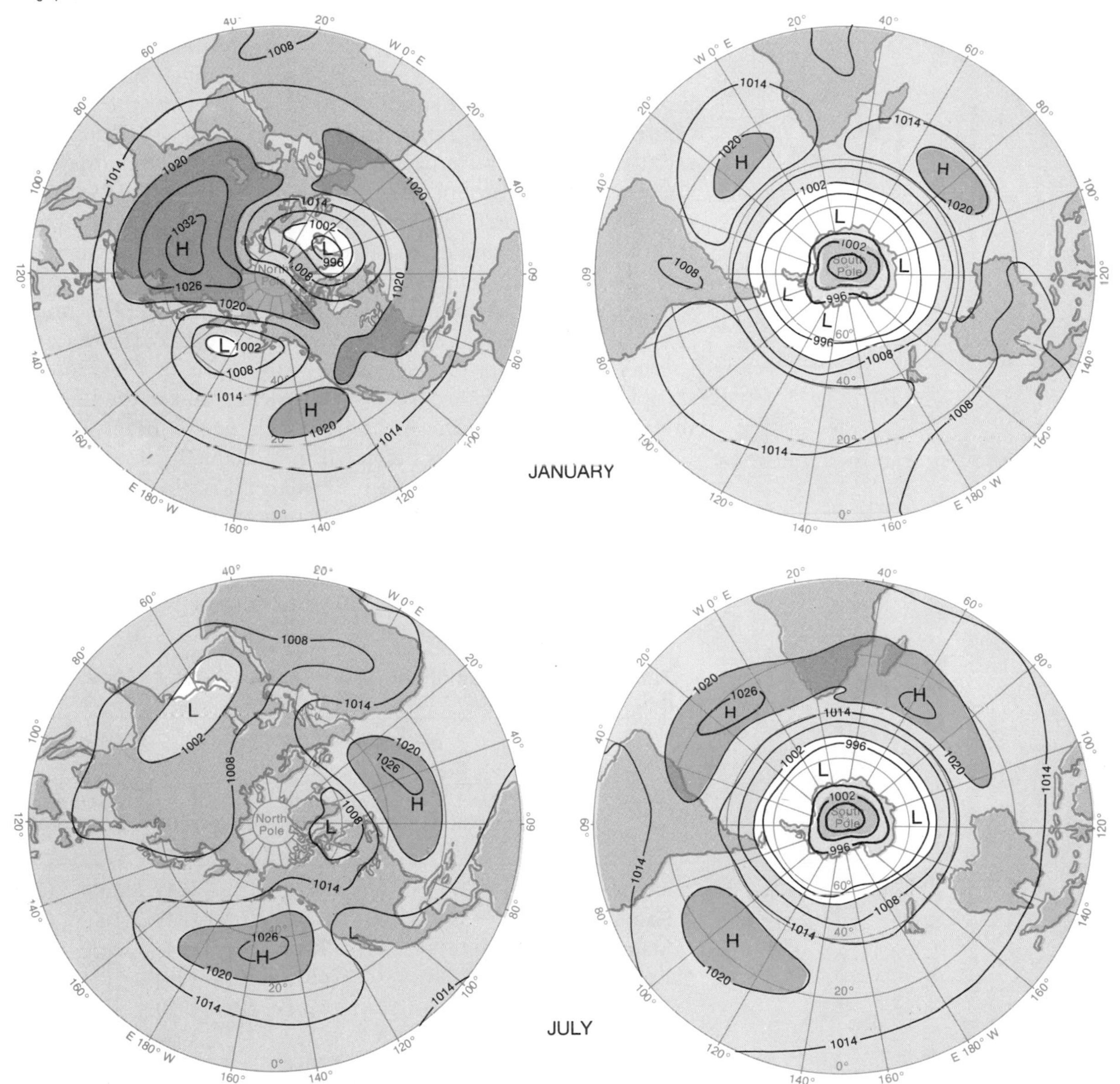

Figure 5.8 Global maps show the average direction of surface winds and the average atmospheric pressure at the surface for January and July. Pressure readings are in millibars. Near the Antarctic Circle, where there is open ocean around the entire globe, the isobars form a continuous belt. Elsewhere, the presence of landmasses causes the regions of high and low pressure to be broken into individual pressure cells. The cells tend to be distributed along bands of latitude that are analogous to the pressure belts that would form on a uniform earth. The formation of pressure cells is affected by temperature differences between the oceans and the continents. (After the United States Weather Bureau)

The average distribution of surface winds closely follows the pressure patterns, with winds tending to flow outward from high-pressure regions and inward toward low-pressure regions. In many areas, the winds approach geostrophic flow modified by friction with the earth's surface.

The distribution of pressure cells generally shifts to the north in July and to the south in January. Over the poles, the high-pressure cells are more diffuse and less numerous in summer than in winter. The resulting seasonal shift in wind patterns generates a seasonal variation in the weather of many regions. The winds on the east coast of central Africa, for example, change from the northeast trades in January to the southeast trades in July. The inset map shows how the mean position of the intertropical convergence zone, which is the meeting place of the trade winds, shifts between January and July. (After Schwerdtfeger, 1970, and Flohn, 1969)

toward the equator. As it sweeps southward, traveling more than 4,000 km (2,500 miles) across the Great Plains and down the Mississippi River valley to New Orleans and the Gulf Coast, the cold air of the polar outbreak is modified only slightly. As Table 5.1 shows, at New Orleans during mid-winter, the mean temperature of the cP air is close to 0°C. And temperatures may be lower than −5°C during severe outbreaks! However, when the cP air moves out over the warm waters of the Gulf of Mexico, a steep lapse rate develops; the cP air is warmed rapidly and gains moisture. Modification is so quick that another "new" air mass is produced within 48 hours or so. This warm, moist air mass is designated maritime tropical (mT), and it will move away from the source region eventually. Table 5.1 gives the air temperature and the dew point temperature for both cP air and mT air at New Orleans during the course of a year. Note that there is much less seasonal variation in the temperature of mT air. This is because the sources of these air masses, namely, the Gulf of Mexico and the Caribbean Sea, are quite close to New Orleans, and so the thermal and moisture properties of the air have little chance to undergo alteration. Dew point temperature is indicated in the table because it is a useful measure of the moisture content of an air mass; the moisture content of the mT air is consistently greater that that of the cP air. There is also less difference between the air temperature and the dew point temperature of mT air than of cP air; the mT air is closer to saturation, and only a small amount of lifting and cooling will produce saturation, condensation, and clouds.

Table 5.1. Mean Monthly Air Mass Temperatures at 6 A.M. During Fair Weather at New Orleans, 1971–1974

	cP Air				mT Air			
	Air Temperature		Dew Point Temperature		Air Temperature		Dew Point Temperature	
	°C	(°F)	°C	(°F)	°C	(°F)	°C	(°F)
January	3	(37)	−1	(30)	16	(61)	16	(61)
February	4	(39)	0	(32)	16	(61)	14	(57)
March	8	(46)	4	(39)	18	(64)	16	(61)
April	10	(50)	7	(45)	20	(68)	18	(64)
May	18	(64)	15	(59)	22	(72)	20	(68)
June	22	(72)	18	(64)	24	(75)	22	(72)
July	23	(73)	21	(70)	24	(75)	22	(72)
August	23	(73)	20	(68)	23	(73)	22	(72)
September	21	(70)	18	(64)	23	(73)	22	(72)
October	16	(61)	13	(55)	22	(72)	21	(70)
November	9	(48)	6	(43)	19	(66)	17	(63)
December	4	(39)	1	(34)	17	(63)	16	(61)

Muller, Robert. 1977. "A Synoptic Climatology for Environmental Baseline Analysis: New Orleans." *Journal of Applied Meteorology*, Vol. 16: 20–33.

Air masses move over the earth's surface according to certain predictable patterns. Figure 5.9 shows typical tracks of air masses over the United States in winter and summer. To a large degree, the movement of these air masses determines weather conditions in the regions

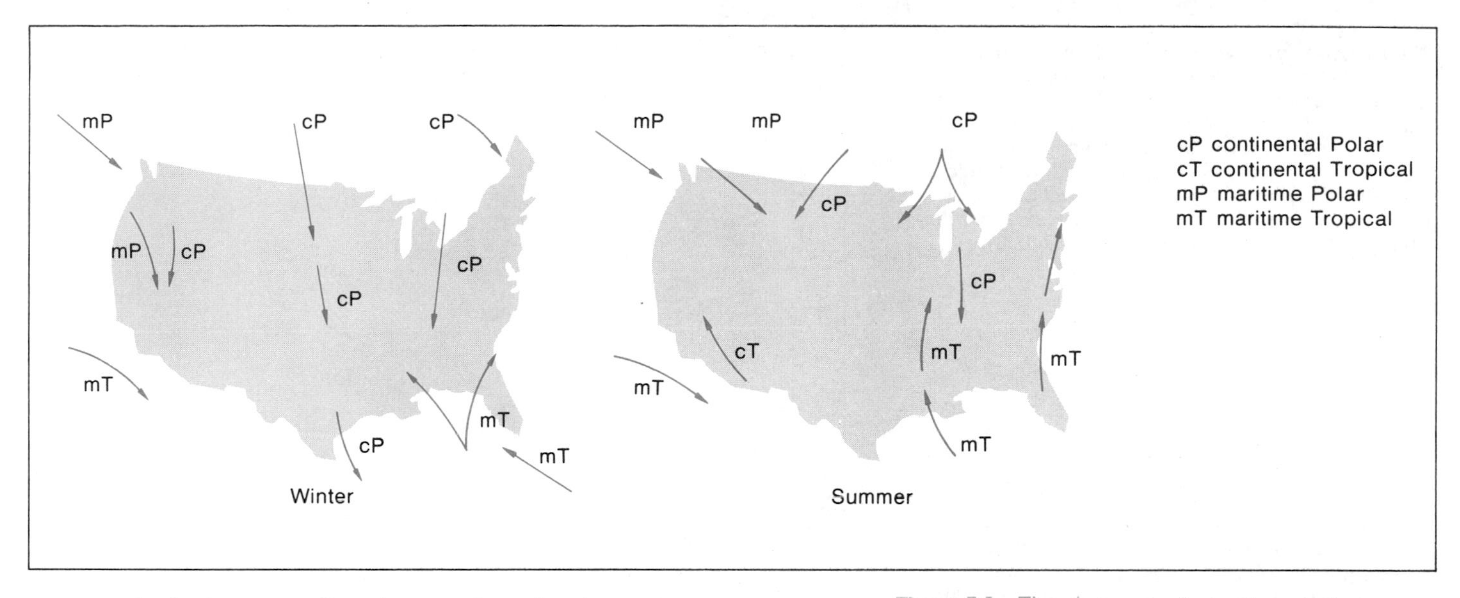

Figure 5.9 The air masses that enter a region can strongly influence weather conditions there by replacing the existing air with air of different temperature and humidity. Maritime air tends to be humid, and continental air tends to be dry. The maps in this figure depict the average movements of air masses over the conterminous United States during winter and summer. During winter much of the northern half of the United States is invaded by cold polar air, whereas during summer the northward flow of tropical air dominates the weather in most regions. (After Brunnschweiler, 1957)

over which they pass. But the weather also depends on the surface characteristics of the regions themselves. Continental polar air, for example, is generally associated with cold, windy, and fair weather over the Great Plains and Mississippi River valley. But when cP air moves down over the Great Lakes, the lower layers receive much heat and moisture from the water; consequently, cP air masses often bring spectacular snow squalls to the eastern and southeastern shores of the Great Lakes. This situation is an exception, however. On the whole, very little precipitation falls from continental air masses. Instead, they tend to gain water vapor from evapotranspiration and return it eventually to the oceans. The opposite is true of maritime air masses; they tend to lose more water vapor by precipitation to the continents than they receive back by evapotranspiration.

Fronts

A front is a boundary between two air masses that differ in temperature, humidity, or both. Much of our knowledge of air masses and fronts comes from the work of a small group of Scandinavian meteorologists who carried out their investigations at the Norwegian School for Meteorology in Bergen, Norway, in the years following World War I. The most distinct fronts are those occurring between air masses whose properties contrast most sharply—namely, those between cold polar air and warm tropical air. Temperature, humidity, cloudiness, and wind can differ markedly from one side of a front to the other, sometimes over a distance of only a few kilometers.

When warm air moves into a region occupied by a colder air mass, the forward edge of the warm air mass is designated as a *warm front*. Figure 5.10a, p. 132, shows the formation of a warm front, both in cross section and as it would be represented on a weather map. The cool air is retreating toward the right, corresponding on the map to the east or northeast, the usual direction in which a warm front is apt to travel in the United States. The warm air moves in from the left (west

Figure 5.10 (a) A warm front is formed when a mass of warm air encounters a denser mass of cool air. The diagram shows a vertical cross section of a warm front, together with the surface symbol as it would appear on a weather map. As the uplifted warm air becomes cooler, condensation and cloud formation may occur. The vertical section shows that the resulting precipitation arrives at a location ahead of the surface warm front itself.

(b) A cold front develops where a mass of cool air pushes into a region of warmer air. The vertical cross section of the front shows the warm air ahead of the front being forced upward over the incoming cool air. Cloud formation and precipitation may occur as the uplifted warm air cools. The vertical section shows that precipitation normally arrives at a location just ahead of the surface front.

(c) An occluded front involves three air masses of different temperatures, and it combines some of the features of both warm and cold fronts. The vertical cross sections show that the mass of incoming cold air forces warm air to rise above both regions of cool air. In the forward vertical section, the occlusion is not complete, and the warm air is in contact with the ground in close proximity to a cold front and a warm front. In the rear vertical section, the occlusion has formed, and the warm air is lifted completely above the ground. Precipitation may occur on both sides of an occluded front, as the diagrams indicate. (After Flohn, 1969)

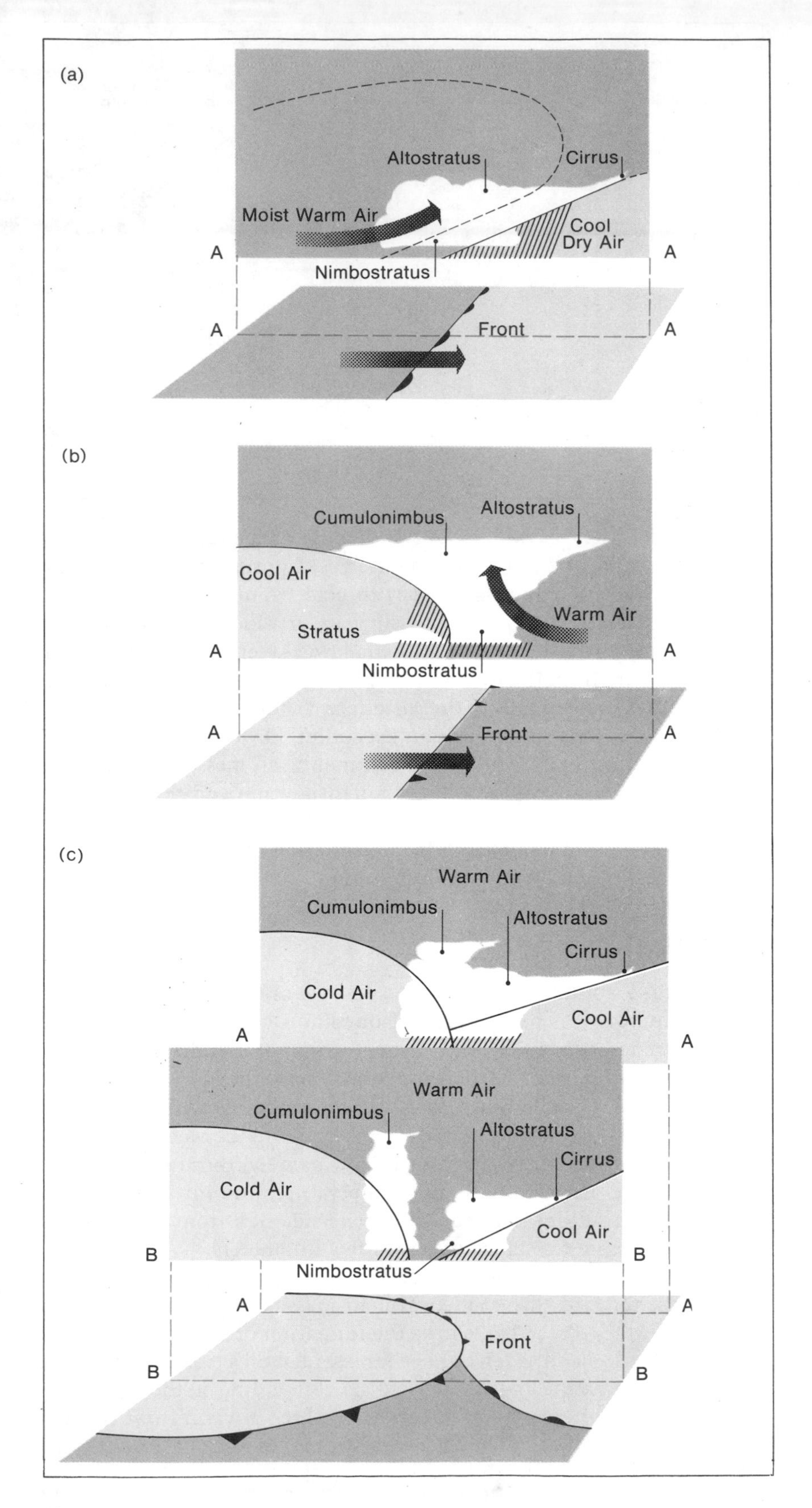

or southwest), replacing the cool air after the front passes over. Although it is not shown on a daily weather map, which indicates only the surface location of a front, the frontal boundary slopes gently up from the earth's surface (toward the right in Figure 5.10a). Thus the front at an altitude of 1 km may be several hundred kilometers ahead of the front at the surface.

Weather conditions in the vicinity of a warm front depend on many factors, such as the properties of the two air masses and the presence or absence of mountains. Nevertheless, the passage of a warm front is usually associated with a characteristic sequence of weather conditions. Condensation and clouds begin where the warm, moist tropical air rises over the cooler air and cools to the dew point. Because a warm front slopes so gently upward, its coming is usually heralded by the appearance of high cirrus clouds a day or more before the surface front arrives. As the surface front approaches, the clouds become thicker and lower, forming sheetlike stratus clouds accompanied by a broad band of precipitation ahead of the surface front; the cloud sequence is also shown in Figure 4.19, p. 109.

A *cold front* develops when cold air moves into a region occupied by warm air. Figure 5.10b shows that the structure of a cold front differs from that of a warm front. The advancing cold air forces the warm air to rise quickly and steeply, creating a slope which is much less gradual than is the case with a warm front. There is rapid development of towering vertical clouds and cumulonimbus, with most of the precipitation occurring just ahead of the surface front. During warm summer weather, the passage of a cold front is usually associated with the sudden appearance of a line of thunderstorms and a rapid drop in temperature.

In those cases where the surface boundary between tropical and polar air masses remains at about the same location for a day or more, the front is designated as *stationary.* More complex *occluded fronts,* involving more than two air masses (see Figure 5.10c), will be treated later in this chapter. All fronts mentioned represent relatively short segments along the larger-scale polar front previously described as a component of the general circulation model. As such, they are associated with moving secondary circulations, such as the midlatitude cyclones, which are discussed next.

Cyclones and Anticyclones

Almost all daily weather maps for the United States show centers of high and low pressure. The isobars around a pressure center usually form asymmetric oval shapes. The converging airflows that form around a central region of low pressure are called *cyclones,* or *lows.* The diverging flows that form around high-pressure regions are called *anticyclones,* or *highs.* The horizontal differences in pressure within large highs or large lows are small and usually amount to less than 10 millibars (mb) over a distance of 100 km (60 miles), only a small fraction of the normal sea level atmospheric pressure of 1,013 mb.

As Figure 5.11 shows, a cyclone, or area of low pressure, is characterized by converging surface air which is forced to ascend and then

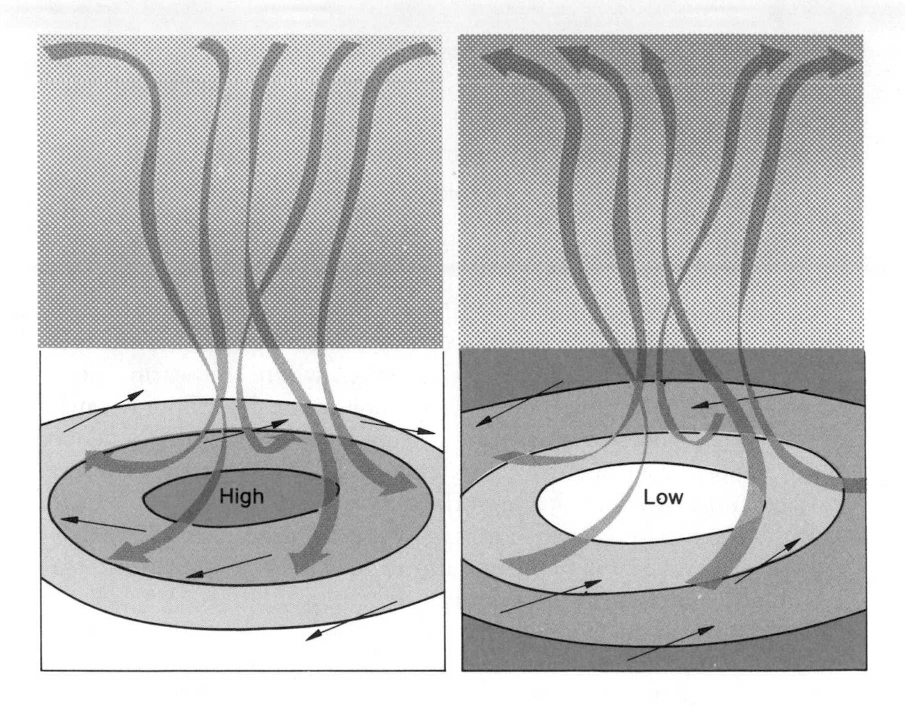

Figure 5.11 In a region of high pressure (left), air descends from the upper atmosphere and diverges outward along the surface of the ground. The air streaming outward from highs therefore tends to be dry. In a region of low pressure (right), air converges inward along the surface of the ground and ascends to the upper atmosphere. If the air streaming into a low is warm and moist, condensation and cloud formation occur as the air rises and cools. The directions of circulation are shown for the northern hemisphere.

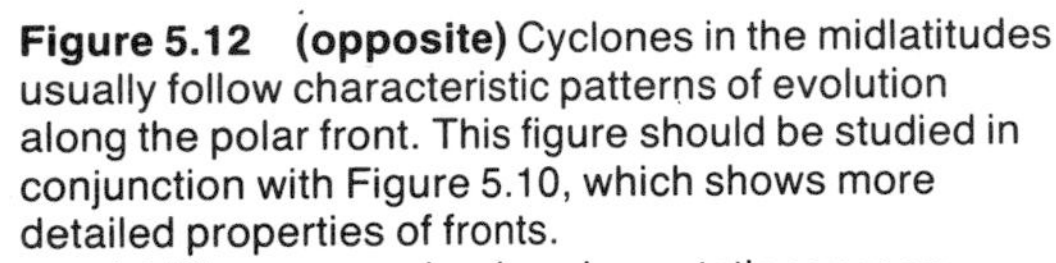

Figure 5.12 (opposite) Cyclones in the midlatitudes usually follow characteristic patterns of evolution along the polar front. This figure should be studied in conjunction with Figure 5.10, which shows more detailed properties of fronts.

(a) The process begins along stationary segments of the polar front where cold and warm air stream in opposite directions.

(b) The stationary front tends to be unstable, and a bulge, or wave, usually develops within one or two days. The waves move along the polar front toward the northeast, and most dissipate within six to twelve hours.

(c) A few waves, however, grow into well-developed circulations with a diameter of 1,000 km (600 miles) or more. In the northern hemisphere the counterclockwise converging circulation of the cyclone includes warm and cold fronts that separate polar and tropical air masses.

(d) The cold front eventually overtakes the warm front, lifting the warm air away from the surface and forming an occlusion. The occlusion eliminates the surface air temperature differences that provide energy for the system, and the circulation then weakens and finally dissipates.

(e) This sequence of development can be traced on the three map sketches for March 5, 6, and 7, 1973. As a cyclone occludes, a new wave disturbance may form farther back along the trailing stationary front. Over the oceans, the polar front often supports a cyclone family of three to five members in various stages of development. Most late fall, winter, and spring precipitation over the United States is associated with midlatitude cyclones and associated fronts.

diverge in the upper atmosphere. An anticyclone, or high-pressure area, on the other hand, is characterized by air subsiding from aloft which diverges at the surface. As air rises in a cyclone, it tends to undergo adiabatic cooling, resulting in cloud formation. For this reason, cyclones are generally associated with cloudy and unsettled weather conditions. The descending air in an anticyclone, by contrast, experiences adiabatic heating, a process which is usually associated with fair weather.

In the northern hemisphere, midlatitude cyclones tend to form along the polar front and migrate toward the east and northeast for as much as 4,000 km (2,400 miles). Most cold anticyclones in North America peel off from the Yukon high and drift southeastward to reach eventually the Atlantic Ocean off the Carolinas. Cyclones are centers of convergence for masses of both warm and cold air, and much of the precipitation over the United States is associated with the ascent of warm, moist air over the cold air within the cyclone circulation system.

Midlatitude cyclones tend to develop along segments of the polar front where warm and cold air come into frequent juxtaposition. Their evolution usually follows a characteristic pattern. Cyclones begin along a stationary segment of the polar front. Figure 5.12a shows a stationary front with a northeast-southwest orientation; such a front may remain over the same region for as much as 24 to 48 hours before a cyclone begins to develop.

Pressure is lowest along the front, increasing toward the northwest and southeast. On the polar side of the front, cold air flows from the northeast; on the tropical side of the front, warm air flows from the southwest. Because such a front tends to slope upward from the surface toward the northwest, there is often warm air aloft over the

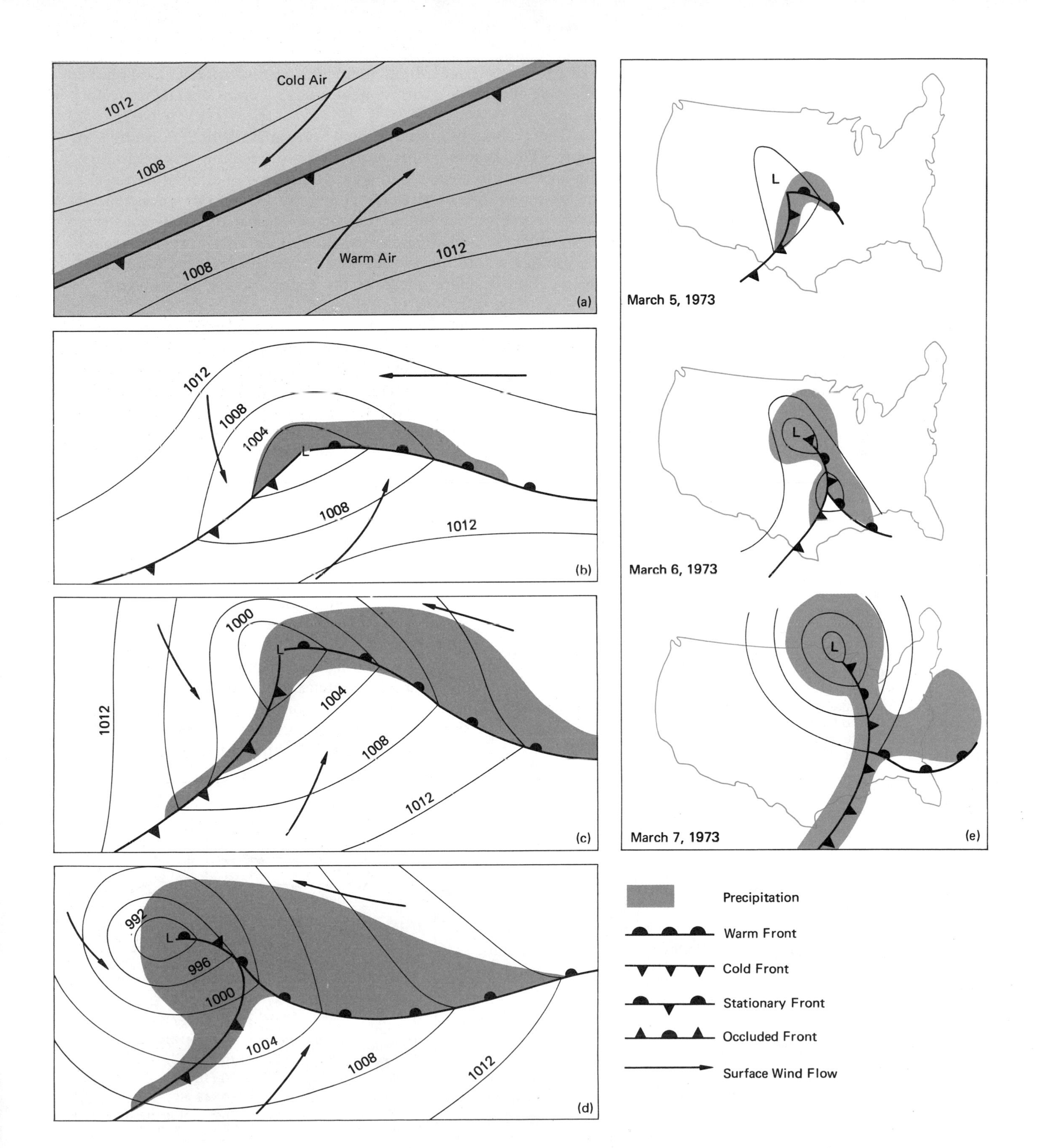
Cold Air
Warm Air
1012
1008
1008
1012
(a)
1012
1008
1004
L
1008
1012
(b)
1000
L
1012
1004
1008
1012
(c)
992
L
996
1000
1004
1008
1012
(d)
L
March 5, 1973
L
March 6, 1973
L
March 7, 1973
(e)
Precipitation
Warm Front
Cold Front
Stationary Front
Occluded Front
Surface Wind Flow

cold air at the surface, which produces a narrow band of cloud on the polar side of the front, with some drizzle and occasional light precipitation.

The cyclone begins as a small wave disturbance, illustrated in Figure 5.12b. At some point along the front, pressure falls even more, and air begins to converge in a counterclockwise circulation toward the low center. Ahead of the low center the warm air advances toward the northeast into the former domain of the cold air, forming a warm front, while the cold air begins to sweep in behind the center, toward the southeast, forming a cold front. A broad shield of warm-front precipitation develops ahead of the center, and a narrow band of showers breaks out along the cold front.

Many waves may move toward the northeast along the polar front and then simply dissipate, usually in 6 to 12 hours. A few waves, however, grow into well-developed cyclonic circulations of 1,000 km (600 miles) or more in diameter. One of the more difficult tasks of the weather forecaster is to predict the development of a fully blown midlatitude cyclone from a small wave disturbance.

Figure 5.12c shows the cyclone at a more advanced stage of development. The central pressure of the cyclone may fall at this time to less than 29.5 in. (1,000 mb), and the converging counterclockwise circulation may expand to a diameter greater than 1,000 km. The cyclone system tends to move in a northeasterly direction with a speed between 25 and 50 km per hour. Meanwhile, the warm front continues to progress northeastward while the cold front moves rapidly east and south. Steady and substantial rains usually occur within the precipitation shield ahead of the warm front, and occasionally there are severe thunderstorms and even tornadoes. (Tornadoes are the topic of the case study following this chapter.) The weather map in Figure 5.13 represents a small, intense midlatitude cyclone out of which several "killer" tornadoes were born.

Figure 5.12d shows the cyclone in its mature phase. The cold front tends to move more rapidly toward the east than the cyclone

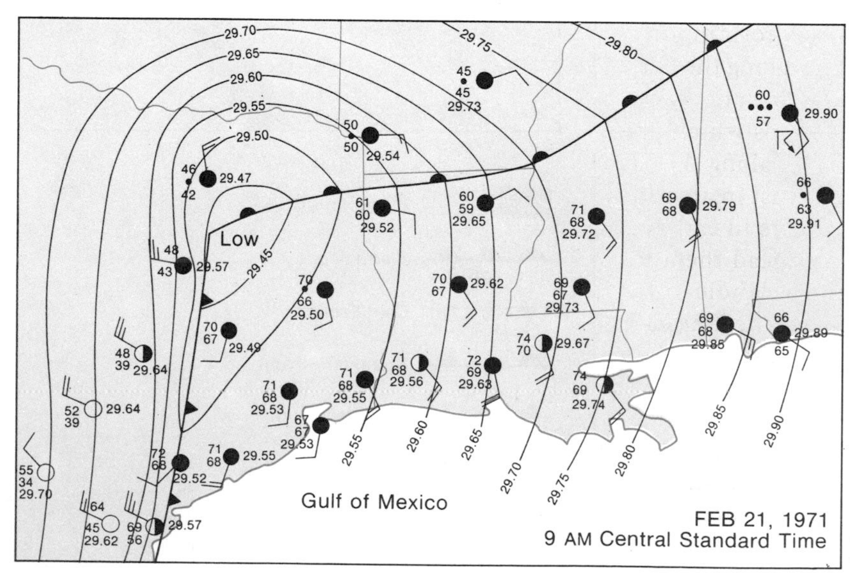

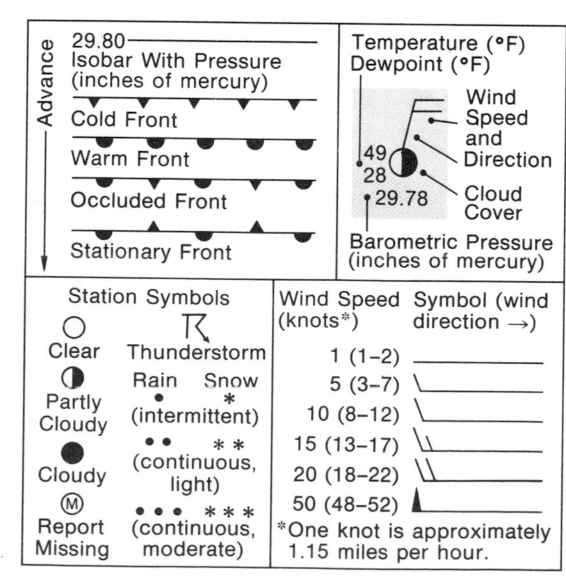

Figure 5.13 The weather map portrays weather conditions in the south central United States on February 21, 1971, based on ground station reports at 9:00 A.M. Central Standard Time. The information on the map can be read with the help of the accompanying symbol table. The map shows, for example, isobars labeled with barometric pressure readings in inches of mercury, a cold front advancing eastward through Texas, a warm front advancing northward, and numerous station reports. The station symbols indicate temperature and dew point in °F, barometric pressure in inches of mercury, wind direction and speed in knots, cloud coverage, and precipitation events. Note that a station in Alabama, near the eastern edge of the map, reports a thunderstorm and heavy rain.

The station symbols in the weather maps prepared by the National Weather Service carry more detailed information than do the station reports in the simplified weather map shown here. Among other additional station data, the Weather Service maps present cloud type, height of the cloud base, change of barometric pressure in the previous three hours, and weather conditions in the previous six hours.

This weather map shows an intense low-pressure system over eastern Texas. Note that most of the surface winds cross the isobars at small angles and converge toward the low center. West of the cold front, winds are strong, with a speed of 25 knots. During the afternoon of February 21, the low-pressure area generated thunderstorms and tornadoes in Louisiana and Mississippi. Damage from the tornadoes was severe, and many people were injured or killed. (After Muller, 1973)

system does; thus the cold front eventually overtakes the warm front, lifting the warm air up from the surface and forming an occlusion. Occluded fronts have some properties of both warm and cold fronts; they are frequently associated with large amounts of precipitation, which falls from the warm, moist air that is lifted above the surface. As Figure 5.10c shows, precipitation may occur on both sides of an occluded front. Occlusions are especially characteristic of cyclone systems over the Pacific Northwest, where the Coast Ranges and Cascades are often inundated by heavy rains and snows at higher elevations during winter and spring.

In the cyclone's final stage, its energy supply undergoes a significant alteration. The developing occlusion eventually eliminates the surface air temperature differences, so that warm, moist air from the warm front no longer feeds into the center of the system. Although the intensity and size of the cyclone are often greatest at the beginning of the occlusion, it is not long before the center of the cyclone begins to fill with cooler, drier air and the cyclone itself loses its energy. The life cycle of a midlatitude cyclone ranges from 24 hours to as much as 5 days; thus it is possible for a particularly long-lived cyclone to travel across most of the United States.

Midlatitude cyclones tend to develop along nearly stationary segments of the polar front where large masses of warm and usually moist air are present on the equatorward side of the front. Favored regions for cyclone development in the United States include the eastern slopes of the Rocky Mountains, from Colorado to Alberta, as well as the Gulf of Mexico and the Atlantic coasts. When a polar outbreak spills southward across the Great Plains, for example, a stationary front forms between the very cold polar air east of the Rocky Mountains and milder Pacific air over the western states. The cold, dense air does not normally flow up and over the mountains to the west; instead, frontal waves on the eastern slopes often develop into winter storms which then sweep eastward. On the southern and eastern margins of the polar outbreak, the cold front sometimes becomes stationary over the warm waters of the Gulf of Mexico or over the Gulf Stream off the Carolinas or Virginia. Storms forming in these regions often become well developed as they sweep along the Gulf and Atlantic coasts toward New England, where they are known as "nor'easters" because of the blustery northeast winds which precede the passage of the storm center. Cyclones that form along the eastern coast of Asia usually reach the Pacific Northwest coast in an occluded stage. Cyclones over the central and eastern states are in earlier stages of development, however, and follow paths which lead them toward the northeast and eventually to the region of the Icelandic low. Most of the late fall, winter, and spring precipitation over the United States is due to the passage of midlatitude cyclones and associated fronts.

Jet streams also play a part in the formation of midlatitude cyclones, although the details of the processes involved are not fully understood. It is believed that the presence of a jet stream aloft may aid in the divergence of surface air that has ascended through the low-pressure center of a cyclone. Cyclones appear to intensify when the surface flow of air coincides with favorable airflow conditions

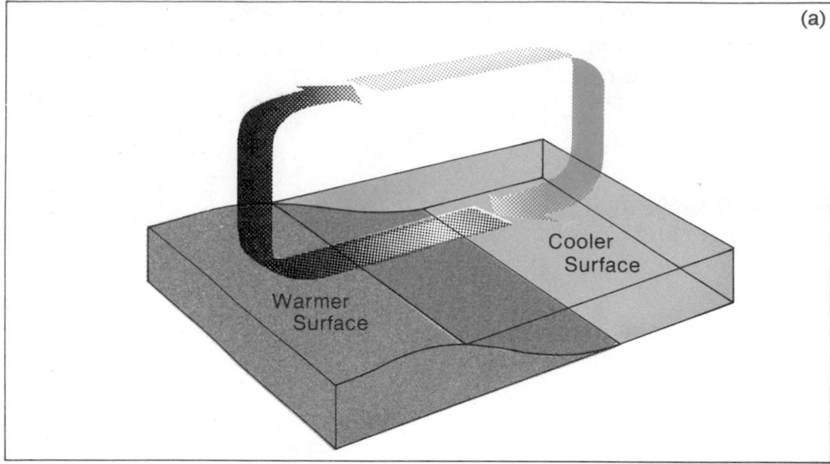

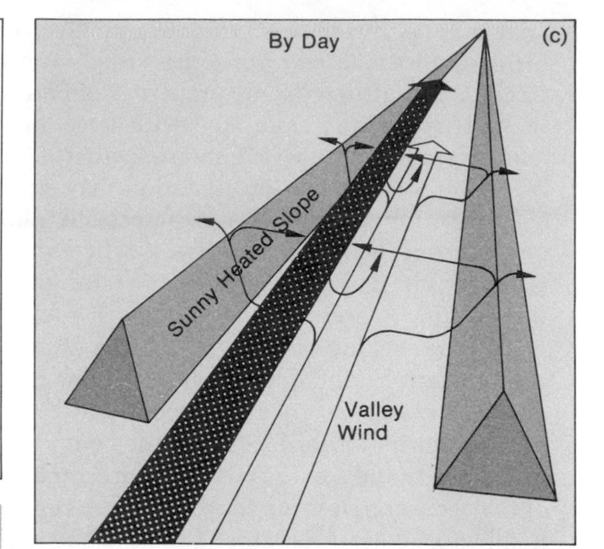

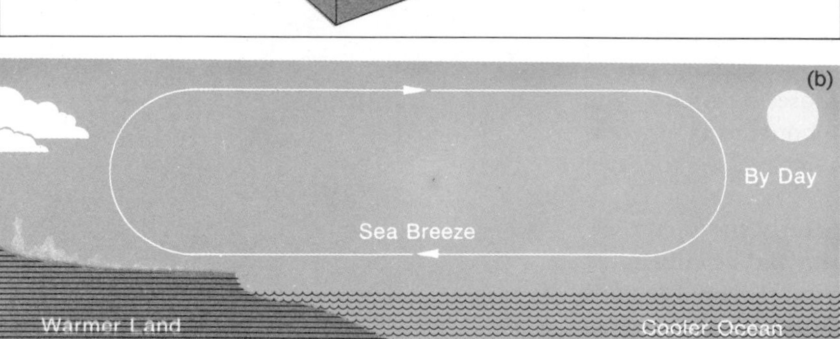

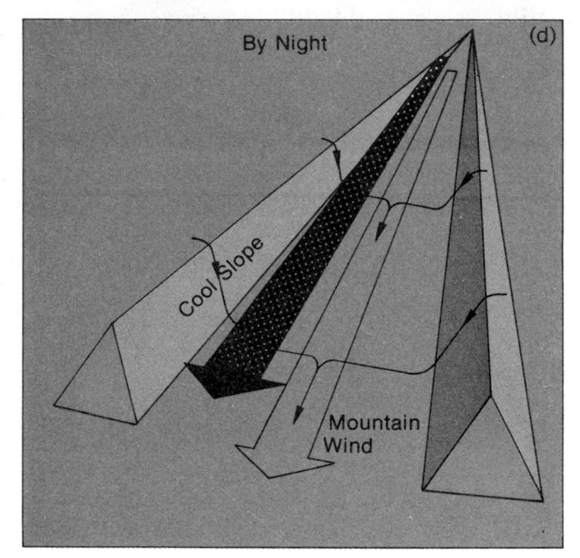

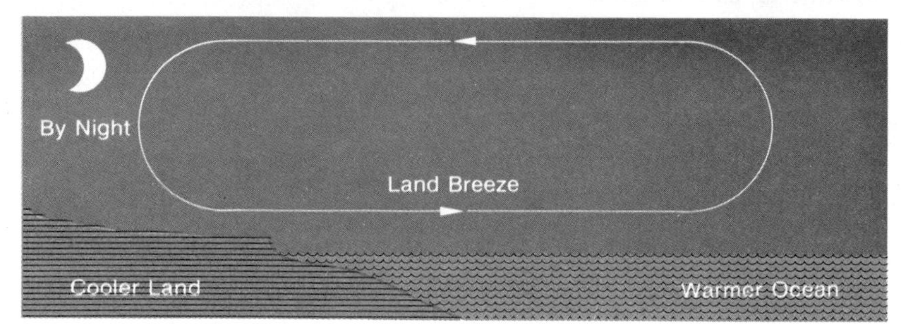

Figure 5.14 Small-scale circulations of air can significantly modify local weather conditions.

(a) This diagram shows a small convective cell generated near the boundary between a warm and a cool region.

(b) Land and sea breezes develop because of the difference in temperature between the ocean and the land. During the day, the land heats more rapidly than the ocean, and a surface sea breeze develops from the ocean toward the land. At night, the land cools rapidly, and the surface flow of air is a land breeze from the land to the ocean.

(c) Mountain and valley winds develop because of the temperature difference between valleys and mountain slopes. During the day, the slopes are warm, and as air rises up the slopes, a surface valley wind flows up the valley to replace the ascending air.

(d) At night, cool air descends the slopes and flows down the valley out of the mountains.

aloft; when airflow aloft is unfavorable, the development of cyclones seems to be suppressed.

Anticyclones are centers of higher pressure with surface air diverging outward—in a clockwise flow in the northern hemisphere. Because the subsiding air in an anticyclone is subject to adiabatic heating, there is little cloud formation, and, hence, anticylones are usually associated with fair weather. In North America during winter, cyclones and anticyclones often cross the continent in rapid succession. Stationary polar front segments, which often give birth to cyclones, frequently carry the beginnings of an anticyclone in their wake. The very changeable winter weather of the Northeast is due to the progression of polar outbreaks and cyclones. Some winter weeks may see the progress of as many as three pairs of cyclones and anticyclones down the St. Lawrence River valley toward the North Atlantic. Terrain configurations, such as coastlines and mountain slopes, significantly affect local circulations and weather, as shown in Figure 5.14. This is especially true during periods of weak anticyclonic flow.

TROPICAL SECONDARY CIRCULATIONS

Much less is known about tropical weather patterns than about mid-latitude weather patterns. The region between the tropic of Cancer and the tropic of Capricorn has few observing stations, and weather data from the extensive tropical ocean surface are especially sparse. The Coriolis force is small near the equator; thus winds are not geostrophic there. Rotating cyclones and anticyclones do not form near the equator, and frontal activity is absent. In the middle latitudes, the interaction of different air masses dominates the weather. In the continental tropics, the daily cycle of heating and cooling of a comparatively homogeneous mass of warm air is the principal determinant of the weather. The temperature variation during a single day in the tropics often exceeds the variation in the average temperature through the year. In spite of a lack of variety in weather patterns, the tropics do exhibit characteristic surface circulations not found elsewhere in the world.

Easterly Waves

Aside from the intertropical convergence zone—that band of persistent clouds and precipitation which marks the meeting place of the northeast and southeast trades—weather in the tropics varies little from day to day. Nevertheless, certain dynamic atmospheric instabilities do occur to interrupt the long stretches of hot, sunny days.

Sometimes a weak trough of low pressure many hundreds of kilometers long and running roughly north and south forms in the path of the trade winds and drifts slowly westward with them (see Figure 5.15). These disturbances, called *easterly waves,* form most often during the summer or high-sun season. In the tropics, a layer of warm air usually overlies the surface layer, forming a weak tempera-

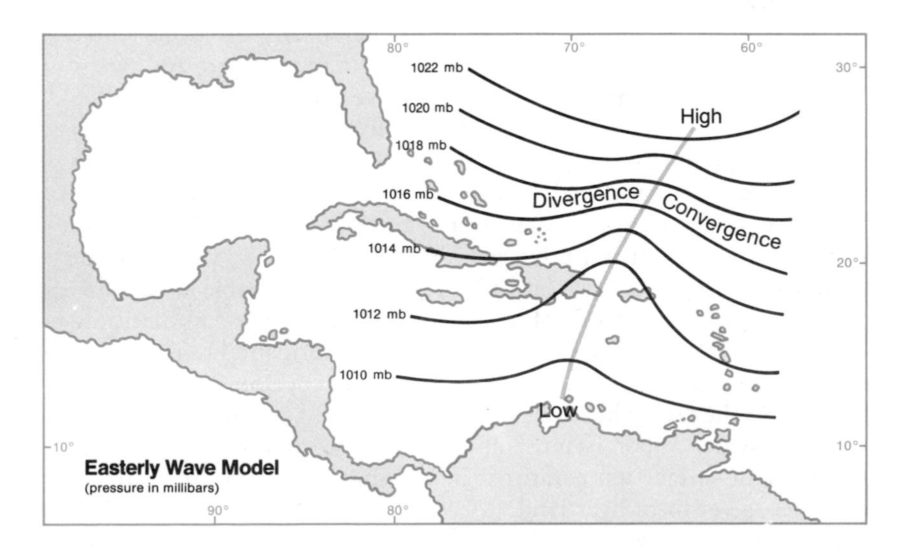

Figure 5.15 The kink in the isobars denotes the presence of an easterly wave. Easterly waves move toward the west with the trade winds. Ahead of an easterly wave, to the west of the trough, warm, dry air descends from the upper atmosphere, and the weather is fair. Behind an easterly wave, to the east of the trough, moist air from the surface ascends to great heights, and severe thunderstorms may be generated. Each year a few easterly waves increase in intensity and develop into severe tropical cyclonic storms such as hurricanes. (After Flohn, 1969)

ture inversion and preventing the surface air from rising to higher altitudes. This inversion is often temporarily destroyed by an easterly wave, resulting in weather disturbances which vary from mild to extremely violent. Ahead of a wave, to the west of the trough, the surface winds tend to diverge, or fan out, horizontally. Warm, dry air descends from the upper atmosphere, and the weather is fair. Behind the wave, to the east of the trough, the winds tend to converge horizontally. Moist air from the surface can ascend to great heights, and severe thunderstorms may be generated. Each year a few easterly waves increase in intensity and develop into severe tropical cyclonic storms or hurricanes.

Tropical Cyclones

A surprising feature of the usually monotonous tropical weather is the development of intense cyclonic disturbances. If its wind speed exceeds 120 km per hour (75 miles per hour), the disturbance is classified as a *tropical cyclone*. Tropical cyclones are called by different names in various regions of the world. Those that form near the east or west coast of North America are called *hurricanes*, and those forming in the western Pacific are called *typhoons*. Both words mean "big wind"—in Arawak and Chinese, respectively.

A tropical cyclone is a compact and intense low-pressure center, often with a diameter of only a few hundred kilometers, yet with pressure gradients that can exceed 30 mb per 100 km. The winds that whirl around the center of a tropical cyclone may have velocities greater than 200 km per hour (120 miles per hour); the accompanying rain and thunderstorms can be severe.

Many hurricanes begin as a rotating tropical storm associated with an intense easterly wave. Why some storms die out and others continue to build to hurricane strength is not known. The characteristic structure of a fully formed tropical cyclone consists of rapidly rotating walls of cloudy, moist air surrounding a nearly calm, relatively cloudless central "eye," as illustrated in Figure 5.16. The height of a tropical cyclone is typically 10 to 15 km (6 to 9 miles), and the eye may be several tens of kilometers in diameter. Dry air from above descends into the eye and becomes warmer than the surrounding air because of adiabatic heating, keeping the eye open and relatively free of clouds. The moist air in the cloud walls is warm and follows an ascending spiral pattern.

Tropical cyclones rarely form within 5° latitude of the equator because of the weakness of the Coriolis force at low latitudes. Nor do they usually form at latitudes greater than 30°N or S. Warm ocean surfaces between these two latitudes provide the favorable conditions needed for the formation of tropical cyclones. Unlike midlatitude cyclones, which are powered primarily by the horizontal temperature differences between different air masses, tropical cyclones derive their force from the latent heat released by condensing water vapor in the cyclone. Water vapor is most abundant in warm tropical air over oceans with surface temperatures greater than 27°C (81°F). Tropical cyclones never form over land.

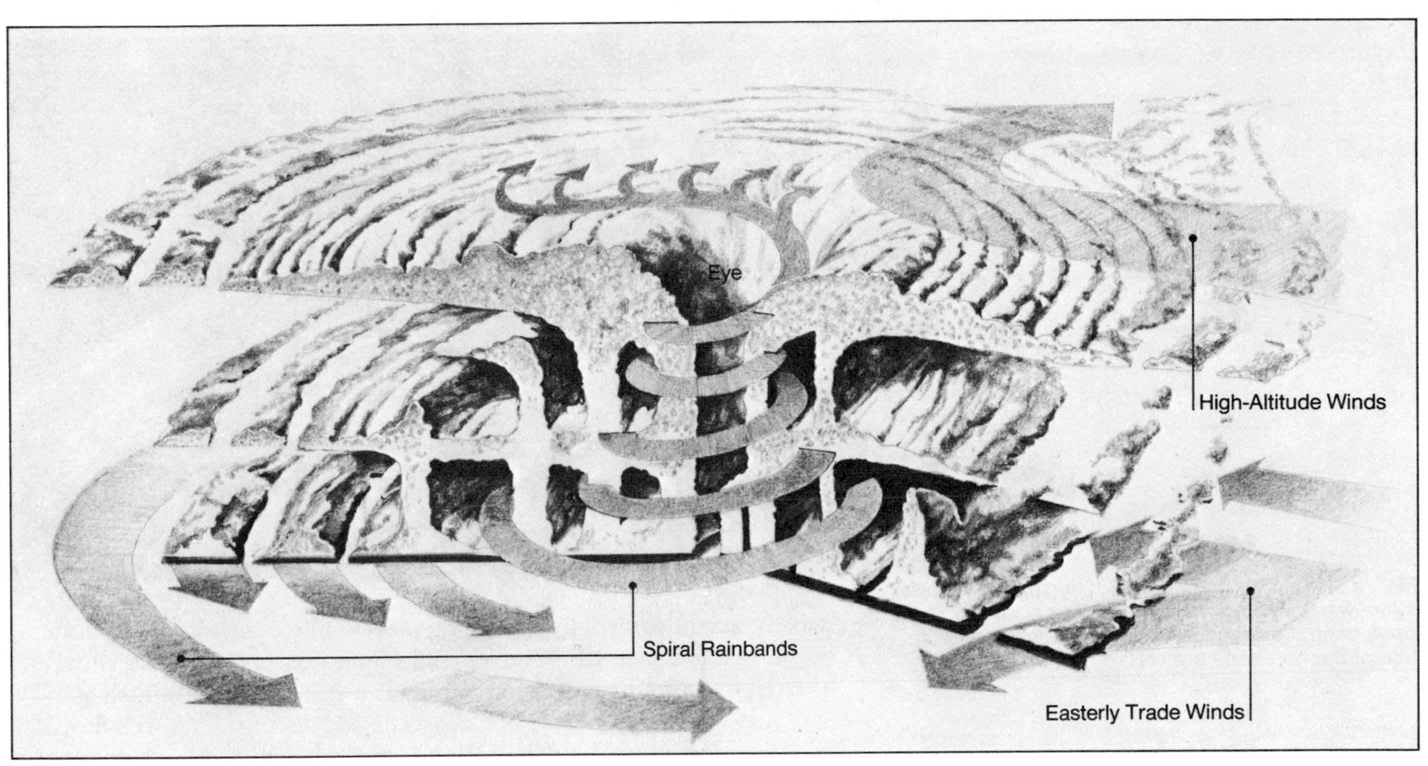

Figure 5.16 This cross section of a hurricane shows the central column, or eye, and the swirling cloud bands that give a hurricane its characteristic appearance as seen from above. A tropical cyclone in the Northern Hemisphere has an intense counterclockwise rotation. Wind speeds near the center may exceed 200 km per hour (120 miles per hour). The central eye is a region of low atmospheric pressure, and the dry air descending into the eye from above makes weather in the eye clear and calm.

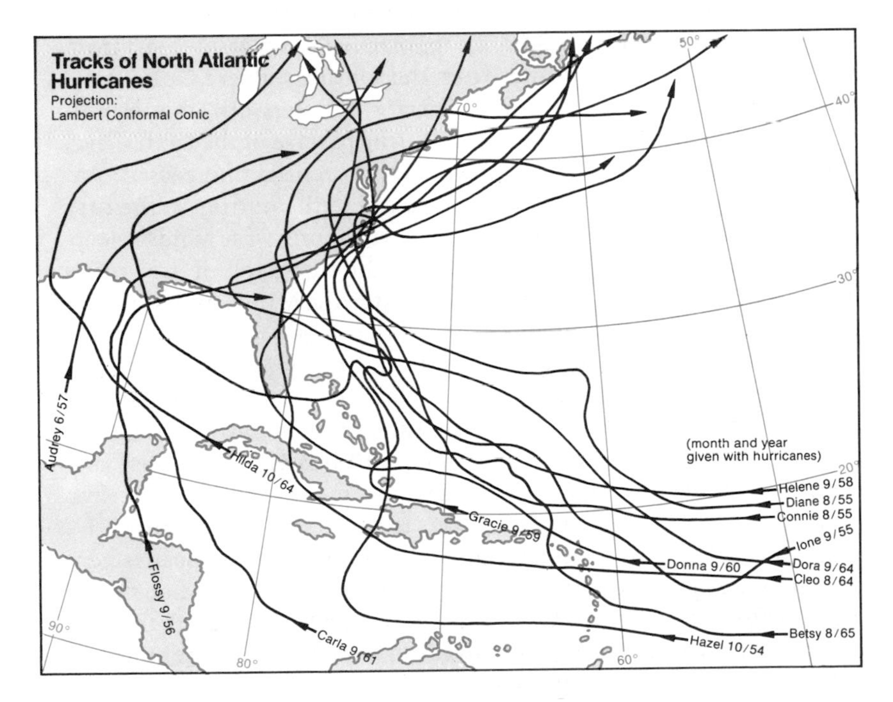

Figure 5.17 This map shows the tracks of some devastating North Atlantic hurricanes for the years 1954 through 1965. The path of an individual hurricane tends to be erratic, but as the map indicates, many of the hurricanes generated in the Caribbean Sea follow the same general course westward and northward. Some hurricanes turn northward early and skirt the East Coast of the United States; others enter the Gulf of Mexico before heading north. Hurricanes tend to lose strength over land because of friction and lack of water vapor, but some hurricanes, such as Hazel, 1954, have retained their destructive intensity across the Middle Atlantic States. A hurricane traveling over the land often brings heavy rains that cause rivers to overflow. (After *The National Atlas of the United States of America,* 1970)

Atlantic hurricanes usually follow one of several broad paths, as shown in Figure 5.17. Over the open ocean in the lower latitudes, they travel only a few tens of kilometers per hour but increase in speed as they invade the middle latitudes and are steered forward by the circumpolar vortex of westerlies aloft. However, this increase in speed is

Figure 5.18 This photograph of hurricane Gladys, 1968, was taken from the Apollo 7 spacecraft. The open central eye and the counterclockwise circulation of the hurricane are visible.

usually accompanied by a loss of power. Once a tropical cyclone begins to travel over land or over cold water, it is cut off from a supply of energy in the form of water vapor, and its strength diminishes.

At sea or along a low-lying coastline, a direct hit by a well-developed hurricane can be devastating to an inhabited area. Along the southwestern coast of Louisiana, Hurricane Audrey, 1957, is still a terrible, tragic memory. As Audrey approached the marshy coast from the southwest, hurricane winds from the southeast drove Gulf waters nearly 50 km (31 miles) inland; seawater 3 meters (10 ft) high battered the coastal village of Cameron. During passage of the central eye, the winds abated for 10 to 20 minutes, the torrential rain ceased, and the sky was partly cloudy. As Audrey passed to the northeast, the fury of the winds and rain was renewed, but with northwest winds sweeping coastal waters away from the shore. The tragic toll of drownings and destruction became apparent only after the winds and seas subsided as Audrey moved inland.

Tropical cyclones are dangerous storms: when forceful hurricane winds cause the ocean to surge up onto low coasts, the water may rise several meters above the normal high-water mark and cause extensive flooding. A rainfall of more than 25 cm (10 in.) is not unusual as a hurricane passes by, and if the hurricane moves over land, heavy runoff from the rains soon swells rivers to the flood stage. Fortunately, tropical cyclones are relatively rare. The number of hurricanes generated in the North Atlantic and the Caribbean varies from one to two per year to ten or eleven. Only three hurricanes formed off the east coast of North America in 1972, a year in which ocean water temperatures were much below normal.

Predicting the precise track that a tropical cyclone will take is often very difficult because hurricanes are steered by the airflow in the upper atmosphere. Hurricanes tend to move westward within the trade wind regime in tropical latitudes and eastward along the equatorward margins of the circumpolar westerlies in midlatitudes. The high altitude currents responsible for steering a hurricane between the easterlies and westerlies in subtropical latitudes, however, are especially weak and erratic.

But while it may be difficult to predict its future course, it is possible to keep track of a hurricane's progress using present-day satellite technology. Because the distinctive cloud formations of a tropical cyclone are easy to see on weather satellite photographs (see Figure 5.18), a moving storm can be accurately tracked in time to give advance warning to threatened areas. Improved warning and communications systems have steadily reduced the loss of life resulting from hurricanes in the United States, even though those who are in danger do not always take the warnings seriously. In less developed areas of the world, large numbers of people may be unable to benefit from advanced methods of hurricane tracking. When a typhoon in the Bay of Bengal struck East Pakistan in 1970, the almost total lack of warning, together with ineffective communications and transportation, resulted in the loss of an estimated 250,000 lives.

Monsoons

Monsoons are secondary circulations which involve a reversal in the direction of surface winds. Seasonal shifts in the direction of trade winds over India and the Indian Ocean strongly affect weather there. In summer, the southwest wind picks up moisture as it crosses the sea toward India, and precipitation from this air accounts for 70 percent of India's annual rainfall. The mountains of India help induce orographic precipitation from the moist southwest winds. The mountain town of Cherrapunji, for example, averages 1,168 cm (460 in.) of rain a year; between 85 and 90 percent normally falls during the five months from May through September, and almost 570 cm (224 in.) was recorded for June 1956.

The change between the northeast winter winds and the southwest summer winds occurs gradually across India. The summer monsoon begins to migrate into southern India late in May but is not felt in northwest India until July. In northern India, the northeast winter winds begin in September but do not arrive in southern India until January.

Upper air phenomena in the tropics also seem to be connected with the monsoon. A jet stream blowing from the west lies at an altitude of 16 km (10 miles) above the south slope of the Himalaya Mountains in northern India. At the start of the summer monsoon, it shifts north of Tibet while another jet stream blowing from the east appears over northern India. The precise relationship between the monsoon and the jet streams, however, is not yet understood. While the monsoon is at its most extreme in India, weaker monsoonlike circulations are characteristic of other locations, such as Southeast Asia, northern Australia, and the Guinea coast of Africa.

GENERAL CIRCULATION OF THE OCEANS

The circulations of the atmosphere and the oceans are closely linked through a series of important interactions. Air moving over the surface of the water drives nearly all the principal ocean currents, and the oceans in turn, in their capacity as reservoirs of heat energy, affect the general circulation of the air. Like atmospheric circulations, ocean currents contribute to the redistribution of energy by carrying warm waters toward the poles and cool waters toward the tropics.

The Oceans: Energy Banks

In Chapter 2, we saw that the heat capacity of ocean water is two or three times greater than that of continental surfaces, and very much greater still than that of the atmosphere. The great heat capacity of the oceans makes them truly enormous storage banks of energy for powering both global and local weather phenomena. The greater heat capacity of water causes it both to gain and lose heat more slowly than other substances in the environment. Consequently, ocean temperatures are "conservative," changing slowly through the seasons. Conservative ocean temperatures have a significant effect on global climates, as will be seen in Chapter 7.

We can appreciate the great effect that the oceanic heat reservoirs have on climate if we examine instances in which relatively small changes in ocean temperatures have produced dramatically large changes in weather patterns. In recent years, oceanographers have discovered the existence of large and persistent "pools" of oceanic surface water which differ in temperature from that of the water surrounding them. These pools may be only 1° or 2°C warmer or cooler than the average water temperature, but they seem to cause far-reaching climatic variations, probably by producing small changes in atmospheric circulation. For example, unusually warm water off the East Coast of the United States during 1971 and 1972 was associated with very wet weather along the coastal states and probably helped to provide energy for Hurricane Agnes, a tropical cyclone that was particularly destructive in Virginia and Pennsylvania. Because of interactions with the atmospheric circulation aloft, the same pool of warm water was also associated with freezes, droughts, and grain failures in the Soviet Union in 1972. Similarly, warmer than normal surface water north of the Hawaiian Islands and cooler than normal surface temperatures off the West Coast of the United States between 1957 and 1970 were associated with colder winters across the northern and eastern parts of the country, especially the bitterly cold winter of 1962–1963. Then, in 1971, the pattern of surface temperatures in the North Pacific reversed itself, and the northeastern United States began to enjoy relatively mild winters. More recently, however, temperatures in the Pacific have again assumed the pattern characteristic of the 1960s. This change may be related to the extremely cold weather in the eastern United States during the fall and winter of 1976–1977 and to the devastating drought in the West during the same period.

Surface Currents

Wind blowing along the surface of the ocean exerts a push on the water, but because friction forces are much stronger in water than in air, the speed of the water is only a small fraction of the wind speed.

Typical ocean currents have speeds ranging upward from a few kilometers per day to a few kilometers per hour. Friction causes the speed of fast ocean currents to decrease rapidly with depth below the water's surface. Most fast currents are confined to the upper hundred meters or so of the ocean.

The response time of large ocean currents to the atmospheric circulation is many months. The motion of a large ocean current cannot respond rapidly to changes in the wind, and even the waves that strong winds produce on the surface of the ocean require many hours to reach their full heights. Because ocean currents reflect average wind conditions over a period of a year or more, the general circulation of the oceans is closely related to the general circulation of the atmosphere.

If water entirely covered the earth, the winds would form well-defined belts, and the ocean currents would move around the earth in belts under the influence of the prevailing winds. However, only the ocean around Antarctica has relatively unhindered passage around the globe. The largest oceans, the Atlantic and the Pacific, are mostly confined to basins bounded by continents on the east and west; thus free flow of their currents around the earth is impossible.

The presence of continents strongly influences the motion of the wind-driven currents. Because a current cannot easily leave its basin, the idealized general circulation in the ocean basins consists of closed loops of circulation, or *gyres*. These gyres correspond to the major global wind circulations; both the Atlantic and the Pacific basins have three gyres north of the equator and three to the south. To understand the ideal gyre pattern shown in Figure 5.19, think of the wind directions. The trade winds, for example, drive the low-latitude currents of the subtropical gyres, and the prevailing westerlies drive the high-latitude return currents from the west. The actual circulation of the oceans is shown in Figure 5.20. The gyres are marked by stronger currents on the perimeter and relatively little movement internally. The main gyre patterns can be seen in the Atlantic and Pacific oceans, except perhaps near Antarctica, where the westerly winds establish a westerly global ocean current.

The major ocean currents shown in Figure 5.20 are driven by the wind; however, due to the Coriolis effect, the relationship between the direction of prevailing winds and ocean currents is not one of complete parallelism. The drift of water impelled by seasonal wind regimes tends to be turned by the Coriolis effect, to the right north of the equator and to the left in the southern hemisphere. The surface water actually drifts off at a 45° angle to the wind direction, this angle increasing downward to a depth of 60 to 100 meters (200 to 330 ft), where motion effectively ceases. Thus, along the west coasts of continents, for example, where winds blow equatorward around the east sides of oceanic anticyclones, the water beneath tends to be driven obliquely toward the centers of the anticyclones and away from the adjacent coasts.

Ocean currents play an important role in redistributing heat over the globe. A general northward current of warm equatorial water flows at the western edge of each ocean basin in the northern hemisphere,

Figure 5.19 The idealized oceanic circulation in an ocean basin consists of loops, or gyres, which are driven by prevailing winds. (After Weyl, 1970)

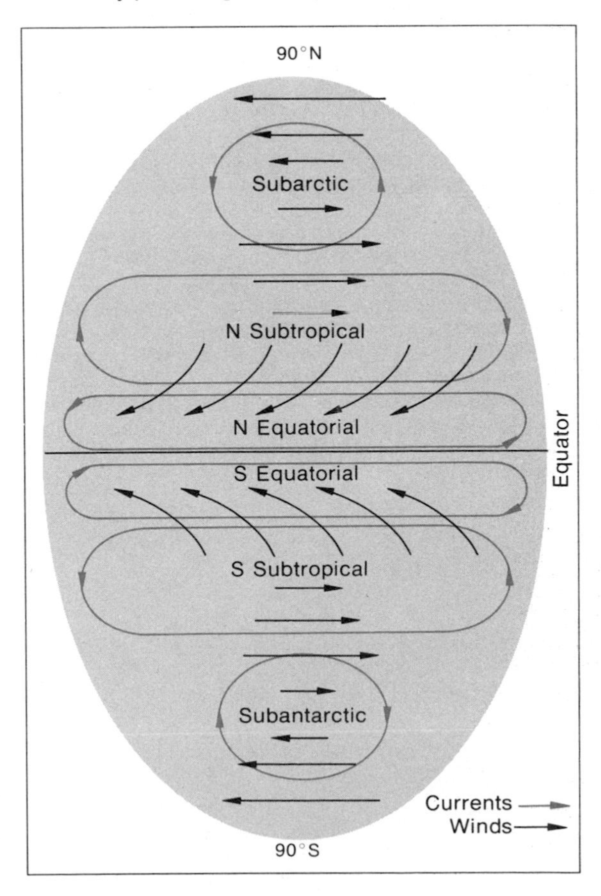

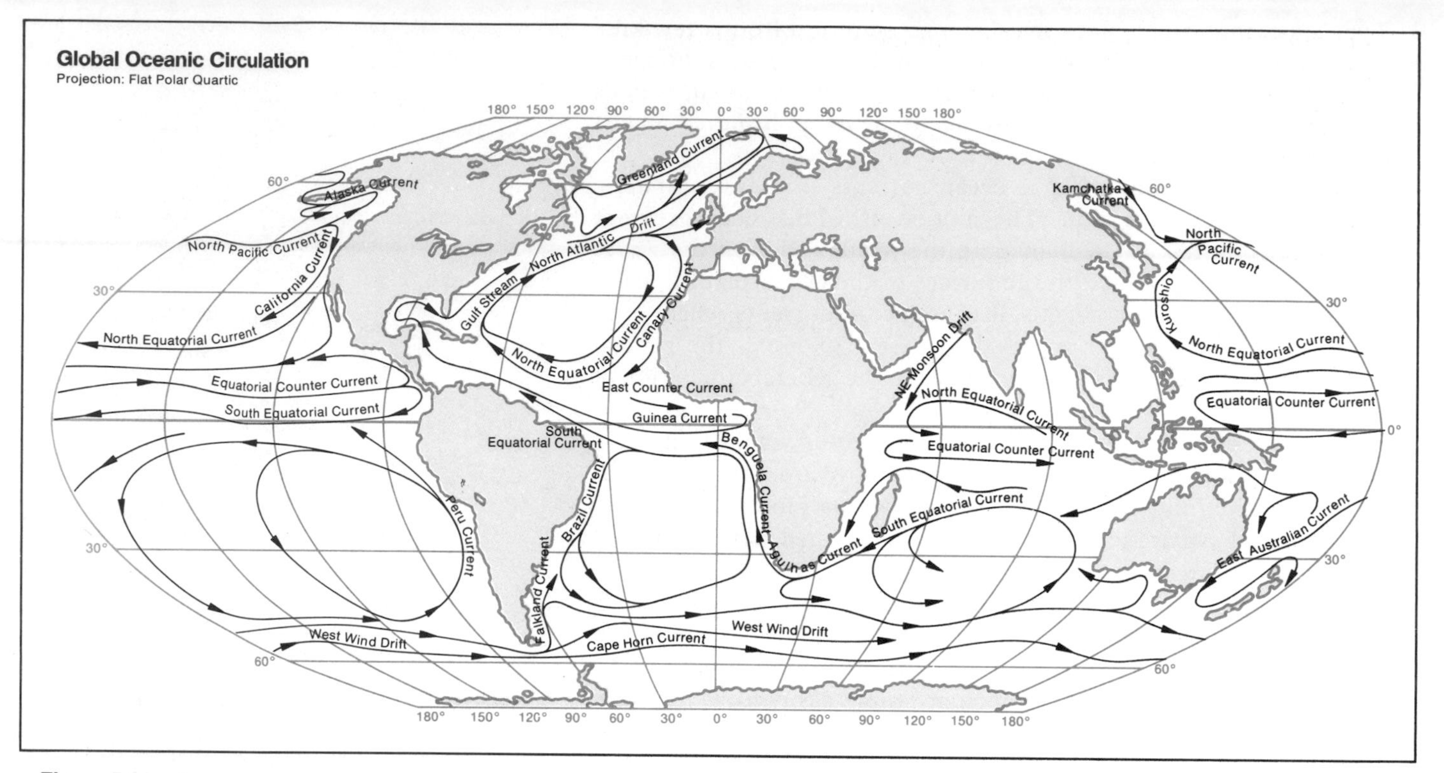

Figure 5.20 This map shows the principal oceanic currents in the surface layer of the oceans. The currents have a tendency to form loops of circulation that are marked by strong currents on the perimeters and relatively little movement internally (compare with Figure 5.19). Ocean currents move warm water poleward and cold water toward the tropics, which helps to equalize the distribution of radiant energy over the earth.

The principal ocean currents along the east coast of the United States include the Florida Current and Gulf Stream, which carry warm waters northward, and the Labrador Current, which carries cool waters southward past the coast of New England. (The Florida Current is the name given to the portion of the Gulf Stream from Florida to Cape Hatteras, North Carolina.) The principal ocean current along the West Coast of the United States is the southward-flowing California Current. (After Leet and Judson, 1965)

and cooling southward currents flow at the eastern edge. The Gulf Stream is the warm northward current in the Atlantic basin; its counterpart in the Pacific basin is the Kuroshio off Japan. Both are fast, narrow currents moving several kilometers per hour. In contrast, the southward currents on the eastern edges of the basins are more sluggish and less well defined.

The amount of water the Gulf Stream transports is more than 30 times as great as the amount of water that flows into the oceans from the continents. The Gulf Stream proper runs from Cape Hatteras, North Carolina, to near the Grand Banks of Newfoundland, where it merges into the broader, slower North Atlantic Drift. The warm water that the North Atlantic Drift carries to western Europe significantly raises the average winter temperature there.

Upwelling

Along the west coasts of continents in the subtropical latitudes, winds blowing toward the equator, together with the Coriolis force, can cause surface water to move away from the coast and out to sea. A flow of cool water from a depth of a few hundred meters then moves up to replace the water that has moved seaward. This phenomenon is known as *upwelling*.

Coastal upwelling of cool water occurs on the west coasts of all continents, but it is particularly pronounced off California, Peru, Senegal, and southwest Africa. Upwelling is extremely important ecologically because it brings chemical nutrients up into the sunlit surface waters, enabling photosynthetic plankton to thrive and support an exceptionally rich food chain. Off the west coast of South America, seaward deflection of surface water and upwelling of cold water occasionally slacken due to weakening of the normal equatorward-moving winds. This has disastrous effects. When upwelling weakens, warm waters from the equatorial region drift southward. Without their supply of nutrients, the plankton die off, and consequently the entire food chain collapses. Fish and other marine organisms die in great numbers, and birds that feed on them must migrate or starve. Decomposing fish litter the beaches and the waters, giving off so much hydrogen sulfide that the white lead paint used on ships turns black. Thus the phenomenon is often called the "Callao painter," after the port of Callao in Peru. Since this effect usually occurs around Christmas time, the southward drifting current is called El Niño, the Christ Child.

The Deep Circulation

In the deep ocean, below the surface currents, there is a slow circulation of water driven primarily by density differences. The density of seawater varies with both temperature and salt content. The densest water is cool and highly saline; such water is produced off Antarctica when surface water freezes, leaving its salt content in the remaining unfrozen water. The ocean water from the region around Antarctica slowly slides down toward the ocean floor. Because the deep circulation is slow, water masses tend to move with little mixing of different waters. According to carbon 14 dating, water in the deep sea may remain there for hundreds of years before returning to the surface.

The deeper waters of the oceans also move horizontally in streams known as *undercurrents.* Some deep undercurrents travel as fast as or even faster than the surface currents above them, and commonly in the opposite direction. Such currents have been discovered as deep as 2,000 meters (6,500 ft) below the surface. It may be that no part of an ocean is truly at rest, not even its deepest waters.

PRECIPITATION DISTRIBUTION

A knowledge of atmospheric and oceanic circulations and their interactions allows us to understand and predict the distribution of precipitation over the globe—an important factor in global climatic patterns. Recall from Chapter 4 that in order for precipitation to occur, there must be rising air cooling to saturation, condensation, and the formation of thick clouds. These processes are usually associated with moist surface air converging into low-pressure systems and rising aloft. On the other hand, in areas where air is subsiding and diverging outward from high-pressure centers near the surface, cloud formation and precipitation are unlikely. It follows, therefore, that in equatorial and tropical latitudes, precipitation patterns should mirror the distribution of moist, unstable tropical air, particularly along convergence zones. In higher latitudes, the greatest amount of precipitation should occur along fronts and throughout the tracks of midlatitude cyclones.

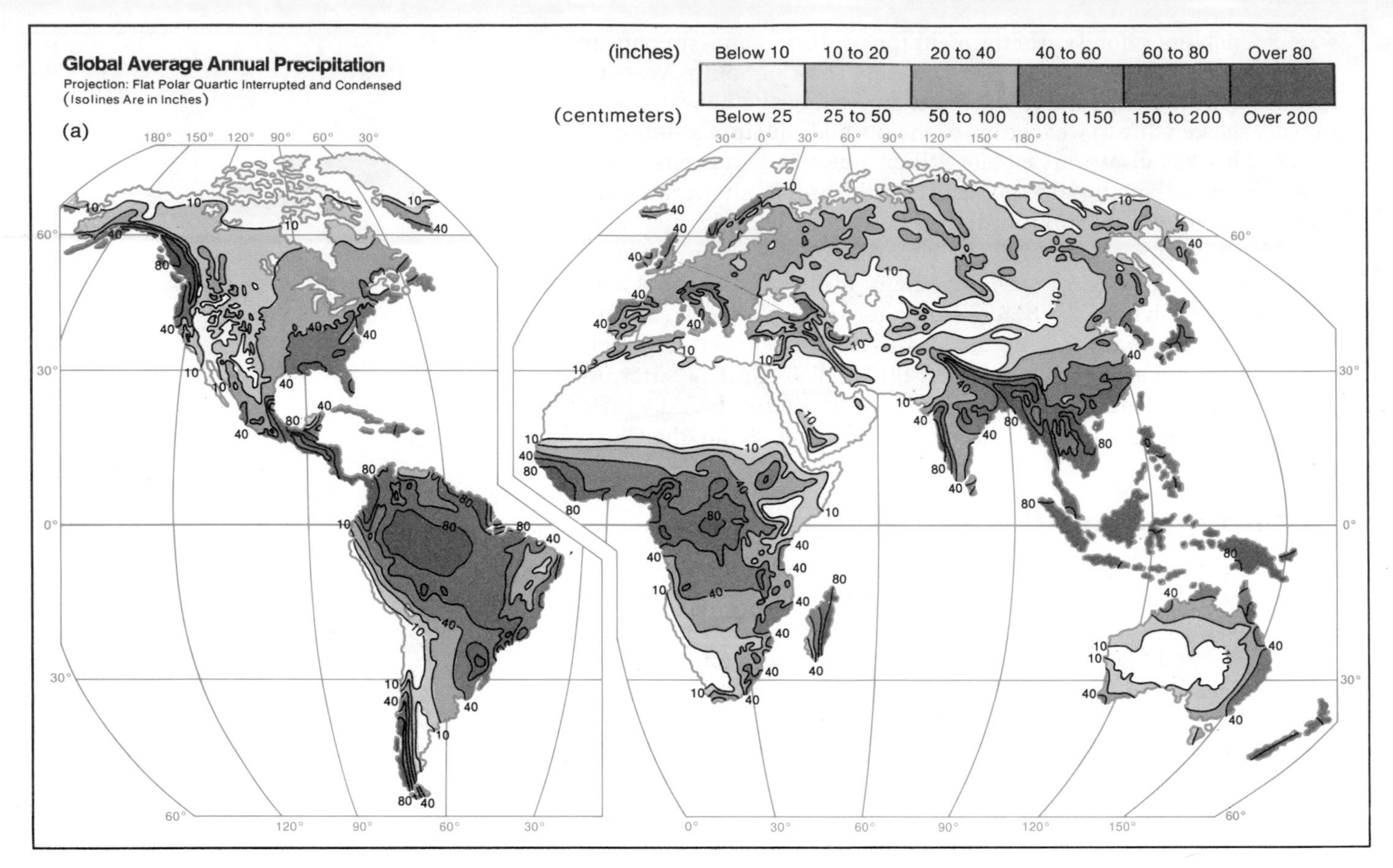

Figure 5.21 (a) This map of average annual precipitation illustrates the complex distributions associated with the positions of continents, ocean basins, and mountain barriers. In spite of the complexity, however, global patterns of high and low precipitation are evident, and these should be related to the global pressure and wind systems shown in Figure 5.8, pp. 128–129.

(b) The pattern of average annual precipitation over the United States displays some broad regularities. The East tends to be moist because of atmospheric moisture from the Gulf of Mexico and the adjacent Atlantic. Precipitation decreases westward from the Mississippi River valley. Moist air from the Gulf of Mexico seldom flows westward, and moist air from the Pacific releases most of its moisture as orographic rainfall on coastal mountain slopes.

(c) The Great Plains are dry in winter because of the influx of cP air from the north. The Far West receives most of its precipitation in winter, when midlatitude cyclones sweep the coast; a strong subtropical high over the eastern North Pacific inhibits precipitation during summer. The East receives precipitation during both winter and summer. (After *Goode's World Atlas*, 1970)

The map of average annual global precipitation distribution shown in Figure 5.21a, above, bears out these expectations. Precipitation is especially great in equatorial regions, where the intertropical convergence zone is present most of the year, and across much of the monsoon regions of India and Southeast Asia. Where midlatitude cyclones recur frequently along the polar front, such as western Europe and eastern North America and Asia, precipitation is relatively high. Precipitation is also very high where mountain ranges intercept moist air regularly, such as along the Pacific coast of Canada. In contrast, precipitation is low over most subtropical regions, the Sahara Desert, for example; over the interior of Asia and North America, where oceanic moisture sources are distant; and over polar latitudes, where the capacity of the atmosphere for water vapor is small. Figures 5.21b and 5.21c show details of the annual and seasonal distributions for the United States.

SUMMARY

Solar radiation provides energy to drive the motion of the earth's atmosphere. Differential heating of the surface gives rise to horizontal differences in atmospheric temperature and hence pressure. Air tends

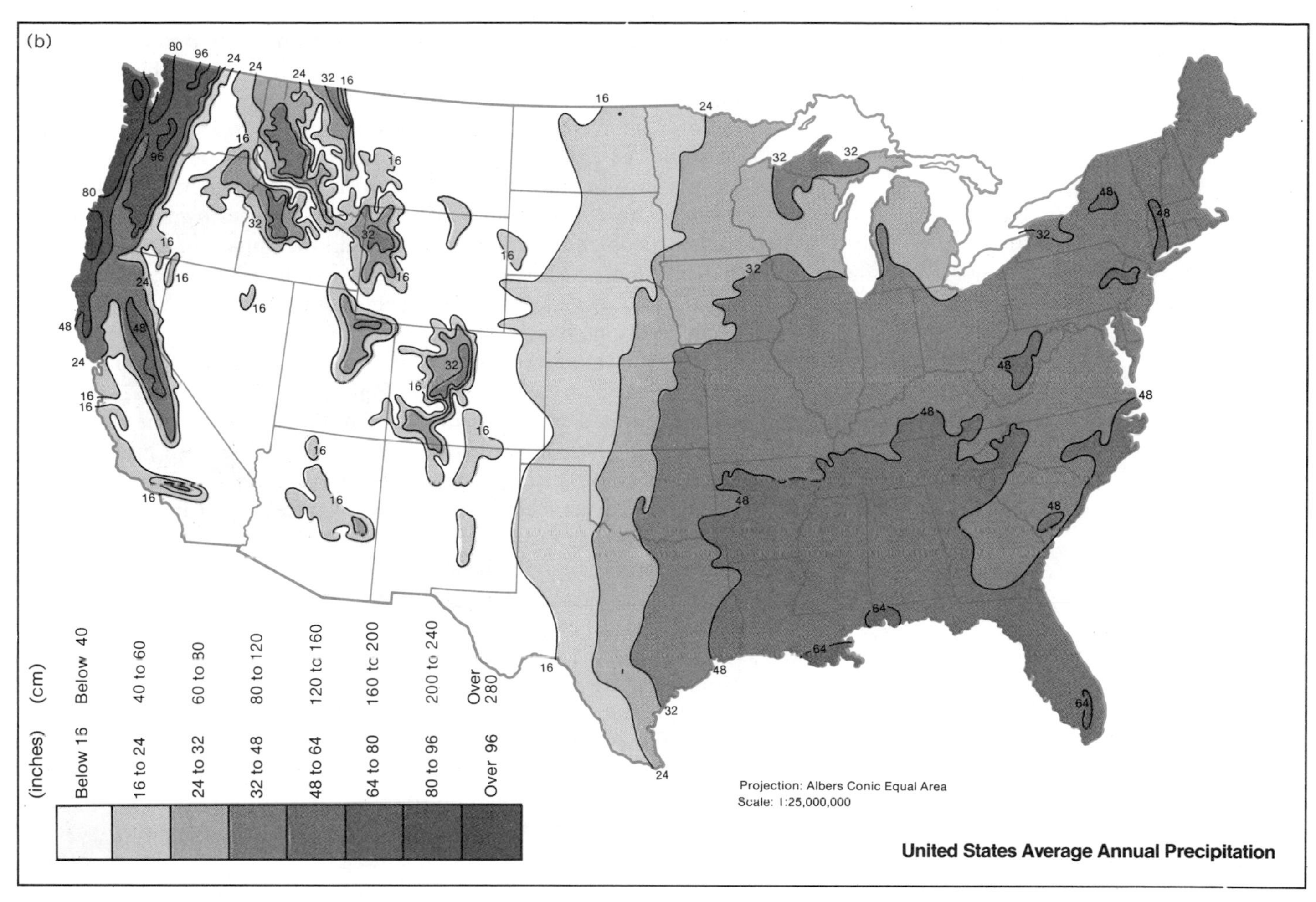

(b)
(inches)
(cm)
Below 16
Below 40
16 to 24
40 to 60
24 to 32
60 to 30
32 to 48
80 to 120
48 to 64
120 tc 160
64 to 80
160 tc 200
80 to 96
200 to 240
Over 96
Over 280
Projection: Albers Conic Equal Area
Scale: 1:25,000,000
United States Average Annual Precipitation

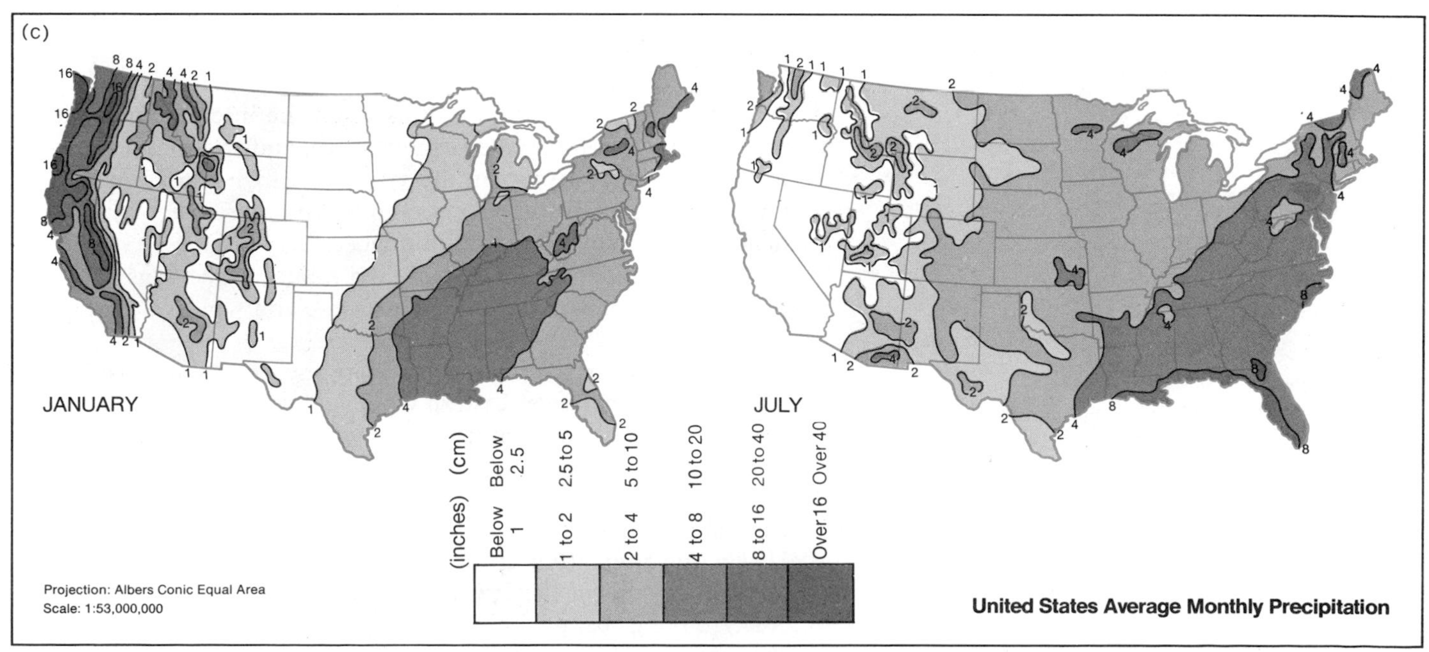

(c)
JANUARY
JULY
(inches)
(cm)
Below 1
Below 2.5
1 to 2
2.5 to 5
2 to 4
5 to 10
4 to 8
10 to 20
8 to 16
20 to 40
Over 16
Over 40
Projection: Albers Conic Equal Area
Scale: 1:53,000,000
United States Average Monthly Precipitation

to move down the pressure gradient, but it is deflected to the right in the northern hemisphere and to the left in the southern hemisphere. The deflection is due to the Coriolis force. Above the friction layer, the Coriolis force is equal and opposite to the pressure gradient force, and air flows parallel to the isobars, with low pressure to the left and high pressure to the right in the northern hemisphere; this flow is known as the geostrophic wind. Near the surface, however, air flows across the isobars at small angles toward lower pressure. In the northern hemisphere again, a counterclockwise flow converges into low pressure, and a clockwise flow diverges out of high pressure.

In terms of heat and moisture, the air tends to take on the properties of the surface when it "sits" over a large homogeneous region for several days. Instead of mixing, mT and cP air masses remain separated by fronts in the middle latitudes of the northern hemisphere, where most precipitation is associated with midlatitude frontal cyclones which move generally northeastward along the polar front zone. Air mass contrasts and fronts rarely exist in tropical and equatorial regions, and precipitation in the form of showers and thunderstorms generally develops along the intertropical convergence zone and is associated with easterly waves.

The general circulations of the atmosphere and oceans describe average flows which together equalize net radiational gains in lower latitudes and net losses to space in higher latitudes. The details of the global distribution of precipitation are also related to the pressure and wind systems of the general atmospheric circulation, but the distribution is even more complex because of the irregular pattern of continents, ocean basins, and mountain barriers. The following chapter analyzes the local utilization of precipitation and its return to the atmosphere and oceans.

Review Questions

1. What does Buys Ballot's law tell us about the interaction of the pressure gradient force, the Coriolis force, and friction?
2. How does surface airflow in the vicinity of high- and low-pressure centers differ in the northern and southern hemispheres? Why do these differences exist?
3. How do continents, ocean basins, and mountain barriers alter the hypothetical zonal pattern of pressure and wind systems over the globe?
4. Why might air mass contrasts be greater over the southeastern United States than over eastern China?
5. Outline the stages of a midlatitude cyclone and associated weather over the eastern United States during winter.
6. Contrast the form, processes, and properties of midlatitude and tropical cyclones.
7. How is the circulation of the oceans related to wind and pressure systems of the atmosphere?
8. Analyze the global precipitation map in Figure 5.21a in terms of the general and secondary circulations of the atmosphere.

Further Reading

Anthes, Richard A., et al. *The Atmosphere.* Columbus, Ohio: Charles E. Merrill Publishing Co., 1975. (339 pp.) This introductory text has an interesting historical perspective. Other unusual features include synoptic and seasonal analyses of midlatitude weather.

Dunn, Gordon E., and **B. I. Miller.** *Atlantic Hurricanes.* Baton Rouge, La.: Louisiana State University Press, 1964. (326 pp.) This is a most readable account of tropical storms and their consequences.

Eagleman, Joe R., Vincent U. Muirhead, and **Nicholas Willems.** *Thunderstorms, Tornadoes, and Building Damage.* Lexington, Mass.: D. C. Heath & Co., 1975. (317 pp.) This provocative analysis of tornado damage emphasizes safety of individual rooms and structures relative to the approach of tornadoes.

Fujita, T. Theodore. "Graphic Examples of Tornadoes." *Bull. Amer. Meteor. Soc.,* Vol. 57 (1976): 401–412. Focus is on the record outbreak of 148 tornadoes during April 3–4, 1974. Closely related articles on severe storms are also in this issue. Up-to-date survey and position papers on meteorological and climatological research are regular features of this bulletin.

Hidore, John J. *Workbook of Weather Maps.* 3rd ed. Dubuque, Iowa: Wm. C. Brown Publishers, 1976. (81 pp.) This paperbound compilation of official U.S. weather maps for 11 series of weather events includes the record polar outbreaks of January and February 1962 and Hurricane Agnes in June 1972.

Lehr, Paul E., R. Will Burnett, and **Herbert S. Zim.** *Weather.* New York: Golden Press, 1965. (160 pp.) This excellent paperback is one of the well-known Golden Nature Guide series. The color diagrams of fronts and midlatitude cyclones are especially informative.

Muller, Robert A. "Snowbelts of the Great Lakes." *Weatherwise,* Vol. 19, No. 6 (1966): 248–255. Focus is on weather events which produce persistent and deep snowfalls over small areas of the Northeast. Anyone interested in weather and its consequences should enjoy each issue of this journal.

Neiburger, Morris, James Edinger, and **William Bonner.** *Understanding Our Atmospheric Environment.* San Francisco: W. H. Freeman & Co., 1973. (293 pp.) Basic fundamentals are presented in a nonmathematical framework. Includes a chapter on modern forecasting techniques.

Riehl, Herbert. *Introduction to the Atmosphere.* 2nd ed. New York: McGraw-Hill Book Co., 1972. (516 pp.) This nonmathematical basic text includes an especially informative chapter on tropical weather.

Stewart, George R. *Storm.* New York: Random House, 1941. (349 pp.) This fictional classic follows the evolution of a midlatitude cyclone, named Maria, across the Pacific and then over the United States. Much of the perspective is through the eyes of forecasters at San Francisco and managers and workers of transportation and communications networks. This novel is a must for students interested in interactions between weather and human endeavors. Reprinted in paperback in 1974 by Ballantine Books, New York.

Weems, John Edward. *A Weekend in September.* New York: Henry Holt & Co., 1957. (180 pp.) This is a careful journalistic account of the 1900 hurricane which swept Galveston, Texas, with a great loss of life.

Case Study:

The Tornado Hazard

Of all the earth's winds, tornadoes are undoubtedly the most feared and awesome. Along the track of a severe tornado, destruction is nearly total, while just a few meters away on either side, there may be no damage at all. Tornadoes are especially frightening because of their sudden and almost unpredictable appearance. At night or in densely forested country, it is difficult to see the funnel cloud of an approaching tornado; the sound, frequently compared to the roar of an express train, is sometimes the first warning. The National Weather Service has an effective program of tornado watches for alerting the public to areas where tornadoes may break out, but there is no way to predict the precise time and place of local occurrence.

A *tornado* is a narrow vortex of rapidly whirling air. It is almost always associated with severe thunderstorm activity. The rotating vortex extends downward from a cumulonimbus cloud, and it becomes visible when water vapor condenses, producing a light gray cloud, the familiar *funnel.* Dust and debris swept up from the ground create a much darker and more ominous-looking funnel. Occasionally, several funnels may dangle from the same cloud, and many funnel clouds aloft never reach the ground. As the mother cloud moves on, the funnel is often retarded at the surface by friction, so that it becomes tilted, crooked, and sometimes even broken. Tornado funnels over coastal waters and seas are often called *waterspouts.*

Tornadoes usually advance with a forward speed ranging between 32 and 48 km/hour (20 and 30 miles/hour). Thus a tornado will normally pass a given point in much less than a minute. The tornado and its track along the surface are usually only a few meters wide, although occasionally extending up to several hundred meters. Some tornadoes skip across the landscape, leaving a broken or interrupted track of destruction below. Most tornado tracks are about 5 to 10 km (3 to 6 miles) in length, but a few exceed 300 km (185 miles); on May 26, 1917, a single tornado tracked more than 400 km (250 miles) across Illinois and Indiana in a little less than eight hours.

Although tornadoes are short-lived and their tracks short, they are nonetheless extremely violent. The winds that whirl around the tornado vortex have been estimated to reach speeds of up to 600 km/hour (370 miles/hour). Within the funnel cloud, atmospheric pressure is as much as 50 mb lower than the adjacent air, and pressure drops of up to 100 mb have been estimated from damage patterns. As a tornado passes over a building, the strong winds of the vortex rip at the exterior, and the abrupt pressure drop causes the building literally to explode, with the roof and walls blown out. The debris is then caught by the rotating winds and strewn along the tornado's path; each piece of debris becomes a flying missile of destruction and potential death. Cellars, interior halls, and bathrooms offer greater safety than rooms adjacent to outside walls. When severe tornadoes pass over modern slab homes, sometimes only the plumbing fixtures, such as bathtubs and toilets, remain. House trailers are especially vulnerable to tornado winds.

In the United States, weather situations most favorable for tornado development occur most frequently over the Great Plains, the Mississippi River valley, and the Southeast. The region extending from the Texas Panhandle northeastward across Oklahoma and eastern Kansas is sometimes called "tornado alley." Tornadoes are very infrequent west of the Rocky Mountains and across northern New England and the upper Great Lakes.

The accompanying map shows the distribution of tornadoes in the United States. The mean annual number of tornadoes reported has increased significantly in recent decades, with the total more than 1,000 during 1973. Since 1970 the annual average is about 900. Although it is recognized that the atmospheric circulation patterns over the United States produce many more or less tornadoes during some years, the recent increase is attributed mainly to much improved observations and detection.

Tornadoes break out most commonly along thunderstorm or squall lines ahead of cold fronts, where temperature contrasts are large and instability great. Instability is usually associated with a deep layer of warm, dry cT air from the Southwest above warm, moist mT air from the Gulf of Mexico. These conditions usually occur in the warm sector of a midlatitude cyclone, immediately ahead of the leading edge of much colder polar air behind a vigorous cold front. The thunderstorms are steered forward by the wind flow in the middle to upper troposphere; in the warm sector, these upper winds are usually from the southwest, so most tornadoes track from the southwest toward the northeast.

The seasonal distribution of tornado outbreaks tends to follow the geographical distribution of air mass contrasts among mT, cP, and cT air masses, of frontal positions and midlatitude cyclones, and of maximum instability through a deep layer of the troposphere. Therefore, tornadoes are most frequent in the South during late winter and early spring, with the pattern of maximum frequency migrating northward into central states, such as Kansas and Missouri, and finally to Iowa and Nebraska by June; at this time, there are few tornadoes near the Gulf Coast. A large proportion of tornadoes occur during late afternoon, when instability tends to be greatest, but unfortunately, occurrences remain relatively common at night, when visual detection is difficult.

A close view of the tornado which swept across sections of Dallas, Texas, on April 2, 1957.

Note the especially narrow band of damage associated with a small tornado during Hurricane Edith in Baton Rouge, Louisiana, in September 1971.

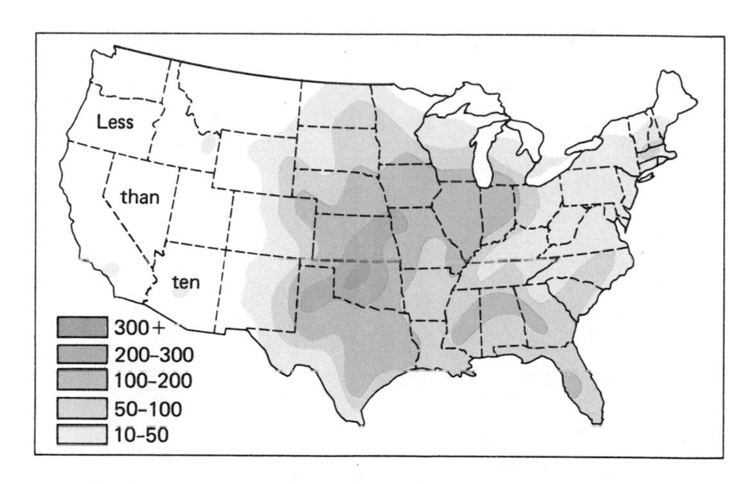

This map shows the number of tornadoes that were reported between 1955 and 1967 in the United States. The main features of the map include "tornado alley" over the Great Plains and the very small number of tornadoes westward from the Rocky Mountains.

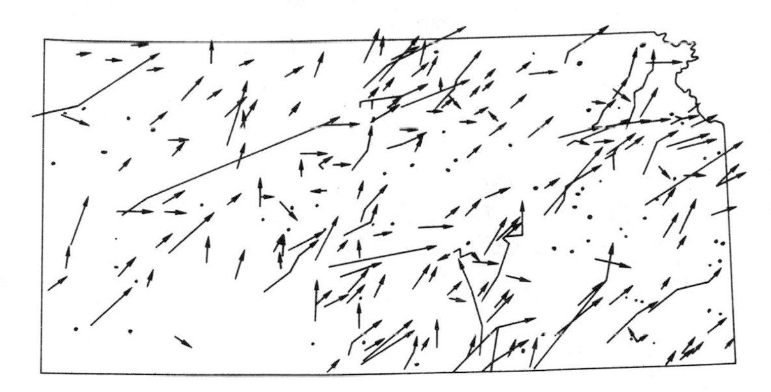

This map shows the direction and path lengths of tornadoes that occurred in Kansas between 1950 and 1970. Most tornadoes are dragged along within thunderstorms by the upper winds from the southwest that steer the thunderstorm cells. (From Eagleman, Muirhead, and Willems, 1975)

Rainy Season in the Tropics by Frederic E. Church, 1866. (Fine Arts Museum of San Francisco)

The sun's radiant energy provides the power for the atmospheric circulation to deliver energy and moisture to the earth's surface. On the land the water interacts with the rocks, soils, and vegetation. In time, most of it returns to the atmosphere in a never-ending cycle of renewal.

SIX
The Hydrologic Cycle and Local Water Budget

INLAND FROM THE GENTLY rolling lowlands surrounding Puget Sound in Washington, the landscape changes abruptly into a spine of high mountains that bisects the state. This is the Cascade Range, and it marks a region of contrasts. To the west of the mountains, rain is present much of the year, falling from storm clouds that sweep in from the North Pacific. Here on the western slopes of the Cascades, the waters accumulate in streams and rivers that drain toward the seacoast. To the east lies a vast region of dry, rolling grasslands, protected by the mountains from much of the heavy rain characteristic of the western part of the state. Here are the sun-drenched orchards and fields of the Yakima, Wenatchee, and Grand Coulee areas. And here man must turn to irrigation practices in order to sustain his crops. These contrasting regions illustrate only a few of the variables at work in determining what happens to water when it returns to the land, and how man adjusts his activities to accommodate the variation.

Average daily use of water in the United States amounts to 6 cu meters (1,500 gal) per person, not including water that is momentarily diverted for hydroelectric use. Of this, the public uses only about 10 percent directly for such things as bathing, washing the car, or watering the lawn. Industry uses another 10 percent, while the remaining 80 percent is divided about equally between fossil fuel electric power plants, which use water for cooling, and agriculture.

Obviously, if the water used for these purposes were permanently removed from the environment, an acute water shortage would result in a relatively short time. But water is a renewable resource. Nearly all of the water man uses at home, on farms, and in industry is returned to the hydrologic cycle. Often, however, water is taken from a convenient supply and discharged to a less convenient one. The water used in irrigation may come from a river or from groundwater, but evapotranspiration returns most of it to the atmosphere. The water used in homes may come from a convenient supply of groundwater, but it may be returned to streams by the sewage system and ultimately flow to the sea. The total amount of fresh water received by the earth as precipitation each year is more than sufficient for human needs, but the water is not always conveniently available.

Because water is an essential resource, man throughout history has engaged in extensive construction to ensure a supply of water adequate to meet continuing demands. The irrigation projects which characterize agriculture in eastern Washington are by no means a new idea. Five thousand years ago in Egypt, simple clay dams were built to retain water from the Nile River. These early efforts have their modern counterpart in the immense Aswan High Dam on the Nile. The amount of water stored in the lake formed behind the Aswan Dam exceeds the annual flow of the Nile. The dam was built to generate power, to control floods, and to provide water for irrigation throughout the year. The aqueducts in California that carry water hundreds of miles to Los Angeles have historical parallels in the aqueducts that brought ancient Rome a water supply that was ample even by modern standards. More ambitious water control projects are being contemplated; there are plans to dam the Zaire (Congo) River in Africa and the Amazon River in South America, which would create huge freshwater lakes. In other areas, attempts have been made to desalinate ocean water and transport it inland. The high cost of such an operation, however, makes it uneconomical in many situations.

The need for fresh water, which has always been acute, grows even more pressing as populations increase and expand into areas where natural water supplies are scarce. Because of the growing world-wide need to manage available water, there is increasing emphasis on the study and analysis of water on and below the surface of the land—a field known as *hydrology*. Hydrology concentrates on an aspect of the hydrologic cycle which we have thus far touched on only briefly. We have already discussed the atmospheric part of the cycle: water's evaporation from the earth's surface, its storage and transport in the atmosphere, and its subsequent return to the surface by precipitation. This chapter reintroduces the entire hydrologic cycle but then focuses on the pathways of water on the continents. A key concept in this discussion will be the idea of the local water budget as a tool for evaluating the availability of water in a particular area. In later chapters, the water budget model will be used to describe local and global interactions between water and solar energy, vegetation, soils, even the shaping of the land surface.

THE GLOBAL WATER BUDGET

The first step in drawing up any kind of budget should be to take an inventory of available resources, and the global water budget is no exception. As Figure 6.1 shows, the earth's water supplies are stored in several different reserves, and in amounts that differ markedly from one another. The oceans are by far the largest single water source. They cover about 70 percent of the earth's surface and contain nearly 98 percent of the earth's total water supply. Ocean water contains dissolved minerals—primarily salt—and is unfit for human consumption or for agriculture. However, when water evaporates from the oceans, the dissolved salts are left behind; thus the water that condenses from maritime air masses is fresh. Man relies almost entirely on the fresh water produced by this natural desalination process.

About three-fourths of all the world's fresh water is stored as ice and snow, primarily in the ice sheets covering Antarctica and Greenland. The ice sheets receive an annual snowfall equivalent to about 10 cm (4 in.) of liquid water. Some of this water returns to the oceans when coastal ice melts or when icebergs break off. For the most part, however, ice sheets represent water in long-term storage and are not important in the year-to-year utilization of water.

The amount of water stored in the atmosphere at any given time is relatively small: if all the water in the atmosphere were to fall uniformly over the earth as rain, it would form a layer only a few centimeters deep. In the course of a year, however, the atmosphere actively transports and recycles enough water to cover the earth with a layer of precipitation 95 cm (37 in.) deep.

For the global water budget to be in balance, the amount of water that enters the atmosphere by evapotranspiration must be equal to the amount that leaves as precipitation. This balance, however, is not apparent unless we view the system in its entirety (see Figure 6.2, p. 156–157). Considering the oceans alone, we find that the average quantity of water evaporated exceeds the quantity precipitated. Over the continents, on the other hand, precipitation exceeds evapotranspiration. If no mechanism existed to compensate for these imbalances, it would not be long before the continents became inundated with moisture and the level of the oceans began to sink. Obviously, this does not occur, because water is continually exchanged between the oceans and the continents. Over the oceans, precipitation is less than evaporation because a large portion of the evaporated moisture is transported through the atmosphere by maritime air masses hundreds or thousands of kilometers in diameter, and this moisture eventually falls as precipitation on the land. The land redistributes its excess moisture in two ways: by contributing to precipitation over the oceans through the moisture transported in continental air masses and by adding to the oceans' reserves through runoff from rivers and

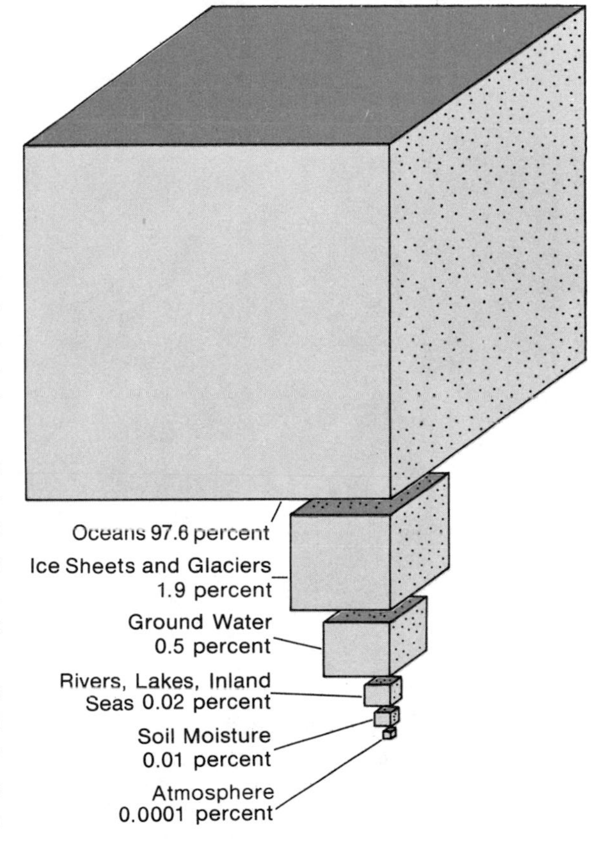

Figure 6.1 The volumes of the cubes show the relative amounts of free water in storage on the earth. Nearly 98 percent of the water is stored in the oceans, which contain an estimated volume of 1.3×10^{18} cu meters (46×10^{18} cu ft) of water. Ice sheets and glaciers contain the largest store of fresh water, but the turnover is too slow to be important except on the geologic time scale of ice ages. Most of the readily available fresh water is stored in porous rock beds as groundwater. The amount of water stored in the atmosphere is relatively small, but because it is actively transported, it plays a key role in the hydrologic cycle. (After Nace, 1969)

Transport of Water
Vapor from the Oceans
(94)
(12)
Storage as Ice and Snow
Evapotranspiration
from the Continents
(69)
Temporary Surface Storage
Surface Runoff
Percolation
Storage in Rivers and Lakes
Groundwater Storage
Groundwater Runoff to Streams

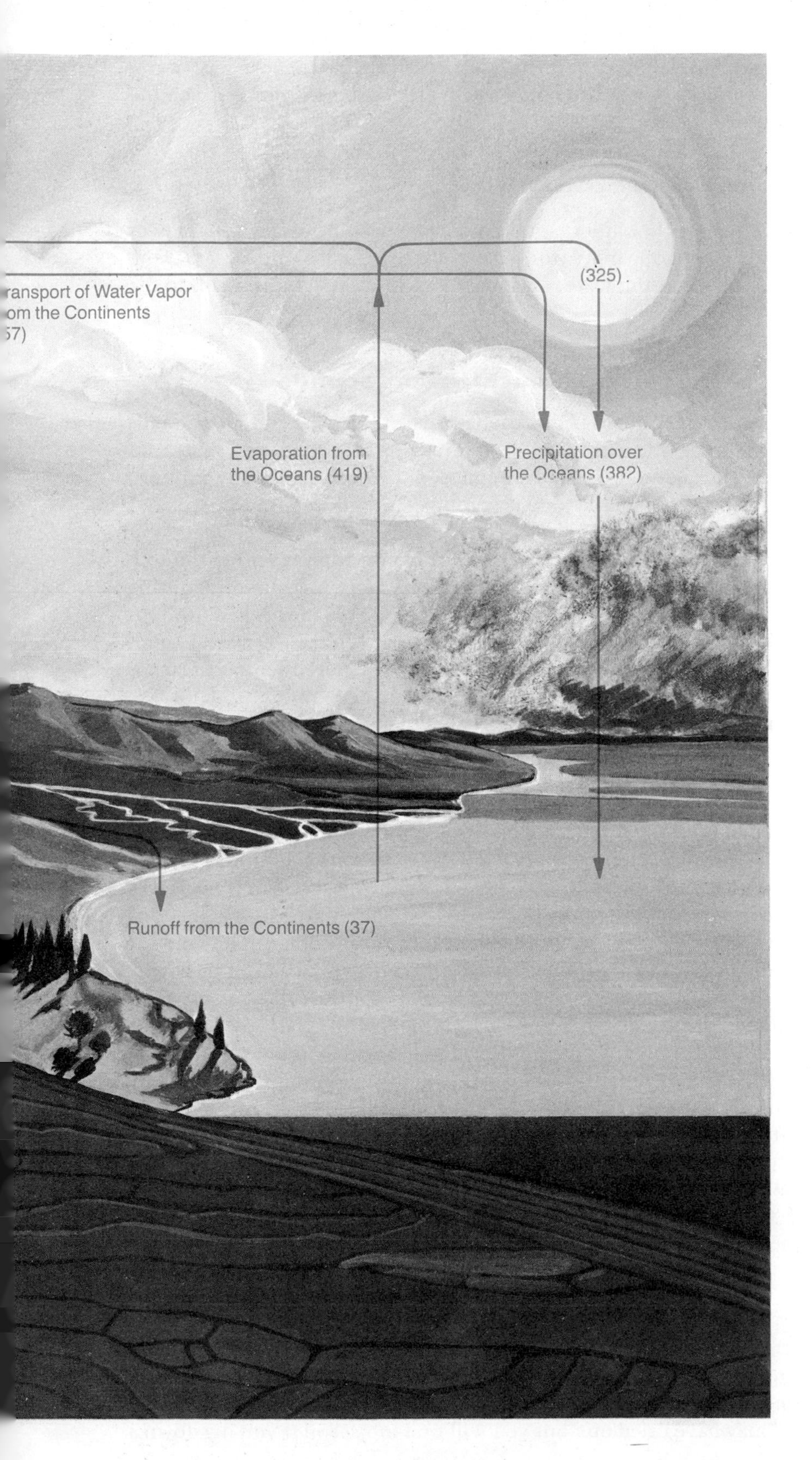

Figure 6.2 The movement of water through the hydrologic cycle involves numerous interactions and storage processes. Each year about 419,000 cu km of water evaporate from the oceans into maritime air masses, and evapotranspiration from the continents into continental air masses amounts to an additional 69,000 cu km (values in parentheses in the figure are water volumes in 1,000 cu km). This total volume of 488,000 cu km is equivalent to a mean annual precipitation over the globe of about 95 cm (37 in.).

Precipitation back to the oceans amounts to only 382,000 cu km, with 325,000 cu km originating from maritime air masses and 57,000 cu km supplied by continental air masses which move over ocean areas. Over the continents, precipitation (106,000 cu km) is greater than evapotranspiration (69,000 cu km); the difference, 37,000 cu km, represents runoff which returns in streams and rivers eventually to the oceans. Note that precipitation over the continents is supplied largely by water vapor from the oceans.

When precipitation falls on the land, a portion of the moisture is intercepted by vegetation and evaporates from temporary storage on leaves. The moisture that reaches the ground either infiltrates the soil, runs off across the surface, or evaporates from temporary storage in pockets and depressions. Some of the water that infiltrates the soil is stored as soil moisture, and a portion percolates deeper into the ground and enters groundwater storage. The flow of streams is maintained both by direct surface runoff and by underground flow from groundwater.

streams. But before its return to the oceans or the atmosphere, water can be involved in a number of transport and storage processes on the land.

WATER ON THE LAND

What happens to water that falls on the land? Some of the principles of hydrology are apparent from everyday experience and observation. Consider, for example, the various routes rain can take as it falls to the earth.

Interception, Throughfall, and Stemflow

Not all of the rain that falls reaches the soil. In cities, some rain falls on buildings and paved areas. If the rain is heavy, sidewalks and parking lots become flooded, and most of the rain eventually runs off. The water that remains on pavements and buildings evaporates and returns to the atmosphere.

Vegetation cover also prevents some rain from reaching the soil. Except in arid regions, plant growth covers most of the ground during the growing season. Figure 6.3 shows how leaves *intercept* raindrops falling on plant cover. At the beginning of a rainstorm, leaves catch and hold a considerable amount of water, and not much rain reaches the soil. For the first few minutes of a rain, the ground under a tree remains fairly dry, but if the rainfall is prolonged, the capacity of the leaves to retain water will be exceeded and water will start to drip from the leaves to the soil as *throughfall.* As the rain becomes heavier and more prolonged, the tree becomes less effective as a shelter. A small amount of water also reaches the soil as *stemflow* by trickling along branches and down the trunk or stem to the ground.

After rain stops falling, water may continue to drip from leaves onto the ground for a time. Nevertheless, some water always remains on the leaves and evaporates back into the atmosphere. The amount of water that actually reaches the ground is less than the amount that falls as precipitation. Because of interception, the amount of water reaching the soil depends not only on the total rainfall but also on its pattern and timing. Compare the effects of interception on a continual, steady rain totaling 1 cm with the effects on ten brief showers falling at the same intensity and totaling 0.1 cm each. In the steady rain, the leaves' capacity to retain water will soon be exceeded, and throughfall will carry water to the soil. But with the brief showers, the leaves will have a chance to dry in between, and a significant portion of the rain from each shower will be intercepted and will subsequently return to the atmosphere; very little will reach the soil.

Infiltration and Soil Moisture Storage

When you water your garden after a dry spell, the water readily soaks into the soil. Puddles begin to collect on the surface only after prolonged heavy watering. A few days after watering, the surface of the soil may have dried out, but you will find moist soil if you dig down a

Figure 6.3 Some of the precipitation that is intercepted by vegetation and held in temporary storage by leaves returns to the atmosphere by evaporation and does not reach the soil.

few centimeters, which demonstrates that soil can store water.

Water enters, or *infiltrates,* soil from the surface of the ground. The maximum rate at which a given soil can absorb water is called its *infiltration rate,* and it depends on the soil's porosity, permeability, surface condition, and moisture content.

Soil consists of fine grains of mineral and organic matter, which vary in size from one type of soil to another. For instance, clay soils have much smaller grains than sandy soils. Water moves through and is stored in the spaces, or *pores,* between the grains. If the pores are interconnected, the soil is *permeable.* Water can move downward through permeable soil because of gravity or upward by capillary action.

The infiltration rate also depends on the condition of the surface layer of the soil. If the pores are open, infiltration proceeds easily. But if the soil grains have been compacted by vehicles or animal hooves, the soil may not be able to absorb moisture at all. Bare soil can also be compacted by rain, which falls with enough force to pack it and decrease its infiltration rate noticeably. A layer of vegetation absorbs the impact of raindrops and helps the soil maintain a high infiltration rate. If animals such as sheep have grazed an area too heavily and the vegetation cover is sparse, it will be difficult to restore the land to agricultural productivity because the compacted soil will not absorb moisture readily.

If water arrives at the surface at a rate less than the infiltration capacity of the soil, all of the water will be absorbed. Once the upper layers of soil are saturated with moisture, additional water can enter only as the water already in the soil slowly seeps down to lower levels. As Figure 6.4a illustrates, dry porous loam can absorb water at an initial rate of over 6 cm per hour (2.4 in. per hour), but as the soil becomes wetter, infiltration rapidly decreases until it levels off at a slow constant rate.

The water in a saturated soil drains, or *percolates,* down to lower levels under the force of gravity, but some water resists the pull and remains stored in the pores between grains. This water is held in small pores by capillary action and surface tension. These forces are too weak to support the greater volumes of water in large pores, and that water drains away over several days. The moisture that remains in storage after the excess has drained is known as the *field capacity* of the soil (see Figure 6.5).

Moisture stored in the soil returns to the atmosphere, either by direct evaporation from the soil or by transpiration through the leaves of vegetation. Plants withdraw soil moisture for transpiration throughout their rooting zone, but some water is unavailable to them because it is too tightly bound to the surfaces of soil particles by molecular attraction. Most plants begin to wilt when soil moisture is depleted to this level, known as the *wilting point.* If soil moisture is depleted to the wilting point too often, crop yields will decrease significantly, and eventually the plants will die if soil moisture is not replenished by rains or irrigation. With the exception of swamp and marsh vegetation, too much soil moisture is also harmful. Most plants

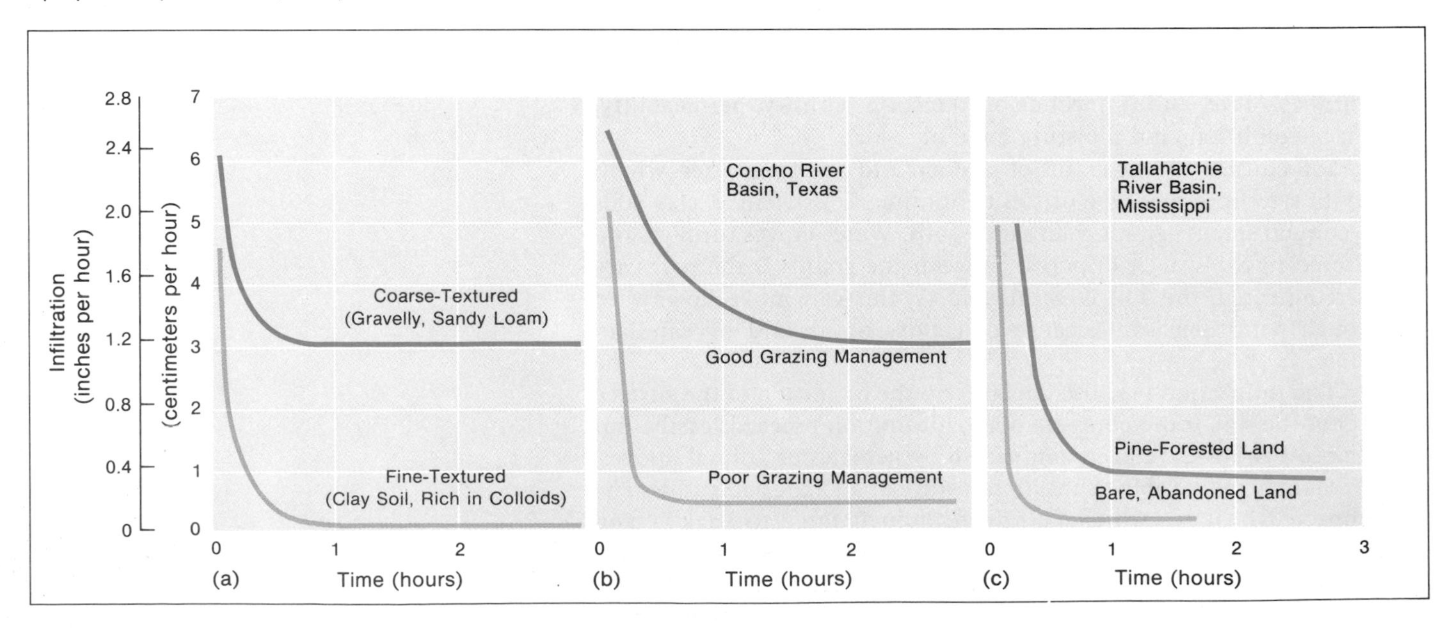

Figure 6.4 The infiltration rate of water into a soil depends on such factors as the texture, porosity, permeability, and moisture content of the soil and on the condition of the surface layer. The graphs show the infiltration rates of various soils measured from the time at which water is added to their surfaces. The infiltration rate falls sharply at first in all cases as the top layer of soil becomes well moistened; then a constant rate of infiltration is attained.

(a) Water infiltrates a coarse, permeable soil more easily than it does a dense clay soil, in which the water passages are small.

(b) The surface of well-managed grazing land retains an open texture and has a high infiltration rate. The infiltration rate is lower on poorly managed land because overgrazing exposes the soil, which allows the bare surface to become compacted by raindrops and animal hooves.

(c) Soil has a low infiltration rate when there is no vegetation cover to protect the surface from the impact of raindrops. Infiltration rates under natural forest tend to be higher than for any other vegetation cover. Infiltration rates for all land uses in the Tallahatchie River basin are unusually low because of soil properties. (After Foster, 1949)

begin to suffer when soil moisture remains greater than field capacity for several weeks or more. The soil pores must contain air as well as water because roots require oxygen and space to eliminate the carbon dioxide waste from root metabolism.

When the ground is covered by vegetation, transpiration is normally far greater than evaporation from the soil surface. Several factors work to reduce evaporation from the soil. Once the top few millimeters of soil have dried out, evaporation decreases because water vapor diffuses upward through soil very slowly. Furthermore, plant cover reduces both the temperature of the soil and the wind speed at the surface, thereby reducing evaporation. Transpiration from plants produces a blanket of humid air near the ground, which further reduces the rate of direct evaporation from the soil.

In Figure 6.5, the space between the field capacity and wilting point curves represents the stored soil moisture available for evapotranspiration, and we shall make use of this concept in the water budget analysis later in this chapter. Note especially that the field capacity, wilting point, and available soil mixture vary according to the size of the soil particles.

Groundwater and Aquifers

The subsurface portions of the continents contain a vast and accessible store of fresh water equal to 15 percent of the amount of fresh water stored in the ice sheets. The water is stored in underground deposits of gravel or porous rock, such as sandstone. These deposits, called *aquifers,* supply the water for wells and springs (see Figure 6.6, p. 162). The water in the fully saturated portion of an aquifer is known as *groundwater.* Groundwater constitutes some 30 times more fresh water than is found in all the rivers, lakes, and streams of the world. Groundwater is found within 100 meters (330 ft) of ground

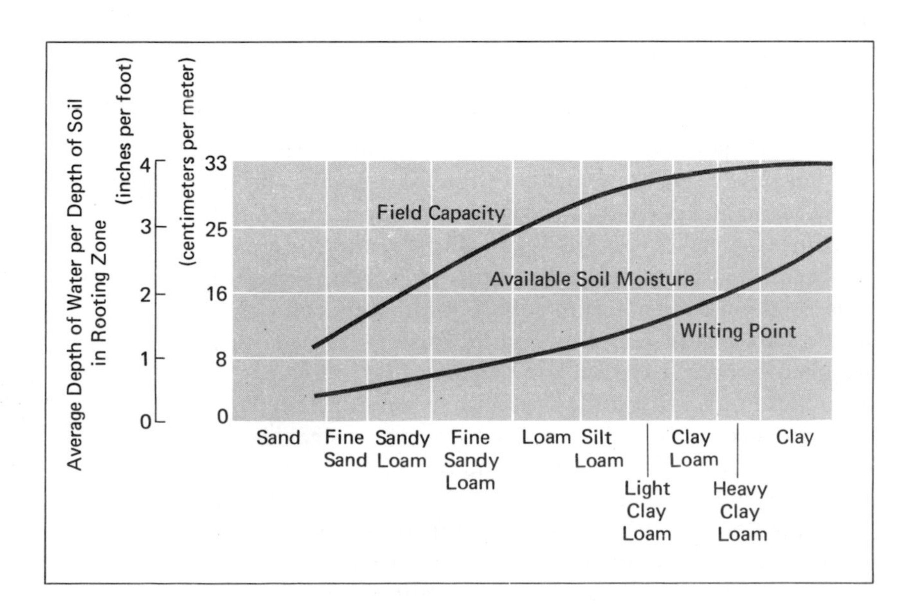

Figure 6.5 The field capacity is the amount of soil moisture stored in the pores between grains after drainage of excess water by gravity. The wilting point represents the amount of soil moisture still retained by surface tension after plant roots have taken up all the water available to them for transpiration. Both the field capacity and the wilting point are related to the texture, or size, of the soil particles. For example, nearly 33 cm of water can be stored in a layer of clay 1 meter deep (or 4 in. in a layer 1 ft deep), but only a little more than 8 cm in sand 1 meter deep. The soil moisture available in the rooting zone averages about 16 cm per meter (2 in. per ft) in silt loam, a little more than 8 cm per meter (1 in. per ft) in clay, and less than 8 cm per meter (1 in. per ft) in sand. (After Smith and Ruhe, 1955)

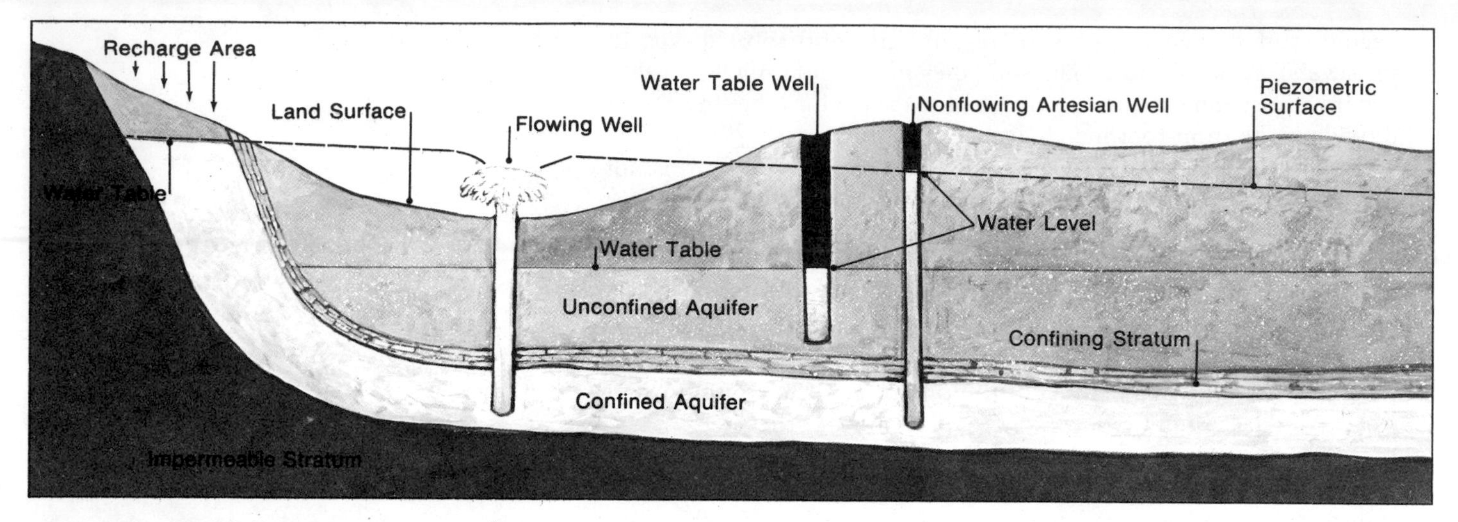

Figure 6.6 This diagram shows the principal features of aquifers in schematic form. If the rock above an aquifer is permeable enough to allow the vertical movement of water, the aquifer is said to be unconfined. The water table, or the water level in an unconfined aquifer, is the level to which water will rise in a well sunk into the aquifer. If the rock above an aquifer is impermeable, the aquifer is confined and must be replenished from a recharge area that is permeable to water from above. When a well is sunk into a confined aquifer, the level to which the water rises is called the *piezometric surface.* The piezometric surface can be a considerable height above a confined aquifer, particularly in the lower portion of a sloping aquifer, where the water pressure is high. The piezometric surface shown here slopes to the right, toward a region where the aquifers drain slowly into surface streams. (After Kazmann, 1972)

level just about everywhere—even in arid regions, such as Arabia, where there is little surface water. The thin layer of soil moisture is *not* considered to be groundwater.

The quality of groundwater is usually excellent, except in some arid and coastal regions where it may be contaminated by mineral salts. The porous rock in the aquifer filters the water and removes suspended particles and bacteria, although contaminants seep into aquifers occasionally. Natural springs, frequently noted for the purity of their water, are fed by groundwater. Many aquifers are extensive enough to serve as permanent water supplies for large cities, enabling some cities to rely on comparatively pure groundwater rather than artificially purified water from a badly polluted river. The temperature of groundwater taken from depths of 10 to 20 meters (30 to 60 ft) in midlatitude locations is usually only 1° or 2°C higher than the average annual temperature at the surface pumping station, which means a supply of cool water in summer. The relatively constant cool temperature of groundwater makes it suitable for manufacturing processes and industrial cooling.

Several different kinds of subsurface materials are sufficiently porous and permeable to make suitable aquifers. The majority of aquifers now being drawn from in the United States are underground beds of sand and gravel. Some of these beds were deposited long ago by receding glaciers; others are the products of ancient erosion. Sand and gravel aquifers are typically 50 meters (160 ft) thick and cover an area as large as several thousand square kilometers. The cross section of aquifers in Figure 6.6 shows sand and gravel beds. Clay deposits, although porous, have such fine pores that water can be extracted from them only with difficulty. Also, removing water from clay may cause the clay to compress and become impermeable. Sandstone and limestone form deep, extensive aquifers. A limestone aquifer 300 meters (1,000 ft) thick underlies a large area of the southeastern United States, and a limestone aquifer in England is being tapped to supply water for London. Limestone is fissured with long channels and cracks that allow water to flow through it easily.

If an aquifer rests on a layer of relatively impermeable rock or compacted clay, water is prevented from draining out below it. The top of the saturated portion of the aquifer is called the *water table;* it rises and falls in response to precipitation that seeps through the soil and upper unsaturated zones of the aquifer and to slow drainage from the aquifer into nearby streams and rivers. The depth of the water table below the surface of the ground is the minimum depth a well must be sunk in order to draw water from the aquifer.

The water table is not necessarily horizontal, but tends to follow the slope of the land. Water in an aquifer need not reach a horizontal level, as free water in a pond does, because the water in a sloping aquifer is not in stationary equilibrium but slowly flows both laterally through the porous rock and vertically from higher to lower elevations. Unlike the swift flow of a mountain stream down a steep slope, the flow in a sloping aquifer is gradual because the fine pores of the aquifer bed offer resistance to water movement.

Almost all of the water stored in aquifers comes originally from precipitation. After field capacity is reached, rain water that infiltrates the soil percolates slowly downward through the ground to the underlying groundwater. Similarly, water may seep through lake bottoms or stream beds into aquifers. The surface region over which water for an aquifer collects is called its *recharge area.* Because water travels horizontally through aquifers, the principal recharge area for an aquifer may be hundreds of kilometers away from the point at which water is being withdrawn. Local precipitation at the point of withdrawal does not necessarily govern the recharge rate of an aquifer.

Sometimes part of the upper surface of an aquifer is bounded, or *confined,* for some distance by a layer of impermeable rock. Because the impermeable upper layers may slope downward through the ground, the difference in height between the higher and lower parts of a confined aquifer may result in a considerable pressure differential. If a well is sunk into a low section of a confined aquifer, where the water pressure is high, the water in the well rises to a considerable height above the aquifer bed (see Figure 6.6); this height is known as the *piezometric surface.*

The speed at which water flows through aquifers varies from a few meters a year to several kilometers a day. Water is stored for a long time—many hundreds of years—in large aquifers that have a low rate of flow. Careful management of water entails ensuring that the amount of water withdrawn from an aquifer does not exceed the net flow into it. Many aquifers are pumped at rates greater than the rate of recharge, which depletes the amount of stored water and lowers the water table.

Approximately 50 percent of all the groundwater extracted in the United States is used for irrigation in Texas, Arizona, and California. Excessive withdrawal has lowered the water table in northwest Texas so seriously that artificial recharge has been tried as a remedial measure to restore some water to the region's aquifer (see Figure 6.7, p. 164). During the rainy season in Texas, water collects in shallow depressions. Some of this water, which would ordinarily evaporate during the dry season, has been pumped to the recharge area of the

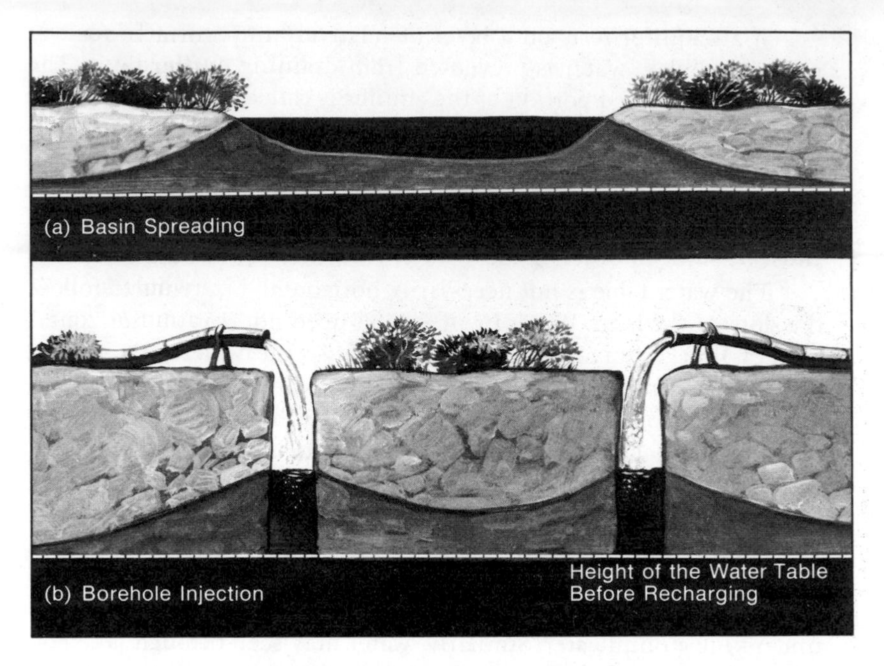

Figure 6.7 This diagram illustrates two methods that have been used to recharge aquifers and raise the level of the water table in regions where supplies of groundwater have been depleted by excessive withdrawal.

(a) Water pumped into shallow surface depressions in the natural recharge area of an aquifer seeps through the permeable rock into the aquifer.

(b) Water pumped into boreholes sunk into an aquifer seeps into the permeable rock to recharge the aquifer.

aquifer to help replenish it. Artificial recharge has also been used in coastal regions when extensive mining of water from an aquifer has allowed seawater to seep in and contaminate wells.

Runoff and Streamflow

Rainfall that reaches the surface of the ground initially infiltrates the soil, but as soon as the rate at which water reaches the surface exceeds the infiltration rate of the soil, some of the excess water collects on the surface. Surface irregularities and vegetation, such as the dense underbrush of a forest, store some of the water for later infiltration. Once the capacity of the surface to hold water is exceeded, water begins to flow across the ground in trickles or sheets under the influence of gravity. This surface runoff increases from small rivulets at the beginning of a rainstorm to a steady torrent if the rain lasts long enough to saturate the storage mechanisms in and on the soil. Toward the end of a prolonged rain, nearly all of the water that reaches the ground joins the surface runoff. A rain of long duration produces more surface runoff than several brief showers of the same intensity separated by dry periods.

Moist regions are laced with a network of small streams fed by surface runoff. The small streams carry water into a few large streams, which in turn supply major rivers. The water that rivers carry from the land to the sea represents precipitation which did not return to the atmosphere by evapotranspiration. (The organization of river systems is discussed in more detail in Chapter 12.)

Hydrologists are especially concerned with the variable flow of streams and rivers over time, particularly in response to runoff from precipitation. *Runoff* is a measure of the average depth of water which flows from a drainage basin during a specified time period. Consider, for example, an impervious parking lot with well-engineered storm drains. Assume now an intense 2.5 cm (1 in.) rainstorm lasting an hour. The storm drains are efficient, and 15 minutes after the storm there is only an average depth of .05 cm of water left on the parking lot in a few depresssions and cracks; later this water will return to the atmosphere by evaporation. The runoff from the parking lot amounts to 2.45 cm, or 98 percent of the rainfall; it is assumed that the runoff comes equally from each area of the parking lot. Incidentally, the runoff proportion from an equal area of nearby park would undoubtedly be far smaller because much of the rainfall there would go into soil moisture storage, to be returned eventually to the atmosphere by evapotranspiration.

Streamflow, on the other hand, represents the volume of water moving down a stream channel during a very short time period, and it is usually expressed in cubic feet per second (cfs) or cubic meters per second. A plot of streamflow, or discharge through time, is called a *hydrograph,* and Figure 6.8 shows a hypothetical example for a stream of intermediate size in a humid region. Between rainstorms, the stream's flow is maintained by an almost steady supply of groundwater slowly seeping into the channel from surrounding uplands. This discharge of the stream is known as the *base flow.* During or shortly after a rainstorm, surface runoff from the drainage basin reaches the stream, and the figure illustrates the characteristic rapid rise and slower recession of the flow associated with the storm runoff and recharged groundwater supplies.

The annual regimes of daily streamflow for three small rivers are shown in Figure 6.9, p. 166. For all three rivers, streamflow is highest in winter and spring, when evapotranspiration rates are low and soil moisture and groundwater supplies are greatest, and lowest in late summer and autumn, when evapotranspiration rates are high. The peaks superimposed on the base flow are due to surface runoff.

The rivers are located in different climatic regions. Bundick Creek in Louisiana is representative of a warm and humid climate. Streamflow is greatest during winter and spring; the peaks in summer and fall are associated with runoff from heavy thunderstorms. Independence Creek is in the humid but cold climate of the Adirondack Mountains region of New York. There runoff normally peaks in early spring due to snowmelt and spring rains; a secondary peak in late autumn occurs because of decreasing evapotranspiration demands. The minimum flow in winter is associated with the winter snowcover and only brief winter thaws. No water flows in Pope Creek, located in a warm, dry region of California, during summer and fall, but in winter and early spring, groundwater contributes to streamflow.

Figure 6.8 This schematic diagram shows the effect of an upstream rainstorm on the amount of water carried by a stream, measured from the time of the storm. From *A* to *B*, water from the rain has not reached the downstream region, and the discharge, or rate of flow, is due to the base flow supplied by groundwater. From *B* to *C*, the discharge rises rapidly as direct surface runoff from the upstream drainage basin reaches the downstream region. From *C* to *D*, the discharge falls slowly as the last of the surface runoff, including the runoff retarded by the vegetation cover, makes its contribution. From *D* to *E*, the discharge is due primarily to groundwater supplies, which have been newly recharged by the rain. The discharge of the stream gradually decreases as the aquifers become depleted. (After Ward, 1967)

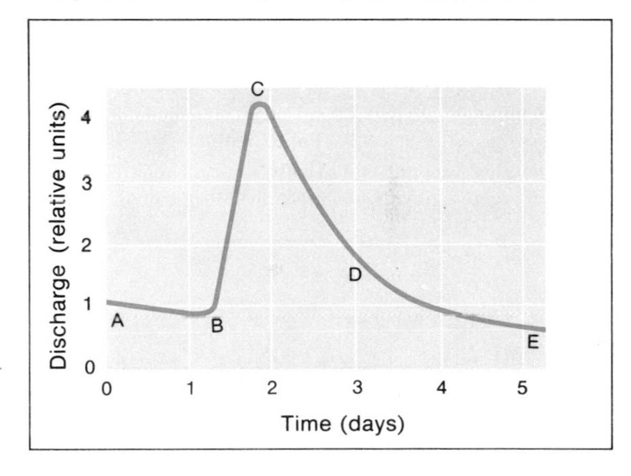

THE LOCAL WATER BUDGET: AN ACCOUNTING SCHEME FOR WATER

Water is not always available in the desired amounts when and where we need it. The principles of hydrology discussed earlier in this chapter show how soil moisture and runoff depend on the amount and timing of precipitation. But precipitation is a specialized and localized

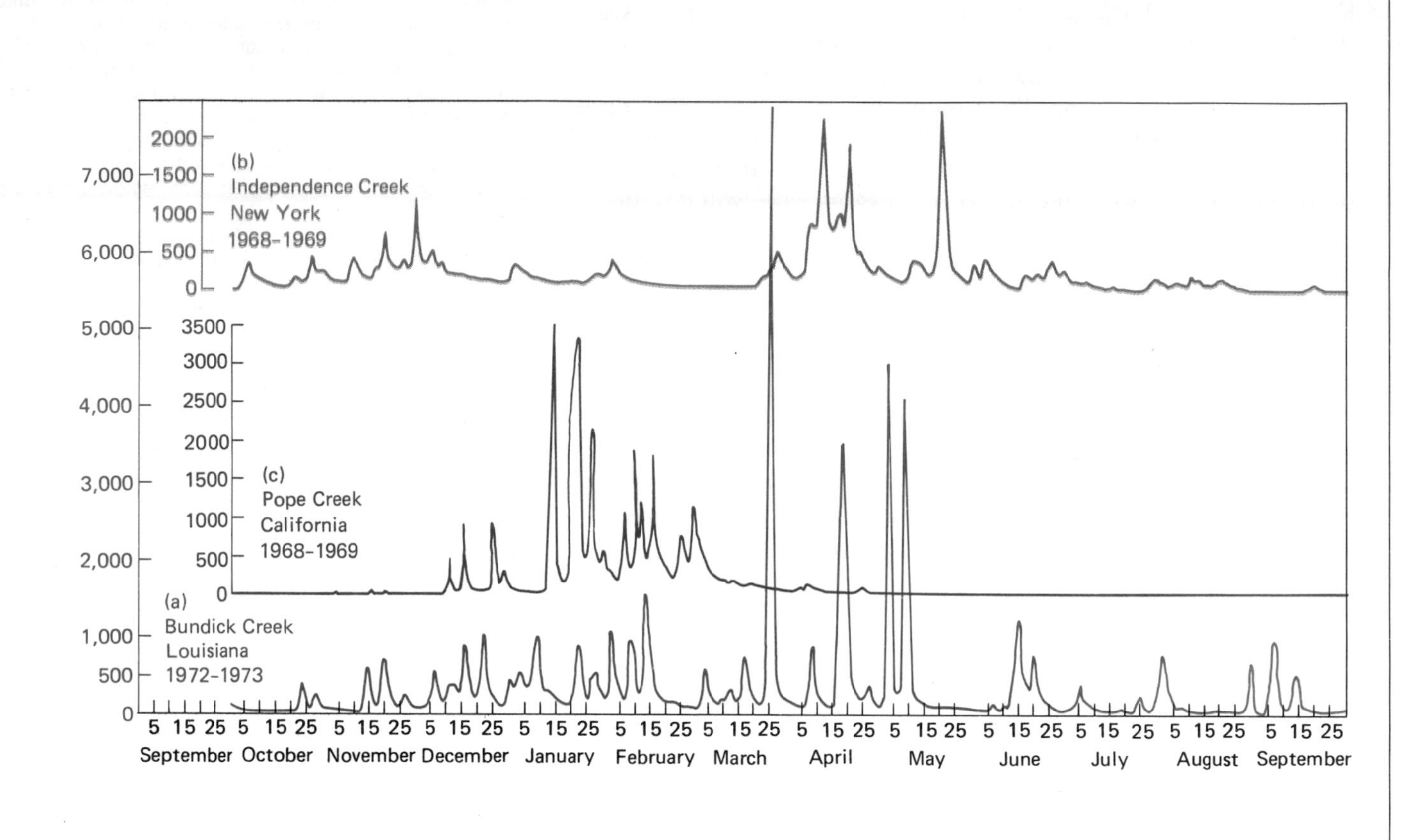

event, subject to marked variations from month to month and from place to place. The variability of precipitation shows up in the variability of water available for soil moisture and for runoff to streams.

To estimate the availability of moisture, hydrologists make use of the water budget. Essentially, the water budget is the local version of the hydrologic cycle. Or, to put it another way, the hydrologic cycle represents the sum of the local water budgets for all regions of the earth. The water budget is an extemely useful tool with a wide variety of applications. For example, water budget methodology may be used to estimate the moisture content of the soil, and this information may then be employed to evaluate irrigation needs or to gauge the extent to which sands and clays will support the frequent passage of wheeled vehicles. The water budget concept also plays a part in calculating evapotranspiration, an essential step in estimating potential vegetation types and crop yields. Estimates of runoff and streamflow are possible with the water budget, even in regions where only temperature and precipitation measurements are available. Finally, the water budget system can be used to evaluate local environments in terms of water-resource potentials and climatic classification.

The local water budget takes into account four principal components of water distribution: precipitation, soil moisture, evapotranspiration, and runoff. When appropriate, snow accumulation and melting may also be included. Of all these components, evapotranspiration is the most difficult to treat properly because of its complicated dependence on biological and meteorological factors. In the 1940s, the American climatologist C. Warren Thornthwaite developed a way of treating evapotranspiration. To overcome the difficulties in estimating the actual rate of evapotranspiration, Thornthwaite introduced the concept of *potential evapotranspiration* and developed a formula for calculating it (see Appendix III). Potential evapotranspiration, the key element in the water budget, is worth examining in detail.

Potential and Actual Evapotranspiration

Potential evapotranspiration is the rate at which water is lost to the atmosphere from a dense homogeneous vegetation cover that has been supplied with all the soil moisture it can use. If the soil moisture is partly exhausted, the actual rate of evapotranspiration from a field may fall below the potential rate because plants may be unable to withdraw the full amount of water they could use.

Evapotranspiration is expressed as the depth of liquid water converted to vapor in a given time over an area. We usually think of precipitation in terms of centimeters of water; the amount of water involved in evapotranspiration is expressed in the same units. A typical value for potential evapotranspiration over a 30-day period in a warm, sunny region is about 15 cm (6 in.).

Of all the factors that determine evapotranspiration, the most important is the amount of solar energy the plants absorb. Both evaporation and transpiration of water require the expenditure of about 590 calories per gram of water. Evapotranspiration therefore depends on the duration of daylight, the degree of cloud cover, and the albedo of the plant cover, because these are the principal factors that determine the amount of energy the plants absorb.

Potential evapotranspiration from a field is essentially independent of the type of plant cover. Imagine looking down on a forest or a field during the middle of the growing season. The plants present a uniform, dense cover of overlapping green leaves. The albedo of almost all green plants is about 15 to 20 percent. Therefore, an acre of forest and an acre of cotton field, both densely covered, absorb about the same amount of energy under the same external conditions.

It is significant that potential evapotranspiration depends only on external conditions, such as solar energy input and weather, and not on the detailed characteristics of the plants. The *potential* evapotranspiration for a given region at a given time can in principle be calculated from astronomical and meteorological data alone. By contrast, *actual* evapotranspiration, or the amount of evapotranspiration that actually occurs, varies with the amount of soil moisture and the condition of the vegetation cover as well as with external conditions. The

Figure 6.9 (opposite) Hydrographs for three small rivers in different climate regions, each draining an area between 194 and 310 sq km (75 and 120 sq miles). The hydrographs are organized by water years, which begin October 1 and end September 30. Discharges are in cubic feet per second (cfs). The water year is a useful calendar for water-resource management because streamflow tends to be lowest in late September, with minimum carryover of groundwater contributions to later base flow. Typically, storm runoff is superimposed as spikes, or peaks, on the base flow contributed by groundwater. Storm runoff tends to rise quickly and to recede more slowly.

(a) Bundick Creek is located in the warm and humid climate of southwestern Louisiana. Groundwater provides for some base flow year round, but streamflow is largest on the average during winter and spring. A considerable proportion of the annual flow is produced in just a few days by the very large spring flood flows, and the maximum daily discharge of 7,980 cfs on March 25, 1973, was the highest daily flow in 17 years of measurements.

(b) Independence Creek is located on the western flanks of the Adirondack Mountains in northern New York. This watershed is representative of climates where persistent low temperatures during winter allow most of the precipitation to accumulate as snowpack. Much of the annual flow occurs during spring as the snow melts, but there is usually a secondary discharge peak in autumn before the winter snows begin to accumulate.

(c) Pope Creek drains a low mountainous region of the coastal ranges of northern California west of Sacramento. The climate is hot and dry during summer with no precipitation. Groundwater contributes to streamflow only during winter and spring, and normally there is no water in the creek from June through October.

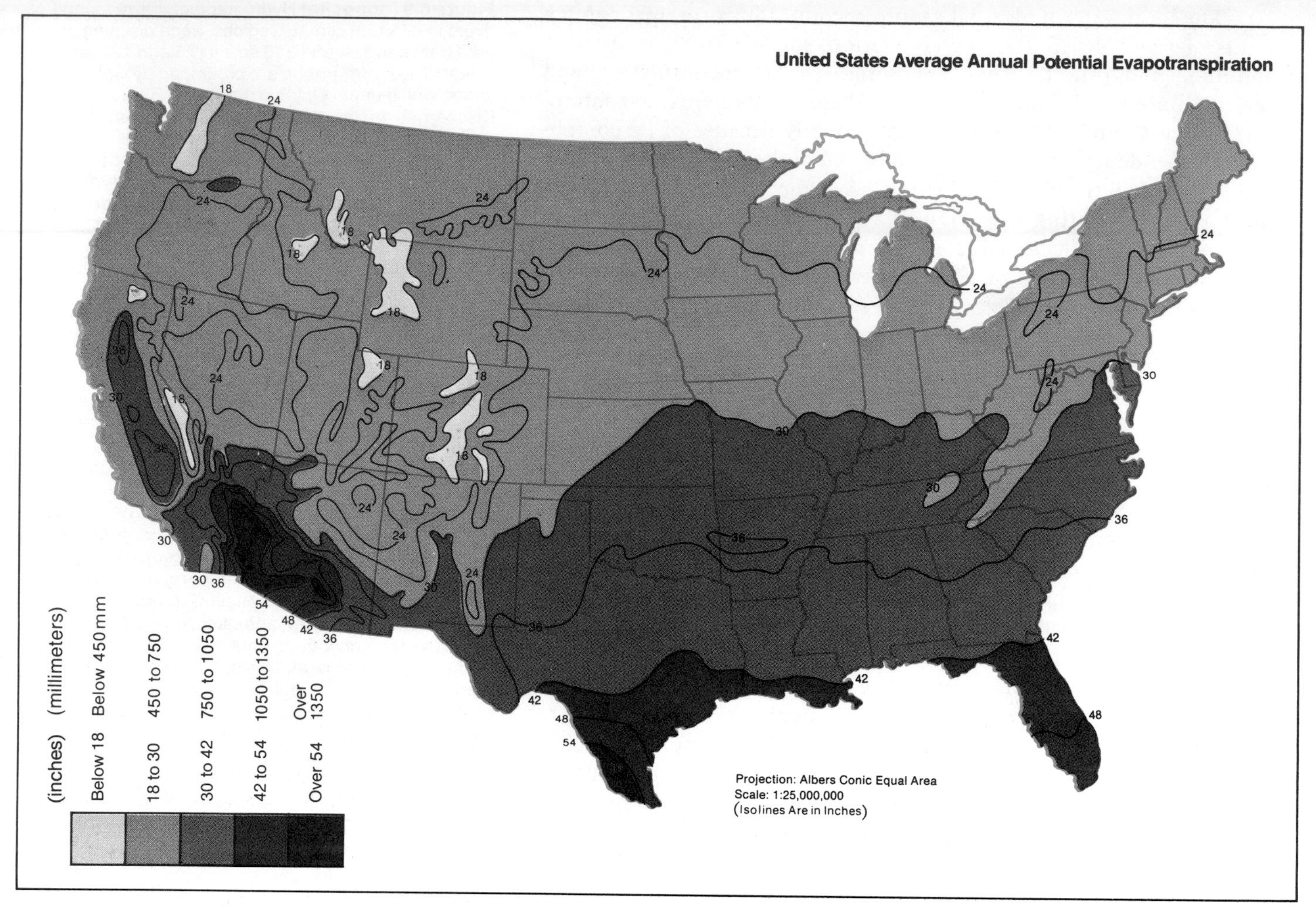

Figure 6.10 This map shows the average annual potential evapotranspiration, or *PE* for short, for the coterminous United States, calculated from Thornthwaite's formula. The Southwest, Texas, and Florida have high values of *PE* because of greater solar radiation income and very warm weather. Along the Canadian border, the *PE* is much lower because of the decreased solar radiation income in winter and much cool and cloudy weather. Across the high mountain areas of the West where temperatures are low, the *PE* is also low. There is also a strong seasonal regime of *PE* throughout the United States: *PE* tends to be low in winter and high in summer, although this seasonality is least marked along the West Coast and across the south.

average annual potential evapotranspiration for the United States, calculated from Thornthwaite's formula, is shown in Figure 6.10.

Measuring Actual and Potential Evapotranspiration

Potential evapotranspiration at a particular location can be measured in a device called an *evapotranspirometer,* which consists essentially of an open tank sunk flush with the ground. The tank, perhaps 60 cm (2 ft) in diameter and 90 cm (3 ft) deep, is filled with soil and planted with a dense cover of grass. If the grass is well watered, the soil moisture remains sufficient and constant so that the difference between the amount of water added to the tank and the amount draining from the bottom represents the amount of water used by the plants and the amount lost by direct evaporation from the soil, or the potential evapotranspiration in this case. Actual evapotranspiration is measured by weighing the tank; changes in weight represent changes in the amount of water stored. A weight increase represents water added to the tank by precipitation or irrigation, and a weight decrease represents evapotranspiration losses or percolation, which is caught in a false bottom and measured directly. In an attempt to simulate the natural environment, some tanks are built large enough to carry a stand of small trees. Setting up and operating an evapotranspirometer is a major undertaking that requires the time and attention of skilled personnel. Measurements of potential evapotranspiration have therefore been restricted to a few locations over limited periods of time, and for most practical applications a calculated estimate of potential evapotranspiration, such as Thornthwaite's, must suffice.

Thornthwaite's calculation of the monthly potential evapotranspiration at a particular location depends mostly on two factors: (1) the average air temperature of the given month and (2) the length of daylight for the given month and latitude. The average monthly air temperature and the day length represent solar energy input; the same conditions that cause temperature to be high also increase potential evapotranspiration.

Although more complex formulas have been developed to estimate potential evapotranspiration, Thornthwaite's formula is one of the most widely used for regional analyses. It has the advantage of being based on standard meteorological data, and it requires no additional measurements. Knowing the potential evapotranspiration of a region is essential to calculating its water budget. Graphs and tables for calculation of potential evapotranspiration are provided in Appendix III.

The Water Budget Equation

The local water budget simply represents a systematic accounting of input, output, and storage of water at a particular location (see Figure 6.11). All of the incoming precipitation (P) during a given time period is allocated to evapotranspiration (AE), soil moisture recharge (ΔST, where Δ represents changes), or "leftover" water, called surplus (S). The water budget equation can then be written:

$$P = AE + S + \Delta ST$$

Surplus represents the surface runoff and groundwater recharge that are largely unavailable to local vegetation; these surpluses ultimately find their way to streams and rivers, where they become the focus of attention for hydrologists and water-resource managers.

Another way to calculate the local water budget is in terms of solar energy input rather than precipitation. Remember from Chapter 2 that about 590 calories are needed to evaporate and transpire a layer of water 1 cm deep from each square centimeter of moist lawn (or about 1,500 calories to produce an inch of evapotranspiration from the same square centimeter of lawn). Potential evapotranspiration (PE) therefore is a measure of energy income. Incoming climatic energy is utilized for evapotranspiration, and any "leftover" energy—not used for evapotranspiration—goes into additional heating of the earth's surface or atmosphere. In water budget terms, the extra energy is called deficit (D) because there is not enough water to meet the climatic demand. Thus the simple energy budget equation is:

$$PE = AE + D$$

All of these water budget components are explained further in the following section.

Figure 6.11 This schematic diagram illustrates the principal components of the local water budget. The input to the system of soil and vegetation is the amount of moisture supplied by precipitation. A large portion of the input returns to the atmosphere by evaporation and plant transpiration. Some of the input moisture is stored in the soil. However, the storage capacity of the soil is limited; when the storage is full, a moisture surplus becomes available to supply surface runoff and groundwater recharge.

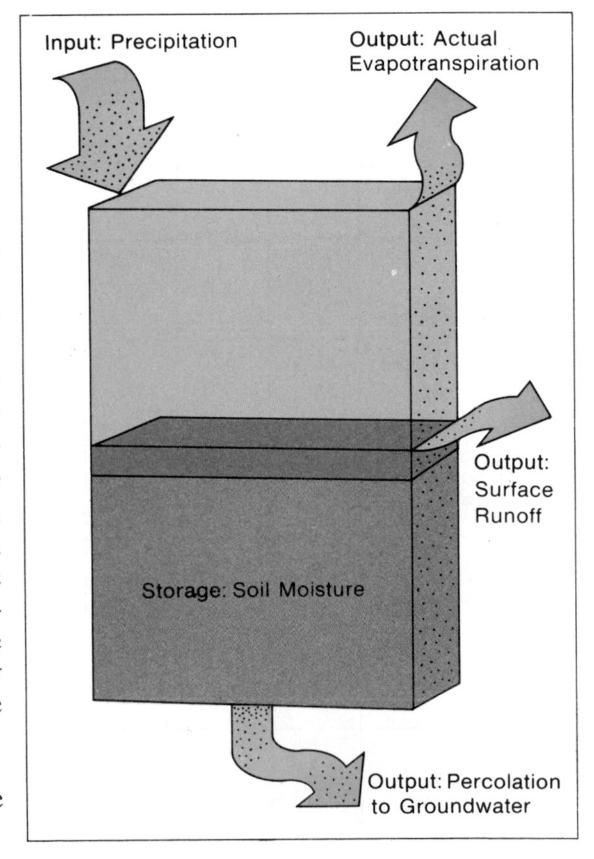

A Local Water Budget: Baton Rouge

To illustrate the calculation of a local water budget, consider the example of Baton Rouge, Louisiana, in 1962. In an average year, Baton Rouge receives more than 50 in. (125 cm) of rainfall, a comparatively large amount. Nevertheless, the water budget analysis shows that for several months in 1962 Baton Rouge was "dry" and crops needed watering.

In this example, income, output, and storage are considered month by month. Weekly or even daily periods could also be used. The first step in constructing a monthly local water budget is to use meteorological records to find the monthly precipitation at Baton Rouge, which is listed in the first row of the chart in Figure 6.12. The values are given in inches because weather data in the United States are recorded in English units. The total precipitation for the year was 52.9 in. (134.4 cm), almost all of which fell as rain.

Involved computations are required to calculate monthly potential evapotranspiration using weather records and Thornthwaite's formula. The results have been worked out and are listed in the second data row of the table. As we can see, potential evapotranspiration is much smaller in the winter months than it is in the summer because *PE* tends to follow the seasonal regime of temperature.

Figure 6.12 This is the local water budget for Baton Rouge, Louisiana, during 1962. See the text for a discussion of how the budget is calculated. The water budget equation is $P = AE + S \pm \Delta ST$. For 1962 the equation works out: $52.9 = 39.3 + 15.7 - 2.1$. Similarly, the energy budget equation is $PE = AE + D$, and for 1962 this is $41.9 = 39.3 + 2.6$. Each equation balances, so we can be quite confident that we have not made any calculation errors.

	JAN	FEB	MAR	APR	MAY	JUN	JUL	AUG	SEP	OCT	NOV	DEC	Total
1. Precipitation (*P*)	6.4	0.7	3.3	9.7	1.6	11.4	2.0	4.5	4.3	5.2	0.9	2.9	52.9
2. Potential Evapotranspiration (*PE*)	0.4	1.7	1.3	2.6	5.3	6.1	7.4	6.8	5.2	3.4	1.1	0.6	41.9
3. Precipitation minus Potential Evaporation (*P-PE*)	6.0	−1.0	2.0	7.1	−3.7	5.3	−5.4	−2.3	−0.9	1.8	−0.2	2.3	11.0
4. Change in Stored Soil Moisture (ΔST)	0.0	−1.0	1.0	0.0	−3.7	3.7	−5.4	−0.6	0.0	1.8	−0.2	2.3	−2.1
5. Total Available Soil Moisture (*ST*)	6.0	5.0	6.0	6.0	2.3	6.0	0.6	0.0	0.0	1.8	1.6	3.9	
6. Actual Evapotranspiration (*AE*)	0.4	1.7	1.3	2.6	5.3	6.1	7.4	5.1	4.3	3.4	1.1	0.6	39.3
7. Deficit (*D*)	0	0	0	0	0	0	0	1.7	0.9	0	0	0	2.6
8. Surplus (*S*)	6.0	0	1.0	7.1	0	1.6	0	0	0	0	0	0	15.7

Precipitation is the source of water income. In months when rainfall exceeds potential evapotranspiration, the excess of moisture allows plants to transpire at the maximum rate. The monthly potential evapotranspiration is subtracted from monthly precipitation and the result entered in the third data row. The six negative values show that for 6 months of 1962, precipitation was less than potential evapotranspiration.

Soil moisture is the principal storage mechanism considered in the local water budget. Because 1961 had been a wet year, 1962 began with the soil moisture in Baton Rouge at its full capacity of 6.0 in. (The soil moisture storage capacity of 6 in. is representative of the Baton Rouge area; on a global basis, the storage capacity ranges from about 2 to 12 in.) The simple water budget considered here assumes that all of the soil moisture is equally available to plants, even if the soil moisture is below field capacity; the model assumes that if potential evapotranspiration exceeds rainfall, plants will use soil moisture as needed until the supply is exhausted. If rainfall exceeds potential evapotranspiration, the model also assumes that the soil will absorb water up to the soil moisture storage capacity of 6.0 in.

Data row 4 shows the change in soil moisture, and row 5 shows the total soil moisture at the end of each month. These quantities must be calculated together, as a few examples will indicate. In January, rainfall exceeded potential evapotranspiration, but the excess was not absorbed because soil moisture was already at full capacity. In February, a rainfall deficiency of 1.0 in. caused 1.0 in. of water to leave soil moisture, which reduced the stored total to 5.0 in. In March, an excess precipitation of 2.0 in. made water available for soil moisture, but only 1.0 in. could be absorbed because this amount brought the available soil moisture back to its full capacity of 6.0 in. At the end of July, the total soil moisture was 0.6 in.; a deficiency of rainfall in August caused the remaining 0.6 in. of soil moisture to be used, which left the soil moisture at 0.0.

Potential evapotranspiration is equal to actual evapotranspiration only when the soil contains some moisture. Row 6 shows actual evapotranspiration equal to potential evapotranspiration for all months in which total soil moisture is greater than 0.0. For September, when soil moisture was exhausted, actual evapotranspiration was equal to the amount of rainfall.

The difference between potential and actual evapotranspiration, called the *deficit*, is listed in row 7. A deficit represents the minimum amount of irrigation water required to restore crops to full transpiration; it does not represent an actual loss of water. The deficit represents the amount of unavailable water that plants could have used. Baton Rouge suffered small deficits in August and September of 1962.

Whenever precipitation exceeds actual evapotranspiration, there is an excess of water, part of which recharges the depleted soil moisture supplies. Any excess water that remains after soil moisture is brought to full capacity is called the *surplus*. Row 8 shows the calculated monthly surpluses for Baton Rouge in 1962. Surpluses were present only in January, March, April, and June. In March, rainfall exceeded actual evapotranspiration by 2.0 in., but of this excess, the

(a)	Jan	Feb	Mar	Apr	May	Jun	Jul	Aug	Sep	Oct	Nov	Dec	Year
1. Precipitation (*P*)	4.4	4.8	5.1	5.1	4.4	3.8	6.5	4.7	3.8	2.6	3.8	5.0	54.0
2. Potential Evapotranspiration (*PE*)	0.7	0.9	1.6	3.0	4.7	6.2	6.7	6.3	4.9	2.8	1.2	0.8	39.8
3. Precipitation minus Potential Evapotranspiration	3.7	3.9	3.5	2.1	−0.3	−2.4	−0.2	−1.6	−1.1	−0.2	2.6	4.2	14.2
4. Change in Stored Soil Moisture (ΔST)	0	0	0	0	−0.3	−2.4	−0.2	−1.6	−1.1	−0.2	2.6	3.2	0
5. Total Available Soil Moisture (*ST*)	6.0	6.0	6.0	6.0	5.7	3.3	3.1	1.5	0.4	0.2	2.8	6.0	
6. Actual Evapotranspiration (*AE*)	0.7	0.9	1.6	3.0	4.7	6.2	6.7	6.3	4.9	2.8	1.2	0.8	39.8
7. Deficit (*D*)	0	0	0	0	0	0	0	0	0	0	0	0	0
8. Surplus (*S*)	3.7	3.9	3.5	2.1	0	0	0	0	0	0	0	1.0	14.2

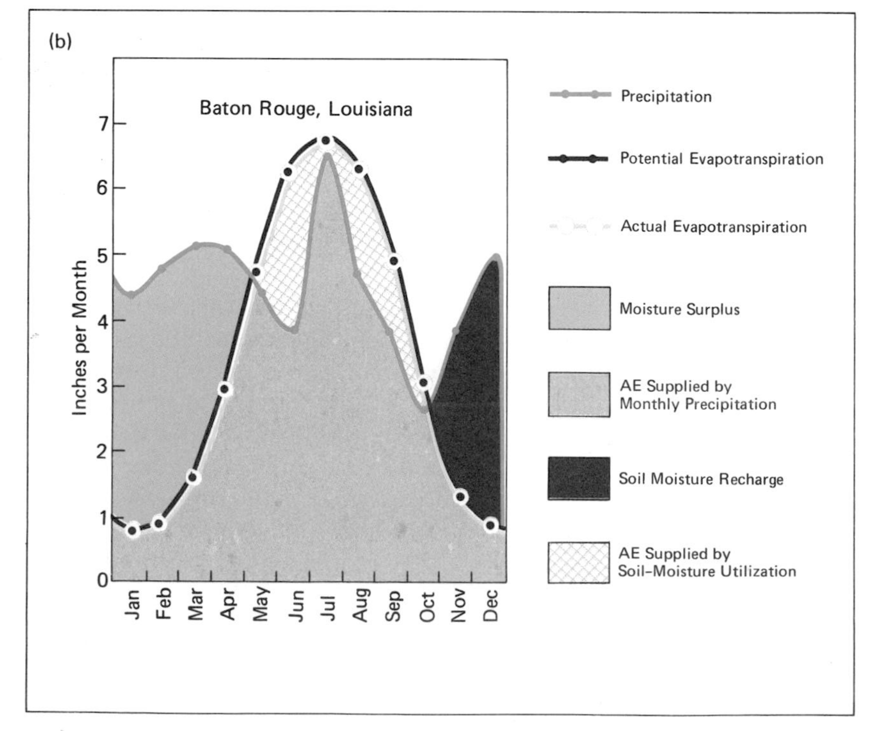

Figure 6.13 Average water budget for Baton Rouge, Louisiana, based on standard climatological data for 1941–1970.

(a) This average budget is based on 30 years of temperature and precipitation data. Average monthly precipitation is much less variable than monthly precipitation in individual years, such as 1962 in Figure 6.12, p. 170. Annual precipitation averages 137 cm (54 in.), with the least rainfall during autumn and late spring on the average. In this particular average water budget, *AE* is always equal to *PE*, and there is no deficit *(D)*. The distinction between the components of an average water budget and the budget of an individual year (Figure 6.12) should be kept clear.

(b) This graph of the average water budget for Baton Rouge is an example of standardized graphs that have appeared in research publications. In graphical form it shows the seasonal regimes of *P*, *PE*, *AE*, and *S*, as well as soil moisture withdrawal and recharge.

soil absorbed only 1.0 in., which left a surplus of 1.0 in. Surplus water is important because it is available for surface runoff and groundwater recharging. It is this water that enters streams and rivers and that eventually shapes the landscape by erosion and by sediment deposition.

The long-term average water budget over a 30-year period for Baton Rouge is shown in Figure 6.13. Calculations for the long-term average are similar to those for an individual year. We have to be careful about where to begin soil moisture storage calculations, however. Because precipitation exceeds potential evapotranspiration by more than 6 in. for November and December together, the soil will be fully recharged to its capacity of 6 in. during December. Storage will remain at capacity until May, when precipitation falls below potential evapotranspiration.

Individual wet months compensate for dry months, which makes long-term average monthly precipitation much less variable than monthly precipitation in individual years, such as 1962 shown in Figure 6.12. In this particular long-term water budget, the soil moisture is never fully depleted; thus *AE* is always equal to *PE* and there are no deficits. In the graph, the seasonal regimes of *P*, *PE*, *AE*, and *S* are evident.

In the next chapter, we will focus on regional differences in global climate, but this is an appropriate time to stress how much energy and moisture regimes vary seasonally and from year to year in the middle and higher latitudes. To illustrate this variability, a more complex model was used in Figure 6.14, p. 174, to calculate the monthly water budget components for Baton Rouge for 1960 through 1967. Precipitation is not shown, but its interaction with *PE* in terms of *AE* and *D* is evident. Deficits occur each year, but they range from small deficits with little environmental consequence to relatively large deficits during 1962, 1963, and 1965. Surpluses, on the other hand, are even more variable than precipitation because they represent the "leftover" water within the budget. Surpluses are normally restricted to the winter and spring months, but note especially the large surpluses during the winter-spring of 1961–1962 compared to the small surpluses for 1962–1963. This variability has considerable environmental impact, and it is due to small changes in the upper atmosphere circulation patterns which tend to persist for a few months or years.

APPLICATION OF THE WATER BUDGET MODEL

The water budget model presented in this chapter is a simple representation of a very complex environment. For example, it does not account for the fact that potential evapotranspiration can be better estimated by measurements of solar radiation input and the energy balance. The model assumes for simplicity that soil moisture is readily available to plants, regardless of the moisture content of the soil. In reality, the smaller the amount of moisture stored in the soil, the more difficult it becomes for plants to utilize it. The model also assumes that all rainfall infiltrates the soil whenever there is a storage opportunity; it makes no allowance for intense rainfalls which exceed infil-

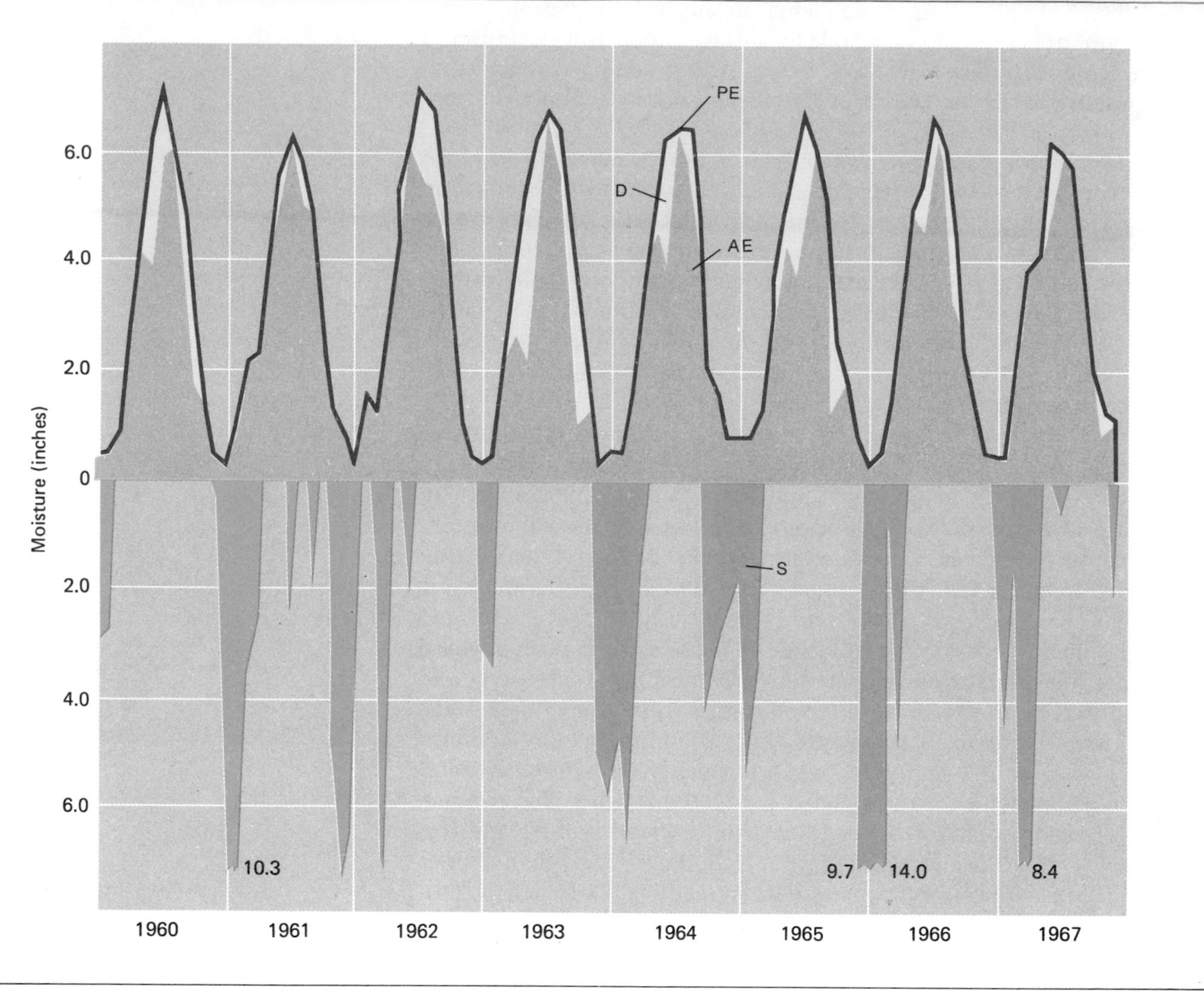

Figure 6.14 Monthly regimes of calculated water budget components for Baton Rouge between 1960 and 1967. This graph emphasizes seasonal consequences of the variability of precipitation, which itself is not shown.

A more realistic, but at the same time more complex, model of soil moisture storage and depletion was used for these calculations. In the water budgets for Figures 6.12, p. 170, and 6.13, p. 172, it was assumed that soil moisture was "equally available" to meet the climatic demand of *PE* during months when $P - PE$ was negative. In the model used for this figure, it was assumed that the soil moisture would become "decreasingly available" as it was depleted; in other words, the vegetation would not be able to withdraw all of the needed soil moisture even though the soil still contained some available moisture. The decreasing availability model of soil moisture depletion requires special tables or equations.

The water budget graphics in the following chapters are based on the decreasing availability model. Note especially that this model shows that small deficits recur each year at Baton Rouge. (After Muller, 1976)

tration rates. In addition, the model does not partition the surplus to runoff and groundwater recharge.

For certain purposes, the monthly water budget involves too long a time unit. For irrigation and flood forecasting, for example, a daily water budget becomes a necessity. Modern computer facilities allow for very complex analysis of daily water budgets for research and water-resource management. When Thornthwaite first worked out the water budget concept during the 1940s, computers were not readily available. Although he worked with daily water budgets for agricultural research (see Figure 6.15), most of his textbook examples were within a framework of average monthly budgets.

A further difficulty is that the precipitation data required for a water budget analysis are not always sufficiently accurate or available. Precipitation is conventionally measured by noting the accumulation of water in a standard bucket, or rain gauge, exposed under standard conditions. The gauges, however, do not catch a fully representative sample on windy days or when the precipitation falls as snow. In the

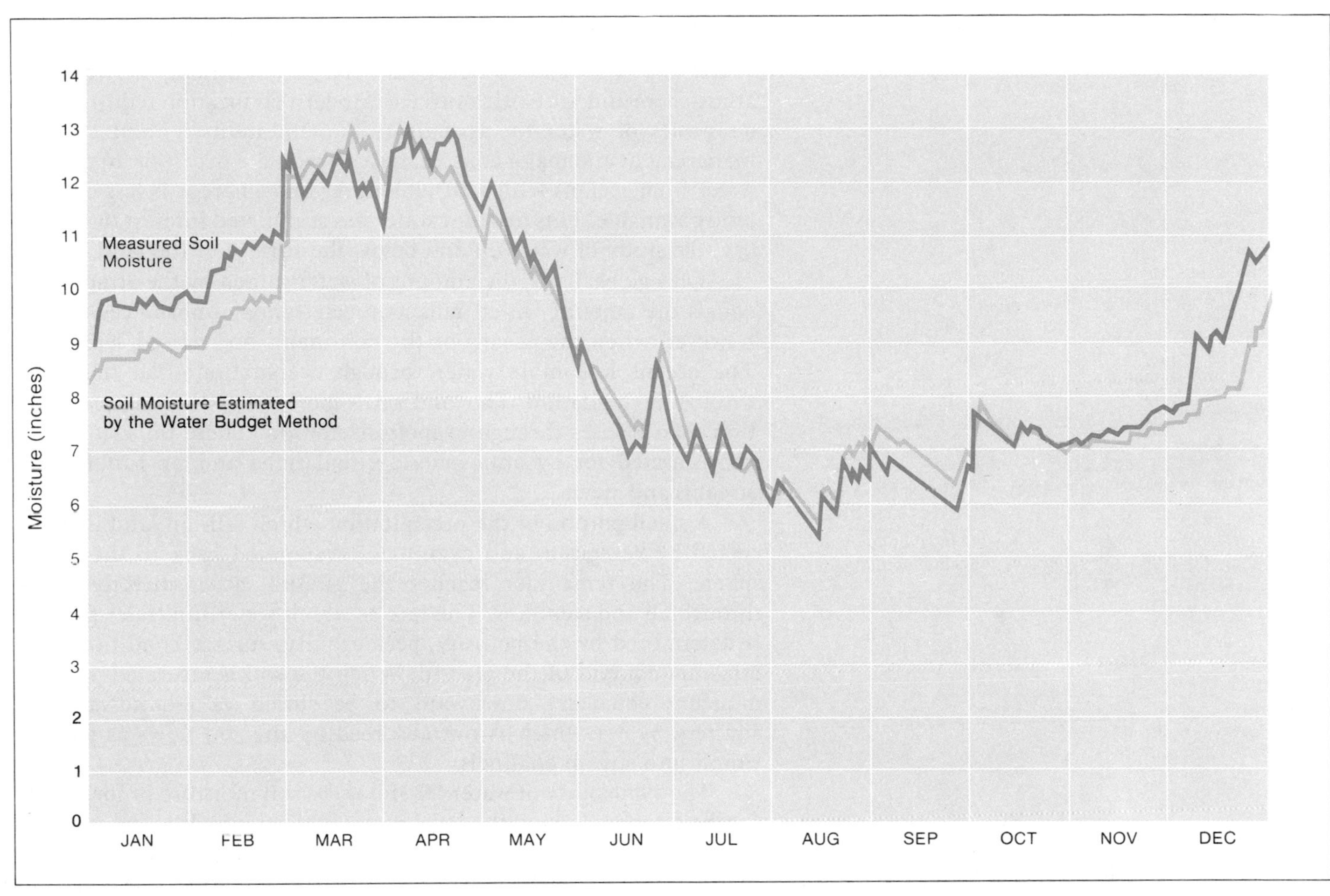

Figure 6.15 This graph compares the amount of stored soil moisture estimated by the daily water budget method to the measured amount. The excellent agreement indicates the suitability of the local water budget approach to agricultural problems.

The variation in soil moisture through the year reflects seasonal conditions. The amount of stored soil moisture is maximum in late winter and spring, when precipitation is heavy and transpiration is small due to lack of plant cover. Soil moisture decreases in April and May and reaches a minimum during summer, when precipitation is low and high temperatures and dense plant cover cause evapotranspiration to be high. The local peaks in the soil moisture are caused by rapid recharge during rainstorms. Can you explain why the average level of soil moisture increases in late autumn? (After Thornthwaite and Mather, 1955)

United States, there is an average of only about one rain gauge for every 500 sq km, or 200 sq miles; all of the official rain gauges could fit between the 40-yard lines of one standard football field! Because storm rainfall is often very localized, the rain gauge network rarely can provide accurate precipitation data for an entire drainage basin.

Despite some deficiencies, the water budget model is a valuable tool for a wide range of objectives. In more complex form, it can be utilized for research in fields ranging from agriculture to flood hydrology. Different water budget components can be applied to particular objectives or relationships. For example, evapotranspiration is related to plant growth and even the yield of certain crops; deficit is an index of irrigation needs for some crops; and surplus represents the water available to recharge groundwater supplies or to enter stream channels as runoff. Calculations of surplus in various locations have revealed that streamflow, or water yield, is significantly affected by land use and the vegetation cover of a drainage basin: for the same precipitation input, urban areas and some croplands produce the largest water yields, while heavily forested regions contribute the smallest proportions of precipitation as streamflow over the year.

SUMMARY

Water is a renewable resource that travels continually between the atmosphere and the earth's surface. Modern civilization requires large quantities of water for agriculture and for industry, and efficient management and major construction projects are necessary to transfer water from regions with a surplus to regions where it is less conveniently available. This need for water has stimulated interest in hydrology, the study of water on and below the surface of the land.

On a global basis, the amount of water gained by the atmosphere equals the amount which falls as precipitation; imbalances appear, however, when the oceans or the continents are viewed separately. The oceans lose more water through evaporation than they gain through precipitation. The land gains more water through precipitation than it loses through evapotranspiration. These imbalances are compensated for by atmospheric circulations and by runoff from streams and rivers.

A small portion of the precipitation which falls on land is intercepted by vegetation and eventually evaporated back to the atmosphere. The remainder reaches the ground either directly or as throughfall and stemflow. The rate at which water infiltrates the soil is determined by the porosity, permeability, surface condition, and moisture content of the ground. When the soil is saturated, surplus moisture percolates downward to be stored as groundwater in aquifers. Water which is not absorbed by the soil flows as surface runoff into stream channels.

The availability of water for storage as soil moisture or for runoff can be estimated using the local water budget, which is the regional version of the hydrologic cycle. Potential evapotranspiration represents an estimate of the maximum rates of evaporation and transpiration from a vegetation-covered landscape with no water shortage. The water budget method compares potential evapotranspiration, which is a measure of the climatic demand for water, with precipitation and soil moisture reserves. The water budget model is an extremely useful tool which can be employed in a variety of situations ranging from agriculture to roadway engineering to the study of floods.

In most midlatitude locations, the precipitation supply from year to year is much more variable than the energy income. Water-resource managers must cope with large variations in the surplus water available for groundwater and streamflow. In the following chapters, we will use the water budget from time to time to assess the availability of water for various purposes and processes in different climates. In Chapter 7, the water budget will be reformulated to evaluate global differences in climatic elements.

Review Questions

1. Discuss some of the environmental problems which the water budget model can be used to study.
2. How does the pattern and timing of rainfall, as distinct from quantity, affect the amount of moisture absorbed by the ground?
3. How does land use affect infiltration and surface runoff?
4. Compare field capacity, wilting point, and available soil moisture for soils of different textures.
5. What is groundwater, and how is it recharged? What is the water table?
6. Explain the terms "runoff" and "streamflow."
7. Describe the operation of an evapotranspirometer.
8. Compare potential evapotranspiration (*PE*) and actual evapotranspiration (*AE*).
9. What is meant by the term "deficit" in regard to the local water budget, and what is its significance for irrigation procedures?
10. How is the surplus (*S*) of the local water budget related to a hydrograph of a river in a humid region?

Further Reading

Chorley, Richard J., ed. *Water, Earth, and Man.* London: Methuen & Co.,1969. (588 pp.) This unusual book is a well-organized collection of essays on various aspects of the hydrologic cycle and their applications in hydrology, geomorphology, and socio-economic geography. Many of the contributors are British, giving the book a perspective from the British Isles.

Kazmann, Raphael G. *Modern Hydrology.* 2nd ed. New York: Harper & Row, 1972. (365 pp.) Emphasis in this work is on critical evaluation of hydrologic variables for water-resource management.

Leopold, Luna B. *Water: A Primer.* San Francisco: W. H. Freeman & Co., 1974. (172 pp.) This excellent little book stresses the interrelationships between hydrologic principles and environmental responses.

Mather, John R. *Climatology: Fundamentals and Applications.* New York: McGraw-Hill Book Co., 1974. (412 pp.) Chapters 3 through 7 focus on the water budget and its applications.

Muller, Robert A. "Water Balance Evaluation of Effects of Subdivisions on Water Yield in Middlesex County, New Jersey." *Proc. Assoc. of Amer. Geog.* (1969): 121–126. This brief article illustrates the utilization of the water budget model for evaluation of the effects of land use and land cover on streamflow and water yield.

Thornthwaite, C. W., and **John R. Mather.** "The Water Balance." *Publ. in Climatology,* Vol. 8 (1955):1–86. This is a summary of the early water budget research and applications by Thornthwaite and his colleagues at the Laboratory of Climatology.

———, and **John R. Mather.** "Instructions and Tables for Computing Potential Evapotranspiration and the Water Balance." *Publ. in Climatology,* Vol. 10, 3 (1957): 185–311. This instruction manual provides explanations and tables for calculation of daily and monthly water budgets using the decreasing availability model of soil moisture depletion.

Ward, R. C. *Principles of Hydrology.* 2nd ed. London: McGraw-Hill Book Co., 1975. (367 pp.) Various components of the hydrologic cycle and the water budget are treated systematically. This text incorporates a geographic perspective, and it also includes an extensive bibliography organized by topic.

Case Study:

The California Water Plan

The management of fresh water is often a controversial subject. According to one view, there is plenty of water on the earth; it simply needs to be redistributed. Others argue that humans should not plan their activities where there is a shortage of water; they should live where water is naturally available. However, the fact is that man has been moving water about for thousands of years, turning barren earth into fertile farmland.

(below left) *This simplified map shows the existing and planned channels of the California state water project. Now under construction, the project will extend nearly the length of California, bringing water from the north to the more densely populated southern portion.*

(below right) *In this photograph, a portion of the California Aqueduct is shown winding through the San Joaquin Valley.*

A case to consider is southern California, which has become greener and more productive by the importation of water. In the early 1900s, it became clear that the growth of Los Angeles—then a small coastal town—would be stimulated by irrigating the surrounding land. For a source of water, Los Angeles turned to the Owens Valley, 400 km (250 miles) to the north at the eastern foot of Mount Whitney. The farmers of the Owens Valley had hoped for the construction of an irrigation system along their own valley, but the rights to the water were purchased by Los Angeles. One reason Los Angeles could buy the water was that the climate in the Owens Valley could not support the highly profitable citrus groves that grow well around Los Angeles. Southern California has a subtropical climate that is conducive to productive agriculture if adequate water is available.

The Los Angeles Aqueduct from the Owens Valley to Los Angeles was completed in 1913. Soon more water was needed, however, and aqueducts from the Colorado River, to the southeast, were built to supply Los Angeles and San Diego. Now more than 100 cities rely on water from the Colorado—a supply that has recently been supplemented by water provided by a state project. Rather than continuing to rely on local and regional initiatives, the state of California has prepared a water resources development plan for the entire state.

The California Water Plan, known earlier as the Feather River Project, is founded on the recognition that

most of the state's population and irrigated land are found in the south, whereas most of the water is in the north. The plan details the construction of the aqueducts, canals, dams, reservoirs, and power stations needed to transport water from the north to the south. The project begins at massive Oroville Dam on the Feather River north of Sacramento. The principal source of water for the present stage of the plan is the Sacramento delta, 100 km (60 miles) east of San Francisco. The delta is formed by a confluence of streams from the west slope of the Sierra Nevada as they drain toward San Francisco Bay through the only opening in the coastal range.

A major project in the plan, Perris Dam, has been completed. The dam is located southeast of Los Angeles and is one part of the effort to move an immense amount of water from one area of the state to another. Perris Lake traps water that has flowed from the Oroville Dam more than 1,000 km (600 miles) away. Never before has one of man's projects sent so much water flowing so far. Some sections of the concrete-lined aqueducts are as wide as a major river. At one point in its journey southward, the water is lifted nearly 600 meters (2,000 ft) through the Tehachapi Mountains—a record lift for so much water. Five huge dam-fed power plants provide the energy to lift the water in stages from sea level at the delta to its highest point, the Tehachapi crossing.

In addition to supplying water for southern regions, the aqueducts provide irrigation for the dusty, windblown San Joaquin Valley. Eventually the new land made available for agriculture will amount to 1 million acres.

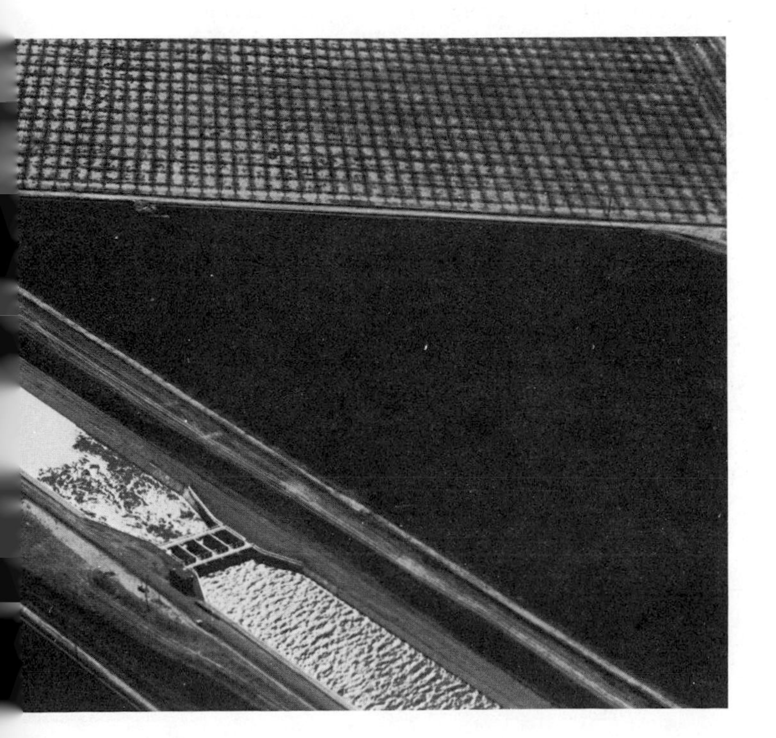

The water development program, which has spanned some 30 years, has been one of the most controversial programs ever undertaken in California. Adoption of the plan instigated an angry sectional feud between the moist "north" and the dry "south." Southern California was accused of trying to steal northern water, but the population concentration in the southern part of the state carried the vote and the issue was approved. Environmentalists still claim that the project has scarred the countryside irreparably, upsetting natural balances of streams, estuaries, vegetation, and wildlife. It is also argued that providing water will promote population growth, which will lead to further urbanization and land development. For instance, perhaps the availability of water in Los Angeles will lead to a larger population, more congestion, and more smog.

In such a controversy, sound arguments come from both sides. Proponents of the water plan say that a rapid population increase is *not* expected. They also contend that a canal can be built from the Sacramento delta to provide water to maintain the estuarine area near San Francisco Bay.

Many of the questions and controversies over the water plan center on whether the water is really needed. Ninety percent of the water used in southern California is for irrigation; there is evidently abundant water for domestic and industrial use. If the water used for irrigation were cut to 80 percent, the amount available for other uses would double from 10 to 20 percent. However, if the irrigation supply were reduced, what effect would that have on agriculture and food supplies for the entire country? California is now one of the nation's most productive agricultural areas.

The allocation of water resources is a matter of economics as well as a matter of technology. Water demand is, in reality, often a demand for water at a sufficiently low cost to make irrigation profitable. The emphasis on higher priced specialty crops, such as fruit and nuts, in California agriculture reflects the need for profitable crops to offset the cost of irrigation water. The California water project has been criticized for using public funds to increase the value of privately held farmland. Furthermore, technological advances in desalination plants may give a new dimension—unforeseen when the water plan was devised—to the problem of water resources.

Although more aqueducts, canals, and pumping plants are planned for the late 1970s and the 1980s, it is uncertain when and if they will be completed. The pace of the water plan is influenced by projected water needs, the development of new technology, and ecological impacts.

Sky Above Clouds II by Georgia O'Keeffe. (Collection of Mrs. Potter Palmer, Lake Forest, Illinois; with permission of Georgia O'Keeffe)

Dynamic processes in the atmosphere cause the weather patterns that give each region of the earth a distinctive climate. Climate can be viewed in two ways: as the delivery of energy and moisture by the atmosphere or as the interaction of energy and moisture at the earth's surface with systems of soils, vegetation, and landforms.

SEVEN

Global Systems of Climate

THE SOUTHWESTERN PART OF the United States is a region where potential evapotranspiration far exceeds precipitation. The annual deficit of moisture at Phoenix, Arizona, for example, is greater than 102 cm (40 in.) in an average year, and extensive irrigation is needed to support the growth of crops. A plan has been advanced recently to increase radically precipitation in the Southwest, so that moist woodlands would grow where only cactus and desert shrubs can exist today. The idea is to build a stupendous dam across the Bering Strait between Siberia and Alaska. The dam would have to be 100 km (60 miles) long, far more massive than any dam existing on the earth.

How could a dam at the Arctic Circle affect weather in the southwestern United States 5,000 km (3,000 miles) away? If such a dam could be built, it would block the circulation of cold arctic waters into the Pacific Ocean. Advocates of the plan anticipate that warming the Pacific Ocean would increase rainfall in the Southwest, and the Russian proponents hope that it would indirectly improve the cold, dry climates over much of Siberia as well. But meteorologists opposed to the scheme argue that it is impossible to predict the full consequences of such a drastic global scheme to modify climate. Changing the temperature of the Pacific Ocean would change the temperature of the air above the Pacific basin and affect the pattern of the general circulation of the atmosphere. And because the general circulation of the atmosphere exerts controls on the climates of the earth, the effects of such a dam could be far-reaching indeed.

Climate is often said to be the average condition of the atmosphere near the earth's surface over a period of years, and it is generally described in terms of such meteorological conditions as temperature, precipitation, wind, degree of cloudiness, and so forth. But most geographers have a more comprehensive perspective of climate and include such surface conditions as evapotranspiration and soil moisture storage as well. The climate of a region has profound influence on its soils, vegetation, and landforms; it determines the suitability of a region for agriculture, and it influences the processes that shape the landscape. Understanding the global distribution of the earth's climatic regions is the key to understanding many of the characteristics of those regions. This chapter discusses how solar energy input and the circulation of the atmosphere determine the climate of a region. Two widely used methods of climate classification are described, and the local water budget is used to show why climate is the active factor controlling other systems.

CLASSIFICATION OF CLIMATE

The atmosphere delivers energy in the form of warm air and water vapor around the world. The general circulation of the atmosphere is an important factor in determining the variety and timing of weather events that can occur at each global location. The comparative stability of the general circulation produces characteristic regional weather patterns that make each region distinctive. The winter vacationer in Florida, for example, has every reason to expect a succession of warm, sunny days during his stay. Because the distribution of energy and moisture has implications for other systems in addition to weather, it is useful to classify the regions of the earth according to the type of climate experienced. Climate classification helps reveal the general patterns in other environmental systems and serves to organize a wealth of information about the earth's surface.

To be most useful, a classification scheme should have enough categories and distinctions to account for a system's principal features, but not so many that a general understanding is lost in a welter of detail. The classification schemes discussed later in this chapter arrange the earth's climates according to about five main categories and one or two dozen subcategories. No two places on earth have exactly the same climate, however, so any classification scheme inevitably submerges individual details. Because climatic characteristics do not change abruptly, the boundaries of climatic regions should usually be interpreted as broad zones of transition rather than as sharp divisions. On a global scale, however, the transition zones are frequently narrow compared to the size of the climatic regions; thus it is meaningful to speak of boundaries if the concept is not pushed too far.

Measures of Climate

Classification systems for climates generally employ one or more measures, or *climatic indexes*, to distinguish one climatic type from another, according to rules devised by the inventor of the system. What constitutes an appropriate index of climate depends on the intended use of the classification system. Someone interested in constructing buildings would want to emphasize wind speed and direction to assess the possibility of wind damage. Someone concerned with transportation would be especially interested in the frequency of fog and icing conditions in various regions. The timing and duration of temperature inversions must be known to assess a region's potential for air pollution.

Climate classification systems in physical geography generally employ temperature and precipitation as indexes of climate because these are easily measured and clearly influence how vegetation is distributed and what potential there is for agriculture in a particular region. Indexes of wind, temperature inversions, or other atmospheric phenomena are not ordinarily used because too many parameters can encumber a system and because for many regions the only available data are temperature and precipitation records. Nevertheless, temperature is only an indirect measure of energy inputs to systems. Other measures, such as potential evapotranspiration, may be more suitable indexes of climate, as discussed later in this chapter.

Time and Space Scales of Climate

Weather conditions change constantly. The movement of air masses into a region can change the local air temperature significantly within a few hours. Weather conditions also vary markedly from place to place in the same local region. Passing storm clouds can bring heavy rain to isolated sections while other locations nearby remain dry. Given that the principal indexes of climate are temperature and precipitation, geographers must still decide when, where, and how often to analyze these factors. The following example shows how important the time scale—the "when"—of measurement is.

The average annual temperature at Aberdeen, Scotland, is 8.3°C (46.9°F), and at Chicago, Illinois, it is 9.7°C (49.5°F), less than 2°C (3°F) higher. The average annual precipitation at both locations is the same, 84 cm (33 in.). Despite these similarities in annual averages, the climates at the two locations differ significantly. The monthly averages give a much more accurate indication of the climate in each city. During January, the coldest month, Aberdeen averages 3.9°C (39°F) while Chicago averages –3.9°C (25°F), a difference of nearly 8°C (14°F). During July, the warmest month, Aberdeen averages 14°C (57°F) and Chicago averages 23°C (73°F), which is 9°C (16°F) higher. The variation of temperature at the two places is markedly different through the year; the climate at Chicago is more extreme than the climate at Aberdeen. The example shows that, in general, the timing of temperature and precipitation, as well as the average annual values, is important in characterizing the climate of a region.

The areal scale (size) of climatic regions is also important in classification systems. Figure 7.1 shows temperature readings made at 11 P.M. at various points in a backyard located in Baton Rouge, Louisiana. The yard has its own microclimate, with a wide range of temperatures over a small area. The microclimate of the yard is too small to appear on even a county-wide climate survey, but climatic conditions in different parts of the yard are important to someone trying to protect subtropical vegetation from a hard freeze.

The best areal scale for a climate classification system depends on the intended use of the system. On a global scale, most of the West Coast of the United States would be considered to be in a single climatic region. But someone concerned with water resources on the West Coast would need to see a much more detailed picture that revealed the gradation from the abundant rainfall in the Pacific Northwest to the dryness of southern California.

This chapter focuses on climate classifications on a global scale to give an idea of the general distribution of climate on the earth. Excessive detail is not desirable on such a large scale. The Köppen system of global climate classification discussed later in this chapter assigns New York City and Nashville, Tennessee, to the same climatic region. On a nationwide scale, the climates of the two cities are obviously different, but on a global scale, these climates resemble each other more than they do the climates of tropical rainforests or Asian steppes.

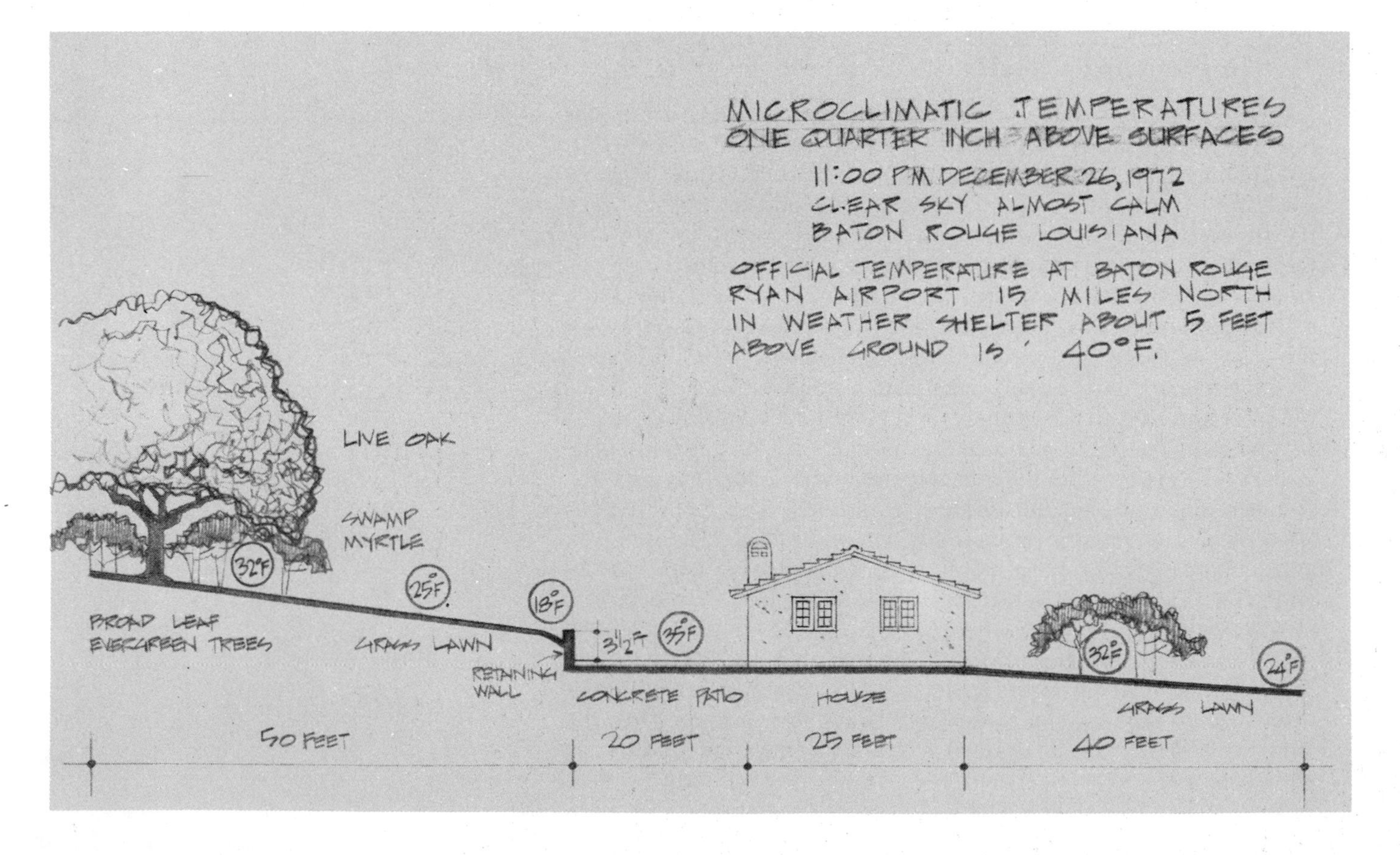

Figure 7.1 The nighttime temperatures in this Louisiana backyard show a large degree of variation from place to place because of such factors as local movements of air, differences in the amount of heat stored in the ground and buildings, and differences in cooling rates. Climate measured on such a small scale is called the *microclimate* of the area. Such details are lost when the climates of large regions are classified into broad categories. (After Muller, 1973)

ONE VIEW OF CLIMATE: THE DELIVERY OF ENERGY AND MOISTURE

The delivery of energy and moisture to different regions of the earth follows global patterns imposed by the distribution of solar radiant energy and the general circulation of the atmosphere. To understand how solar energy input and the general circulation control the gross features of the earth's climates, we can analyze the climate regions of an idealized hypothetical continent.

Distribution of Climatic Regions on a Hypothetical Continent

The hypothetical continent shown in Figure 7.2, p. 184, is flat and featureless with no interior mountains, seas, or gulfs. Nevertheless, it embodies many of the features of the actual continents: it is surrounded by oceans; it is broad at the north and extends to high latitudes, like North America and Asia; and it tapers toward the south and ends near latitude 60°S, like South America. Also like actual continents, it is divided into several different climatic regions. This climatic division of the hypothetical continent is determined by certain variables—solar energy input, global patterns of pressure and winds, and the seasonal meridional migration of the general circulation.

The average distribution of temperature on the hypothetical continent is directly related to the pattern of incoming solar radiation. Tropical and subtropical regions, for example, receive the greatest annual amounts of solar energy and tend to be the warmest year-round (see Figure 3.17, p. 79). Regions at higher latitudes receive less solar energy and have lower average temperatures. In addition to these annual temperature differences, there are variations in the range of temperature over the year according to latitude. The tropical regions receive a nearly uniform seasonal input of solar energy and, as a result, maintain comparatively constant temperatures throughout the year. But because of the tilt of the earth's axis, the higher latitudes receive a much greater amount of solar energy in summer than in winter. As a result, the difference between average summer and average winter temperatures on the hypothetical continent tends to be greater the farther regions are from the equator.

Air masses, wind patterns, and pressure systems control the distribution of atmospheric moisture and precipitation to the various climatic regions on the hypothetical continent. Air masses reaching the continent after crossing the ocean are moisture-laden as a result of evaporation. This humid maritime air may then become a source of precipitation if it is cooled sufficiently to cause condensation. Cooling generally takes place when the air is forced to rise, either by convergence into a center of low pressure or by undergoing initial heating at the earth's surface. Such movements of air do not occur haphazardly, but rather follow regular patterns in accordance with the general atmospheric circulation. The distribution of moisture over the hypothetical continent takes place in response to these patterns.

The intertropical convergence zone and the equatorial trough of

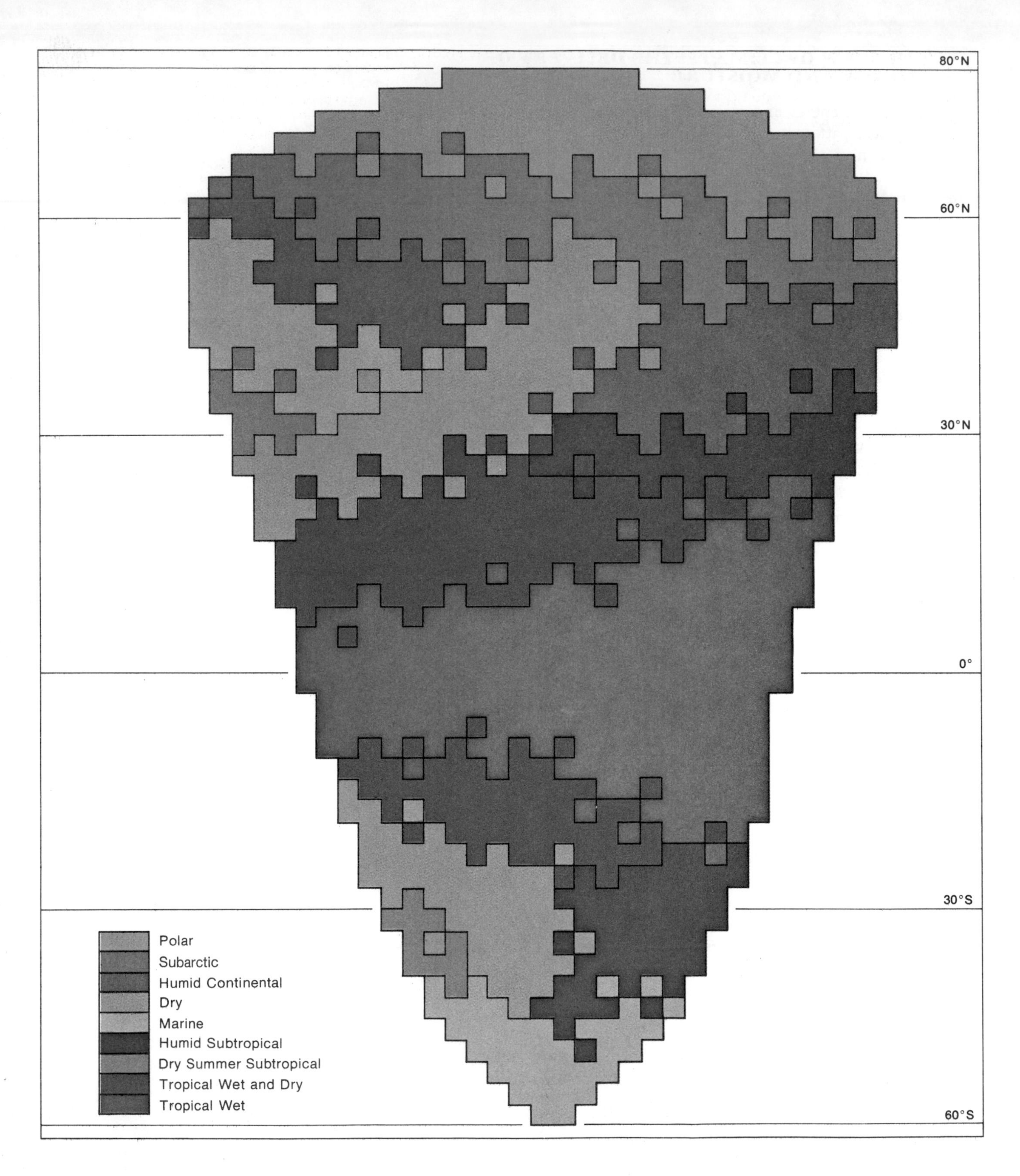
80°N
60°N
30°N
0°
30°S
60°S
Polar
Subarctic
Humid Continental
Dry
Marine
Humid Subtropical
Dry Summer Subtropical
Tropical Wet and Dry
Tropical Wet

low pressure occupy the region near the equator, as shown in Figure 5.8, pp. 128–129. The easterly trade winds that converge into this trough are frequently moist due to evaporation from the oceans, and as the converging air rises up into the circulation of the Hadley cells, condensation and precipitation occur. Precipitation tends to be frequent year-round near the equator; thus the climatic region there is classified as *tropical wet*. Because the trades sweep onshore and across the eastern side of the continent between about 15°N and S, the eastern side tends to receive more moisture than the western side at these latitudes.

The *dry* climates near 30°N and S are located in the zone of subtropical highs, where dry air from the upper atmosphere subsides almost to the surface. As it subsides, the air warms adiabatically, resulting in a persistent inversion and stable environmental lapse rates at middle altitudes of the atmosphere. Despite intensive surface heating, the dry air and stability aloft combine to inhibit most precipitation processes. In both hemispheres, dry regions extend poleward across the interior of the continent. These midlatitude dry climates are not directly associated with the subtropical highs but occur because they are far from oceanic moisture sources. Weak midlatitude cyclones are unable to produce much rainfall from the relatively dry air masses here. The midlatitude dry regions are much cooler during winter than their subtropical counterparts.

Along the western margins of the continent in the subtropical latitudes, the delivery of moisture to the *dry summer subtropical* and *marine* climatic regions is governed mostly by subtropical highs centered over the oceans to the west. In winter, the cells are at lower latitudes in each hemisphere, allowing midlatitude cyclones to bring moist maritime air and precipitation to both the subtropical and marine climatic regions. In summer, the subsidence associated with the high-pressure cells moves poleward and stabilizes the atmosphere over the subtropical climatic region. Cold ocean currents offshore further strengthen the inversion and stability (see Figure 5.20, p. 145). The subtropical climatic region has a wet winter and a very dry summer, while the marine climatic region farther poleward receives moisture throughout the year.

On the eastern side of the continent at subtropical latitudes, atmospheric moisture is available year-round for precipitation in the *humid subtropical* regions. During summer, the subsidence associated with the subtropical highs is less well developed and persistent than on the western side of the continent. The oceanic circulation brings warm water to the eastern coasts at these latitudes, in contrast to the cold water along western coasts. Maritime tropical air formed over the warm ocean off the eastern coasts is swept around the western margins of the subtropical highs and over the coastlines toward the interior. Summer showers and thunderstorms occur frequently in the unstable, warm, humid air. Precipitation is also heavy during winter, when it is the result of midlatitude cyclones along the polar front.

Between the tropical wet zone astride the equator and the drier subtropical areas are the *tropical wet and dry* climatic regions. In the

Figure 7.2 (opposite) The climatic regions represented on this hypothetical continent are determined by the input of solar radiant energy and the delivery of air masses and moisture by the atmospheric systems. The regions are depicted with overlapping boundaries to suggest that climate changes gradually from one location to another across the surface of the earth. The distribution of the climates should be studied in conjunction with the text and with the general circulation maps in Figure 5.8, pp. 128–129.

northern hemisphere summer, the intertropical convergence zone and the belt of rising moist air shift to the north, bringing precipitation to the tropical wet and dry climatic region in the northern hemisphere and relative drought to its counterpart in the southern hemisphere. In the winter low-sun season, the ITC migrates to the south and the situation reverses. The subtropical highs in both hemispheres also play a role: as the ITC shifts equatorward, the highs move in behind, and the subsiding air and temperature inversions effectively shut off precipitation processes for many months.

In the higher middle latitudes of the northern hemisphere where the continent is broad, there are *humid continental* climatic regions both east and west, separated by the interior dry climate. Radiational heating of the land in summer and cooling in winter lead to marked seasonal temperature differences; summers are warm to hot and winters cool to cold. Because the moderating effects of the oceans on temperature are minimal, these climatic regions are characterized by very large annual temperature ranges, a feature known as *continentality.* The westerly wind flow, on the average, carries the continentality effect from the interior to the east coast and at the same time prevents the extension of continental climates to the west coast. In the southern hemisphere, the continent is narrow and the moderating effects of the oceans greater; thus there is no development of humid continental climate there. During winter and summer, precipitation is associated mainly with midlatitude cyclones, but precipitation over most areas is greater during summer, when the warmer air masses are supplied with abundant moisture. Midlatitude cyclones tend to become more intense during winter and spring, but the stormy periods are interspersed between stretches of cold, fair weather.

In the northern hemisphere, the *subarctic* climatic region is poleward of the humid continental climates and extends along a northwest-southeast axis from coast to coast. Winter weather is dominated by the polar high displaced southward over the colder continent: the air is always cold, and the precipitation which falls as snow is light except near the coasts. Most precipitation falls during the mild summer, when it is associated with midlatitude cyclones.

On the hypothetical continent, *polar* climates are found only in the northern hemisphere where the wide continent extends poleward of latitude 60°N. This region is dominated by the polar high much of the year and also strongly influenced by nearby cold and often ice-covered ocean surfaces. Winters are cold with meager precipitation. Much of the annual precipitation falls as rain during the short summer, when temperatures are normally above freezing over all but extensive ice-covered areas.

Distribution of Climatic Regions on the Earth

A global map of the earth's climates indicates that many of the climatic regions described for the hypothetical continent are apparent on the actual continents (see Figure 7.3, pp. 188–189). Near the equator are tropical wet regions, with tropical wet and dry regions immediately to the north and south. Africa and Australia show the expected dry regions near latitude 30°. Dry summer subtropical climates are located on the western sides of the continents in subtropical latitudes: California and Oregon, central Chile, much of the Mediterranean coastline, the tip of South Africa, and small areas of western and southern Australia. The midlatitude west coasts of North and South America and of Europe show the expected marine climates, and the interiors of the United States and Europe are humid continental climatic regions. Subarctic and polar climates occupy the northern portions of Asia and North America.

The effects of continents, oceans, and mountain ranges on the distribution of climates on the earth are also apparent on the map. The mountain ranges in the high latitudes of western North and South America, for example, prevent the west coast marine climates from extending as far inland as they do in Europe. As a further consequence, the regions to the east of the west coast mountain ranges are dry because moist air is blocked from reaching them. Similarly, the west coast ranges in the low latitudes of South America block the easterly trades, which prevents moisture from reaching the west coast between the equator and latitude 30°S, making the coastal regions there dry. The monsoon circulation greatly influences the climatic regions of India and Southeast Asia. The complex causes of the monsoons are not fully understood, but the high-pressure cell over Siberia in winter, low pressure over the Indus Valley and Tibet in summer, the presence of the Himalaya Mountains, and the seasonal changes in the position of the intertropical convergence zone and jet streams over Asia all play a role.

ANOTHER VIEW OF CLIMATE: ENERGY AND MOISTURE INTERACTIONS AT THE SURFACE

The preceding view of climate emphasizes the atmosphere as a delivery system and minimizes the interaction of energy and moisture with systems on the surface of the earth. However, systems on the earth's surface share the energy and moisture that the atmosphere delivers; plant growth, for example, depends on the amount of soil moisture available to vegetation and not directly on total precipitation. So to bring out the relationships between climate and other systems, the view of climate can be extended to include the interactions of delivered energy and moisture with the systems on the earth's surface.

Penck's Three Functional Realms of Climate

In the late nineteenth century, the German geographer Albrecht Penck recognized that in terms of the interaction of climate with other systems, three distinct *realms of climate* can be designated. The criteria that distinguish Penck's three natural realms of climate can be stated in terms of potential evapotranspiration. The *frozen* realm embraces regions where potential evapotranspiration equals zero. Plants cannot transpire when the temperature is so low; thus the realm in which potential evapotranspiration equals zero is characterized by low temperatures. The *dry* realm includes all regions where

Figure 7.3 This global map classifies climates according to solar radiation input and atmospheric delivery of energy and moisture to the earth's surface. Compare it with the distribution of climates on the hypothetical continent shown in Figure 7.2. The landmasses of the high northern latitudes possess polar and subarctic climates like the hypothetical continent because of the low input of solar radiant energy. Like the hypothetical continent, Africa and South America possess tropical wet climate regions flanked by tropical wet and dry climate regions. Africa and Australia exhibit dry climate regions near latitude 30°. The effect of mountains on climate distributions is apparent in the way the tropical wet region of South America does not cross the Andes to the west coast.

The distribution of continents and oceans on the earth affects the movements of air masses. The southeastern United States, for example, possesses a humid subtropical climate instead of a humid continental climate because of the northward movement of moist tropical air from the Gulf of Mexico and the Caribbean. The United States and much of Europe lie in the belt of westerly winds. Parts of western Europe and the west coast of the United States possess marine climates. In Europe the marine climate gradually merges into a humid continental climate region, but in the United States the west winds from the Pacific Ocean are blocked by mountains. The region immediately inland from the mountains is desert. The north-central and northeastern United States possess a humid continental climate, however, partly because of cold winds from the north in winter and warm winds from the south in summer. (After Finch and Trewartha, 1949)

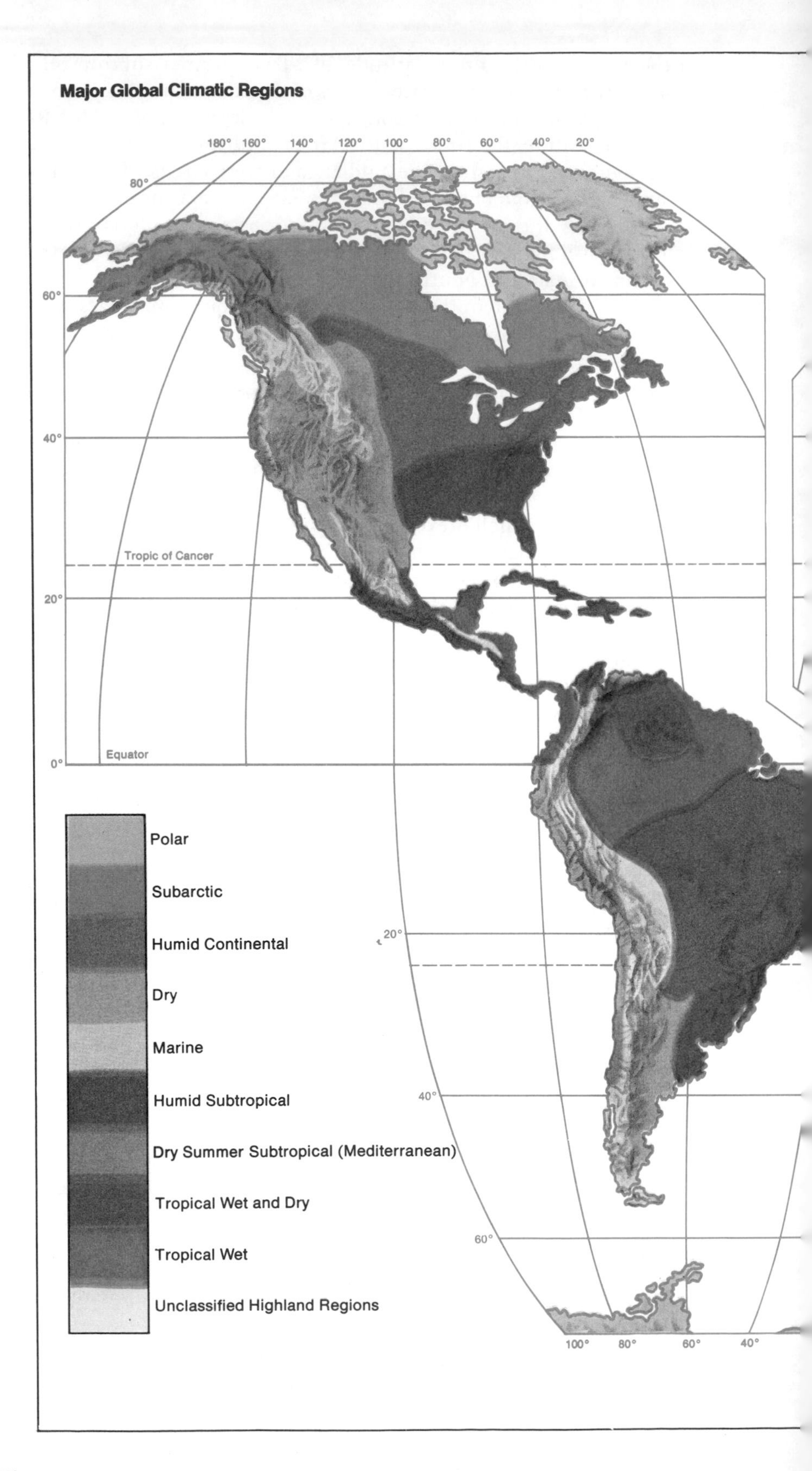

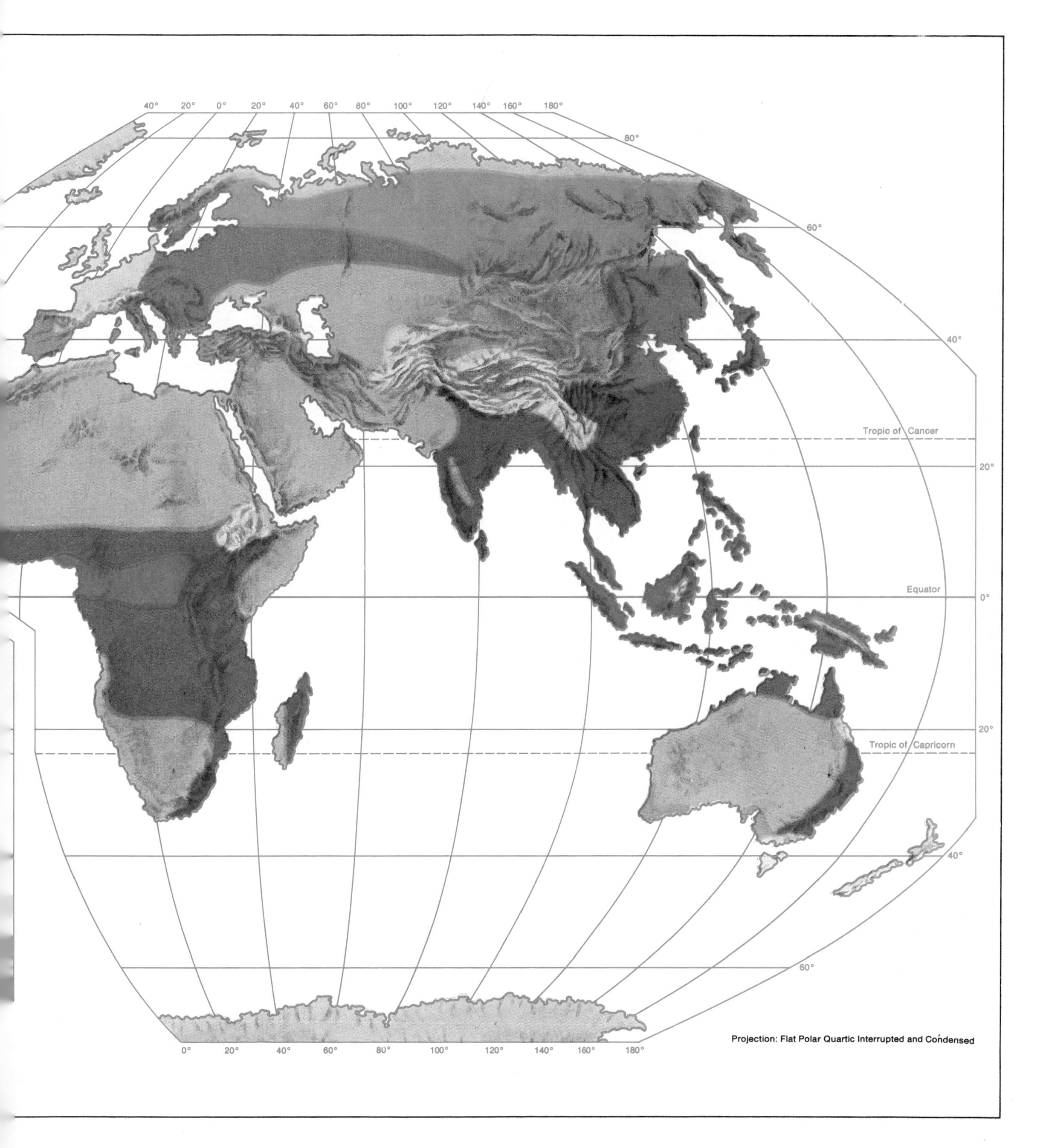
40°
20°
0°
20°
40°
60°
80°
100°
120°
140°
160°
180°
80°
60°
40°
Tropic of Cancer
20°
Equator
0°
20°
Tropic of Capricorn
40°
60°
0°
20°
40°
60°
80°
100°
120°
140°
160°
180°
Projection: Flat Polar Quartic Interrupted and Condensed

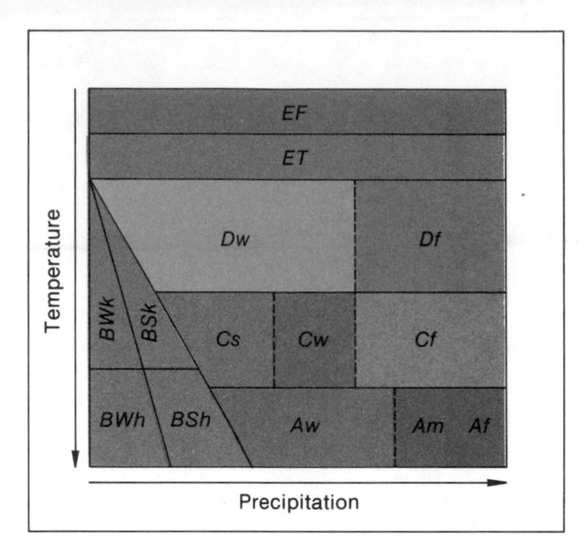

Figure 7.4. This schematic diagram represents the principal climatic regions of the Köppen classification system in terms of average annual temperature and precipitation. The *E* climates, near the top of the figure, are the coldest, and the *A* climates, near the bottom, are the warmest. Reading from left to right in the direction of increasing precipitation, the *BW* climates are the driest and the *Af* climates are the wettest. The precipitation scale in particular is highly schematic. In the *Cs* (dry summer) type, for example, a smaller annual precipitation can support forest vegetation because most of the rain falls during winter when *PE* is low. The precipitation scale shows general trends in each temperature zone, but it does not indicate that the annual precipitation in *Df* regions is much less than in *Af* regions.

precipitation is less than potential evapotranspiration. Characteristic of the dry realm are dry grasslands and deserts, where deficiencies of soil moisture are common. The *moist* realm, in which forests are located, includes all regions where precipitation exceeds potential evapotranspiration.

Each of the foregoing realms can be subdivided by imposing criteria connected with energy, most often using some measure involving temperature. However, the resulting climatic types, such as hot-moist, cool-moist, and so forth, prove in practice to be inadequate to characterize the world's climates in sufficient detail, even on a global scale. The dry and moist realms require further division.

Different systems of climate classification divide the dry and moist realms in different ways. Two widely accepted systems that take this perspective of energy-moisture interaction are discussed next.

The Köppen System of Climate Classification

Wladimir Köppen, a German botanist and climatologist, developed and refined a system of climate classification in the early decades of this century. Köppen's system was strongly influenced by the work of the nineteenth-century plant geographers who mapped the vegetation of the world on the basis of extensive field studies.

Köppen's classification employs five principal climatic types, which he labeled *A, B, C, D,* and *E.* The types *A, C,* and *D* are divisions of the moist realm, *B* is the dry realm, and *E* is a polar climatic type that includes the frozen realm. The *B* type is subdivided into the *BW,* or arid desert climatic type, and the *BS,* or semiarid steppe climatic type. Köppen subdivided the *E* climatic type into the *ET,* or tundra type, and the *EF,* or perpetual ice type. His definition of the *ET* climatic type illustrates his use of vegetation as an indicator of climate. The *ET* climatic type is represented by climates in which the average temperature of the warmest month falls between 0°C (32°F) and 10°C (50°F). Such a climate can support limited growth of vegetation, and the 10°C limit for the warmest month corresponds approximately to the poleward boundary of forests. By choosing this limit, Köppen ensured that regions with *ET* climates would be essentially treeless.

Köppen devised definite criteria based on temperature and precipitation to distinguish between principal climatic types. A schematic diagram of his method of division is given in Figure 7.4. The *A* climatic types are hot and moist, the *C* types are warm and moist, and the *D* types are cool and moist. The *B* climatic types, however, include a wide span of temperatures and a range of moisture. The boundary line of the *B* climatic types moves to greater precipitation as the temperature increases. The reason for this shift is that the boundary between the dry realm and the moist realm is intended to fall where precipitation and potential evapotranspiration are equal, and potential evapotranspiration increases with temperature.

The Köppen system was an outstanding achievement for its time, and it still provides an orientation to the distribution of the world's climatic regions. One of the strong points of the Köppen system is its flexible terminology, which is outlined in greater detail in Appendix III. The flexibility of the system comes from the use of additional symbols to describe special features of climatic regions. The symbol *f*, for example, means that adequate precipitation falls in every month of the year. Therefore, the symbol *Af* specifies a tropical wet climate, as in the equatorial rainforests of South America. The symbol *Cf* describes the warm, moist climate of the eastern United States. The symbol *w* means that little precipitation occurs in winter, and *s*, that little precipitation occurs in summer. Additional symbols are added to characterize special conditions, such as frequent fog.

However, Köppen did not possess an accurate means of estimating potential evapotranspiration. He assumed that the transition between the dry *B* climatic type and the moist *A*, *C*, and *D* climatic types occurred at forest boundaries. By studying temperature and precipitation data for typical forest boundaries, Köppen established empirical relations involving temperature and precipitation that could be used to define the boundary of the *B* climatic types. His assumption that the forest boundary is where precipitation equals potential evapotranspiration is incorrect; precipitation exceeds potential evapotranspiration at most forest boundaries. Because his empirical relations for the *B* climatic type boundary were not very precise, he sometimes placed the boundary hundreds of kilometers from the true boundary between the dry and moist realms.

The Köppen system has been criticized because many of its boundaries were developed to correspond to the boundaries of certain vegetation regions. Therefore, the climatic regions defined by the Köppen system usually show some correlation with regions of vegetation, and an evaluation of the relationship between climate and vegetation cannot be made objectively and independently. Another criticism of the system is that the boundary between *Cs* and *Cf* climatic regions, for example, is based solely on precipitation, whereas the boundary of *B* regions depends on both precipitation and temperature in a crude measure of potential evapotranspiration.

Nevertheless, the Köppen system is comprehensive and rather flexible. Most of the climatic types can be related to features of solar energy input and the general atmospheric circulation which produce the seasonal regimes of temperature and precipitation. The terminology has been recognized by most geographers and environmental scientists, and the system has been used in recent years primarily to introduce the regional aspects of global climates. Figure 7.5, pp. 192–193, shows the climates of the world according to Köppen's system, and a few examples of representative landscapes are illustrated in Figure 7.6, pp. 194–195.

The Thornthwaite System of Climate Classification

In an attempt to overcome the deficiencies of the Köppen system, C. Warren Thornthwaite devised in 1948 a method for classifying climates in which climatic types are defined according to water budget evaluations of energy and moisture. In Thornthwaite's classification system, energy is specified by potential evapotranspiration, and mois-

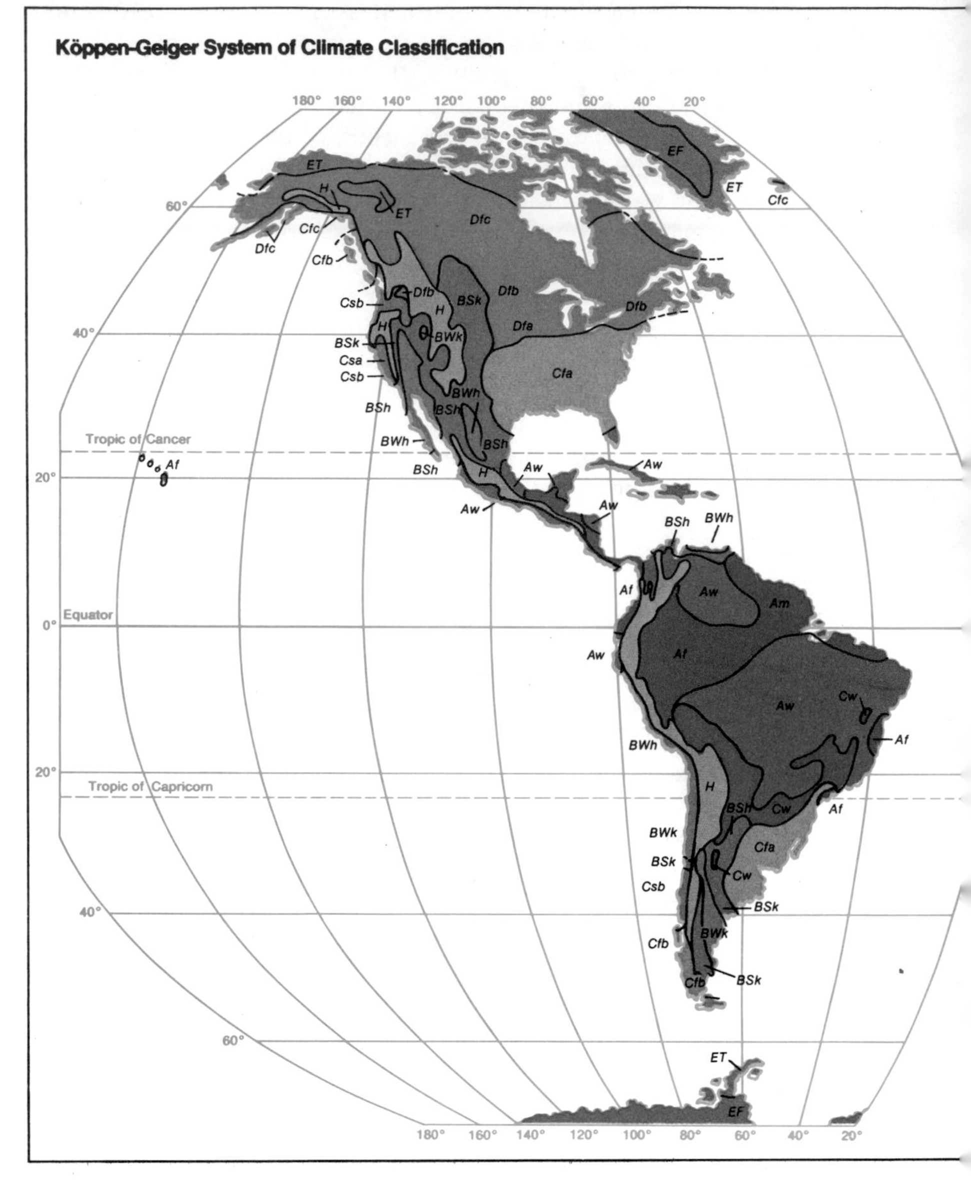

Figure 7.5. This map depicts the broad global classification of climates according to the Köppen system of climatic classification. The definite boundary lines shown between the principal climatic regions are assigned by the Köppen system; in actuality, climatic regions shade gradually into one another on the average. The map of climates according to Köppen differs in an important respect from the map of climates presented earlier in this chapter (Figure 7.3, pp. 188–189). The earlier map shows a classification of climates based on the delivery of energy and moisture. The Köppen system of classification takes into account very general relationships among seasonal temperatures, precipitation, and vegetation. Although Köppen's estimates of evapotranspiration are far less accurate than more recent formulations, his classification of global climates is useful for a historical perspective and for a relatively simple global regionalization. (See Appendix III for a more detailed explanation of Köppen's criteria for climate classification.) (After Köppen-Geiger-Pohl map [1953], Justes Perthes; and Köppen-Geiger in *Erdkunde*, volume 8; and Glenn T. Trewartha, *An Introduction to Climate*, fourth edition. New York: McGraw-Hill, 1968)

ture is specified by a *moisture index*, which is defined in Figure 7.7, p. 196.

Thornthwaite's moisture index depends essentially on the difference between precipitation and potential evapotranspiration. The moisture index has the value −100 when precipitation is 0, and it may exceed +100 where rainfall far exceeds potential evapotranspiration. At the boundary between the dry and moist realms, the moisture index is 0.

Figure 7.8, p. 196, shows schematically how basic climatic types are defined in Thornthwaite's classification system. Each climatic type corresponds to a well-defined range of energy and moisture values, which change proportionately from one type to the next.

One of the principal differences between the Köppen and the Thornthwaite classification systems is in the way the dry and moist

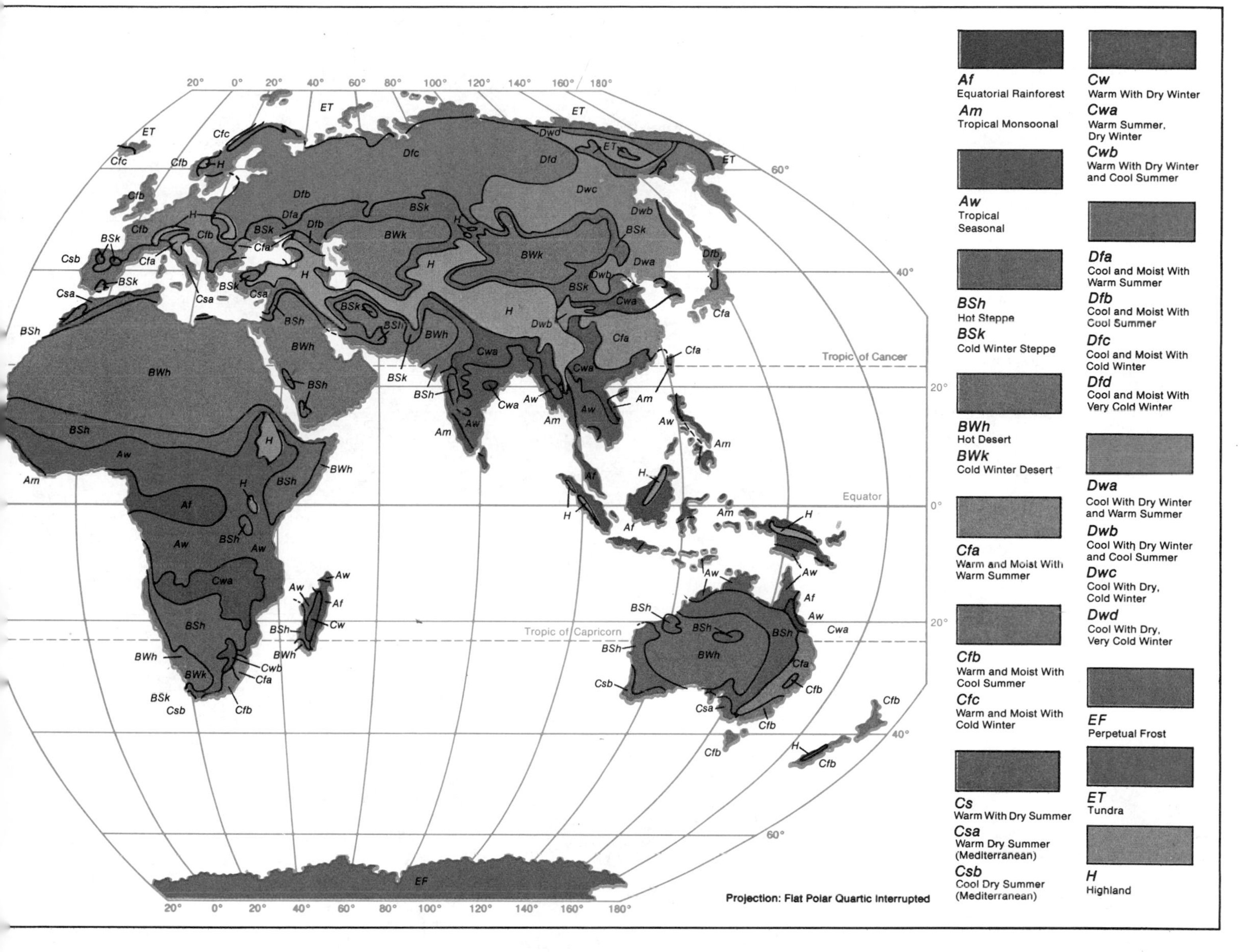

realms are divided. Köppen divided the moist realm into three principal climatic types, *A*, *C*, and *D*, and the dry realm into two types, *BW* and *BS*. Thornthwaite also divided the dry and moist realms into five principal climatic types: two in the moist realm, two in the dry realm, and one, the subhumid, that spans the boundary between the dry and the moist realms. The subhumid climatic type contains the boundary where precipitation and potential evapotranspiration are about equal. When both classification systems are applied to find the boundaries of climatic regions for the United States, the difference between them emerges clearly. For example, the soil and vegetation in the dry grasslands of the central United States are so distinctive that they appear to occupy a separate climatic region between the moister areas to the east and the drier ones to the west. When Köppen's criteria are used, the dry grasslands do not emerge as a distinct climatic region. However,

Figure 7.6 (a) England has a marine climate because of moist westerly winds from the Atlantic Ocean. The climate is moist throughout the year, with moderate temperatures. This early morning mist in Devon, England, formed in a layer of cool air near the wet ground.

(b) Orographic lifting of persistent easterly trade winds creates a cloud canopy over the Hawaiian island of Kauai. These clouds produce frequent precipitation that makes the summit area in the background of this view one of the wettest places on earth.

(c) This snowy cornfield in Iowa exemplifies the seasonal temperature variations of a humid continental climate region. Snow is frequent in Iowa during the cold winter months, but sunshine and high temperatures make the summer an excellent growing season. Moisture is available throughout the year.

(d) Precipitation in arid climates seldom exceeds a few centimeters per year. This oasis in the Sahara depends on groundwater to supply the needs of humans and animals.

(e) The Mediterranean climate of the coastal regions of central California is moist during the winter and fosters the growth of vegetation. During the rainless summer season, the vegetation dries out and brush fires are common. The fires in this view burned for several days in August 1977 in the hills near Oakland, California.

(f) The climate near San Francisco, California, is moderate and supports the growth of subtropical vegetation. A rare occurrence of freezing temperatures during the winter of 1972 killed these eucalyptus trees in the Berkeley Hills, to the east of San Francisco.

(g) In subarctic climate regions, snow and ice may persist through the year. The photograph shows icebergs breaking off, or *calving,* from a glacier in northern Iceland.

$$\text{Moisture Index} = 100\ \frac{P - PE}{PE}$$

$$\text{Moisture Index} = 100\ \frac{S - D}{PE}$$

P = Precipitation

PE = Potential Evapotranspiration

S = Surplus

D = Deficit

Figure 7.7 The value of Thornthwaite's moisture index can be calculated from either of these two equations. The value of the moisture index is positive at places where mean annual *P* is greater than *PE*, and it is negative where *P* is less than *PE*. The moisture index is also positive at places where *S* is greater than *D*. Even though mean annual *P* is greater than *PE*, deficits can occur where there is a strong seasonality of precipitation, such as in dry summer subtropical regions.

Figure 7.8 This schematic diagram represents the principal climatic regions in the Thornthwaite classification system. Potential evapotranspiration is used as a measure of moisture utilization by environmental systems on the earth's surface. The climates run from cool and dry at the upper left corner of the diagram to hot and wet at the lower right. The climates are divided according to orderly ranges of values for potential evapotranspiration and moisture index. In the higher latitudes where energy is in short supply the emphasis is on energy availability. In lower and middle latitudes where large amounts of energy in terms of radiation and heat are normally available, the emphasis is on moisture availability. Five principal moisture regions, *A* through *E*, are recognized for middle and lower latitudes. The subhumid climatic region, *C*, is centered where the moisture index equals 0. (After Carter and Mather, 1966)

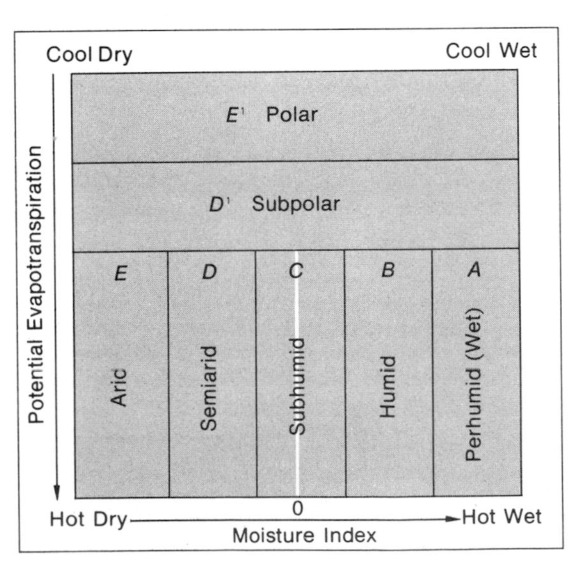

the dry grasslands correspond closely to the subhumid climatic type of Thornthwaite's system. Figure 7.9, p. 197, shows the regions of the United States divided according to Thornthwaite's moisture index alone, and the subhumid region occupies the central United States.

Thornthwaite originally subdivided the seven basic climatic types in Figure 7.8 into 65 combinations of energy and moisture units which were then related to global distributions of vegetation, soils, precipitation, and runoff—the interactions of which help to shape the surface of the continents. Perhaps because it is too complex for most general purposes, the system is not often used to define climatic regions on a global basis. Thornthwaite's calculations of potential evapotranspiration on a monthly basis represent much improvement over Köppen's temperature and precipitation relationships, but his *PE* estimates are still not accurate enough for daily or weekly assessments of irrigation requirements during dry spells.

WATER BUDGET CLIMATES: INTERACTIONS WITH ENVIRONMENTAL SYSTEMS

Climatic systems are, as we have seen, the products of interactions between energy and moisture. A useful tool for understanding climate in terms of energy and moisture is the local water budget discussed in Chapter 6. The water budget deals with energy by using the concepts of potential and actual evapotranspiration, and it deals with moisture in terms of the availability of water for plant growth and the amount left over in the form of runoff for streamflow and erosion. The individual components of the water budget which can be used as indexes of climate include potential evapotranspiration, precipitation, soil moisture, evapotranspiration, deficit, and surplus or runoff. By analyzing these components, geographers are able to understand the ways in which climate controls various systems in the environment. In particular, studies of water budget components are useful for making environmental comparisons from place to place, as well as revealing the magnitude of climatic variability over time.

Consider the average local water budgets at two locations—Cloverdale, California, and Manhattan, Kansas. Cloverdale receives an average of about 100 cm (39.4 in.) of precipitation annually, and Manhattan receives 80 cm (31.5 in.). Both locations have a total annual potential evapotranspiration of nearly 80 cm (31.5 in.); thus their energy endowments are essentially equal.

On the basis of annual precipitation, Cloverdale appears to be more humid than Manhattan. But as discussed in Chapter 6, the utilization of moisture at a place depends on the timing of precipitation. Figure 7.10, p. 198, shows the water budgets for the two towns. At Cloverdale, large deficits of soil moisture are generated in the summer months, when potential evapotranspiration is high and precipitation is low. In terms of interactions, Cloverdale has a dry climate for vegetation during summer. In winter and spring, however, heavy precipitation results in a large surplus of water for runoff, making Cloverdale moist for purposes of streamflow.

At Manhattan, the water budget shows that the timing of precip-

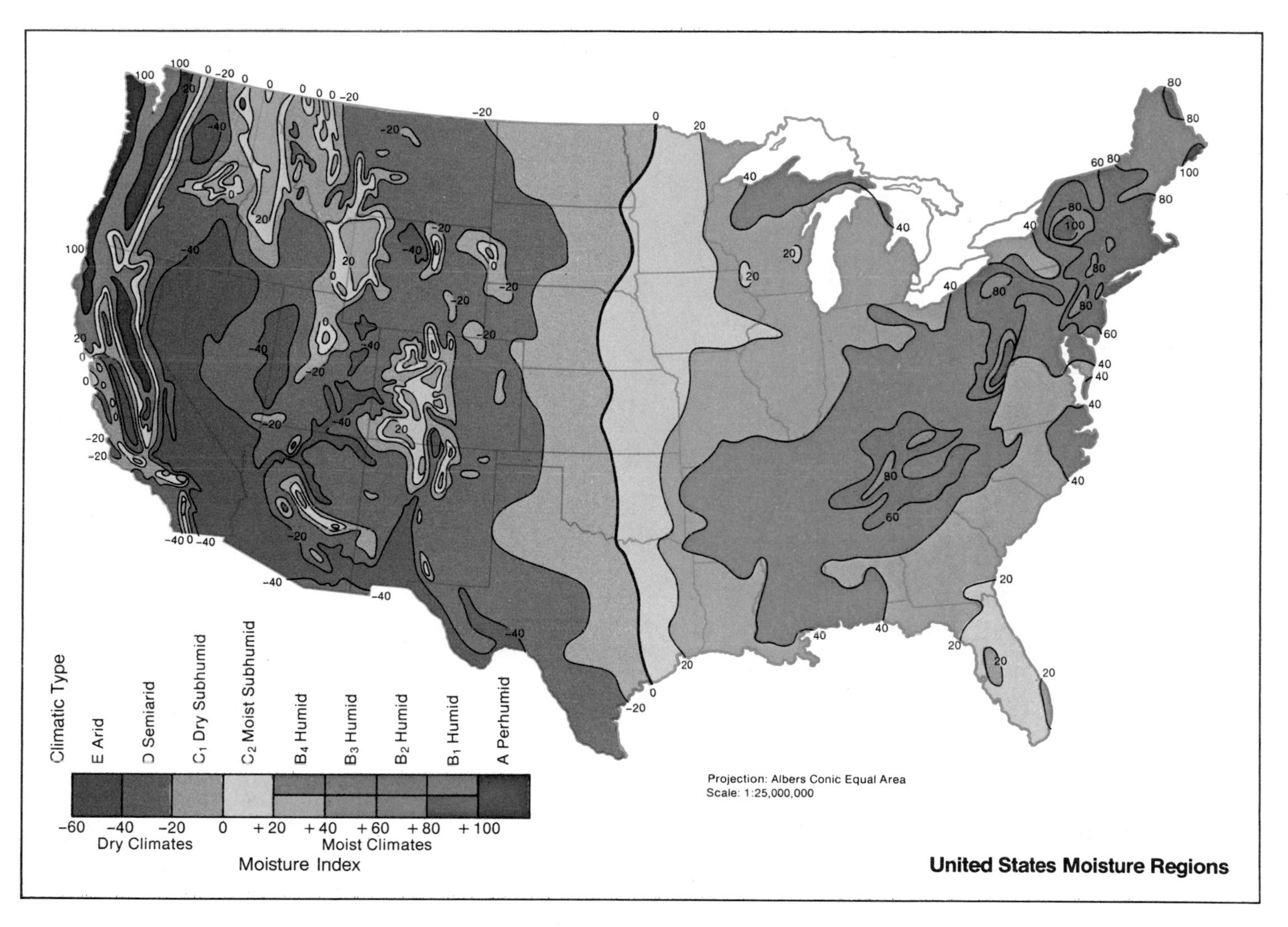

Figure 7.9 This map of the coterminous United States shows the principal moisture realms as classified according to the Thornthwaite moisture index. Precipitation exceeds potential evapotranspiration where the index is positive, and it is less than potential evapotranspiration where the index is negative. The line where the moisture index is 0 extends from western Minnesota south to the Gulf of Mexico. Immediately to the east of this line, there is a surplus of moisture, on the average, and immediately to the west, there is a deficiency, on the average. The moisture index of a region bears a close relation to the form of its vegetation. Forests, for example, can maintain themselves only where the moisture index is positive and has a value of at least 20 or so. Only grasslands and desert vegetation can survive where the moisture index is negative. (After Thornthwaite and Mather, 1955)

itation tends to follow the variation of potential evapotranspiration through the year. Therefore, only small deficits and surpluses are generated. For vegetation purposes, Manhattan is moist, but for streamflow, it is dry.

The examples of Cloverdale and Manhattan show that when a geographer considers the relationship of climate to other systems, it is not sufficient to consider only the amounts of energy and moisture delivered; the distribution of energy and moisture to systems must also be taken into account. As in the case of Cloverdale—moist for streamflow but dry for vegetation—a given location may exhibit different types of climate according to which component of the water budget is considered.

The surplus in the water budget is a good index of the spatial and temporal variability that can exist within a given climate. Figure 7.11a, p. 199, shows, for example, the mean winter-spring surplus across Louisiana for a 30-year period, 1941–1970. Although Louisiana has the highest average precipitation of the 50 states, the average seasonal surplus shown on the map ranges from more than 55 cm (20 in.) to less than 35 cm (14 in.) along the Gulf Coast. Even in this wet state,

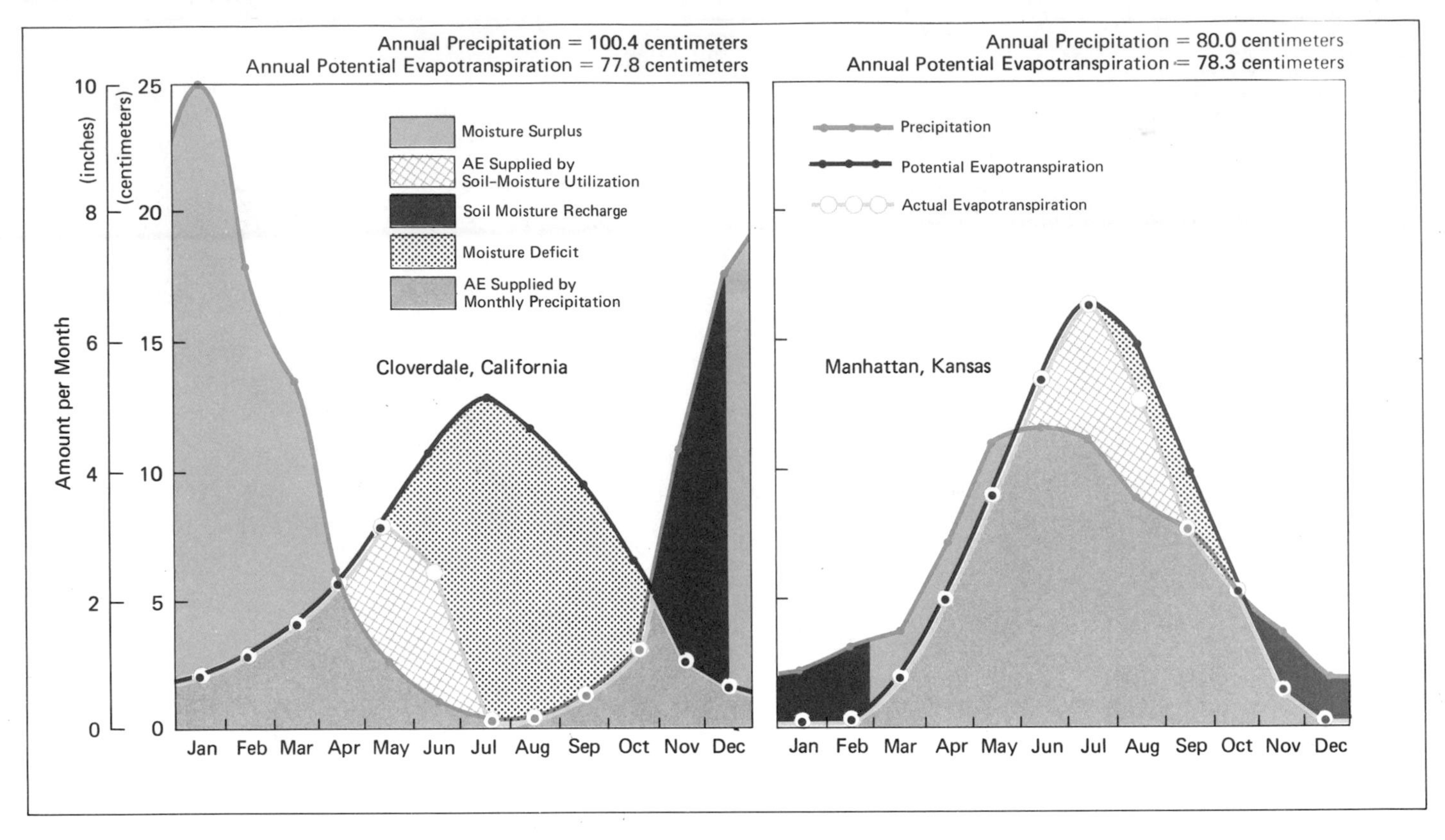

Figure 7.10 Cloverdale, California, and Manhattan, Kansas, receive similar amounts of precipitation annually and have nearly the same total annual potential evapotranspiration. The climates at these locations appear to be similar on the basis of the annual endowments of moisture and energy. But their local water budgets indicate that the timing of precipitation compared to the demand for moisture at each location makes the climate different for systems on the earth's surface. At Cloverdale, heavy precipitation and a low demand by vegetation during the winter generate large surpluses of moisture. Deficits occur during the dry summer. Thus Cloverdale is moist for runoff and dry for vegetation. At Manhattan, the timing of precipitation through the year is in close accord with the demand by vegetation, making surpluses and deficits small. Manhattan is therefore dry for runoff and moist for vegetation. (After Carter, Schmudde, and Sharpe, 1972)

the spatial variability is considerable; this is especially so because surplus water is the "extra" water within the local water budget.

Temporal variability from one year to the next, which depends on small changes in the upper atmospheric circulation, is frequently even more important than spatial variation. The surpluses across Louisiana for three winter-spring seasons are shown in Figure 7.11b–d. These examples represent a season that is "wet" everywhere, a season with extreme spatial variability, and finally, a "dry" season everywhere. It is obvious that the range between extreme seasons is very large, and we can suspect that this has some impact on a great number of environmental interactions and responses.

Figure 7.12a, p. 200, shows the annual surplus generated at seven locations in the United States. The data are presented in terms of frequency distributions over a 30-year period to indicate the degree of variability that can occur at each place. At Blue Canyon, California, for example, a moisture surplus greater than 135 cm (53 in.) was generated half of the time. But in individual years, the surplus ranged from over 229 cm (90 in.) to under 64 cm (25 in.). In 70 percent of the years, an annual surplus greater than 110 cm (43 in.) occurred at Blue Canyon. From the standpoint of moisture surplus, Blue Canyon should be classified as exceptionally humid, Morgan City, Louisiana, as humid, and Phoenix, Arizona, as arid.

Recall from Chapter 6 that the deficit *(D)* in the water budget represents the difference between potential evapotranspiration *(PE)* and actual evapotranspiration *(AE)*. The deficit is a measure of the

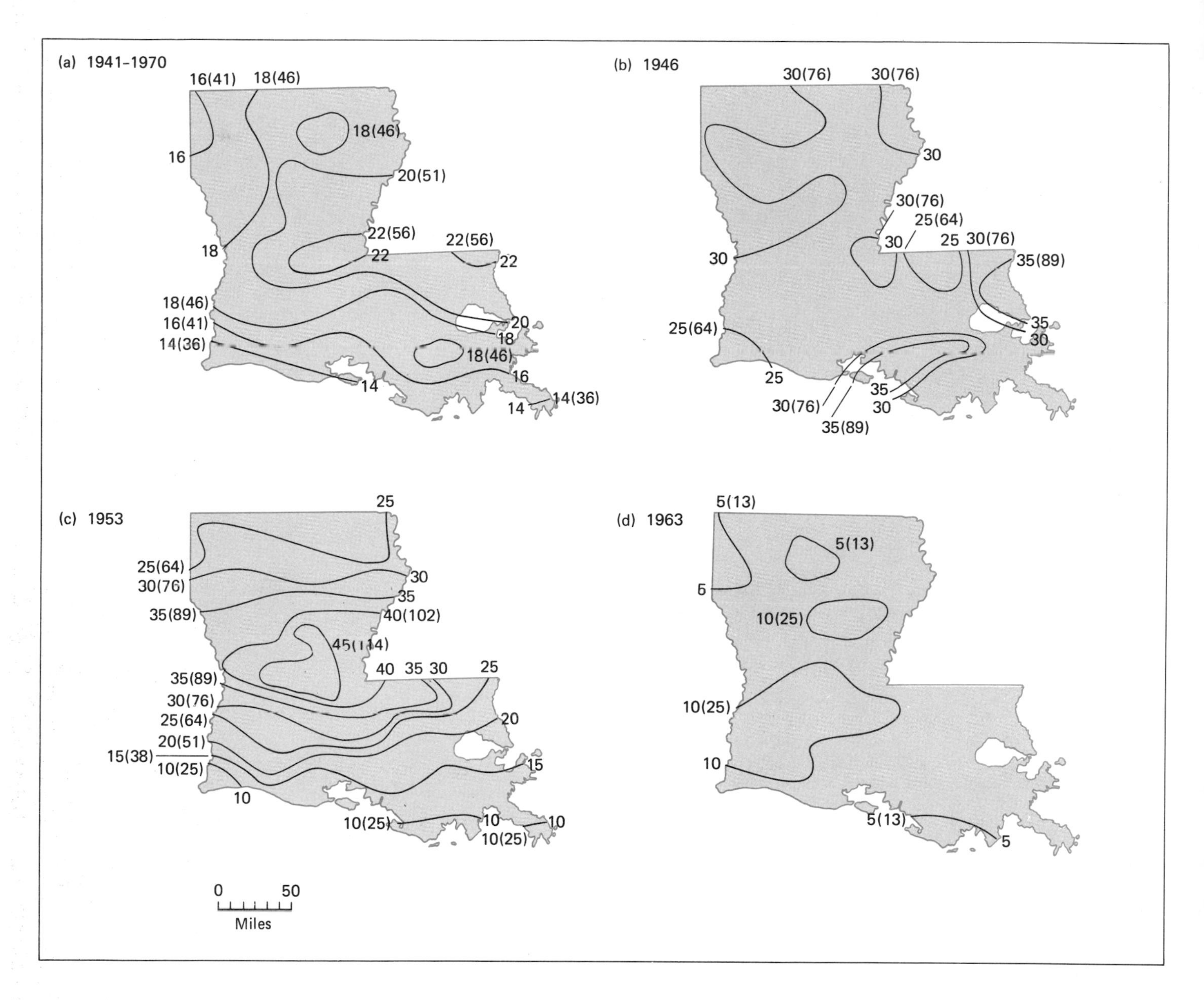

degree to which precipitation and soil moisture reserves do not meet the climatic demand for water *(PE)*, and it can be used to calculate the irrigation potential for crops and the potential water shortage for most native vegetation. In Figure 7.12b, the same seven locations are arranged according to the ratio of actual to potential evapotranspiration expressed as a percentage. Vegetation that is native to places with high values of *AE/PE*, such as Morgan City, will not do well without irrigation in drier climatic regions, such as Brownsville, Texas, where this ratio exceeds 60 during only about 10 to 15 percent of the years. On this climatic scale, Morgan City and Miami are highest, Blue Canyon is only intermediate, and Phoenix is again lowest. Depending upon which water budget component is selected, these locations change their relative positions along a climatic scale.

Figure 7.11 These maps illustrate in inches and centimeters the spatial and temporal variability of the surplus for Louisiana during the winter-spring season (November through May).

(a) The mean surplus over a 30-year period shows considerable variability across the state.

(b) During 1946 a wet winter-spring season occurred everywhere.

(c) Extreme spatial variability existed in the winter-spring of 1953, when record floods occurred across south-central Louisiana.

(d) Very small surpluses were recorded during the extremely dry winter-spring of 1963. (After Muller and Larimore, 1975)

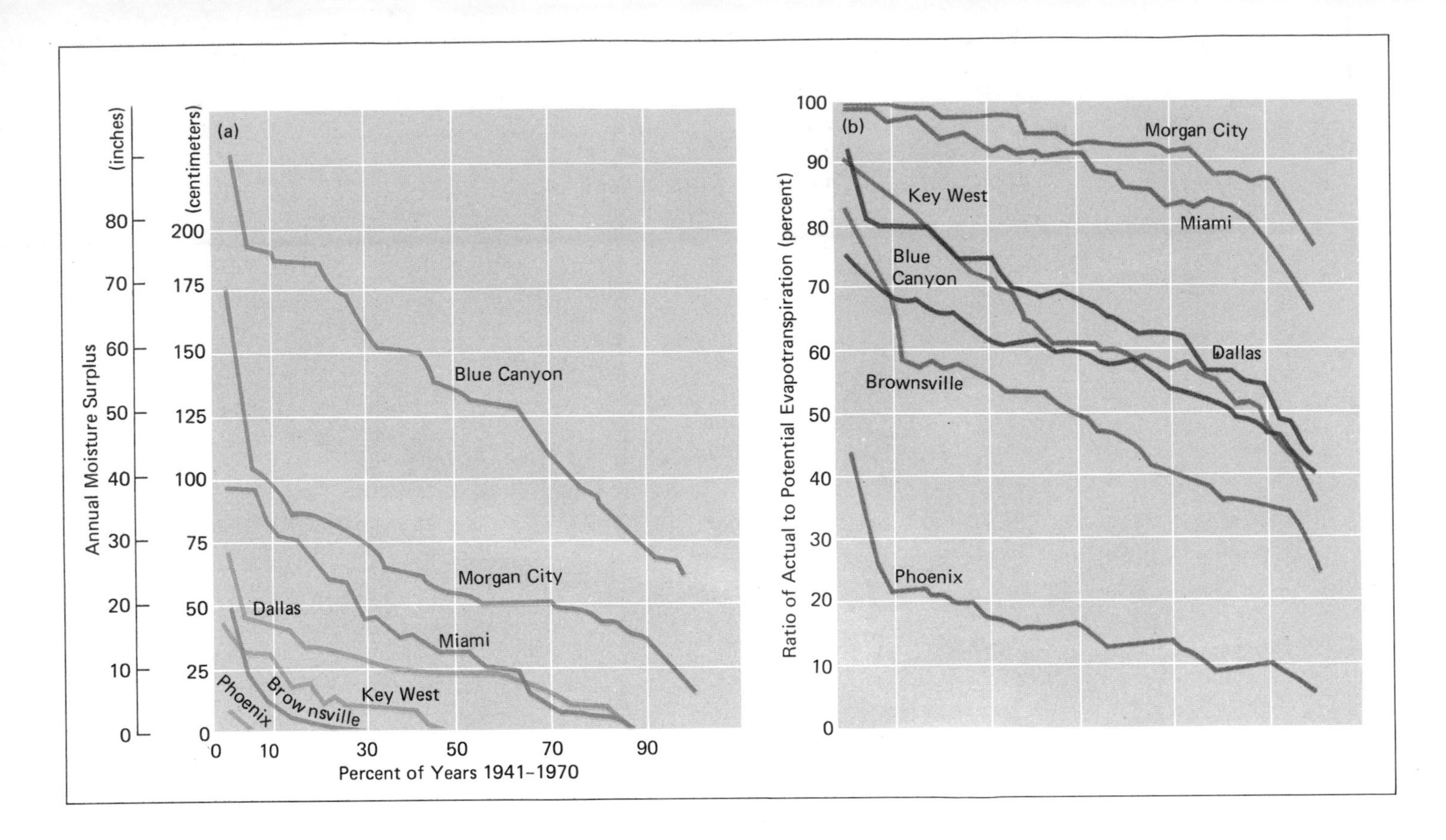

Figure 7.12 Components of the local water budget can be used as climatic indexes. These graphs emphasize variability of the climate at seven locations across the United States for the 30-year period between 1941 and 1970. The annual values of the components are plotted with the largest value to the left down to the smallest value on the right.

(a) The annual surpluses are shown here. For example, the largest annual surplus over the 30 years at Blue Canyon, California, amounted to more than 229 cm (90 in.) and the second largest surplus to about 190 cm (75 in.); the smallest surplus was only about 64 cm (25 in.). Over the 30-year period, the average surplus was about 140 cm (55 in.). On the basis of the cumulative frequency curve, we can see that the surplus amounted to 157 cm (62 in.) or more during 30 percent of the years. Similarly, during 90 percent of the years, the surplus was equal to or greater than 76 cm (30 in.); we can transpose the perspective and also say that during 10 percent of the years the surplus amounted to less than 76 cm (30 in.). The largest surpluses were at Blue Canyon; Morgan City, Louisiana, ranked second of the seven.

(b) The ratio of actual to potential evapotranspiration is a measure of the degree to which precipitation and soil moisture meet the climatic demand for water. Most crops and forest vegetation do poorly when the ratio is low over the growing season. For this index, Morgan City ranked first and Blue Canyon only fifth! (After Muller, 1972)

SUMMARY

The combined processes of solar energy input and the circulation of the atmosphere deliver energy and moisture to the earth's surface. Variations in the distribution of energy and moisture to particular locations create the pattern of different climates around the globe. Many factors combine to produce these differences in moisture and energy distribution, and the climatic patterns which result show significant spatial and temporal variations, whether they are viewed on the scale of a continent, a region, or even a microclimate within the confines of a suburban backyard. Out of the infinite number of possible climate classifications, scientists attempt to abstract a small number of typical climates in order to represent geographical areas in terms of their dominant climatic patterns. Different schemes of climate analysis and classification are useful for different purposes.

Climates may be viewed in terms of the response of particular geographical locations to solar energy input and the general atmospheric circulation. This perspective can be understood by analyzing the distribution of climates on a hypothetical continent which has a smooth surface without mountain barriers, interior seas, or gulfs. The effects of mountain ranges, seas, and continental landmasses become apparent on a global map of actual climatic patterns.

An alternate approach to climate, and one that is particularly useful for many of the objectives of physical geography, is the analysis and classification of energy and moisture interactions at the surface of the earth. These interactions directly affect the form and distribution of vegetation, soils, and landforms.

The climate classifications of Penck, Köppen, and Thornthwaite represent progressive attempts to evaluate energy-moisture interactions at the surface on a global scale. Penck stressed the distinction between the moist and dry realms, and Köppen and Thornthwaite developed increasingly complex criteria for evaluation.

The local water budget, which analyzes interactions between energy and moisture in terms of precipitation, soil moisture, evapotranspiration, and runoff, is a useful tool for understanding the ways in which climate controls various environmental systems and for comparing the spatial and temporal variability within climates. By evaluating climates in terms of the various water budget components, we can see that areas which receive similar amounts of moisture and solar energy may in fact be very different climatically. In subsequent chapters, we will use the water budget concept as an analytical tool for understanding environmental interactions at the earth's surface.

Review Questions

1. What climatic indexes are generally used in physical geography?
2. What factors determine variations within a microclimate? How do they differ from those which determine global climates?
3. Why is the frequency with which solar energy input and moisture availability are measured important in characterizing the climate of a particular area?
4. Why are different types of climate developed on the east and west coasts of the hypothetical continent at subtropical latitudes?
5. What are the main differences in the climatic pattern over the northern and southern hemispheres of the hypothetical continent?
6. Why does the marine climate of Western Europe extend further inland than the marine climates on the west coasts of North and South America?
7. According to Penck's system of climate classification, what conditions characterize the boundary between the moist and dry realms?
8. Why was Köppen's dependence on vegetation as an indicator of climate a drawback to his system?
9. What are the principal ways in which Thornthwaite's system of climate classification differs from Köppen's?
10. Where would you expect farmers to depend more on irrigation—Manhattan, Kansas, or Cloverdale, California? How would a comparison of the water budgets of both areas show this difference?

Further Reading

Carter, Douglas B., and **John R. Mather.** "Climatic Classification for Environmental Biology." *Publ. in Climatology,* Vol. 19, No. 4 (1966): 305–395. The evolutionary sequence of the several Thornthwaite climatic classifications and some of their relationships to vegetation distributions are presented.

———, **Theodore H. Schmudde,** and **David M. Sharpe.** "The Interface: As a Working Environment: A Purpose for Physical Geography." *Tech. Paper No. 7,* Comm. on College Geog., Assoc. of American Geog., 1972. (52 pp.) This monograph stresses the relationships between water budget components and environmental responses.

Chang, Jen-hu. *Atmospheric Circulation Systems and Climates.* Honolulu: Oriental Publishing Co., 1972. (326 pp.) This intermediate level text focuses on the dynamics of the general circulation.

Hare, F. Kenneth. *The Restless Atmosphere.* Rev. ed. London: Hutchinson Publishing Co., Ltd., 1956. (192 pp.) This brief introductory classic contains outstanding regional and continental chapters which focus on the dynamics of the general circulation.

Muller, Robert A., and **Philip B. Larimore, Jr.** "Atlas of Seasonal Water Budget Components of Louisiana." *Publ. in Climatology,* Vol. 28, No. 1 (1975): 1–19. This monograph includes more than 120 maps which illustrate the spatial and temporal variability of water budget components.

Thornthwaite, C. Warren. "An Approach Toward a Rational Classification of Climate." *Geographical Review,* Vol. 38, No. 1 (1948): 55–94. Thornthwaite first set out the potential evapotranspiration and water budget systems in this classic paper.

Trewartha, Glenn T. *An Introduction to Climate.* 4th ed. New York: McGraw-Hill Book Co., 1968. (408 pp.) Temperature and precipitation properties of climatic regions over the globe are stressed in this introductory text.

Wilcock, Arthur A. "Köppen After Fifty Years." *Annals Assoc. of American Geog.,* Vol. 58, No. 1 (1968): 12–28. This article presents a concise review of Köppen's development of his climatic classification and the subsequent modifications by other climatologists.

Case Study:

How Much Do Climates Change?

Although the weather seems to be relatively the same from year to year, one summer may be hot and dry, and another cool and wet. Climatic fluctuation on this scale is very costly to technological civilizations, especially in terms of energy consumption and agricultural production. An early season freeze on Labor Day weekend in 1974 devastated crops across much of the upper Midwest. The unusually cold winter of 1976–1977 in the eastern United States was discussed briefly in Chapter 5; its impact was probably especially great in the Northeast because it had been preceded by several unusually mild winters. Similarly, the record-breaking heat in the Southeast during the early summer of 1977 took its toll in terms of increased energy use for air conditioning, and withered corn and peanut crops. Fluctuations in water budgets are especially significant. The greater-than-normal deficits shown for New Jersey in Figure 8.21, p. 223, and for Kansas in Figure 8.24, p. 226, significantly reduced crop yields, and along the Mississippi River, alternations in the surplus have resulted in both floods and low water problems (see Case Study following Chapter 12).

Are climates more variable now than in the past? Are there signs of recent trends toward warmer or colder, wetter or drier conditions? It was not until the nineteenth century that geologists and other scientists began to appreciate the extent of glaciation and the associated climate changes during the earth's history. By analyzing debris moved about by the glaciers and sediments laid down by glacial meltwater, they concluded that during the Pleistocene epoch, beginning about 1 million years ago, there were four glacial stages and three warm interglacials. From most recent to oldest, the classical glacial chronology is as follows:

North America		Europe (Alps)	
Glacial	Interglacial	Glacial	Interglacial
Wisconsinan		Würm	
	Sangamonian		Riss-Würm
Illinoian		Riss-	
	Yarmouthian		Mindel-Riss
Kansan		Mindel	
	Aftonian		Günz-Mindel
Nebraskan		Günz	

More recent investigations suggest that this is an oversimplification; the Pleistocene probably began as much as 2 million years ago, and there were many more glacial stages. A computer model has been utilized recently to characterize global climates about 18,000 years ago, during the Wisconsinan glacial stage. Sea level is estimated to have been 85 to 130 meters (280 to 430 ft) lower than at present, but surface water temperatures in summer in the Gulf of Mexico and Caribbean Sea may have been only about 2°C cooler.

Only in the last decade have scientists begun to unravel the detailed outline of the earth's climatic history for the past million years, using studies of tree rings, fossil plant pollen, ancient soils, lake and deep-sea sediments, mountain glaciers, and cores of ice caps on Greenland and Antarctica. The accompanying figure illustrates the major trends of global climate based on these analyses. Perhaps the most astounding features are the glacial-interglacial cycles that last about 100,000 years, with shorter interglacials separated by long glacial climates, as shown in graph *a*. Glacial conditions appear to develop slowly and irregularly, with interglacial climates evolving suddenly. Recent studies indicate that climatic variability on time scales of 100,000 and 20,000 years may be related to systematic variations of the earth's orbit about the sun, which produce seasonal and latitudinal changes in the distribution of incoming solar radiation. Graph *b* shows the rapid warming trend that began about 15,000 years ago. During the very sudden and brief Younger Dryas cold interval about 10,500 years ago, much forest in Europe was destroyed. Graph *b* also shows that the mildest temperatures since the Pleistocene occurred about 5,000 to 7,000 years ago.

Temperature variations in Eastern Europe over the past 1,000 years are shown in graph *c;* the temperature reconstruction is based on historical records of the forward positions of the snouts of alpine glaciers and of the wine production at various monasteries. The lower temperatures associated with the "Little Ice Age" between 1400 and 1850 undoubtedly resulted in social and economic dislocations that are not yet well understood. The thermal maximum of the 1940s represents the highest average temperatures of the past 1,000 years. Graph *d* shows the cooling which began about 1945, but recent data suggest that the global cooling may have leveled off since the mid-1970s.

Much shorter-term climatic variability has an especially direct impact on environmental processes and economic activities. The next figure shows monthly temperature and precipitation variability for southeastern Louisiana from 1911 through June 1977. Each month's deviation from its respective average for the 30-year period 1931–1960 is shown in graphs *a* and *c*. For example, January 1940 was 7°C (12°F) below normal. Graphs *b* and *d* show short-term variations in terms of six-month "running averages"; for the six months ending December 1931, the temperature averaged about 3°C (5°F) above normal, and for the six months ending December 1976, about 4°C (8°F) below normal.

There are two outstanding features of these graphs for southeastern Louisiana. One is the downward step of temperature beginning in the fall of 1957 and continuing, with the exception of several winters, to the present. The other is the tendency for clusters of months or of years to be warmer or colder, or wetter or drier, than normal. Especially obvious are the dry periods of 1915–1918, 1933–1938, and 1951–1955, all associated with droughts on the Great Plains; other intense dry periods include 1924–1925 and 1962–1963. Climatic variability on these time scales significantly affects agricultural yields and water-resource management.

Explaining why climates change is extremely difficult because many interactions of the general circulation are not understood in detail. Furthermore, slight climatic changes can produce disproportionate effects in the environment. If the average global temperature were to drop a few degrees, ice and snow would cover a greater proportion of the earth's surface. Because of its high albedo, the ice would reflect more radiant energy back to space, which would cause further cooling and further spread of ice and snow.

A number of different explanations have been proposed to account for climatic variability over geologic time, but none is entirely satisfactory. Glacial climates appear to have been associated with continental drift and plate tectonic processes (see Chapter 11), which caused continents to move into polar regions, allowing ice sheets to develop. In addition to Pleistocene glaciation, there is evidence of at least three widely separated glacial periods over the last billion years. Within these glacial periods, it is now believed that systematic variations in the earth's orbit account for the 100,000-year glacial-interglacial cycle. Furthermore, mountain uplift is believed to be connected with colder temperatures and increased snowfall over uplands. Stratospheric dust from intense volcanic activity has been linked to periods of somewhat diminished solar radiation income and lower temperatures at the earth's surface. The early summer freezes and snows in northern New England in 1816 followed the massive eruption of the volcano Tambora in the Dutch East Indies in 1815, and several years of spectacular red sunsets followed the eruption of Mt. Agung in Indonesia in the early 1960s.

Numerous attempts have been made to relate sunspot activity to climatic variation, especially drought periods in the Great Plains. Although there is some correlation between sunspots and dry periods on a time scale of several years, no cause-and-effect relationship has been demonstrated. Climatic variability on shorter time scales, such as those shown for southeastern Louisiana, are obviously related to changing patterns of the circumpolar vortex of westerlies in the upper atmosphere. Meteorologists and geophysicists continue to debate the causes of these circulation changes and the eventual return to more "normal" patterns. Precise forecasting of the circulation patterns a season or more in advance remains an elusive objective at present.

In predicting the climates of the near future, most concern centers on the effect of human activities, particularly the use of the atmosphere as a dumping ground. The atmosphere is a sensitive controller of climate; cloudiness and dust in the air directly affect the amount of energy reaching the earth, and carbon dioxide in the air affects the amount of energy that is radiated away. Large-scale industry and agriculture add smoke and soil particles to the air, increasing the albedo of the atmosphere and decreasing the amount of solar radiation received at the surface, and possibly lowering surface temperatures as well.

The interrelatedness of the earth's systems makes it difficult to predict the net effect from a particular change. The amount of carbon dioxide in the atmosphere is expected to increase by at least 20 to 30 percent in the last half of this century, primarily because of industrial activities. The direct effect of the increase in carbon dioxide will be a decrease in longwave radiation losses from the earth's surface, which will raise the average temperature. The higher temperatures will increase evaporation, perhaps leading to increased cloudiness. Because it is difficult to determine the degree to which each of these modifications will affect other environmental properties, the final temperature change caused by the increase of carbon dioxide cannot be predicted.

Any small, long-term change in the atmosphere is critical. The present state of the environment is a function of its past history; there is no fresh start each year because the environment accumulates the effects of minor changes. Orbit wiggles, volcanic dust—these and other small and complicated interactions influence climatic trends. Now human activities can be added to the list.

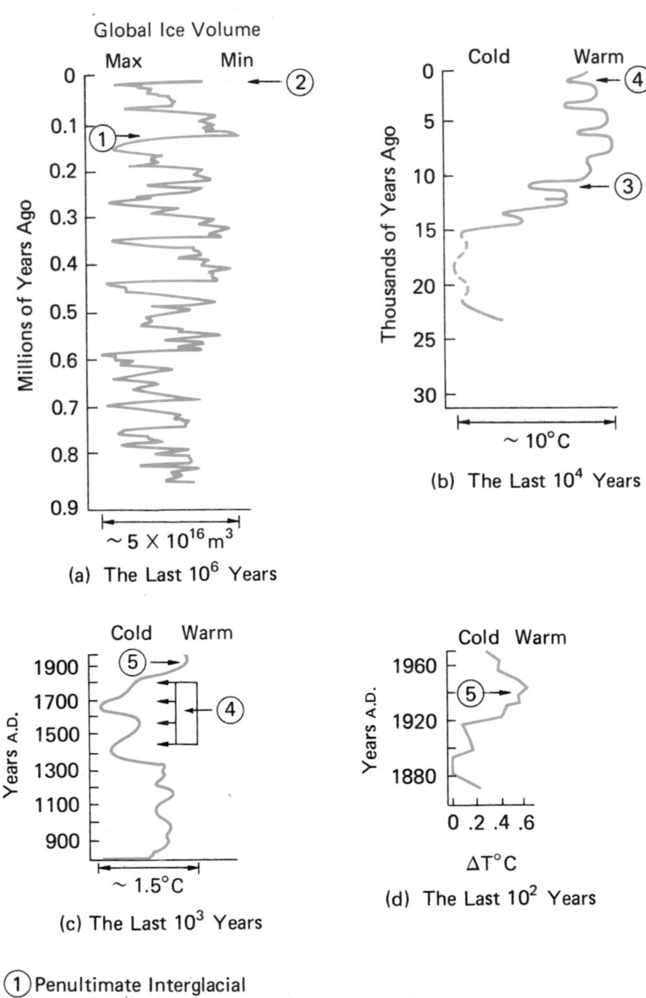

The main trends of global climate during the past million years. Graph (a) is based on isotope analyses of deep sea floor sediments; (b) on alpine tree lines, fluctuations of alpine glacier snouts, and fossil pollen; (c) on historical records of alpine glacier snouts and wine and grain production; and (d) on measured temperature data. (From National Academy of Sciences, 1975)

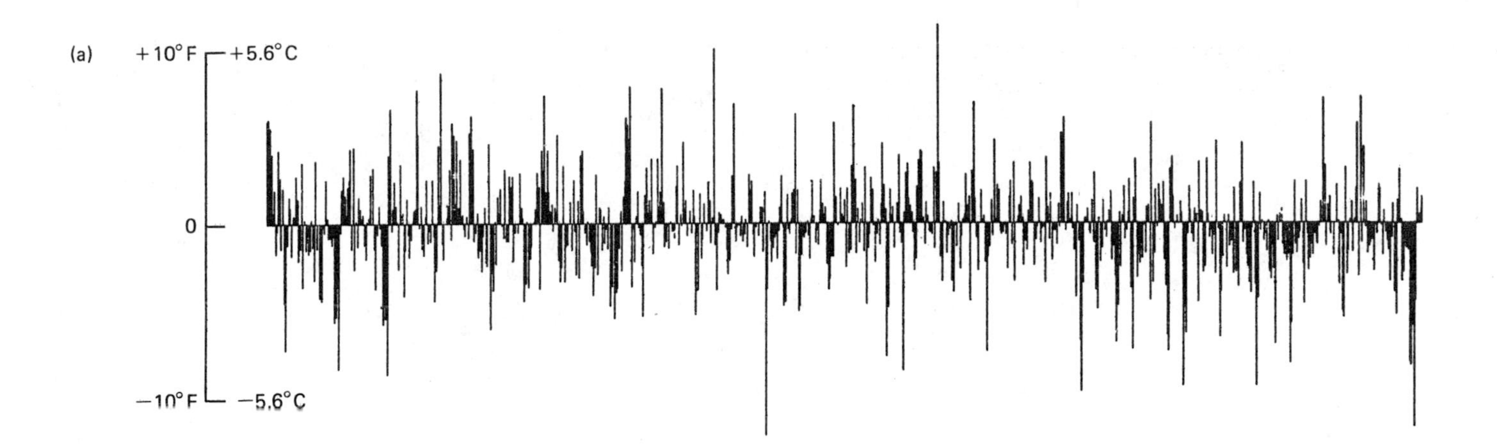

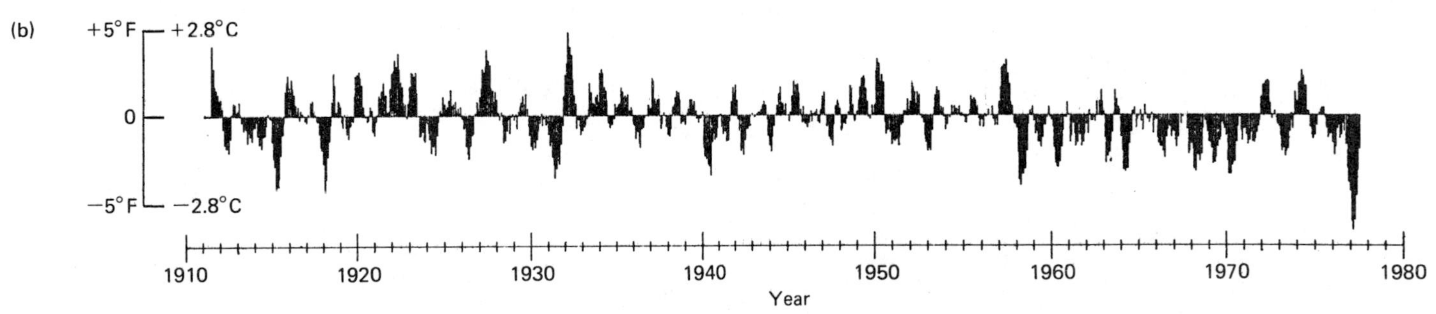

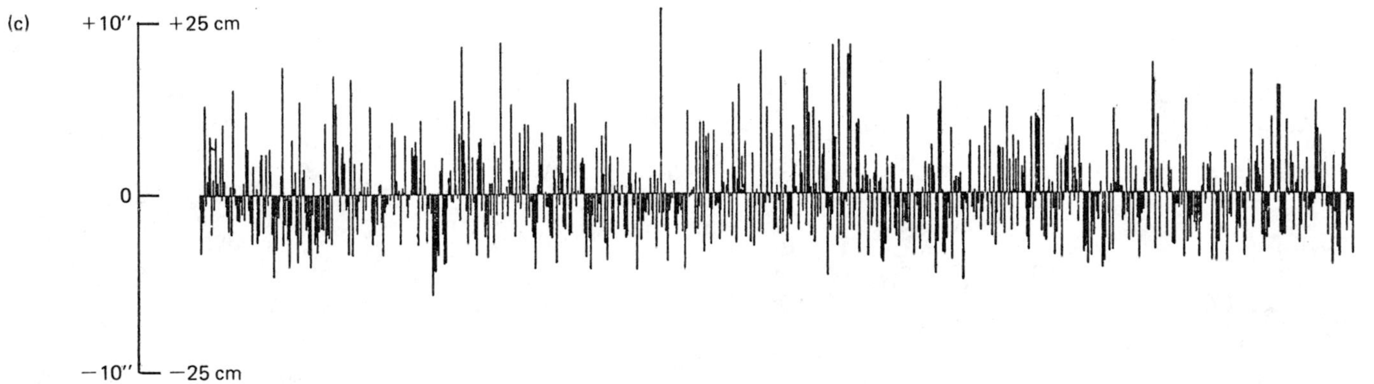

Temperature and precipitation variability in southeastern Louisiana. The temperature deviation of each month from the averages of the 30-year period between 1931 and 1960 is shown in graph (a), and precipitation deviations are shown in graph (c). For example, January 1940 was 7°C (12°F) below normal, and the rainfall during October 1937 was 28 cm (11 in.) above normal. Six-month running averages of temperature deviations are used in graph (b) to illustrate a "smoothed" interpretation of short-term temperature variation, and graph (d) is a similar smoothed interpretation of precipitation variation. Note especially the colder temperatures beginning in late 1957 and the fluctuations between warmer and colder, and wetter and drier conditions. (From Muller and Willis, 1978)

The Park by Gustav Klimt, 1910. (Collection, The Museum of Modern Art, New York; Gertrud A. Mellon Fund)

EIGHT

Climate and Vegetation

Sunlight and rain showers give an arboretum its colorful, luxuriant foliage. If the climate in a particular region were to change, the plants in that region would also change because of the close interaction of vegetation with climate.

ONE OF THE WONDERS of our planet is its vegetation, the living blanket of green that covers most of the land. Green plants are unique to earth; elsewhere in the solar system, temperatures do not remain within the narrow but critical range that allows water to be present in liquid form, rather than being boiled away or permanently frozen. It is water that makes possible the chemical reactions that permit living cells to form and multiply, and it is the balance between precipitation and potential evapotranspiration that largely determines the nature and abundance of plant life on our planet.

When land plants evolved on earth, some 400 million years ago, they quickly spread over the land surface, except for those regions covered by glacial ice. The ability of plants to flourish under a wide variety of energy and moisture conditions is the result of a multitude of adaptations.

Although plant life varies in form from lichens to giant forest trees, most plants are similar in having a root system to gather chemical nutrients dissolved in soil moisture, green leaves to convert solar energy into chemical energy for plant maintenance and growth, and stems or branches to support the energy-converting leaves and to channel nutrients and chemical energy to and from other parts of the plant.

However, these basic features vary greatly in order to operate efficiently over a wide range of environmental conditions. Roots can be deep where moisture and nutrients are far below the land surface, shallow where moisture and nutrients are concentrated near the surface, or far-spreading where moisture is scarce. Leaves may be large to permit high rates of transpiration or to intercept the dim light within dense forests. Where water supplies are limited, leaves tend to be small or needlelike, with waxy coverings to minimize water loss. Leaves may be retained by the plant all year so that food can be manufactured continuously or as soon as conditions are appropriate, or they may be shed during drought or when soil moisture is frozen in order to stop transpiration and thereby retain moisture. Where moisture is extremely limited, leaves may be dispensed with altogether, with green stems and branches taking over the manufacture of food, as in the case of cacti. Stems may be thinly protected where there is no temperature hazard or danger from water loss, or covered by thick insulating bark in cold regions, by fine hairs or dead leaves where shade is required, by waxy coatings where water loss must be minimized, or by needles or thorns where predation by animals is a danger. Sometimes the stems (or trunks) are equipped to store moisture in aqueous tissues (for example, desert cactus and the baobab of the savannas).

In terms of their relationship to moisture supplies, plants may be classified as *hydrophytes*, which grow in water; *xerophytes*, which are structurally adapted to survive in extremely dry soils; or *mesophytes*, which occur where the water supply is neither scanty nor excessive. Likewise, plants may be *perennial*, enduring seasonal climatic fluctuations, or *annual*, dying off during periods of temperature or moisture stress but leaving behind a crop of seeds to germinate during the next favorable period for growth. The seeds may have hulls that must be dissolved away chemically, or hard coverings that must be ground off or broken by floods or fires, or pulpy edible covers that are consumed by animals, who thus carry the seed to new ground in their waste products.

By these and many more adaptations, plants have developed the variety allowing them to exist from the edge of wind-whipped snowfields to black-water swamps and sun-baked deserts. This chapter describes some of the ecological factors that are generally responsible for the development of the particular vegetation in a region. Some properties of the principal vegetation regions of the globe and their associated climates are also described. Analysis of the precise interactions of solar radiation, atmospheric gases, and soil water that produce plant tissue is reserved for the following chapter.

ECOLOGY OF VEGETATION

Over long periods of earth history, vegetation in widely separated regions with similar climatic endowments and stresses has evolved toward similar forms. Cactus species are prominent in the dry climatic realms of the Americas, for example, but across many of the dry regions of Africa, various species of euphorbia have developed similar spines and water-storing mechanisms. Adaptations to climate are similar although the individual species are quite different. This type of *convergent evolution* of different species toward similar forms is seen in nearly all climates that involve periods of moisture or temperature stress.

Plant Communities

In any midlatitude forest, there are various kinds of trees, shrubs, ferns, grasses, flowers—perhaps fifty or more different species of vegetation—all living together in one *plant community.* A forest, a grassland, or a prairie consists of numerous species of plants living in association with each other and with bacteria, fungi, insects, and larger burrowing, seed-collecting, grazing, and browsing animals. A plant community affects its local environment by its modification of soils and moisture storage conditions. The individual plants interact with one another and respond to changes in a way that gives the community a definite pattern of organization.

The different species in a plant community do not usually compete for energy and moisture because each species has its own special *niche* in the community. The niche includes not only the precise microenvironmental setting of each plant species but also the particular way in which the species utilizes energy and moisture. In a broadleaf forest, for example, low shrubs use the subdued light that filters through the high leafy canopy and do not compete with the towering trees for direct sunlight. Such plants are physiologically adapted to grow in partial shade and cannot tolerate long periods of direct sunlight. Mosses growing on rocks do not rely on the same moisture supply drawn on by ferns growing on the forest floor. In central Africa, farmers often successfully grow several dissimilar crops—such as bananas, a tall open plant, and cassava, a low shrub—in the same field.

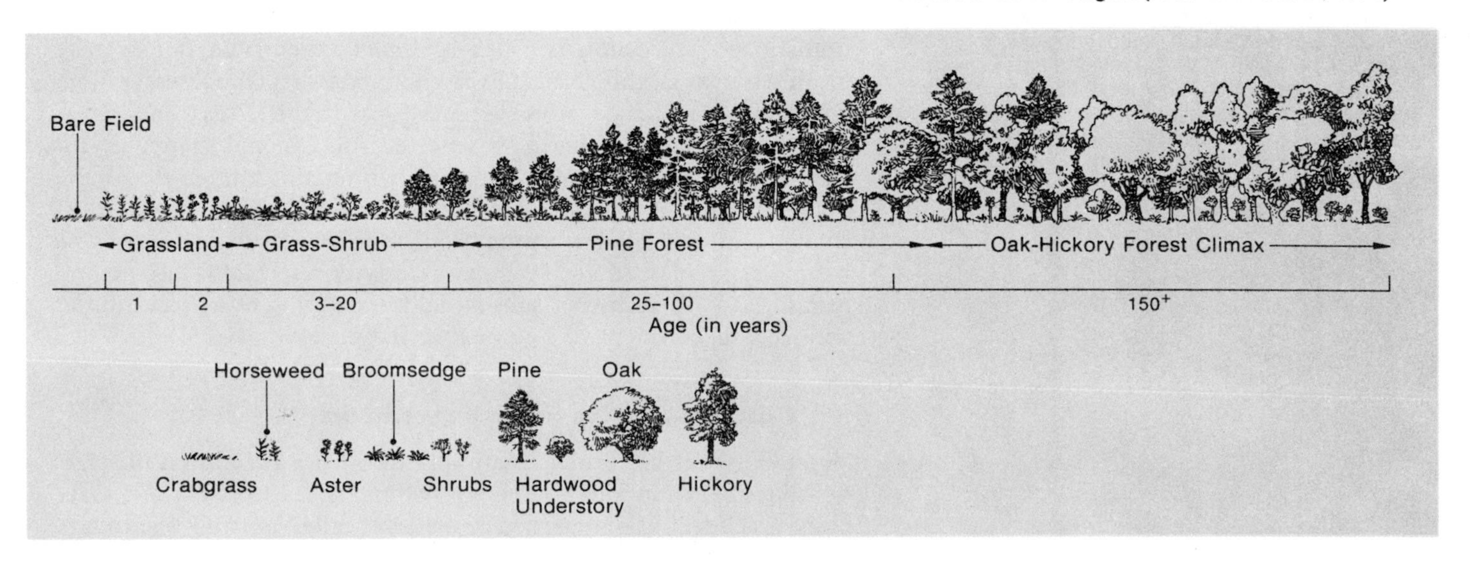

Figure 8.1 This diagram illustrates a typical plant succession from open land to forest in the middle latitudes. In the initial stages of the succession, low, fast-growing grasses and shrubs dominate. Each stage alters the soil and microclimate of the land, enabling other species to establish themselves and become dominant. The final stage consists of high trees that overshade and force out some of the low shrubs of earlier stages. (After E. P. Odum, 1971)

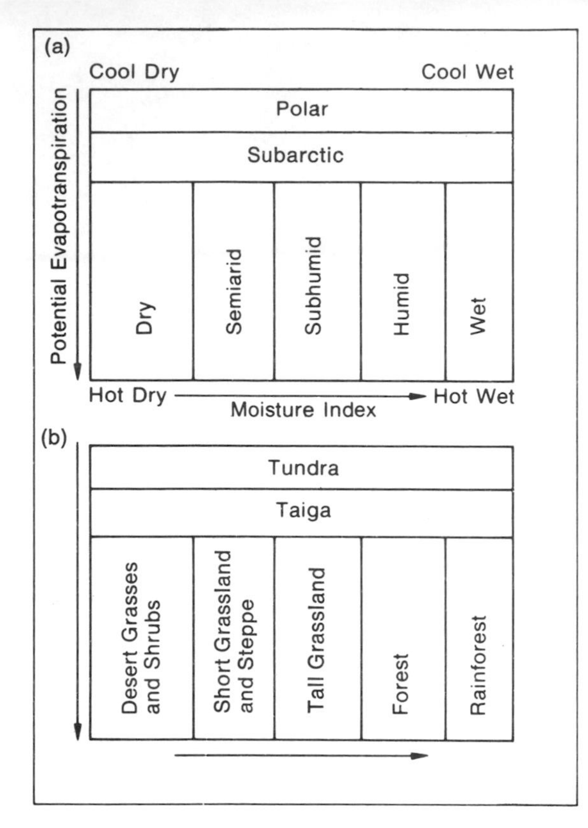

Figure 8.2 (a) The schematic diagram shows the principal climatic regions according to the Thornthwaite system of classification. (After Blumenstock and Thornthwaite, 1941)

(b) The principal vegetation regions are diagrammed according to the same *PE* and moisture index scales. At higher latitudes, energy availability determines the presence of tundra or forest; at middle and lower latitudes, the vegetation is determined primarily by the availability of moisture. (After Blumenstock and Thornthwaite, 1941)

(c) **(opposite)** This figure shows the actual relationships between *PE*, the moisture index, and selected natural vegetation in North America and the tropics. Only some of the vegetation types discussed later in the chapter are included. The figure shows that the distribution of natural vegetation is quite similar to the simple theoretical scheme in (b) and that there is some overlap among the types. (Modified from Mather and Yoshioka, 1968)

As long as the crops do not compete for the same moisture and sunlight, the total crop yield per acre can be greater than if the crops were planted separately. If two species in a plant community happen to compete for essential requirements, one may eventually be eliminated by the other because two strongly competitive species cannot occupy the same niche indefinitely.

Plant Succession and Climax

A hiker walking through the woods of New England sometimes comes upon the foundation stones of farm buildings that were abandoned more than a hundred years ago. Although the land was once cleared for agriculture, it is now forest again. A forest does not immediately establish itself when cleared land is left to nature. A sequence or *succession* of plant communities occurs, one plant community appearing and then giving way to the next, until a stable community is attained and there is no further change in its composition. Each community alters the local microclimate and surface condition, making possible the appearance of more demanding species, which subsequently become dominant. The particular order in which communities succeed one another depends on whether the succession begins in cleared land, filled-in marsh, burned-over forest, or some other condition. It also depends on the climate and the kind of stable plant community characteristic of the region as a whole. In a succession that changes from cleared land in wooded country to forest, the early communities are dominated by grasses and low shrubs, as shown in Figure 8.1, p. 205. These give way to communities containing higher shrubs and small trees. The general trend in succession in all plant communities is toward *taller vegetation,* toward *more diverse vegetation,* and toward *greater stability.* As more advanced communities become established, the pace of succession slows because the individual plants have longer life cycles. For example, the larger trees of deciduous forests may live several hundred years and not produce seeds for their first two decades.

The stable plant community that results at the end of a long, undisturbed succession is called a *climax vegetation.* It may take hundreds or even thousands of years to establish a climax vegetation. Environmental change, moreover, may disturb the climax once it has developed, thereby initiating a renewed successional sequence. For example, accumulation of ice during a winter storm might break the upper branches of a forest; a fire might burn over thousands of acres of timberland; or a flood or severe drought might result in the loss of a particular species. Intensive human activity, especially logging and farming, usually results in substantial interruption of a succession or complete replacement of a climax vegetation.

Climate and Natural Vegetation

Vegetation that has evolved naturally through a succession of stages without significant human interference and in response to the average climate conditions of a particular region is called *natural vegetation.*

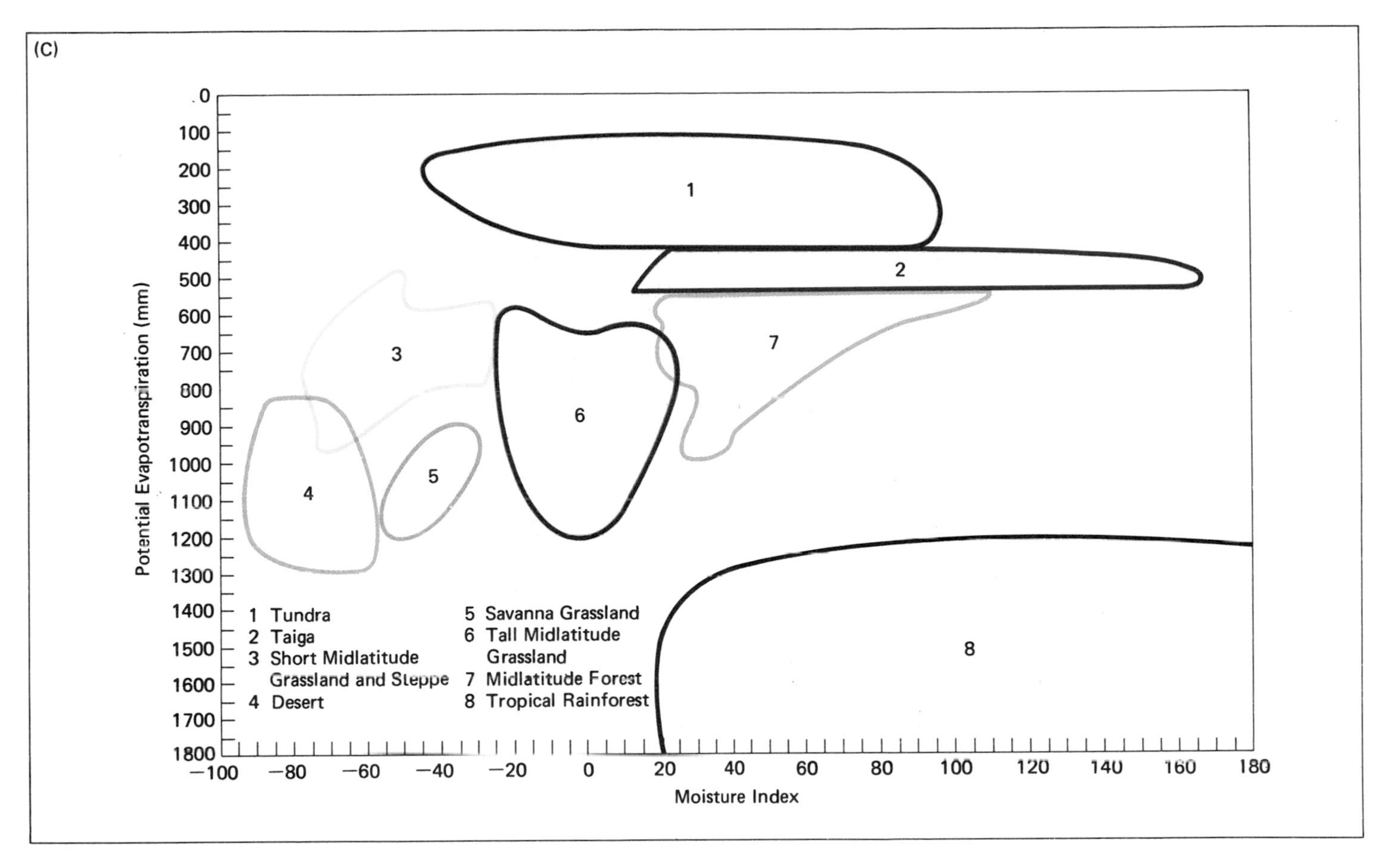

Figure 8.2 shows the close interrelationship between global climate and natural vegetation patterns. This schematic analysis was originally worked out by Thornthwaite and David Blumenstock in the 1930s. As diagram *b* demonstrates, energy availability is crucial in determining natural vegetation at higher latitudes, whereas moisture supply is more important at lower latitudes. In subhumid climates where the moisture index is near zero, grasslands are the predominant vegetation form. In areas where the moisture index is somewhat greater than zero, the predominant natural vegetation is forest. At higher latitudes, characterized by polar and subarctic climates, solar energy input is more significant than moisture in determining natural vegetation, giving rise either to tundra or to subarctic coniferous forests known as taiga.

PRINCIPAL VEGETATION REGIONS AND THEIR CLIMATES

A global map of natural vegetation types is shown in Figure 8.3, p. 209. The map represents a much simplified distribution because it assumes that climate is forever unchanging and that there are no widespread disturbances. Also the earth's many diverse plant communities have been grouped into only nine major types, sometimes called *biomes*, a term that includes not only vegetation but associated animals as well.

Botanists have produced numerous versions of this map according to various and complex criteria. The problems are similar to those encountered in climate classification: How should the various vegetation formations be grouped, and how many types should be shown? Unlike the boundaries between climatic regions, which are based on long-term averages of weather data, the pattern of vegetation distribution may depend on short-term fluctuations as well as on averages. If a dry region has a year with above-average moisture from time to time, plants from neighboring moister regions may establish themselves in the region classified as dry. Some overlap at regional boundaries is therefore to be expected.

Many of the major vegetation regions described in this section correspond closely to climatic regions discussed in Chapter 7. In the following pages, we outline the nature of the relationships between the elements of climate and the vegetation formations on the earth, beginning at the equator and progressing poleward. Each region's characteristic plant growth is described and discussed in terms of the amount and distribution of energy and moisture it receives.

Tropical Rainforests

The tropical rainforest is the vegetative response to a climate that exerts no limits on vegetation growth—a climate that provides abundant energy and moisture in every month of the year. The resulting vegetation consists mainly of broadleaf evergreen trees that shed some leaves throughout the year. The large leaves permit photosynthesis at low light levels and also high rates of transpiration, which removes latent energy and thus prevents the leaves from heating excessively. Trees that drop all of their leaves in one season are seldom found in the rainforest because a tree that remains dormant for part of the year has a competitive disadvantage against evergreens that maintain photosynthesis and growth in all seasons.

Perhaps due to the absence of climatic constraints, biological variation, in terms of the number of plant and animal species per unit of area, is far greater in the tropical rainforest than elsewhere. The diverse vegetation of a rainforest presents a crudely layered appearance as plant species adapted to differing light intensities reach varying heights. The tall tree species form a comparatively dense canopy about 30 meters (100 ft) above the ground. A few still higher light-seeking trees protrude above the canopy here and there. A dense tangle of smaller trees and vines commonly makes a wall along sunlit riverbanks and around forest clearings. Ferns are often found in the dim light and abundant moisture near ground level (see Figure 8.4, p. 210), and the forest floor is relatively open and free of underbrush.

In rainforests we usually encounter an abundance of vines, some rooted in the ground and twining upward around the trunks of larger trees in their quest for sunlight, and some, with no roots at all, hanging down from the branches of the forest giants. The latter types extract moisture from the air and nutrients from the litter of plant debris that lodges in the branches of the trees. Such vines and other plants of this

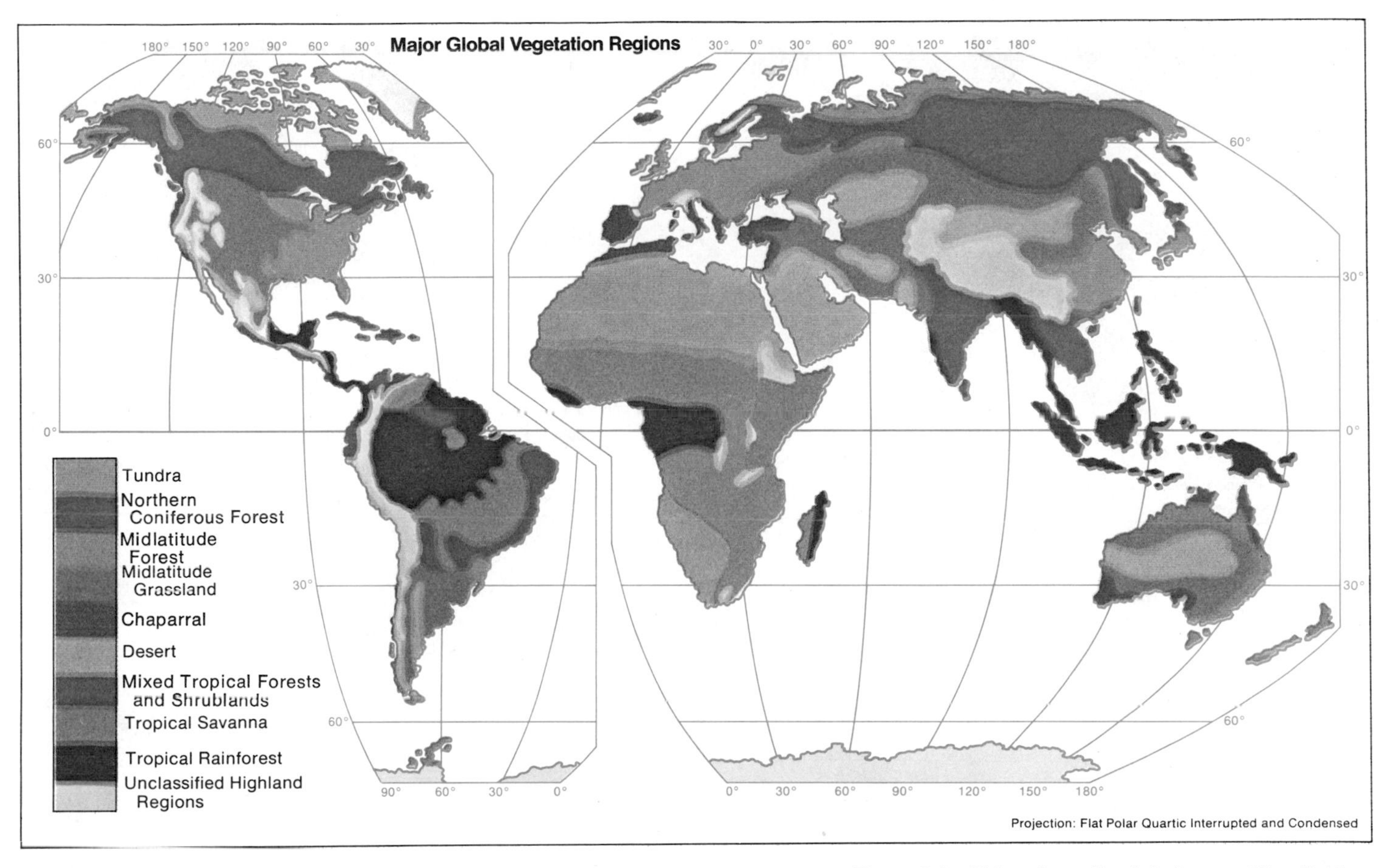

Figure 8.3 This schematic global map of the distribution of major vegetation types should be compared with the global map of climates (Figure 7.3) and the global map of soils (Figure 10.17). Climate, soils, and vegetation are strongly interacting systems, and their global distributions bear marked similarities. The regions occupied by tropical rainforest, for example, are found in tropical wet climates, and the underlying soils in the rainforest tend to be Oxisols. Similar broad global generalizations can be made about most of the major vegetation types. (After Finch and Trewartha, 1949)

type, known as *epiphytes*, depend on rapid decay of organic material and are widespread only in the constantly hot, damp climates of the tropics.

The major limitation in the environment of the wet tropics is the poverty of the soils, which usually are lacking in mineral nutrients. In Chapter 10, we shall see that the heavy rainfall of the wet tropics dissolves soluble mineral matter in the soil and flushes it away, leaving behind a very infertile residue. The most fertile part of the soil is the surface layer, which receives nutrients from the plant litter (leaves, branches, fruit, etc.) that accumulates there. The roots of rainforest trees are generally shallow because of the nutrient distribution and especially because roots will not penetrate perpetually wet soils. Consequently, the tall trees often have buttressed trunks that flare widely at the base to support the shallow-rooted but massive structure that reaches high toward the sunlight.

It may seem hard to believe that the survival of the great rainforests of the wet tropics is not at all certain, but such is the case. Human activities, such as logging and clearing for agriculture, are destroying the rainforest at a rapid pace. Regeneration of a mature rainforest seems next to impossible due to the poverty of tropical soils, which deteriorate very rapidly and irreversibly as soon as the forest cover is removed.

In tropical rainforest areas, most of the precipitation falls as brief but intense showers and thunderstorms between late morning and early evening, when the humid tropical air is most unstable because of solar heating at the surface. It is often said that the showers occur like clockwork each day, but this precise regularity seems to be restricted to a few places where terrain features help to promote instability at certain times. Elsewhere, there appear to be clusters of wetter days followed by days with only a few widely scattered showers. The wetter periods are believed to be associated with the westward passage of weak low-pressure disturbances along the intertropical convergence zone and within the trade wind systems.

Tropical rainforest areas are generally characterized by air temperatures that vary little either daily or seasonally. Singapore, for example, is located at latitude 1°N, in a region classified as having a tropical rainforest climate. The average monthly temperature shows little variation from season to season, and rainfall exceeds 15 cm (6 in.) in every month (see Figure 8.5, p. 212). Another way to describe such a climate is by a *thermoisopleth* diagram, which shows the variation in air temperature through the day and through the year. Air temperature through the year at a given time of day can be read along a horizontal line; air temperature through the day, along a vertical line. The pattern of the temperature contour lines, or *isotherms*, on a thermoisopleth diagram reflects the climatic characteristics. When the isotherms run primarily in a horizontal direction, temperature variations during a day are greater than variations from season to season. In Figure 8.6, p. 213, the diagram for Belém, Brazil, located in a tropical rainforest near the mouth of the Amazon River, exhibits such a pattern. The temperature changes by 5°C (9°F) or more during each day but varies by no more than 2°C (3.6°F) at any given hour through the year.

The local water budget can be used to provide another perspective on climate. The water budgets presented in this chapter are based on a more realistic model of soil moisture availability than the one presented in Chapter 6. It is assumed that plants can withdraw only a portion of the total soil moisture in any month; thus both a deficit and a utilization of soil moisture may occur in a given month or a succession of months. Figure 8.7, p. 213, gives the water budget for Kribi, Cameroon, located in the western section of the tropical rainforest region of Africa. This water budget shows no deficits, on the average, and large surpluses for nearly every month. The extraordinary heavy rainfalls and large surpluses have great impact on nearly all environmental responses in tropical rainforest regions.

Mixed Tropical Forests and Shrublands

In spite of the overall wetness of the climate, some regions of the tropics experience a very short dry season, largely because of the seasonal migration of the ITC and its associated showers. The vegetation of the mixed tropical forests and shrublands must adapt to the moisture deficiency of this short dry season. Many of the trees are deciduous, shedding their leaves for a month or two when soil mois-

Figure 8.4 (opposite) The dense vegetation in this rainforest in Ecuador often reaches great heights and may have specially adapted forms such as climbing vines.

ture is unavailable for evapotranspiration (see Figure 8.8, p. 214). The canopy of a mixed tropical forest is lower and less dense than that of a rainforest, and more light penetrates to ground level. As a result, the ground layer of vegetation tends to be more varied. In some areas where the dry season is accentuated, the forest gives way to low thorny trees and shrubs (see Figure 8.9, p. 215) with small hard-surfaced leaves that resist water loss. The global map in Figure 8.3, p. 209, shows that this vegetation type corresponds to Köppen's tropical monsoon climatic regions *(Am)* and to portions of the tropical wet and dry climatic zones. Mixed tropical forest and shrubland vegetation is widespread in the monsoon regions of Southeast Asia and India, with smaller areas in Central and South America.

Figure 8.5 Singapore is a station in the tropical rainforest climate (Köppen *Af* region). As the graph shows, the average monthly temperature and precipitation are high and nearly constant through the year. (After Nelson, 1968)

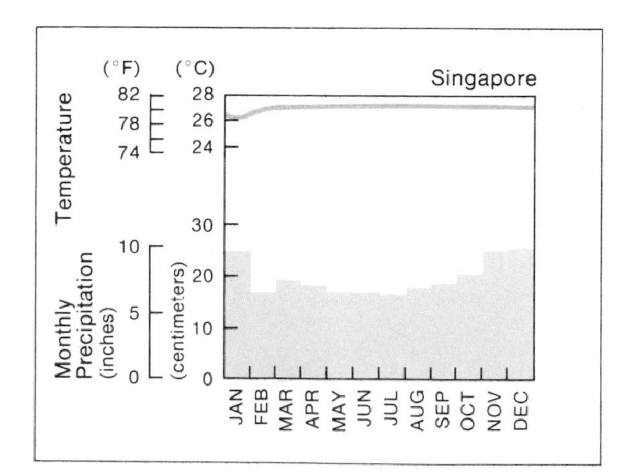

Tropical Savannas

A *savanna* is a tropical grassland, generally with scattered trees. Although not all savannas correspond to a single climatic region, most are located within the tropical wet and dry climatic zone, equivalent to Köppen's *Aw* climate, between the tropical rainforests and the drier regions near latitudes 30°N and S. The climate is one of heavy summer rainfall, associated with the poleward margins of the ITC during the high-sun season, and nearly complete winter drought, when the dry air of the subtropical highs subsides during the low-sun season. Extensive savannas are located between the rainforest and the Sahara Desert in Africa, in the region south of the Sahara known as the Sahel, across much of southeastern Africa, and over most of the interior of Brazil south of the Amazon basin.

Savannas vary from open woodland with a ground cover of grass to open grassland with a few isolated trees, such as baobab and acacia (see Figure 8.10, p. 215). Most of the trees shed their leaves during the dry season. Savanna trees commonly have thick fire-resistant bark and small drought-resistant leaves. The grasses are well adapted to alternating wet and dry seasons; during the long winter drought, the grass dies off above the surface, but the root systems survive and send up new shoots when moisture conditions improve. The lack of trees in some savanna regions may result from widespread grass fires that destroy young trees, rather than from insufficient soil moisture alone. Human use of savanna land for livestock grazing also tends to prevent forests from becoming established.

The African savannas are famed for their enormous herds of grazing (grass- and herb-eating) and browsing (twig- and leaf-nibbling) hoofed animals, including zebras, giraffes, buffalo, and many varieties of gazelle—always followed by predators, such as lions, cheetahs, and, of course, man. This host of short and tall creatures is collectively equipped to chew at everything from the lowest herbs to the leaves and twigs of the spreading treetops. Strangely, no such collection of herbivores is present in the extensive savanna regions of South America. Due to hunting and rapid agricultural encroachment, the great herds of game animals in Africa seem doomed and may shortly be restricted to a few protected reservations.

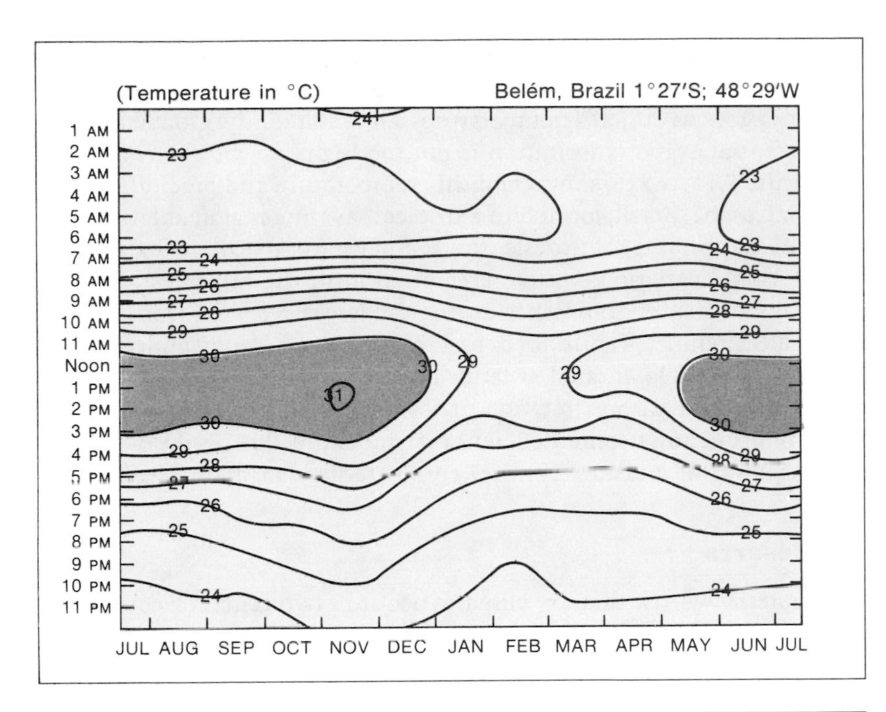

Figure 8.6 Belém, near the mouth of the Amazon River, is located in a tropical rainforest climate (Köppen *Af* region). This thermoisopleth for Belém gives the average hourly temperature through a year. The scale of months begins with July because Belém is in the southern hemisphere, so that January and February are summer months. The noon temperature through the year, which can be read by tracing across the diagram from left to right, is between 30° and 31°C (86° and 88°F) until December, when it falls 1° or 2°. The diagram can also be used to trace the temperature through a given day. On January 1, for example, the temperature after midnight remains at about 23°C (73°F) until the early morning, then rises to nearly 30°C (86°F) at midday. The temperature falls again in the evening. In a tropical rainforest, the temperature range through a day is greater than the range of temperature through the year. (After Troll, 1958)

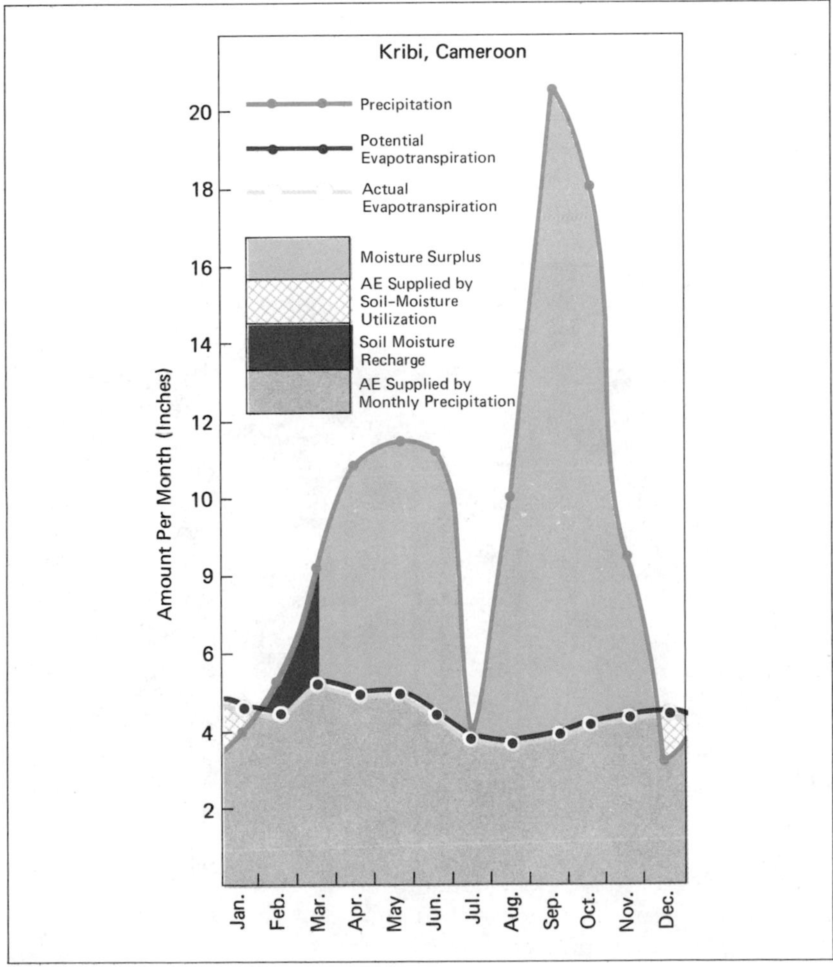

Figure 8.7 Kribi, Cameroon, is located at 3°N latitude in the tropical rainforest of Africa (Köppen *Af* region). Monthly *PE* is high throughout the year, but during most months precipitation is even greater and large surpluses are generated. The ITC and associated showery precipitation tend to follow the seasonal migration of the sun; thus the two very rainy seasons at Kribi occur when the sun is overhead at noon near the equator. (After Carter and Mather, 1966)

Savannas are found in regions with annual precipitations as high as 150 cm (60 in.), although precipitation as low as 50 cm (20 in.) can support a tree savanna if temperatures are sufficiently moderate so that potential evapotranspiration is not too high.

Figure 8.11, p. 216, shows monthly temperature and precipitation data for Cuiabá, Brazil, located in a tropical savanna region at latitude 16°S. As in tropical rainforests, the temperature at Cuiabá exhibits little seasonal variation. Unlike precipitation in the rainforest, however, rainfall greatly diminishes during the low-sun season from May through September. Figure 8.12, p. 217, shows the water budget for Caracas, Venezuela, located at latitude 10°N in the savanna of Venezuela. The annual precipitation at Caracas is 80 cm (32 in.). The water budget shows some deficiency of moisture during the winter months, but soil moisture is never completely exhausted.

Deserts

In Chapter 7, we saw that dry climates occur in two general locations; one is in subtropical latitudes in association with the subtropical highs and the other is in the middle latitudes in interior continental locations distant from oceanic moisture sources.

The deserts of the subtropical dry realms are the most extensive on the earth. The Sahara (Arabic for "the desert") stretches across Africa from the Atlantic coast to the Red Sea; its eastern sections are sometimes referred to as the Libyan and Egyptian deserts. The same dry realm extends across the Arabian peninsula, Iran, and Afghanistan, extending even farther east into the Thar Desert of Pakistan. The

Figure 8.8 These photographs of ancient Maya sites in Honduras show a mixed tropical forest during the dry season **(left)** and the wet season **(right)**.

Figure 8.9 (above) This tropical shrub region in Australia is analogous to the shrub regions of South America and Africa. The low trees are a species of acacia. The shrubs and trees in tropical shrublands generally have small leaves.

Figure 8.10 (left) The grass savanna in Amboseli Park in Kenya, Africa, is an open grassland dotted with acacia trees.

desert zone stops here due to the effect of the rain-bearing summer monsoon of southern Asia. Subtropical deserts are also located in the southwestern United States and northern Mexico, and in the southern hemisphere in Australia, in Chile and Peru in South America, and southern Africa. On the equatorward margins of the subtropical deserts, the modest rainfall maximum usually occurs during summer; on the poleward margins, in contrast, the precipitation maximum tends to occur during winter in association with weak wave disturbances along the polar front.

The deserts of the middle latitudes are concentrated in the northern hemisphere where massive continents reach into the higher latitudes. The deserts between the Caspian and Aral seas in Soviet Central Asia and the Gobi Desert in Mongolia come quickly to mind. But there are also vast dry regions in the western United States in the basins and plateaus between the coastal mountain ranges and the Rocky Mountains. Although these midlatitude deserts experience cold winter temperatures, summers are hot.

Desert regions also occur along cold water coastlines in both the tropics and subtropics where the oceanic circulation produces upwelling of cold subsurface waters. These coastal deserts are especially well developed in Peru and northern Chile, South-West Africa and Angola, Baja California, Morocco, and the Somali Republic. The coastal deserts tend to be cloudy, foggy, cool, and damp, but substantial precipitation is very uncommon because of the atmospheric inversions and stability associated with the subtropical highs aloft over the cold coastal waters.

The vegetation cover developed in areas of dry climate consists of some plants that are drought-resistant and some that are drought-evading. Drought-resistant species have specialized structural and morphological adaptations that allow them to conserve moisture during long rainless periods, whereas drought-evading plants germinate and complete their life cycles quickly during those brief periods when rain dampens desert soils.

The drought-resisting plants are mainly shrubs. These are spaced widely apart, with extensive root systems to gather moisture, small and often waxy-surfaced leaves to conserve moisture, and in some cases tissues that can store moisture. In a few instances, there are no true leaves at all but only a much enlarged green stem that takes over the function of leaves in the photosynthesis process by which plants manufacture their food. The giant saguaro cactus of southern Arizona is the prime example (see Figure 8.13, p. 218). Since the fleshy tissue of cacti is attractive to animals, most cacti are armed with sharp spines. Some desert shrubs, such as the common creosote bush and sagebrush, send their roots many meters downward in search of moisture; others, such as the saguaro cactus and Joshua tree, have shallow, wide-spreading roots that take advantage of the surface moisture produced by light rain. Some remarkable shrubs can undergo almost complete dehydration without injury, becoming "resurrected" from a dormant state only when rains fall.

The ephemeral drought-evading vegetation consists of grasses and flowering herbaceous plants that resist germination until guaranteed a moisture supply sufficient to take them through their life cycles. This seems to be accomplished by chemical germination inhibitors that must be washed away by water before germination can proceed. Certain other plants produce seeds surrounded by a tough hull that can only be worn away by mechanical abrasion or cracked by mechanical force. These seeds germinate after being battered by stones carried in desert floods. Following damp periods, the desert surface may be covered by flowers as the annuals spring into activity. Once their seeds are produced and distributed, the annuals wither away, leaving the ground bare until the next wet period causes the new crop of seeds to germinate. While the perennial shrubs remain, the cover of annuals varies enormously from season to season and from year to year. This creates great difficulty for animals dependent upon vegetation for food, and also for desert nomads whose sheep, goats, and camels are at the mercy of the unreliable rains.

The most common characteristic of desert regions is that precipitation is so much less than potential evapotranspiration. Almost all precipitation goes into soil moisture storage, and much of that is quickly returned to the atmosphere by evapotranspiration. There is some local surface runoff from time to time during heavy rainshowers. Most of this water drains into local basins and returns to the atmosphere by evaporation, but some percolates to the groundwater table.

Figure 7.12, p. 200, shows that at Phoenix, Arizona, for example, surpluses are indeed rare climatic events. Where soil texture and nutrients are favorable, irrigation agriculture can be very productive, but irrigation water must be obtained from upland and mountain areas where orographic precipitation is much greater than *PE*. In Figure 7.12, Blue Canyon, California, is an example of an upland area

Figure 8.11 Cuiabá, in the Mato Grosso section of western Brazil, has a tropical wet and dry climate (Köppen *Aw* region). The graphs of average monthly temperature and precipitation show a dip during the winter months of May through September. Precipitation is nearly zero during June and July. (After Nelson, 1968)

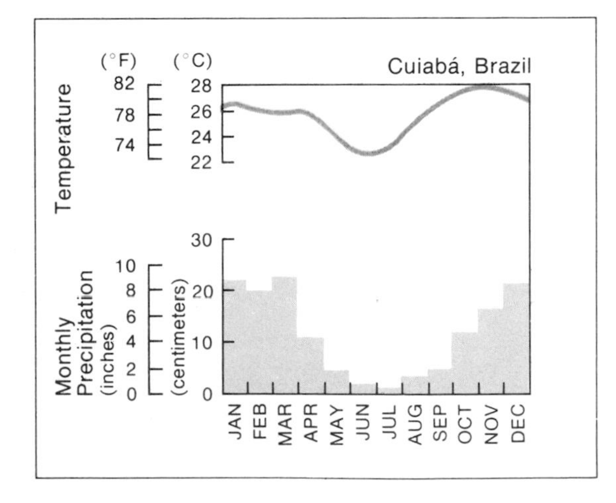

with a very large production of surplus water. Because there is little water available for evapotranspiration, most of the net radiation gain in desert regions goes into heating. The subtropical deserts become especially hot during most summer days. At Phoenix, afternoon summer temperatures typically exceed 40°C (104°F).

Chaparral

In the dry summer subtropical climatic regions along the western coasts of continents between about 30° and 45° latitude in both hemispheres, an almost unique vegetation assemblage has evolved in response to mild, rainy winters and hot, dry summers. Both the climate and associated vegetation are sometimes described as Mediterranean because relatively similar climates and vegetation border all of the shores of the Mediterranean Sea except in Libya and Egypt. This climate type is also found in smaller coastal areas of California, Chile, South Africa, and Australia.

The vegetation characteristic of the dry summer subtropical climate is commonly known as *chaparral*. Chaparral consists of an almost impenetrable mat of brush ranging from knee-high to twice the height of a man. The shrubs are small-leaved and deep-rooted to survive the summer drought period and are generally evergreen (see

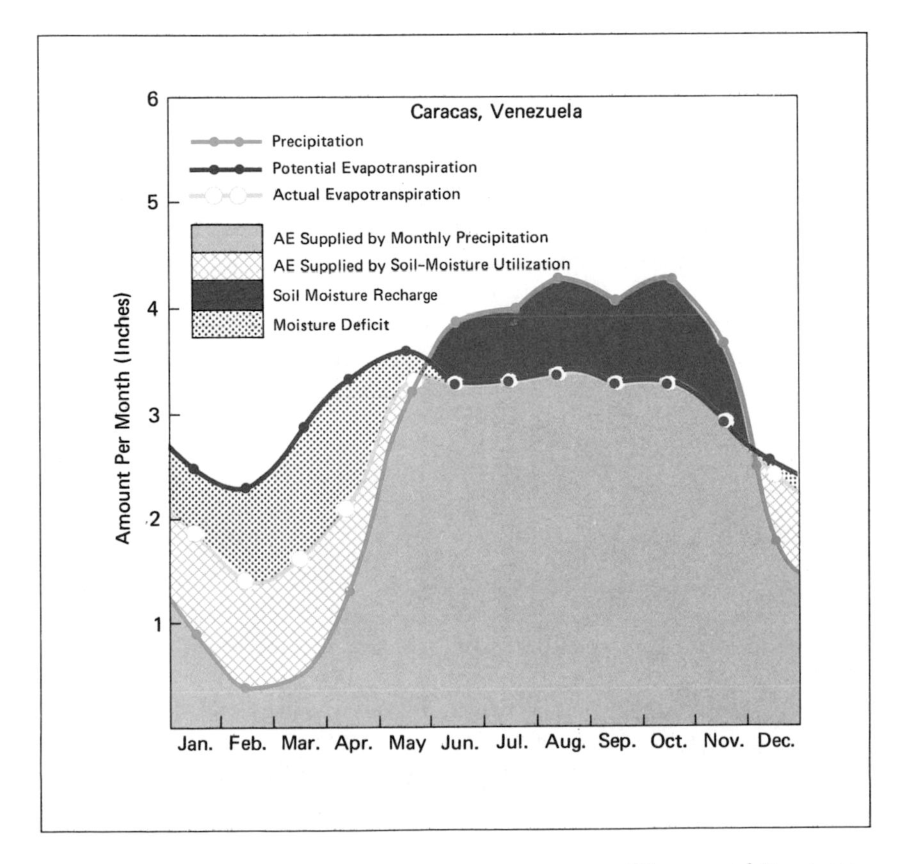

Figure 8.12 Caracas, Venezuela, has a tropical wet and dry climate (Köppen *Aw* region). The average annual precipitation is somewhat less than the annual potential evapotranspiration, but as the water budget shows, the seasonality of precipitation causes moisture deficits to occur during the months of December through May. Soil moisture stores are recharged during the remaining months. (After Carter and Mather, 1966)

Figure 8.13 (above) Little or no plant life can survive in this portion of Death Valley because it receives so little moisture.
(below) These 5 to 8 meter (15 to 25 ft) tall saguaro cactus near Phoenix, Arizona, have no leaves, a waxy skin, and are able to store water within their tissues, enabling them to survive in a desert despite their great size.

Figure 8.14). Since summer drought increases the danger of fire, most chaparral species have evolved the capability of resprouting from subsurface roots after being burned off above ground. Damp north-facing slopes may be covered by oak thickets, and flat areas with good drainage and deep soils may be virtual savannas of large oaks rising from grass or scrub. As in the case of deserts, chaparral vegetation on different continents and in different hemispheres is remarkably similar in appearance, although the plant species included may be quite different. Convergent evolution of plants is especially clear in this distinctive climatic realm.

The chaparral, or brush, vegetation in southern California sometimes receives national publicity during late summer and fall when fires sweep quickly up the steep slopes of the mountains, as seen in Figure 7.6, p. 194. Very low humidity and strong gusty winds combine to make fire fighting in chaparral dangerous and often ineffective. The chaparral serves as watershed protection on steep slopes, and if a heavy winter rain occurs before the vegetation has resprouted, water, mud, and boulders choke stream channels, and often flood through homes.

The monthly temperature and precipitation regime for Palermo, Italy, shown in Figure 8.15, is representative of dry summer subtropical climatic regions. Winter rainfall is associated with midlatitude cyclones, and the mild temperatures at this site are due mostly to the protection from cold polar continental air offered by neighboring mountain barriers and by maritime air masses from adjacent sea waters. Because of subsidence associated with the subtropical highs, summers are hot and dry. In other locations, such as San Francisco, California, cool upwelling ocean water along coasts keeps winter temperatures mild and summer temperatures cool (see Figure 8.16, p. 220).

The average water budget for San Francisco (Figure 8.17, p. 221) shows a large moisture deficit during summer and fall and a small surplus during spring after soil moisture storage has been recharged. In most dry summer subtropical climatic regions, local water surpluses are not great enough to sustain irrigation agriculture during the summer, and winter surpluses from mountain regions must be stored in reservoirs to be delivered to irrigation projects later in summer.

Midlatitude Forests

Midlatitude forests are dominant across the eastern United States, western Europe, eastern China and Japan, and much smaller areas of South America, Australia, and New Zealand. These forests tend to be located within humid continental climatic regions, but they also occur in most of the humid subtropical and marine west coast climatic regions. Except in upland areas, most of these regions have been cleared for cropland and pasture so that very little "original" forest remains.

Figure 8.14 This chaparral vegetation is located in the Rif Mountains of northern Morocco, with the Straits of Gibraltar in the distance. The plants in such regions are adapted to Mediterranean climates and are protected from prolonged drought by small, leathery leaves. Nevertheless, climate is not the sole factor involved in the establishment of chaparral. Many areas were once forested, and it is believed that land clearing, overgrazing, and fire altered the ecological balance and made it difficult for trees to survive.

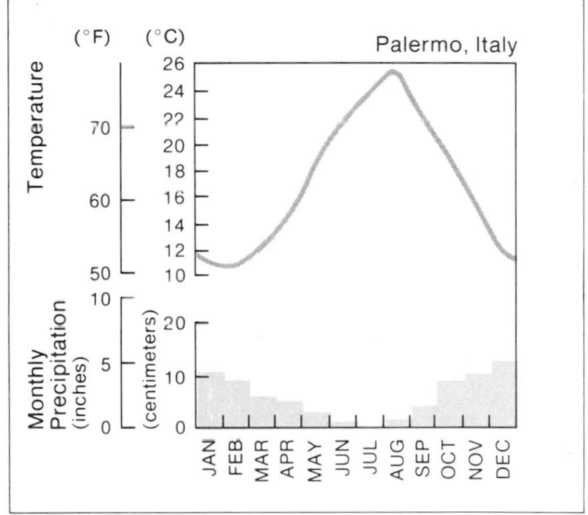

Figure 8.15 Palermo, Italy, has a Mediterranean climate characterized by moderate temperatures and a dry summer (Köppen *Cs* region). As the graphs show, winter in Palermo is cooler and more moist than summer. The average annual precipitation is nearly 80 cm (31 in.), but June, July, and August combined receive only 4 cm (1.5 in.). (After Nelson, 1968)

Because of their spatial separation on each of the continents, midlatitude forests include many different combinations of species. In the eastern United States, for example, oak, hickory, maple, and beech each tend to be dominant in various areas. The canopy of the handsome multistoried forest is usually about 30 meters (100 ft) or more above the ground. Some shrubs and shade-tolerant annuals occupy the ground surface, but the forest tends to be relatively open below the crowns. Most of the trees drop their leaves during winter, so the appearance of the forest changes dramatically through the seasons (see Figure 8.18, p. 222). Winter defoliation is a response to physiological drought. In most midlatitude forest regions, soil water in the root zone is frozen in winter and therefore unavailable to the trees. Continued transpiration of moisture by plant leaves, with no uptake of moisture from the soil, would surely be fatal to the plant. Therefore plants with a leaf-dropping mechanism have an evolutionary advantage under these conditions when compared to broadleaf evergreen trees. Recall that the deciduous habit is similarly favored in areas of winter drought, such as the mixed tropical forests.

In the eastern United States and western Europe, the broadleaf deciduous forests become mixed with coniferous evergreens on the northern margins and with broadleaf evergreens on the southern margins; broadleaf evergreens are dominant in the southern hemisphere, where winters are mild. Pines tend to be dominant on the coastal plain of the southeastern United States, where sandy soils store

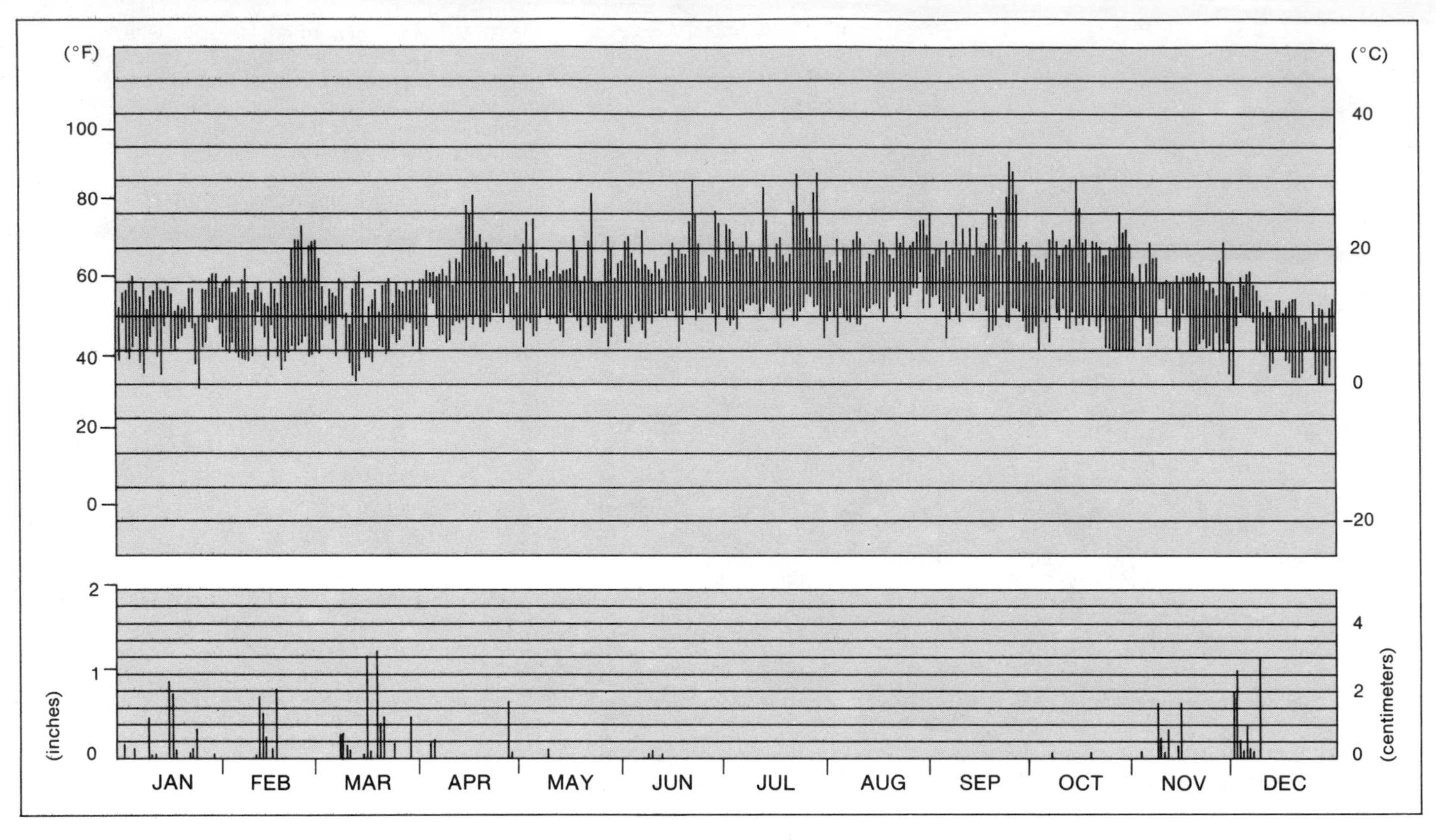

Figure 8.16 The graphs show the daily temperature and precipitation at San Francisco during 1954. Each vertical line on the temperature graph indicates the maximum and minimum temperatures recorded for a given date. Temperatures remain moderate at San Francisco through the year because of the prevailing winds from the Pacific Ocean. The temperature reached or exceeded 30°C (86°F) on only five days in 1954, and freezing temperatures occurred only twice. Note the almost complete lack of precipitation during the summer months. Precipitation during the rainy season tends to occur in episodes of a few rainy days separated by periods of dry weather. (After Hendl, 1963)

only limited reserves of soil moisture. The pines are also more susceptible than deciduous trees to the fires that sweep areas of the coastal plain from time to time; some scientists believe that the pines are only an intermediate successional stage toward the broadleaf deciduous forest climax.

Most midlatitude forest regions are characterized by a large temperature range between winter and summer and a relatively even distribution of precipitation, or slight summer maximum. Pittsburgh, Pennsylvania, is near the boundary between Köppen's humid subtropical (Cfa) and humid continental (Dfb) climates; its monthly temperature and precipitation regimes are displayed in Figure 8.19, p. 222. In general, summers are hot, but temperatures well below freezing are common during winter months. Precipitation varies little. Klagenfurt, Austria, is representative of midlatitude forest regions with humid continental climates; its large seasonal temperature range is shown in the thermoisopleth in Figure 8.20, p. 223. The graph should be compared with Figure 8.6, p. 213, which illustrates the thermoisopleth for a tropical rainforest location.

The average water budgets of most locations with midlatitude forests show relatively small summer deficits and relatively large winter and spring surpluses. Figure 8.21, p. 223, illustrates, however, the variability of seasonal deficits and surpluses for Middlesex County, in central New Jersey. There are small deficits just about every summer, but this figure shows the large deficits and small surpluses during the

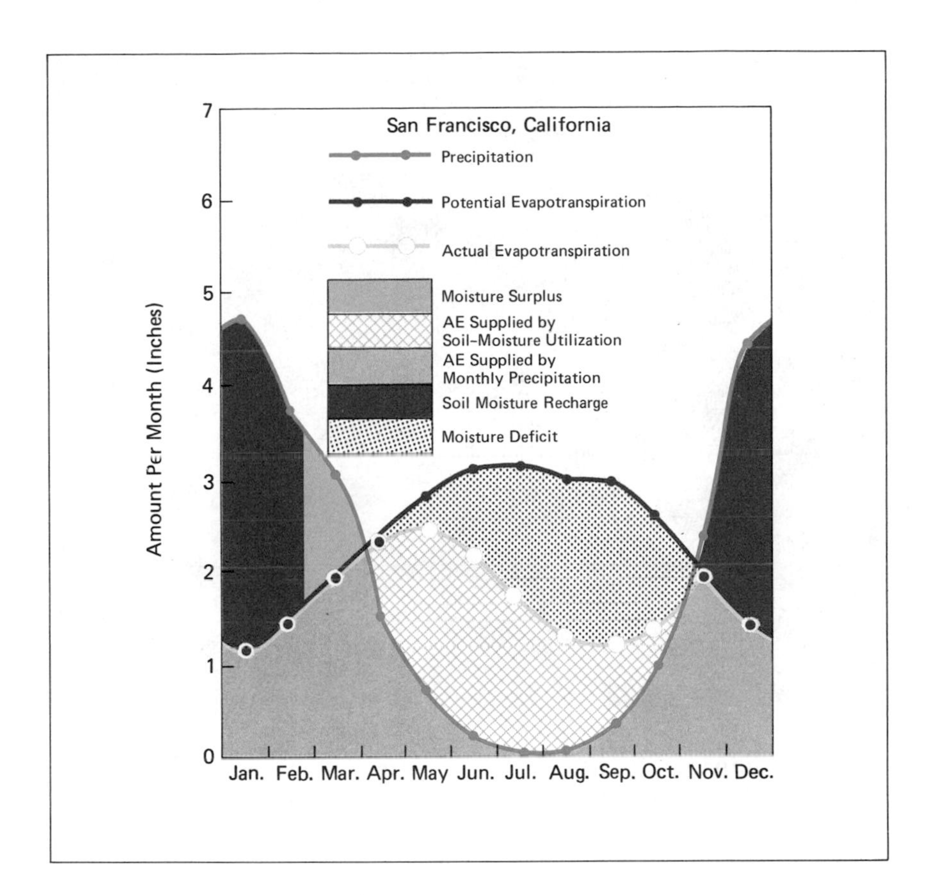

Figure 8.17 The local water budget for San Francisco, California, is characteristic of stations with a Mediterranean climate (Köppen *Cs* region). A deficiency of moisture persists through the dry summer months, and plants that live through the summer must draw upon fog drip and stored soil moisture. From October through March, when most of the precipitation is received, soil moisture stores are replenished, and a surplus is generated. (After Carter and Mather, 1966)

great drought of the early 1960s which extended throughout most of the northeastern United States.

Midlatitude Grasslands

Grasses dominate the vegetative cover in areas where precipitation does not fully meet the needs of trees and shrubs, or where periodic fires prevent tree regeneration from seedlings. Vast areas of continuous grasslands once extended across central North America from Texas to Alberta and Saskatchewan; extensive grasslands, known as *steppes,* were also found from the Ukraine in the western Soviet Union across Asia to Manchuria. These grasslands included both *tall-grass prairies* and *short-grass prairies.* Similar extensive grasslands were also present in South America, particularly in Argentina and Uruguay. Due to intensive agricultural exploitation, there are very few areas of natural grassland left in any of these regions.

Midlatitude grasslands contain a large number of plant species that are different from the grasses of the tropical savannas. The dominant grasses are usually perennials that lie dormant during winter and continue their growth in the next growing season. Near the boundary between forest and grassland, where moisture is comparatively abundant, the natural grassland vegetation is usually tall prairie grass 1 to 2 meters in height. Most prairie areas throughout the world are now utilized for intensive agriculture or wheat farming. Where moisture is

Figure 8.18 The mixed broadleaf deciduous forests of the United States are alive with color for a few weeks in autumn; this example is from southern Wisconsin. Each species progresses through a sequence of color changes and leaf fall, with some species regularly ahead or behind others. The variable temperature and moisture conditions of each fall season cause the period of maximum color to vary by as much as four weeks or more.

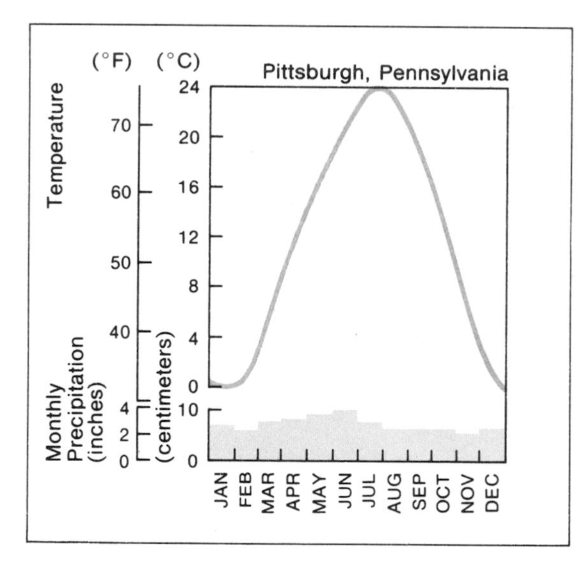

Figure 8.19 Pittsburgh, Pennsylvania, is near the boundary between a humid subtropical and a humid continental climate (Köppen *Cfa* and *Dfb* regions). The graphs of temperature and precipitation show the great range of the average monthly temperature between summer and winter. The monthly precipitation is nearly constant through the year, however. and does not exhibit seasonal behavior. (After Nelson, 1968)

less, the dominant vegetation is short prairie grass, generally less than a meter high (see Figure 8.22, p. 224), and in the driest grasslands, the grass becomes bunched and tufted, with bare ground often visible between the clumps. In all types of grassland, patches of trees may occur here and there, especially along streams.

Like the tropical savannas, the temperate grasslands of Eurasia and North America were once immense pastures, supporting vast numbers of grazing animals. Upwards of 40 million bison and 50 to 100 million antelope pastured on the American grasslands, with single herds of 100,000 to 2 million animals having been seen. All modern breeds of horses appear to have descended from the vast herds that formerly dominated the Eurasian grasslands.

The tall-grass prairies in Illinois and Iowa were located near the western margins of Köppen's humid continental climatic type. Almost all this land is now utilized for corn, small grain, and hog farming, but many ecologists believe that the climate of this region could have supported forests if the fires that swept the grasslands prior to modern agricultural land uses could have been eliminated.

Farther west on the high plains of Alberta, Montana, the Dakotas, Wyoming, Colorado, and Texas, short-grass prairies are present. Here tree growth is precluded (except along stream courses) by complete drying out of the subsoil during periodic droughts. This phenomenon is very clear in the morphology of the soils developed in this area, as will be seen in Chapter 10. These grasslands are used predominantly as unfenced pasture for cattle. The tall-grass prairies grade into the areas of short bunched grass through a broad transition zone centered just east of the 100th meridian from central Texas to the Dakotas, with a westward swing into Saskatchewan and Alberta. This transition zone is tempting but dangerous for agriculture; its cycles of wet years lure wheat farmers in only to meet disaster when the inevitable dry years follow, transforming the land into a "dust bowl."

Most of the world's grassland areas are located within interior dry continental climates, specifically Thornthwaite's subhumid continental type (see Figure 7.8, p. 196) where precipitation is, on the average, just barely equal to potential evapotranspiration. In the grasslands of continental interiors in the northern hemisphere, summers tend to be hot and winters very cold and dry. Most of the rains fall in summer. In the less continental climates of the southern hemisphere, grasslands develop mainly in thc rainshadow of the Andes Mountains in Argentina and Uruguay.

The average water budget for Huron, South Dakota, located near the diffuse boundary between the tall and short grasslands, is shown

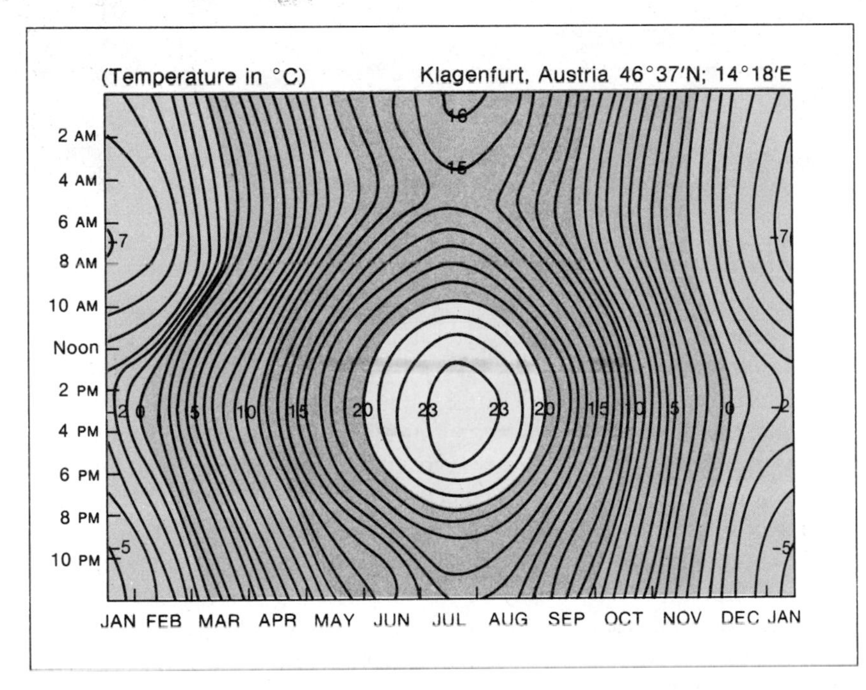

Figure 8.20 Klagenfurt, Austria, has a humid continental climate (Köppen *Dfa* region). As the thermoisopleth shows, the range of temperature through the year exceeds the range through the average day. In January the noon temperature is approximately −4°C (25°F), whereas in July the noon temperature is 22°C (72°F). Note that in November the temperature through the day varies by only a few degrees, partly because of the moderating effect of cloudy weather. (After Troll, 1965)

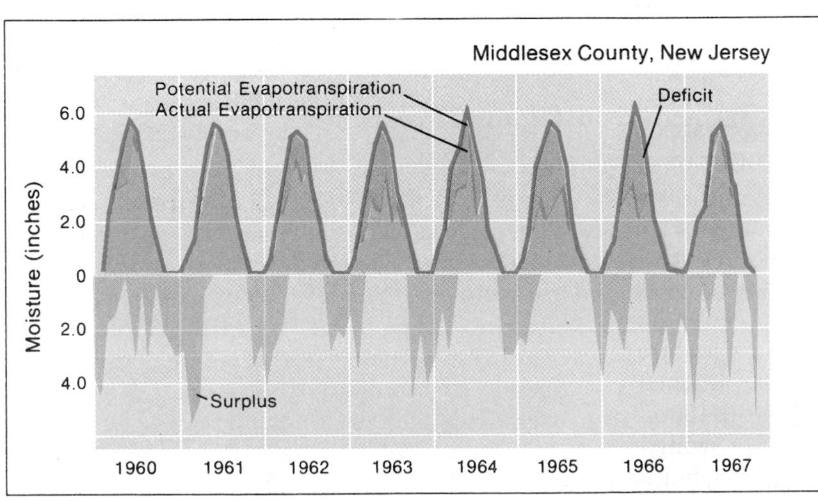

Figure 8.21 These local water budgets for Middlesex County, in central New Jersey, illustrate climatic variability in water budget terms at the equatorward margins of the humid continental climatic region with a midlatitude forest. During the drought years of 1962 through early 1966, deficits were large, and winter-spring surpluses were not great enough to meet the water resource needs of the region. Water restrictions were common, and there was some loss of shrubs and trees in suburban areas. (After Muller, 1969)

Figure 8.22 This tussocked grassland is in South Island, New Zealand. Radiocarbon dating of wood fragments indicates that extensive forests occupied some of this region 500 years ago. The conversion to grassland may have occurred as a result of human use of fire.

in Figure 8.23. Despite the rainfall maximum in summer, the deficit is still relatively large, and precipitation during winter is not great enough to generate a significant surplus. Some climatologists and ecologists believe that clusters of drier than normal years are mainly responsible for the absence of forests within the wetter margins of the subhumid climatic regions. Figure 8.24, p. 226, shows the annual ratio of actual to potential evapotranspiration *(AE/PE)* at Dodge City, Kansas, near the drier western margin of the midlatitude grasslands. This graph clearly illustrates the spells of wetter and drier years that tend to be characteristic of the Great Plains and even the more humid regions immediately to the east. These fluctuations are apparently associated with changes in the upper atmospheric circulation of the winds of the circumpolar vortex. In the past, and especially during the "dust bowl" years of the 1930s, this variability has had considerable impact on the economic and social structure of the United States.

Northern Coniferous Forests

Coniferous forests dominated by spruce, fir, and pine extend in a broad band across North America, Europe, and Asia between about 50° and 65°N latitude. These forests, also known as *taiga* or northern boreal forests, have developed in the subarctic climatic region where extremes of continentality occur; that is, these forests experience the largest annual temperature ranges encountered on earth. Winters are very cold and dry, with only light snowfall. These areas are dominated by the Canadian and Siberian highs and are the source regions for polar continental air masses. Although the summer is very brief, there are many daylight hours, and temperatures are mild, even warm to hot, occasionally exceeding 25°C (77°F).

The conifers of the taiga are tall, slim, and tapered upward, as in Figure 8.25, p. 227. Most conifers are evergreen and do not lose their needles during winter. Their small thick-surfaced needles and thick bark resist moisture loss and help them withstand the long, cold winters. Since they do not need to produce a set of new leaves before they can begin manufacturing food in the spring, they can start growth as soon as temperatures rise sufficiently. This is a clear advantage where the growing season is very short, as it is in these latitudes.

Coniferous forests contain a comparatively small number of plant species. The dense layers of needles allow little light to reach the ground, and the cool temperatures also limit plant growth. While spruce is common in the coniferous forests of North America, larch, which drops its needlelike leaves in winter, is found in eastern Siberia, where conditions may be too severe even for needleleaf evergreens. In the southern margins of the coniferous forests, the trees are tall and densely packed. Farther north, the trees tend to be smaller and increasingly separated from one another.

Figure 8.26, p. 227, shows the mean monthly regimes of temperature and precipitation at Moose Factory, in central Canada. Temperatures range from −20°C (−3°F) in winter to 16°C (60°F) in summer. Much of the annual precipitation falls as rain during summer, when

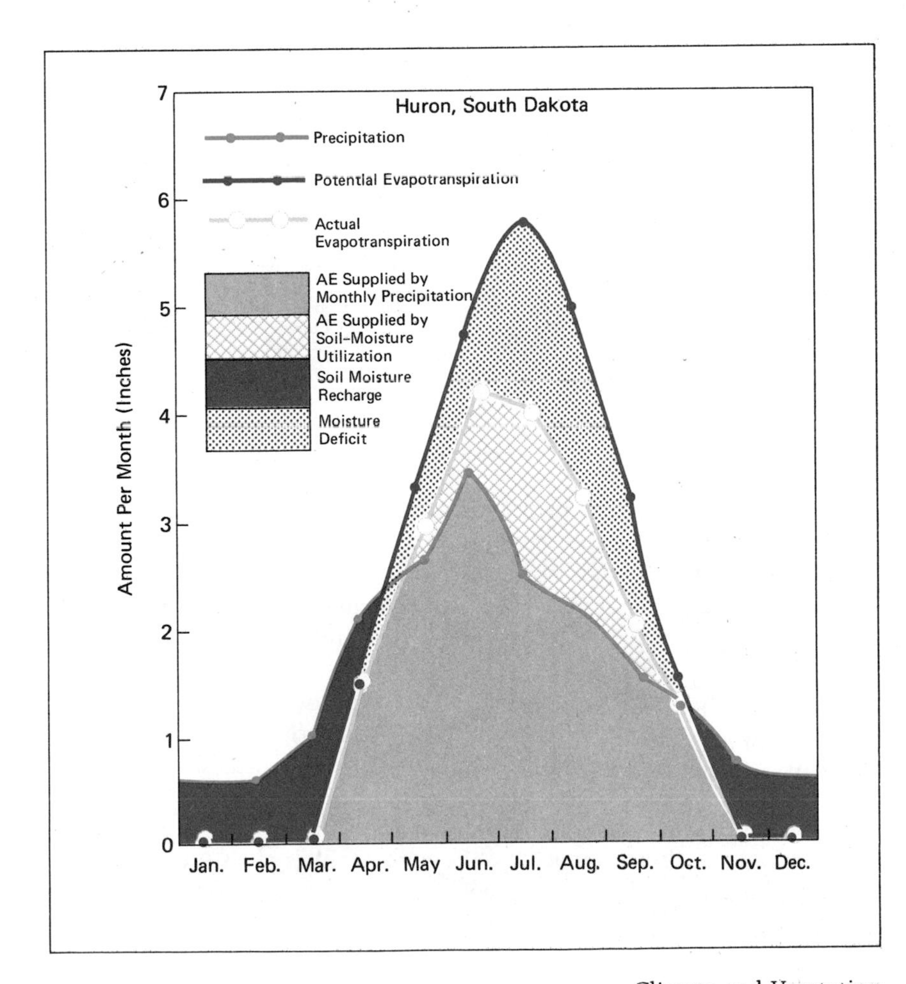

Figure 8.23 Huron, which is located in the prairie grasslands of South Dakota, has a humid continental climate with moderate moisture and cold winters (Köppen *Df* region) and a subhumid climate *(C)*, according to Thornthwaite. The local water budget for Huron shows that precipitation occurs in all months, and is most plentiful during early summer. However, high summer temperatures cause potential evapotranspiration to be high then as well, and moisture deficits normally occur. The soil moisture is recharged during the cooler months, when the vegetation's demand for moisture is small, but winter precipitation is not large enough, on the average, to generate surpluses for runoff. (After Carter and Mather, 1966)

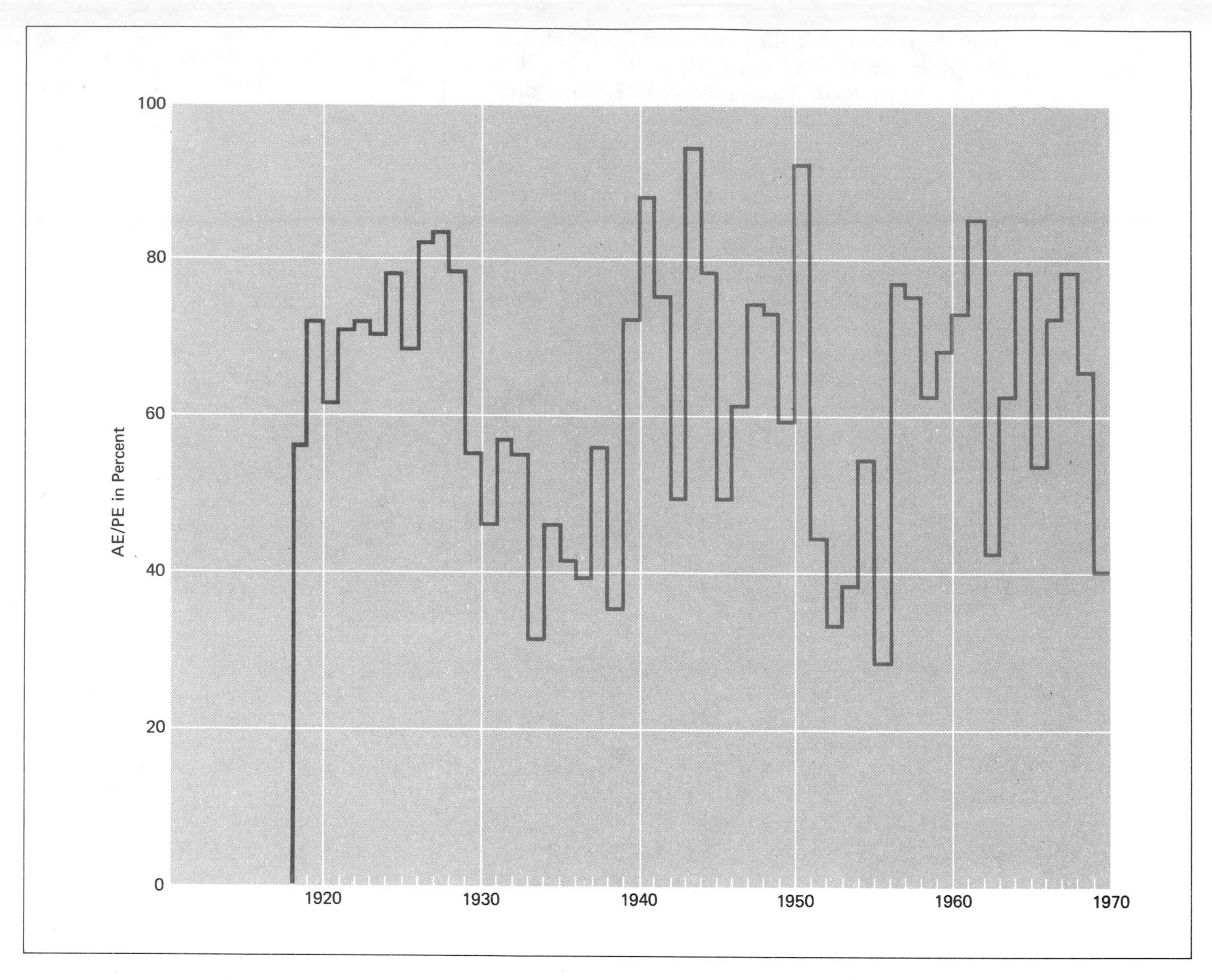

Figure 8.24 This graph shows the annual ratio of actual evapotranspiration to potential evapotranspiration *(AE/PE)* in percent for Dodge City, Kansas, which is located near the drier western margin of the midlatitude grasslands. The clusters of drought years in the 1930s and 1950s, "dust bowl" years, are remarkable features of the climate. In 1934, for example, *AE* amounted to only 33 percent of *PE,* but in 1944, the ratio climbed to 94 percent. These climatic fluctuations have been characteristic of the Great Plains and most other subhumid climatic regions, and they are likely to have impacts on most natural response systems and human settlement and economic activity.

PE is just about as high as in the midlatitudes. Hence, most summer rainfall is utilized for evapotranspiration, and nearly rainless periods occasionally result in small deficits; some summers are dry enough for the danger of fire to be serious. Local water budgets show that most of the surplus is produced in late spring during and just after the snowmelt season.

In North America, the coniferous forests of the subarctic climatic region spread southeastward from the Mackenzie River valley in northwestern Canada across the Canadian border into sections of Michigan, New York, and New England. The border between the taiga and the tundra to the north is much farther south in eastern than in western Canada; this orientation oblique to parallels of latitude can be attributed mostly to shorter and cooler summers over much of eastern Canada.

Figure 8.25 Coniferous evergreens, commonly spruce, fir, and pine, occupy large regions of the northern middle latitudes and the subarctic. The uniform appearance of this coniferous forest in Washington's Cascade Range shows how only a few species of conifers form the dominant vegetation. Mt. Rainier is in the background.

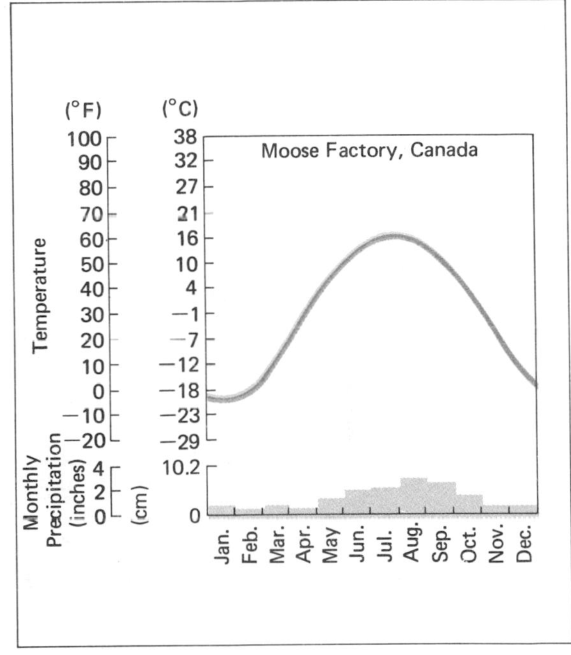

Figure 8.26 Mean monthly temperature and precipitation regimes at Moose Factory, in central Canada. The range between winter and summer temperatures in the subarctic climate type is the largest of all global environments. The northern coniferous forest is well adapted to short but warm summers, when much of the annual precipitation falls.

Along the northwestern coast of North America, coniferous forests extend southward into Washington, Oregon, and northern California. These magnificent forests consist of different species from those of the taiga. Redwoods (Sequoia) and Douglas fir are perhaps the best known of these giant trees, but other species of needleleaf evergreens are also common. These forests appear to be adapted mostly to cool, very wet winters and mild, foggy summers; the term "rainforest" is often applied to the more luxuriant stands.

Tundra

Trees cannot endure unless the average temperature of the growing season exceeds about 10°C (50°F) for a period of two to three months. Near the Arctic Ocean, the trees of the northern coniferous forest give way to low shrubs, grasses, flowering herbs, mosses, and lichens (see Figure 8.27). This vegetation association is known as *tundra.* Mosses and lichens live on rock surfaces without soil, and in sheltered locations they sometimes form layers several centimeters thick.

Figure 8.27 Low shrubs, grasses, and flowering herbs are the dominant vegetation types in tundra regions, such as shown here in Norway. Accumulations of water are visible in the background; tundra regions tend to be wet during the period of thaw because water cannot drain through the permafrost layer below the surface.

Although winter temperatures in the tundra regions are not as low as those in the more continental taiga of eastern Siberia, they are nevertheless extremely severe and are commonly accompanied by galelike winds from which there is no shelter. A 30 kph (20 mph) wind at −18°C (0°F) produces the equivalent temperature of −40°C (−40°F); at the same wind speed −30°C (−24°F) is equivalent to −55°C (−68°F). This combination of wind and temperature is dangerous to water-bearing tissues and produces extreme moisture stress because the evaporation induced by wind cannot be offset by water intake from the ground, which is frozen. As a consequence, tundra plants are small-leaved, like desert plants, and low-growing so that they

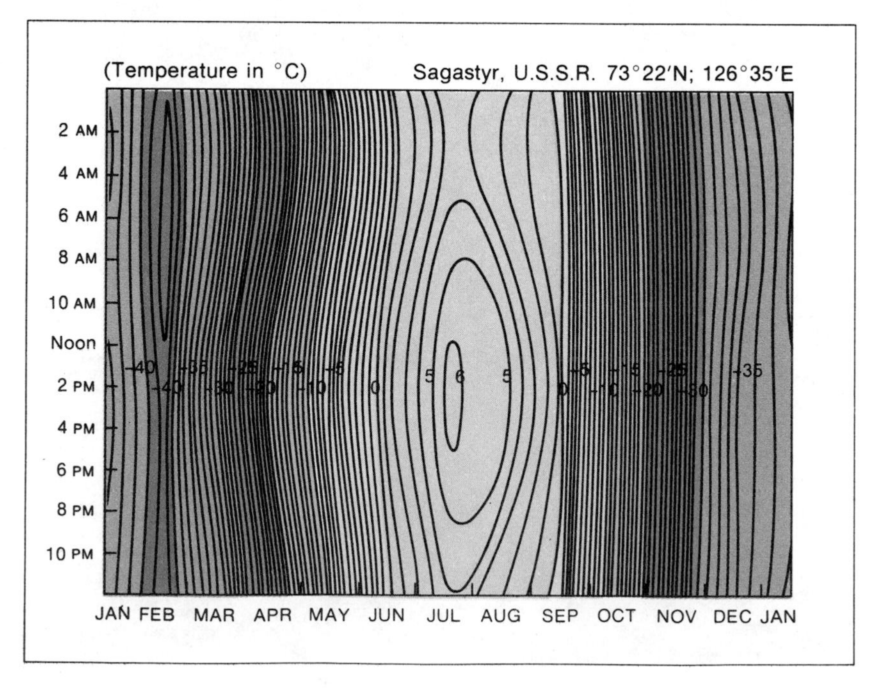

Figure 8.28 Sagastyr, U.S.S.R., is a station in the tundra region of Siberia north of the Arctic Circle (Köppen *ET* region). The temperature during the year varies through an extreme range because of the lack of sun near the time of winter solstice and the continual daylight at summer solstice. The nearly vertical temperature contour lines imply that the temperature during any given day is essentially constant. (After Troll, 1965)

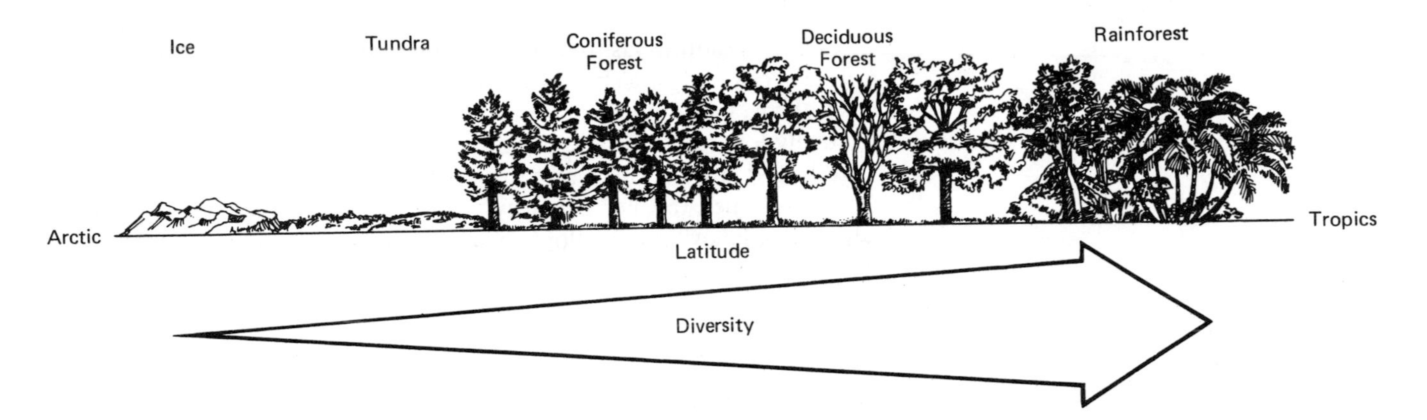

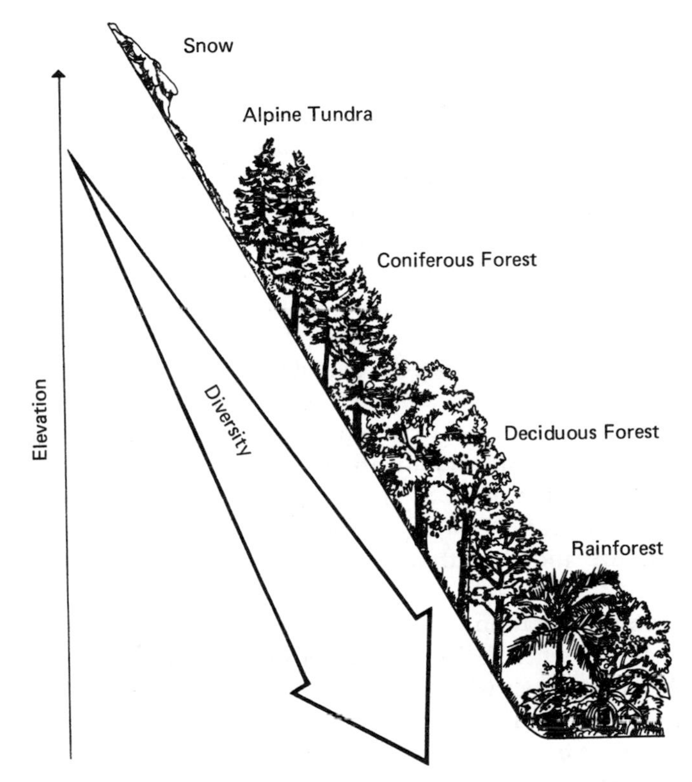

Figure 8.29 Relationship of latitude and elevation on vegetation distribution and community diversity. Increasing latitude or elevation results in a shift to cooler vegetation types with a decrease in community diversity. Climbing 4,500 meters (15,000 ft) up a tropical mountain will reveal shifts in communities analogous to what could be observed in a trip from the tropics to the pole.

are covered by snow when icy winter winds sweep over the land surface. At the forest margin, tundra persists on exposed wind-swept ridge crests, which suggests that evaporative windstress combines with the length of the growing season to determine the poleward limit of tree growth.

Figure 8.28 is a thermoisopleth diagram for Sagastyr, at latitude 73°N, in the tundra of northern Siberia. The nearly vertical pattern of isotherms shows that the daily variation of temperature is only a few degrees because of the long daylight periods in summer and the absence of sunlight during midwinter. Seasonally, the average midday temperatures vary from 6°C (43°F) in July to –40°C (–40°F) in February.

Precipitation tends to be low throughout the year. In winter, the tundra is covered by a shallow snowcover for six to eight months, but most of the sparse precipitation falls as rain in summer. Despite the relatively low summer precipitation, the tundra is usually moist and waterlogged during the warm months because water cannot drain downward into the permanently frozen subsoil, called *permafrost,* which lies below the shallow surface layer of soil that thaws each summer.

Highland Vegetation

Average temperatures tend to decrease with increasing elevation almost everywhere so that one can progress through climatic and vegetation zones as one climbs upward in mountainous regions.

Figure 8.29, p. 229, shows that in humid climatic regions the vertical zonation of vegetation with increasing elevation is somewhat like the zonation of vegetation from equatorial latitudes toward the poles. In dry climatic regions, increasing precipitation with elevation produces spectacular vegetation zonation within short horizontal distances. In the San Francisco Peaks region of northern Arizona, where the vertical zonations of North America were first studied, the upward progression is from desert shrubs at 1,200 meters (4,000 ft) elevation to a spruce-fir forest at 3,000 meters (10,000 ft) that is not unlike the spruce-fir forests of subarctic Canada. Due to elevation differences, the desert shrubs and spruce-fir forest, which normally grow thousands of miles apart, here exist within a few miles of each other. In tropical upland regions, the diversity of contrasting vegetation is even greater. In such areas as the island of Hawaii or the highlands of Papua, in eastern New Guinea, nearly all of the world's latitudinal temperature and vegetation zones can be recognized in vertical succession.

Local climates and vegetation in highland areas are very complex, however, and analogies with latitudinal zonation should not be pushed too far. Adjacent slopes with different orientations differ dramatically in terms of solar radiation, temperature, and precipitation; some north-facing slopes of deep valleys in midlatitudes may receive little direct sunlight over the year, while nearby south-facing slopes may bake in many hours of sunshine each day. On windward slopes, precipitation tends to increase sharply with elevation; thus for at least the first 1 or 2 km of elevation, upland areas tend to produce much larger water surpluses than surrounding lowlands. A flourishing vegetation cover in upland regions retards water runoff, reducing soil erosion on steep slopes and decreasing flood hazards in adjacent lowlands. At the same time, forested uplands release water slowly to streams, sustaining their flow through rainless periods. Thus the management of upland vegetation, the control of soil erosion, and the stabilization of streamflow all go hand-in-hand as matters of critical concern to people within the highland regions and far beyond them on the great river flood plains of the earth.

Further Reading

Bennett, Charles F., Jr. *Man and Earth's Ecosystems: An Introduction to the Geography of Human Modification of the Earth.* New York: John Wiley & Sons, Inc., 1975. (331 pp.) This unique text is organized by world regions in order to focus on the ecological impacts of human use of the earth. The book specifically analyzes the geographical and historical background of the environmental crisis.

Billings, W. D. *Plants, Man, and the Ecosystem.* 2nd ed. Belmont, Calif.: Wadsworth Publishing Co., 1970. (160 pp.) This brief paperback, part of the Fundamentals of Botany series, includes many succinct sections that supplement Chapters 8 and 9 of this text.

Eyre, S. R. *Vegetation and Soils: A World Picture.* 2nd ed. Chicago: Aldine Publishing Co., 1968. (328 pp.) This book, written from the perspective of the British Isles, focuses on relationships between vegetation and soils. It is organized by global vegetation types and is especially useful for Chapters 8 and 10.

Kellman, Martin. *Plant Geography.* London: Methuen & Co., 1975. (135 pp.) This more advanced introduction is especially useful for students undertaking field projects.

Seddon, Brian. *Introduction to Biogeography.* New York: Barnes and Noble Import Division of Harper & Row, 1973, (220 pp.) The factors that control vegetation distributions are studied, as well as the distribution changes that have occurred during recent earth history.

Tivy, Joy. *Biogeography.* New York: Longman, 1971. (394 pp.) Plant distributions and vegetation formations are explained in terms of environmental factors.

SUMMARY

Supplies of solar radiation and moisture determine the type of vegetation that can grow in a particular location. Botanists have classified the exceedingly complex mosaic of vegetation formations into groups, or plant communities. When disturbance of natural vegetation occurs, or when agricultural land is abandoned, the open land progresses in time through an orderly and largely predictable succession of plant communities until a mature, or climax, vegetation is attained. The climax vegetation is assumed to be in equilibrium with the environment and thus does not change further in composition. Although climax vegetation is determined chiefly by the availability of moisture and energy, climatic variability and many other factors also have a significant impact from time to time.

In this chapter, natural vegetation formations have been combined into nine major global types. Forests are restricted to the humid climatic realm, but various types have evolved in response to different seasonal energy and moisture regimes. Grasses and specialized xerophytic vegetation dominate the dry climatic realm; a subhumid climatic zone astride the boundary between the humid and dry realms contains elements of both humid- and dry-land vegetation. Energy availability is so restricted toward the polar margins of the humid climatic realm that forests give way to various forms of low vegetation collectively called tundra. In highland areas, the distributions of local climates and vegetation are complex, but there are some similarities to the poleward gradations of climates and vegetation regions. The next chapter focuses more specifically on the energetics of climate and plant interrelationships within a community framework.

Review Questions

1. Describe a typical plant succession from open land to climax community in a midlatitude region.
2. There is far more diversity of plant species in moist equatorial regions than in subarctic latitudes. Speculate about the reasons for this difference.
3. Compare the daily and annual temperature cycles for the tropical rainforest and tundra. How are these cycles related to the characteristic vegetation in those regions?
4. Why is the impact of logging and other human activities more devastating in tropical rainforests than in subarctic or midlatitude regions?
5. What factors limit extensive tree growth on savanna lands?
6. Compare the adaptations of drought-resistant and drought-evading plants.
7. Compare the environmental factors that have determined the physical appearance of cacti and coniferous evergreens.

Walter, Heinrich. *Vegetation of the Earth in Relation to Climate and the Eco-Physiological Conditions.* Translation of the 2nd German edition. New York: Springer-Verlag, 1973. (237 pp.) This work stresses relationships between climate and vegetation and also includes short descriptions of the vegetation formations of each natural vegetation type.

Watts, David. *Principles of Biogeography.* New York: McGraw-Hill Book Co., 1971. (402 pp.) This text emphasizes environmental science within a systems framework.

Whittaker, Robert H. *Communities and Ecosystems.* 2nd ed. New York: Macmillan Co., 1975. (385 pp.) This is another outstanding paperback on ecology, especially pertinent to Chapters 8 and 9.

Case Study:

Slash-and-Burn Agriculture: An Efficient or Destructive Land Use?

In vast areas of the tropics, inhabitants use a method of agriculture largely unknown to midlatitude populations. Slash-and-burn agriculture—also known as shifting cultivation, *swidden* cultivation, *milpa* cultivation (Latin America), and *ladang* (Southeast Asia)—constitutes the principal method of farming practiced on one-third of the total land area currently used for agriculture in Southeast Asia. It has been estimated that in some countries, among them the Philippines, up to 10 percent of the population depends on shifting cultivation for its food.

In Latin America, typical crops produced by shifting cultivation are maize, manioc, and squash. In Southeast Asia, rice, cucumbers, and maize are often planted the first year, followed by second-year crops of cassava, sugarcane, and squash.

The actual technique of slash-and-burn agriculture is essentially just what the name implies. Several months before the region's rainy season begins, the trees and undergrowth on a selected plot of forested land are cut. The remaining vegetation is left on the ground to dry and is later cleared away by burning. At the beginning of the rainy season, crops are planted, usually by using sticks or hoes to dig holes for the seeds.

The same plot is replanted until after a few harvests the decrease in soil fertility or the growth of weeds and grasses reaches a point at which cultivation is no longer worthwhile. Cultivators then abandon the field, leaving it to fallow for several years, and move on to another plot to repeat the process of cutting, burning, planting, and harvesting.

Shifting cultivation is not feasible for a high-density population because crop yields are low and because the long fallow period required for reestablishment of the cleared vegetation means that only a fraction of the arable land can be productive at any given time.

The overall effect of slash-and-burn agriculture on the ecology of a region is difficult to evaluate. Many experts argue that shifting cultivation is an efficient and, if properly practiced, ecologically sound method of utilizing tropical land that would be difficult to cultivate by most other methods. They are also quick to point out that in underdeveloped countries, slash-and-burn agriculture may be necessary to support the people who eke out a subsistence living by practicing it.

Other experts, however, argue that this method of agriculture is a reflection of cultural level rather than of any unavoidable or unalterable physical limitations of the environment, and that on the whole the method is wasteful and inefficient. They point out that the clearing of vegetation exposes the already thin, infertile topsoil of the slopes to the powerful erosive force of tropical rainfall. The heavy erosion results in an increased silt load in stream channels, which decreases their depth and causes flooding. In some regions, slash-and-burn agriculture also is responsible for the depletion of various species of timber.

Most scientists would agree, however, that any damage wrought by slash-and-burn agriculture is minimal—perhaps even negligible—under certain conditions and if properly practiced. The amount of available land must be sufficient to allow the abandoned plots adequately long fallow periods, and there must be sufficient seasonal variation in the climate and rainfall of the region so that cut vegetation will dry and newly planted crops will be watered. The method can be used only to support a low-density population with a subsistence economy that does not require a large surplus of food for trade, and there must be minimal influence or pressure from external culture groups, such as loggers, that practice economic systems in conflict with the system of shifting cultivation.

In addition, certain practices must be followed: a safety path must be made around the clearing to prevent the spread of fire during the burning process; the same plot must not be used over and over again for the same crop in order to avoid depletion of certain soil nutrients; and secondary rather than primary forests must be used as cultivation sites. If these practices are not followed, and if the proper conditions are not present, slash-and-burn agriculture can indeed do great damage to soil, biotic, and water resources, reducing vast areas to economic uselessness in a short period of time under the climatic extremes of the tropics.

(top) *The natural vegetation on this Honduran slope has been cut, dried, and burned, and replaced by the* milpa *crops shown here. The farmers who cultivate this plot will abandon it for a new area after only a few harvests.*

(bottom) *Most of this formerly forested area near Oaxaca, Mexico, has been transformed into a wasteland as a result of* milpa *agricultural practices.*

The Waterfall by Henri Rousseau. (Helen Birch Bartlett Memorial Collection; courtesy of the Art Institute of Chicago)

All interactions among the parts of an ecosystem are united by the energy flow through the system. This intimate interdependence of living things implies a certain stability, a certain dynamic reciprocity, and a certain harmony of nature.

NINE
Ecologic Energetics

ALL LIFE REQUIRES ENERGY, and the ultimate energy source is the solar radiation that green plants convert to chemical energy. It is this chemical energy stored in plant tissues that animals and humans use for food. Indeed, even the fossil fuels, such as coal and petroleum, represent concentrated forms of solar energy derived from plants and microscopic animals which interacted with the solar radiation of the past millions of years of earth history. Physical geographers therefore are concerned with how environmental inputs, such as solar energy and precipitation, control plant productivity and the resultant interactions among all life forms, and how these interactions vary from place to place.

In the preceding chapter, we saw that climate, in particular energy and moisture, largely determines the distribution of natural vegetation formations over the earth. We also saw that plants live in groupings, or communities, that appear to be mutually beneficial to the individual members. Animals also live in association with the plants and rely directly or indirectly on them for their food. These interactions of plants and animals with one another and with their environment give the communities definite patterns of organization. However, the task of tracing and understanding all of the interactions is extraordinarily difficult. For example, a typical midlatitude deciduous forest contains perhaps fifty species of plants and thousands of species of animals, primarily insects; within a Douglas fir forest in Oregon, there are about fifty species of lichens and mosses alone.

One way to simplify the study of these interactions is to focus on the basic ecological unit, the ecosystem. An *ecosystem* is a community of plants and animals generally in equilibrium with the inputs of energy and materials in their particular environment. An ecosystem can be as small as a tidal pool or a single tree or as large as a lake or a forest. The interactions within an ecosystem can be analyzed in terms of energy flows and materials cycles. Solar energy enters and cascades through the ecosystem, and is partitioned among physical processes, such as longwave radiation exchanges and evapotranspiration, and biological processes, such as photosynthesis and the production of chemical food energy. Materials and nutrients, on the other hand, are continually recycled within the system. In this chapter, we will analyze ecosystems by means of energy flows, similar to the energy budgets used in Chapter 3. All of the energy entering an ecosystem can be accounted for, and the pathways through the system can be traced. The accounting begins with the solar radiation that green plants convert to chemical energy.

ENERGY USE BY GREEN PLANTS

We can begin a study of the energy flow in an ecosystem by looking at the factors that determine how a plant utilizes energy. Of the solar radiant energy falling on a leaf, a small amount is immediately reflected, and approximately 80 percent is absorbed, as Figure 9.1

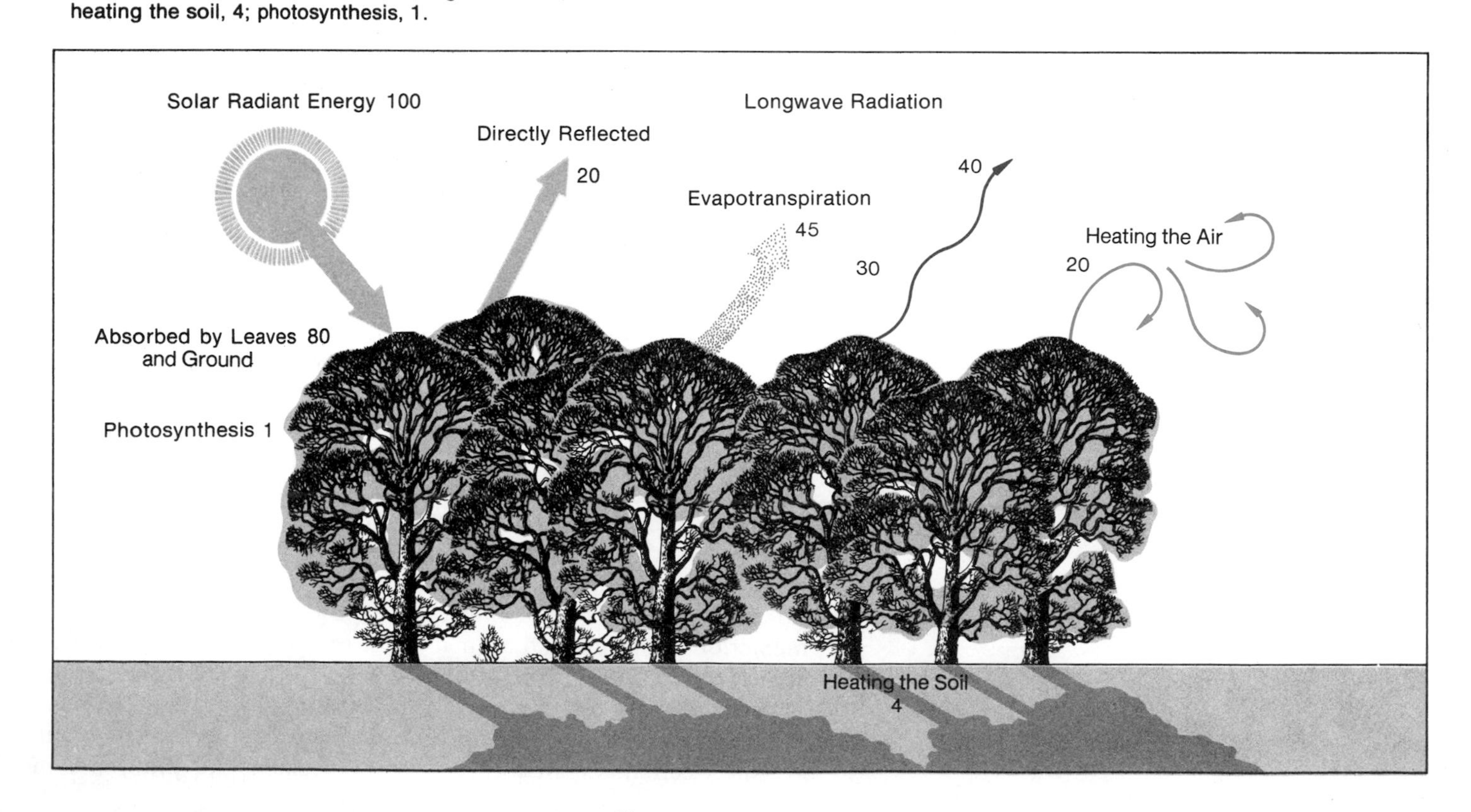

Figure 9.1 For every 100 units of solar radiation incident on a forest in a humid climatic region, only about 1 unit becomes stored chemical energy through photosynthesis. For this hypothetical midday situation, about 80 percent of the incoming solar radiation is absorbed by the vegetation and the ground. However, the forest crowns also receive longwave radiation redirected downward from the atmosphere. This diagram shows how the 130 units of energy received (100 solar radiant, 30 longwave) are utilized for various components of the local energy budget: reflection, 20 units; evapotranspiration, 45; longwave radiation, 40; heating the air, 20; heating the soil, 4; photosynthesis, 1.

illustrates. Most of the absorbed energy functions to warm the leaf and is then given off as longwave radiation, while some of the absorbed energy is used in the evaporation and transpiration of water stored in the plant. Only about 1 percent of the solar radiation is used in the process of photosynthesis to produce the chemical energy the plant requires for growth and maintenance.

The detailed chemical processes involved in photosynthesis are quite complex. In terms of energy, however, photosynthesis may be thought of as a process in which simple molecules, water (H_2O) and carbon dioxide (CO_2), are joined with the aid of solar radiant energy to form a more complex carbohydrate (sugar or starch) molecule (CH_2O):

$$H_2O + CO_2 + \text{solar energy} \rightarrow CH_2O + O_2$$

The solar radiant energy is stored in chemical form as carbohydrates. Further chemical reactions use carbohydrates and nutrients from the soil to produce the more complex protein molecules that the plant's cells require for growth. In addition to synthesizing carbohydrates, photosynthesis produces free oxygen gas. The formation of oxygen is incidental to photosynthesis, but it is important to the ecosystem as a whole. Without photosynthesis, there would be little oxygen in the atmosphere, and animal life as we know it could not have evolved (see Figure 9.2).

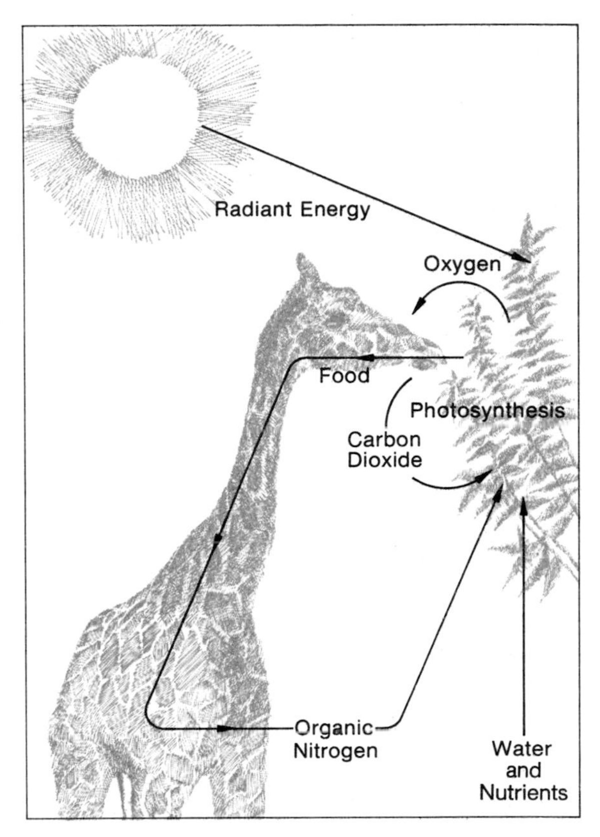

Figure 9.2 Photosynthesizing plants combine carbon dioxide, water, and nutrients with the aid of radiant energy to produce stored food energy. Animals rely on plants for food, and they breathe the oxygen released as a by-product of photosynthesis. Carbon dioxide is returned to the atmosphere by the respiration and decay of plants and animals; nutrients such as organic nitrogen compounds also cycle between organisms and the environment.

Photosynthesis and Plant Growth

A significant question for agriculture is this: How much chemical energy, or carbohydrate, can a plant produce photosynthetically for a given amount of incident solar energy? In 1926 Edgar Transeau measured the carbohydrate-producing efficiency of a cornfield. He recognized that all the carbon in a plant comes from the carbohydrate produced during photosynthesis, and by measuring the amount of carbon in corn plants, he was able to estimate the amount of carbohydrate produced during the growing season. Because the amount of energy required to produce 1 kg of carbohydrate by photosynthesis had already been determined in laboratory studies, Transeau was able to conclude that of the total solar energy incident on a field of corn during the 100-day growing season, only 1.6 percent becomes stored chemical energy through photosynthesis. This percentage is the energy conversion efficiency of corn; all plants have energy conversion efficiencies of only a few percent.

The rate of photosynthesis in a leaf varies with the intensity of the incoming light; photosynthesis is dependent on only visible light portions of the total solar spectrum. If the intensity of light is increased, the rate of carbohydrate production will also increase, up to the maximum value for each plant species. Further increases of light intensity beyond this point will not result in increased photosynthesis. A leaf in full sunlight receives more than enough light to carry on maximum photosynthesis. Even the leaves that are partially shaded or that receive only indirect light are able to carry on photosynthesis near the maximum rate. Because plants adapted to the tropics receive

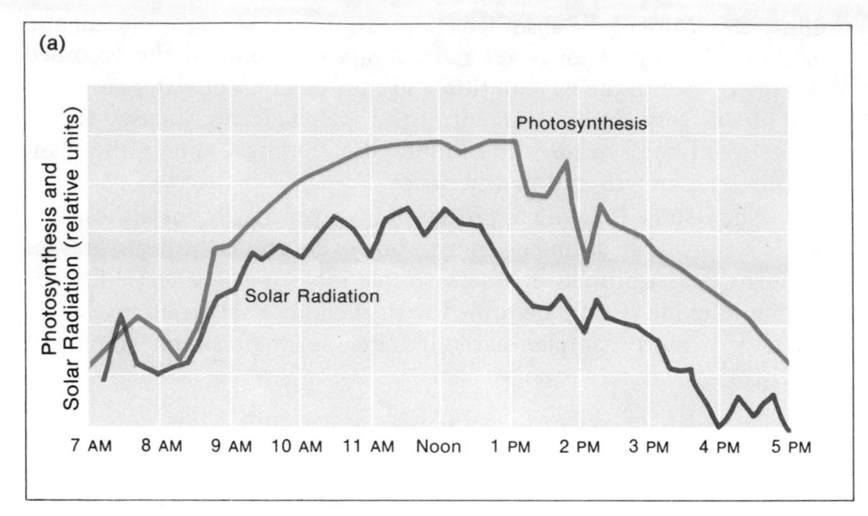

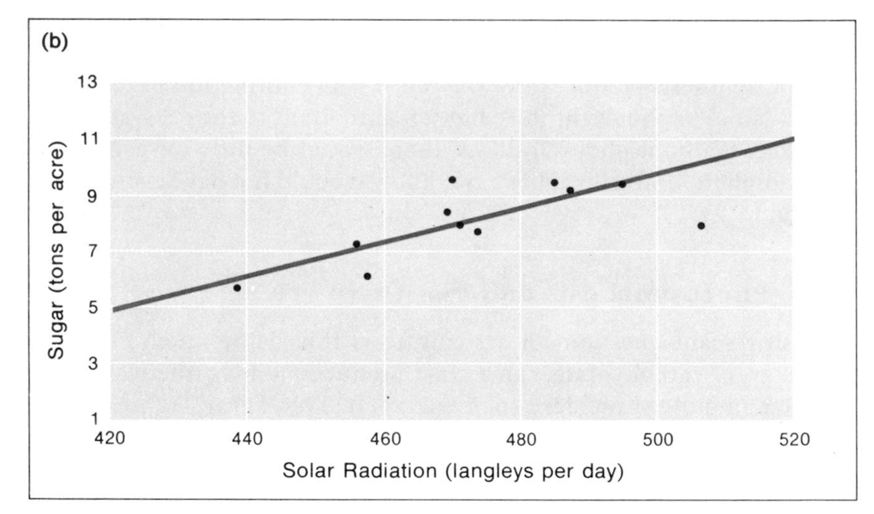

Figure 9.3 (a) The course of photosynthesis through the day rises and falls with the variations of incident solar energy.

(b) This graph shows the yield of sugar from a Hawaiian sugarcane plantation during different growing seasons. Each dot shows average daily solar radiation input and associated yield of sugar during a season. The solid line represents the general trend; it shows that the yield is high during seasons when fair weather causes the average radiant energy input to be high and is low during cloudy seasons. (After Jen-hu Chang, 1968)

more solar energy, they generally have higher maximum rates of photosynthesis than plants native to midlatitude regions, where less solar energy is available. Graphs depicting the relationships between photosynthesis and solar radiation are shown in Figure 9.3.

The rate of photosynthesis also depends on the temperature of the leaf, which may differ from the air temperature. For plants in midlatitude regions, photosynthesis for a given light intensity reaches a maximum at a leaf temperature of about 25°C (77°F). For arctic plants, maximum photosynthesis occurs at lower leaf temperatures. The rate of carbohydrate production decreases both above and below a plant's optimum leaf temperature, and production stops if leaf temperatures rise above 40°C (104°F) or so. If there is no wind to cool the leaves, leaf temperatures may rise so high that photosynthesis stops during the middle of the day, when solar energy input is maximum. Under such conditions, carbohydrate production is limited to a period in the morning and to a brief period in the late afternoon. Partly for this reason, the warmest regions of the tropics tend to have lower agricultural potential than midlatitude regions.

A number of other factors also influence the rate of photosynthesis. Adequate supplies of water and carbon dioxide are necessary for efficient photosynthesis. The availability of nutrients from the soil, particularly nitrogen and phosphorus, also affects the rate of carbohydrate production. Nitrogen is required for the production of the plant proteins necessary for cell growth. Phosphorus, a comparatively rare element in the earth's crust, is required for the synthesis of chemical compounds important in the photosynthetic process. Phosphorus deficiency is often the limiting factor for plant growth in moist climatic regions, lakes, and coastal areas.

Plant Respiration

Not all of the carbohydrate a plant produces in photosynthesis is available to animals as food. The plant's own cells require a continuous supply of energy, day and night, for maintenance. In the process of respiration, carbohydrate combines with oxygen and is reduced to carbon dioxide and water, which releases stored chemical energy for use by the living cells. Respiration is a necessary process which provides energy to the plant to maintain its vital functions. Photosynthesis stops in the absence of light or when leaves become too hot. Respiration, however, continues both night and day, although it is greater during the day when the temperature is higher and the leaves are exposed to light (see Figure 9.4). Increased temperature increases respiration, and as respiration rises, the amount of energy available for use in photosynthesis declines. Therefore, cool nights help plants conserve the chemical energy produced during the day.

One way to find the amount of energy a plant uses in respiration is to seal a plant in a dark box and measure the amount of carbon dioxide it produces by respiratory activity. Studies show that a wide variety of plants, including phytoplankton (floating microscopic green plants), use approximately one-fourth of their stored energy for respiration.

Figure 9.4 The upper portion of the graph shows the trend of net photosynthesis (that is, photosynthesis minus respiration) in a leaf for different leaf temperatures and different rates of solar energy input during daylight hours. Note that photosynthetic activity is small at both low and high temperatures and that it reaches a maximum at an intermediate temperature. For a leaf at a given temperature, photosynthesis increases with increased solar energy input, but not proportionately. Doubling the solar energy from 0.8 to 1.6 langleys per minute increases photosynthesis by only 30 percent or so.

The lower part of the graph shows the energy consumed by respiration at different temperatures for a leaf that is kept in the dark. (Note that in order to emphasize the relation of respiration to net photosynthesis, the lower graph shows respiration in negative quantities that increase *down* the scale.) Respiration increases with increased temperature until the leaf is too warm to function. High temperatures at night therefore promote the use of stored energy for respiration and tend to decrease the net productivity of the plant. (After Gates, 1968)

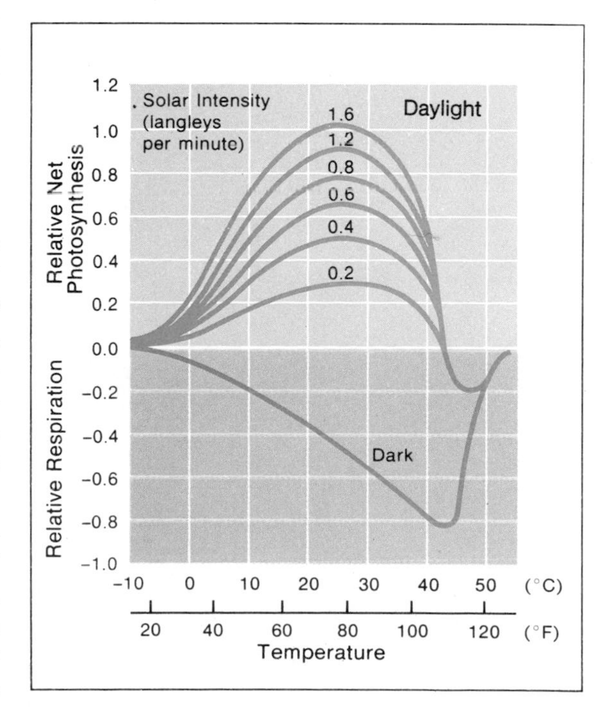

Productivity Measures

The rate at which a plant converts light to chemical energy is called its *gross productivity.* Since the plant uses some of the chemical energy for respiration, only a portion of the total chemical energy remains stored in plant tissues. The net rate at which a plant stores energy, exclusive of the energy it uses for respiration, is called the *net productivity.* Stated another way, net productivity equals gross productivity minus respiration. Net productivity is the quantity important to the consumer of a plant because it represents the amount of organic material produced by the plant.

The *standing crop,* or *biomass,* is the amount of energy present in plants at any given time, expressed in grams of dry matter per sq meter of ground surface. Biomass should not be confused with productivity, which is the rate at which a plant produces food. A mature, unlogged redwood forest has a large biomass but a comparatively low productivity. A large amount of energy is stored in the trees, but not much

Table 9.1. Net Productivity of Major Ecosystems

Type of Ecosystem	Area (millions of square kilometers)*	Net Productivity per Unit Area (dry grams per square meter per year) †		World Net Productivity (billions of dry tons per year)
		Normal Range	Mean	
Tropical Forest	24.5	1,000–3,500	2,000	49.4
Temperate Forest	12.0	600–2,500	1,250	14.9
Boreal Forest	12.0	400–2,000	800	9.6
Woodland and Shrubland	8.5	250–1,200	700	6.0
Savanna	15.0	200–2,000	900	13.5
Temperate Grassland	9.0	200–1,500	600	5.4
Tundra and Alpine	8.0	10–400	140	1.1
Desert and Semidesert	42.0	0–250	40	1.7
Cultivated Land	14.0	100–3,500	650	9.1
Swamp and Marsh	2.0	800–3,500	2,000	4.0
Lake and Stream	2.0	100–1,500	250	0.5
Total Continental	149.0		773	115.0
Open Ocean	332.0	2–400	125	41.5
Continental Shelf, Upwelling	27.0	200–1,000	360	9.8
Algal Beds, Reefs, Estuaries	2.0	500–4,000	1,800	3.7
Total Marine	361.0		152	55.0
World Total	510.0		333	170.0

* One square kilometer is equal to about 0.39 square mile.
† One gram per square meter is equal to about 0.0033 ounce per square foot.

Source: Whittaker, Robert H. 1975. *Communities and Ecosystems*, 2nd ed. New York: Macmillan.

additional energy is added to storage each day. The phytoplankton in the ocean have a high productivity, but their biomass is small because they are continuously consumed by predators. *Yield* is the amount of energy stored during the growing season in the desired portion of a crop, such as the fruit. It is usually determined by weighing the harvested portion of the crop.

The productivity of a particular plant is expressed in terms of the amount of energy stored during a given time interval, such as a day or a year. Both gross and net productivity can be measured. Suppose a plant is placed in a transparent sealed container. When no light is allowed to enter the container, the rate at which the plant gives off carbon dioxide is a measure of the rate at which it is using stored energy for respiration. When the plant is exposed to light, photosynthesis and respiration both occur. Photosynthesis causes carbon dioxide to be absorbed at the same time that respiration causes carbon dioxide to be given off. The net rate at which the plant absorbs carbon dioxide is therefore a measure of its net productivity. Gross productivity can be found by adding the rate of energy usage for respiration to the net productivity. For instance, if the net productivity of a carrot plant is 4 grams of carbohydrate in a given time, and it uses the equivalent of 1 gram of carbohydrate for respiration, the gross productivity of the carrot is 5 grams of carbohydrate.

Because it is very difficult to carry out direct measurements of energy flow, productivity is usually measured indirectly. The net productivity of agricultural crops is often estimated by harvesting the complete plants, roots and all, at the end of the growing season. The plants are then dried and weighed, and the productivity is expressed in terms of the dry matter that a given area produces during a specified period of time. Net productivity measured by the harvest method can be expressed as grams of dry matter per sq meter of field per year, as in Table 9.1. When the harvest method is applied to natural plant communities, measurements of dry matter must be corrected for losses caused by the shedding of leaves, the depredations of insects, and the death of some plants during the growing season.

The energy content of a gram of dry matter differs among plant species depending on the relative amounts of proteins and carbohydrates in the plant tissues. To correct for this difference, productivities expressed in terms of dry matter can be converted to energy units, as in Figure 9.5, by burning the dry matter in a device that measures the heat energy released in complete combustion.

The average global net productivities of all important food crops, such as wheat, rice, corn, and potatoes, are of the order of 500 to 700 grams of dry matter per sq meter per year. Surprisingly, mean agricultural productivity appears no greater than that found in many "natural" ecosystems (see Table 9.1). Productivities two or three times greater are attained in areas where intensive mechanized agriculture is practiced. The world average net productivity of sugarcane is high—about 1,500 grams per sq meter per year. Productivities of more than 9,000 grams have been measured for sugarcane in some tropical regions, where the growing season lasts the entire year and the soil is moist and rich in nutrients.

The global distributions of net productivity in Table 9.1 correspond approximately to the patterns of climate and vegetation types shown in Figure 8.3, p. 209. Productivity is greatest in the tropical

Figure 9.5 This diagram compares the productivity of various environments in terms of energy units (kilocalories per sq meter; 1 kilocalorie equals 1,000 calories). The deep oceans and deserts, which cover approximately 80 percent of the earth's surface, have low productivities; the deep oceans lack nutrients, and the deserts lack moisture. The most productive areas include tidal estuaries. If all agricultural areas had a productivity as high as that of estuaries, the annual agricultural productivity of the earth would approach 10^{18} kilocalories. The present world productivity is estimated to be 10^{14} kilocalories per year. (After E. P. Odum, 1971)

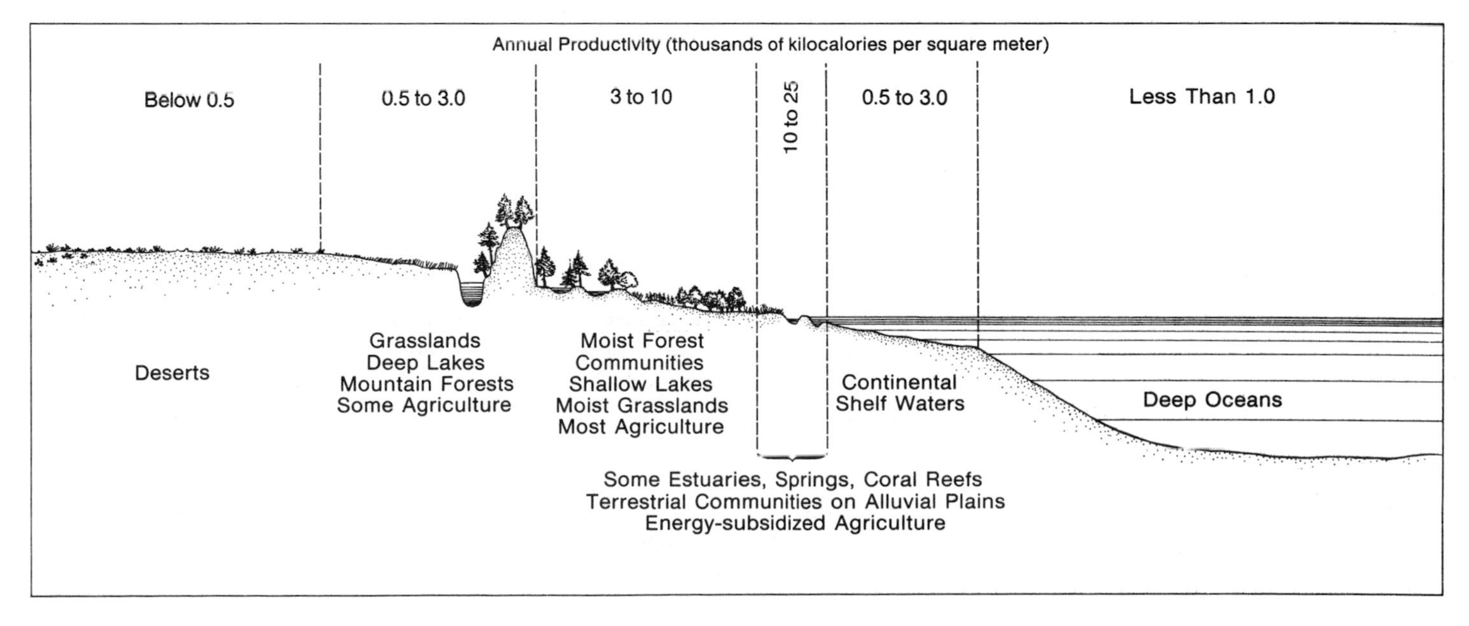

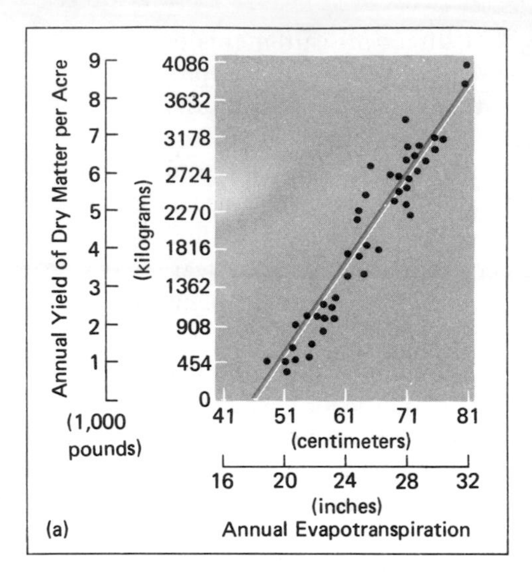

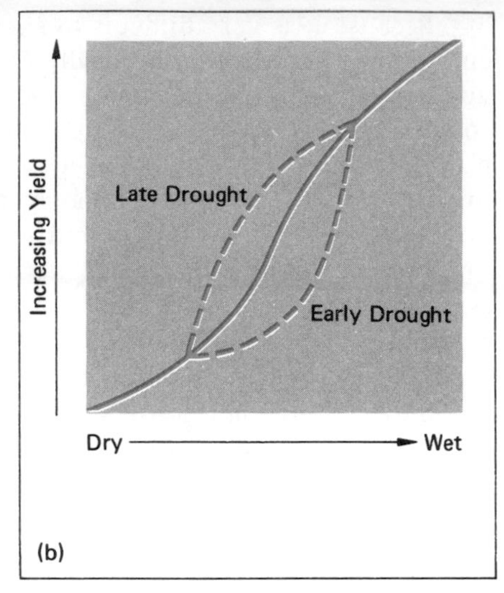

Figure 9.6 (a) This graph shows the relationship between annual evapotranspiration and the yield of several crops grown in evapotranspirometers near Columbia, South Carolina. Significant yields begin only after evapotranspiration is greater than 46 to 51 cm (18 to 20 in.).

(b) This graph illustrates the relationship between soil moisture availability and sugarcane yields in Hawaii. The curve shows that moisture deficits early in the growing season tend to decrease yields, but that deficits late in the growing season tend to increase yields. (After Jen-hu Chang, 1968)

rainforest, and it decreases as both moisture availability and potential evapotranspiration decrease. Average productivity of the coniferous, or boreal, forest is less than half that of the tropical rainforest, and annual rates for desert regions are less than 5 percent of those of the rainforest. Because of the large areal extent and high productivity rates of the tropical and temperate (midlatitude) climates and forests, these two together account for more than half of the total productivity of the continents. Annual productivity of the continents averages four to five times more than that of the oceans, even though the oceans make up more than two-thirds of the surface area of the globe. Figure 9.5 shows that the most productive regions of the oceans are the waters of the continental shelves and especially estuaries, where nutrient-rich streamflow mixes with ocean waters across broad expanses of brackish and saline marshes.

Climate and Crop Growth

A plant's basic requirements include solar radiant energy, carbon dioxide from the atmosphere, and water and nutrients in solution from the rooting zone of the soil. Insufficiency of any of these factors inhibits a plant's growth. Normally, each plant is adapted to the regional solar energy regime, but unusual spells of cold weather or freezes can retard plant growth or "kill back" tender vegetation intolerant of freezing temperatures. Of course, factors other than climate, such as disease or insects, sometimes reduce plant growth drastically.

Because evapotranspiration, photosynthesis, and absorption of nutrients from the soil all depend on soil moisture availability, components of the local water budget represent climatic indexes of crop growth and yield. For example, the general relationship between crop yields and evapotranspiration is shown in Figure 9.6a; once evapotranspiration exceeds an initial base level of 46 to 51 cm (18 to 20 in.) per year, crop yields tend to increase with greater evapotranspiration. The relationship is not simple, however. Figure 9.6b shows how moisture deficits affect yields of sugarcane. When soil moisture is low during the growing season, the yield is low, and when soil moisture is high, the yield is high. However, the dashed lines show that if a drought occurs early in the growing season during the period of active growth, the yield declines, but if the same degree of drought occurs late in the season, the yield increases. This seasonal drought relationship is characteristic of many but not all crops.

Potential Photosynthesis and Crop Productivity

An important concern for world agriculture is the evaluation of the potential productivity of a given region, assuming that enough water is supplied and that good agricultural practices are followed. The maximum value of a plant's net productivity, given sufficient water and nutrients, is referred to as its *potential photosynthesis.* Because photosynthesis depends on light intensity and temperature, these two factors set a natural limit to the expected rate of photosynthesis of a particular plant species in a region. If photosynthesis is occurring at its maximum rate, additional water or fertilizer will not increase productivity.

Potential photosynthesis, therefore, is a measure of a plant's maximum net productivity if the only limiting factor is solar radiant energy input. Potential photosynthesis for a particular crop can be estimated from the records of temperature and solar radiation alone. Figure 9.7, p. 242, shows estimates of the global distribution of potential photosynthesis for a crop of sugar beets. Photosynthesis in sugar beets is especially adapted to the solar radiation levels and light intensities of middle and higher latitudes. As the figure indicates, potential photosynthesis for sugar beets is lowest in the tropics and polar regions. Since high temperatures limit the rate of photosynthesis, the higher intensities of solar radiation and light in tropical latitudes do not result in increased rates of photosynthesis in sugar beets, while crops such as sugarcane are better adapted to the tropics and have higher productivity there. In the tropics, the lowest values of potential photosynthesis for sugar beets occur near latitude 10°N. Lowland areas at that latitude have the hottest climate on the earth, which increases the proportion of chemical energy allocated to respiration. Productivity at latitude 10°S exceeds productivity at latitude 10°N because temperatures for a given latitude are somewhat lower south of the equator. Tropical highlands have better agricultural potential than land near sea level because the cooler temperatures at higher altitudes favor a higher rate of photosynthesis.

According to estimates of potential photosynthesis, the world regions with the greatest potential productivity for sugar beets probably lie in the middle latitudes and have a Mediterranean climate. Regions with the potential for a highly productive 8-month growing season include the central United States, California, southwestern Europe, southern Australia, and New Zealand, all of which are excellent agricultural regions in actual practice. The high estimated productivity near latitude 60°N during the 4-month growing season is of particular interest. The long days of the northern summer and the

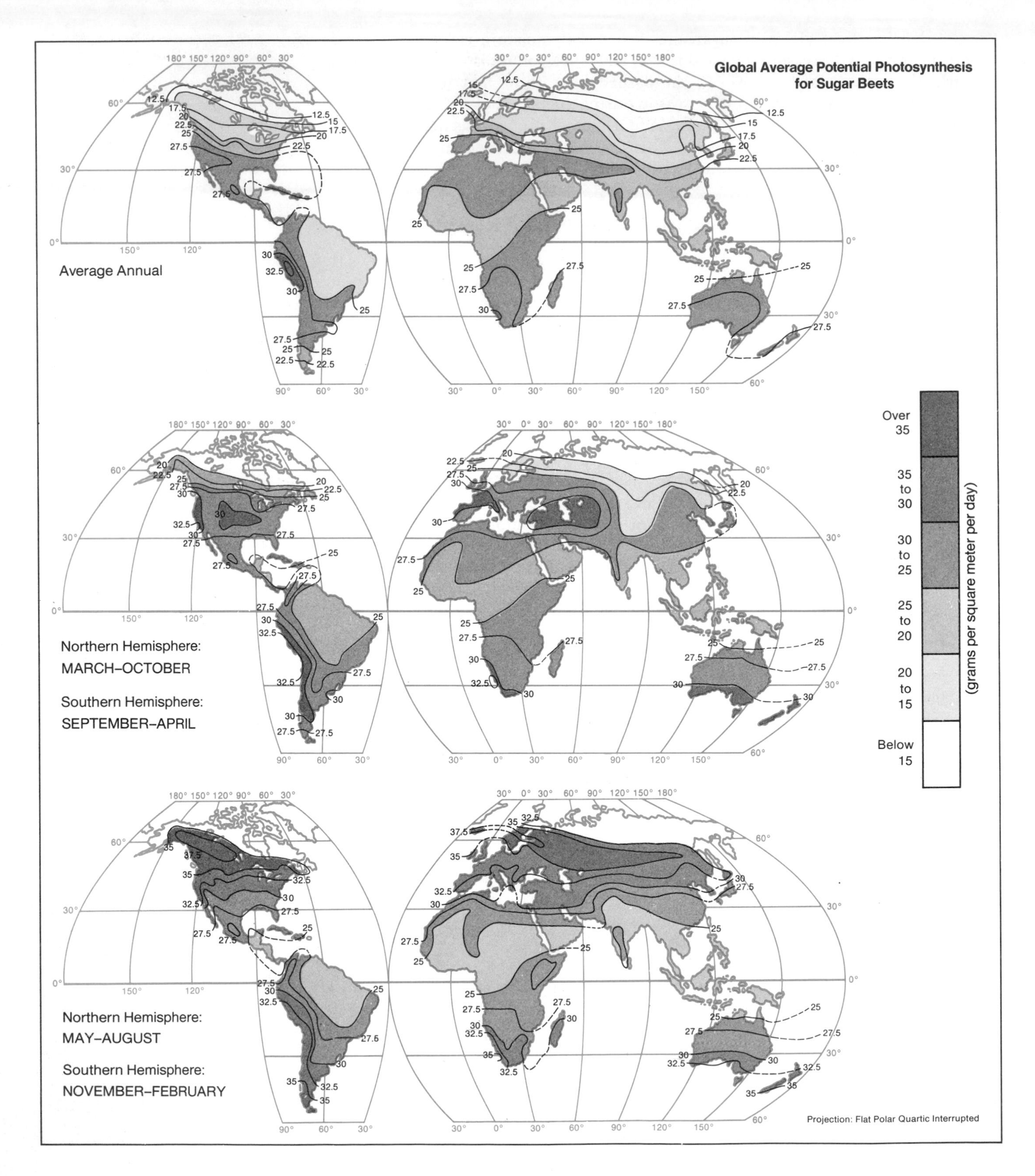
Global Average Potential Photosynthesis for Sugar Beets
Average Annual
Northern Hemisphere:
MARCH–OCTOBER
Southern Hemisphere:
SEPTEMBER–APRIL
Northern Hemisphere:
MAY–AUGUST
Southern Hemisphere:
NOVEMBER–FEBRUARY
Over 35
35 to 30
30 to 25
25 to 20
20 to 15
Below 15
(grams per square meter per day)
Projection: Flat Polar Quartic Interrupted

moderate temperatures favor a high level of plant activity. The Matanuska Valley in southern Alaska, for example, is known for its giant cabbages and other luxuriant crops produced during the short summer growing season. Unfortunately, the shortness of the growing season severely limits the types of crops that can be grown.

ENERGY FLOW THROUGH ECOSYSTEMS

In any ecosystem, there are producers of energy, consumers of energy, and decomposers of waste. Green plants, the energy producers, are the only organisms capable of producing food energy directly from inorganic matter and sunlight by photosynthesis. Organisms such as animals and bacteria and fungi that rely on organic matter for food are the energy consumers. Food energy flows from producers to consumers in an ecosystem. Ecologists view this flow of energy (or food) as a pyramid in which each level supports a smaller number and mass of organisms. Once the food energy has been spent, it cannot be replenished except by further exposure of green plants to solar radiation.

Food Chains and Food Webs

The animals that ingest plants are called *herbivores,* and these animals may in turn be ingested by other animals, called *carnivores.* Some animals, such as man, are both herbivorous and carnivorous. A *food chain* is a sequence of consumption and represents energy transfer through the environment. There are two basic types of food chains. The *grazing food chain* begins with green plants and goes on to herbivores and then to carnivores. A grazing food chain can be symbolized as a linear relationship: Plant→herbivore→carnivore→second carnivore→top carnivore. The *decay food chain* begins with dead organic matter and goes on to microorganisms and to their predators, bacteria and fungi. These decomposers live within and on organic materials, especially dead tissues, breaking them down and returning minerals to the soil. Usually much less than half of the plant material on the continents is consumed directly by animals; the rest is recycled by decomposers.

Some food chains involve only a few links. In an agricultural ecosystem, for example, cattle eat grass and grain and are in turn eaten by carnivores. However, most feeding relationships in nature do not take the form of simple, isolated chains. Many food chains are interconnected, forming complex *food webs* (see Figure 9.8, p. 244). For this reason, tracing feeding relationships is not a simple task. In a midlatitude forest, for example, there are numerous species of herbivores, each of which may feed on several plant species. Carnivores in the forest may also have a varied diet, feeding on herbivores and other carnivores. One way to analyze a food chain is to take samples of all relevant animal species in the ecosystem and examine the contents of their digestive tracts. An owl's recent diet, for instance, is revealed by the bits of bone and hair in the pellet the bird spits up every few days. The complexity of the food web is apparent when we consider

Figure 9.7 Potential photosynthesis is a measure of a plant's maximum net production if the only limiting factor is solar radiant energy input. The global maps in this figure show calculated potential photosynthesis in units of grams of production per sq meter per day for sugar beets, which are especially adapted to light intensities of the middle and higher latitudes.

(opposite top) Average annual potential photosynthesis is given on this map. The highest values occur in regions that are classified as *Cs* or *Aw* climates according to the Köppen system.

(opposite center) Potential photosynthesis is shown for an 8-month growing season (March through October in the northern hemisphere, September through April in the southern hemisphere). Many of the highest values are reached in what are now desert regions because of the clear skies prevailing in such areas. White areas on the map indicate regions where data are unavailable.

(opposite bottom) Potential photosynthesis is given for the 4-month principal growing season (May through August in the northern hemisphere, November through February in the southern hemisphere). Note the high values of potential photosynthesis that occur in the higher latitudes because of the long duration of daylight at low intensities during the summer months. Cool nights lessen respiration, permitting high net productivity values to develop. (After Jen-hu Chang, 1970)

Figure 9.8 This diagram shows the principal members of a food web in a salt marsh near San Francisco, California, during winter. The energy producers are various salt marsh and marine plants, which grasshoppers, snails, fish, and marine invertebrates feed upon. Birds and rodents are intermediate carnivores, and the top carnivores of the web are hawks and owls. (After Smith, 1966, and Johnston, 1956)

that a few acres of grassland may harbor several hundred different species of insects and numerous other animals.

Laboratory Ecosystems

Because a simple, natural ecosystem contains so many species of plants and animals, quantitative measurements of energy flow are difficult to accomplish. Ecologists have instead turned to laboratory ecosystems made up of a clearly defined plant-herbivore-carnivore chain. Although laboratory ecosystems are artificial, they have the advantage that conditions in them are easily varied. Careful study of the response of laboratory ecosystems to a variety of conditions has led to insight into ecologic principles.

Lawrence B. Slobodkin, an American ecologist, studied a laboratory ecosystem comprising a species of one-celled green algae *(Chlamydomonas)* as the primary producer (energy producer), water fleas *(Daphnia)* as the herbivores, and Slobodkin himself as the carnivore. Slobodkin divided his colonies of water fleas into five groups. He gave each colony in the first group a standard amount of energy each day in the form of algae. Successive groups were given proportionately more food, so that colonies in the fifth group received five times the standard portion. In the absence of predation, each colony grew to a size proportional to the rate at which food energy was supplied (see Figure 9.9a).

Slobodkin then acted as a predator on each colony. Every fourth day he removed a number of adult water fleas equal to a given percentage of the new offspring in the colony. He then compared the energy output, or the yield of water fleas that could be extracted, from each colony with the amount of energy supplied as food. As Figure 9.9b shows, the energy output of the colonies dropped when predation rates were high. The reason is that if a colony is "overfished," the few remaining animals are unable to make full use of the food supplied.

An intelligent predator's most efficient strategy for maintaining its own food supply is to eat only enough animals so that the right number remain to use all the available food. Slobodkin found that, in order to maintain the energy efficiency of the colonies, the energy output (the amount he could extract by predation) must be no more than 12 percent of the energy input (food), irrespective of the feeding level.

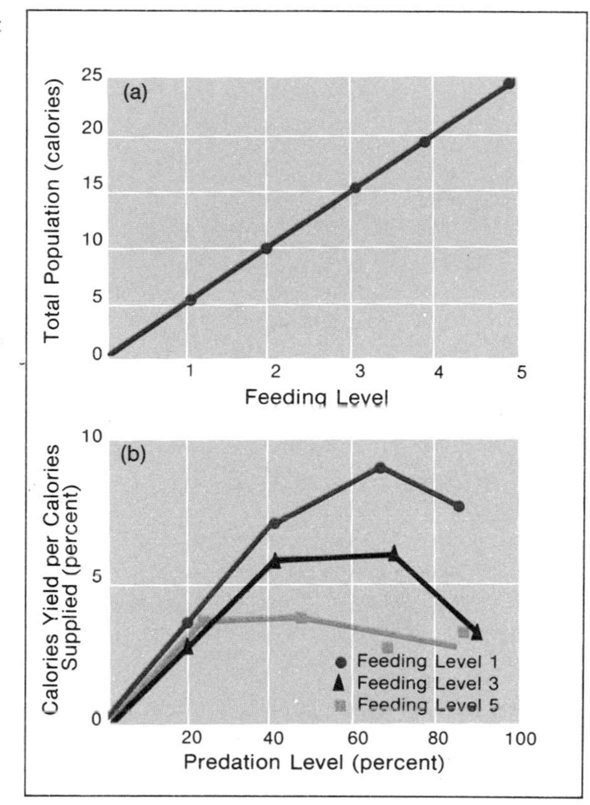

Figure 9.9 These graphs summarize the results of L. B. Slobodkin's experiments on a laboratory ecosystem consisting of algae, water fleas, and a predator. Five groups of water flea colonies were fed algae, such that the group with feeding level 5 received five times as much food as the group with feeding level 1.

In the absence of predation, each colony reached a population proportionate to its feeding level.

The yield of water fleas that could be extracted from each colony per unit of food energy supplied decreased at high levels of predation because the number of water fleas remaining could not use all of the food. (After Slobodkin, 1959)

Energy Transfer Along Food Chains

Slobodkin's results indicate that at each step of a food chain there is a significant loss of useful energy. Only a fraction of the net production of one stage becomes available as food for the next stage. Food chains in natural ecosystems exhibit approximately the same fractional transfer of useful energy as the laboratory ecosystems show. In a grazing food chain, for example, only about 10 percent of the energy absorbed during one stage is available to be absorbed during the next stage. This is because no organism can convert the food it eats into an equal amount of stored energy. An adult person uses most of the energy obtained from food for body heat, motion, and work. Only a small amount of energy is stored for growth, and only the stored energy is available to a predator. Another reason for the low efficiency of the energy transfers along food chains is that the chemical reactions required for life are always accompanied by the transformation of energy into forms of heat that cannot be utilized. In addition, at each step in the grazing food chain, energy in the form of waste products is lost to decay food chains.

Consider a simple plant-herbivore-carnivore food chain consisting of grass plants, mice, and snakes. The mice obtain approximately 10 percent of the energy absorbed earlier by the grass plants, and the snakes obtain approximately 10 percent of the energy absorbed by the mice. Thus the carnivore receives only 1 percent or so of the energy originally absorbed by the plants. The fraction of the original energy available to a succeeding carnivore stage—a hawk, for instance—is still less. The rapid decrease of available energy along a food chain limits such chains to four or five links. Large carnivores, such as lions, which are the last natural link in a food chain, obtain only a small fraction of the energy absorbed by the plants in their habitat. Lions must roam over large areas to obtain their food, and one region cannot support many of them.

STUDIES OF NATURAL ECOSYSTEMS

Despite the difficulties involved, the energy flows in several natural ecosystems have been studied in detail. The results of such studies can be conveniently summarized in energy flow diagrams, which show the main energy pathways through the system.

The energy flow diagram in Figure 9.10a represents a typical ecosystem. The diagram illustrates how the energy input to each stage in a food chain is divided among various energy flows and outputs. The principal outputs include the energy used in life processes, or respiration, the energy transferred to the next stage of the food chain, the energy transferred to the decay food chain, and the energy utilized for

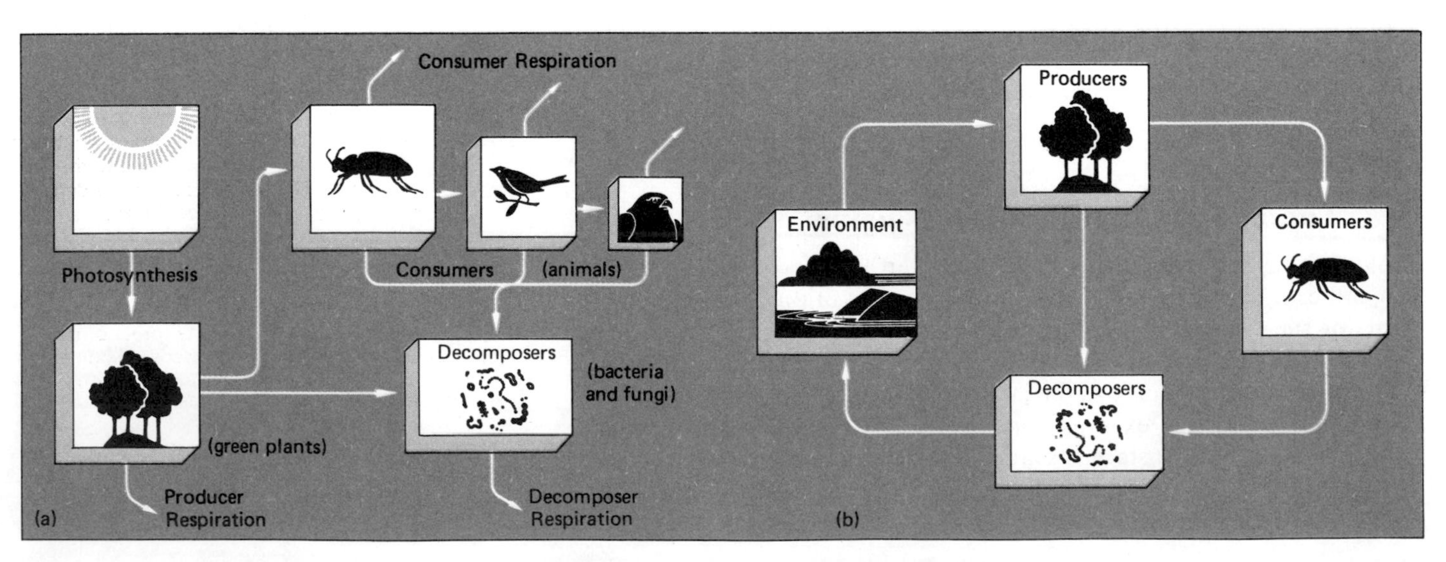

Figure 9.10 (a) This schematic diagram illustrates the flow of energy in a natural ecosystem. Following the flow from left to right, solar radiant energy input is converted to food energy by green plants. The plants use some of the energy for respiration, herbivores consume some of it, and organisms of decay feed on some of it. Part of the energy in the herbivores is transferred to successive carnivores. However, the amount of transferred energy decreases at each step because energy is used for respiration and other purposes.

(b) This diagram indicates schematically how materials circulate between the principal units of an ecosystem and between the ecosystem and the environment. (After Whittaker, 1970)

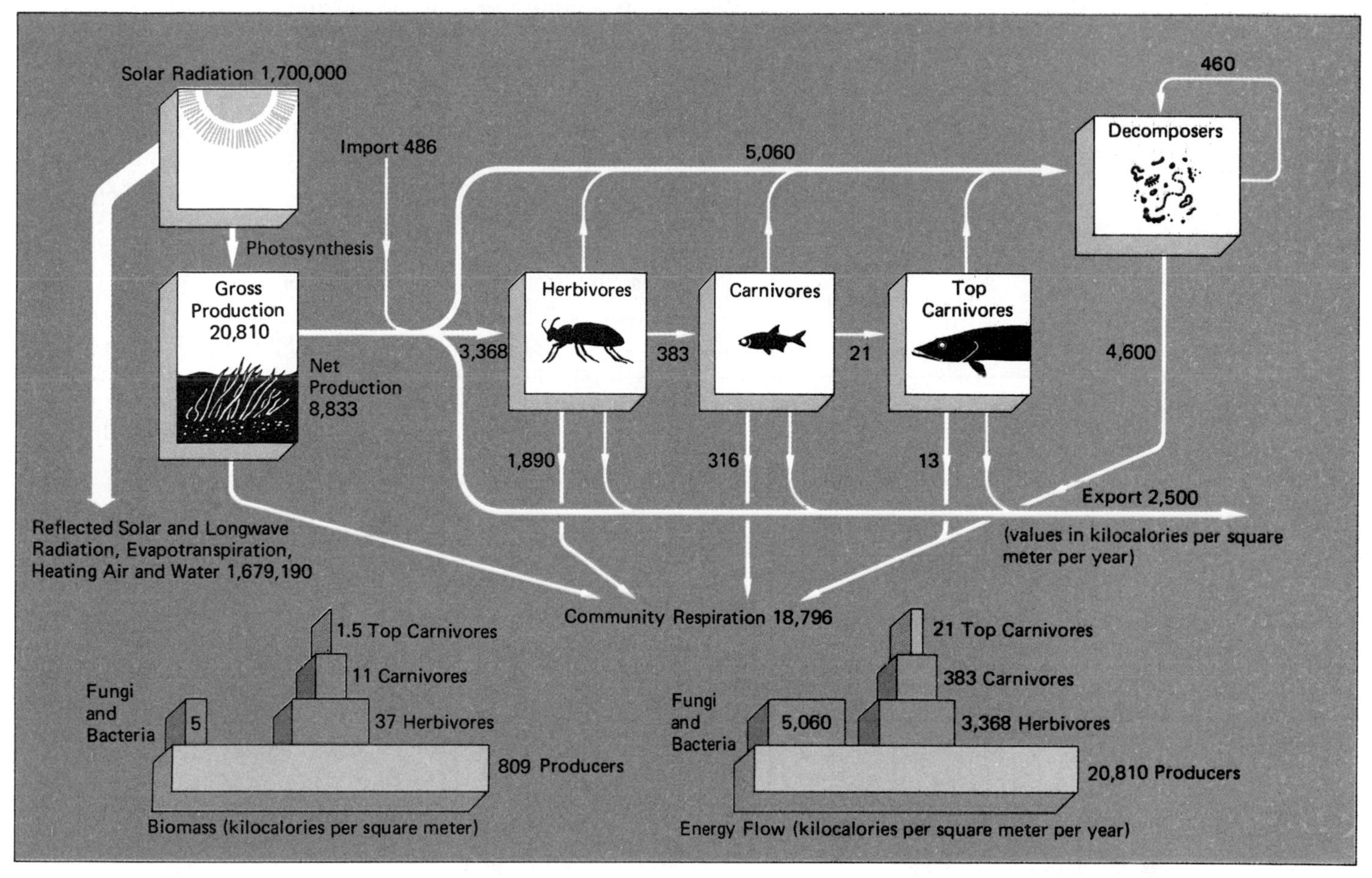

Figure 9.11 As this energy flow diagram for the Silver Springs ecosystem shows, the energy input to the ecosystem is principally solar radiant energy, with a small input of organic matter that has entered the section of stream under study. The energies are expressed in kilocalories per sq meter per year. Most of the solar energy incident on the system is utilized for respiration and energy budget components, such as reflection, evapotranspiration, and heating the air and water. Only a small fraction is converted to food energy by various water plants. Some of the net production is absorbed by herbivores, some is exported in the form of plant matter that is carried downstream out of the system, and some is transferred to the chain of decay. The small energy loop in the decay chain represents the decomposition of decay organisms. Note the rapid decrease in energy transferred along the main food chain. The top carnivores receive only 0.6 percent of the energy absorbed by the herbivores. The *biomass pyramid* for Silver Springs shows that at any given time, most of the biomass of the ecosystem is in the form of primary producers, with only a small fraction in the form of herbivores and carnivores. The *energy flow pyramid* summarizes the energy flow data from the main diagram. (After H. Odum, 1957, and E. P. Odum, 1971)

other purposes. In a pond, for example, the energy in plant and animal materials that sink into the bottom sediment is exported from the system. Figure 9.10b is a schematic materials flow diagram for an ecosystem consisting of several stages, including a decay food chain.

One of the classic energy flow studies of a natural ecosystem was carried out at Silver Springs, Florida, by the American ecologist Howard Odum and his colleagues. Their results are summarized in the energy flow diagram of Figure 9.11. Energy is expressed in kilocalories per sq meter per year.

In the Silver Springs ecosystem, radiant energy from the sun provides almost all the energy input to the producer forms of life in the spring. A small additional contribution of energy is imported through plant and animal debris that enters the system. Only a small fraction of the absorbed radiant energy is employed in photosynthesis.

There are four links in the Silver Springs grazing chain: plant, herbivore, carnivore, and top carnivore. The significant reduction of energy from stage to stage is evident from the flow diagram. Note that the decay chain is somewhat more important in terms of amount of energy transferred than the grazing chain: the energy input to the herbivores is 3,368 kilocalories, but the energy passed to the decay chain is 5,060. Plant and animal materials moving downstream out of the region of study account for the exported energy flow.

When plant communities have undergone succession toward the stable, comparatively unchanging system called *climax vegetation*, very little energy is used for the production of new plant material. The energy released by respiration and by the decay of dead plant and animal tissues balances the energy stored in new growth, so that the total energy stored in a climax forest remains effectively constant. Although an individual tree in a climax forest uses energy for growth, the forest ecosystem as a whole uses essentially all the input energy for respiration. Researchers who studied the energy flow for an oak and pine forest near the Brookhaven National Laboratory on Long Island, New York, found in the course of their work that of the annual gross production of dry matter per sq meter, only about one-fifth was stored as new growth, litter, and humus. The major part of the gross production was used in respiration (see Figure 9.12). Respiration accounted for more than 80 percent of the energy used, which indicates that the Brookhaven forest is in a late stage of succession.

AGRICULTURAL ECOSYSTEMS

Early human societies followed a hunting and gathering mode of existence; the eventual domestication of animals and plants represented a more effective way of assuring an available food source. Both patterns persist today, and in both it is man who is the top carnivore. However, a hunting or grazing economy cannot support a dense population of top carnivores because so much useful energy is lost at each stage of the long food chains involved. Societies that depend on hunting or grazing are therefore small in size and widely dispersed.

Figure 9.12 This diagram traces the energy flow in an oak and pine forest on Long Island, New York, in units of grams per sq meter per year. Less than 20 percent of the gross productivity is used for net forest production, indicating that the forest is nearly at the climax stage, the point at which most of the new growth in the forest is balanced by the death and decay of old growth. The gross productivity in this forest is used primarily for the respiration of plants, animals, and decomposers. (After Woodwell and Whittaker, 1968)

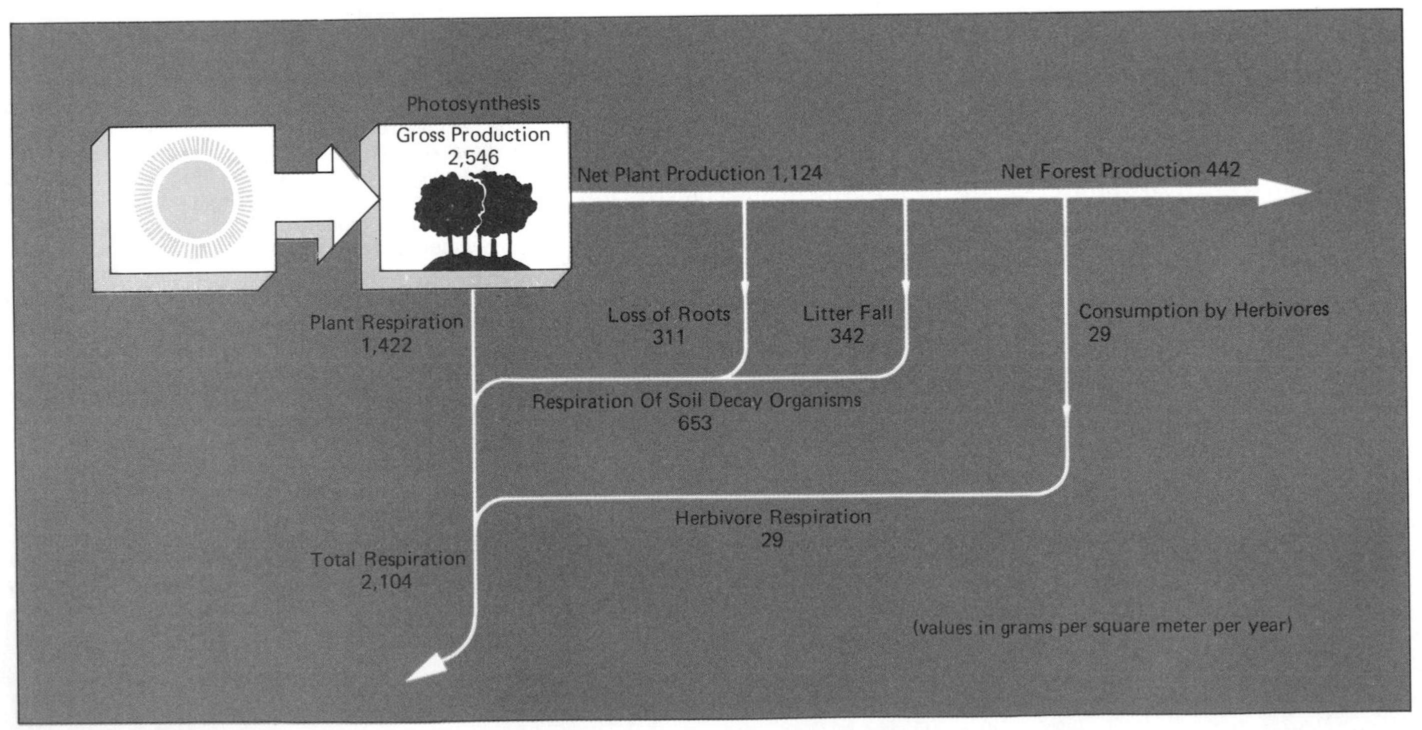

The cultivation of selected plant species in agriculture is more efficient than hunting or grazing because the food chain is much shorter. In an agricultural ecosystem, man often plays the part of herbivore, although some agricultural output becomes food for domestic animals. Although people need protein in their diets, only members of affluent societies can afford the cost in energy of eating meat every day—meat is at the end of a long food chain. In the United States, much of the energy available from cereal grains is consumed indirectly as meat because grain is the diet of livestock. The average American consumes directly or indirectly more than four times the amount of grain available to the average person in developing countries.

The efficiency of agricultural ecosystems has been remarkably improved in the past hundred years or so. In the Middle Ages in Europe, a return from a cereal crop of eight or ten grains for every grain planted was considered good. Today the average return from most cornfields in the United States is 500 grains for each grain planted. There are several reasons for the marked improvement in food production. The principal factors are the development of improved plant varieties; the use of irrigation; the extensive use of fertilizers, weed-killers (herbicides), and insecticides; and the increased mechanization of agriculture by utilization of a large energy input from fossil fuels.

Figure 9.13b shows that from 1800 to 1940, corn yields in the United States averaged 25 bushels per acre. Average present-day yields exceed 60 bushels per acre. A primary reason for the improvement was the introduction of high-yield varieties, principally hybrid corn. The recent development of highly productive strains of wheat and rice for use in tropical and subtropical regions has contributed greatly to the food supplies available to countries such as India. The new strains mature early and respond well to fertilizer applications. They also are relatively insensitive to seasonal variation in the duration of daylight; so a new crop can be planted almost any time during the year. This characteristic allows three crops a year to be harvested regularly in tropical countries, which increases the annual productivity of the land.

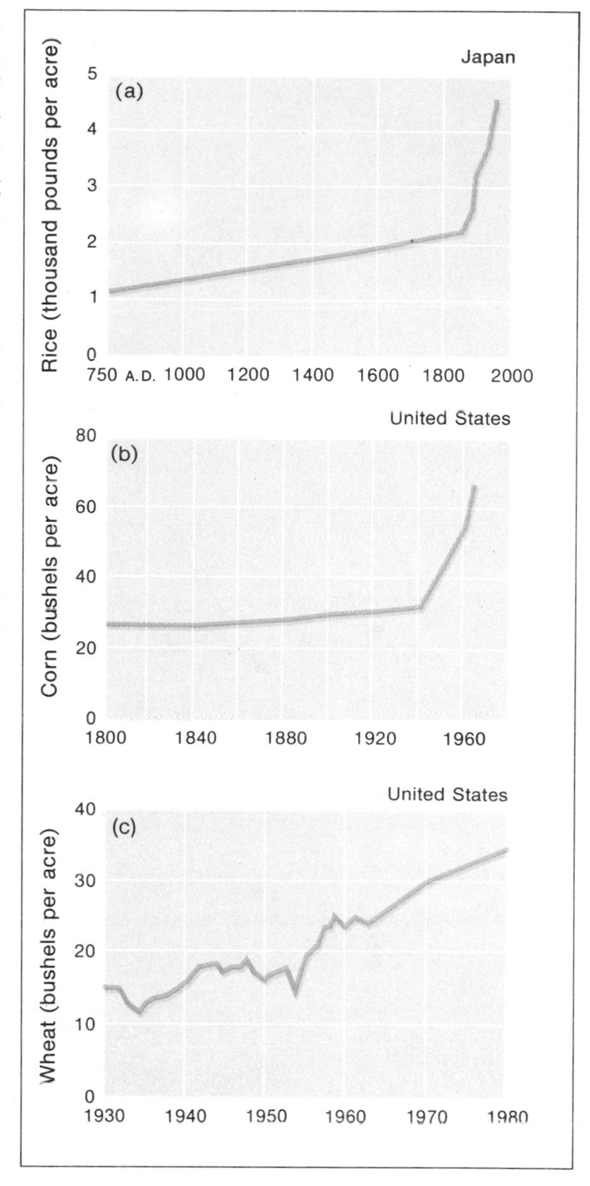

Figure 9.13 Estimated (a) rice yields in Japan and (b) corn yields in the United States have shown rapid increases in the past few decades because more productive species have been developed and because efficient, energy-intensive agricultural practices have been used. Such a rate of increase cannot be sustained indefinitely for a given crop.

(c) Wheat yields in the United States increased by 3.5 percent per year from 1950 to 1965, but the projected rate of increase from 1965 to 1980 is 2 percent per year.

PROBLEMS OF WORLD AGRICULTURE

An agricultural ecosystem is not stable in the way that a climax forest is, a fact well known to anyone who tries to maintain a beautiful green lawn. Man has artificially stabilized agricultural ecosystems by supplying them with large amounts of external energy. In industrialized nations, the stored energy of fossil fuels drives the machinery required to increase agricultural productivity. So the true energy input for food production includes not only radiant energy but also the cost of building tractors, producing fertilizer, training agricultural field agents, and so forth. In highly mechanized agriculture, man may supply more than 1 kilocalorie of energy for each kilocalorie of food energy produced.

The rapid increase in agricultural yields experienced in the past few decades cannot be expected to continue indefinitely because there are natural limits to agricultural performance. A plant may respond well to moderate applications of fertilizer, but applying ten times as much fertilizer will not increase the yield tenfold. Although further improvements in productivity will undoubtedly be made, a given plant species cannot be improved indefinitely.

The climatic limits set by potential photosynthesis also affect yields. In recent years, the amount of land devoted to agriculture has declined in the world because of its unsuitability for raising crops. The 100 million acres of virgin land in Central Asia put to agricultural use by the Soviet Union are often too dry to produce good harvests, and much of this land is now being released from agricultural activity.

Industrialized nations, such as Japan and the United States, now grow more food on less agricultural land than they did 20 years ago. It appears that if advanced agricultural methods were introduced into less developed countries, agricultural performance could be significantly improved there. Many social and economic problems are involved, however. There are social and political implications when a society changes from subsistence farming, in which one farm produces barely enough to support its own workers, to modern productive agriculture, in which one farm produces surpluses for other sectors of society. Fewer farm workers are needed, and those forced to leave the land usually migrate to cities to seek new employment. Marketing and transportation systems are needed to distribute farm products. Modern agriculture requires costly equipment and supplies, and the new methods involved require intensive educational programs for farmers.

It is expected that the population of the world will almost double by the end of this century. Agricultural production will not be able to keep pace with such growth indefinitely. The food shortage in the Soviet Union and a number of other countries after the poor harvests of 1972 shows that food production is not high enough even now. The problem at hand is to feed people adequately until a stable population is attained. Lack of protein is one of the most serious nutritional problems in the world today. The Food and Agriculture Organization of the United Nations has developed a world plan for agriculture to meet food needs through 1985. The plan emphasizes cereal production by intensive agriculture and calls for increased production of animals, particularly hogs and poultry, which are relatively efficient to raise. The United Nations plan calls for cereal production to increase by 3.6 percent each year. Reports at the end of 1976, however, indicate that the growth rate has not increased appreciably from 2.6 percent, the value that has been characteristic of the past decade. World grain production and reserves are shown in Figure 9.14.

ECOSYSTEMS AND MAN

Humans have always interacted with natural ecosystems, and have set up ecosystems of our own to fit our needs. Throughout most of history, and in many countries today, our thrust in dealing with nature has been to obtain the immediate necessities for survival. Sometimes, however, these activities have made survival more difficult over time.

Our interaction with nature has been comparatively successful, nevertheless. A large number of people are well-clothed and well-fed today, at least in comparison with earlier ages. It is becoming increasingly clear, however, that human activities in the drive for survival have important repercussions in nature. As we attempt to turn a greater proportion of the earth's productivity to our own uses, we sometimes act in ways that will make that productivity less available to us in the future.

The traditional method of farming in the rainforests of Central and South America and Southeast Asia has been to clear a section of forest by cutting and burning, to farm the land for a few years, and then to move on to a new section. In a rainforest, nutrients are stored in the natural vegetation rather than in the soil; thus after a few years cleared land becomes less productive. This slash-and-burn, or *swidden* or *milpa,* cultivation was carried on for thousands of years with no apparent long-term ill effects, as long as population densities were low and the forest was allowed time to reestablish itself in a region before clearing was repeated.

Today, however, increased population pressures and the need for food have led the people to shorten the cycle between successive rotations. Furthermore, the cleared land is often used to graze cattle after it has been farmed for a few years. The grazing destroys secondary growth and prevents the reestablishment of the forest. The cattle pack the earth to a nearly impervious layer that seriously diminishes infiltration of rainfall and increases runoff and erosion. The accelerated pace of swidden cultivation may make large tracts of land in the rainforests economically useless and difficult to restore to productivity (see Case Study following Chapter 8).

The search for food has caused many countries to turn to the sea. The sea as a whole is not a highly productive region (see Figure 9.5, p. 239). Photosynthesis is limited to the upper 50 to 100 meters (160 to 330 ft) of the ocean, and growth in this layer is usually limited by lack of nutrients in the water. An acre of midlatitude grassland is several times more productive than an acre of ocean; the net production of the ocean is less than half the net production of the land, and much of this productivity is concentrated in areas of upwelling (such as off the western coasts of North and South America and off South Africa), which represent less than 1 percent of the ocean surface.

The strategy of many nations is to maximize the investment made in modern fishing fleets and methods. Often fishing pressure intensifies as the catch declines. For this reason, the ocean is being overfished, and many species, including some types of whales, have almost disappeared. In 1935 commercial fishermen in California landed 60,000 tons of Pacific sardines. The catches decreased with time, and by 1961 the sardine industry in California no longer existed. Similar overfishing is evident with Pacific perch, mackerel, and tuna. As Slobodkin showed with his laboratory food chain, the best long-term strategy for an intelligent predator is to leave enough animals to make good use of the available food. However, the desire of many

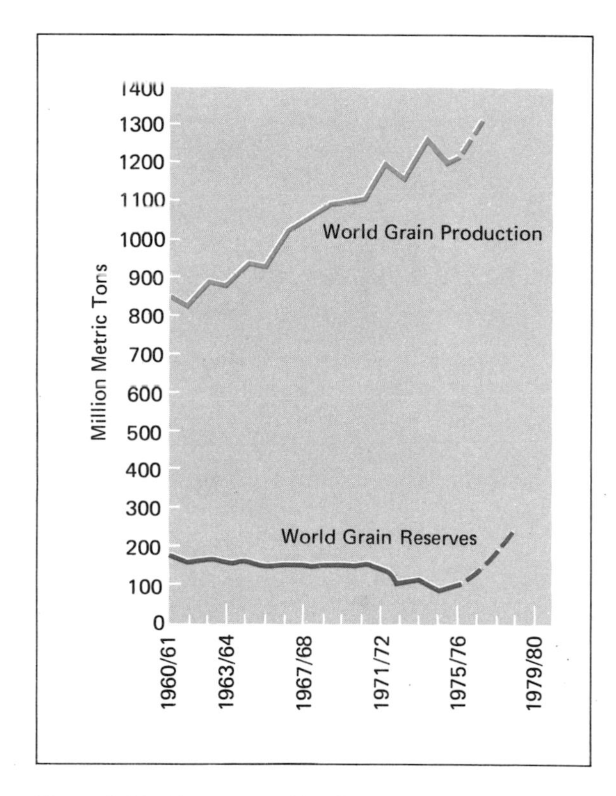

Figure 9.14 Estimates of total world grain production beginning with the marketing year 1960–1961 are shown by the upper curve. The upward trend is due largely to technology rather than increasing acreage, but the variability in the 1970s has been associated mostly with climatic stress, such as drought in the Soviet Union in 1972. The lower curve shows world grain reserves, which have decreased substantially during the 1970s because of population growth. Reserves during the middle 1970s fell to only about 10 percent of annual world production, but reserves have increased rapidly during 1976–1977 because of favorable climatic conditions. (Modified from *World Agricultural Situation,* U.S. Dept. of Agriculture, 1976)

nations for an immediate increase in their living standard becomes more important than the inevitable decline in ocean productivity.

The misuse of rainforest land and the overfishing of the oceans are only two examples that illustrate how immediate benefits can be accompanied by large hidden costs. The farmers and the fishing fleets gain access to an inexpensive source of energy, but restoring the ecosystems that produced the energy will eventually be costly to society as a whole. Because the true costs of human actions are seldom considered, mankind may eventually suffer the consequences of bankrupt ecosystems.

With the rapid increase in world population, man is far from being in a stable long-term relationship to natural and agricultural ecosystems. Andrei Sakharov, a leading Russian physicist, has predicted famine among millions in the next 20 years if production continues to fall further behind population, and he has called upon the affluent nations of the world to use one-fifth of their annual production to help other nations reach higher standards of living. Research can provide the basis for informed actions, but the time available for building a foundation of knowledge is shrinking rapidly.

SUMMARY

Ecosystem analysis is the study of the relationships between organisms and their surroundings. Plants and animals interacting together with the environment are frequently evaluated within a systems framework; one useful theme is energy transformation and storage.

In the photosynthesis process, green plants make use of carbon dioxide from the air, water and nutrients in solution from the soil, and solar radiation to produce complex molecules of plant carbohydrates (sugars and starches). Only a very small proportion of incident solar radiation is utilized directly; usually much less than 5 percent is converted to stored chemical energy.

Photosynthesis is restricted to daylight hours, but plants use energy day and night in respiration to maintain life processes. The net rate at which a plant stores energy—or photosynthesis minus respiration—is called its net productivity. Productivity tends to be greatest in tropical rainforests and to decrease toward drier and colder climatic regions. In water budget terms, productivity is related directly to evapotranspiration. Oceanic productivity is generally low except for regions of cold water upwelling, some shallow coastal areas, and especially estuaries, where nutrient-laden freshwater runoff mixes with ocean waters. The productivity of agricultural ecosystems is about equal to that of the more productive natural systems, but more of the agricultural productivity is available, directly or indirectly, for human consumption.

Energy flows through ecosystems can be evaluated in terms of transfers of energy along food chains and webs. Because of complex interactions, only about 10 percent of the energy absorbed at one stage of a food chain is available to the succeeding stage, which severely restricts the population density of carnivores at the highest level of the chain.

Population growth has exerted great pressure on the agricultural lands of the earth. Until early in this century, agricultural yields had remained almost constant for hundreds of years. Recently, yields of many grains and other crops have increased spectacularly, but the energy input in terms of fossil fuels has been very costly. Agricultural lands will have to be managed carefully and wisely if food production is to keep pace with population growth and the rising expectations of people everywhere.

In the following chapter, we turn to the soils of the earth. Soils physically support plants and also provide a medium for the storage and recycling of materials essential for plant growth. In turn, soil properties strongly reflect climatic inputs and the responses of plant and animal communities.

Review Questions

1. Compare and contrast the diurnal cycles of photosynthesis and respiration.
2. Describe the global patterns of net productivity.
3. Compare the productivity of oceanic and continental ecosystems.
4. How is plant growth related to water budget components?
5. Discuss the various ways plants and animals interact within an ecosystem.
6. How is photosynthesis related to the energy budget of a forest stand?
7. Describe energy flows through food chains.
8. Discuss factors which affect global food production and consumption.

Further Reading

Eckardt, F. E., ed. *Functioning of Terrestrial Ecosystems at the Primary Production Level.* Proceedings of the Copenhagen Symposium, July 1965. UNESCO, 1968. (516 pp.) This collection of technical papers by some of the world's leading scholars includes several studies which are especially useful at the introductory level.

Golley, Frank B., and **Ernesto Medina, eds.** *Tropical Ecological Systems: Trends in Terrestrial and Aquatic Research.* New York: Springer-Verlag, 1975. (398 pp.) This is Volume II of a series which focuses on analysis and synthesis of ecological research. It includes a wide range of papers on tropical environments, which probably present the most difficult resource-management dilemmas.

National Academy of Sciences. *Productivity of World Ecosystems.* Washington, D.C., 1975. (166 pp.) This short book is a collection of papers presented at a symposium held in Seattle during 1972. The papers, written at beginning and intermediate levels, analyze the world's major ecosystems.

Odum, Eugene P. *Fundamentals of Ecology.* 3rd ed. Philadelphia: W. B. Saunders Co., 1971. (574 pp.) This is a comprehensive work by an outstanding American ecologist.

———. *Ecology: The Link Between the Natural and Social Sciences.* 2nd ed. New York: Holt, Rinehart and Winston, 1975. (244 pp.) This paperback introduces ecological principles for students and laymen. More emphasis is placed on man and ecosystem interactions than in the preceding standard work.

Phillipson, John. *Ecological Energetics.* New York: St. Martin's Press, 1966. (57 pp.) This very small book is often considered a classic; most of the basic themes of ecology are introduced succinctly.

Case Study:

The Tragedy of the Whale

Just below the surface of the sea live some of the most immense creatures on earth—whales. These air-breathing mammals are found in all oceans of the world. Most species are social animals, and they move about the oceans in groups that vary in size from small families of a few animals—some of them coming together only during part of the year for pairing and mating—to large groups that number in the thousands. All species of whales have been severely reduced in number by centuries of intensive whaling. Yet mounting evidence warns us that we must consider the whale not as an unlimited resource but as an integral part of the ocean ecosystem.

In the cooler waters of northern and southern oceans, energy is most abundant. The polar seas are more productive than the tropic ones, partly because in the tropics the more dense, nutrient-rich underlying water cannot easily mix with the very warm, less dense surface water that has been largely depleted of nutrients. In the polar seas, life is so abundant that during the short growing seasons the ocean surface is often discolored with plankton blooms for many square miles. Plant plankton, which carry on the major portion of photosynthesis that occurs in the ocean, use only carbon dioxide, seawater, and certain nutrients to synthesize proteins, fats, and carbohydrates. These plant plankton, which lie at the base of the energy pyramid, are eaten by a variety of animals ranging in size from single-celled protozoa to crustaceans resembling small shrimp. These are in turn devoured by other, usually larger, animals, some of which are eaten by other animals, and so on. In general, 80 to 90 percent of the energy is lost at each step in the food chain.

Baleen whales, which constitute the most abundant group of whale species, feed almost exclusively on the largest plankton animal, krill. Their food chain can be represented as follows:

Plant plankton → animal plankton → whales
(primarily krill)

Because these whales obtain their food from animals that are near the base of the energy pyramid, the amount of energy lost in the food chain is relatively small. Thus whales can harvest an important part of the ocean's food relatively efficiently, and they store their harvest in neat, easily collected multi-ton packages: themselves. The whale represents the best access that we presently have to the abundant plant plankton of the northern and southern seas.

But the activities of the whaling industry have significantly depleted the whale population in much of the Arctic during the last century and in the North Pacific and the Antarctic during this century. Whale oil has been used in household paint, in margarine and soap, and in cosmetics. Outside of Japan where it constitutes 10 percent of the meat supply, whale meat is fed primarily to domestic animals—dogs and cats in Europe and North America and ranch fox and ranch mink in the United States, Canada, and Norway. The remarkable fact about the various whale products other than meat is that all of them could be made from synthetic materials or other, more abundant natural raw materials. The business of whaling, far from being an outdated practice from our history, has until very recently been more extensive than it ever was in previous centuries. The total of whales killed in the 1960s was the largest ten-year kill ever made. Nevertheless, large-scale whaling operations continue to be pursued only by Japan and the Soviet Union. The Japanese justify their whaling industry as a necessity to supply protein to an increasingly meat-hungry population. Despite its own harvest, Japan imports additional whale meat from the U.S.S.R., Korea, and Peru. Were Japan to terminate its whaling industry, in which it has a heavy financial investment, it would be forced to import vast quantities of more expensive beef, contributing to the already troublesome balance of payments problem created by the rising cost of imported petroleum.

At present, only four species of the "great" whales (fin, sei, Bryde's, and sperm) remain in commercially profitable numbers. These should remain at or rebuild to optimum population sizes if current practices continue. (Of these four species, only fin whale stocks are currently below optimum size.)

How many whales can be taken from a population without endangering the species? Scientists are attempting to determine this by studying the rate at which a species can replace itself and thus maintain its population. Under dense population conditions, whales, like all animals, produce more young than are necessary to maintain the whole population, and a fairly high percentage of newborn whales presumably die of natural causes—predators, disease, inadequate food supply, failure to compete, and so on—during the first year. When the population of a region is substantially reduced for any reason, a larger proportion of the young survives. Also, the number of offspring produced in a depleted population may increase because of a lowering of the average age of sexual maturity and an increase in the pregnancy rate. Female fin whales, for example, mature sexually at about six years in a depleted stock, as opposed to about ten years in a stable, unexploited stock, and the pregnancy rate increases from about 33 percent to about 50 percent.

From a commercial viewpoint, a viable number of whales would be the maximum number of animals that can be harvested from a species each year without preventing the species from making up that loss. This number is known as the *sustainable yield.* In order to calculate the sustainable yield of a whale species, one must know, among other things, the size of the overall population, how many young a female bears each year, how long it takes the young to reach reproductive age, how long an individual's reproductive age lasts, and what the life expectancy of each age group is.

Some scientists have claimed that the main schools of whales in the Antarctic have been brought to near extinction. But just what does "near extinction" mean? The whaling industry is concerned only with *commercial extinction,* in which the number of remaining animals is too small to be harvested profitably. Other scientists studying whales are quick to point out that there is little danger that any species will become totally extinct—there will always be at least a handful of surviving members of a species. In terms of ecology, however, neither of these interpretations seems relevant; the few whales that remain at the point of commercial extinction can no longer play a very significant part in the ecosystems of the oceans. In this sense, the concept of *significant extinction* seems a more relevant one. And it seems apparent that it is the significant extinction of whales that we should strive to prevent.

Whale species are conveniently placed in two categories, the toothed *whales and the* baleen *whales. Most toothed whales—porpoises, dolphins, narwhales, and killer whales, for example—are relatively small. The sperm whale, the largest of the toothed whales, is perhaps the best known of all whales; of the five species that sustained the nineteenth-century whaling industry, sperm whales were most frequently killed.*

The larger baleen whales are so named because of the comblike structure that descends from the roof of their mouths. It is made of closely spaced parallel triangular plates of hornlike substance called baleen. The inner edges of these long, thin slats fray out into a thickly matted net of hair that lines the entire inner surface of the comb so that as the whale swims, it sieves out and swallows whatever food the seawater contains.

The throat and belly of most baleen whales, such as the humpbacks, are deeply furrowed by long pleats from chin tip to navel. These pleats allow the whale's throat to expand so that it can engulf great quantities of water, from which the whale strains its food. Baleen whales without these pleats—the right whale, for example—have longer baleen strips and strain their food by swimming, with their mouths partially open, directly into masses of small marine life, thus collecting the tons of krill and other animal plankton they feed on.

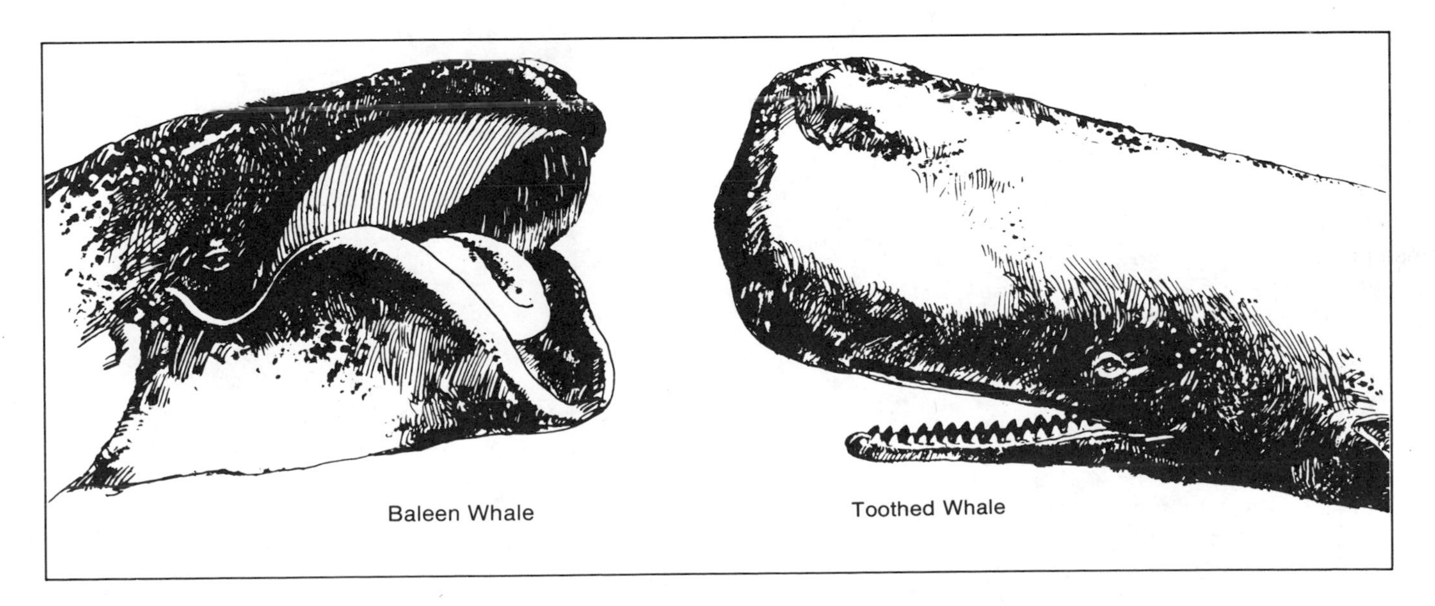

Lime Banks by Andrew Wyeth, 1962. (Collection, Mr. Smith W. Bagley; copyright © by Andrew Wyeth)

A thin layer of soil cloaks the earth's surface, softening its contours and supporting its vegetation. The soils we see today are products of the interactions of living and nonliving materials with energy and moisture over hundreds or thousands of years. The development of characteristic soil types is associated particularly with climate and plant cover.

TEN
The Soil System

AN ESSENTIAL FACTOR IN the productivity of any ecosystem is the nature of the soils that support it. Soils are indispensable for the higher forms of life that have evolved on our planet, for they are the medium in which plants grow and, as we have seen in the preceding chapter, plants are the basis of virtually all terrestrial food chains. Soil is itself a much more complex phenomenon than most people realize; it is certainly not just fine material—or "dirt," as it is often called. The "soil" we see taken from excavations or being pushed around by earth-moving equipment is generally not soil at all in the scientific sense. True soil is the rather thin veneer of organic matter and fine rock particles, differentiated into layers of contrasting physical and chemical characteristics, which forms over a greater depth of decomposed rock that is the raw material for soil development but is not soil itself.

A true soil is the product of a living environment and has particular features that reflect the influence of climate and the activity of living organisms. Where there is no life, as on the moon, there is no true soil. Since the earth is teeming with life, soils are present almost everywhere, although their degree of development and state of preservation may vary greatly. As soon as water begins percolating downward through fine mineral material and some form of vegetation roots among the mineral particles, the development of soil begins. The less water and vegetation present, the weaker the development of the soil. In many desert regions, it is truly hard to say whether the loose surface material is indeed soil or merely finely divided rock debris. Elsewhere, human activities have led to such massive losses of soil by erosion that it is difficult to reconstruct the original nature of the soil. Let us consider what is required to produce this crucial resource.

SOIL-FORMING PROCESSES

Soils are found covering nearly all types of terrain. In every case the soil represents a surface modification by organic activity and percolating water of a mass of fine mineral particles ultimately derived from solid rock. Thus the story of soil development begins with the fragmentation of rocks by chemical and mechanical action, a process known as weathering.

Weathering: The Disintegration of Rocks

The earth's surface configuration is not static. The vast tonnages of sediment fed into the seas daily by large rivers indicate ceaseless removal of material from the land surface. This removal begins with the fragmentation of rock by weathering processes and continues as the fragmented material is transported away from its point of origin by the agents of *erosion:* running water, wind, waves, glacial ice, and the force of gravity itself. Subsequent chapters are concerned with how this removal occurs and the surface forms that result from it. But there would be no removal, no change in the land surface, without prior fragmentation of solid rock by the various weathering processes.

Both mechanical and chemical stresses can fragment solid rock. Downward erosion into the land surface initiates the mechanical breakup of certain massive types of rocks, such as granite; as erosion reduces the confining pressure on the subsurface rock, the rock expands upward by developing internal cracks. This seems to be the only disintegration process that does not involve the action of water in some form. All rocks (which are discussed more fully in Chapter 13) contain openings by which water can enter them. These may be microscopic pores between the particles forming the rock or they may be wider partings, principally the joint systems that divide most rocks (see Figure 10.1). Water splits rocks when it freezes in such openings, just as it bursts pipes when heating systems fail in winter. This occurs because the change of state from liquid to crystallized (frozen) water produces nearly a 10 percent increase in volume. Where water fills a crack, it freezes first at the surface and then progressively downward; thus the force of the expanding water acts against the confining rock rather than simply causing the ice to move upward. The expansion of freezing water in the tiny pores of rocks like sandstone may break loose all of the grains in the freezing layer. Similarly, shale is a common rock type that is easily split into small flakes by freezing water. Wherever below-freezing temperatures are frequent and sustained for long periods, this so-called *frost weathering* is a very important agent of rock disintegration. In dry climates, the crystallization of salts contained in evaporating water can also cause the detachment of small particles of rock. Where rainfall is more abundant, soluble salts are flushed away and play little or no part in rock disintegration. The prying action of growing roots and the burrowing activities of animals also enlarge natural openings in rocks. The major products of mechanical weathering are chemically unchanged rock particles of various sizes: boulders, gravel, sand, and silt.

Figure 10.1 (top) Granite rock usually breaks into blocklike masses along joints when it is exposed at the earth's surface. The joints in this granite at Pike's Peak, Colorado, have been enlarged by physical and chemical weathering processes that slowly transform such rock into sand and clay.

(bottom) This photo illustrates the effect of frost weathering in cold climates in high latitudes and at high altitudes. The angularity of the forms produced here in the Teton Range of Wyoming is characteristic of frost weathering.

Chemical decay tends to play a greater role than mechanical stress in most rock fragmentation. The positively charged hydrogen (H^+) and negatively charged hydroxyl (OH^-) ions in water are quick to react with the chemical compounds (minerals) of which rocks are composed (see Chapter 13). In addition, water that percolates into the ground usually becomes a weak acid through absorption of the carbon dioxide and acidic decay products produced by vegetation (see Figure 10.2, p. 258), increasing its capacity to cause chemical reactions. Some of the products of these reactions are quite familiar, such as the orange rust that develops on iron tools left out in the rain or the green deposit that appears on copper roofs and leaky copper plumbing. Both of these are products of the chemical process known as *oxidation*, in which oxygen ions in the water combine with other ions. Oxidation affects

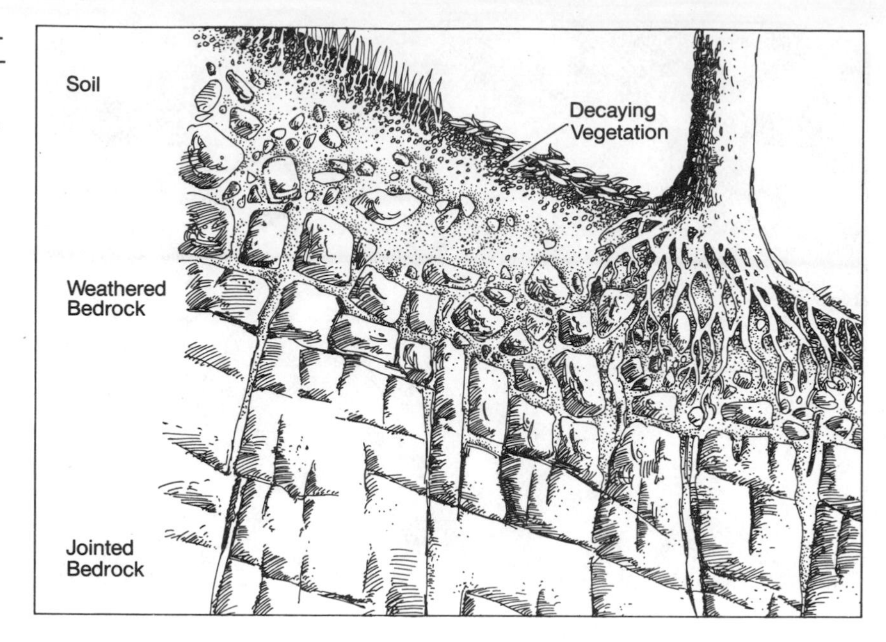

Figure 10.2 Chemical and physical weathering processes break down massive rock into the small particles that form the inorganic component of soil. In warm, moist climates, weathering proceeds actively along the exposed surfaces of jointed rock. Some of the mineral nutrients supplied to the soil by the parent rock are incorporated into growing vegetation. The nutrients are returned to the surface of the soil when plant litter decays, releasing nutrients, which rain water subsequently washes downward into the soil.

rock-forming minerals as well as metals. Most soils, and the decomposed rock under them, are red-brown, yellow-brown, gray-brown, or plain brown due to the oxidation of iron-bearing minerals.

Oxidation seldom acts alone and is but one of several chemical reactions that decompose rocks by swelling and softening some minerals, dissolving others, and causing the complete alteration of still others to form new secondary minerals. Chemical weathering is, of course, most effective in warm, damp environments, where it can produce a *weathered mantle* of fine particles tens of meters deep over the solid bedrock. The ultimate product of chemical weathering is clay, a substance almost as common as air and water and all of it produced by rock decay. Clay results from thorough chemical alteration of the most abundant of the rock-forming minerals, and it varies considerably in its physical characteristics. Although not all minerals weather to clay, the proportion of clay minerals present in a soil can often be used as an index of the soil's relative age or degree of development, while the specific types of clay minerals are a clue to the environment in which the weathering occurred.

Translocation

The mantle of fine material created by weathering is a vital resource, but it is not yet a true soil. Many changes are required. When rain water sinks downward into the weathered mantle, it produces both mechanical and chemical translocation of material. The process of *eluviation* is the flushing of fine particles or dissolved substances to lower levels in the soil. Most soils tend to develop a surface layer that is relatively porous due to the downward translocation of the finest particles. Conversely, the deposition of these fine particles at a lower level, known as *illuviation,* makes the soil there more compact than either the eluviated zone above or the weathered mantle below. Thus the soil begins to be differentiated into layers, or *horizons,* by vertical changes in *texture* and *density*—the size of the soil particles and their degree of compaction.

Downward percolating water also causes the translocation of chemical components in the weathered mantle. Like the finest particles, these dissolved substances are removed from the surface layer. The solutes may be taken up by plant roots, removed by groundwater outflow, or deposited at a lower level in the soil. Complete chemical removal is known as *leaching.* In a later section, we will see that soil moisture can be acidic enough to remove normally inert iron and aluminum oxides from the upper soil. When this occurs, the oxides are deposited in the compact zone deeper in the soil, giving the illuvial layer a yellow or red color. Under semiarid conditions, calcium compounds are eluviated from the upper layer and deposited as lime (calcium carbonate) in the deeper part of the soil in forms varying from scattered white flecks to concrete-like masses. The latter are vivid demonstrations of the effectiveness of eluviation in the development of soil horizons.

Organic Activity

A mass of mineral particles alone, even differentiated into contrasting layers by translocation, does not constitute a true soil. No real soils existed until life appeared on the land. Organic activity is vital to soil development, and organic matter is a crucial component of soils. Several types of life are necessary in the development of soil: plants to contribute organic material, fungi and bacteria to reduce this organic matter to a semisoluble chemical complex called *humus,* and larger soil organisms, such as earthworms and termites, to mix the humus into the mineral matter of the soil.

Humus is the chemical substance that makes the upper part of the soil so much darker than the subsoil below it. Humus is itself dark brown to black and is difficult to see or study in isolation because it becomes intimately mixed with the mineral matter of the soil. The properties of humus are all beneficial: through its own water-absorbing capacity, it increases the soil's ability to store moisture and to retain soluble nutrients against the eluviation process; it is an important source of the carbon and nitrogen required by plants; and it maintains a favorable soil structure (neither too compact nor too porous) for plant growth.

In addition to their role in the creation of humus, soil bacteria are crucial in converting gaseous nitrogen to forms that can be utilized by plant life, a process known as *nitrogen fixation.* Many forms of bacteria can extract free nitrogen from air in the soil. Some of these are parasitic on plant roots; others live in the soil itself. Bacteria also liberate nitrogen locked in organic soil proteins so that it can be absorbed by plants. Most plants require nitrogen; thus air is a vital component of soils, both as a source of this essential plant nutrient and as a source of oxygen for beneficial soil organisms. Where air is excluded, as in waterlogged soils, the organic activity vital to soil formation is greatly diminished and soils are minimally developed.

The degree of organic activity in soil is evident from the fact that 1 cu cm of soil may contain over 1,000,000 bacteria, and a normal hectare (2.47 acres) of pasture land in a humid climate usually contains a million or more earthworms and close to 25 million insects. Earthworms and insects are very important in mixing and aerating the soil as well as in producing a part of its humus content in their own digestive systems.

Vegetation affects soil formation by providing the plant debris that organisms subsequently decompose to produce humus. Plant root secretions in the soil assist in the chemical weathering process, and plant roots wedge apart rock masses, contributing to mechanical weathering. Different vegetation types influence soil development in different ways through their varying root depths, contributions of organic litter, and effects on soil chemistry. Some of these influences will be discussed in later sections.

PROPERTIES OF SOILS

The soil-forming processes outlined above give soils varying physical and chemical properties. These properties reflect a number of influences to be discussed more fully in later pages. Soils are generally classified according to several attributes, which include texture, structure, chemical characteristics, color, and profile development. Some of these are a consequence of the soil-forming environment; others reflect the soil's *parent material*—the raw material from which the soil is formed.

Soil Texture

The *texture* of a soil refers to the size distribution of the mineral particles composing the soil. Particle sizes are conventionally classified into three main groups: sand, silt, and clay. In the United States, particles with diameters between 62 and 2,000 microns (1 micron = .001 mm) are called *sand,* particles from 2 to 62 microns are called *silt,* and those smaller than 2 microns are called *clay,* as shown in Figure 10.3 and Table 10.1. The designations sand, silt, and clay refer only to the size of the particles and not to their chemical or mineralogical composition.

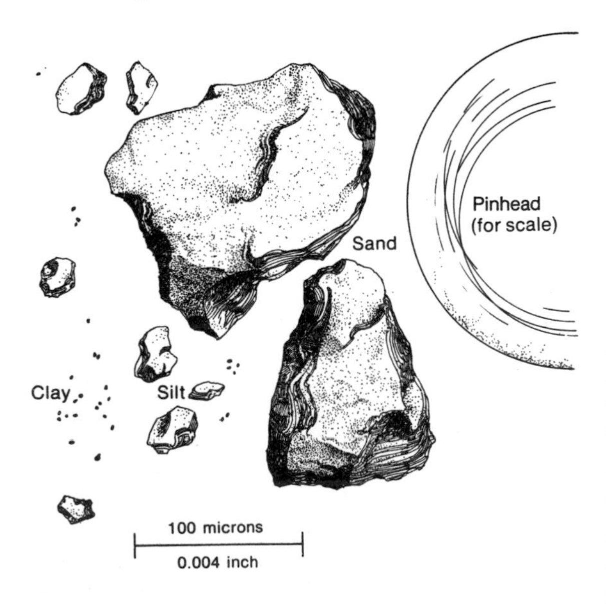

Figure 10.3 The particles that constitute the inorganic component of soil are classified according to size. Particles smaller than 2 microns in diameter are called clay, particles from 2 to 62 microns in diameter are called silt, and larger particles are considered sand or gravel.

The United States Soil Conservation Service has set up a broad classification of soil textures. The texture class to which any soil belongs can be determined by measuring the proportions of sand, silt, and clay in the soil. Figure 10.4 shows the standard classification of soil textures. Many soils are classified as *loams* of different types, which means that they contain a mixture of sand, silt, and clay. Texture influences a soil's ability to absorb and store moisture. Coarse-grained sandy soils absorb water easily but cannot retain water through long periods of drought. The small pores in fine-grained silt and clay soils are efficient in retaining moisture but make infiltration difficult. Texture also influences the ease of working the soil, as well as its rate of heating in spring and cooling in fall.

Table 10.1. Particle-Size Classification

Class	Millimeters	Microns
Gravel	greater than 2.0	greater than 2,000
Very Coarse Sand	2.0 – 1.0	2,000 – 1,000
Coarse Sand	1.0 – 0.5	1,000 – 500
Medium Sand	0.5 – 0.25	500 – 250
Fine Sand	0.25 – 0.125	250 – 125
Very Fine Sand	0.125 – 0.062	125 – 62
Very Coarse Silt	0.062 – 0.031	62 – 31
Coarse Silt	0.031 – 0.016	31 – 16
Medium Silt	0.016 – 0.008	16 – 8
Fine Silt	0.008 – 0.004	8 – 4
Very Fine Silt	0.004 – 0.002	4 – 2
Clay	less than 0.002	less than 2

Source: Ruhe, Robert V. 1969. *Quaternary Landscapes in Iowa.* © Iowa State University Press, Ames.

Soil Structure

Grains of soil join together in larger clumps called *peds.* The shape and size of the peds characterize the *structure* of a soil. The peds can be plate-like, granular, blocky, or prismatic, as shown in Figure 10.5, p. 262. Plants generally grow best in a soil with a surface structure consisting of peds 1 to 5 mm (0.04 to 0.2 in.) in size. Smaller peds reduce the infiltration of air and water into the soil, and larger peds leave too much space around small plant roots. One of the reasons for plowing soil before planting is to make the soil structure as suitable for plant growth as possible.

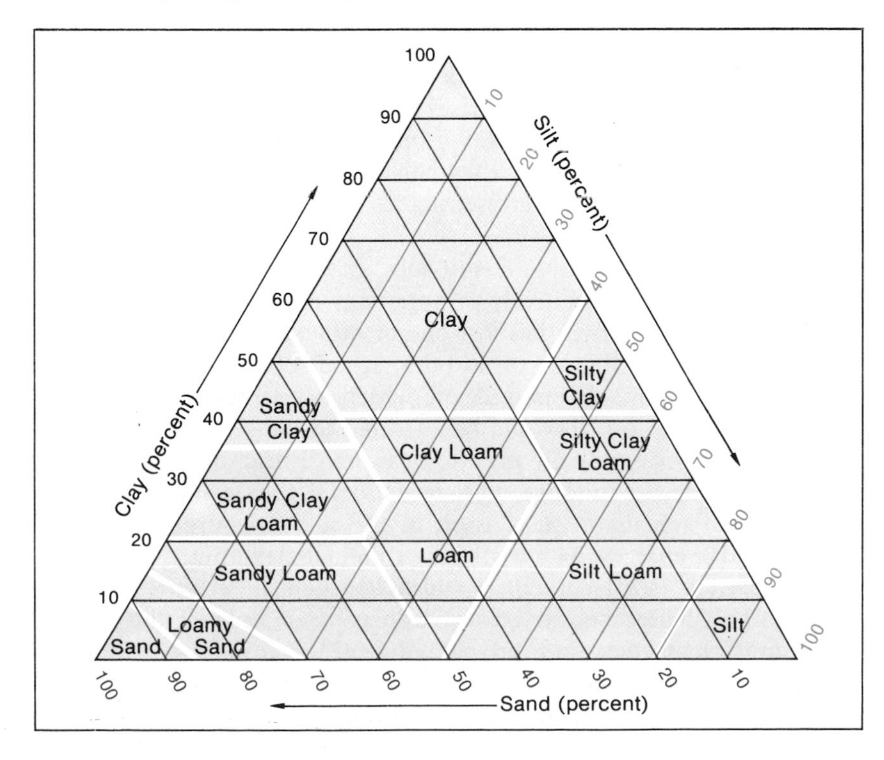

Figure 10.4 The texture of a soil is determined by measuring the proportions of clay, silt, and sand in the inorganic part of the soil. Texture is measured by sifting the soil sample through a series of screens graded from coarse to fine. The soil texture triangle shown in the figure can be used to classify the texture of a soil sample once the percentages of the components are known. If a soil sample contains 30 percent clay and 40 percent sand, for example, it would be classified as a clay loam. (After Bridges, 1970)

Figure 10.5 Four common soil ped structures are shown here. Scales are in inches.

a. Platy
b. Prismatic
c. Blocky
d. Granular

In addition to ped form, soil *bulk density* (mass per unit of volume, including pore space) is an important property. Bulk density normally increases with clay content and is a measure of soil compactness. Soil bulk density may be increased artificially and unintentionally through compacting by heavy agricultural equipment. Both natural and artificial compaction present problems in soil utilization since they reduce the rate of water infiltration into the soil.

Soil Chemistry

The chemical behavior of soil is closely connected with its physical characteristics. The fine clay particles in soil become attached to small particles of humus, forming a *clay-humus complex* composed of particles so small that chemically they behave like large molecules. The clay-humus complex plays an essential part in maintaining soil fertility.

When chemical compounds dissolve in soil moisture, they *ionize,* or dissociate into positively or negatively charged atoms. A positively charged ion, which has lost one or more electrons, is called a *cation* (since it would move toward the cathode of an electrolytic cell or battery). A negatively charged ion, which has gained one or more electrons, is an *anion* (it would be attracted to the anode of an electrolytic cell). The extremely minute (less than 1 micron in size) clay and humus particles in soils tend to be *colloidal,* that is, they are solid particles that remain in suspension in the soil moisture; they also carry negative charges. As a result, the colloidal clay-humus complex attracts cations (see Figure 10.6) and holds them by electrical force (adsorption). These bound cations, consisting mainly of the bases calcium (Ca), magnesium (Mg), potassium (K), and sodium (Na), are essential plant nutrients. They go into solution very easily and would be quickly leached from the soil (see Figure 10.10, p. 266) were it not

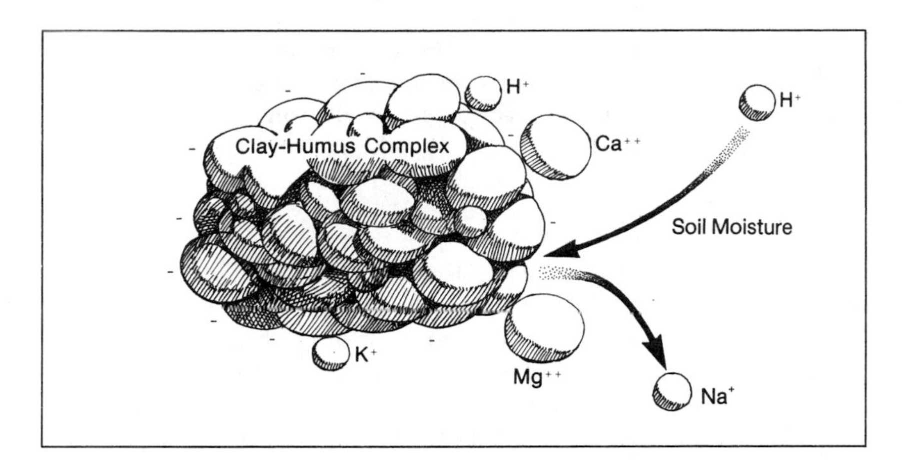

Figure 10.6 The colloidal inorganic particles and humus in a soil bind together to form the clay-humus complex. On a submicroscopic scale, the particles of the clay-humus complex act like giant molecules with the power to attract cations electrically. The complex performs an important function in a soil by preventing chemical nutrients needed by plants from washing out of the soil. Acid soil moisture contains numerous hydrogen ions that can replace basic cations on the surface of the complex, as the figure shows. Hence acid soil moisture removes basic inorganic nutrients from the soil.

for their electrical bonds to the clay-humus complex. The colloidal state of soils is evident in several characteristics, including plasticity, cohesion, and volume changes.

In a productive soil, the clay-humus complex maintains a delicate balance. Its hold on the essential cations must be strong enough to prevent the cations from being washed away, but weak enough to allow plant roots to absorb cations as required. The clay-humus complex holds different kinds of cations with different degrees of strength. Calcium is most strongly bound, sodium least, with magnesium and potassium intermediate. A cation weakly bound to the complex can be replaced by a new cation that will be more strongly bound. Basic cations can also be replaced by metallic cations, or by hydrogen ions from acidic water. This process of cation exchange can cause the soil properties to vary with time as one type of bound cation gradually replaces another. The capacity of various soil colloids to acquire and retain cations is a definite quantitative value, expressed as the *cation exchange capacity* (measured as milligram equivalents per 100 grams of soil material). The degree to which a soil is saturated with exchangeable cations other than hydrogen is its *base saturation,* expressed as a percentage of the cation exchange capacity.

Not all soils have a functioning clay-humus complex. The soils of arid regions contain little or no humus and a smaller percentage of clay than most humid region soils. Thus while the soils of arid regions are initially rich in soluble plant nutrients, the lack of a clay-humus complex permits them to be rapidly leached of these nutrients when they are artificially irrigated. This is a major problem of desert irrigation agriculture. On the other hand, to utilize some desert soils it is necessary to leach them purposely to remove toxic salts.

The *acidity* of a soil is one of the measures used to characterize soil chemistry. Acidity is measured in terms of the concentration of hydrogen ions, known as the pH of the soil. Pure water, which has a pH value of 7, contains 10^{-7} gram of hydrogen ions per 1,000 cu cm. A weak acid with 10^{-6} gram of hydrogen ions per 1,000 cu cm has pH 6. The lower the pH, the more acidic the soil and the lower its content of plant nutrients. Values greater than pH 7 indicate basic, or alkaline, soils. Natural soils have pH values between 3 and 10; values between 5

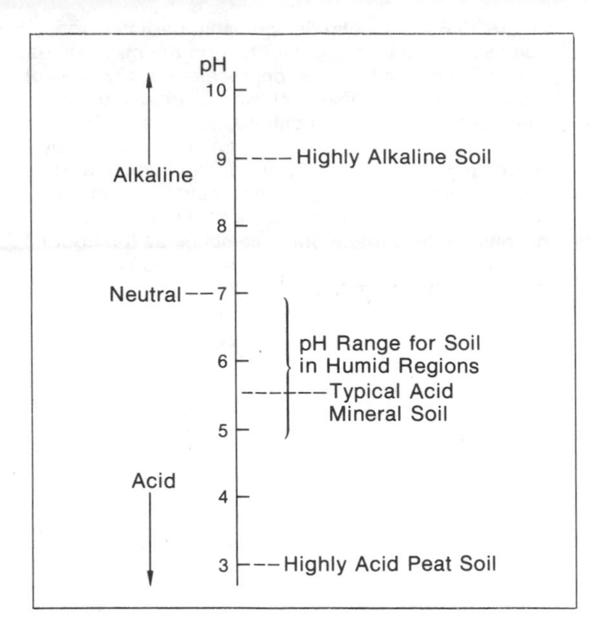

Figure 10.7 The pH value of a soil, or its hydrogen ion concentration, is one of the measures that can be used to estimate a soil's suitability for agriculture. A low pH indicates that a soil is acid and may have lost many of its nutrients by exchange with hydrogen ions. A high pH indicates that a soil contains strong alkalis, which may be damaging to plant root tissues. (After Lyon and Buckman, 1943)

Figure 10.8 V. V. Dokuchaiev, the father of soil science, used this drawing to illustrate the soil profile of chernozem soil, a productive agricultural soil of the Mollisol type found in Russia and the United States. The *A* horizon at the top is rich in organic material, which soil animals and percolation of water rapidly carry downward. The *C* horizon in a chernozem soil often contains the burrows of soil animals, lined with lime from the calcium salt concentrations in the soil.

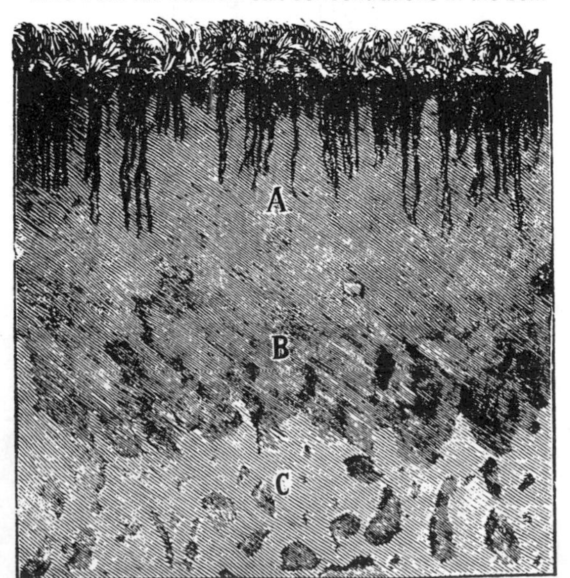

and 7 are the most common and seem best for most crops (see Figure 10.7). Soil pH can be modified artificially by the application of fertilizers: lime is used to decrease acidity of low pH soils and ammonium sulfate to increase acidity of high pH soils.

Soil Color

One of the most obvious characteristics of a soil is its color. Local soil colors vary both laterally and vertically. On a larger scale, soil color varies geographically by environment. Soils of the humid tropics are generally red; those of the temperate grasslands tend to be black; those of northern coniferous forest regions may be gray. Soil color is almost entirely a consequence of the amount and state of iron and organic matter present. Iron oxides, as we have seen, produce the characteristic reds, yellows, and browns of soil. Soils that have been leached of iron are gray. Iron may also be present in reduced (deoxidized) form, giving soils greenish and gray-blue hues. This occurs where air has been excluded from the soil environment, as when the soil is permanently waterlogged. Organic matter colors the soil black. Thus it is the combination of iron oxides and organic content that gives most soils their brown color. Other coloring materials are sometimes present, particularly white calcium carbonate, black manganese oxides, and black carbon compounds.

Soil color is generally included in soil descriptions, color identification being based on the standardized Munsell color notation system, which classifies all colors according to hue (position in the color spectrum), value (relative lightness), and chroma (color purity or strength). Thus a yellowish-red soil might be described as 5YR 5/6—midway in the range of yellow-red hues, with a value of 5 (halfway between absolute black and absolute white) and a chroma of 6 (moderate color purity). These identifications are made with detailed color charts designed especially for the purpose of soil classification.

Soil Profiles

Every soil has distinctive characteristics that are expressed in the nature of its *profile,* or the sequence of horizontal layers resulting from the soil-forming processes of eluviation and organic activity (see Figure 10.8). Five major layers are generally present, termed the *O, A, B, C,* and *R* horizons (see Figure 10.9). Each of these is further subdivided using numbers (*A0, A1, A2, B1, B2,* etc.). The *O horizon* is the layer of undecomposed plant debris or raw humus at the soil surface. Below it is the *A horizon,* which has two characteristics: it is the layer in which humus and other organic materials are mixed with mineral particles, and it is the zone of translocation from which eluviation has removed certain fine particles and soluble substances, both of which may be deposited at a lower level (see Figure 10.10, p. 266). Thus the *A* horizon is dark in color and usually light in texture (sandy or silty) and porous. The *A* horizon is commonly differentiated into a darker upper *A1* horizon of organic accumulation and a lighter lower *A2* horizon showing loss of material by eluviation.

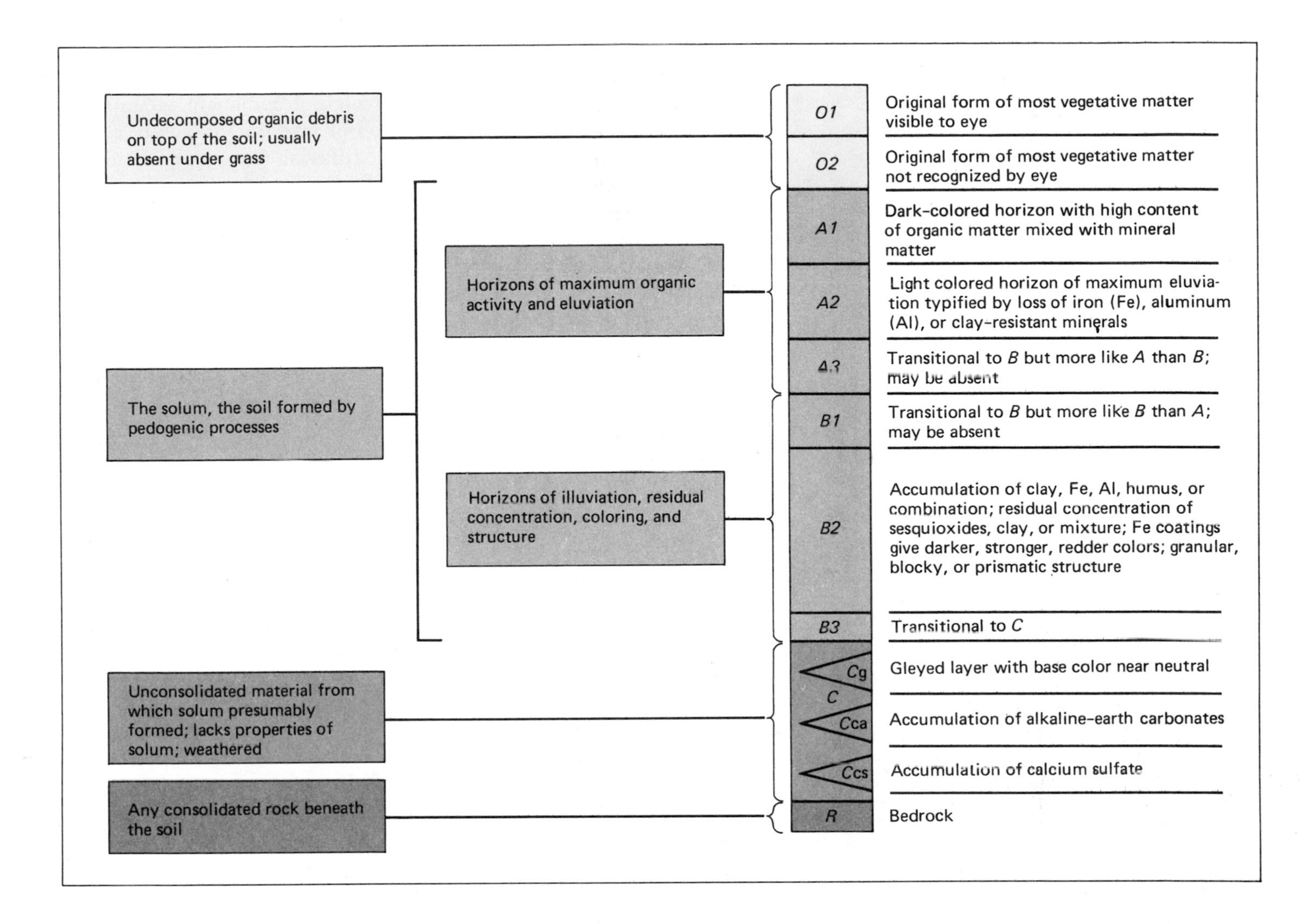

Figure 10.9 Standard horizons in soil profiles. No single profile contains all of the horizons shown. Additional subhorizons similar to those indicated for the *C* horizon include: *B2t*—illuvial clay; *B2ir*—illuvial iron; *B2h*—illuvial humus; *B2m*—strong cementation; *Csi*—cementation by silica; *sa*—enriched by salts; *f*—permanently frozen; *x*—hardpan composed of sand and/or silt (fragipan). (Modified from Ruhe, 1975, and Soil Survey Staff, 1951, 1962)

The *B horizon* is the *illuvial* layer. It receives material eluviated from the *A* horizon. Generally it is compact, with a higher bulk density than the *A* horizon due to its greater clay content. A clay-rich *B* horizon is called a *textural B horizon.* Here is where we encounter blocky and prismatic soil structures and sometimes dense clay hardpans as much as a meter thick. Clay hardpans occur only in semiarid areas and probably mark the maximum depth of moisture penetration and clay formation within the soil. Thus they are not entirely illuvial in origin. The *B* horizon may be vividly colored by oxides of iron and aluminum or by calcium carbonate leached from the *A* horizon.

The *A* and *B* horizons together are known as the *solum,* the soil generated by pedogenic (soil-forming) processes. The *C horizon* is composed of weathered parent material that has not yet been significantly affected by the pedogenic processes of translocation and organic modification. This layer also includes unconsolidated parent material, such as deposits left by streams (alluvium), wind (sand and loess), and glaciers (till). The *R horizon* consists of unweathered bedrock.

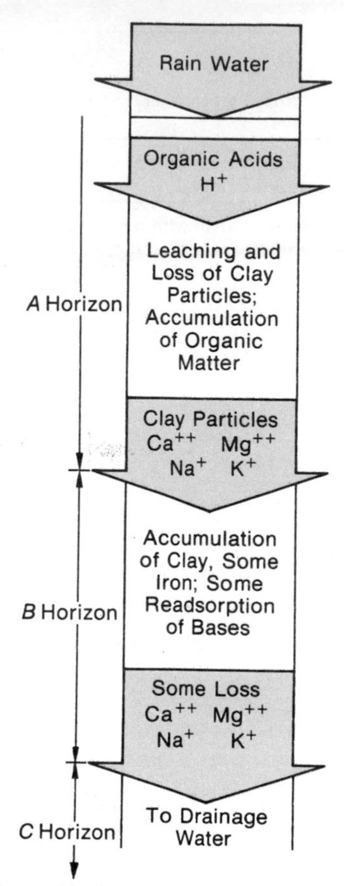

Figure 10.10 In the eluviation process shown here schematically, rain water, which has become acidic from dissolved carbon dioxide or from organic acids in the humus on the ground, infiltrates the soil. The hydrogen ions in the acidic water displace basic cations from the clay-humus complex, causing a downward movement of soluble nutrients. If the acidity of the soil becomes high, the clay particles disintegrate chemically, usually into compounds of aluminum and iron. (After Bridges, 1970)

These horizons vary greatly in thickness, depth below the surface, and degree of development. Special characteristics are indicated with additional symbols, such as *ca* (calcium carbonate accumulation), *g* (for gleying, meaning reduction of iron due to lack of aeration), *ir* (illuvial iron accumulation), *t* (illuvial clay accumulation), and so on. Figure 10.11 shows a typical soil description according to the common notation system.

Many of the important characteristics of soil profiles are not visible to the naked eye but have to be ascertained by tests in the field or laboratory. These include measurements of bulk density and the specific content of organic matter, carbonate, clay, oxides, and silica at different levels. When all the properties of a soil are known, the soil can be classified in terms of its morphology, potential productivity, and probable genesis. Many soil profiles reveal changes in the pedogenic processes over time; thus a soil profile is a good indicator of both present and past environmental conditions in an area.

FACTORS AFFECTING SOIL DEVELOPMENT

The present study of soils, the science of *pedology,* is largely an outgrowth of work begun by Russian soil scientists more than a hundred years ago. Over the vast expanse of Russia, the nature of soils seemed to be very closely associated with large-scale patterns of vegetation and climate. These relationships were outlined in 1886 by V. V. Dokuchaiev, who was the first to analyze soil horizon development and who is recognized as the founder of soil science. The studies of Dokuchaiev and his associates indicated that soil formation was influenced by a number of factors, including (1) parent material, (2) climate, (3) site, (4) organisms, and (5) time. Present-day soil scientists generally call these the *factors of soil formation.*

Parent Material

As we have seen, the inorganic material on which a soil develops, known as the parent material, determines what chemical elements are initially present in the soil. The parent material may be either rock that has decayed in place or unconsolidated material deposited by the action of streams, glaciers, or wind or by massive gravitational transfer (see Chapter 11). Some parent materials have an abundant supply of the chemicals most needed by plants, others are lacking in plant nutrients, and a few contain substances that are actually toxic to many plants. Of course, the elements that we ourselves require for proper nutrition come indirectly from this parent material through both the plant and animal material in our own diets.

In addition to the relatively inert common elements silicon, aluminum, and iron, we have seen that soils usually contain the soluble bases calcium, potassium, magnesium, and sodium. From an agricultural standpoint, all of these are desirable components of soils. Acidic water moving through the soil tends to remove soluble bases by exchanging hydrogen ions for basic cations. The soil therefore becomes more acidic and less suitable for agriculture. However, if the parent material is rich in bases, they can be resupplied to the soil and taken up by plant roots, in effect replacing those leached from the soil. For this reason, soils developed on soft limestone bedrock in areas of humid climate are especially fertile, for limestone is composed almost entirely of calcium carbonate (see Chapter 13). On the other hand, soils derived from sandstone are often infertile, having few soluble nutrients to begin with and being coarse in texture, which facilitates the leaching process.

Many other essential plant nutrients, such as phosphorus and sulfur, are found in soils. Trace amounts of such elements as boron and copper are likewise necessary for plant nutrition because they enable a plant to produce the chemicals required for photosynthesis and growth.

Although the parent material of a soil largely determines which chemical elements are present in the soil, the original rock does not necessarily exert a dominant effect on soil formation. Due to the

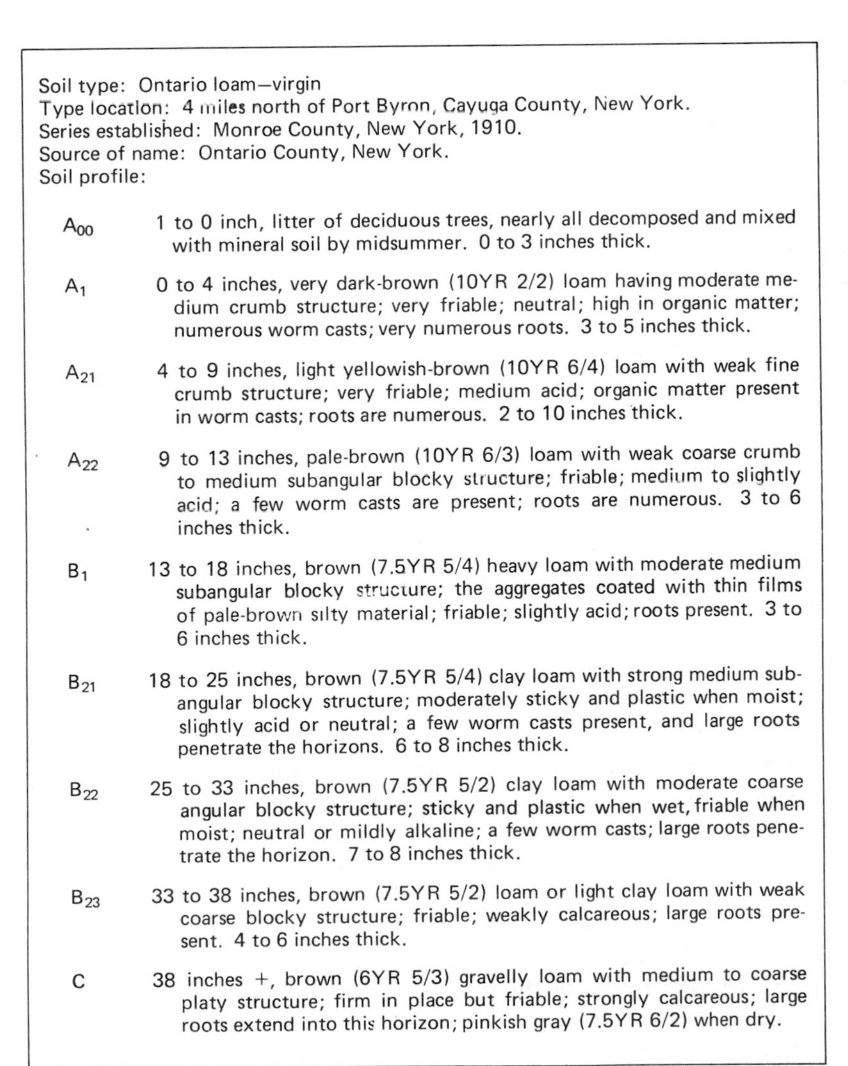

Soil type: Ontario loam—virgin
Type location: 4 miles north of Port Byron, Cayuga County, New York.
Series established: Monroe County, New York, 1910.
Source of name: Ontario County, New York.
Soil profile:

A_{00} 1 to 0 inch, litter of deciduous trees, nearly all decomposed and mixed with mineral soil by midsummer. 0 to 3 inches thick.

A_1 0 to 4 inches, very dark-brown (10YR 2/2) loam having moderate medium crumb structure; very friable; neutral; high in organic matter; numerous worm casts; very numerous roots. 3 to 5 inches thick.

A_{21} 4 to 9 inches, light yellowish-brown (10YR 6/4) loam with weak fine crumb structure; very friable; medium acid; organic matter present in worm casts; roots are numerous. 2 to 10 inches thick.

A_{22} 9 to 13 inches, pale-brown (10YR 6/3) loam with weak coarse crumb to medium subangular blocky structure; friable; medium to slightly acid; a few worm casts are present; roots are numerous. 3 to 6 inches thick.

B_1 13 to 18 inches, brown (7.5YR 5/4) heavy loam with moderate medium subangular blocky structure; the aggregates coated with thin films of pale-brown silty material; friable; slightly acid; roots present. 3 to 6 inches thick.

B_{21} 18 to 25 inches, brown (7.5YR 5/4) clay loam with strong medium subangular blocky structure; moderately sticky and plastic when moist; slightly acid or neutral; a few worm casts present, and large roots penetrate the horizons. 6 to 8 inches thick.

B_{22} 25 to 33 inches, brown (7.5YR 5/2) clay loam with moderate coarse angular blocky structure; sticky and plastic when wet, friable when moist; neutral or mildly alkaline; a few worm casts; large roots penetrate the horizon. 7 to 8 inches thick.

B_{23} 33 to 38 inches, brown (7.5YR 5/2) loam or light clay loam with weak coarse blocky structure; friable; weakly calcareous; large roots present. 4 to 6 inches thick.

C 38 inches +, brown (6YR 5/3) gravelly loam with medium to coarse platy structure; firm in place but friable; strongly calcareous; large roots extend into this horizon; pinkish gray (7.5YR 6/2) when dry.

Figure 10.11 An example of a complete soil description according to U.S. Department of Agriculture practice.

operation of the other soil-forming factors, a single parent material can give rise to soils that have dissimilar properties, and soils from two different kinds of parent material may develop similar properties.

Climate

On a global scale, major soil types show a close relationship to climatic zones. The energy and moisture delivered by the atmosphere influence the vertical movements of soil moisture and dissolved substances in soils and the rates of chemical reactions, weathering, and erosion. In addition, soils often show a close association with vegetation types, which are, in turn, influenced by climate.

The nature and intensity of the soil-forming processes vary significantly in different energy and moisture environments. Clearly, eluviation is most important where rainfall is abundant; cation exchange is favored by warm, wet conditions; organic activity is most vigorous in moderately warm and moist climates.

The rates of production and destruction of organic matter in the soil are largely a function of climate. In short-summer, high-latitude regions, vegetative production is relatively low, but so is the activity of soil bacteria and larger organisms, such as earthworms. Thus organic material decays very slowly, much of it forming acidic peat consisting of partially decayed vegetation rather than adding to a clay-humus complex. At the opposite extreme in the humid tropics, vegetative production is at a maximum. However, here the destruction of organic matter by insects and bacteria is so rapid and complete that humus cannot form, and the soil has only a minimal organic horizon.

Energy and moisture considerably affect the mobility of iron, silica, and soluble bases. In many cool, moist, coniferous forest regions with acidic soil moisture, iron becomes mobile and is lost from the upper layer in the eluviation process while silica remains fixed in the soil, which produces a gray *A* horizon. As we shall see, the coniferous vegetation plays a role in this. In warm, moist regions with neutral soil moisture, iron remains in the soil and silica is leached, which produces a red *A* horizon. In dry regions, moisture evaporates from the soil rather than percolating down to the water table; thus soluble substances often become concentrated at the downward limit of moisture penetration. In the soils of semiarid grasslands, a white layer of calcium carbonate frequently accumulates in the *B* or *C* horizon. In hot, dry climates where irrigation or natural runoff into closed basins raises water tables, soluble salts may move upward from the elevated water table by capillary action, producing a white powdery deposit on the ground surface.

The major pedogenic regimes, to be discussed later in the chapter, are all related to particular climatic conditions and the associated patterns of vegetation cover and drainage. Figure 10.12 illustrates the general correspondence between climate and the major soil types recognized in the present United States Department of Agriculture Soil Classification System. In the top part of the figure, the major climatic regions are arranged according to moisture and energy availability. Below, the major soil types are similarly arranged according to the moisture and energy conditions of the regions in which they are found. The association of particular types of soil with specific climatic regions is apparent.

Site

The soils encountered in such varying situations as hilltops, slopes, and valley floors have particular characteristics determined by the site itself. Since water drains downward, the soils at the foot of a slope evolve in a wetter environment than those on the slope or hilltop. The weathered mantle available for soil development is likewise thinner at the top of the slope, due to erosion, than either on the slope, where the rock debris, called *colluvium*, is slowly moving downward (see Chapter 11), or at the base of the slope, where the colluvial material may be accumulating. The material at the slope foot generally is finer than that higher up the slope, since fine material is flushed down and deposited by flowing water; likewise, the greater dampness at the slope foot favors further chemical decomposition. North-facing slopes are damper and often have deeper soils than south-facing slopes that receive more solar energy.

The plant cover on slopes is often different from that on level ground due to differences in insolation, temperature, soil depth, and moisture conditions. Thus organic activity and the organic content of soils vary from one topographic setting to another.

In general, soils developed on slopes are thinner, stonier, lower in organic matter and cation exchange capacity, and less well developed than those of level or low-lying land. Soils at the bases of slopes can range from deep and well drained to waterlogged, depending upon the specific form of the slope foot. If water collects at the slope foot, acidic peat may form, for plant material decomposes slowly in an oxygen-poor, waterlogged environment. On level surfaces where the rate of erosion is slow, weathering may produce clays that create heavy soils or hardpan layers within the soils.

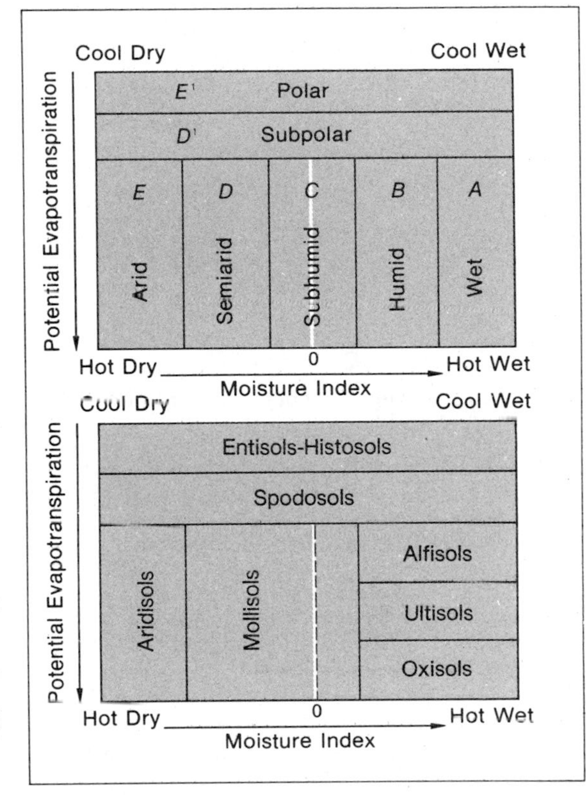

Figure 10.12 Climate is one of the important factors in soil formation. The top diagram shows the principal climate regions according to the Thornthwaite system. Potential evapotranspiration, a measure of energy, increases downward from cool to hot on the vertical scale. The moisture index, a measure of the moisture available for systems, increases from dry to wet along the horizontal scale. The bottom diagram classifies the principal soil types according to the same method.

The similarities between the two diagrams show that some soils form primarily in a particular climatic region. Waterlogged Entisols and Histosols are found in polar climates, where subsoils are frozen. Mollisols rich in humus and bases form under grassland in the subhumid climate region, where the moisture index is small or zero. Aridisols with little profile development occur in the arid climatic region. In the humid and wet climate regions, the soils arrange themselves primarily according to energy rather than to moisture, partly because of the important effect of vegetation type on soil development. Thus soils range from thoroughly leached and acidic Spodosols developed under coniferous forest, through more neutral Alfisols, to successively less fertile Ultisols and Oxisols formed under subtropical and tropical humid climates. The soil terminology is discussed in a later section. (After Blumenstock and Thornthwaite, 1941)

Organisms

We have seen that it is the presence of organic activity that distinguishes a true soil from simple weathered parent material. The intensity and nature of organic activity vary geographically, and much of this variation is due to climate or microclimate. Thus the climatic factor and the organic factor are hard to separate. We have previously discussed the general importance of organic activity to soil formation, but there remains a particular aspect that is of special importance to soil development: the *nutrient cycle*.

Wherever organic activity exists, there is a constant cycling of material and energy between the life forms and their environment. This is beneficial to both. Organisms need nutrients present in soils to carry out their life-sustaining processes, but they reciprocate by returning the same nutrients to the environment in their waste products, or as litter, or in the form of their own bodies when they die. In the absence of organically controlled nutrient cycling, soluble compounds would soon be leached out of the soil in humid environments. But in a life-filled soil, nutrients, rather than being leached away, are picked up and utilized almost as soon as they go into solution (see Figure 10.13). The same atoms that are part of the soil one day may be part of a microorganism or plant the next; then perhaps plant litter or part of a herbivore; then part of a carnivore; then the dead carcass of a herbivore, carnivore, or microorganism; then one of its decomposition products; finally reaching the soil to repeat the cycle again.

Whatever nutrients are not used in the cycle are leached away; thus in certain environments some compounds are lost but others not, which affects the life-sustaining capacity of the soil and determines what can survive in it. Without organic nutrient cycling, most soils of the wet tropics and moist midlatitudes would soon be leached of soluble bases, leaving few nutrients to support the forest vegetation, which would soon deteriorate.

Figure 10.13 The nature of the nutrient cycle helps determine soil fertility.

(left) Plant species with high nutrient demands prevent soluble compounds from being leached from the soil. The plant extracts nutrients and then returns them to the soil in the form of plant litter. The constant two-way exchange between the plant and the soil maintains a high base saturation in the soil.

(right) Plant species with low nutrient demands permit unused soil chemicals to be leached away and return few nutrients to the soil. Thus the soil's base saturation remains low.

Time

The role of time in soil formation is clear in some respects but less so in others. Obviously, it takes time for leaching, eluviation, and organic activity to produce distinct horizonation in soils, and in some

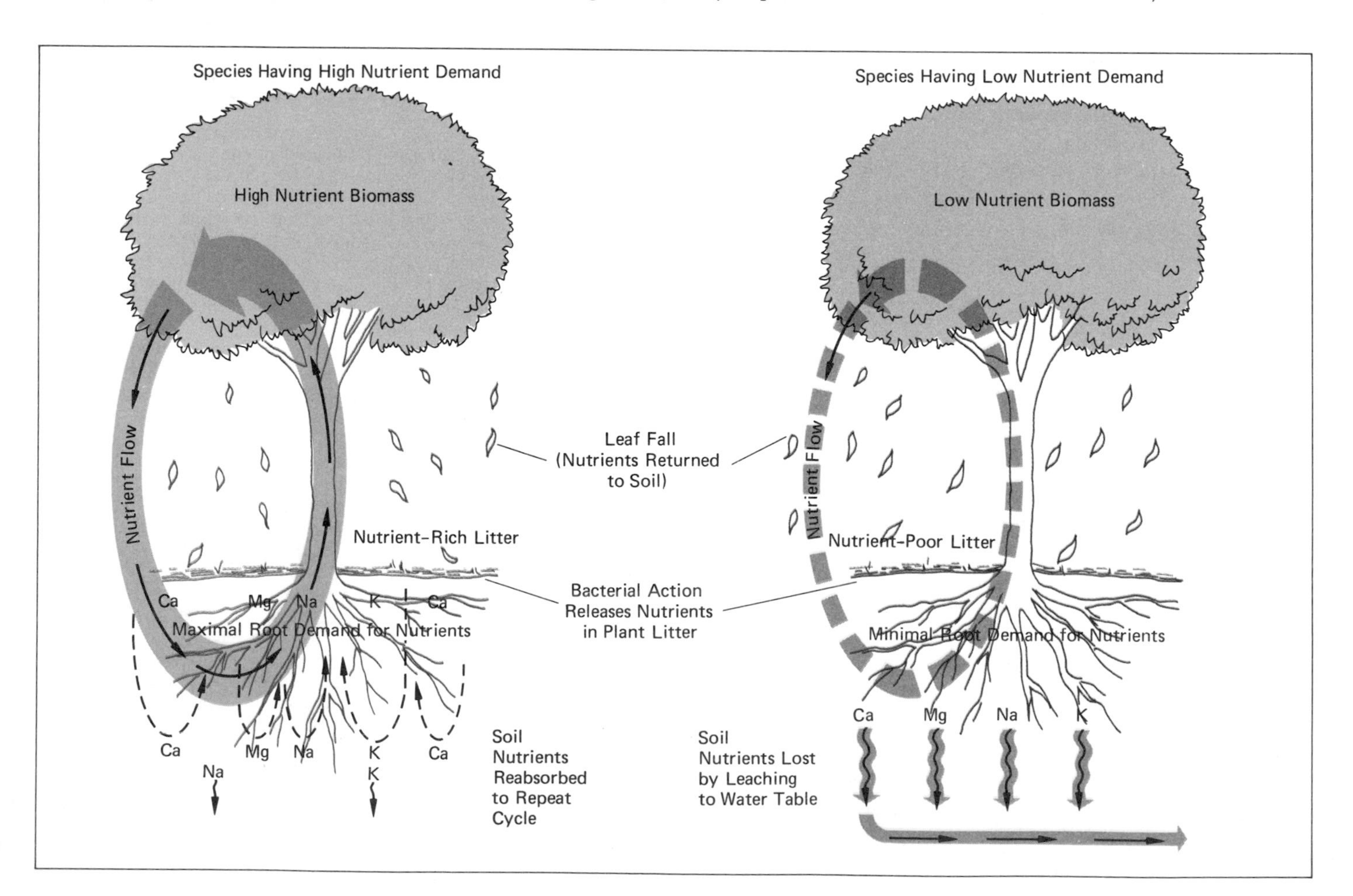

cases we can see these processes modifying parent materials of a known age. The question is: do soils reach a certain equilibrium with their environment, after which they cease to change, or is change slow but continuous through time?

In a few instances, the rate of soil formation can be determined from knowledge of geologic events. After the explosive eruption of the volcano Krakatau in Indonesia in 1883, soil formation in the fresh volcanic ash was rapid—the soil thickening at a rate of about 1 cm (0.4 in.) per year. This was a consequence of the region's moist tropical climate. In somewhat drier areas in Central America, volcanic ash has taken a thousand years to develop a soil about 30 cm (1 ft) deep. The development of a visible calcium carbonate layer or clay hardpan in the southwestern United States probably requires 5,000 to 10,000 years.

Many of the world's most productive soils are fertile because they are geologically young. Thus the soils of river flood plains developed on recent (younger than 10,000 years old) stream deposits, or *alluvium*, are generally very productive, as are soils on volcanic ash and on those glacial deposits that are rich in calcium carbonate. These recent deposits and the soils forming on them have not had time to be greatly eluviated. By contrast, the soils covering flattish land surfaces of great age have strong horizonation, with eluviated *A* horizons overlying massive textural and chemically rich *B* horizons. The soils of some of these old land surfaces have been eroded down to the *B* horizon, which forms a hard crust, known as *duricrust*, over weathered bedrock. Duricrusts may be strongly cemented by iron and aluminum oxides (ferricrete), silica (silcrete), or calcium carbonate (calcrete). An example of a calcrete duricrust is shown in Figure 10.14.

It is generally believed that soils tend toward a state of equilibrium with local climate and environmental conditions. However, it requires thousands of years for such equilibrium to be attained. Moreover, significant climatic and environmental changes—changes in the soil-forming factors—may occur more rapidly than the soil can respond. So it is probable that soils rarely, if ever, reach a completely stable condition; instead, they continually undergo slow modification in one direction or another with occasional interruptions due to deposition of new material or truncation by erosion.

Many soils contain relict features inherited from past conditions that no longer prevail today. Such features are important clues to past environments in the regions where they are found. Soil phenomena have provided evidence of periodic shifts in the forest-tundra boundary in the higher latitudes and the forest-grassland boundary in the mid-continent regions, the great expansion of desert climates in subtropical regions prior to Pleistocene time, and recurring climatic fluctuations in nearly all parts of the earth where soils have been studied. Still more important are ongoing human effects on the soil system, which are causing world-wide changes in soil quality. This is a consequence of interference with the soil system by removal of natural vegetation, interruption of nutrient cycles, addition of fertilizers, and acceleration of erosion rates. We shall return to this topic at the conclusion of this chapter.

Figure 10.14 A calcrete layer, called *caliche* in the southwestern United States, forms a resistant caprock and produces a bold ledge along the valley of the Virgin River in southwestern Nevada: **(top)** general view; **(bottom)** detail of the calcrete caprock. In tropical savanna regions, ferricrete (laterite) crusts produce similar landforms. Such crusts form within the soil and are exposed by later erosion of the upper part of the soil profile.

MAJOR PEDOGENIC REGIMES

Soil properties change gradually from one area to another, and many gradations exist in the transition zones between different soil types. Nevertheless, several distinctive *pedogenic,* or *soil-forming, regimes* have been recognized. They are related to climate both directly and indirectly through the influence of climatically controlled vegetation cover and moisture conditions. Each major pedogenic regime produces a highly distinct soil type that largely reflects latitudinal variations in energy and water budgets (see Figure 10.15).

Laterization

The process of *laterization* (also known as *ferrallitization* or *desilication*) produces the red iron-rich soils that are widely encountered in the permanently or seasonally wet portions of the tropics. The distinctive feature of laterization is the solution and removal of silica

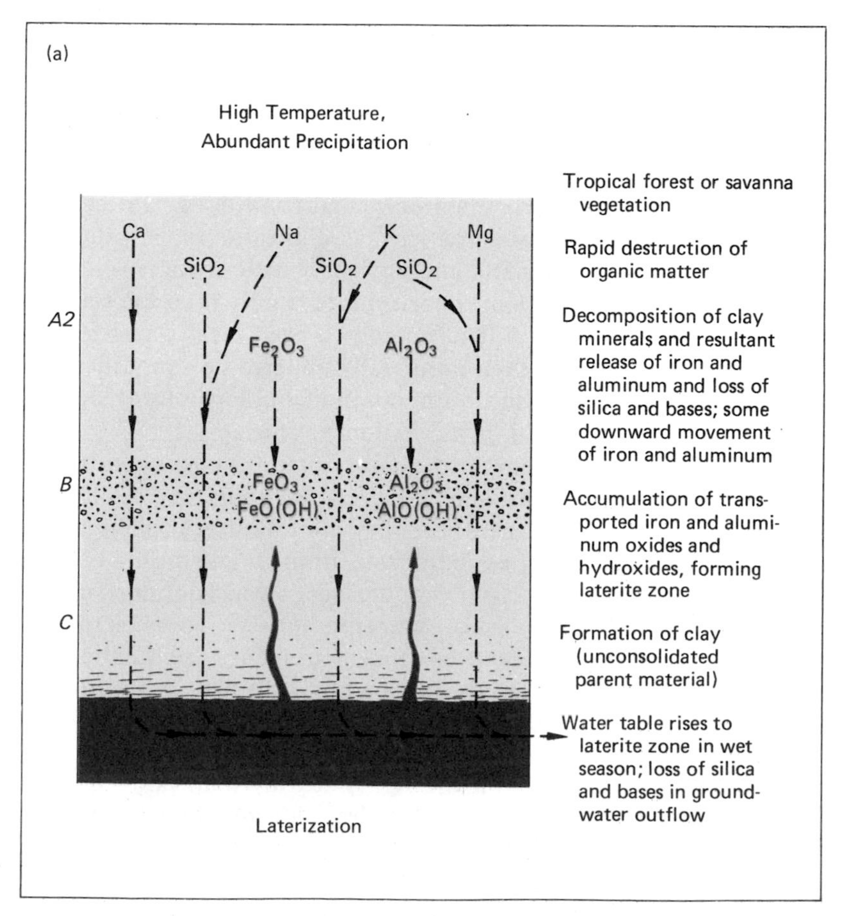

Figure 10.15 Some of the major soil-forming regimes, showing movements of chemical substances. Substances in red in this figure are accumulating in place as residuum or by illuviation.

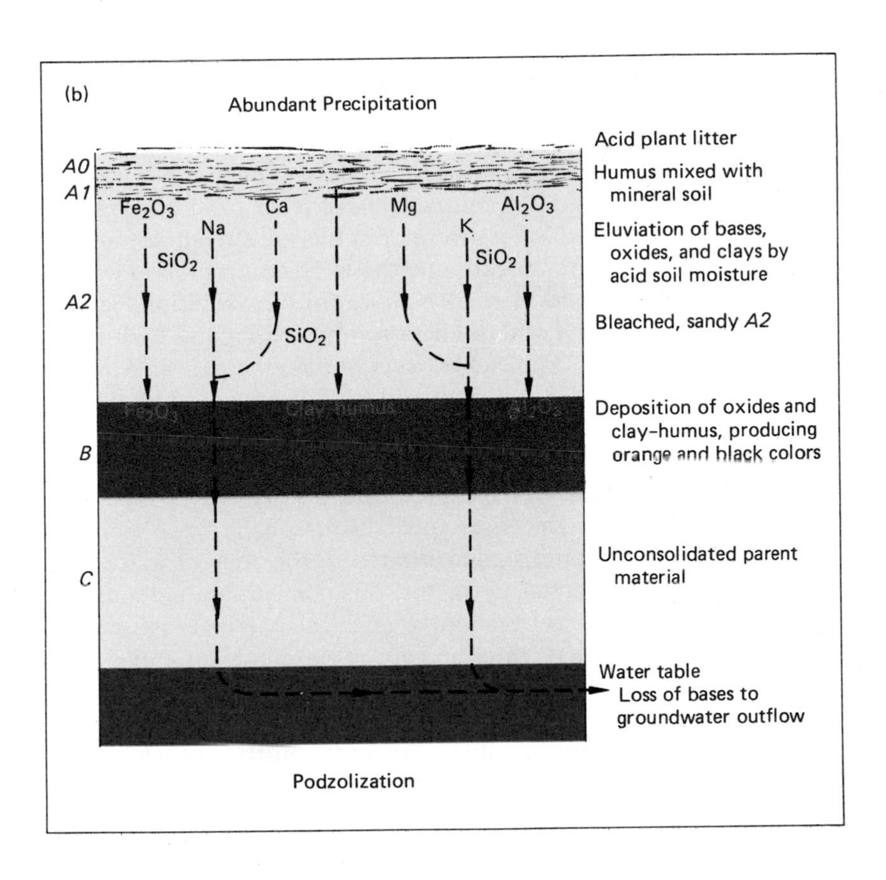

Podzolization

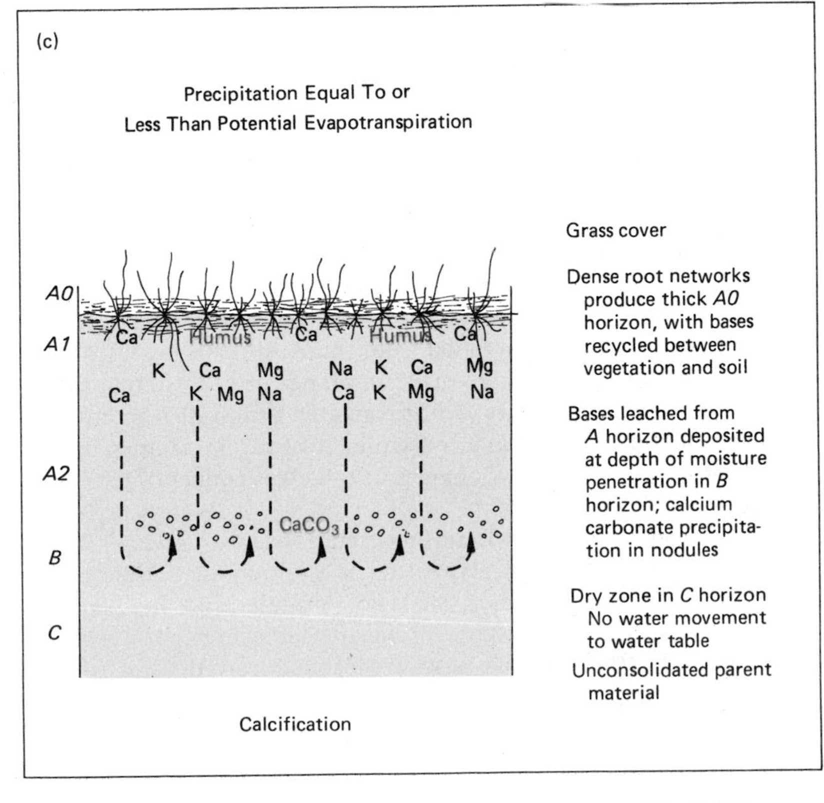

Calcification

(SiO_2), the most common mineral found in both solid and decomposed rock material (see Figure 10.15a). Silica is highly resistant to chemical weathering, and its removal indicates a weathering and leaching regime of maximum aggressiveness. Under moist tropical conditions, rapid bacterial decomposition of plant litter prevents the formation of humus, which is essential in the soil's retention of soluble nutrients. Carried to an extreme, the laterization process decomposes clays and removes virtually all cations in solution, leaving behind only quartz sand and the least soluble oxides and hydroxides of iron and aluminum. The end product of the process is a residue of earthy aluminum oxides and hydroxides known as *bauxite*, which, where found in commercial quantities, is the major source of metallic aluminum. The laterization process has a peculiar effect in those tropical areas that have wet summers and dry winters. Under such conditions, the water table rises and falls seasonally, with iron and aluminum oxides becoming concentrated in the zone of water table fluctuation. When exposed to the air, this concentrate hardens to a brick-like crust known as *ferricrete* or *laterite*. Laterite crusts may be as much as 6 meters (20 ft) thick and are a common feature in areas of savanna vegetation (see Chapter 14).

Laterization may proceed without total impoverishment of the soil as long as vegetation maintains an active nutrient cycle so that bases are not completely lost. But clearing of forest vegetation may lead to a loss of the soluble bases brought up by plant roots and to complete desilication and formation of soil crusts. This is a hazard in the utilization of tropical forest regions for the expansion of food production.

Podzolization

Podzolization involves the breakdown of clay and the eluviation of iron along with other cations and even humus, with silica left behind as a residue (see Figure 10.15b). Iron can be eluviated only by soil moisture that includes organically produced acids that decompose clay minerals and dissolve basic substances but do not affect silica. This eluviation results in a bleached gray *A* horizon that has a silty or sandy texture. The iron eluviated from the *A* horizon is precipitated in the *B* horizon of the soil, giving it an orange or yellow color that is in sharp contrast to the gray *A* horizon. Deposition of iron and aluminum oxides at this level is a consequence of several factors, including the presence of electrical charges in the clays concentrated in the *B* horizon.

Podzolization occurs wherever vegetation does not have heavy nutrient requirements and plant litter is itself acidic. These conditions are most widespread in the cool, middle- and high-latitude coniferous forests. Many types of needleleaf evergreens have low nutrient needs so that soluble bases are leached from the soil, causing it to become increasingly acidic, with acidic decay products from slowly decomposing needles contributing to the tendency. Since soil organisms do not thrive in the cool, acidic environment of the soil, humus production is retarded. As in moist tropical regions, the lack of a functioning clay-humus complex is an important factor in the leaching of nutrients from the soil. On a smaller scale, podzolization occurs wherever the parent materials are acidic and lacking in soluble bases. Thus acidic, nutrient-poor, podzolized soils develop on quartz sand and chemically acidic rock types in all climates.

Calcification

In arid and semiarid climates, there is insufficient movement of moisture through the soil to leach away soluble bases. What is usual in such areas is for some leaching to occur in the *A* horizon of the soil and for the downward-moving moisture to evaporate from the dry subsoil, where its chemical load is precipitated (see Figure 10.15c). The principal mineral involved is calcium carbonate ($CaCo_3$), which may be deposited in various forms in the *B* horizon of the soil, where it forms a subsidiary *calcic horizon*. This subsoil carbonate-enrichment process is called *calcification*.

Soils subject to calcification show varying degrees of horizon development in addition to their carbonate concentrations. In mid-latitude grassland areas, dark organic horizons normally occur above *B* horizons that contain only scattered flecks or nodules of calcium carbonate. In semidesert areas, there may be almost no organic horizon but instead strong textural *B* horizons that include conspicuous veins of carbonate. The depth at which the calcic horizon forms depends on the depth to which the soil moisture penetrates before evaporating and precipitating its carbonates; the greater the annual rainfall, the farther below the surface the calcic horizon.

Over long periods of time, the accumulation of calcium carbonate may produce a concrete-like layer (K horizon) a few centimeters to several meters in thickness. In the southwestern United States, such lime crusts have long been called *caliche*. In scientific terms, they are known as *calcrete*. The development of a layer of calcrete requires either a parent material rich in carbonates or a constant influx of carbonates, as in wind-blown dust, and a length of time probably exceeding 10,000 years.

Salinization

In dry regions, water that runs off rocky hillslopes is often trapped in depressions that have no drainage outlet (see Chapter 14). This may produce a temporary lake that soon evaporates. However, in such depressions the water table may be close to the surface for long periods. Since the subsurface water is constantly evaporating but leaving its chemical load behind, the remaining water may become saline. Above the water table will be a *capillary fringe* of saline water moving upward toward the surface through capillary action. This water also evaporates, causing the salts dissolved in it to be precipitated either in the soil or as a white powder over the land surface. The deposit consists of various chlorides, sulfates, and carbonates of calcium, magnesium, and sodium, including common salt (NaCl), gypsum

($CaSO_4$), and lime ($CaCO_3$). In high concentrations, many of these chemicals are toxic to most plants and soil organisms, making the land essentially useless. This process is known as *salinization*.

Unfortunately, agricultural irrigation without the provision of artificial drainage facilities to keep the water table low produces the same result and has often destroyed vast areas of potentially useful land. We have written records of man-induced salinization occurring nearly 5,000 years ago in southern Mesopotamia, where the ancient Sumerians built networks of canals to channel the waters of the Tigris and Euphrates rivers into their fields. The resulting salinization eventually caused a shift in economic and political power away from Sumeria in the south to Babylonia farther north. Salinization has continued to take land out of production in Mesopotamia (Iraq) in the present century. With the recent establishment of year-round irrigation in Egypt's Nile Valley and delta, the same problem is becoming increasingly serious there. Artificially induced salinization has also beset various irrigation projects in the American West, such as those of the lower Colorado River region.

Gleization

In regions of high rainfall, a persistent waterlogging of low-lying areas also causes soils to develop distinctive properties. Here the water is not saline, and vegetation is able to grow in the muck. Thus the soils of wet depressions have a dark organic *A* horizon, in which plant litter decays only slowly due to the lower level of bacterial action in an oxygen-poor environment. This surface layer often takes the form of peat. However, as the vegetative matter decays in oxygen-poor stagnant water, organic acids are released that react with the iron in the soil and convert it to a chemically reduced (instead of oxidized) condition. Rather than producing red colors, iron in the reduced (or ferrous) state produces black to blue-gray colors. Wetland soils of these colors are called *gley soils* and are produced by the process of *gleization*.

Gley soils generally have a pH below the tolerance of most crop plants, and bases must be added (in the form of lime) to make them agriculturally productive. Liming of gley soils raises their pH levels to a point at which soil microorganisms can thrive in them, assisting the decomposition of organic matter and creating humus in the soil.

CLASSIFICATION OF SOILS

The classification of soils into a limited number of categories is even more difficult than the classification of climates. This is because there are more variables to take into account. Whereas a climate may be characterized by its annual temperature and moisture regimes, soils vary in many respects: texture, structure, parent material, organic matter, cation exchange capacity, and the details of horizon development. Furthermore, unlike climates, soils can differ considerably from point to point within an area the size of a city block. Additionally, human activities can greatly modify the soil: two farms a mile apart that have the same temperature and rainfall regimes may have noticeably different soils due solely to varying agricultural practices.

Explanatory Classification: Great Soil Groups

To be of real value, a system of soil classification must be detailed enough to describe precisely the significant characteristics of an almost infinite number of specific soils. We will discuss such a classification system shortly. However, to reveal the general influence exerted on soil development by climate and vegetation, a more generalized explanatory type of classification is useful. Climate and vegetation do indeed tend to foster the development of widely distributed general soil types having similar characteristics from one continent to another. Thus the soils developed under coniferous forests in Canada are much like those found in a similar environment in the Soviet Union, regardless of the soil parent material.

In 1938 the United States Department of Agriculture published a soil classification system utilizing the concepts of zonal, azonal, and intrazonal soils. These formed the three main *orders* of soils. Well-developed, or "mature," general soil types with good horizonation were called *zonal soils,* reflecting the zonation of climates and major vegetation types around the world. In this broad classification, soils that for some reason have not developed typical profiles, or lack profile development altogether, were *azonal soils.* Azonal soils have not had time to come into equilibrium with the climate—as in the case of soils developing on recent alluvium or volcanic ash or soils that are eroded away as fast as they form. A third group consisted of soils whose outstanding characteristics are determined by local nonclimatic factors, such as drainage conditions or parent material. These were called *intrazonal soils.*

The zonal soils fell into two broad groupings that were related to the pedogenic regimes discussed in the preceding pages. Soils subject to the laterization and podzolization processes, which cause them to accumulate iron and aluminum at some level in the soil profile, were called *pedalfers* (*ped,* soil; *al,* aluminum; *fer,* iron). Soils of humid regions are of this general type, as shown in Figure 10.16. In drier areas, calcification becomes important, producing a group of soils known as *pedocals* (*ped,* soil; *cal,* calcium). As Figure 10.16 indicates, the boundary between pedalfers and pedocals in the United States nearly coincides with the boundary between the moist and dry regions defined by Thornthwaite's moisture index.

Figure 10.16 Soils in the coterminous United States can be divided into two major classes, the pedalfers of the East and the pedocals of the West. Pedalfer soils are leached and contain aluminum and iron compounds from the breakdown of clay by acids. Pedocal soils are alkaline and contain calcium compounds. The dividing line between pedalfers and pedocals nearly coincides with the line where Thornthwaite's moisture index is 0. Pedalfers are formed in regions where moisture for leaching is abundant, and pedocals form in dry regions. (After Marbut, 1935)

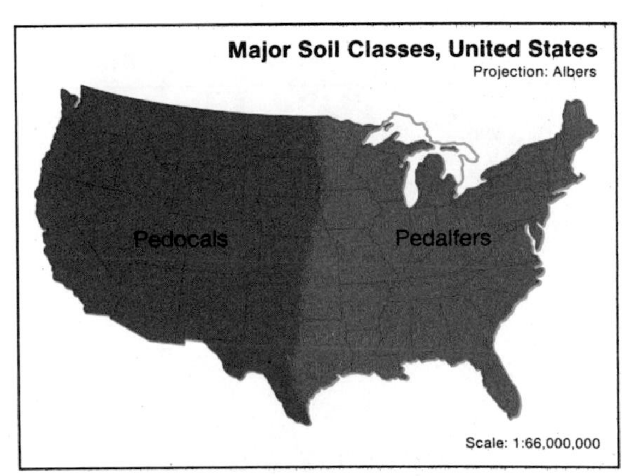

Each of the main soil orders in the 1938 classification was divided into suborders, reflecting the influences of climate, vegetation, and other soil-forming factors. Further subdivision related to soil morphology resulted in the more than forty *great soil groups* shown in Table 10.2. Each of the great soil groups has very specific characteristics that differ from those of other groups, reflecting differences in the environments in which the soils developed. The great soil groups were subdivided into *families* and the families into *series* bearing the name of the location in which the specific soil type occurred.

Over the years, soil scientists became increasingly dissatisfied

Table 10.2. 1938 U.S. Soil Classification System and Approximate Equivalents in the 1960 Comprehensive Classification

1938 SYSTEM				1960 SYSTEM
Order	*Suborder*	*Great Soil Groups*	*Suborder*	*Order*
ZONAL	Pedalfer	Soils of forested warm-temperate and tropical regions	Red-brown Lateritic Red-Yellow Podzolic Low-humic Latosol Latosol Humic Latosol Laterite	Humult, Udalf, Udult Udult Tropept, Ustox Tropept, Humult, Andept, Ustult Humult, Humox, Andept Orthox	Oxisol, Ultisol, Alfisol, Inceptisol
ZONAL	Pedalfer	Soils of cold regions	Polar Desert Arctic Brown Tundra Alpine Turf	Aquept, Ochrept, Umbrept, Andept	Inceptisol
ZONAL	Pedalfer	Soils of forested cool-temperate regions	Podzol Brown Podzolic Gray-Brown Podzolic Gray Wooded Sols Bruns Acides Western Brown Forest	Orthod, Humod Orthod, Andept, Ochrept Udalf, Udult Boralf Ochrept, Umbrept	Ultisol, Spodosol, Alfisol, Inceptisol
ZONAL	Pedalfer	Soils of forest-grassland transition	Degraded Chernozem Noncalcic Brown	Boralf, Boroll Xeralf, Ochrept	Alfisol, Mollisol, Inceptisol
ZONAL	Pedalfer / Pedocal	Dark-colored soils of semiarid, subhumid, and humid grasslands	Reddish Prairie Prairie (Brunizem) Chernozem Reddish Chestnut Chestnut	Ustoll Udoll, Boroll, Xeroll, Ustoll Boroll, Ustoll, Xeroll Ustalf, Ustoll Xeroll, Ustoll, Boroll	Mollisol, Alfisol
ZONAL	Pedocal	Light-colored soils of arid regions	Reddish Brown Brown Sierozem Red Desert Desert Polar Desert	Ustalf, Orthid, Argid Ustoll, Xeroll, Argid, Orthid, Boroll Argid, Orthid Argid, Orthid Argid, Orthid Argid, Orthid	Aridisol, Mollisol, Alfisol

with the 1938 classification system, which did not prove to be precise or comprehensive enough to permit exact description of the enormous variety of soils actually existing on the earth. Nevertheless, the 1938 classification remained useful because of its emphasis on the broad influence of climate and vegetation type upon the development of soil characteristics. Thus its terms are still widely used.

Morphological Description: The U.S. Comprehensive System

In 1960 a new soil classification scheme was developed by the United

1938 SYSTEM			1960 SYSTEM	
Order	*Suborder*	*Great Soil Groups*	*Suborder*	*Order*
INTRAZONAL	Hydromorphic soils in areas of imperfect drainage or high water table	Humic Gley	Aquoll, Aquept, Aquult, Aqualf	Inceptisol, Mollisol, Alfisol, Spodosol, Ultisol, Entisol
		Low-Humic Gley	Aquult, Aquent, Aquept, Aqualf	
		Alpine Meadow	Aquod, Aquoll, Umbrept	
		Bog	Suborders of Histosol	
		Half Bog	Aquept, Aquoll, Aqualf	
		Planosol	Aqualf, Alboll	
		Ground-Water Podzol	Aquod	
		Ground-Water Laterite	Aquult, Udult, Ueult	
	Halomorphic soils (saline and alkali) in areas of imperfect drainage in arid and coastal areas	Solonchak	Orthid, Aquept	Inceptisol, Aridisol, Mollisol, Alfisol
		Solonetz	Natric great groups of Alfisol, Mollisol, Aridisol	
		Soloth	Natric subgroups of Mollisol and Alfisol	
	Calcimorphic soils formed from calcareous parent materials	Brown Forest	Ochrept, Xeroll, Udoll	Inceptisol, Mollisol
		Rendzina	Rendoll	
AZONAL		Lithosol		Entisol, Inceptisol, Mollisol
		Regosol		
		Alluvial		

Source: P. W. Birkeland, 1974. *Pedology, Weathering, and Geomorphological Research.* New York: Oxford University Press, pp. 44–45.

States Department of Agriculture. Officially termed the *United States Comprehensive Soil Classification System,* the new scheme is often called the *7th Approximation,* since it was actually the seventh revision of the original proposal submitted to soil scientists for test applications. In the new system, the focus is strictly upon the specific properties of a soil as it presently exists, rather than upon its environment, genesis, or virgin condition. An entirely new terminology has been created, in which mainly Greek and Latin roots are combined to form terms that precisely describe a soil's properties (see Appendix III).

In the 7th Approximation, ten *soil orders* are recognized. Each order includes a number of *suborders* based on varying influences of climate, vegetation, or drainage conditions. The world-wide distribution of the orders is shown in Figure 10.17, pp. 280–281, and the formative influences on the major types are illustrated in Figure 10.18, p. 282. The suborders are themselves subdivided into *great groups* on the basis of horizon development. The great groups are differentiated into *subgroups,* and then, as in the 1938 classification, into *families,* and finally into *soil series* named after the place in which a specific homogeneous soil type occurs. The general nature of the new classification is apparent when we encounter the nomenclature for the distinctive types of soil horizons (see Table 10.3, p. 283) whose presence consigns a soil to one or another of the new soil orders.

Figure 10.17 World-wide distribution of soil orders according to the U.S. Comprehensive Soil Classification System (7th Approximation). (Adapted from USDA, Soil Conservation Service, 1972)

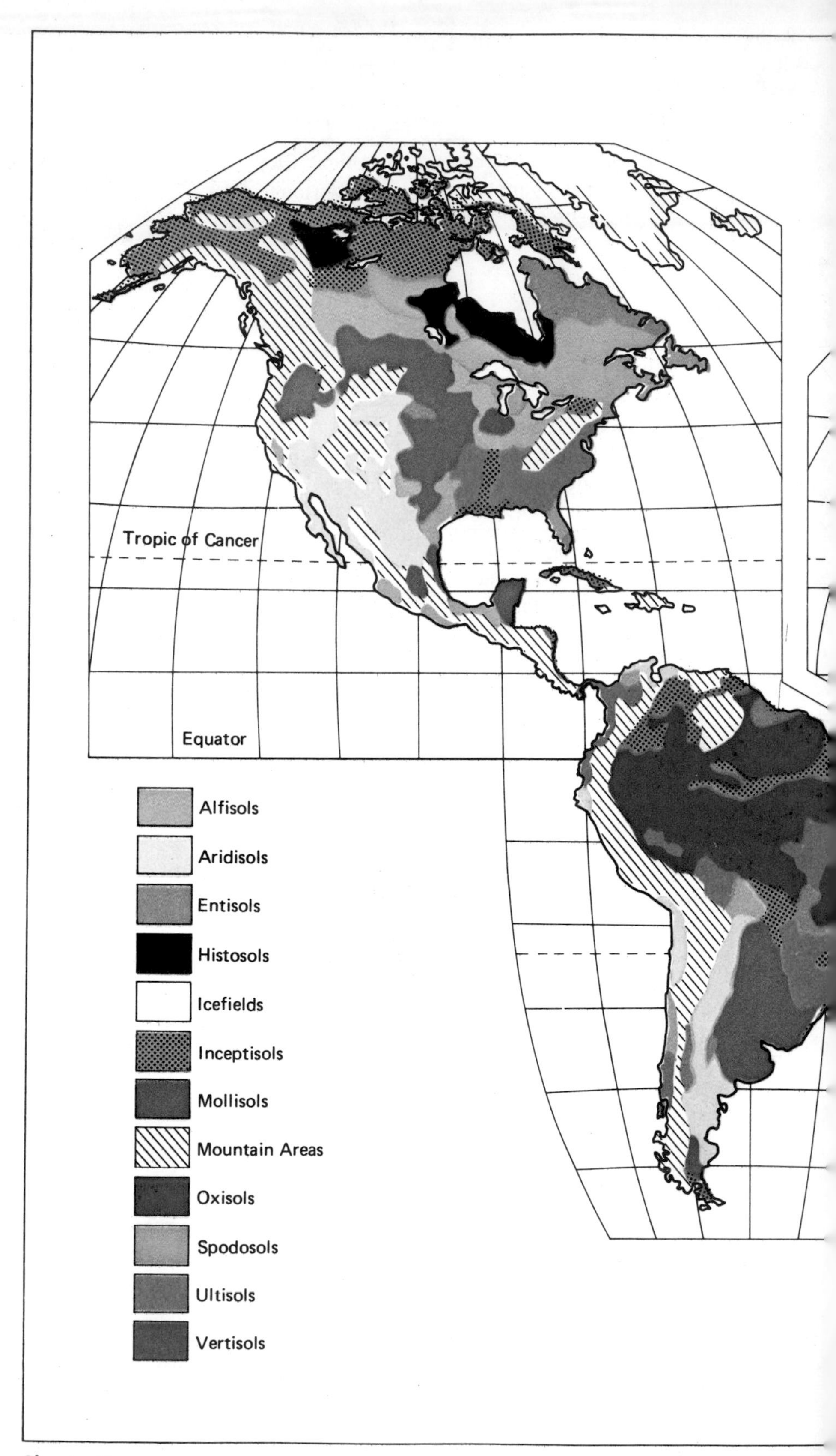

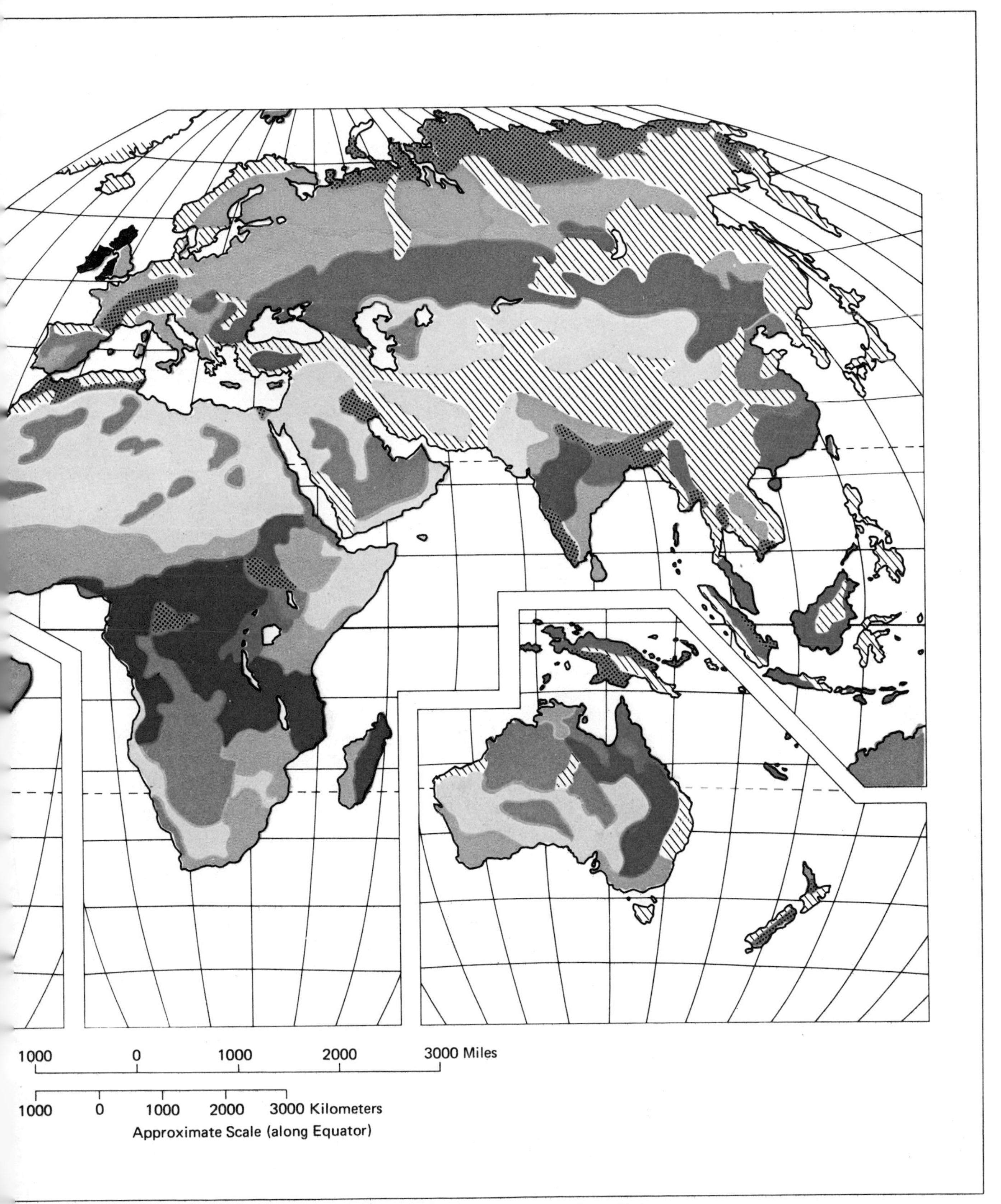
1000 0 1000 2000 3000 Miles
1000 0 1000 2000 3000 Kilometers
Approximate Scale (along Equator)

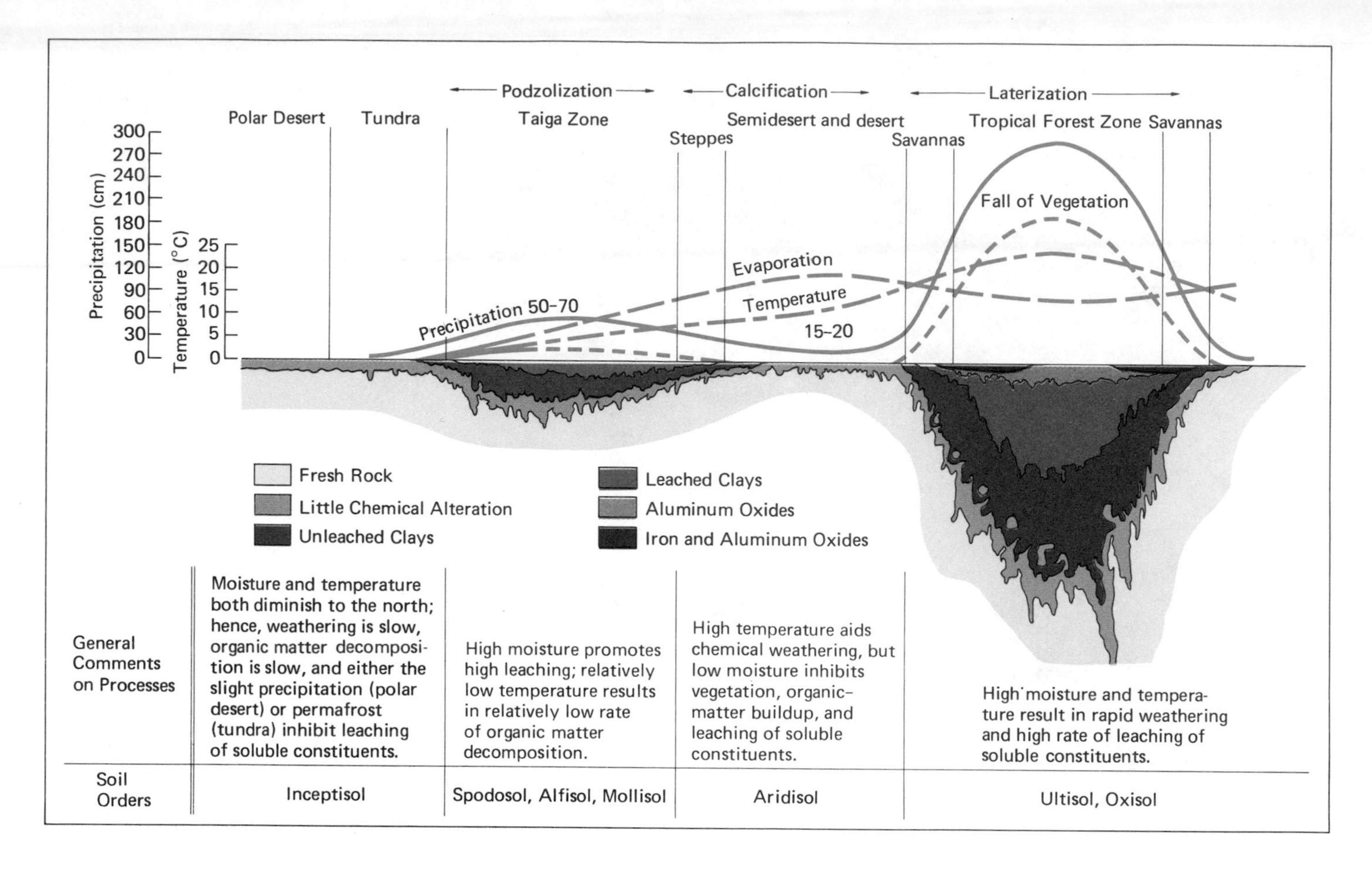

Figure 10.18 Latitudinal distribution and formative influences on major soil orders. This diagram summarizes how precipitation, energy, evaporation, and vegetation affect rock weathering and soil formation. (Modified from Strakhov, *Principles of Lithogenesis*, and Birkeland, 1974)

There is some similarity between the major soil orders in the new Comprehensive System and the most common of the great soil groups in the 1938 classification (see Table 10.2, pp. 278–279). The great groups of the new classification are much more specific than the great soil groups of the older system, their names being very descriptive once the formative elements are learned: for instance, a *Plinthaquult* has a well-developed textural *B* horizon (order Ultisol), including a zone of concentration of iron and aluminum compounds (plinthite) in a wet setting (aqu). In the following pages, we shall summarize the nature of the soil orders in the Comprehensive System (the suborders are summarized in Appendix III). The soils are described here in order of increasing development.

Entisols (from the word "recent") are soils with poor horizonation due to one of several factors: youth of the soil, rapid erosion as the soil forms, waterlogging, or human interference, such as plowing. Obviously, these conditions differ considerably, and thus the range in Entisol types is great. Suborders make the necessary distinctions. Entisols are equivalent to the various azonal soils in the older classification.

Histosols (from Greek *histos*, tissue) are composed primarily of plant material. They develop locally in waterlogged environments where organic matter decomposes slowly due to the lack of oxygen required by bacteria (see Figure 10.20, p. 286). Suborders relate to drainage conditions and the degree of decomposition of the plant material. Histosols are usually, but not always, acidic, and their horizons are based on the degree of compaction and state of decomposition. They can develop almost anywhere from the high arctic to equatorial forests, being excluded only from desert regions, but they are most common in glaciated landscapes of higher latitudes. Histosols can be very productive when drained; however, the subsequent oxidation and drying of the organic material lead to compaction and subsidence of the land surface and increase the hazards of both fire and wind erosion.

Vertisols (from Latin *verto*, to turn) are soils in which horizon development is impeded by the churning effects of repeated changes in their volume due to alternating wetting and drying. Such soils are dominated by certain clay minerals that have exceptional water-

Table 10.3. Nomenclature for Principal Soil Horizons: U.S. Comprehensive System*

Epipedon: Surface horizon darkened by organic matter; includes eluvial horizons.

Mollic epipedon (from Latin *mollis*, soft; Greek *epi*, outer; Greek *pedon*, earth)
Dark colored; at least 1 percent organic matter; more than 50 percent base saturation (that is, 50 percent of total exchange capacity saturated by base cations).

Umbric epipedon (from Latin *umbra*, shade)
Dark with humus, but less than 50 percent base saturation.

Ochric epipedon (from Greek *ochros*, pale)
Light in color and low in organic matter.

Subsurface horizon

Albic horizon (from Latin *albus*, white)
Bleached due to removal of oxides and clays.

Argillic horizon (from Latin *argilla*, clay)
Enriched by illuvial or authigenic (formed in place) clay.

Natric horizon (from Arabic *natrun*, sodium carbonate)
Having columnar or prismatic structure and a high sodium content.

Spodic horizon (from Greek *spodos*, wood ash)
Enriched by dark organic matter or iron and aluminum oxides.

Oxic horizon (from *oxygen*)
Strongly weathered, with bases leached to leave a residue of hydrated iron and aluminum oxides and permeable clays.

Cambic horizon (from Latin *cambiare*, to exchange)
Little changed from weathered parent material, but having minor illuvial carbonate or clay and/or greater coloring by iron and aluminum oxides than higher or lower horizons.

* This table includes only the most common of the many horizons possible.

Figure 10.19 The environment is an important factor in soil formation, and many characteristic types of soil are associated with certain climatic, vegetation, and drainage conditions, as these illustrations suggest.

(a) *Tundra Region.* Arctic tundra vegetation consists of mosses, lichens, low shrubs, grasses, and flowering herbaceous plants. Tundra soils are generally developed over ground that is permanently frozen and thus impermeable to water. Repeated disturbances of the soil by freeze and thaw, as well as seasonal waterlogging, are characteristic. Networks of ice wedges are also common in low-lying areas. The soils have poor horizonation and are mainly Entisols and waterlogged Histosols.

(b) *Coniferous Forest Region.* Coniferous forests have relatively low nutrient requirements. Thus they produce acidic litter and do not recycle soluble bases in large amounts. Consequently, the soils beneath such forests are acidic, with clay, humus, bases, and even the oxides of iron and aluminum eluviated by acidic soil moisture. The result is the distinctive Spodosol, in which a bleached sandy *A2* horizon overlies an oxide-colored illuvial *B* horizon. Such soils are infertile unless neutralized by the addition of lime.

(c) *Short-Grass Prairie Region.* Grassland soils are generally high in nutrients due to low precipitation, rapid nutrient cycling, and a rich clay-humus complex. The organic horizon is well developed and dark in color due to its humus content. Short-grass prairies are pedocals and develop shallower soils than tall-grass prairies because soil moisture does not penetrate as deeply. Their soil structure is excellent for agricultural use. Thus they fall into the Mollisol soil order. The chernozem soils, transitional between short- and tall-grass prairies, are among the most fertile of all soils. Short-grass prairies have a visible zone of lime accumulation in the *B* or *C* horizon due to shallow penetration of moisture.

(d) *Tall-Grass Prairie Region.* Tall-grass prairies receive more precipitation and develop soils that are deeper and even richer in organic matter than the soils of short-grass prairies. They, too, are Mollisols with high base saturation and a rich nutrient cycle. However, they are somewhat more leached than the short-grass prairie soils and are pedalfers.

(e) Midlatitude Forest Region: Alfisols

(f) Tropical Forest Region: Ultisols, Oxisols

(g) Desert Region: Aridisols

(h) Bogs and Meadows: Histosols

This fact suggests that at times these soils have been occupied by forests, which have affected the soil's development. They are the most fertile soils located within a zone of reliable rainfall.

(e) *Midlatitude Forest Region.* Beneath the broadleaf deciduous forests of the moist midlatitudes, soils tend to be relatively well supplied with mineral nutrients due to nutrient cycling where deep roots penetrate the parent material and nutrients stored in the vegetation are returned seasonally to the soil in leaf fall. In the midlatitude region, bacterial action supplies abundant humus. The resulting soils are Alfisols.

(f) *Tropical Forest Region.* Plant litter decomposes so rapidly under humid tropical conditions that humus is not formed. Nevertheless, the steady decay of plant litter on the forest floor returns nutrients to the soil, where they are immediately taken up by vegetation. Thus the soil nutrients are largely locked up in the vegetation itself, with the subsoil being eluviated of clays, bases, and even silica. What remains in the soil are mainly oxides and hydroxides of iron and aluminum, which color the soil orange or red. These leached red soils are Ultisols and Oxisols.

(g) *Desert Region.* Desert soils are low in clay due to the lack of water for chemical weathering and low in organic matter as a consequence of the sparse plant cover. Thus they lack a functioning clay-humus complex. They are high in basic nutrients since there is little eluviation, but are easily leached when irrigated artificially. Desert Aridisols are pedocals and commonly have calcrete (caliche) layers.

(h) *Bogs and Meadows.* Where soils are waterlogged, they may support grasses, herbs, mosses, and low shrubs. Trees cannot subsist in perpetually wet ground. Decaying plant material makes most waterlogged soils strongly acidic, and the upper horizon may be dark peat. Such soils are known as Histosols. Changes in water level produce color changes as iron is oxidized by contact with air or reduced by the exclusion of air.

absorbing capacity. When dried, Vertisols shrink, harden, and develop systems of cracks as much as 2.5 cm (1 in.) wide and 50 cm (20 in.) deep. These cracks collect organic debris that falls into them. Wetting causes the soil to swell, closing the cracks and forcing sticky soil masses to squeeze and churn against one another. In much of California, Vertisols of the suborder xererts are the normal soil type and are known as "adobe" from their use in the making of sun-dried brick in the Spanish period of California history. Vertisols are widespread and have no exact counterpart in the older soil classification system.

Inceptisols (from Latin *inceptum*, beginning) are young soils that have developed in humid regions on recent alluvium, glacial or aeolian (wind-produced) deposits, or volcanic ash. Although some soluble compounds have been removed from the *A* horizon, these soils show no clear illuvial horizon.

Aridisols (from Latin *aridus*, dry) normally have the thinnest profiles of any regional soil type due to shallow and infrequent penetration of water. They contain a minimum of organic matter, have maximum stoniness, and often include concentrations of calcium carbonate (lime), calcium sulfate (gypsum), and sodium chloride (salt), either in discrete masses or in surface or subsurface layers (see Figure 10.21). Clay hardpans and iron and silica crusts may also be present, the latter inherited from prior pedogenic regimes. Aridisols are pedocals in the older classification system, specifically red desert and sierozem (gray) desert soils. Calcification and salinization may both be factors locally.

Figure 10.20 This waterlogged Histosol in southern Michigan consists of a layer of muck over undecomposed peat. (7th Approximation: limnic medisaprist)

Mollisols (from Latin *mollis*, soft) have dark, humus-rich *A* horizons (mollic epipedons) with high base saturation (see Figures 10.22 and 10.23). They develop in the transition zone between arid and humid climates under semiarid to subhumid grasslands. This vegetation cover generally maintains a rich clay-humus complex due to the extremely dense mass of roots, which yield copious decay products. The grasses demand an abundant supply of bases, which are thus kept in the nutrient cycle. Mollisols show varying degrees and depths of calcification and are pedocals in the 1938 system. Unlike Vertisols, which may also be dark and rich in organic matter, Mollisols do not experience major volume changes, nor do they harden when dry. Thus, provided there is sufficient moisture, the best Mollisols are the "cream" of agricultural soils. Their natural cover of grass has been replaced everywhere by crop grasses: wheat, barley, rye, corn, and sorghum. Unfortunately, they extend into areas that experience periodic droughts, leading to crop failures and recurrent "dust bowl" conditions. Mollisols rarely, if ever, form on bedrock. They are derived from alluvium, glacial deposits, and loess (fine material deposited by the wind). Mollisols include the prairie, chernozem, and chestnut-brown soils of the older classification system; thus they vary significantly in character moving westward from Illinois to the much drier high plains of eastern Colorado. This is reflected in the soil suborders.

Spodosols (from Greek *spodos*, wood ash) are soils in which a leached and eluviated light-colored *A* horizon (albic horizon) overlies an illuvial *B* horizon (spodic horizon) that is colored by iron or aluminum compounds or relocated organic carbon (see Figure 10.24, p. 288). Sometimes an iron-cemented clay hardpan is present. Spodosols are most common in forested cold-winter areas and wherever nutrient-poor sandy horizons are present. The major soil-forming process is podzolization, and these soils are the podzols and brown podzols in the older classification. Spodosols are notoriously infertile, most of their soluble bases having been replaced by hydrogen ions. They are acidic and not retentive of moisture. To be productive for grains, they must be neutralized by treatment with calcined lime (quicklime), which may increase grain yields by as much as one-third and pasture production by 300 percent. However, some root crops, such as potatoes, thrive in the acidic environment of these soils.

Alfisols (from aluminum, *al*, and iron, *fe*) have yellowish-brown *A* horizons (ochric epipedons) that have been partially leached of bases, causing the upper soil to be colored by iron and aluminum compounds (see Figure 10.25, p. 289). A clay hardpan is usually present in the illuvial zone. Alfisols are transitional between the Mollisols of the pedocal zone and the Spodosols and Ultisols of the pedalfer region. In the United States, they are found mainly in the southern Great Lakes area; however, Alfisols are widely distributed, occurring from middle to tropical latitudes. The retention of a significant proportion of bases allows the better Alfisols to be very productive agriculturally, and the intensively farmed American "corn belt" is developed partly on them. These soils occupy a position somewhat similar to the gray-brown podzols in the older classification but also

Figure 10.21 (left) This Aridisol is located in southern New Mexico and has developed on sediments washed into a basin from neighboring mountains. (7th Approximation: petrocalcic paleargid)

Figure 10.22 (center) This Mollisol has developed under tall-grass prairie in central Iowa and is used for growing crops such as corn and soybeans. The parent material is loess, which is fine dust transported long distances by wind. (7th Approximation: typic argiudoll)

Figure 10.23 (right) This Mollisol has developed under short-grass prairie in eastern Colorado. The organic content and the soil depth are much lower than those of the Mollisol in Iowa. This soil is used primarily to grow small grains. (7th Approximation: abruptic paleustoll)

include areas formerly assigned to several others of the 1938 great soil groups.

Ultisols (from Latin *ultimos*, ultimate) are similar to Alfisols but more thoroughly leached of bases (see Figure 10.26). Often they are found on land surfaces older than those occupied by Alfisols. Thus Ultisols may be an advanced state of Alfisol development. Ultisols occur in climates that are warmer and wetter than those of Alfisols; they are redder in color due to a greater proportion of iron and aluminum oxides in the *A* horizon; and they are significantly more acidic, poorer in humus, and less productive than Alfisols. The bases present in the upper soil have been brought up by deeply penetrating tree roots and are fed into the *A* horizon by way of plant litter. Forest cutting interrupts this nutrient cycle and results in rapid leaching of the bases remaining in the upper soil. In North America, Ultisols are found throughout the southern Atlantic states and lower Mississippi Valley. They are the soils so severely damaged by 150 years of intensive cotton farming, which resulted in erosional loss of nearly the entire solum in some localities. Ultisols are more or less equivalent to the red and yellow podzols of the older classification.

Oxisols (from the word "oxide") are even more thoroughly leached than Ultisols and are the soils characteristic of the wet tropics (see Figure 10.27). The diagnostic feature of Oxisols is a subsurface oxic horizon consisting of a residue of clay and iron and aluminum oxides and hydroxides, with virtually all bases removed. Oxisols are the consequence of the laterization process. They are found on long-exposed land surfaces in the tropics that receive heavy precipitation either seasonally or all year around. Oxisols, formerly known as latosols (the only direct equivalents in the old and new systems), are more widely distributed than the climates presumably required for their development. Some Oxisols presently occur in relatively dry regions, indicating that climatic change has occurred in these areas. Oxisols are primarily encountered in the equatorial forest regions of South America, Africa, and Southeast Asia, but they have a considerable latitudinal spread, extending well beyond the tropical rainforests into the regions of tropical savannas and scrub forests. As in the case of Ultisols, the bases present in Oxisols are a consequence of nutrient cycling by the natural vegetation; thus clearing the natural vegetation results in rapid leaching of soluble plant nutrients, leaving a severely impoverished soil.

Figure 10.24 This Spodosol has developed on sandy glacial outwash from granitic rocks of the Adirondack Mountains of northern New York. The bleached eluvial *A2* horizon and the iron-enriched illuvial *B* horizon are well shown. (7th Approximation: typic haplorthod)

SOIL MANAGEMENT

Replacement of natural vegetation with crop plants greatly alters the earth's soil resource. Although undisturbed soil tends toward a state of equilibrium, it is a constantly functioning dynamic system. In the equilibrium state, the nutrients returned to the soil annually by the natural debris from forest or grassland balance the uptake of nutrients from the soil. Clearing the land for agriculture interrupts the process of nutrient cycling because food and fodder crops return a much reduced proportion of organic litter to the soil; so, in time, the base content of the soil decreases. Wherever rainfall is generous enough to permit large crop yields, clearing the land for agriculture has accelerated the leaching process. At the same time, clearing the soil exposes it to the impact of rain and to runoff, and plowing it loosens its cohesion and lays it bare to the effects of wind. The result is accelerated erosion of the most fertile and moisture-retentive part of the soil, its humic *A* horizon. This reduces the soil's permeability and water-holding capacity, which, in turn, increases storm runoff and flooding in lowlands. Plant roots formerly occupying loose, humus-rich *A* horizons increasingly must search for nutrients in dense clay subsoils. As soils become thinner and less fertile, plants become more sensitive to the periodic moisture variations that are a part of all climates. Thus many "droughts" are actually normal events which have a disastrous impact because the soil system has been artificially weakened by human exploitation.

Artificial compaction and erosional loss of humus and nutrient-rich surface soil horizons have made many soils less permeable, less moisture retentive, less fertile, and less able to exchange their nutrients with the plants growing in them (see Figure 10.28, p. 291). Erosional transport of soil particles into river valleys has choked streams with sediment, causing stream beds to rise, which further increases flood hazards. This is a world-wide phenomenon. The silt moving into valleys is filling reservoirs at two to three times the anticipated rates, threatening future water supplies. For want of drainage facilities, salinization has destroyed the soils of many irrigation projects. Perhaps 75 percent of all irrigated land is affected by salinization,

Figure 10.25 (left) This Alfisol is located in northern Michigan and has developed on sandy glacial outwash. The soil is relatively infertile because of the excessive drainage of water through the sandy parent material. Alfisols developed on other parent materials formerly supporting deciduous forests may be quite productive. (7th Approximation: typic udipsamment)

Figure 10.26 (center) This Ultisol has developed on sandy coastal plain sediments in central North Carolina. The soil is strongly acidic and over 200 cm (80 in.) thick. (7th Approximation: plinthic paleudult)

Figure 10.27 (right) This Oxisol has developed on deeply weathered rock in central Puerto Rico. The soil is low in nutrients but is permeable to water and easily worked in agriculture. (7th Approximation: tropeptic haplorthox)

and every year some 200,000 to 300,000 hectares (500,000 to 750,000 acres) of irrigated land are abandoned as a consequence of salinization and waterlogging.

What can be done to save our soil resource? In some regions, remedial measures, such as fertilization and crop rotation, can restore exhausted soils. In one test, the soil in a plot used solely to grow corn for 30 years experienced a loss of two-thirds of its base content and a corresponding loss in its nitrogen. Similar plots that used a 3-year crop rotation of corn, wheat, and clover showed little loss of nutrients. The year devoted to clover pasture allowed organic matter to enter the soil and restore its nutrient status. At the same time, clover, like alfalfa, is a legume that supports bacterial parasites that can extract nitrogen from soil air, converting it into forms that can be utilized by plants. Adding manure to the soil has a similar effect, but is practical only on a small scale. Major agricultural operations depend upon artificial phosphate and petrochemical fertilizers, which are becoming increasingly expensive. Nevertheless, these fertilizers are indispensable for the high-yield crops that most commercial farming now depends upon.

To reduce surface erosion, stubble and litter from harvested crops may be left in the fields, providing some protection against rain impact and wind erosion. In rolling country, plowing should be along the contour rather than up and down slope. This considerably retards runoff and reduces soil erosion to about a third of the rate that prevails where this practice is ignored. Erosion can be reduced still more by strip cropping (see Figure 10.28), in which different crops, such as grains and fodder, are planted in successive bands along the contour. In upland areas, massive tree plantings can regenerate former woodlands, protecting slopes against further erosional losses. In dry regions, irrigation drains are as essential as canals and must be built into the system in order to stave off artificial salinization and waterlogging. All of these procedures are being carried out in many parts of the world, but often only after centuries of neglect.

Unfortunately, in some regions it is already too late. Salt covers vast expanses, looking like permanent frost on the ground. Reservoirs have already lost much of their water-storage capacity, and flood frequency cannot be diminished in the absence of water-absorbent soils. The red soils of Georgia and the rocky slopes of Palestine are spectacular examples of truncated remnants of much deeper virgin soils that have been eroded to their *B* horizons and even to bedrock by agricultural pressure on the land.

In parts of the warm, wet regions where Ultisols and Oxisols have been cleared of forest and exploited agriculturally, the soil resource seems damaged beyond repair, leached of nutrients on level surfaces and eroded away on slopes. Meanwhile, as cities and suburbs continue to expand over ever-larger areas, some of our most productive soils are blotted out, and the crops formerly grown on them are forced onto new lands—lands previously idle due to inferior soils, less suitable climate or topography, or uneconomic distance from markets.

It is a double irony that the world-wide population drift to cities, accelerating their sprawl, is in many areas partly a consequence of the failure of agriculture on lands devastated by farming practices that ignored the limits to exploitation of the soil. While this has at times been a factor in North America, as during the "dust bowl" years of the 1930s, it is a phenomenon of frightening proportions in the Third World, where population pressures and land tenure systems have led to intense "mining" of the soil, resulting in the exhaustion and erosional destruction of the basic resource needed to feed millions who are already undernourished and poverty-stricken.

Figure 10.28 (top) Gullying of this cornfield in Missouri resulted from the improper practice of plowing and harrowing up and down the slope. The damage was done in only a few days by several heavy rains. At the time of the photo, this land, which had a natural cover of grass, had been used agriculturally for only three years.

(bottom) This scene in southern Wisconson illustrates contour strip cropping which retards runoff and greatly reduces soil erosion. According to the Soil Conservation Service, agricultural yields have doubled where contour strip cropping has been introduced in this area.

SUMMARY

Soil is a complex living system intimately combining mineral and organic matter both physically and chemically. The mineral matter is produced by rock decomposition, and the organic matter results from bacterial decomposition of plant and animal material. Soil development begins with the fragmentation of rock by mechanical and chemical weathering processes. The nature of weathering, and of the resulting raw material for soil development, varies with changing energy and moisture conditions. Eluviation and organic activity convert decomposed rock to soil; soil differs from weathered rock by having a profile, which consists of separate horizons of varying texture, structure, bulk density, color, and chemistry. The measures of a soil's chemistry are its pH, or concentration of hydrogen ions, and its cation exchange capacity. The clay-humus complex of a soil is of particular importance in the retention of basic cations, which are the essential plant nutrients that, in the absence of a clay-humus complex, tend to be leached from the soil. Soluble bases are also kept in the soil by the nutrient cycle, in which vegetation returns to the soil the same nutrients it has extracted.

Parent material, climate, site, organic activity, and time all affect soil development. The climatic factor, together with the influence of vegetation, produces three different soil-forming regimes: laterization, podzolization, and calcification. Drainage conditions produce two additional soil-forming regimes: salinization and gleization.

Due to the great number of variables in soil development and morphology, soil description and classification are extremely difficult. Nevertheless, because of similarities in soil types in different parts of the world, there is a need for a comprehensive classification system. The United States Comprehensive Soil Classification System, often called the 7th Approximation, is designed in such a way that every type of soil can be described in detail and still be related to world-wide great soil groups.

Soil management is an extremely difficult task since any interference with the natural soil system tends to disturb soil equilibrium and increase erosion and nutrient loss. Agricultural use of soils demands careful attention to these problems and may involve specific procedures to combat soil erosion and costly artificial fertilization to prevent soil exhaustion. In many areas, soil loss has reached crisis proportions, causing masses of people to abandon their farms and migrate to cities. Thus soil deterioration is a factor in population shifts and urban growth.

Further Reading

Bennett, Hugh H. *Soil Conservation.* New York: McGraw-Hill Book Co., 1939. (993 pp.) This abundantly illustrated classic, dealing with the processes and results of soil erosion in the 1930s, was written by the first Chief of the U.S. Soil Conservation Service. It is useful for its regional treatment of soil erosion problems in the United States and foreign areas and its wealth of detail concerning specific localities.

Birkeland, Peter W. *Pedology, Weathering, and Geomorphological Research.* New York: Oxford University Press, 1974. (285 pp.) This work provides a thorough treatment of weathering, the factors of soil formation, and the study of soils as geological deposits that convey considerable information on recent environmental history. It is not concerned with soil-plant relations or agricultural applications.

Black, C. A., ed. *Methods of Soil Analysis.* 2 vols. Madison, Wis.: American Society of Agronomy, 1965. (1,572 pp.) This is a complete manual of the procedures used in the laboratory study of soils.

Bridges, E. M. *World Soils.* New York: Cambridge University Press, 1970. (89 pp.) This brief work is a well-illustrated (with color) outline of soil development and soil types in major natural regions of the world.

Bunting, B. T. *The Geography of Soils.* Chicago: Aldine Publishing Co., 1965. (213 pp.) While concise, this book is filled with interesting facts on the soil-forming processes and major soil types—rather zestfully written.

Carter, V. G., and **T. Dale.** *Topsoil and Civilization.* 2nd ed. Norman: University of Oklahoma Press, 1974. (292 pp.) The authors interpret the rise and fall of great world civilizations as due to man-induced deterioration in their resource base—particularly the soils that supported their agricultural systems.

Clarke, G. R. *The Study of Soil in the Field.* 5th ed. Oxford: Clarendon Press, 1971. (145 pp.) This is an excellent manual outlining in very readable fashion the exact techniques used in analyzing soils in the field, with information on soil mapping and the use of aerial photographs.

Review Questions

1. Throughout most of the earth's history, the land surfaces have been absolutely free of vegetation. Were soils of any type present under such conditions? If so, analyze their character in hot, cold, wet, and dry regions.
2. The soils of Java are considerably more productive than those of Brazil, yet both areas lie in similar latitudes in the humid tropics. What is the explanation for the difference in fertility?
3. How can the water budget concept be applied to the various degrees and types of soil development?
4. Termite nests are generally high in calcium. Would you expect the soil under a termite nest in a tropical savanna landscape to differ from the surrounding soil? If so, how?
5. How could a soil develop a profile "in equilibrium" with its environment? Do soils ever reach a completely stable condition?
6. What changes in soils occur where a midlatitude forest advances into a former grassland area? where a grassland expands into a former forest area?
7. How is it possible for a calcrete layer 1 meter thick to be found 1 meter below the surface in a desert region where the heaviest rainfalls only wet the soil to a depth of half a meter?
8. What conditions triggered the great American "dust bowl" of the 1930s?
9. What are the various ways in which human activities can directly or indirectly cause deterioration of the soil resource?
10. How could natural changes in the environment duplicate the effects of human activities on soil systems in different types of settings?

Eckholm, Erik P. *Losing Ground: Environmental Stress and World Food Prospects.* New York: W. W. Norton, 1976. (223 pp.) The author presents an ominous view of the problem of current man-induced soil deterioration in varying environments, with special stress on future problems of food production, especially in the Third World—should be on every geographer's reading list.

FitzPatrick, E. A. *Pedology: A Systematic Approach to Soil Science.* New York: Hafner Publishing Co., 1976. (306 pp.) This complete and copiously illustrated treatment of soils offers a highly regarded alternative to the U.S. Comprehensive Soil Classification System.

Russell, E. W. *Soil Conditions and Plant Growth.* 10th ed. New York: Longman, Inc., 1973. (849 pp.) This text provides a very complete exposition of all aspects of soils in relation to plant growth.

Soil Survey Staff, U.S. Dept. of Agriculture. *Soil Taxonomy: A Basic System of Soil Classification for Making and Interpreting Soil Surveys.* U.S.D.A. Handbook 436. Washington, D.C.: U.S. Government Printing Office, 1975. (754 pp.) This is the updated official presentation of the U.S. Comprehensive Soil Classification System (7th Approximation), providing the system's rationale and detailed application.

Steila, Donald. *The Geography of Soils.* Englewood Cliffs, N.J.: Prentice-Hall, Inc., 1976. (222 pp.) This valuable paperback is a good introduction to the U.S. Comprehensive Soil Classification System (7th Approximation), including material on the uses and management of soils of the different orders in the classification scheme.

United States Department of Agriculture. *Soils and Men.* Yearbook of Agriculture, 1938. Washington, D.C.: Dept. of Agriculture, 1938. (1,232 pp.) Although dated, this work contains a wealth of information on soils, their classification, characteristics, uses, and maintenance.

Rocky Landscape by Paul Cezanne, 1898. (Reproduced by courtesy of the Trustees, The National Gallery, London)

Like an artist's composition, the natural landscape of a region is made up of many parts. And each of these parts—mountain, boulder, or pebble—was shaped by processes that continue to change the face of the earth. Energy from the sun, acting through climate, soils, and vegetation, combines with energy from within the earth to create unique and inspiring scenery.

ELEVEN
Sculpture of the Landscape

CERTAINLY ONE OF THE MOST IMPORTANT parts of the "personality" of any place is the configuration of the land surface, or *topography.* Mention Switzerland and picturesque mountains instantly come to mind; most people's concept of Kansas involves vast featureless plains; Arabia, Tibet, Norway, New England—all these and many more place-names conjure up scenes of a specific nature. Virtually everyone, even the stay-at home, has a large inventory of real or imagined landscapes tucked away in his or her mind and is aware of the fascinating variety of surface types present on our planet.

The earth's scenic diversity is enormous and far exceeds that of any other planet in our solar system. This is a consequence of the earth's unique atmosphere, hydrosphere, and unusually active interior. As with all physical systems, the earth's surface configuration can be explained in terms of varying inputs and transformations of energy as they are applied to the materials composing the system. Three sources of energy are involved: solar radiation, gravitational force, and radioactive decay that generates heat within the earth itself. Solar radiation drives the atmospheric circulation, producing wind, rain, snow, and ice. On the earth's surface, these combine with gravitational force to move fragmented rock material from high to low elevations in the never-ending process of *gradation* by which projections on the earth's surface are worn down and depressions are filled in.

Combating this leveling tendency is the earth's internal energy, which periodically disturbs the planet's surface, warping, crumpling, twisting, breaking, uplifting, and depressing it to provide new work for the gradational forces.

There is unmistakable evidence that the earth's surface configuration is perpetually changing. Sometimes the changes are easily seen, as when a severe flood alters the channel of a river or a landslide tears away a hillside (see Figure 11.1). Other changes are too slow to be perceptible in a human lifetime. Rivers slowly alter their courses; valleys deepen and expand; waves pounding against shores gradually gnaw away the land. Hills and mountains seem enduring, but after each spring thaw or heavy rain, the streams running off steep slopes are brown with debris removed from the land. The hills, large and small, are being washed to the sea, slowly, grain by grain.

The interrelation between the earth's surface forms and the gradational processes modifying them becomes directly apparent as soon as we try to alter terrain for our own purposes, whether to increase an area's productivity or to protect ourselves from natural hazards, such as floods. If a dam is built to regulate a river for irrigation, flood control, or power development, the flow of sediment down the river is interrupted. This may cause ocean beaches miles away to disappear. Why? The answer is that beaches require constant sand arrivals to counterbalance steady losses of material due to wave erosion. The

Figure 11.1 Although many processes of geomorphic change are slow, some changes are rapid and dramatic. This photograph shows some of the effects of the Hebgen Lake, Montana, earthquake of August 1959 near Yellowstone National Park, Wyoming. The strong shock caused a portion of the mountainside shown in the photograph (left background) to slide downward into the valley in the center of the photograph, leaving a broad scar devoid of vegetation. The lake was created when the rockslide dammed the stream in the valley. Nineteen campers are buried under this rockslide.

output of sand from the river may be the input required to sustain the beaches. In previous chapters, we have seen that human alterations in one process or form frequently trigger a chain of unforeseen reactions, some of which may be quite undesirable. This is as true of terrain features as it is of the atmosphere, the hydrosphere, and the biosphere.

The remainder of this book analyzes the earth's surface configuration and the processes that produce individual terrain features, which are known as *landforms.* This chapter provides a general introduction to landform development and discusses the various scales of the earth's relief features, the factors differentiating landforms, the energetics of erosion and deposition, and general theories of landform development. The study of landforms, a science known as *geomorphology,* draws upon many of the facts in the earlier discussions of energy, the hydrologic cycle, climate, vegetation, and soils—all of which play important roles in landform development. In turn, terrain features affect climates by channeling movements of air, determining surface elevation, causing orographic precipitation, and producing rainshadow regions. Similarly, surface form often determines the nature and amount of streamflow, the location of water supplies, and the local character of vegetation and soils. Thus landforms are a functional part of the human environment. Succeeding chapters will consider in greater detail the most important processes that produce the distinctive characteristics of individual landscape types.

ORDERS OF RELIEF

Before discussing how the earth's landforms develop, it is first necessary to establish that variations in surface configuration occur on a variety of scales, which have been termed *orders of relief.* When we speak of *relief,* we are referring to vertical differences in elevation between high and low places on a surface, but all relief features have horizontal as well as vertical dimensions. Relief features on the earth's surface occur on all scales, ranging from raindrop-impact craters a fraction of 1 mm deep and only 1 or 2 mm across to the continents and ocean basins themselves, which extend over millions of square kilometers and have a maximum difference in elevation of about 20 km (12.4 miles). The continents and ocean basins, which constitute the largest scale relief features of the earth's surface, both vertically and horizontally, are referred to as the *first order of relief.*

First Order of Relief: Continents and Ocean Basins

The earth's outer layer, or crust, is composed of a mosaic of adjoining plates some 70 to 100 km (40 to 60 miles) thick. These plates are great "floating" rafts of relatively low-density rock supported from below by the denser, semiplastic rock that forms the upper portion of the earth's mantle. The continents and ocean basins are embedded in these plates.

The geologic record shows that plants, animals, and landforms have changed with time, and recent evidence indicates that the con-

Figure 11.2 This satellite photograph shows a portion of the east coast of Africa looking toward Saudi Arabia, with the Red Sea at the left and the Gulf of Aden at the right. The opening of the Red Sea is believed to have begun 5 to 10 million years ago when the lithospheric plates carrying Africa **(bottom)** and Arabia **(top)** began to drift apart.

tinents themselves have shifted restlessly over the earth's surface (see Figure 11.2). In the first decades of the twentieth century, the German meteorologist Alfred Wegener and others suggested that portions of the earth's landmasses have drifted long distances over geologic time. Wegener argued that the presence in Pennsylvania of extensive coal deposits formed from tropical vegetation implied that the region was once located near the equator. Similarly, ancient glacial deposits in South Africa, Brazil, and Australia seemed to indicate that these areas had formerly been close to one of the poles. Under the glacial deposits in Africa and South America are rock structures that also seemed to match up across the Atlantic. The fossilized remains of ancient plants likewise indicated a former connection between all of the southern hemisphere continents.

More obviously, the close fit between the facing coasts of Africa and South America, which is very apparent on a globe, could hardly be an accident. Wegener was not the first to notice the similarity in these two coasts or even to propose the movement of continents, but he was the first to analyze the evidence in detail and to present a strong case for continental "drift." Wegener agreed with several predecessors that Africa and South America had probably been part of the same landmass and further proposed that all of the world's continents had once been united in a vast supercontinent, which he called Pangaea, as shown in Figure 11.3. More recently it has been shown that all of the continents fit together fairly well if their actual submerged "edges" at a depth of 2,000 meters (6,500 ft) are used, rather than their shorelines. Wegener proposed that the breakup of Pangaea produced the present continents, which have since been moving differentially with respect to one another.

For almost half a century, Wegener's ideas were scoffed at by most geologists, largely because there seemed to be no conceivable force that could displace continent-sized masses of the earth's crust. The fact that the continental slabs project downward into the denser material of the earth's interior made their displacement seem all the more impossible. Nevertheless, Wegener's ideas about continental drift turned out to be partially correct.

In 1944, the geologist Arthur Holmes suggested an alternative idea—that it is the sea floors that are moving, expanding in some places, thus widening the separation between some of the continents and elsewhere forcing continents together. Holmes proposed that the driving force was thermal convection currents in the upper part of the earth's mantle. Holmes's theory was refined by the oceanographer Harry Hess in 1962 when he demonstrated the relevance of several newly discovered features of ocean floor topography. During World War II, oceanographers had found that all of the oceans are divided by continuous undersea ridge systems, such as the Mid-Atlantic Ridge, and that the margins of some oceans include deep trenches, such as those that ring the western Pacific. Hess suggested that new ocean floor material is created by volcanic eruptions along the ridge systems and that old ocean floor material disappears by descending into the trench systems in the process of *subduction* (Figure 11.4, pp. 300–301). According to this theory, the continents ride passively on ocean

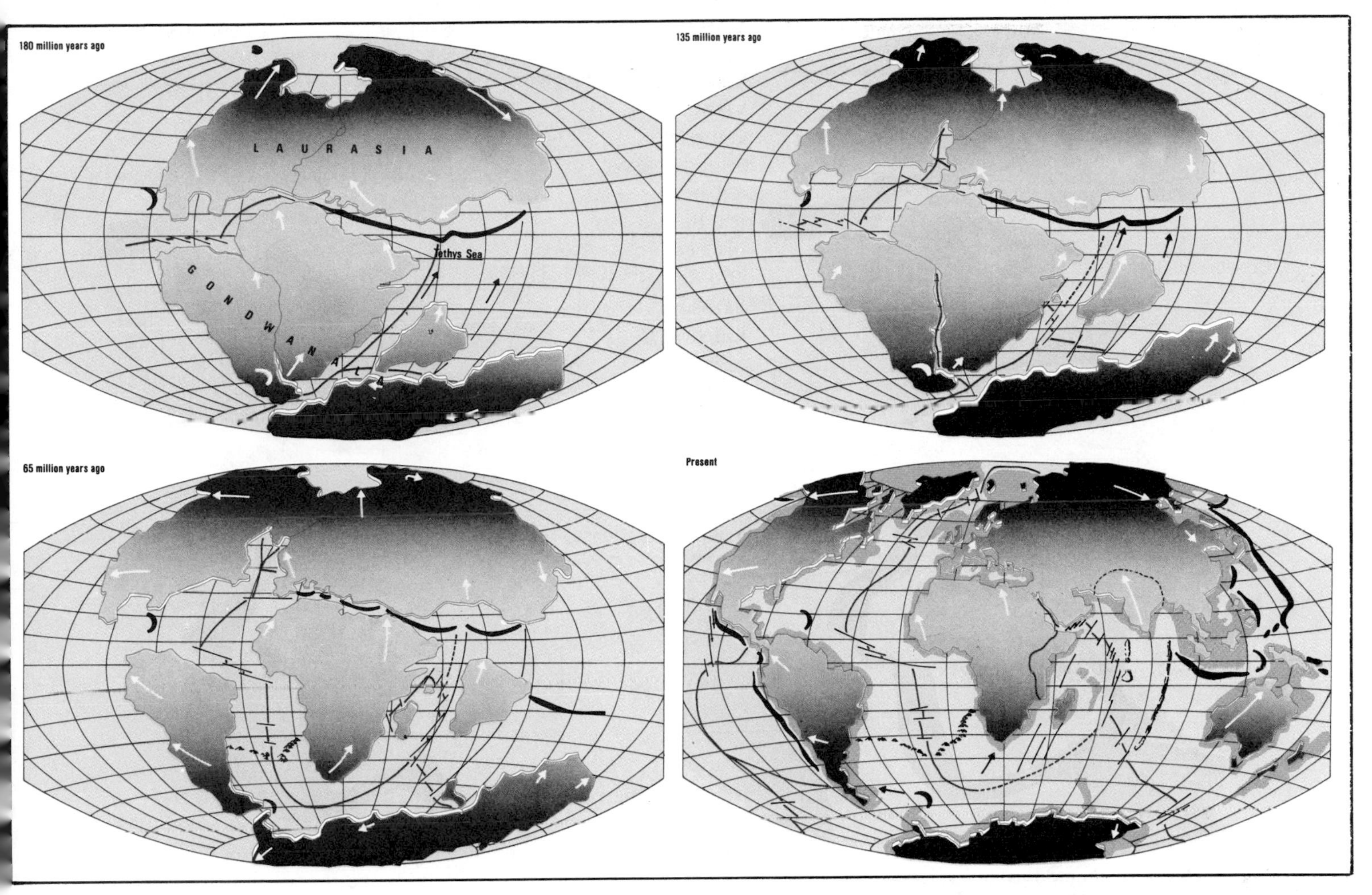

Figure 11.3 Alfred Wegener suggested in 1912 that about 200 million years ago all of the landmasses of the earth were grouped together, forming a supercontinent, Pangaea. Pangaea subsequently separated into two continents, Laurasia in the northern hemisphere and Gondwanaland in the southern hemisphere. The rifting has continued, with crustal plates and associated continents moving in the directions indicated by the arrows. The plate that carries India, for example, separated from Antarctica and drifted northward until it collided with the Asian plate. The drift of the continents is now thought of in terms of plate tectonics, seafloor spreading, and subduction. In this figure, subduction zones are shown in black and spreading centers in red.

floor material that is moving laterally from the ridges toward the trenches in the process now known as *sea floor spreading*. The global pattern of lithospheric plates, spreading centers, and subduction zones is shown in Figure 11.5, p. 302. The movement of lithospheric plates and the interactions between them are known as *plate tectonics*.

Sea floor spreading has been investigated experimentally. Research ships, such as the *Glomar Challenger*, have drilled into the floor of the ocean basins with hollow coring devices to bring up samples of sediment for analysis. More than 500 such holes have been drilled into sea floors around the world since 1968. Although the earth itself is about 5 billion years old, no oceanic sediment older than 170 million years has been found, which suggests that the ocean floors are relatively young. Furthermore, the thickness and age of the sediments increase with distance from the oceanic ridges. This reflects the fact that the sea floors nearest the ridges have formed most recently and have had the least time to collect sediment.

In a few places, isolated hot convective plumes of volcanic activity in the mantle remain fixed in position, and, as the plates creep by, the hot spots feed surface volcanic eruptions, producing volcanic

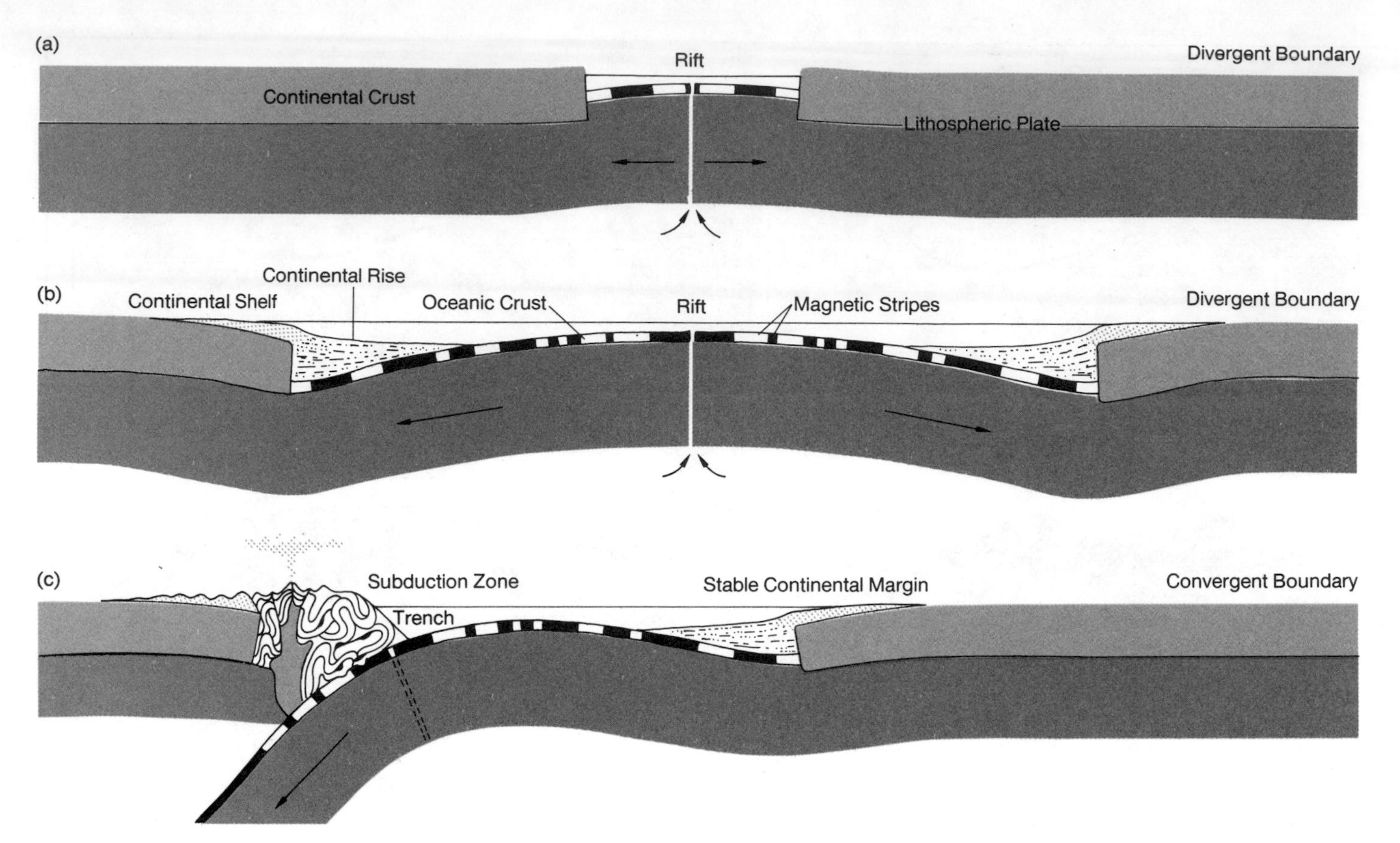

Figure 11.4 (a) A newly developed spreading center causes continental plates to rift and move apart (divergent boundary).

(b) The development of new crust at a spreading center causes a sizable ocean basin to form over the course of tens of millions of years. Note that the sediment deposits are thickest near the continental plates, where the oceanic crust is oldest. This schematic diagram represents the Atlantic Ocean basin. The magnetic stripe pattern indicates the record of the earth's magnetism that is frozen into new crustal rock as it cools. The pattern is the same on both sides of the spreading center; such patterns occur in all ocean basins.

(c) Deep ocean trenches, such as those in the western Pacific Ocean, occur where a plate of oceanic crust plunges downward into the mantle beneath either a continental plate or another oceanic plate (convergent boundary). Crustal deformation at a convergent boundary leads to mountain uplift, to the formation of island arcs, and to earthquakes and volcanic activity.

(d) **(opposite)** A collision between two continents—India and Eurasia, for example—produces high mountain ranges such as the Himalayas. The continental crust is doubled in thickness and the surficial sediments are compressed to form contorted geologic structures. (From Robert S. Dietz, "Geosynclines, Mountains, and Continent Building." Copyright © 1972 by Scientific American, Inc. All rights reserved)

islands on the moving plates. The Hawaiian Islands appear to have been formed in this way because they exhibit a definite progression in age from one end of the chain to the other (see Figure 11.6, p. 303).

Perhaps the most convincing of all evidence for sea floor spreading is the record of ancient magnetism, or *paleomagnetism*, in the volcanic rocks of the sea floor. When molten rock extruded at the oceanic ridges solidifies, certain of its iron-bearing minerals become oriented in accordance with the earth's magnetic field. Because the earth's magnetic field changes in strength and orientation (polarity) through time, the magnetic imprint in older volcanic rocks farther from the oceanic spreading centers differs from the imprint in younger rocks nearer the spreading centers. The volcanic rocks on the floors of the various seas exhibit similar patterns of magnetic imprinting, except for variations resulting from differing rates of sea floor spreading. These patterns are strikingly symmetrical on the opposite sides of each of the undersea ridge systems. The study of paleomagnetism in ocean floor rocks reveals that in the last few million years the East Pacific basin has been widening at a rate of 10 to 12 cm (4 to 5 in.) per year, compared to an annual spreading rate of only 2 cm (less than 1 in.) in the North Atlantic basin. The midoceanic ridges, the spreading sea floors, the subduction zones and associated deep oceanic trenches, and the mantle therefore represent the recycling of crustal and mantle materials on an awesome global scale at an almost imperceptibly slow rate.

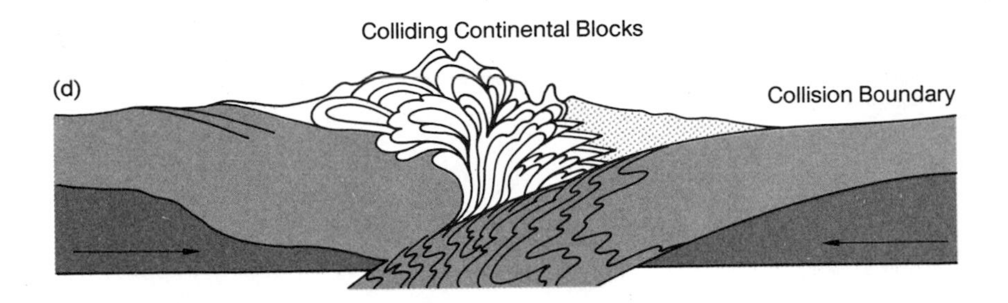

Second Order of Relief: Continental and Oceanic Mountain Systems, Basins, and Plateaus

After the continents and ocean basins, the most extensive relief features on the earth are continental mountain ranges, such as the Andes, Alps, and Himalayas; subcontinental plateaus—for example, the Tibetan and South African plateaus; and large depressions such as the West Siberian Lowland or the Mississippi Valley. Such features constitute the *second order of relief.* The ocean basins also possess a second order of relief in the form of the oceanic ridge systems and the deep submarine trenches (Figure 11.7, pp. 304–305). The oceanic ridges, which are the source regions for new ocean floor material, extend tens of thousands of kilometers and rise to heights of 3,000 meters (nearly 10,000 ft) above the ocean floor. They form undersea mountain chains with a continuity unrivaled by any mountain system on the land. The ocean trenches, however, are discontinuous, occurring only where subduction of lithospheric plates is presently active. All second-order relief features have been produced by either vertical or horizontal displacements of the earth's crust and may be explained in terms of plate tectonics.

Most of the great continental mountain systems were created as a result of convergence of lithospheric plates (see Figure 11.5, p. 302). The young, rugged mountains bordering the west coast of South America are currently being uplifted as the South American plate expands westward against the eastward moving Pacific plate. A similar relationship in the past accounts for the mountain systems of the Pacific coast of North America. The Himalaya Mountains and Tibetan Plateau north of India were generated when the lithospheric plate that carries India spread northward into and under the plate of continental Asia. Along the Atlantic coasts of the Americas, Europe, and Africa, the continents ride on lithospheric plates that are diverging, and since there are no plate boundaries at these coasts, they are low-lying and geologically inactive at present.

Third Order of Relief: Landforms

The individual hills and valleys of landscapes constitute the *third order of relief.* It is these sculptural details of the second-order relief features that we specifically regard as landforms (Figure 11.8, p. 306). Landforms are on the human scale of magnitude—they are features that can often be seen in their entirety in a single view and are

Global Lithospheric Plates

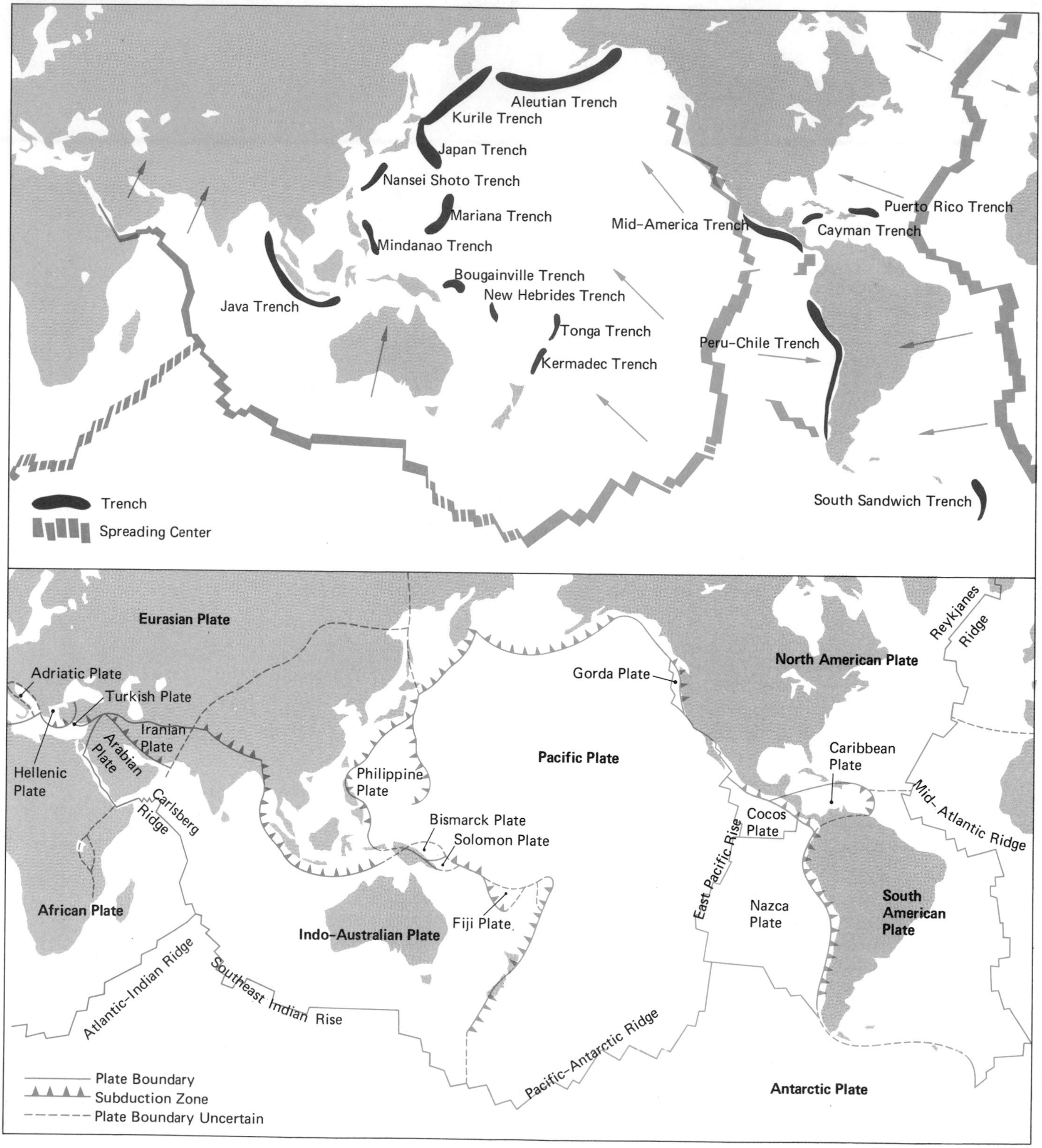

comparable in size to the distance a person can hike or climb in a few hours or a day. Most landforms have been produced by the gradational processes of erosion or deposition rather than by motions of the earth's crust. Man's more ambitious engineering works, such as canals, dams, and reservoirs, are also on the scale of third-order relief features. The science of geomorphology is concerned largely with explaining the processes that create relief features on this scale.

The variety of landforms is enormous, and many attempts have been made at their classification. Some classification schemes are purely descriptive and based on form (morphology) alone. For example, stream systems in drainage basins may be described morphologically as *dendritic* (branching like a tree) or *pinnate* (featherlike) (Chapter 12). Other classifications are genetic or explanatory and are concerned with the geomorphic processes responsible for the creation of the landform. Genetic classifications refer to forms such as *glacial troughs, fluvial terraces,* and *wave-cut benches.* Genetic systems occasionally identify landforms by the local name used in the area where the phenomenon was first described or is most characteristic—as the Gaelic word "drumlin," used in reference to small, distinctively streamlined hills of debris deposited by glaciers, or the Icelandic "jokulhlaup," meaning an enormous and sudden flood *(hlaup)* of meltwater issuing from beneath a glacier *(jokul).*

Many geomorphologists believe that the origin of any landform is best discussed in terms of the mechanics of gravitationally produced stresses acting on materials of varying resistance. The outcome of the contest between stress and resistance is seen in the angles of *slopes* that develop on the materials. This approach emphasizes mathematical measurement of landforms and the processes modifying slopes. However, due to the complex histories of many landscapes, a complete understanding of all process-form relationships remains a distant prospect.

Figure 11.5 (opposite) These maps indicate the probable locations of lithospheric plate boundaries in the earth's crust and the relative motion of the plates. In the upper map, the blue lines represent spreading centers where new crust is being formed, and the purple lines are regions where plates are descending into the mantle in the subduction process. The arrows indicate the general relative directions of plate motion. The lower map shows plate boundaries. The seven major plates are identified in bold type and several of the smaller plates in lighter type. Note that the Pacific basin is largely rimmed by subduction zones.

Many features of the earth's crust can be understood in terms of the motion and interaction of plate boundaries. Note, for example, that mountain uplift is active at the plate boundaries along the west coasts of North and South America, where the mountains are known to be young. Along the east coasts, however, there are no plate boundaries and no mountain-building activity. Earthquake and volcanic activity is associated with converging plates such as those around the rim of the Pacific Ocean basin. (top, from John F. Dewey, "Plate Tectonics."

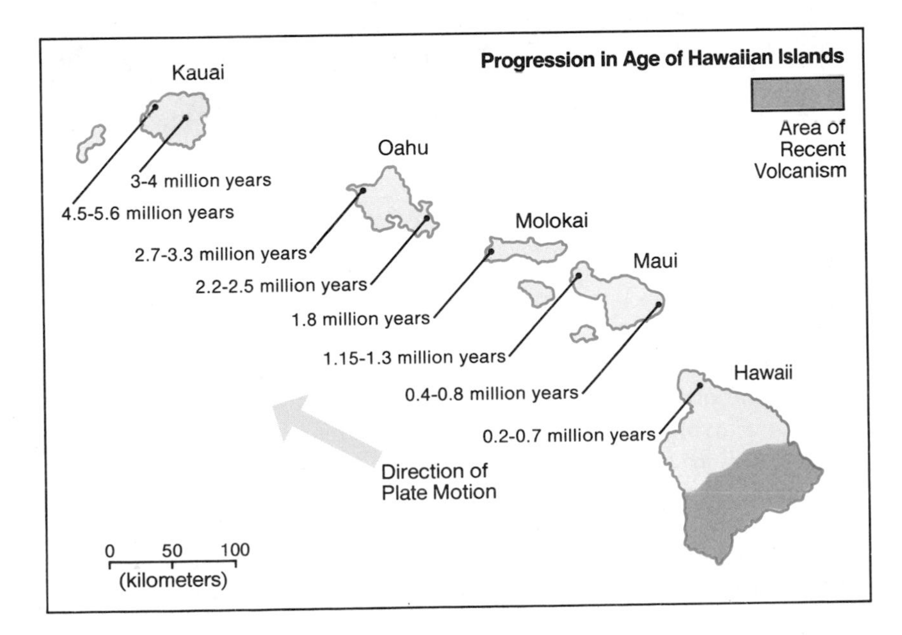

Figure 11.6 The ages of the Hawaiian Islands, as determined by the potassium/argon radiometric dating method, show a general progression from old to young moving southeastward along the chain. These observations were not explained until the development of the theory of plate tectonics, which attributes the formation of the Hawaiian Islands to a plume of volcanic activity fixed in the earth's mantle. As the Pacific plate drifts over the location of the plume, periods of volcanic activity cause the formation of islands of volcanic rock. Can you verify from the map that the measured ages of the Hawaiian Islands are consistent with a drift rate of several centimeters a year? (After McDougall, 1971)

Figure 11.7 The ocean basins, which are first-order relief features, contain second-order relief features, such as deep trenches and extensive undersea mountain ranges. This physiographic diagram of the North Pacific Ocean basin shows the plains, trenches, and seamounts characteristic of the ocean floor. Heights above and below sea level are given in feet. Note in particular the island arcs along the continental side of the great trenches. Such arcs are located at the boundaries where oceanic crustal plates are plunging downward into the mantle under continental plates, and they are sites of volcanic and earthquake activity. (Courtesy of the National Geographic Society)

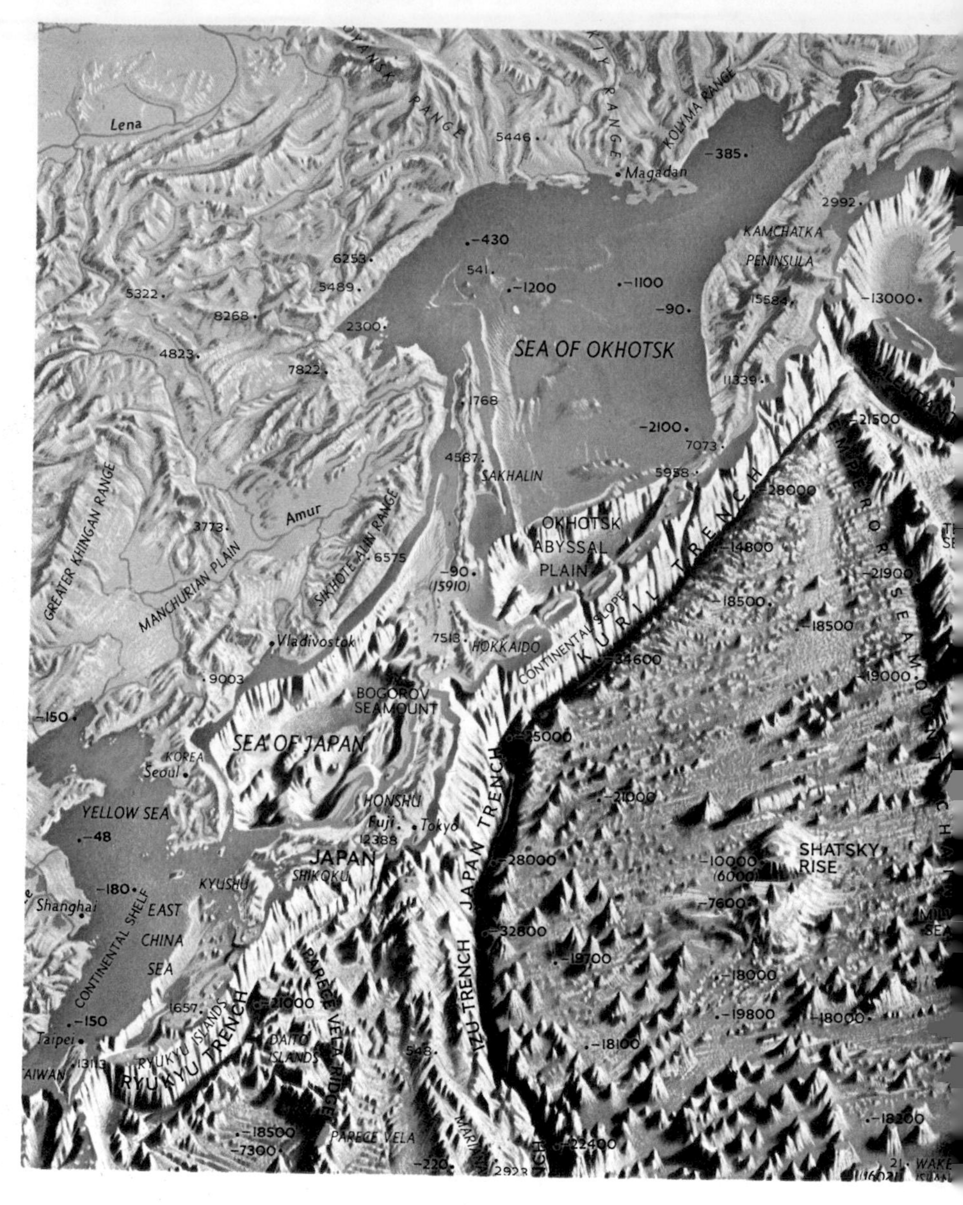

THE DIFFERENTIATION OF LANDFORMS

A limestone hill rising out of the forests and swamps of Malaya or Vietnam certainly has a different appearance from a hill of the same material in Kentucky or Missouri. Why is this? And why are certain highly unusual landforms in widely separated parts of the world similar to one another but different from features only miles away? Four factors are involved: (1) the geologic structure, which includes both the type and arrangement of the materials composing the landform; (2) tectonic activity, or the nature of local movements of the earth's crust resulting from the earth's internal energy; (3) the gradational processes at work, which are largely controlled by climate; and (4) the period of time the tectonic and gradational processes have operated in

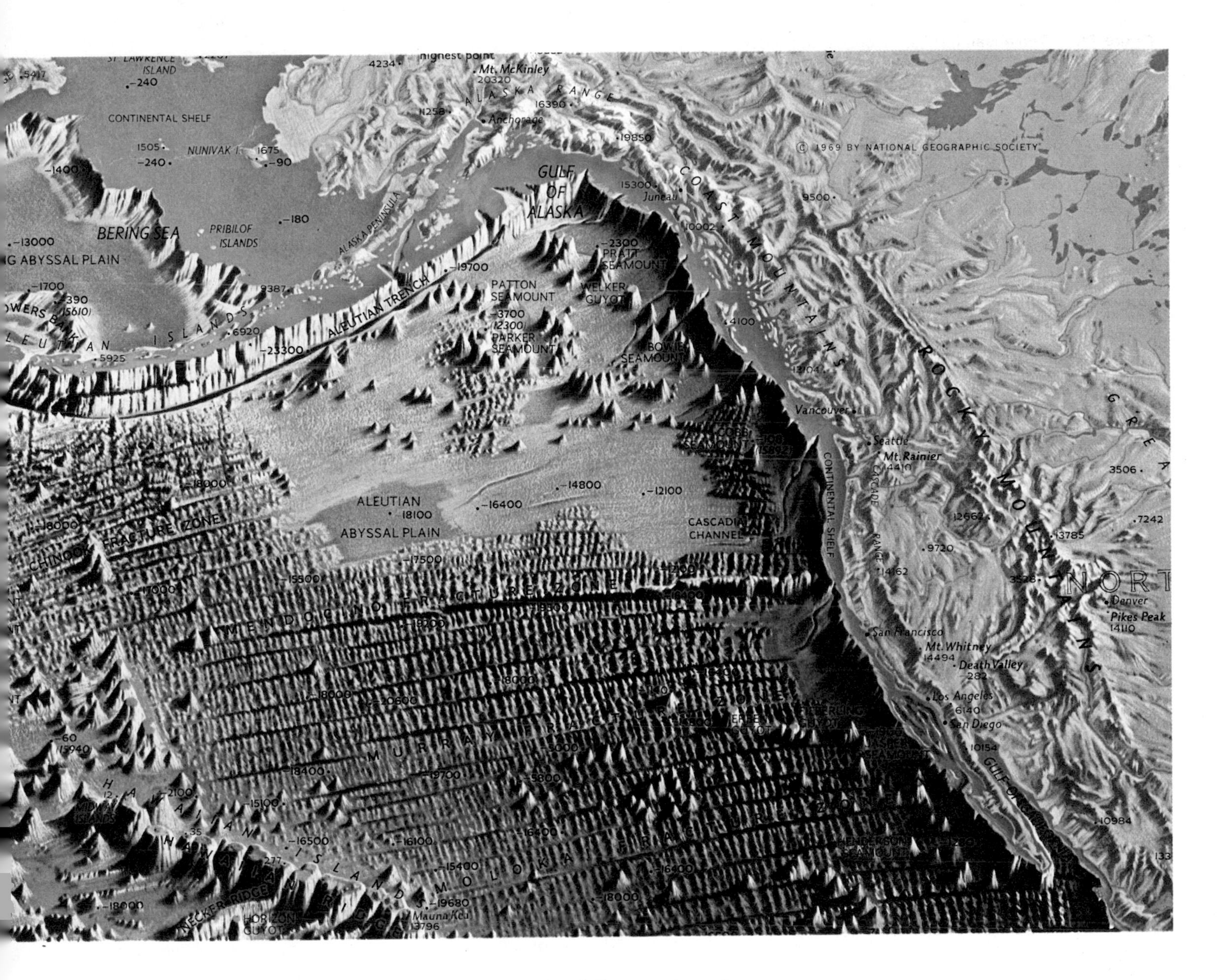

a given manner. The roles these factors play in the differentiation of landforms around the world are considered in the remaining chapters of this book.

Geologic Structure and Landforms

Geologic structure refers to both the nature and the arrangement of the materials from which landforms are built up or in which they are sculptured. The geologic materials encountered at and below the earth's surface are extremely diverse, ranging from thin blankets of dust that have settled from the atmosphere to kilometer-deep masses of uniform hard rock crystallized from a molten condition. Some of these materials are loose deposits; others are coherent enough to give

Figure 11.8 Landforms of the type shown here are third-order relief features. Such features are produced by a variety of erosional and depositional processes. The valley occupied by the stream was created by river erosion and enlarged by glacier scouring. In the right background is the summit of Mt. Rainier, a volcano in Washington's Cascade Range—a constructional landform composed of lava and volcanic ash. The peak to the left is on a rim created by the collapse of the summit of an older volcano on the site of Mt. Rainier. In the middle distance are moraines, ridgelike deposits left by the Emmons glacier when it was much larger than at present. In the foreground are cobbles washed out of the glacier by meltwater. The larger boulders and some of the cobbles were brought here by a rockslide in 1963. The rockslide covers the surface of the Emmons glacier in the center of the photo.

rise to the expression "rock hard." Some are chemically and physically simple; others are extremely complex. These materials may be arranged in layers, in solid lumps, or in vast slabs. The layers may be horizontal, tilted, or wrinkled (see Figure 11.9); the lumps and slabs may be rigid or shattered by fractures. All of these variations produce effects on the landscapes seen on the earth's surface and are important enough to merit separate treatment in Chapter 13.

Tectonic Activity and Landforms

The crustal motions that create the specific arrangements of rock masses that provide the framework for landscape development are known as *tectonic activity*, or simply *tectonism* (from the Greek word *tekton*, meaning "builder"). Tectonic activity is an expression of the earth's internal energy, and it influences scenery in a gross sense merely by elevating and depressing the earth's crust. All of the earth's second-order relief features, and also many local landforms, are produced directly by tectonic activity. The earth's internal energy can produce all imaginable types of stresses on the rocks of the earth's crust, including compression, tension, torsion (twisting), and shearing. When these stresses overcome the internal strength of the rock material involved, the rock fails, either by rupturing or by wrinkling.

Ruptures along which vertical displacements of rock masses have occurred are known as *faults.* When carried out over periods of a million or more years, movement along faults can produce awesome relief features, such as the great east-facing escarpment of California's Sierra Nevada range, which rises more than 3,300 meters (11,000 ft). Rocks that react to compressive stresses by bending instead of rupturing along faults produce wrinkled structures known as *folds.* The distinctive landscapes of portions of the Appalachian Mountains from Pennsylvania to Alabama result from the underlying structure of folded rock layers (see Chapter 13). Where the tectonic stress is vertically upward, the crust is forced to rise in a structural *dome,* which may be tens to hundreds of kilometers in diameter and thousands of meters high. An outstanding example is the Black Hills region of South Dakota.

Different types of tectonic movements are often interrelated. For instance, upward doming of the crust produces tension at the crest of the dome, which may result in collapse or faulting. If the dome is steep, the downward pressure of rocks on the slopes may be great enough to compress those on the flanks, causing folds to develop (see Figure 11.10, p. 308). The doming itself may be caused by the upward injection of molten rock material into the crust as part of the process of *volcanism* (see Chapter 13).

It must be stressed that the structures formed by tectonic activity are assaulted by gradational forces as soon as crustal motion begins. Consequently, tectonic structures are rarely seen intact. Nevertheless,

Figure 11.9 This large, symmetrical ridge of rock in the Zagros Mountains of Iran formed when compressional forces wrinkled the earth's crust in this region. The exposed cross section in the foreground shows rock strata that have been folded by intense tectonic activity.

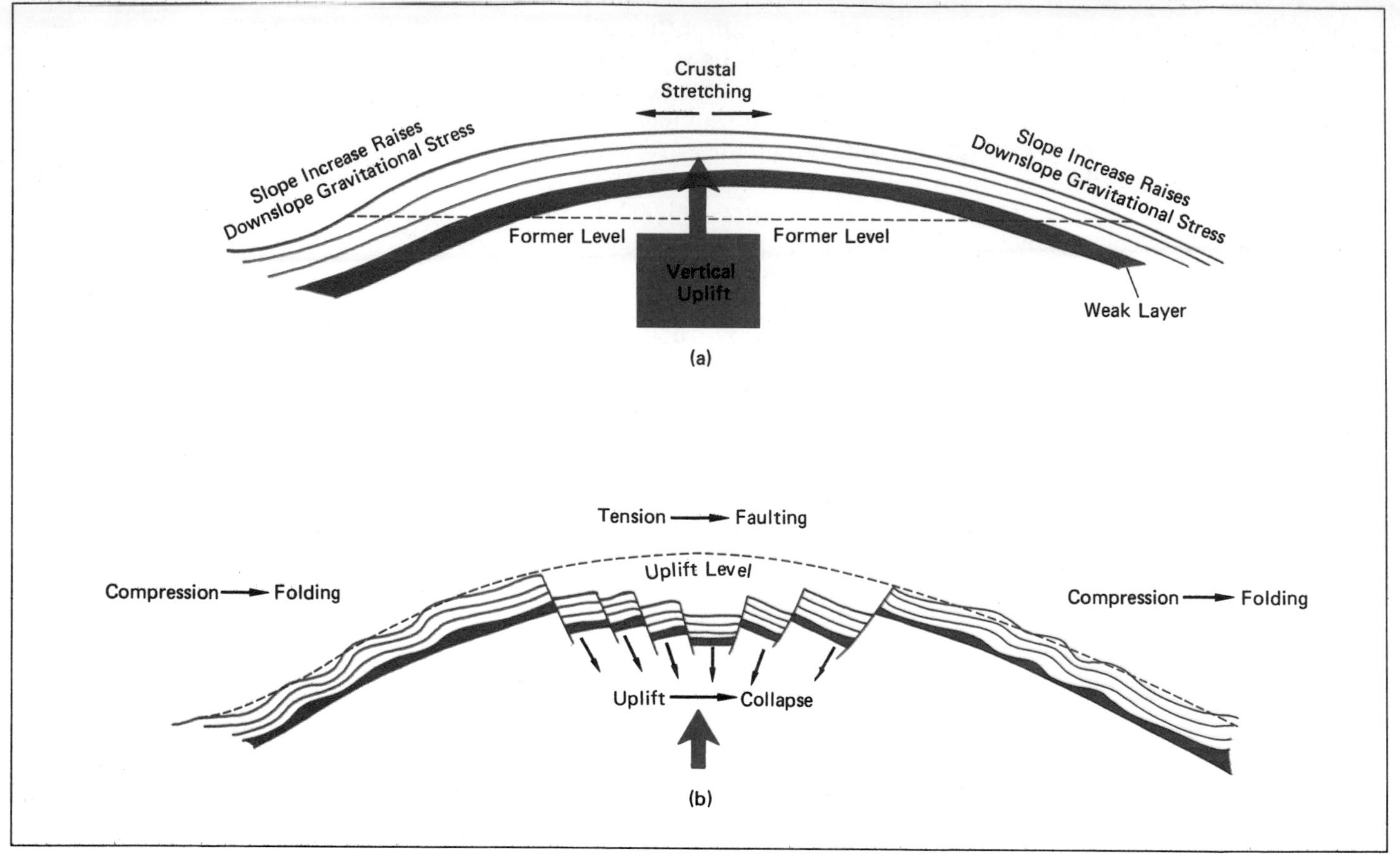

Figure 11.10 Different types of tectonic activity often have a common cause. In this example, broad uplift (a) forming a dome causes both folding and faulting (b). The dome may be hundreds of kilometers wide and thousands of meters high. As the rocks forming the roof of the dome are subject to increasing tension (a), they rupture and gradually subside along faults (b). On the flanks of the dome, rock layers thousands of meters thick slip downward, producing fold structures. This type of folding requires the presence of a weak layer (shown in black), which deforms internally and lubricates the sliding mass above it. Likewise, for collapse to occur, there must be some way the center of the dome can subside. Often the dome collapses into a mass of molten subsurface rock. It may be the rise of this molten rock that lifted the original surface, forming the dome.

by elevating and depressing the crust and by producing the specific arrangements of rock masses of varying physical characteristics, tectonic activity plays a crucial role in the appearance of landscapes.

Gradation and Landforms

Whereas tectonic activity tends to create irregularities on the land surface, gravity, the earth's second internal energy source, tends to smooth the surface by providing the energy for flows of water, ice, air, and loose rock material in the various processes of gradation. *Gradation* includes the fragmentation of rock material by weathering, the detachment (entrainment) and removal (transportation) of fragmented debris by the agents of erosion, and the deposition of material at lower elevations. Thus the general effect of gradation is to wear away projections and fill in depressions on the earth's surface.

The most widespread physical agent of gradation is water (Figure 11.11, pp. 310–311). It is the essential factor in both the chemical and mechanical fragmentation of rock by weathering processes; likewise, the movement of water in its liquid and solid states is the principal means of transporting fragmented rock material from high to low elevations in the processes of erosion and deposition.

While channelled flows of running water are the most widespread single agent of gradation (Chapter 12), other important agents are direct gravitational transfer (discussed later in this chapter), wind (Chapter 14), sheets and streams of frozen water in the form of glaciers (Chapter 15), and ocean waves and currents (Chapter 16). All of these transporting media modify landscapes both by erosion and deposition. Each of these gradational agents may be dominant over vast areas, producing a highly distinctive imprint upon the land. Often they work in conjunction or succeed one another periodically.

It is climate that generally determines which of the gradational processes dominates in an area, the only exception being the action of waves and currents. Where several agents are active simultaneously, the climate usually determines their relative intensity, and where changes in the type and intensity of gradational processes have occurred, a climatic change has usually been the reason (see Chapter 14). This relationship has been somewhat altered since man has become an agent of environmental change. Generally the tempo of geomorphic change has been increased by human activity, mainly through our direct and indirect effects on hydrology and the vegetation cover.

Time and Landforms

Time is clearly an influence on landforms, but its effect is often difficult to evaluate. Theoretically, over periods of tens of hundreds of millions of years, the gradational processes should erase hills and even mountain ranges. Tectonic activity, however, repeatedly sets back the gradational clock, defeating or greatly retarding the leveling tendency. Nevertheless, where vast areas of hard rocks have been eroded to very low relief, as in much of Africa and Australia, it is clear that the landscape has been undergoing erosion without the interference of tectonic activity for an unusually long period of time. Thus, in the absence of tectonism, the ultimate effect of time is to cause gradational processes to reduce landscapes to featureless plains.

Since areas with no crustal movement are few, the influence of time on most landforms must be assessed according to the rates at which the various gradational processes operate. Some processes, such as channelled flows of water or waves breaking on beaches, produce rapid changes in associated landforms. In river and beach systems, variations in the intensity or nature of processes or in supplies of materials cause immediate adjustments in landforms as a new equilibrium between process and form is approached. These adjustments always tend to minimize energy expenditure (or work) in the system. In many quick-response systems, the inputs of energy and materials fluctuate continuously, so form changes may be more or less continuous, with only a quasi-equilibrium condition ever being achieved.

Other gradational processes, such as those affecting hillslopes, require much more time to produce landscape changes. Where the response time of the system is slow, the landform may preserve features resulting from older processes long after different processes have become dominant because there has been insufficient time for current processes to eliminate older inherited forms. Such "left over"

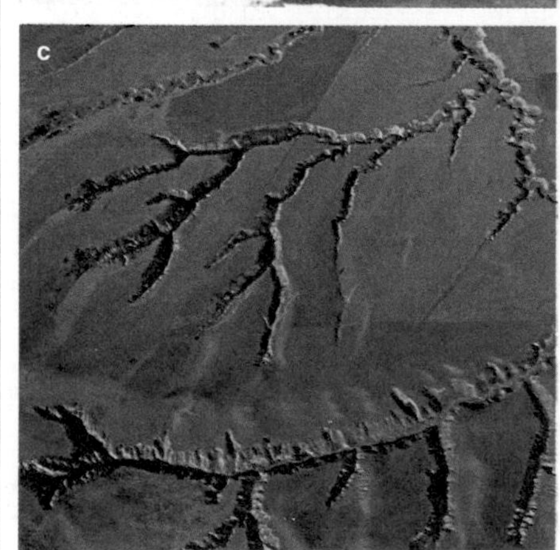

Figure 11.11 Water acting as an agent of erosion and deposition affecting the varied materials and structures of the earth's crust has generated widely different landforms, as these examples show.

(a) Motueka River valley in New Zealand was carved by a river, then filled with sediments washed out of a melting glacier. The present sinuous river slowly changes its course by lateral migration, alternately eroding and depositing a veneer of sediment and producing a wide floodplain between the valley walls.

(b) Water acting on these sedimentary rocks in Red Rock Canyon, California, has cut a variety of shapes and patterns in layers of different resistance to erosion.

(c) Surface runoff and streamflow are carving a dendritic, or branched, drainage pattern into this landscape north of Christchurch, New Zealand. The relatively uniform resistance of the underlying rock favors the development of a treelike drainage pattern as each of the side tributary branches extends by headward erosion away from the main stream channels.

(d) Half Dome, a famous natural feature in Yosemite National Park, California, was originally a complete granite dome, probably shaped by *exfoliation*. When granite is exposed at the surface by downward erosion, the confining pressure on it is released, and concentric sheets split off, or exfoliate, from the main mass. A later rockslide removed half of the dome, with the debris being carried away by a glacier.

(e) This natural arch of sandstone is an erosional remnant formed by collapse and by the action of water, even though it is located in the arid southeast of Utah where the annual precipitation is only 20 cm (8 in.). The lower layers of rock weather more rapidly than the upper layers due to greater dampness at the base of the rock exposure. In time, the entire structure will be weathered away.

(f) The attack of waves at the base of these limestone hills near Dorset, England, has removed the support for the overlying rock. The entire face has become a nearly vertical cliff because of removal of rock from the foot.

(g) The rock spires in Bryce Canyon National Park, Utah, are products of water erosion of weakly consolidated sedimentary rocks. They exhibit a steplike structure because the rock layers have different resistances to erosion.

(h) Pillars have been cut into the soft pumice slopes of an ancient collapsed volcano at Crater Lake, Oregon. Such pillars are formed by rainfall on erodible material containing boulders that prevent portions of the surface from being eroded downward as rapidly as unprotected portions.

features that are not products of current processes are called *relict forms.* The climatic fluctuations of the Pleistocene epoch repeatedly changed both the energy and material inputs of many slow-response landform systems, particularly hillside slopes. These slow-adjusting slopes presently preserve many relict features produced under Pleistocene climates that no longer prevail (see Case Study following Chapter 7). Such slopes are slowly changing at present to come into equilibrium with the process system that prevails today. The most widespread and spectacular of all relict landscapes are those produced by Pleistocene glaciation (see Chapter 15). It will require more than a million years for current gradational processes to erase the evidence of glacial modification of vast areas in North America and Eurasia, and in most of the earth's high mountains, even though the glaciers withdrew from most of these areas some 10,000 years ago.

ENERGETICS OF EROSION AND DEPOSITION

It is possible to analyze the development of landforms in the same way that an engineer analyzes the forces that must be resisted or utilized to allow an aircraft to fly or a bridge to support the load of traffic crossing it. First, we have the material involved, whether solid rock or a mass of loose particles left by some depositional process. This material has many measurable physical properties, among them its internal strength, or resistance to deformation. The mineral particles composing the material may interlock, and they adhere to one another due to electrostatic attraction on the atomic scale. To deform the material or remove particles from it, the cohesion between its constituent particles must be overcome. This requires the application of energy.

It is chemical energy and energy on the molecular scale that initiates change in the land surface. Before solid rock can be moved, it must be fragmented by weathering; this occurs when water is introduced into the rock. Because water contains ions of many elements, its presence eventually causes chemical and mechanical changes. Many of these reactions result from electrical energy, which causes ions of various elements to displace one another and form new combinations that reduce the cohesion of the rock. A molecular energy source is also available when liquids change to solid form, as when water freezes or dissolved substances crystallize. The crystallization of most substances expands them, exerting stress against the confining material. These stresses are frequently enough to shatter weakly coherent rock.

The principal energy source directly responsible for the stresses that move fragmented material is gravity. Gravitational force constantly tugs at everything on and near our planet and puts in motion unconfined fluids, such as water and air, and also plastic substances, such as ice. As we saw in Chapter 2, gravitational energy is also known as potential energy, because it can only cause motion in materials that have the potential to be displaced downward. As fluids move toward lower levels, they exert an oblique downward stress, known as *shear stress*, against the underlying surface. The higher the density and depth of the fluid, the greater its weight (or the gravitational force exerted upon it), and thus the greater the shear stress. A current of air obviously exerts much less stress on the surface than does a current of water moving at the same speed. Shear stress is also influenced by the slope of the surface on which it is acting—the steeper the slope, the greater the downslope gravitational force in comparison with the force tending to hold particles in place on the surface (see Figure 11.12).

When the shear stress applied exceeds the resistance of the material the stress is acting upon, removal of the material commences. Since material of low resistance to shear stress is easily put into motion, removal of such material can occur on moderate slopes. As resistant material can only be detached by high shear stresses, its removal requires steep slopes, which maximize the shear stress (see Figures 11.13, p. 314, and 11.25, p. 328).

The manner in which fragmented material is transported from its point of origin varies with the different gravitationally motivated agents involved, as we will see in the specific chapters pertaining to each. In all cases, transportation of material requires kinetic energy in excess of the downward gravitational force that tends to hold the material against the underlying surface. The greater the kinetic energy, the greater the shear stress producing work in the form of detachment and removal of particles of solid material. However, the energy required to transport material is considerably less than that needed to detach the material and initiate its motion. This is because the particles transported have acquired their own kinetic energy in addition to that of the moving medium carrying them.

When the velocity of the transporting medium is diminished, kinetic energy also decreases, which causes the larger particles to settle out. Thus deposition begins, forming either a temporary or permanent deposit. All deposition occurs in low-energy environments where the

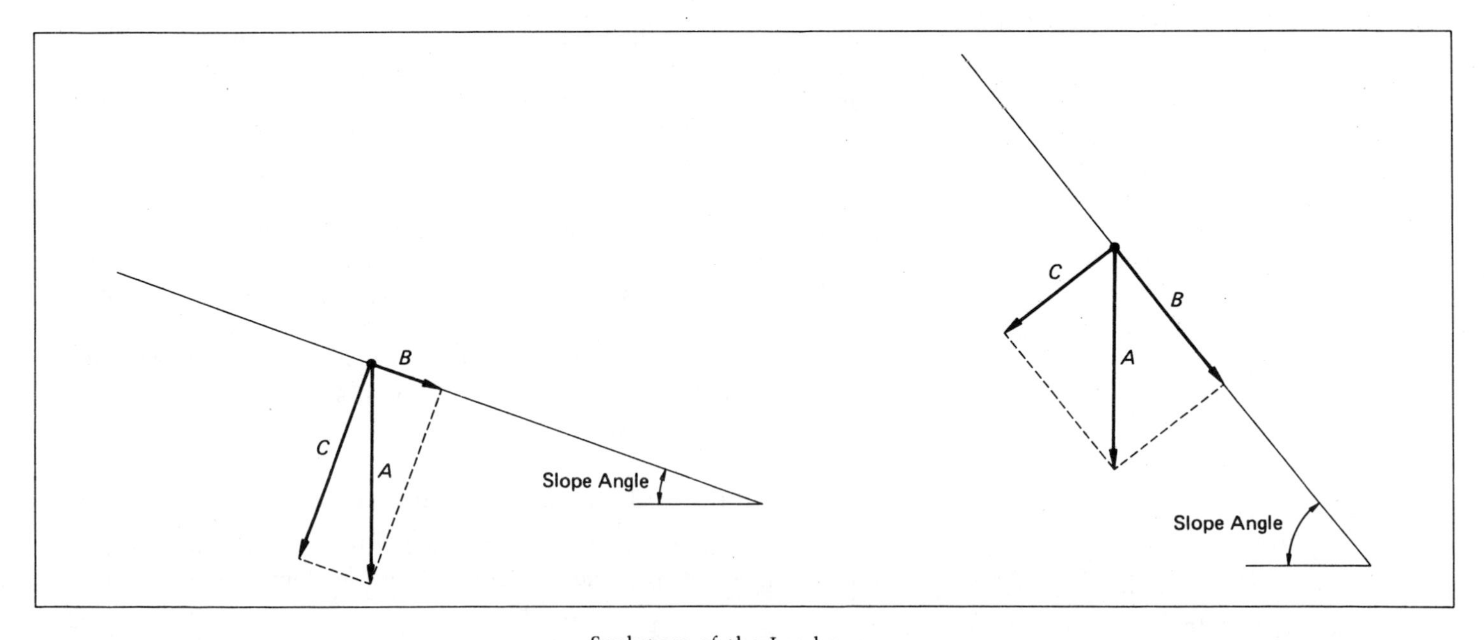

Figure 11.12 Effect of slope on shear stress. Particles of the same weight are shown resting on slopes of differing inclines. The gravitational force, *A*, exerted on the particles can be resolved into a downslope force, *B*, which tends to cause downhill motion, and a force directed into the surface, *C*, which tends to hold the material against the surface. The steeper the slope, the greater the downhill force *(B)* compared to the force resisting movement *(C)*. The particle at the left will remain in place because *C* is greater than *B*. If disturbed, the particle at the right will move downhill because *B* is greater than *C*.

Figure 11.13 In this view over the Watauga River in eastern Tennessee the steep relief on the left and in the background rises high above gentler slopes on the right due to variations in rock resistance to weathering and erosion. Steep slopes are required to effect removal of the resistant Lower Cambrian sedimentary rock on the left, whereas younger and less resistant sedimentary rock in front of the bold escarpment can be removed even on subdued slopes. Such variations in slope angle make it possible for all parts of the landscape to be eroded at similar rates over long periods of time.

velocity of the transporting medium is decreased by friction, a reduction of slope, or loss in volume of the fluid medium itself. The constant gravitational pull on the solid particles carried by the medium then gradually overcomes the declining kinetic energy of the particles, finally holding them immobile against the underlying surface. Thus beaches, sandbars, and fields of sand dunes form where waves, river currents, and air currents respectively lose kinetic energy and the ability to do the work of transporting the material available to them (see Figure 11.14).

SLOPES: THE BASIC ELEMENTS OF LANDFORMS

The basic problem in landform analysis is the explanation of the specific processes producing the slopes of varying lengths and inclinations that make up the earth's landscapes. The photographs in Figure 11.15, pp. 316–317, give examples of the diverse types of slopes encountered in different regions.

Most of the slopes we see around us have been formed by erosion beginning with vertical incision by streams. The same process that cuts gullies in a farmer's field has also excavated the Grand Canyon of the Colorado River. The general effect of stream erosion is to excavate the stream bed downward. However, valleys are expanded by other erosional processes that constantly transfer material down slopes and into streams to be carried away, ever widening the excavation made by vertical stream erosion.

Gravitational Force and Mass Transfer

The form of an erosional slope depends on the relationship between the rate at which the material composing the slope is loosened by weathering and the rate at which the loose material is transported away. The most inescapable force on our planet is gravity, and it follows that it is the gravitational effect on the materials composing slopes that directly or indirectly controls all slope development. Gravity can displace material directly, and the displacement may be rapid or slow, ranging from catastrophic rockslides to the gradual downward creep of soil and rock particles over periods of hundreds of years. The downward transfer of the material on slopes, unassisted by any fluid transporting medium, is known as *gravitational transfer* or *mass wasting*.

Slopes of soil or rock fragments or even solid rock are held together by internal friction and attractive forces between their component particles. Without such cohesive forces, hillsides would collapse under the ever present downward pull of gravity. We know that the steeper the slope, the more effective is gravity in urging materials to move downward (Figure 11.12). A slope can be stable only if the force of gravity acting on each volume of rock or soil is balanced by the cohesive forces acting within the volume. Weathering of jointed rocks on a slope can reduce the cohesive forces to the point where rocks break loose or the whole hillside avalanches downward.

For a given material on a slope—whether a soil blanket or layer of rock debris—there is a maximum *angle of repose* that the material can maintain without slipping downward. At the angle of repose, gravitational stress is just balanced by the internal friction and cohesion of the material involved. If new material is added to a slope that is already at the angle of repose, a portion of the slope may detach and slide downward. Wet soils are particularly likely to become detached from a slope because the absorbed moisture increases the weight of the soil and lessens the friction between the soil particles (see Figure 11.16, a, b, p. 318).

Where a rock cliff appears in the landscape, the slope at the base of the cliff is usually composed of *talus*, rock fragments that have fallen after having been detached from the cliff face by mechanical weathering. Talus slopes are often seen in alpine areas and semiarid climates where rock remains free of vegetation and soil and is exposed to physical weathering processes. Talus accumulates at a steep angle of repose (34 to 39 degrees); thus a slight downward push usually starts the talus sliding. As a consequence, talus can be treacherous to walk over (Figure 11.16c, p. 319).

The scar of a catastrophic landslide or the pile of broken rocks at the foot of a cliff offers clear evidence that slopes can be unstable under gravitational stress. But even the soil of an ordinary pasture slope moves downhill at a fraction of a centimeter per year in a form of mass wasting aptly known as *creep* (Figure 11.17, p. 320). Rates of creep are measured by setting pins or rods into the soil and, after an interval of months or years, observing how far the pins or rods have been displaced. Creep as slow as 1 cm per year can cause fence posts and the tombstones of hillside cemeteries to tilt visibly. Creep probably occurs on all soil-covered slopes steeper than a few degrees.

The usual explanation for the mechanism of creep is that when soil becomes moist or freezes, it expands at right angles to the slope.

Figure 11.14 The formation of desert sand dunes requires two types of transporting agents (water and wind) and two instances of loss of transportational energy. Sand is transported from its point of origin by running water. In dry regions, streams frequently wither away on open plains, allowing the sand to be deposited in the stream bed. As the stream bed dries out, the sand can be picked up and transported to a new site by strong winds. Sand accumulates in dunes when the transporting wind loses power, probably due to loss of velocity as air currents either converge or diverge. This view is in Great Sand Dunes National Monument, Colorado.

Figure 11.15 These photographs show the variety of slope types that result from varying geological materials and climatically controlled slope-forming processes (see also Figure 11.11, pp. 310–311).

(top) Stone Mountain, Georgia, is a core of unjointed granitic rock that rises boldly above jointed, chemically weathered granite in the Piedmont region of the Appalachians.

(bottom) Chemical weathering in jointed granite in arid eastern California creates a jumble of exposed rock. In the distance is the steep wall of the Sierra Nevada, its crest splintered by frost weathering; Mt. Whitney is in the center.

(opposite left) Precipitous slopes have been cut in basaltic lavas in the upland areas of the Hawaiian Islands. Strong vertical incision by torrential streams during heavy rains is accompanied by rapid sliding of the water-saturated soils on steep vegetated slopes (red scar at left).

(opposite right) Smooth rolling slopes in Devon, England, result from chemical weathering of limestone, accompanied by slow downward transfer, or "creep," of the resulting mantle of soil.

When the soil dries or thaws, however, it tends to settle downward in a vertical direction, which results in a small displacement down the slope, as illustrated in Figure 11.17a. The presence of deep-rooted vegetation on a slope tends to reduce the rate of creep because roots have a binding effect on the soil. Rates of creep on grassy slopes are usually less than 1 cm per year, with the rate being proportionate to the slope angle. The effect of the creep process is to produce smooth slopes with broadly rounded summits and a very gradual rise from the base. The creep process is generally regarded as causing hilltops to be lowered and slope angles to diminish with time.

Some slopes are composed of soils that swell and greatly multiply their weight by unusually high absorption of rain water; these soils become unstable as their added weight produces increased gravitational stress at the same time that their cohesion is reduced by waterlogging. Slope failure may then ensue in the form of a *slump*, in which a mass of soil slips downward and tilts backward in a rotational movement (Figure 11.18, p. 321). Slumps often terminate in an *earthflow,* in which a bulge of the collapsed soil is pushed out and downslope by the slump. Slumping is also common along stream banks, particularly during the falling stages of floods, when the stream banks are saturated with water.

Solifluction is a process of gravitational transfer that occurs in cold climates where silty, water-retentive soils cover ground that is permanently frozen, with only the upper meter or so thawing each summer. Water released by the thaw of ground ice cannot drain downward; thus the spaces between the soil particles become satu-

rated and the soil sags slowly downslope. Some parts of the soil move faster than others, creating a surface pattern resembling overlapping scales, as shown in Figure 11.19, p. 322. Each scale is a *solifluction lobe*. Despite their appearance, active solifluction lobes move only a few centimeters per year. Even more than the creep process, solifluction has the general effect of smoothing slopes by filling in depressions and eliminating or discouraging the development of gullies or stream channels because most of the running water moves laterally through the thawed soil rather than over its surface.

Occasionally soil and rock material absorb so much water that the solid particles become separated and flow in a viscous stream known as a *mudflow* or *debris flow*. This can occur in any climate in which all water is not permanently frozen (Figure 11.16a, p. 318). Mudflows are composed entirely of fine particles, while debris flows contain large boulders carried along by the mud. Both are a particular hazard where the vegetation cover is poorly developed, as in arid or alpine regions, or where the vegetation has recently been destroyed by fire. Mudflows and debris flows result from exceptionally heavy rains in such areas and are frequently destructive to roads, residences, and even whole towns.

The most awesome of all gravitational transfers of material are the immense *rockslides* that occur in mountainous country. These may involve several cubic kilometers of rock that fall suddenly and move with incredible speed and destructive force (Figures 11.1, p. 296, and 11.16b). Clearly some triggering force is required to produce such slope movements. Nearly all rockslides are attributed to either

Figure 11.16 (a) Weathered rock that is saturated with water has little cohesive strength, and sometimes large bodies of rock debris break loose and flow rapidly downslope in a debris flow. The debris flow in the photograph occurred in Idaho, near the headwaters of the Pahsimeroi River visible in the foreground. It has a characteristic lobed appearance, and its speed enabled it to spread out into the valley before coming to rest. Debris flows are frequently found in dry regions where sudden heavy rains saturate the soil and where the vegetation cover is too sparse to help bind the soil.

(b) Landslides are sudden detachments of large amounts of material that move downslope. The hillside scar shown in the photograph was produced in 1955 by a landslide that moved tons of soil, rock, and vegetation into Emerald Bay on Lake Tahoe at the border of California and Nevada. (See also Figure 11.1, p. 296.)

(c) **(opposite)** Frost weathering of well-jointed granite rock at elevations above 3,500 meters (11,000 ft) in California's Sierra Nevada loosens rock masses and produces rock falls, particularly during the spring, when ice is melting. This creates cones of rock rubble, known as *talus*, at the base of frost-shattered cliffs.

c

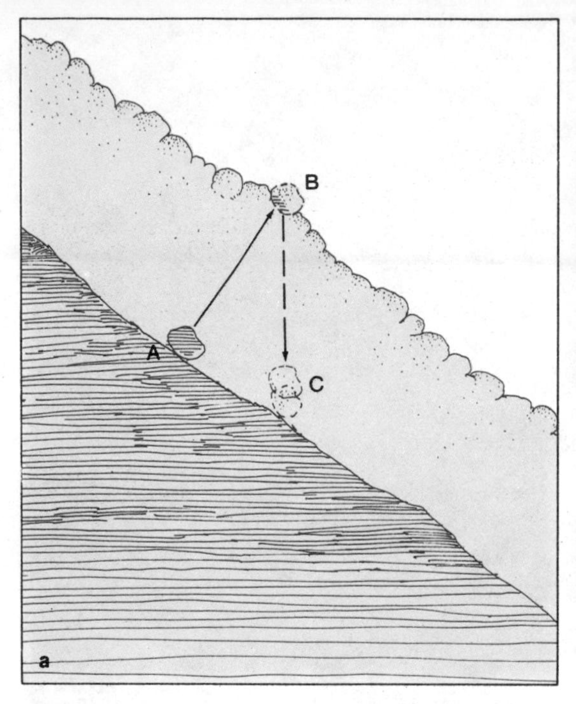

Figure 11.17 (a) Soil tends to creep downslope under the influence of gravity. This schematic diagram illustrates the mechanism responsible for soil creep. When soil becomes moist or freezes, it expands at right angles to the slope, but when the soil dries and contracts, the soil particles tend to move vertically downward, which results in a net downslope migration of the soil surface.

(b) The ice needles, or *pipkrakes,* shown in this photograph are a mechanism for moving soil downslope. When water in the ground freezes and expands outward, it carries soil particles with it at right angles to the surface. Soil particles tend to fall vertically downward or forward when pipkrakes melt, resulting in a net downslope displacement of the particles.

unusual wetting and internal lubrication of the rock or a sudden shock, as from an earthquake. When the two effects occur in combination, a major rockslide is quite likely. Mountain regions are prone to earthquakes, related to the geological movements creating the mountains, and are also periodically sodden with moisture due to heavy orographic rainfall and spring melting of mountain snowpacks. Thus the conditions necessary to produce rockslides are not uncommon in mountain regions, and many large ones have occurred within historic time. Rockslides are more important in terms of their potential destructiveness to man's works than as a creator of landforms. However, landslide scars and deposits are a part of nearly all high mountain landscapes.

Rainsplash and Slopewash

While gravity alone is fully competent to remove material from hillslopes, its effect is greatly multiplied when it sets in motion water that

Figure 11.18 **(left)** A destructive slump and earthflow in Oakland, California, destroyed some 40 homes. The structures visible have moved downslope from street level.

(right) A smaller, more typical slump and earthflow about 50 years old in California's Coast Ranges. Slump and earthflow phenomena are widespread in central California due to the presence of mechanically weak rock and heavy clay soils with high water-absorbing capacity.

can exert shear stress on the particles over which it flows. Water erosion on slopes probably removes 50 to 100 times as much material as mass wasting processes.

When raindrops fall forcefully on bare land surfaces, their impact can knock aside fine soil particles (Figure 11.20, p. 323). *Rainsplash* commonly carries particles several millimeters laterally. On slopes, rainsplash results in a net downward movement of soil particles because the majority of particles splashed into the air land somewhat downslope from their prior position. The steeper the slope, the greater the downhill displacement. Even large rocks are affected indirectly since rainsplash erosion cuts away the finer material beneath them, removing their support. A complete cover of vegetation intercepts rainfall and prevents rainsplash erosion. Thus this process is most effective in arid and semiarid climates where bare soil is widely exposed. Careful field and laboratory experiments have shown that rainsplash erosion can by itself account for the smoothly rounded character of most unforested hillcrests.

During heavy rainfalls, the rate of water arrival at the soil surface may exceed the rate of water infiltration into the soil. In this case, water builds up at the surface and begins to slide down any available slopes, initiating *overland flow* of water. The initial flow is a thin sheet. As it gains speed, the resulting turbulence soon breaks up the flow into a multitude of separate threads, or *rills.* In the initial *laminar* or *sheet flow,* water particles move in parallel paths and are capable of transporting the material already loosened by rainsplash, but channelled turbulent flow in rills is required to detach additional material in the process of *slopewash,* which is the most important process of erosion wherever rainfall is not intercepted by a thick canopy of vegetation. The erosional rills produced during periods of surface runoff commonly fill with debris between periods of heavy rain, only to be re-excavated during later downpours (see Figure 11.21, p. 324).

Many slopes that are covered with vegetation, and even some bare slopes, show little or no evidence of rill erosion. In such instances, it appears that the slopes are able to absorb most rainfall, which percolates into the soil rather than running off on the surface. This soil moisture then moves laterally downslope between soil particles as *throughflow.* Throughflow is capable of removing dissolved substances and also very fine particles that filter between the coarser fragments in the soil. Thus even subsurface throughflow is capable of removing material from slopes, and under average conditions in well-vegetated areas, it may be an important erosional process.

MODELS OF LANDFORM DEVELOPMENT

The interplay of geomorphic processes and the diversity of earth materials and geological structures have generated an immense variety of individual landforms and general landscape types. To systematize this diversity and unify the study of geomorphology, it has seemed necessary to find some common thread or principle in the development of all erosional landforms. As a result, several comprehensive theories of landform development have been created. Three different approaches stand out above the rest, each of undoubted value. These three theories focus, respectively, on the effects of time, tectonics, and the equilibrium tendency in landform development.

The Cycle of Erosion Concept

Figure 11.19 Solifluction lobes, shown here on a slope on the Seward Peninsula, Alaska, develop where silty, water-retentive soils cover permanently frozen ground. The lobes exhibit irregular margins because the rate of downslope movement of the thawed soil varies from point to point.

At the close of the nineteenth century, the theory of organic evolution presented by Charles Darwin in 1859 was finally becoming accepted. At this opportune moment, the soon-to-become-famous American geographer William Morris Davis wrote a series of essays in which he proposed that landforms could be described and analyzed in terms of their evolutionary development. An analogy may help us appreciate Davis's thinking. Imagine a scientist from another solar system visiting earth for the first time, with only a short time to determine the development of human life from infancy to old age. With no time to follow an individual through a full life span, an alternative approach would be to differentiate infants, children, teenagers, adults, and the aged, then to work out a scheme of evolutionary development from one stage to the next. Similarly, geomorphic processes act far too slowly for landform changes to be apparent over the time span of human history, to say nothing of the life of a single observer. Thus Davis examined existing landforms and classified them according to their stage of development, using the terms "youth," "maturity," and "old age." He then inferred how the landforms of one stage would develop into those of the next stage over the course of time.

Davis called his ideal cycle of development the *geographical cycle,* but today it is commonly known as the *Davisian cycle,* or the *cycle of erosion.* To explain the cycle of erosion in the simplest way,

Davis visualized a sudden uplift of the land surface from a lower to a higher altitude and assumed that uplift occurs too quickly for erosion to have a marked effect on landscape form. Once the uplift was complete, however, erosion began shaping the land (Figure 11.22, p. 325). It has subsequently been discovered that the rate of uplift in geologically active regions actually does well exceed the rate of erosion. Thus, in some instances, cycles of erosion probably are initiated more or less as Davis imagined they were.

Davis first concentrated his attention on the erosive action of flowing water. Rainfall on the uplifted block of land initiates erosion as water flows over the surface and down toward the sea. Davis pictured streams cutting downward to form deep, steep-sided river valleys, with larger areas of the original land surface remaining between the new valleys. He called this deeply incised landscape the *youthful* stage of landform evolution.

As rivers deepen their valleys, they decrease their altitude above the sea. This reduces their potential energy and therefore the kinetic energy available for erosional work. Drawing upon the ideas of the first scientific explorer of the arid western United States, John Wesley Powell, Davis called the level below which rivers could not cut downward the *base level of erosion.* Water at base level has no gravitational

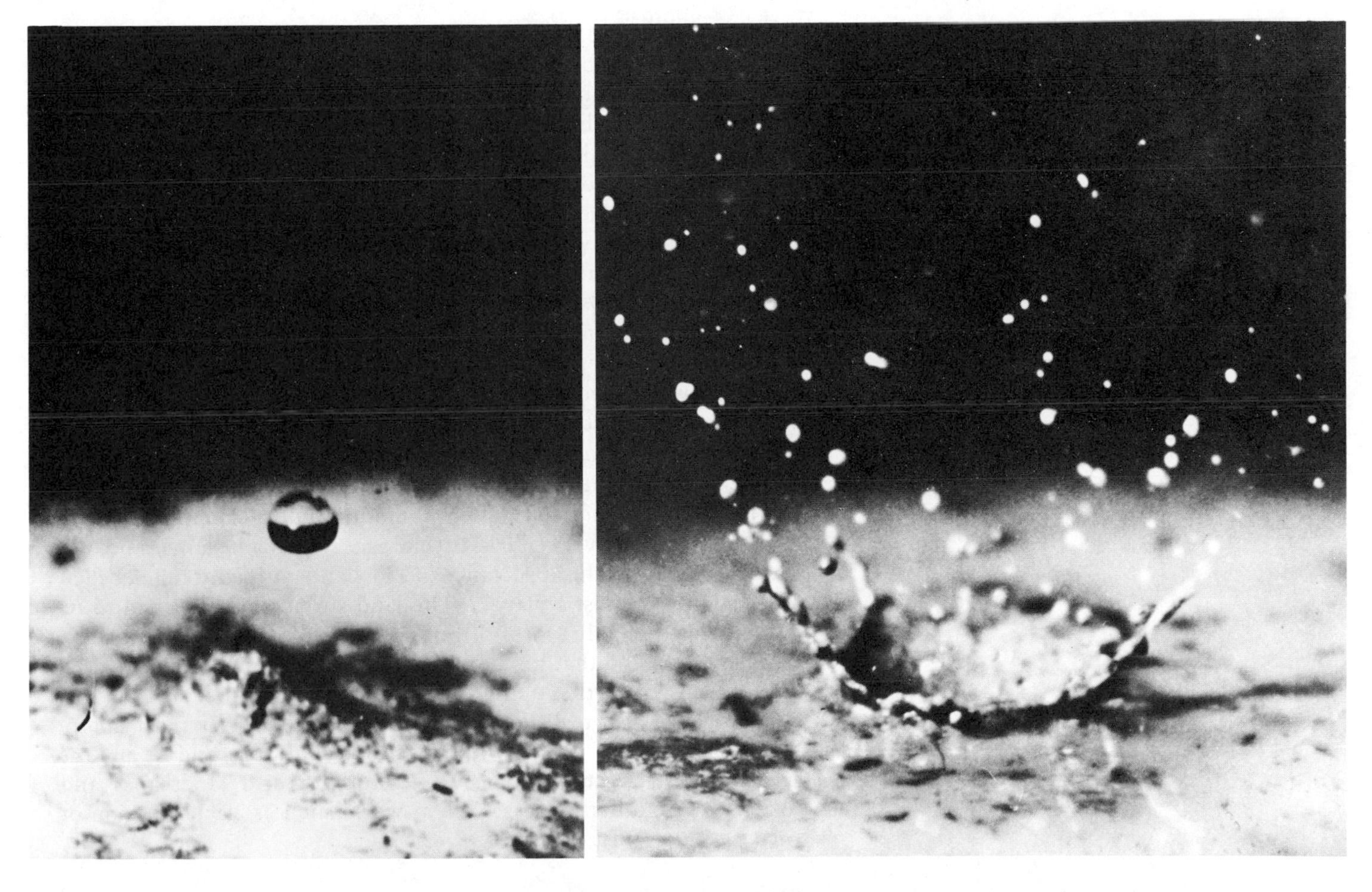

Figure 11.20 These close-up, high-speed photographs of a raindrop splashing into bare soil show graphically the work that can be done by water.

(left) The raindrop is 3 mm in diameter and is traveling at a speed of 11 meters per second.

(right) When the drop strikes the ground, particles of soil are thrown several centimeters by the impact. The impact of raindrops on a bare slope tends to produce a net movement of soil particles down the slope.

Figure 11.21 Rill erosion on bare slopes in Death Valley, California. The transition from rills to gullies is clear at the left of the photo. Rills of this type are rarely seen on slopes having vegetative protection, but they develop in all climates where vegetation is absent or has been artificially removed from slopes.

energy to do the work of erosion. An entire landscape can be eroded down to base level by running water, but it can be eroded no farther than that unless a new uplift occurs. Sea level is the ultimate base level, but streams that drain into closed interior basins may have a base level higher or lower than sea level, depending on the altitude of the basin. In late youth, according to Davis, the vertical incision of rivers would become more gradual as base level was approached, and would finally give way to lateral stream erosion, the expansion of valley floors, and the appearance of flat flood plains.

In the stage Davis called *maturity*, the original uplifted land surface is reduced to rolling hills and flat-floored valleys with flood plains continuously veneered by stream-deposited alluvium. Further evolution would proceed very slowly, requiring tens of millions of years. The *old age* stage, as Davis imagined it, is largely hypothetical; in it erosion has proceeded so far that the land has been worn down close to sea level, with only a few isolated areas of high ground, which Davis called *monadnocks* after a mountainous example in New Hampshire. The river valleys in old age are broad plains, and the landscape presents the appearance of a lowland with faint relief, called a *peneplain* ("almost a plain"). The land surface has returned to its hypothetical pre-uplift form through a complete cycle of erosion.

Davis stressed that his simple model would be modified in actuality, perhaps by further uplift during one of the intermediate stages of evolution. He also realized that this initial model was most appropriate for climates where flowing water was abundant, and he and his students devised other cycles of erosion for arid landscapes and regions where other processes are at work, such as glacial erosion or wave action against coasts.

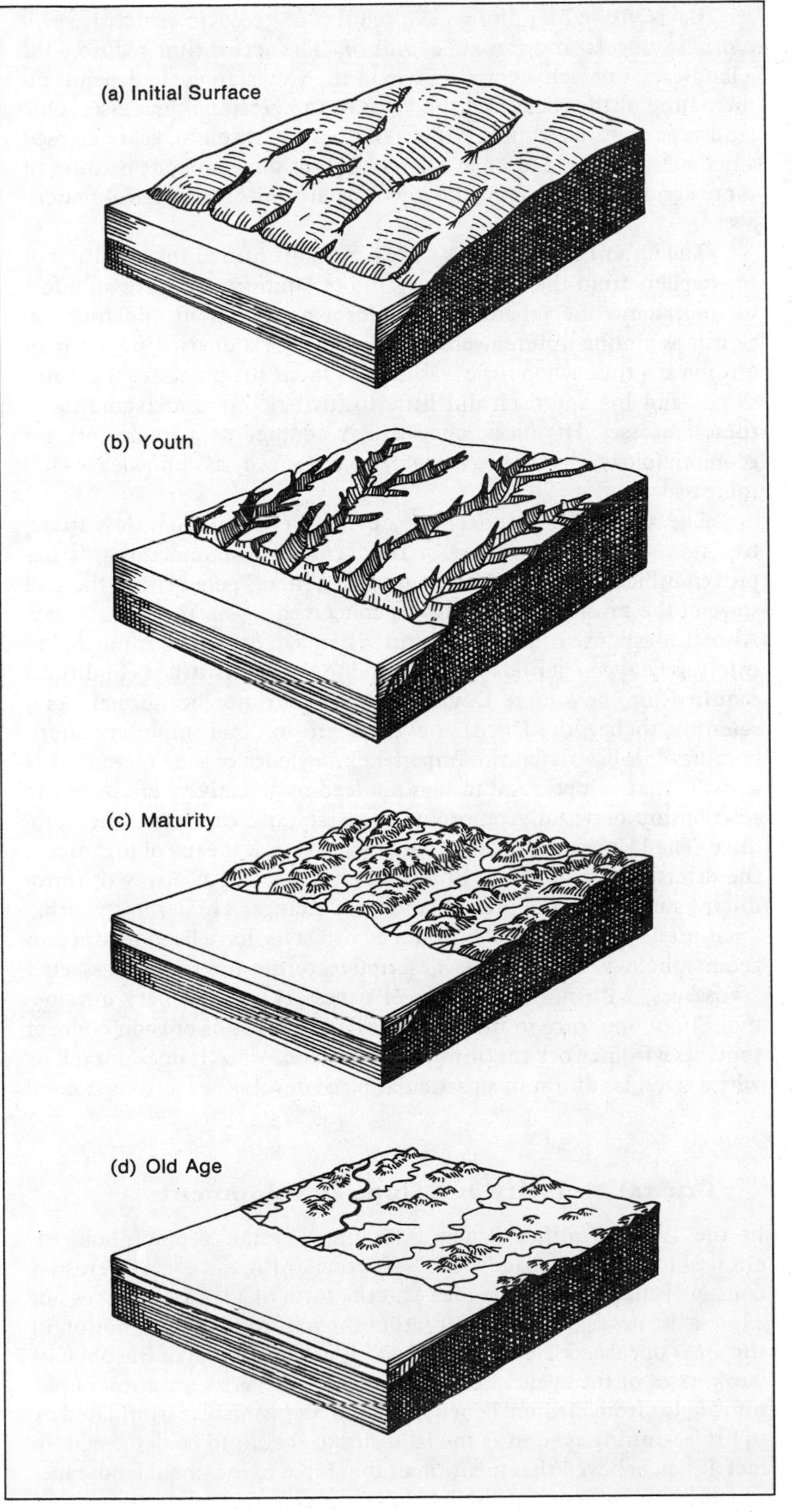

Figure 11.22 This sequence of diagrams illustrates stages of the Davisian cycle of landscape evolution. This cycle applies to moist regions where erosion is accomplished primarily by flowing water. The diagrams assume that the underlying rock is uniform and exerts no controls over landform development.

(a) The initial surface is a landscape of low relief. After uplift of the region, streams have energy to begin cutting downward.

(b) In the stage of youth, the streams have cut comparatively narrow valleys downward, and much of the initial surface is preserved in the high land between the valleys.

(c) In late youth and early maturity, runoff and the creep process have worn the hill slopes to rounded ridges. The streams have widened their valleys, and little trace of the initial plain remains.

(d) In old age, mass wasting and the work of flowing water have worn the region to a plain of low relief *(peneplain)* with isolated hills, or *monadnocks*. (After Longwell, Knopf, and Flint, 1941)

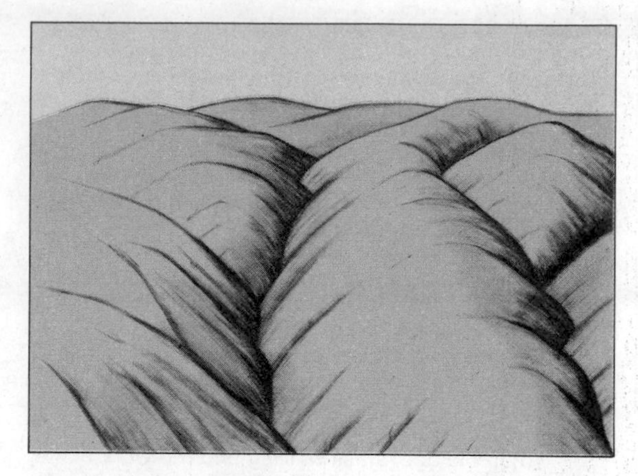

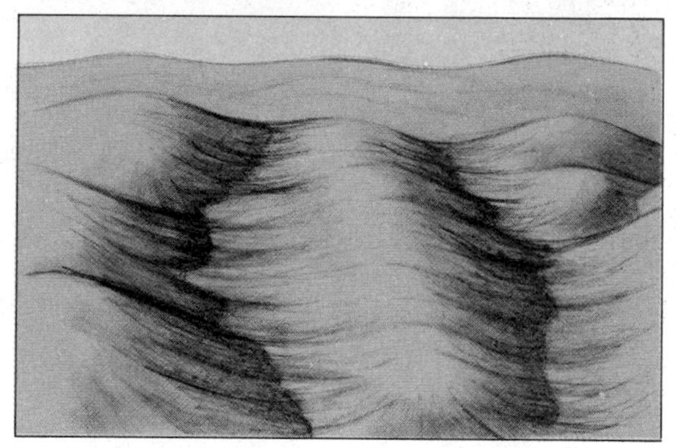

Figure 11.23 According to Walther Penck, these contrasting slope forms indicate varying tectonic conditions in the regions where they occur.

(top) Convex slopes indicate an increasing rate of stream incision due to accelerated tectonic uplift.

(center) Straight slopes suggest a uniform rate of stream incision, either at the maximum possible rate or due to steady uplift.

(bottom) Concave slopes indicate a weakening of stream incision due to a deceleration or cessation of tectonic uplift.

Davis viewed landforms as products of *geologic structure, geomorphic process,* and *stage of evolution.* The actual time required for a landscape to reach a certain stage in the Davisian cycle depends on the nature of the rocks and the vigor of the erosional processes. One landscape may be younger than another in absolute years elapsed since uplift, but it can be at a later stage of development because of more aggressive erosion or the low resistance of its geological materials.

The importance of Davis's work is that it turned the attention of geographers from the pure description of landforms toward an effort to understand the progress of landform development and the relationships among different landforms. But Davis devised his cycle of erosion at a time when little was known about the processes that form slopes, and his approach did little to further our understanding of these processes. His ideas, once widely adopted as a framework for geomorphology, have been severely criticized as knowledge has increased.

The fundamental concept of a cycle of erosion seems less attractive now than it did earlier in this century. For one thing, it has proven difficult to find clear examples of intact peneplains—the end stage of the erosion cycle as Davis conceived it—on the earth today. Also, the response time of landform systems may be long compared to intervals between periods of uplift, and so the undisturbed conditions required for the simple Davisian model may not be normal. As a scientific theory, the Davisian cycle of erosion had important shortcomings. It is based upon an imperfect knowledge of the processes that actually shape slopes, and it does not lead to objective analysis of the relationship between geomorphic processes and the forms they produce. The Davisian cycle is of value mainly as a means of organizing the diversity of landscapes in humid climates and as a way of introducing students to the idea of landscape change. The terms "youth," "maturity," and "old age," introduced by Davis, have been retained in geomorphology, but only as descriptive terms for stream-dissected landscapes, with no implication of necessary evolutionary development from one stage to the next. While the cycle of erosion concept provides a framework for thinking about landforms, it does not tell us why a given landform in a particular place develops exactly as it does.

Crustal Mobility and Slope Development

In the 1920s, Walther Penck, a young German geomorphologist, engaged in a lively debate with Davis concerning the cycle of erosion concept. Penck stressed the idea that the form of a landscape does not express the passage of time, but rather the nature of crustal motion in the area concerned. He concentrated his attack on one of the obvious weaknesses of the cycle of erosion concept—Davis's separation of tectonic uplift from erosion. Penck argued that erosion was stimulated by uplift, beginning as soon as the land surface began to be deformed. In fact, Penck believed that the form of the slopes in erosional landscapes was the best evidence of the local rate of tectonic deformation.

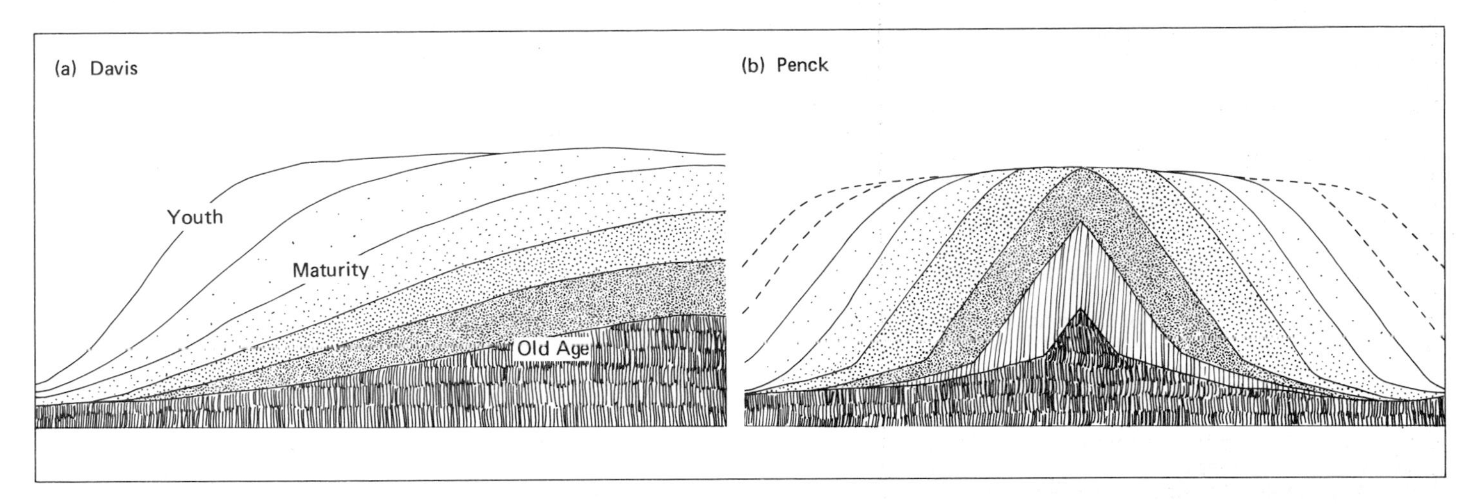

Figure 11.24 (a) According to William Morris Davis, the evolution of a slope proceeds as shown in this diagram. As the slope retreats, it becomes less steep and more rounded. This type of slope development appears to occur when material is transported down a slope more rapidly than it can be removed from the base of the slope by streams or other agents of erosion.

(b) According to Walther Penck, the evolution of a slope through successive stages occurs by parallel retreat, in which the angle of the steep slope remains essentially constant while a new slope of gentle inclination expands upward at the foot of the steep slope. For this type of slope development to occur, the removal of material from the base of the slope must keep pace with the rate of transport of material down the slope face.

Penck pointed out that three slope types are produced by erosion: slopes that are straight, those that are convex, and those that are concave. Penck reasoned that straight slopes indicated steady rates of downward stream incision and removal of material from slopes. This he termed *uniform development,* which he felt was due to a uniform rate of tectonic uplift. Convex slopes become progressively steeper downward. Penck suggested that this indicated an increasing rate of downward stream erosion, or *waxing development,* which was a response to increasing intensity of tectonic uplift. Concave slopes seemed to indicate a decreasing rate of stream incision, or *waning development,* due to diminishing tectonic activity (see Figure 11.23).

A corollary to his interpretation of slope form as an index of tectonic movement was Penck's recognition of the fact that the steep portion of an eroding slope may maintain a constant angle as material is removed from it, rather than becoming less steep as Davis suggested (see Figure 11.24). In Penck's landscape model, steep slopes may be present even in much denuded landscapes that have little erosional energy due to an absence of tectonic activity. This is in sharp contrast to the ideal Davisian peneplain, where the creep process is imagined as wearing summits down and constantly reducing slope angles. In fact, many erosional landscapes, especially those of semiarid and desert regions, accord much better with Penck's model than with the Davisian view of landscape formation. It is now widely accepted that the steep upper portions of slopes do maintain constant angles as they retreat in areas where chemical weathering is weak and erosion by overland flow greatly exceeds removal by mass wasting processes. On the other hand, slopes do seem to become progressively less steep, with steady lowering of hillcrests, in humid areas covered by a creeping mantle of soil that is protected by forest vegetation.

Penck's interpretation of hillslope form as an expression of tectonic activity has been less well received. Concave and convex slope segments may appear temporarily due to changes in erosional intensity, but probably cannot persist in the landscape as Penck suggested they would. The major difficulty is that varying slope angles would have to form and be preserved over long periods of time in rocks that are uniform in resistance to erosional stresses. This is hard to explain

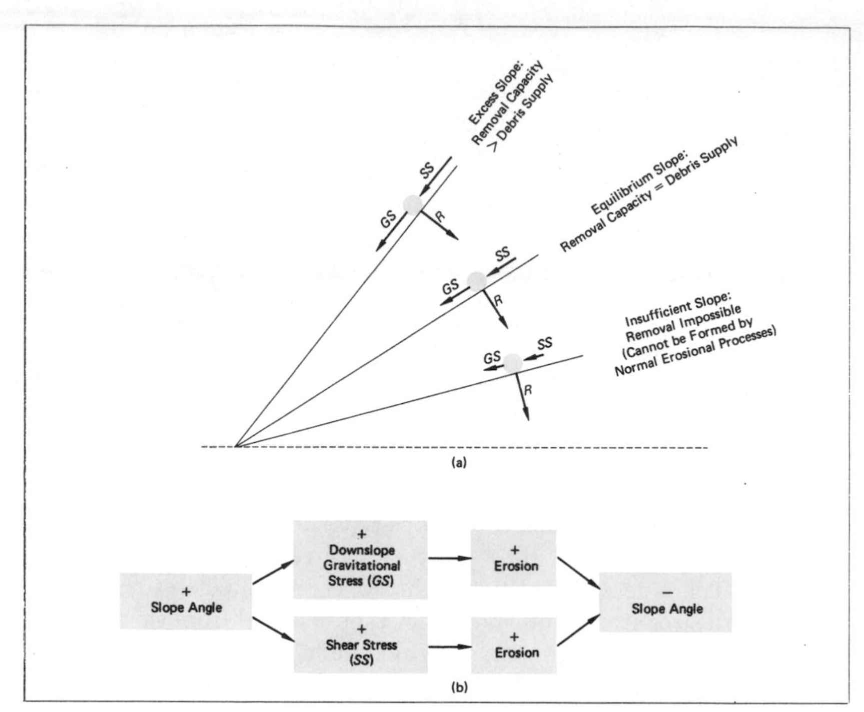

Figure 11.25 Principles of equilibrium slopes.

(a) Hypothetical slope angles indicate downslope gravitational stress *(GS)*, shear stress *(SS)* exerted by the gravitationally driven erosional agent (such as slopewash), and resistance to detachment of slope particles *(R)*. Resistance remains the same at varying slope angles, while *GS* and *SS* diminish with decreased slope. The steepest slope is unstable, as the sum of *GS* and *SS* far exceeds *R*, causing rapid erosional removal. The intermediate slope is stable, as the sum of *GS* and *SS* just balances *R*, so that removal can occur without changing the slope angle. The lowest slope cannot be produced by the stresses shown, as particle resistance *(R)* exceeds erosional stresses, so that erosional removal is impossible.

(b) Diagrammatic representation of negative feedback relationship that causes slopes to be self-adjusting toward angles that equate stress to resistance. Any increase in slope increases the erosional stresses, which in turn decreases the slope angle until equilibrium is reestablished.

in terms of the mechanics of erosion. Unfortunately, Penck's geomorphic model is similar to Davis's in that it is deductive rather than based on careful study of the slope-forming processes themselves. It is a great misfortune that Penck was not able to answer criticisms of his theory, for he died at the age of 35, his major contribution being published posthumously from his notes.

Equilibrium Theory of Landform Development

Since about 1950, geomorphologists have increasingly emphasized the actual mechanics of landform development, focusing attention on the specific relationships between form and process in the landscape. In the previous section on the energetics of erosion and deposition, we saw that most slopes could be explained in terms of applied force versus the resistance of the materials affected. When disruptive forces overcome material resistance, the material is fragmented and set in motion, and erosion commences. But what is the effect of erosion itself on the balance between stress and resistance?

If erosion reduces the slope angle, it has the effect of reducing the downslope gravitational stress on the material forming the slope (see Figure 11.12, p. 313) as well as the shear stress exerted by the erosional agent. How far can erosion reduce a slope angle? Certainly not below the steepness required for removal of the debris produced by weathering on the slope. Conversely, any increase in slope angle will increase the downslope gravitational stress and the energy of the erosional agent, producing more vigorous erosion. This soon reduces the slope to the angle at which the erosional stress is just adequate to remove the specific type of material supplied by weathering (see Figure 11.25). This is a good example of *negative feedback,* in which disturbance of a system that is in equilibrium triggers changes that tend to restore the original system. Negative feedback is the principal means of maintaining equilibrium in physical systems and is a normal feature of most landform systems. Negative feedback makes the slope system a self-regulating one that tends to maintain a steady state through time, with changes in form occurring only as a consequence of changes in the nature or intensity of the removal processes. A slope that tends to maintain just sufficient steepness to equate applied stress to material resistance is an *equilibrium slope.* The tendency of all slope development seems to be in this direction, although an exact balance between stress and resistance is probably rarely achieved.

Whenever there are materials of varying resistance to erosion, there must be varying slopes—steep ones where the material is most resistant to detachment and removal, and gentle ones where the material succumbs readily. Since slope angle tends to equate stress to resistance, it is possible for the whole landscape, tough rocks and fragile ones, to wear away at approximately the same rate over long spans of time.

The equilibrium theory of landform development suggests that slope forms are adjusted to geomorphic processes in a very delicate way, such that the gravitational energy provided is just adequate for the work to be done. Any change in energy or the nature of applied stresses causes an adjustment in form that will reestablish the equilibrium of the process-form relationship. Similarly, any alteration in form unrelated to the normal operative stresses, such as changes produced tectonically or by human interference, causes an increase or decrease in erosional energy, which will in turn tend to reestablish the original equilibrium relationship. Thus it is impossible to modify either landforms or geomorphic processes without triggering a reaction in the natural system, which tends to restore an equilibrium relationship between process and form.

So far we have been speaking of equilibrium in the process-form relationship. Some depositional features are equilibrium forms in another sense: they persist because erosional removal from them is balanced by arrival of new material (Figure 11.26, top). Beaches, river sandbars, talus cones, river deltas, and volcanoes are all examples. Such features enlarge or shrink depending upon the balance between input and erosional loss of material. But some, such as beaches, have achieved semipermanent configurations that over long periods of time balance erosional losses to arrivals of new material. Thus beaches are in a delicate state of equilibrium that is easily upset by modification of any part of the beach system, from the sediment source to the waves and currents that act upon the beach itself and the coast at some distance from it.

The great value of the equilibrium concept of landform development is its focus on the exact relationship between geomorphic processes and surface forms. This approach to landform analysis is more rewarding in terms of understanding the development of landscapes than is either the cycle of erosion concept or Penck's tectonic model

a

b

of landform development. Consequently, the equilibrium model has largely superseded both of these earlier approaches.

In particular, the equilibrium theory has focused attention on landforms that may be in a steady-state condition, that is, not changing progressively in form though constantly being affected by erosion or deposition. A hill region with sharp ridge crests and narrow ravine-like valleys is one of the world's most common landscapes. If equilibrium slopes reach from the ridge crests to the valley floors, stream downcutting will cause removal from the entire landscape with no change in form occurring. Downward valley erosion could go on indefinitely if the landscape is slowly but continuously uplifted, resulting in a landscape in a steady state that is independent of time (see Figure 11.26, bottom). Whenever there is a continuous input of energy or material into a geomorphic system, a time-independent landscape is the result.

Figure 11.26 Two different types of time-independent equilibrium landforms.

(a) The maintenance of the beach at Cadiz, Spain, requires that sand arrivals balance sand losses. Seasonal fluctuations in the sand budget occur, but there is no progressive change over the years, indicating that material income and outgo are in balance.

(b) The sharp-crested ridges in Oregon's Cascade Mountains will not change form as long as downward stream erosion continues due to slow tectonic uplift. In this case, slope erosion causes the sharp ridge crests to be lowered at the same rate as the valley floors.

The Tempo of Geomorphic Change

Landscape form is the outcome of the contest between the opposing tectonic and gradational forces. What are the rates of tectonic movement and gradation today?

Most measurements of the rate of erosion are based on measurements of the total amount of sediment carried by all the streams that flow out of a drainage system. In large drainage basins of moderate relief, such as the basin drained by the Mississippi River, the current rate of erosion is estimated to be about 5 cm (2 in.) per 1,000 years. It is known, however, that this is not the natural rate of erosion that has prevailed over time, but rather reflects human disturbance of the natural landscape in the past century. In high mountain regions with steep slopes, natural erosion rates may be as high as 1 meter (3 ft) per 1,000 years.

In general, the rate of uplift in mountainous regions is much greater than the rate of erosion. Repeated measurements using precise surveying methods indicate that the rate of uplift in regions of mountain growth is about 5 to 10 meters (15 to 30 ft) per 1,000 years. It is interesting that these measurements seem to justify Davis's model of rapid uplift followed by erosional development of the landscape.

Most landscapes are produced by repeated small increments of change that occur frequently over long periods of time. A mountain range is uplifted a meter or less every century or two. Though small in amount, these movements are frequent on the time scale of a million years or more required for mountains to form. On a different time scale, the work of erosion and transportation by flowing water is performed mainly during relatively frequent floods that occur one or more times a year. Measurements of the sediment loads carried by streams under different conditions of flow indicate that most of their load is transported during the five to ten days of highest flow each year. Only in rare high floods can streams loosen the largest boulders in their beds and carry them to new resting places downstream, or erode their beds downward. Such events are so uncommon that more work is actually performed by events of moderate magnitude that occur relatively frequently.

SUMMARY

The surface of the earth displays relief features on several scales. The continents and ocean basins are features of first-order relief, and major mountain ranges, plateaus, and depressions are second-order relief features. Most third-order features, or landforms, such as valleys and hills, are sculptural details of the second-order relief.

Every feature of the earth undergoes change and development. The drift of lithospheric plates and spreading of the sea floor have moved landmasses and opened ocean basins. New oceanic crust is formed by volcanism within the oceanic ridge systems, and old oceanic crust is recycled into the earth's mantle by subduction into oceanic trenches. The convergence of lithospheric plates has raised mountain ranges along the plate boundaries. Landforms undergo change and development as a consequence of gravitational force, the internal energy of the earth, and external solar energy. The earth's internal energy expresses itself in tectonic activity, including displacements of the earth's crust and volcanism (see Chapter 13). Solar and gravitational energy are expressed on the land surface principally by the processes of gradation, including weathering, erosion, and deposition, which tend to wear down projections and fill in depressions on the earth's surface. Gradational agents that shape landforms by erosion and deposition include running water (see Chapter 12), wind (see Chapter 14), glaciers (see Chapter 15), ocean waves (see Chapter 16), and direct gravitational transfer.

The diversity of landforms arises from the complex combination of geological materials, gradational and tectonic processes, and time. Landforms may be studied in terms of energy concepts and the mechanics of their formation, particularly by analyzing the relationship between erosional stresses and the resistance of geological materials. Slopes are the basic elements of landforms, being produced mostly by erosion. The main processes affecting slopes are gravitational transfer (often called mass wasting), rainsplash, slopewash, and throughflow. Gravitational transfer may be slow and imperceptible or rapid and catastrophic.

The importance of structure, process, and stage in the development of landforms was recognized by William Morris Davis, but his cycle of erosion concept is inadequate to deal with detailed questions of form and process in the creation of landforms. Walther Penck stressed the importance of tectonic influences on slope form as an alternative to the cycle of erosion concept, and his theory of slope retreat at constant angles seems to be valid under certain conditions. The equilibrium theory of slope development focuses on the adjustment of landform to geomorphic process, resulting in equilibrium forms that equate erosional stress to material resistance.

The shaping of landforms normally occurs through the accumulation of small increments of change. Although infrequent catastrophic events can produce spectacular results, landscapes are modified principally by events of moderate intensity that occur with considerable frequency.

Further Reading

Bloom, Arthur L. *The Surface of the Earth.* Englewood Cliffs, N. J.: Prentice-Hall, Inc., 1969. (152 pp.) This up-to-date and well-written paperback outlines the processes of landform development—brief but unusually good.

Brunsden, Denys, and **John Doornkamp, ed.** *The Unquiet Landscape.* Bloomington: Indiana University Press, 1974. (171 pp.) This is a magnificently illustrated collection of articles on various aspects of landform development, from a series appearing in the British periodical *The Geographical Magazine.*

Butzer, Karl W. *Geomorphology from the Earth.* New York: Harper & Row, 1976 (463 pp.) Different climatic regions are discussed in this recent textbook on landform development. It is not highly technical and uses a geographical approach.

Calder, Nigel. *The Restless Earth.* New York: Viking Press, 1972. (152 pp.) This is a most interesting exploration of the implications of recent discoveries relating to global tectonics, including the sea floor spreading phenomenon.

Carson, M. A., and **M. J. Kirkby.** *Hillslope Form and Process.* London: Cambridge University Press, 1972. (475 pp.) The authors provide a thorough and somewhat technical treatment of hillslope processes and resulting forms, intended for the serious student.

Davis, William M. *Geographical Essays.* New York: Dover Publications, 1954. (777 pp.) This collection of Davis's seductively written early papers on the cycle of erosion and associated topics was originally published in 1909.

Eckel, E. B., ed. *Landslides and Engineering Practice.* Washington, D.C.: Highway Res. Bd., Spec. Rpt. 29 (NAC-NRC Publ. 544), 1958. (323 pp.) This is a collection of papers, some well illustrated, on problems of slope engineering and various types of downslope gravitational transfers (slumps, earthflows, landslides).

Review Questions

1. What is the difference between the theory of plate tectonics and the continental drift hypothesis conceived earlier by Alfred Wegener?
2. What evidence from the sea floors supports the idea of sea floor spreading?
3. What are the various erosional agencies? How does gravity provide the energy for each of these?
4. One can safely say that all of the world's lakes, wherever they are found, have been created quite recently on the geologic time scale. What tells you that this statement must be true?
5. How is it possible that different gradational processes are dominant in different parts of the earth? Contrast the humid tropics and the arctic tundra regions with the Atlantic seaboard of the United States in terms of variations in gradational processes.
6. What is meant by the response time of a landform system?
7. What energy transformations occur (a) in the case of a rockslide that blocks a valley, damming the stream in the valley? (b) in the case of a wave whose impact against a sea cliff removes enough material from the base to leave the higher part of the cliff without support?
8. It has been noted that for a beach to persist, sand input must be equivalent to erosional loss. What other landforms can you think of in terms of a materials budget?
9. Davis's cycle of erosion concept has been criticized on grounds that a full cycle from youth to old age could rarely be completed. What two different factors could disturb the course of an erosional cycle?
10. According to Walther Penck, what should be the dominant slope form in areas where lithospheric plates are presently converging?
11. What factor allows erosional stress and material resistance to be equated throughout an area of diverse rocks?
12. Why do geomorphic events of moderate magnitude produce more total work in landscape development than events of exceptional magnitude? Do you think this is true of *all* geomorphic systems?

Leopold, Luna B., M. Gordon Wolman, and **John P. Miller.** *Fluvial Processes in Geomorphology.* San Francisco: W. H. Freeman & Co., 1964. (522 pp.) Advanced but very readable, this book offers a modern approach to the study of river and hillslope processes. It is important as a summary of many years of laboratory and field research by the U.S. Geological Survey.

Lobeck, A. K. *Geomorphology: An Introduction to the Study of Landscapes.* New York: McGraw-Hill Book Co., 1939. (731 pp.) Although dated, this text is included here because of its excellent photographs, maps, and diagrammatic illustrations of landforms of all types.

Penck, Walther. *Morphological Analysis of Landforms.* London: Macmillan Co., 1953. (429 pp.) This is the major source on Penck's theories, originally published in German in 1924 from his notes (following his early death), and only partially complete. It is not especially technical but heavy reading nevertheless.

Schumm, Stanley A. "The Development and Evolution of Hillslopes." *Journal of Geological Education,* Vol. 14:3 (1966): 98–104. A leading investigator in the field summarizes the major problems of hillslope formation, with suggestions as to their resolution.

Shelton, John S. *Geology Illustrated.* San Francisco: W. H. Freeman & Co., 1966. (434 pp.) This book is unsurpassed for crisp photographic illustrations of landforms. The text and organization are more imaginative than most—highly recommended, not advanced.

Strahler, Arthur N. "Dynamic Basis of Geomorphology." Geological Society of America, *Bulletin,* Vol. 63 (1952): 923–938. This article was a pioneering statement of the scientific principles of landform development in terms of mechanics and fluid dynamics—a historical landmark following upon the Davisian era.

Wilson, J. Tuzo, ed. *Continents Adrift and Continents Aground.* San Francisco: W. H. Freeman & Co., 1976. (230 pp.) This is a collection of seventeen superbly illustrated articles on plate tectonics originally published in *Scientific American.*

The Pass Through the Mountains by Fukaye Roshu. (The Cleveland Museum of Art, John L. Severance Fund)

Flowing water is one of the most important agents of landscape change, even in arid regions. The energy in moving water empowers it to pick up, carry, and redeposit rock waste. The landscape expresses the contest between gradation by flowing water and uplift caused by geologic processes.

TWELVE
Fluvial Processes and Landforms

THE CHANNELLED FLOWS OF water that move across the land mingle their currents in ever-larger streams that finally produce great rivers. These flows return the land's surplus water to the seas, and, along the way, they serve us in important ways. Rivers provide very cheap transportation, and they are a source of water for our croplands, households, and industries. They can be used to generate power, and they even carry away our sewage, though at increasing cost to their other uses. The utility of rivers is obvious in the fact that ever since the dawn of civilization, the world's greatest cities have grown up beside them: Babylon on the Euphrates, Nineveh and later Baghdad on the Tigris, early Memphis and later Cairo on the Nile, Rome on the Tiber, Paris on the Seine, London on the Thames, New York on the Hudson. It is indeed a challenge to name great inland cities that are *not* situated on sizable streams. The last human works visible through the window of an outward-bound rocket, as the earth falls away below, are the reservoirs created where man has dammed streams, impounding these surplus waters so as to extract every last bit of use from them.

But surplus waters do not only serve mankind. From an altitude of 100 miles, human works are almost invisible, but the effects of fluvial processes (from Latin *fluvius,* meaning river) are conspicuous over all the land. One can discern hills intricately dissected by valleys produced by running water, large streams collecting the runoff of the earth's highlands and moving it to the seas, and the mouths of these streams, with plumes of sediment issuing from them and discoloring the sea, graphically displaying the vital role of rivers as sculptors of the earth and as conduits for the immense volumes of sediment ceaselessly poured from the land into the seas.

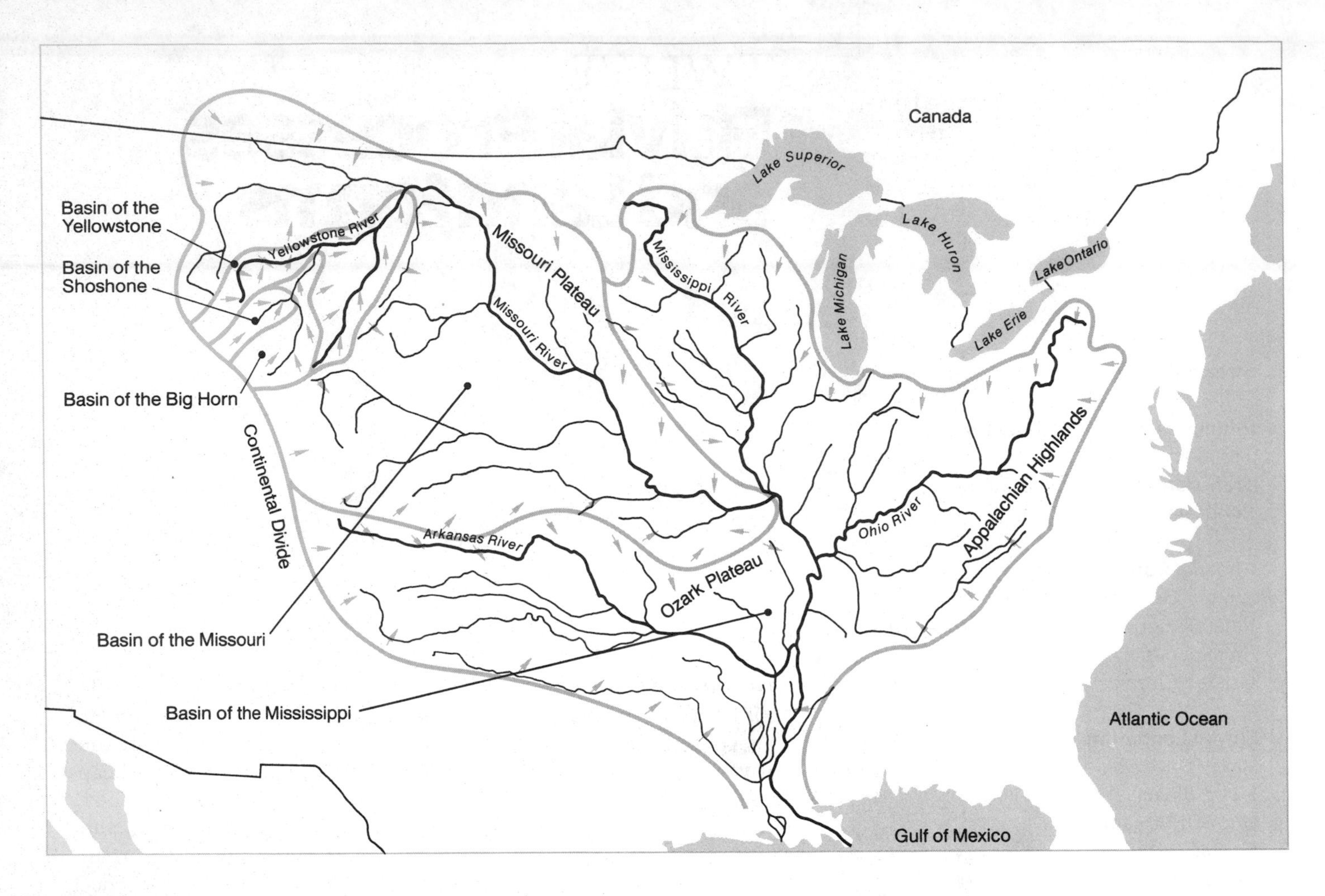

Figure 12.1 The drainage basin of a large river such as the Mississippi contains a hierarchy of smaller nested drainage basins. In this figure, the basin of the Shoshone River is part of the basin of the Big Horn, which is part of the basin of the Yellowstone, which feeds into the Missouri, which is the major tributary of the Mississippi. The arrows indicate the downhill direction of water flow into the various basins. The Continental Divide along the Rocky Mountains separates basins draining eastward toward the Gulf of Mexico from those draining westward toward the Pacific Ocean.

STREAM ORGANIZATION

Only a few streams, such as the Nile River, flow for great distances without being joined by other streams. Most streams are part of a larger hydrographic entity, the *drainage basin,* which is the area drained by a stream and all its tributaries (see Figure 12.1). The scale of a drainage basin may range from a square kilometer or less to a sizable portion of a continent. The pattern of the streams in a drainage basin is called the *drainage net.* Drainage nets differ in texture and pattern. *Drainage texture* refers to the total number or the total length of stream channels per unit of area. Some basins have a very fine (high) drainage texture, indicating that a large proportion of the precipitation runs off the land surface. The *drainage pattern* is largely a reflection of the type and arrangement of the bedrock units composing the drainage basin. The most common patterns are illustrated in Figure 12.2.

A feature common to all drainage nets is that numerous small streams feed into successively larger streams. In 1945 Robert Horton, an American hydraulic engineer, formulated a way to study the orga-

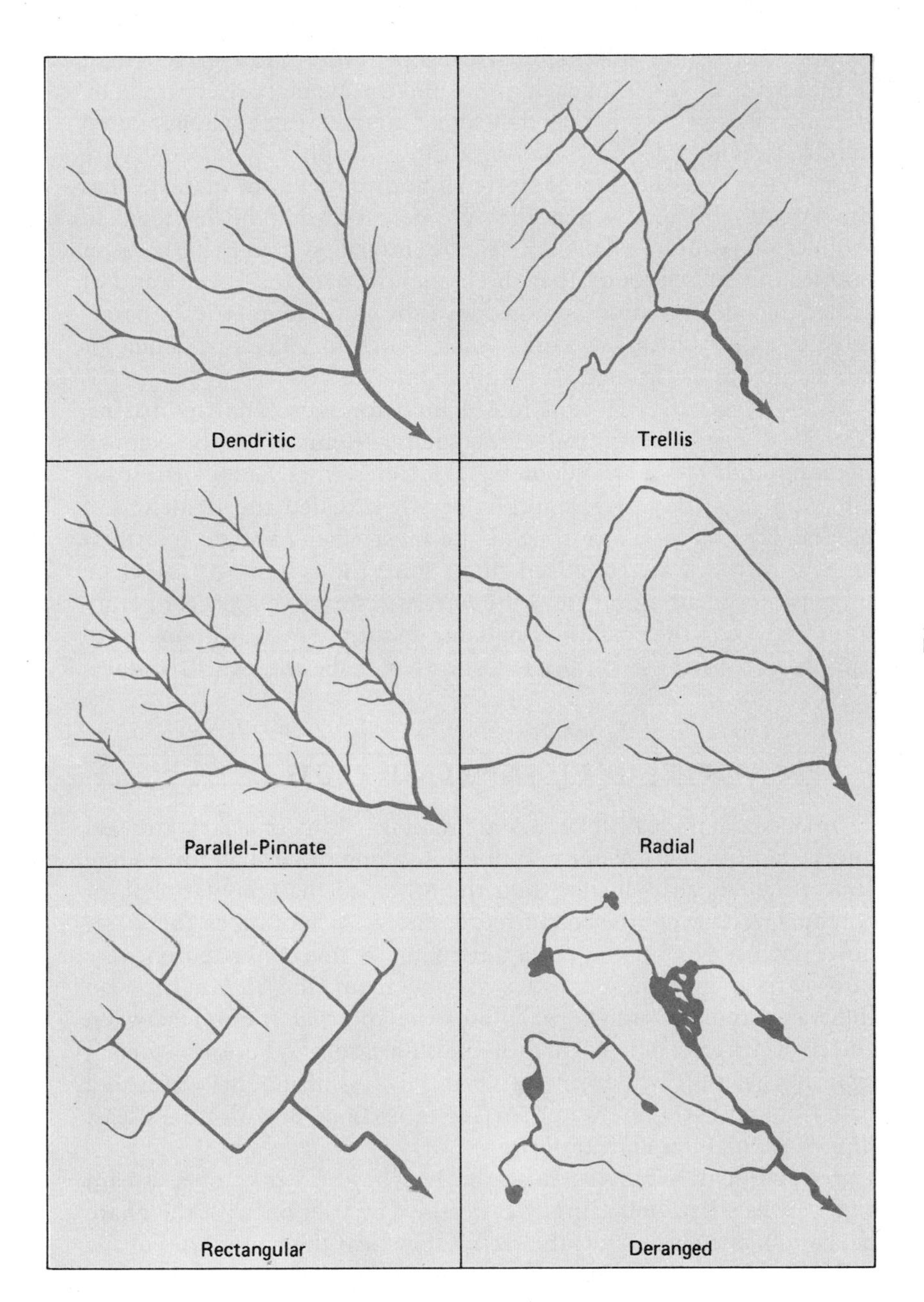

Figure 12.2 These are six of the most frequently encountered drainage patterns. *Dendritic* patterns are found in areas that lack strong contrasts in bedrock resistance, such as flat-lying sedimentary rock or massive crystalline rock that is deeply weathered. *Trellis* patterns develop where inclined layers of sedimentary rock of varying resistance to erosion are exposed at the surface. The parallel segments develop along the outcrops of the erodible layer of rock. *Parallel-pinnate* drainage reflects a topography of long parallel ridges, such as those developed in recently formed rock folds. The short segments drain the flanks of the ridges, and the long segments drain the troughs between. *Radial* drainage indicates the presence of an isolated high mountain area and is common in individual volcanoes or dome-shaped mountainous uplifts. *Rectangular* patterns reflect strong jointing of resistant bedrock, with streams incising along the joint planes. *Deranged* drainage shows no geometrical pattern or constant direction and usually includes numbers of lakes; this pattern indicates destruction of prior drainage by the erosive effects of continental ice sheets, which left behind debris accumulations and deeply scoured out basins, through which runoff makes its way as best it can. Such areas lack true valleys developed by fluvial erosion.

nization of drainage nets using the idea of *stream order.* The smallest streams that have no tributaries feeding them are termed *first-order* streams. According to the most common definition of stream order, a *second-order* stream begins at the junction of two first-order streams, as illustrated in Figure 12.3a, p. 339. Similarly, a *third-order* stream begins at the junction of two second-order streams, and so on.

Horton found significant quantitative relationships between stream order and the *number* of streams represented in each order, the average *area* drained by individual streams of each order, and the average *length* of the stream channels in each order. According to the

Figure 12.3 (opposite) (a) This diagram illustrates the system of stream ordering devised by the engineer Robert Horton and simplified by the geomorphologist Arthur Strahler. A second-order stream arises at the junction of two first-order streams; a third-order stream is produced by the junction of two second-order streams, and so on.

(b) (c) (d) These three graphs represent the Horton analysis of a drainage basin near Santa Fe, New Mexico. The graphs show the regularities that are typical features of drainage development, as revealed by a Horton analysis. (b) The numbers of streams of given order decrease regularly with increased stream order. (c) The average length of streams increases regularly with increased order. (d) The average area drained by a stream of a given order increases regularly as stream order increases. (After *Fluvial Processes in Geomorphology* by Luna B. Leopold, M. Gordon Wolman, and John P. Miller. W. H. Freeman and Company. Copyright © 1964)

Horton analysis, the number of streams per order, the average length of the stream channels in each order, and the average area drained by streams of a given order all fall along straight lines when plotted against stream order on semi-logarithmic graph paper (see Figure 12.3). What does this mean? Horton's findings seem to indicate that drainage development is remarkably systematic and probably tends to produce maximum efficiency in the fluvial system. It has been pointed out subsequently that the branching patterns of trees and of animal blood-circulation systems have the same geometrical characteristics as do stream networks—each being an efficient system for circulating fluids through all points of a larger body.

The regularities Horton found hold for most drainage basins, from small ones covering only a few square kilometers to those covering thousands of square kilometers. Horton's work had a significant effect on the study of geomorphology. It revealed the tendency of natural systems to evolve toward the most efficient state, in which energy and work are equalized throughout the system. At the same time, it turned attention from the purely descriptive aspects of landform analysis to the possibility of using quantitative measurements to gain deeper insight into the processes that shape the earth's surface.

PRINCIPLES OF CHANNELLED FLOW

To understand the nature of fluvial processes, think of clear water in a long cement trough. Water will not flow along the trough unless one end of the trough is higher than the other, so that gravitational, or potential, energy can be converted to energy of motion as the water flows downslope. But the potential energy is not converted entirely into energy of motion; it is also used to overcome the friction between the water and the trough walls and the internal friction between adjacent currents moving within the flow itself. Where no slope is available, friction will eventually bring a mass of moving water to a halt. Consequently, the water surface must slope downward, at least slightly, to maintain streamflow.

Water that is moving near the bed or the banks of a stream experiences a large retarding effect caused by friction with the channel bottom and sides. Thus the rate of flow near the boundaries of the channel is less rapid than that near the center of the channel. Figure 12.4 shows a cross section of a typical stream with flow velocities measured at numerous points. The water moves fastest near the center of the stream, away from the channel boundaries. Boatmen heading downstream attempt to keep their boats in the center of a river, where the water moves fastest, but rowing upstream is easiest near the banks, where the water flows more slowly.

The volume of water a stream carries past a given point during a specific interval of time is called the *discharge* of the stream. Discharge, which is measured in cubic meters (or cubic feet) per second, is equal to the cross-sectional area of the stream times the speed of flow, or velocity (distance traveled per unit of time). If the channel cross section decreases in a downstream direction, the velocity of flow

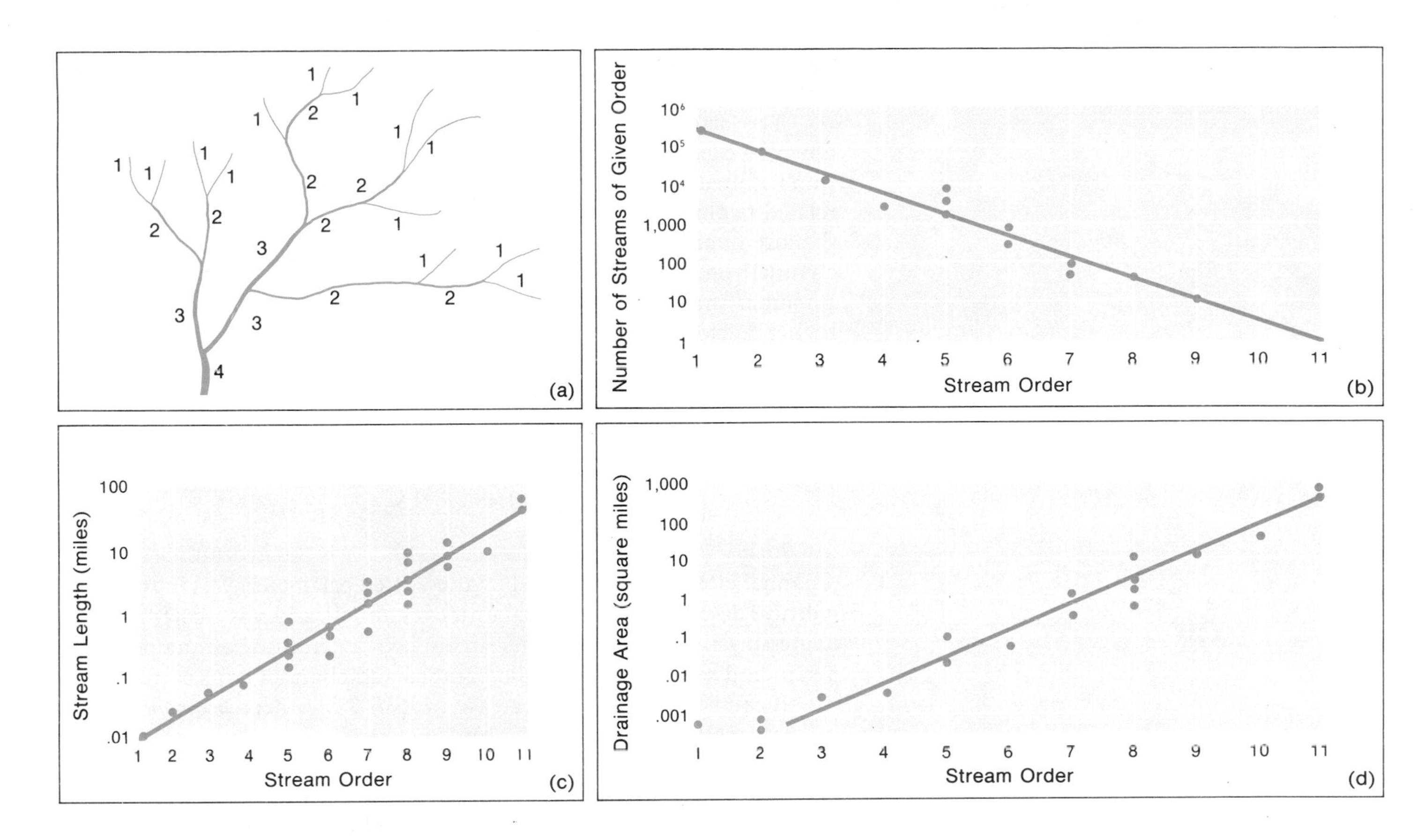

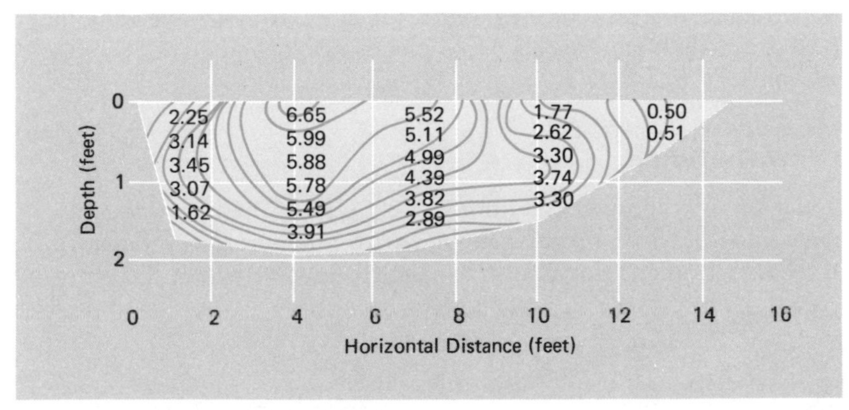

Figure 12.4 (left) This diagram shows the average measured stream velocities, in feet per second, at various points in a cross section of Baldwin Creek, Wyoming. Note that the velocities tend to be lowest near the sides of the channel; the stream cannot, therefore, actively erode its banks in this section. Velocities are highest near the center of the channel. The stream is therefore able to carry material suspended in its waters. Moderate velocities near the stream bed enable some material to be transported over the bed. (After *Fluvial Processes in Geomorphology* by Luna B. Leopold, M. Gordon Wolman, and John P. Miller. W. H. Freeman and Company. Copyright © 1964)

must increase in order to maintain a constant discharge. Figure 12.5, p. 340, illustrates how the velocity of flow must decrease if the cross-sectional area is increased with no change in the discharge.

The United States Geological Survey maintains a network of more than 6,000 gauging stations to measure the discharge of streams in the United States. The data are used to make flood forecasts, to evaluate irrigation water supplies, and for engineering purposes, such as the design of dams, bridges, water intake systems, sewage disposal, and so forth. At most stations, the discharge is inferred from the height of the stream's waters, using a previously determined calibration, or *rating curve,* for the section of stream at the location of the gauging station (see Figure 12.6).

The actual stream velocity reflects the balance between the downslope gravitational stress and the energy lost in overcoming friction within and at the boundaries of the flow. Frictional retardation is related to three independent characteristics of the stream channel: the size of its cross section, the shape of its cross section, and the channel roughness. The larger, deeper (relative to its width), and smoother the channel, the less the energy loss to friction and the higher the flow velocity, other things being equal. Also, the greater the channel slope, the greater the downslope gravitational stress and the higher the resulting flow velocity. These controls of stream velocity are summarized in the Manning equation, which is usually expressed in English units of measurements as follows:

$$v = \frac{1.486\, R^{2/3}\, S^{1/2}}{n}$$

in which v is the velocity in feet per second, R is the hydraulic radius (channel cross section divided by the wetted perimeter), S is the slope (vertical descent per horizontal unit of channel), and n is a coefficient expressing the roughness of the stream bed (ascertained by analogy to a standard series of stream photographs and channel cross sections).

It can be seen that an increase in R or S or a decrease in n will increase stream velocity. Since increases in discharge automatically increase R, normal downstream increases in discharge, as well as those resulting from floods, tend to increase stream velocity. The equation shows that increases in R have a greater effect on velocity than do proportionate increases in S. However, the maximum flow velocity in most streams seldom exceeds 3 to 5 meters per second (7 to 11 miles per hour).

Turbulence

Under very exceptional conditions rarely seen in nature, water can move smoothly over a regular surface, with no point-to-point change in the direction or velocity of flow. Such a flow, in which the water

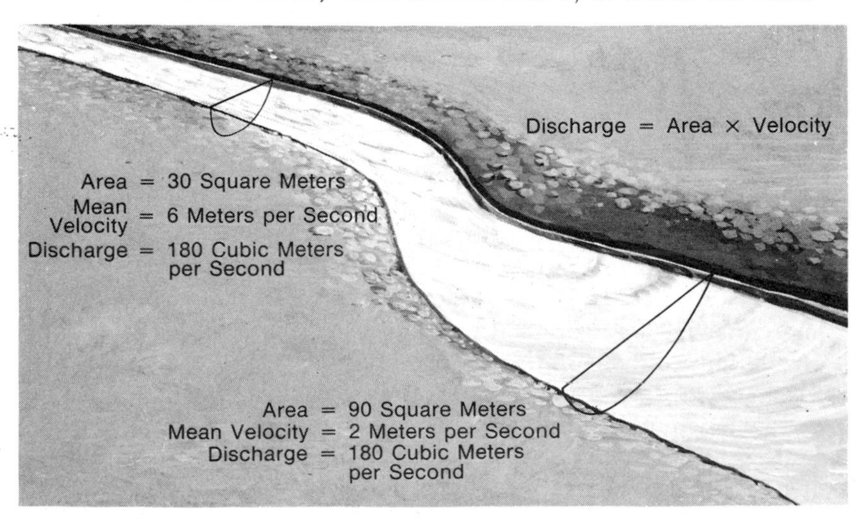

Figure 12.5 As this schematic diagram indicates, the discharge of a stream through any cross section is equal to the area of the cross section, multiplied by the average velocity of flow. If the cross-sectional area of a stream increases downstream, as shown in the diagram, the velocity of flow decreases proportionately, assuming that no water is gained or lost by the stream.

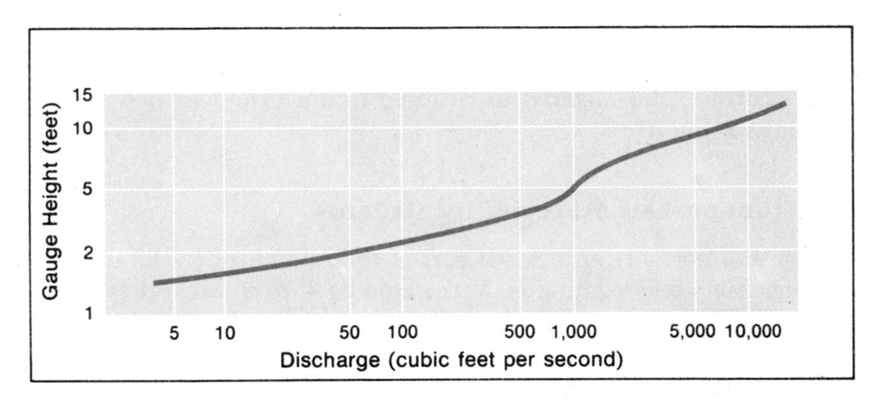

Figure 12.6 This *rating curve* for Seneca Creek, Maryland, relates the height of the stream to its measured discharge and allows the discharge to be inferred from a simple measurement of height. The average discharge of this stream is 100 cu ft per second, but discharges more than 50 times as great have been observed. (After *Fluvial Processes in Geomorphology* by Luna B. Leopold, M. Gordon Wolman, and John P. Miller. W. H. Freeman and Company. Copyright © 1964)

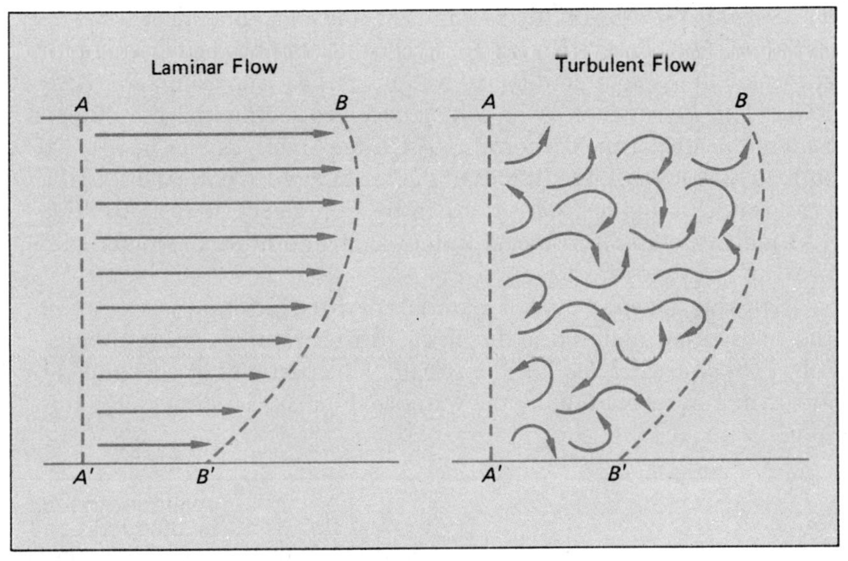

Figure 12.7 These diagrams compare laminar and turbulent flow. In laminar flow (left), the fluid moves like a deck of cards being deformed by internal shearing. The movement is slow and in a single direction. Under these conditions, sediment cannot be incorporated upward into the flow or supported within it. Thus the flow is nonerosive. Such flow is rare in streams but is common in slowly moving groundwater and glacial ice. Turbulent flow (right) is characterized by instantaneous velocities in all directions, which allows sediment to be picked up and supported within the flow, whose overall motion is forward from *A–A'* to *B–B'*.

always moves in strictly parallel trajectories, is said to be *laminar* (see Figure 12.7). This type of slow flow produces little or no erosion because it lacks the kinetic energy required to remove particles from a stream bed or banks. If the speed of flow is increased, however, friction between currents within the flow and between the currents and the channel surfaces also increases, and the flow becomes irregular, or *turbulent*. In turbulent flow, the speed and direction of water movement vary continuously and unsystematically, with momentary currents in all directions, including upward. Nevertheless, the average motion is in the downstream direction. The constantly moving eddies that make a stream so fascinating to observe are evidence of turbulent flow. The rolling, boiling rapids of a swift stream or the water rushing along the gutter of a city street after a heavy rainstorm show the characteristic irregularities of turbulent flow, but even streams that seem placid on the surface usually move fast enough to be turbulent.

In turbulent flow, most of the stream's energy is dissipated as heat, due to the friction generated as internal water currents moving at different velocities and in different directions constantly interfere with one another and with the channel walls. Only a small portion of any stream's energy is available to detach and transport rock debris.

Even so, turbulent flows can be highly erosive. Rapid pressure variations in turbulence near the stream bed and forceful upward-moving eddies permit solid material to be lifted into a turbulent flow and carried away by it.

Transport of Material by Streams

Streams are not only an integral part of the hydrologic cycle, transporting excess precipitation from the land back to the sea; they are also the main agent of gradation. As water moving over the land is channelled into successively larger streams, it carries away rock material contributed by slopewash and gravitational transfer (see Chapter 11). Streams running off the Badlands of South Dakota are milk-white with fine clay. China's Hwang Ho (Yellow River) has been named for the color of its sediment-laden waters, as has the Colorado (Red) River. The Mississippi River, which receives sediment from 40 percent of the area of the conterminous United States, carries nearly 300 million metric tons of sediment to the sea each year, equivalent to the erosional lowering of its entire basin by 1 cm every 200 years. Table 12.1 indicates the discharge, sediment load, and other characteristics of some of the earth's largest rivers.

Streams transport eroded materials in three different ways. Part of the transported material is dissolved in the water, forming the *dissolved,* or *chemical, load* of the stream. Fine particles of clay and silt are carried in suspension by the water as the *suspended load.* It is this

Table 12.1. Characteristics of Selected Rivers

River and Location	Average Discharge at Mouth (thousands of cubic meters per second)	Length, Head to Mouth (kilometers)	Area of Drainage Basin (thousands of square kilometers) ‡	Average Annual Suspended Load (millions of metric tons) §	Average Annual Suspended Load (metric tons per square kilometer of basin)
Amazon (Brazil)	180 (6,400)*	6,300 (3,900)†	5,800	360	63
Congo (Congo)	39 (1,400)	4,700 (2,900)	3,700		
Yangtze (China)	22 (800)	5,800 (3,600)	1,900	500	260
Mississippi (U.S.)	18 (650)	6,000 (3,700)	3,300	296	91
Yenisei (U.S.S.R.)	17 (600)	4,500 (2,800)	2,100		
Irrawaddy (Burma)	14 (500)	2,300 (1,400)	430	300	700
Bramaputra (Bangladesh)	12 (415)	2,900 (1,800)	670	730	1,100
Ganges (India)	12 (415)	2,500 (1,600)	960	1,450	1,520
Mekong (Thailand)	11 (390)	4,200 (2,600)	800	170	210
Nile (Egypt)	2.8 (100)	6,700 (4,200)	3,000	110	37
Missouri (U.S.)	2.0 (70)	4,100 (2,500)	1,370	220	160
Colorado (U.S.)	0.2 (6)	2,300 (1,400)	640	140	210
Ching (China)	0.06 (2)	320 (200)	57	410	7,200

* Numbers in parentheses indicate thousands of cubic feet per second.
† Numbers in parentheses indicate miles.
‡ One square kilometer is equal to about 0.39 square mile.
§ One metric ton is equal to 2,204.6 pounds.

Sources: Holeman, John N. 1968. "The Sediment Yield of Major Rivers of the World," *Water Resources Research,* 4 (August): 737–747.
Fairbridge, Rhodes W. (ed.) 1968. *The Encyclopedia of Geomorphology.* Vol. III, Encyclopedia of Earth Sciences Series. New York: Reinhold.
Espenshade, Edward B. (ed.) 1970. *Goode's World Atlas.* 13th ed. Chicago: Rand McNally.
Curtis, W. F., Culbertson, J. K., and Chase, E. B. 1973. "Fluvial-Sediment Discharge to the Oceans from the Conterminous United States," *U.S. Geological Survey Circular 670.* Washington, D.C.

portion of the load that gives many streams their opaque appearance. Particles too heavy to be carried in suspension are bounced and rolled along the channel bottom as the stream's *bed load.* The bed load cannot be clearly differentiated from the suspended load because small velocity and pressure changes are constantly lifting some particles completely off the bed to join the suspended load for a time, while other previously lifted particles settle back to replace them. The bed load and suspended load together constitute a stream's *solid load.*

The amount of dissolved load carried by a stream under natural conditions depends largely on the composition of the rocks in the area drained by the stream and on the local weathering processes. The latter are controlled in turn by temperature and moisture conditions, the presence or absence of a vegetation cover, and the specific chemistry of the soil moisture. A major portion of the water sustaining streamflow comes from groundwater, which often has a high content of dissolved minerals. Accordingly, the dissolved load of a stream may constitute one- to two-thirds of its total load. Human pollution of streams has added enormously to their chemical load in many areas. Dissolved material, including pollutants, is removed from land surfaces in the conterminous United States at an average rate of about 40 tons per sq km (100 tons per sq mile) per year. The average removal rate of solid material is about 71 tons per sq km (185 tons per sq mile) annually.

To transport its solid load, a stream must maintain forward and upward velocities that exceed the ever-present downward pull of gravity on the particles moving in the flow. The upward components of turbulent motion are important in resisting gravitational force. Thus material carried in suspension glides along in irregular paths within the flow. Heavier material that gravitational force pulls to the base of the flow skips or rolls along the stream bed. Bouncing or skipping transport is known as *saltation,* and rolling and sliding transport is called *traction* (see Figure 12.8, p. 344). As the rocks of the bed load bounce and roll along the stream bed, they collide with one another and with solid rock in the stream bed. Impacts against the stream bed break loose particles of solid rock, causing gradual downward channel incision. This constant battering causes the transported rocks to become smaller and more rounded as they progress downstream—cobbles are reduced to pebbles, and pebbles are disintegrated into sand and silt, which, along with clay particles washed into the stream, are the only solid materials transported far downstream by large rivers.

All types of particle movement increase with an increase in the kinetic energy or velocity of the flow. Any decrease in flow velocity reduces the forces opposing the downward gravitational pull on the transported material. This causes some of the solid material in transit to drop out of the flow. The first particles to settle are the larger ones that are moved by saltation and traction. These coarse materials come to rest in a new deposit of material that may be either temporary or permanent. Thus the process of deposition commences. The lighter particles carried in suspension continue to move despite initial velocity decreases, with only the heavier among them dropping to the bed

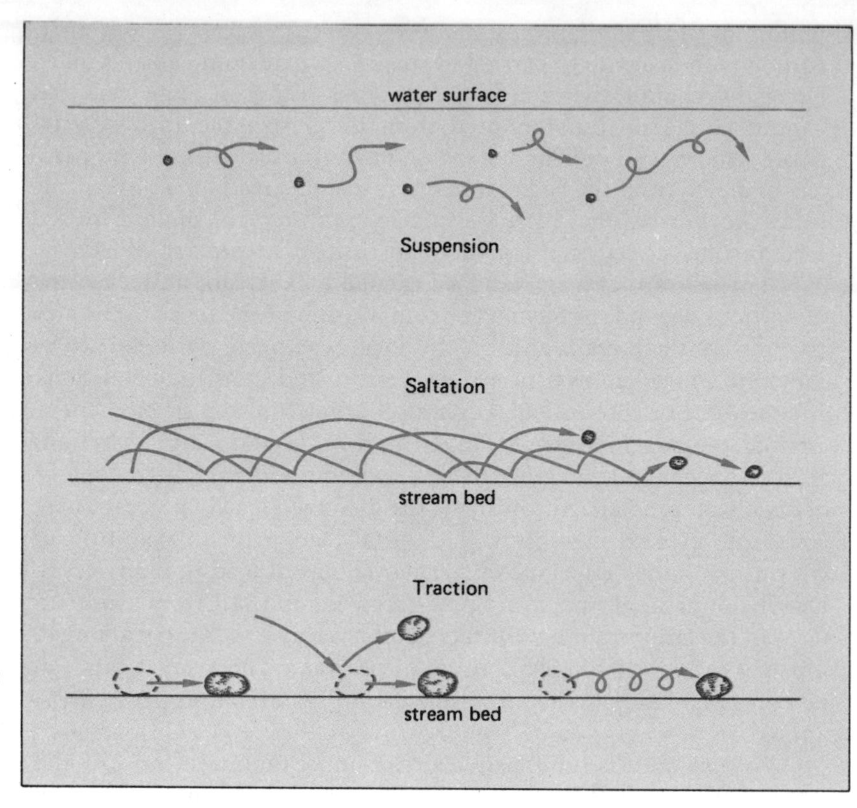

Figure 12.8 Solid material is transported in streams as suspended load and bed load. The finest particles are supported in suspension within the flow (top), while the bed load moves by both saltation (skipping or bouncing) and traction, in which particles are dragged or rolled by fluid shear stress, or are knocked forward by a saltating particle.

of the flow, to move subsequently by saltation and traction. Deposition of the finest suspended material carried in a mass of water occurs only when the water is almost still, as in a lake.

A fast stream has enormous power to transport objects. As a dramatic example of that power, when a train crossing a trestle in the Tehachapi Mountains of southern California was caught in a flash flood during a cloudburst in 1933, the locomotive and tender were carried a kilometer downstream and buried so completely in gravel that a metal detector had to be used to find them.

Scour and Fill

Much of the sediment a stream transports comes to it from smaller streams and unconcentrated runoff from slopes. If the flow near the bed of the stream channel is fast enough, however, the stream can generate sediment by detaching particles from its own bed in the process of stream bed *scouring*. This critical velocity, below which scouring will not occur, depends on the nature of the stream bed because specific amounts of energy are required to put materials of different sizes and shapes in motion. Clay or silt particles, for instance, tend to resist detachment more than does sand. This is due to the greater molecular attraction between the finer particles. Material coarser (larger in diameter) than sand also resists movement more than sand does, due simply to the greater weight of the particles.

Figure 12.9, p. 346, illustrates the flow velocities required to scour particles of given sizes from a stream bed. The most easily eroded particles are those of sand size, having diameters between 0.05 and 1.0 mm. The flow velocity near the bottom of the stream must be on the order of 15 to 30 cm per second (0.3 to 0.7 miles per hour) to initiate scouring of these particles. The graph also shows the range of speeds for which a particle of given size will remain in suspension. A fine particle scoured from the stream bed during high-velocity flow can be carried in suspension even if the speed of the stream subsequently decreases. If the flow velocity continues to diminish, however, progressively finer particles will fall out of suspension and become part of the bed load, then finally part of the deposited sediment in the process of stream bed *filling,* or fluvial *aggradation*.

By scouring and filling, a stream is able to adjust the slope of its bed and the shape of its channel to produce a balance between its energy, or capacity to do work, and the amount of work that must be accomplished. Consider a stream that does not have the energy to transport all of the sediment fed into it. In such a case, sediment is deposited in the stream channel, thereby elevating the stream bed and increasing the stream's potential energy at that point. Any elevation of the stream bed increases the downstream slope, which in turn increases downstream flow velocity. Conversely, a stream whose energy (velocity) exceeds that necessary for sediment transport will flatten its slope, which gradually reduces its velocity. It can do this either by incising downward in the process of *degradation* or by lengthening its course by becoming increasingly sinuous.

CHANNEL FORM AND PROCESS

A stream is a sensitive, dynamic system with the ability to adjust its morphological characteristics in a matter of hours in response to small changes in inputs of energy and material and other environmental influences. The input to a given section, or *reach,* of a stream consists of the discharge of water and sediment from farther upstream. A stream responds to changes in those inputs through adjustments in its channel geometry (cross-sectional shape, size, slope, sinuosity) and bed characteristics, such as roughness, all of which affect the velocity of water flowing through it.

The discharge and sediment load arriving at a reach of a stream for downstream transport require the stream to do work. Exactly how a particular stream reach will adjust to the incoming discharge and load is difficult to predict in detail. The complex interrelationships of stream channel properties can be illustrated by the example of a stream in flood. During a flood, each section of a stream is called upon to carry a greater discharge than it does under average conditions. As the discharge increases, the volume of water becomes too large to be held by the normal channel cross section, and the decrease in friction due to increased flow depth causes the stream velocity to increase. However, the high velocity of flow increases shear stress on the channel boundaries and enables the floodwaters to scour the bed and banks of the stream, which enlarges the cross-sectional area and reduces the

Figure 12.9 The ability of a stream to transport material depends on the local velocity of flow, on the size of the particles, and to some extent on the shape of the particles. This diagram for uniform material relates particle size and flow velocity to mechanisms of erosion and deposition. Combinations of size and velocity located to the right of the settling velocity curve represent the regime of deposition; large particles settle out of a slow-moving stream and become deposited on the bed. Particles with velocities and sizes to the left of the settling velocity curve can be transported either by saltation along the bed or by suspension in the stream. The erosion velocity represents the minimum water velocity for which particles can be loosened from the stream bed; erosion velocity is shown as a band because it depends on particle shape, bed material, and other factors. The graph shows that once a particle becomes suspended it can be transported by water moving more slowly than the erosion velocity. However, particles larger than a few millimeters are transported largely by saltation along the stream bed. (After Morisawa, 1968, from Hjulstrom, 1935)

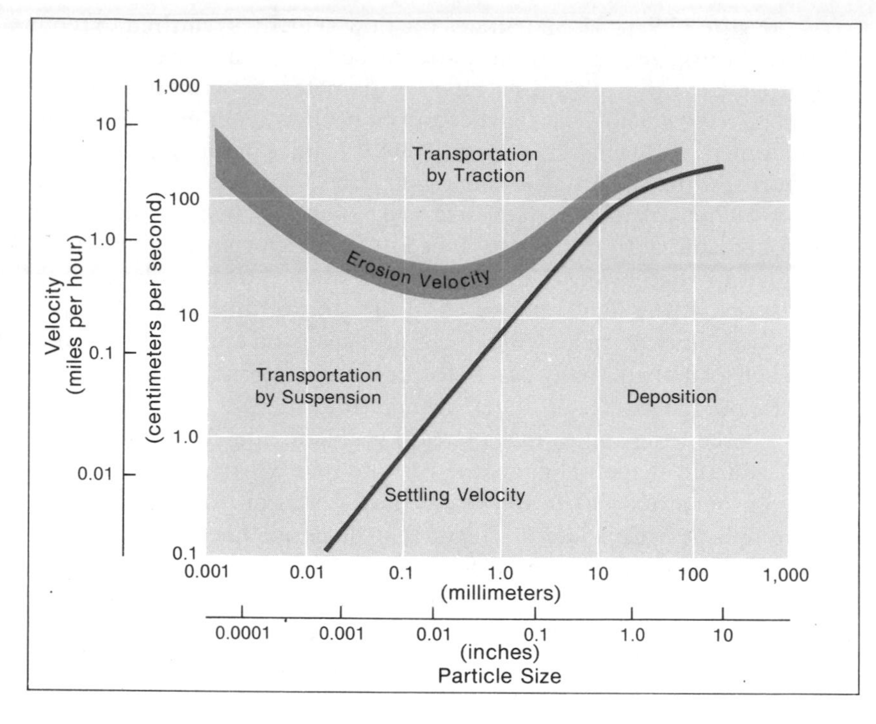

Figure 12.10 (opposite) (a) This schematic diagram illustrates the tendency of the depth and width of a stream channel and the velocity of flow to increase with increased discharge at a given location along the stream. The curves refer to the equilibrium situation in which the discharge has continued long enough for the stream to complete readjustment by scouring or filling.

(b) (c) These channel cross sections for the Colorado River at Lees Ferry, Arizona, show the scouring (b) and subsequent filling (c) that occurred during a high-water period in 1956. Discharge is in units of cubic feet per second. The marked enlargement of the channel by scouring was accompanied by an increase in flow velocity to accommodate the greatly increased discharge. In August, when the discharge returned to almost its initial February value, the cross section of the channel also returned to almost its original size. (After *Fluvial Processes in Geomorphology* by Luna B. Leopold, M. Gordon Wolman, and John P. Miller. W. H. Freeman and Company. Copyright © 1964)

downstream slope (see Figure 12.10). If the flood persists for some time, and if the stream bed is easily eroded, the stream will scour its channel until it reaches a new equilibrium under the conditions of increased discharge.

As the floodwaters subside and the discharge of the stream diminishes, the stream once more finds itself out of equilibrium: its cross section is now larger than required and its slope is too low to provide adequate flow velocity for the diminishing discharge. The decreasing velocity reduces the stream's ability to carry its solid load, and the stream bed begins to fill by deposition of sediment. Eventually, equilibrium is reestablished through the process of deposition, which elevates the stream bed, increasing its slope, and reduces the area of the channel to accommodate the lower stream discharge. Thus the sediment underlying most stream channels is an important factor in the equilibrium process, facilitating channel excavation during floods and the reestablishment of normal channel characteristics after floodwaters pass.

As the example of a stream in flood indicates, several channel properties may be adjusted over a period of time in response to changes in discharge or load. The adjustments always occur in a direction that accommodates altered conditions, with the stream tending toward a state of equilibrium in which stream energy is just adequate for its ceaseless work of sediment transport. Complete equilibrium is impossible since flow conditions change constantly; thus most streams are in a state of constant readjustment, or "quasi-equilibrium." Streams incised in hard rock may not even approach an equilibrium condition, whereas streams in erodible materials can adjust to changes quite rapidly.

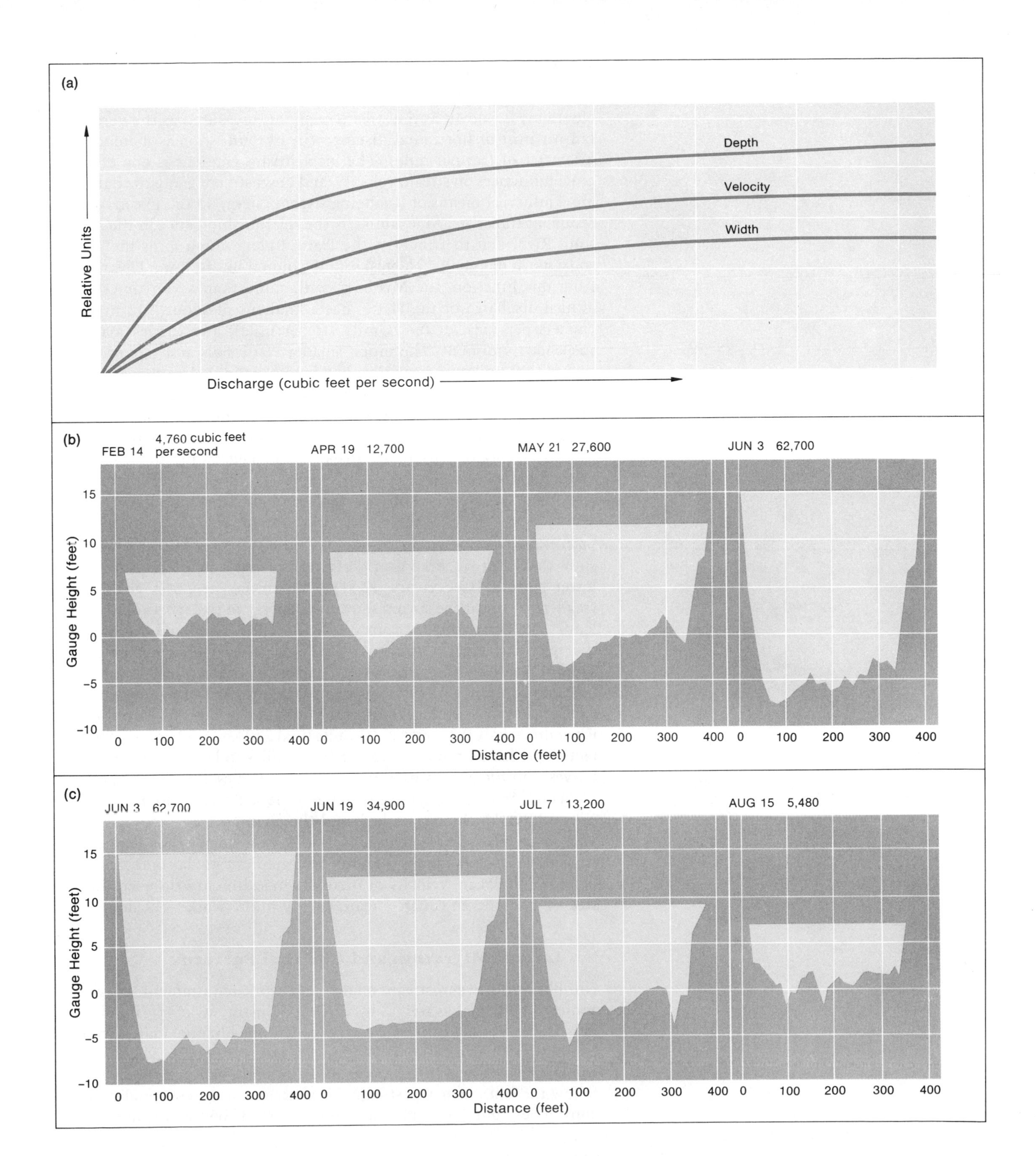
(a)
Depth
Velocity
Width
Relative Units
Discharge (cubic feet per second)
(b)
FEB 14 4,760 cubic feet per second
APR 19 12,700
MAY 21 27,600
JUN 3 62,700
Gauge Height (feet)
Distance (feet)
(c)
JUN 3 62,700
JUN 19 34,900
JUL 7 13,200
AUG 15 5,480
Gauge Height (feet)
Distance (feet)

Stream Gradient

The rate at which a stream channel descends from higher to lower elevations is known as the stream *gradient,* measured as the vertical fall per unit of horizontal distance (in like units, or as meters per kilometer or feet per mile). The local stream gradient is one of the chief influences on stream velocity, and downstream gradient changes are a principal means of balancing stream energy to the necessity of sediment transport. An example is the relationship between the Missouri River and its tributary, the Platte River, which joins the Missouri south of Omaha, Nebraska, and Council Bluffs, Iowa. Upstream from this junction, the Missouri, with more than seven times the annual discharge of the Platte, has a gradient of about 0.13 m/km (meters per km), or 0.7 ft/mile (ft per mile), and carries mainly suspended sediment. The much smaller Platte falls at a rate of 0.76 m/km (4 ft/mile) and carries a bed load of sand. Below the entry of the Platte, the Missouri's gradient increases to 0.27 m/km (1.4 ft/mile), twice its rate of descent above the junction. This sudden gradient increase is required to equate the Missouri River's energy to the increased work of sediment transport produced by the input of the Platte River's bed load.

A graphic portrayal of a stream's gradient from its source to its mouth is called the stream's *longitudinal profile.* Comparisons of profiles of streams that are hundreds to thousands of kilometers long show that most are steep near their sources and much flatter in their lower portions (see Figure 12.11). The initial steepness gives the smaller flows near the stream source the energy to transport coarse bed loads. The flatter downstream gradients reflect increased discharges that produce larger and deeper channels. The gradual decrease in slope also reflects the fact that the caliber of the bed load normally becomes smaller downstream, requiring progressively less energy expenditure for sediment transport. Thus at low water the Mississippi River drops only 0.3 meter (1 ft) in its final 240 km (150 miles). These factors produce a tendency for stream profiles to be concave upward curves. Point-to-point variations in stream profiles are a consequence of changes in discharge, the character of the sediment load, resistance or roughness of the stream bed, and the channel cross-sectional size and shape. Any factor that increases stream velocity or decreases transportational demands permits the gradient to become flatter; decreases in stream velocity or increases in sediment load necessitate steeper gradients to equate stream energy to the work to be done.

Lateral Migration and Channel Patterns

Equilibrium between stream energy and work requirements may be achieved by lateral as well as by vertical adjustments of stream channels. If the channel banks are easily eroded, the channel may migrate laterally a distance of several kilometers over a period of years. This sometimes creates political problems, since many political boundaries follow the lines of major streams. If a stream migrates laterally in a continuous but slow manner, the political boundary normally

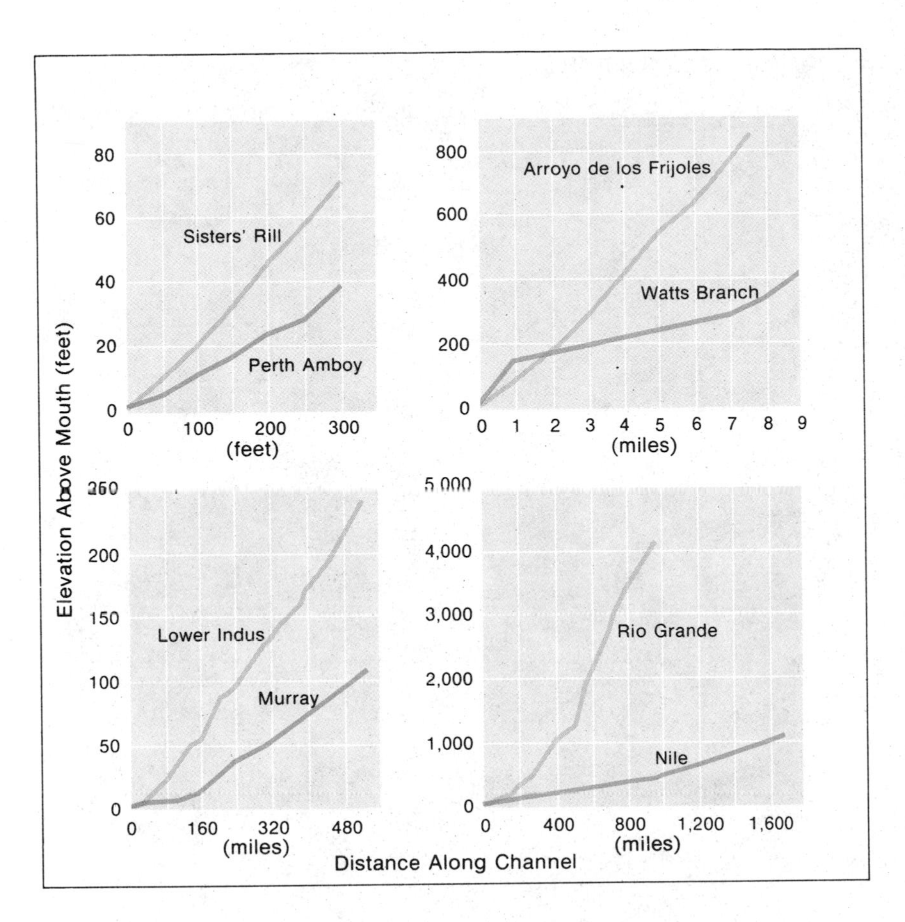

Figure 12.11 The examples of these graphs illustrate the longitudinal profiles of eight streams varying in length from a few hundred feet to several thousand miles. The vertical scales have been exaggerated for clarity. In each case the stream tends to develop a longitudinal profile that is concave upward on the average, with the steepest portion upstream. There are local irregularities in the profiles, perhaps because of local differences in rock structure along the bed. Note that the downstream portions of the Rio Grande and the Nile River develop concave longitudinal profiles even though their discharges are effectively constant because of the lack of tributaries. (After *Fluvial Processes in Geomorphology* by Luna B. Leopold, M. Gordon Wolman, and John P. Miller. W. H. Freeman and Company. Copyright © 1964)

migrates with it. However, a sudden change in channel location, known as an *avulsion*, sometimes occurs during a major flood. In the case of a river avulsion, the political boundary usually does not move, but remains in the former abandoned channel. Figure 12.12 shows how political complications could arise from avulsions along the Mississippi River.

Stream channels are seldom straight and uniform for any appreciable distance. When viewed from above, channels exhibit three broad patterns: meandering, braided, and straight or nonmeandering (see Figure 12.13). Meanders occur in streams of all sizes, from the tiny rills on bare ground to the kilometer-wide Mississippi River. The geometry of the meanders of small streams is strikingly similar to that of large rivers; the average length of meanders, known as meander *wavelength*, on a stream of any size is 7 to 15 times the width of the stream (see Figure 12.14). The development of meanders appears to be normal behavior for streams, particularly for those flowing in homogeneous deposits of easily eroded sediment. Even the meltwater flowing across the surface of glaciers follows a typical meandering path, as do ocean currents such as the Gulf Stream, and even air currents like the jet streams. In the latter examples there are no stream banks or solid load, only boundaries between different types of water and air.

The specific mechanism that generates meanders is yet to be determined. However, even streams without well-developed mean-

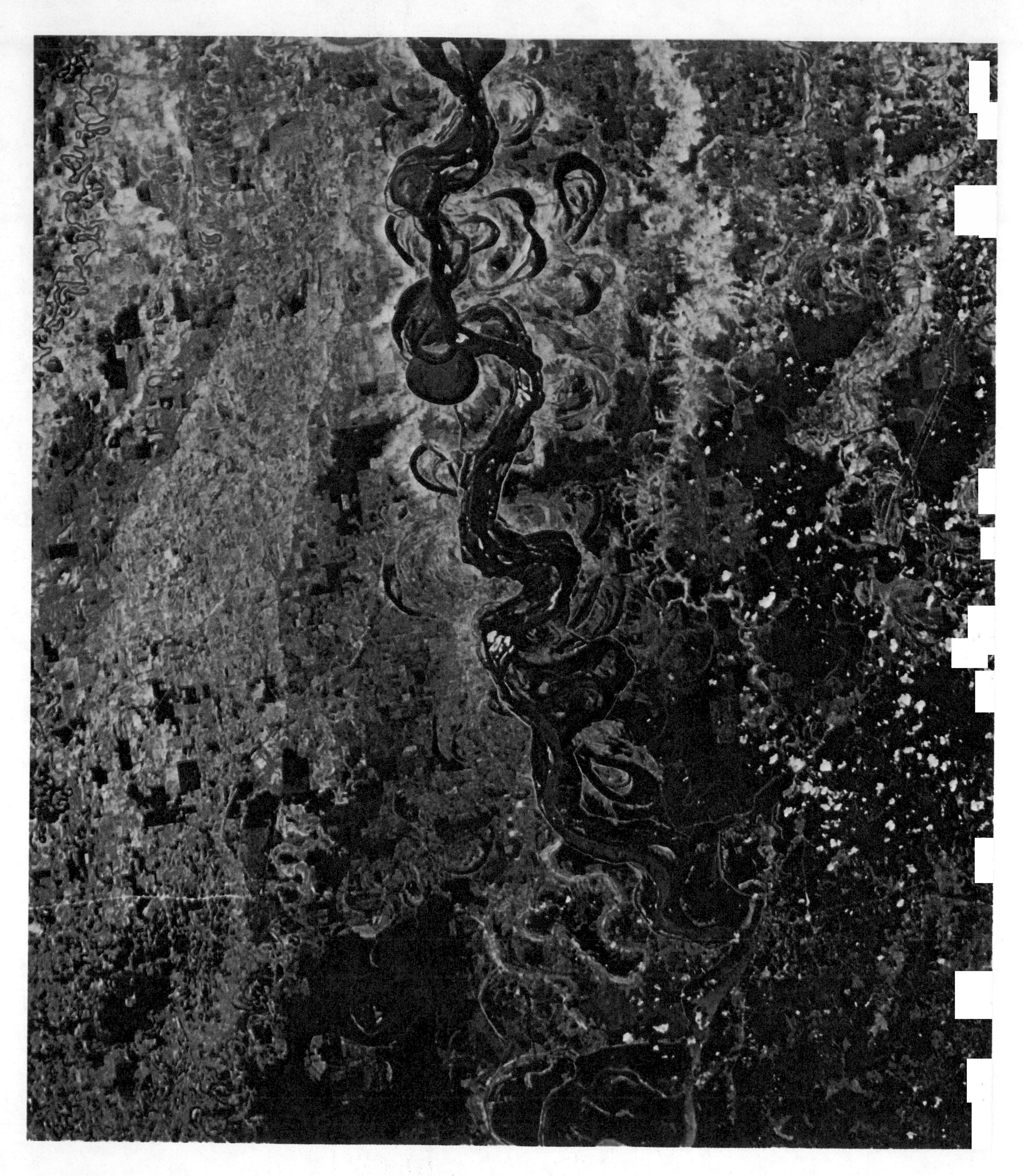

ders often show a succession of deep pools separated by shallow bars, or *riffles*, as shown in Figure 12.15. The scale of pool and riffle spacing is proportional to the width of the stream; the distance from one pool to the next is typically four to seven times the width of the stream, and successive pools tend to be located on alternate sides of the channel. Laboratory experiments with artificial channels indicate that at certain critical flow velocities, smooth channel beds begin to scour and fill, producing pools and riffles. The pools tend to deepen adjacent to the stream banks, which collapse into them, displacing the stream bed laterally. Each pool eventually becomes the site of an outward-pushing loop or meander. The explanation of meandering, therefore, lies in the dynamics of the flow producing the initial irregular channel bed scouring.

The axis of any stream is specified by the line, called the *thalweg*, that follows the deepest part of the stream's cross section. In a meandering stream, the thalweg swings toward the outer bank of each curve and crosses back and forth over the sinuous stream bed. Even in nonmeandering channels, the thalweg wanders from side to side along the stream bed in a manner suggestive of meandering. The maximum flow velocity normally follows the thalweg. Scouring occurs at the outer, or concave, banks adjacent to channel pools, where the flow velocity is greatest, and deposition of sediment occurs in the form of *point bars* on the inner side of each loop at the convex bank, where stream velocity and transporting power are least. Thus the channel shifts laterally with no change in cross-sectional form, undercutting the bank on one side and filling with sediment on the other. Most scouring occurs during flood stages, when stream energy is at its peak, with the greatest deposition of sediment following as flood flows diminish.

The formation of meanders requires an expenditure of energy, just as does vertical stream incision. Why, then, do streams meander? The answer must be that they have excess energy at certain times. The effect of meandering is similar to the effect of vertical incision. In both cases, the result is to flatten the stream gradient, reducing the stream's energy. In the case of meandering, this is done by a lengthening of the stream itself, increasing the horizontal distance required to accomplish each unit of vertical descent. An energetic stream that does not have the tools (bed load) to cut vertically downward into solid rock can thus achieve equilibrium by cutting laterally instead. As we might anticipate, therefore, meandering channels frequently occur where the bulk of the stream load is fine material that is carried in suspension.

A different type of channel results where a stream's load consists of larger particles that can only be moved by traction and saltation. In such a circumstance, a shallow channel is necessary, so that the maximum flow velocity is as close as possible to the stream bed where it can exert shear stress to move the bed load. The resulting channel is the braided type, consisting of many separate intertwining shallow channels (see Figure 12.13b).

Braided channels are associated with the growth of sand or gravel bars in broad, shallow streams that are abundantly supplied with

Figure 12.12 (opposite) This photograph spans a 100-mile section of the Mississippi River north of Vicksburg, Mississippi. It was taken from the Apollo 9 spacecraft in 1969 using color infrared film, which shows vegetation in shades of red (see Appendix II). The sinuous Mississippi has frequently changed channels; remnants of earlier channels are clearly visible. Boundaries between the state of Mississippi and the neighboring states of Louisiana and Arkansas were laid out in the nineteenth century to coincide with the channel of the Mississippi River. Because of subsequent changes in the channel, portions of each state now lie on both sides of the river.

coarse sediment—sand, gravel, and cobble-sized material. The development of these bars tends to produce separate shallow channels with high gradients in which the water thrusts vigorously against the stream bed, enabling the stream to move its bed load of heavy particles. During periods of high discharge, the bars and channels or a

a

b

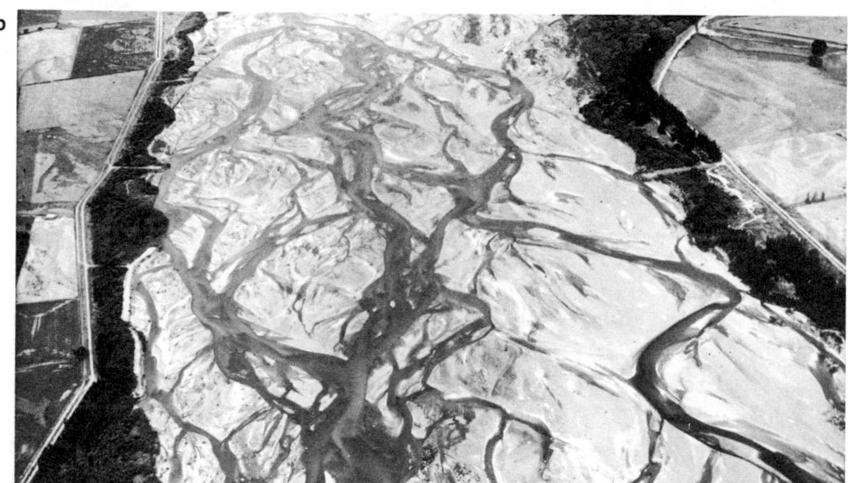

Figure 12.13 (a) The sinuous Animus River near Durango, Colorado, meanders over a broad flood plain produced by alluvial aggradation.
(b) The Waimakariri River on the Canterbury Plains of New Zealand is a typical braided stream. The pattern changes with time as old channels become filled with sediment and new channels are scoured out of the loose sand and gravel of the bed.

braided stream undergo continuous change as sediment is removed from one place and deposited in another. The overall pattern of a braided stream usually undergoes only slow modification, and braiding appears to be a stable way for a stream to maintain the energy required to transport a sand and gravel bed load. Braided streams must have extremely steep gradients to overcome friction at the stream bed, and they have accomplished this by the deposition of the material producing their channel bars; this deposition raises the elevation of the stream bed and increases its downstream slope. Braided patterns tend to develop where stream discharge is highly seasonal, fluctuating very widely, as in semiarid and arid regions, and around the peripheries of glaciers. Wide fluctuations lead to bank collapse, which feeds large amounts of sediment into the channels. Large sediment inputs also occur where the vegetation cover is sparse, where the bank material is sandy or gravelly, and where glaciers are delivering great quantities of coarse waste to streams.

Straight stream channels are uncommon. Where present, they indicate that the pattern of the channel is controlled by the underlying geological structure. Streams incised in solid bedrock are occasionally confined by fractures in the rock, either joints or faults. Even these channels have the pool and riffle sequences found along meandering streams. Some meander patterns are rectangular rather than smoothly curving, indicating that joints control the channel. The channels of some river deltas are quite straight due to clay bed and bank materials, which are resistant to lateral stream erosion (see Figure 12.21, p. 363). The delta *distributaries* of the Nile and Mississippi rivers are examples of straight streams that have clay beds. Distributaries are the opposite of tributaries, flowing out of a larger stream rather than into it; they are normal features of delta landscapes, as will be seen in later pages.

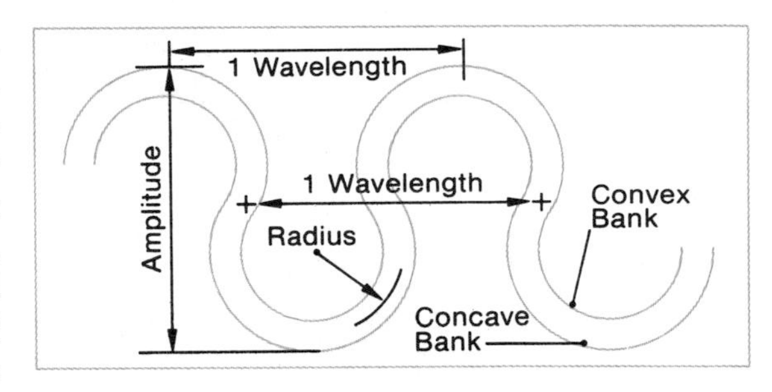

Figure 12.14 The diagram shows how the wavelength, amplitude, and radius of a meander are defined. The radius around a meander is not constant, however, because the form of a meander is more complex than a simple circular arc. The meander wavelength is usually 7 to 15 times the width of the channel. (After Dury, 1969)

Knickpoints

Where waterfalls, cascades, or rapids are present, the stream gradient steepens in an abrupt profile discontinuity called a *knickpoint.* Knickpoints in stream profiles originate in several ways. Some are associated with an abrupt change in bedrock resistance. Other knickpoints may be related to crustal movement, as when an active fault creates an escarpment that the stream must cross.

Knickpoints in stream profiles are extremely common in areas that have experienced severe glacial erosion. The many waterfalls and cascades seen in alpine mountain country usually mark places where glacial erosion has deepened the major valleys, leaving smaller tributary valleys "hanging" above them (see Chapter 15). Often glacial erosion creates steps in the main valleys themselves, each of which will be the site of a waterfall or cascade. Knickpoints can also form in coastal regions because of general lowering of the sea level or local rise of the land over a short period of time. A newly exposed coastal slope is always somewhat steeper than the comparatively flat slope of large rivers in their lower reaches; thus a displacement of the land relative to sea level produces a knickpoint near the original mouth of any river entering the sea.

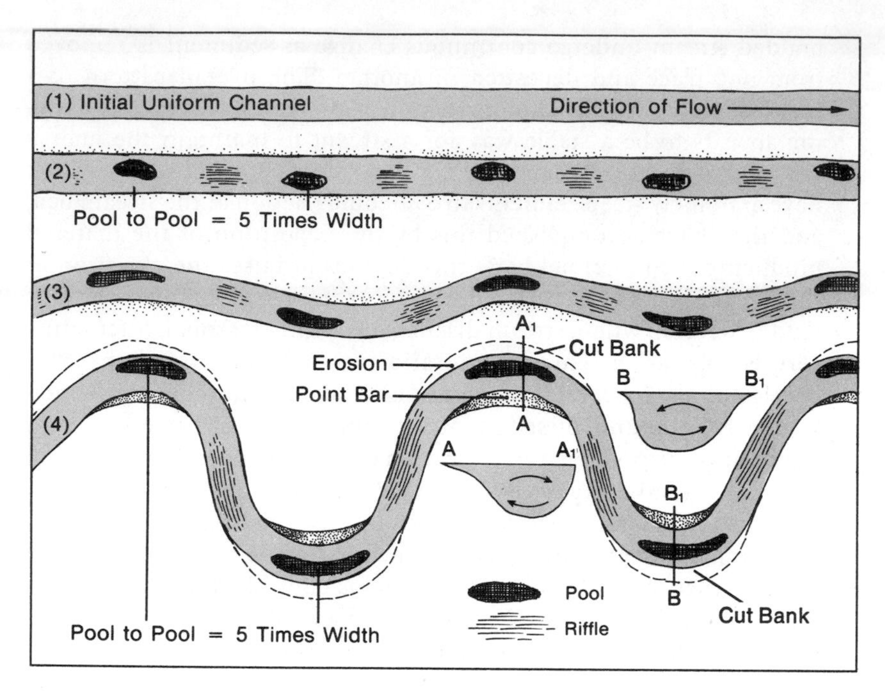

Figure 12.15 Streamflow in a straight channel of uniform cross section (1) seems to be unstable, and if the channel is readily erodible, the stream tends to develop a sequence of alternate deep pools and shallow gravel bars, or riffles (2). The straight channel becomes more sinuous, with pools forming toward the outer concave banks (3). In time, a meandering channel may be developed (4). A sinuous or meandering stream tends to scour its outer concave banks and to deposit sediment against its inner convex banks; the speed of flow is greatest near the concave banks and least near the convex banks. There may also be a lateral flow of water, as the cross-sectional diagram indicates. The lateral flow is thought to move sediment from the cut bank to the point bar. (After Dury, 1969)

In time, most knickpoints are removed by erosional and depositional processes. The stream tends to scour its channel above the knickpoint and to cover the base of the knickpoint with debris. If the stream bed material is erodible, a knickpoint may disappear in only a few years, as illustrated in Figure 12.16a.

Some knickpoints, however, migrate headward (upstream) considerable distances before disappearing. At the immense knickpoint of Niagara Falls, the bed of the Niagara River above the falls is underlain by resistant dolomite that makes downcutting by the recently formed river very slow (see Figure 12.16b). Beneath the dolomite is less resistant shale, which is easily eroded at the exposed face of the falls. As the shale is removed, the dolomite layer collapses and the falls retreat upstream, at an average rate of more than 1 meter (3 ft) per year. The present falls have retreated 11 km (7 miles) since their formation perhaps a hundred thousand years ago. The history of the falls is complex, being related to glacial modification of the landscape, with several phases of waterfall retreat followed by burial beneath glacial deposits.

Graded Streams

Short-term scouring and filling processes constantly adjust stream slopes and channel cross sections to changing inputs of water and sediment. In principle, a stream should eventually reach a state of equilibrium in which, over the long term, it is neither depositing sediment nor scouring its bed, but maintains a gradient and channel cross section that balance stream energy to the transportational work the stream must do. A stream in such a state of adjustment is said to be a *graded* stream. However, the discharges of all streams fluctuate through the year, and from one year to the next, due to variations in

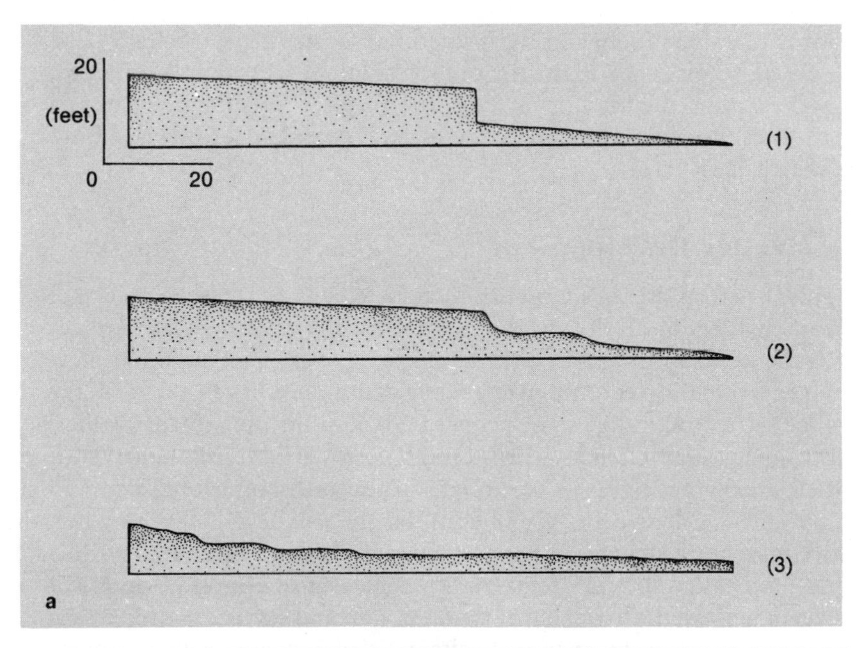

Figure 12.16 (a) Streams generally act in such a way as to reduce knickpoints and to attain a smooth longitudinal profile. The diagram shows a knickpoint (1) that was formed on Cabin Creek, Montana, by an earthquake. The stream began to scour its channel above the knickpoint and to fill below the knickpoint. Because the material on the channel was soft and easily cut, the knickpoint was significantly reduced in only a few months (2). Three years later (3) the profile showed no trace of the knickpoint. (After Morisawa, 1968)

(b) Since the Niagara River began flowing during the Pleistocene era, it has not been able to cut downward significantly into the cap rock of hard dolomite forming Niagara Falls. Instead, the river has cut a long gorge headward by undermining weak shale layers at the base of the falls, causing the collapse of the overlying layers. (After G. K. Gilbert, 1896)

temperature and precipitation; thus a state of permanent equilibrium is impossible. A graded condition really implies that no net scouring or filling occurs over a period of years, despite short-term adjustments reflecting normal seasonal changes. A graded stream is in dynamic equilibrium in the sense that it adjusts itself to slight changes in conditions by scouring or filling for short periods of time to absorb the effects of such changes.

FLUVIAL LANDFORMS

Most hillslopes exist because streams have cut downward, producing valleys. In Chapter 11, we saw that the actual configuration of hillside

Figure 12.17 (opposite) This image taken from an altitude of about 200 km (125 miles) shows the Colorado River's great canyon through the Colorado Plateau in northern Arizona (upper right). The river exits from its canyon at the escarpment of the Grand Wash Cliffs (see arrow), created by uplift along a major fault. Uplift of the Colorado Plateau relative to the area to the west has caused vertical incision of the river to a depth exceeding 1,500 meters (5,000 feet). Healthy vegetation produces the red colors on images such as this. Lake Mead appears in the left (west) part of the image.

slopes is a consequence of mass wasting processes and unchannelled flows of water rather than stream erosion alone. In the following, we shall examine the landforms that are solely a consequence of fluvial processes. But first, it is necessary to understand why valleys should develop at all.

Valley Development

Only when a stream's kinetic energy exceeds the transportational work it must do will the stream incise vertically, creating new surfaces for the slope-forming processes to shape. Streams flowing down new slopes created by tectonic disturbance commonly have excess energy since the tectonic slopes are steeper than equilibrium stream profiles. Streams in such settings will incise, flattening their gradients, until their energy becomes proportionate to the transportational demand upon them. Streams already in equilibrium will begin incising when environmental changes of various types give them excess energy; this has been called stream "rejuvenation." The most common causes of stream rejuvenation are deformation of the land surface, a drop in sea level, an increase in stream discharge, and a decrease in sediment load.

Deformation of the land surface by tectonic activity is the major trigger for erosional rejuvenation that causes valleys to deepen as stream profiles are regraded to restore fluvial equilibrium (see Figure 12.17). Periodic uplift in tectonically active mountain regions results in stream terraces (discussed later in the chapter) and valley-in-valley configurations, in which a steep-walled gorge is cut into the floor of a larger, more open older valley. Elsewhere, slow but continuous uplift produces hilly or mountainous landscapes resulting from steady stream incision. Past sea level changes have produced drops in the base level for all of the earth's largest streams, and the resulting incision of these streams has, in turn, produced a drop in the base level for all their tributaries. Increases in average stream discharge may occur as a consequence of climatic changes or stream piracy, in which an expanding drainage system "captures" the runoff of a neighboring system. Flood discharges may be greatly increased by human effects on surface conditions, such as the cutting of forests and the urbanization process that covers constantly expanding areas with impermeable concrete and asphalt. These tend to cause stream incision and channel enlargement, as well as increasing sediment production. On the other hand, sediment loads are temporarily removed from streams by dam construction, which traps both the bed load and suspended load of streams in artificial reservoirs. Immediately downstream from dams, streams have excess energy since they have no sediment load to transport. Such streams could incise vertically if large volumes of water were released from the reservoirs from time to time. This appears to be occurring along the Colorado River below the Glen Canyon Dam (completed in 1964) and along the Nile below the Aswan High Dam (completed in 1971). Artificial changes in stream behavior are becoming increasingly evident as river control projects and the urbanization process expand.

Flood Plains

When streams attain equilibrium profiles and cease incising vertically, their initially narrow valley floors gradually become wider. The greater part of the valley expansion is produced by the processes of slope erosion, which are constantly at work on the valley walls. However, the initial widening of the valley floor is accomplished by *lateral planation,* in which stream meanders erode valley walls on the outside of each bend, while filling the inside with point bar deposits of sand and gravel. As the stream meanders shift their course, they eventually clear out a continuous flat valley floor that is covered over by stream deposits, or *alluvium.* The stream swings back and forth across this open lowland (see Figure 12.18). Since the stream channel cannot hold the largest discharges that occur at intervals of a year or two, the entire valley floor is flooded periodically; thus it is called a *flood plain.*

Flood plains are a hazardous environment, but they are one of the earth's most valuable landscapes. Each flood that inundates them leaves a covering of fresh silt, often several centimeters thick, that gradually elevates the flood plain. These overbank flood deposits cover the initial sand or gravel point bars and provide nearly level agricultural land whose fertility is renewed by deposits of additional layers of silt every year or two.

River flood plains themselves contain a number of distinctive landforms. During overbank flooding, the maximum deposition of silt occurs just along the edges of the overflowing stream. Here the current is checked suddenly by friction with the flood plain surface, and as much as a meter of new material may be deposited in one major flood. This deposition causes the flood plain to be elevated most rapidly close to the margins of the stream channel, producing a *natural levee.* Natural levees slope almost imperceptibly away from the channel banks toward backswamp areas in which water is trapped between the valley walls and the levee crests. Natural levees along the Mississippi River in Louisiana rise to 5 to 6 meters (15 to 20 ft) above the adjacent

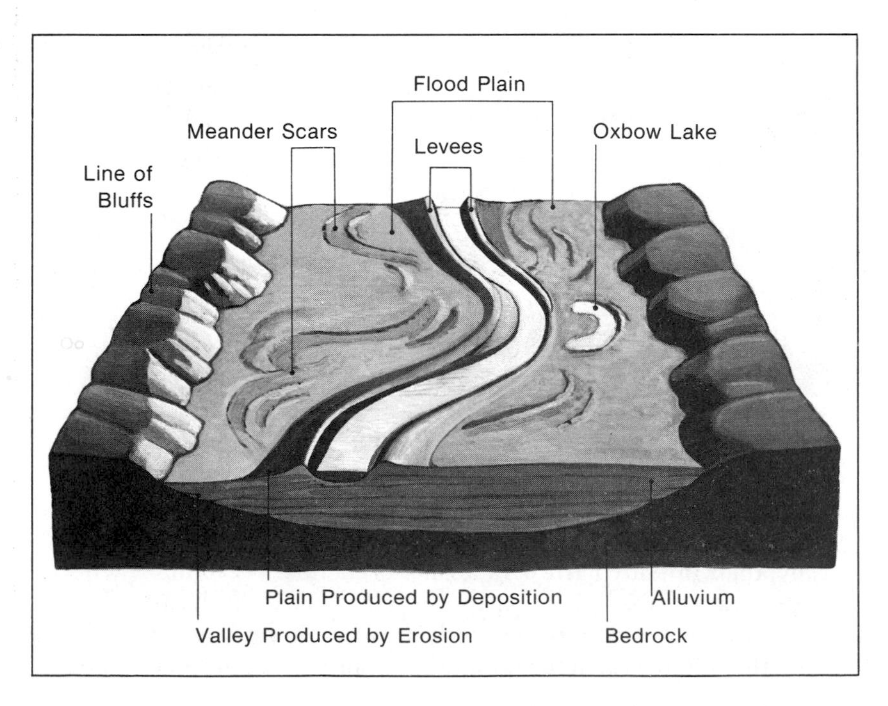

Figure 12.18 Flood plains may result from either fluvial aggradation or lateral stream planation that wears back the valley walls. Oxbow lakes are the remnants of recently abandoned meanders. When the lake is eventually filled with sediment and vegetation, a *meander scar* remains. In times of flood, the overflowing river carries new sediment onto its flood plain and builds raised banks, or natural levees, by deposition of silt close to the river channel. (After Bunnett, 1968)

backswamps, but their slopes are so gradual that they are hardly visible. Natural levees are prime sites for agricultural settlement due to their fertile soils and good drainage characteristics. The highest levees occur where stream channels have been fixed in place for hundreds of years.

If the flood plain stream has excess energy, or if the flood plain itself is composed of easily eroded sand rather than more cohesive silt, meander shifting is more active, and large natural levees cannot form. In such settings, the meanders often expand into flaring "gooseneck" loops that crowd so closely against one another that flood flows break through the narrow intervening necks of land and cut off the meander loops, blocking both ends with sediment. These remnants of the stream are known as *oxbow lakes.* Hundreds of present and past oxbow lakes are visible along the Mississippi River from Cairo, Illinois, to Baton Rouge, Louisiana—many of them created simultaneously by major floods on the river (see Figure 12.12). Oxbow lakes gradually fill with fine sediment and marsh vegetation, and are eventually extinguished, but their traces remain clear in aerial photographs long afterward.

Each stream avulsion that produces meander cutoffs and shortens the stream's length also steepens the stream gradient slightly, giving the stream more energy locally. This triggers the expansion of another loop, both to expend the excess energy and to flatten the stream gradient, eventually leading to another cutoff, and so on—the process is endless, a fine instance of a self-reinforcing feedback effect.

Not all flood plains are produced by erosion. Many of the earth's largest rivers—including the Mississippi, the Sacramento and San Joaquin of California, the Ganges of India, and the Amazon of Brazil—have developed in broad depressions produced by crustal movement. These rivers bring sediment from higher adjacent regions to fill the structural troughs. Such streams do not occupy true valleys, and the crests of their natural levees may be the highest land around for hundreds of square kilometers. Many other flood plains are products of stream aggradation, rather than lateral planation and valley widening.

Underfit Streams

Some streams appear to be much too small for the flood plains and erosional valleys that they occupy. Most such streams are narrow, with closely spaced meanders wandering through a valley that itself meanders on a much larger scale. Streams of this type are said to be *underfit* with respect to their valleys. These conditions suggest a vast reduction in stream discharge. Figure 12.13a shows such a stream. In many cases, the same geometrical relationship exists between valley width and valley meander wavelength as between stream width and stream meander wavelength. Accordingly, the entire valley floor is regarded as a former large stream channel that has become filled with sediment.

There are several explanations for underfit streams. At the end of the Pleistocene, about 10,000 years ago, dwindling continental glaciers poured enormous amounts of meltwater into stream channels, vastly enlarging them; when the glaciers disappeared, the streams were left with their original much smaller discharges. However, many underfit streams are in areas not affected by glaciation. Their underfit character may reflect climatic changes that have greatly reduced precipitation in the last few hundreds of thousands (or even millions) of years. An underfit stream can also be formed by stream piracy, when some of the water formerly entering the stream channel is diverted to another course, diminishing the stream's discharge. The enormous decrease in discharge required to produce some underfit streams is not easily explained, and certain such streams remain something of a geomorphological enigma.

Fluvial Terraces

We have seen that many things can upset stream equilibrium, leading to stream rejuvenation or incision, as well as to stream aggradation. There is much evidence in the landscape that graded streams periodically change their condition, beginning to lower their channel by

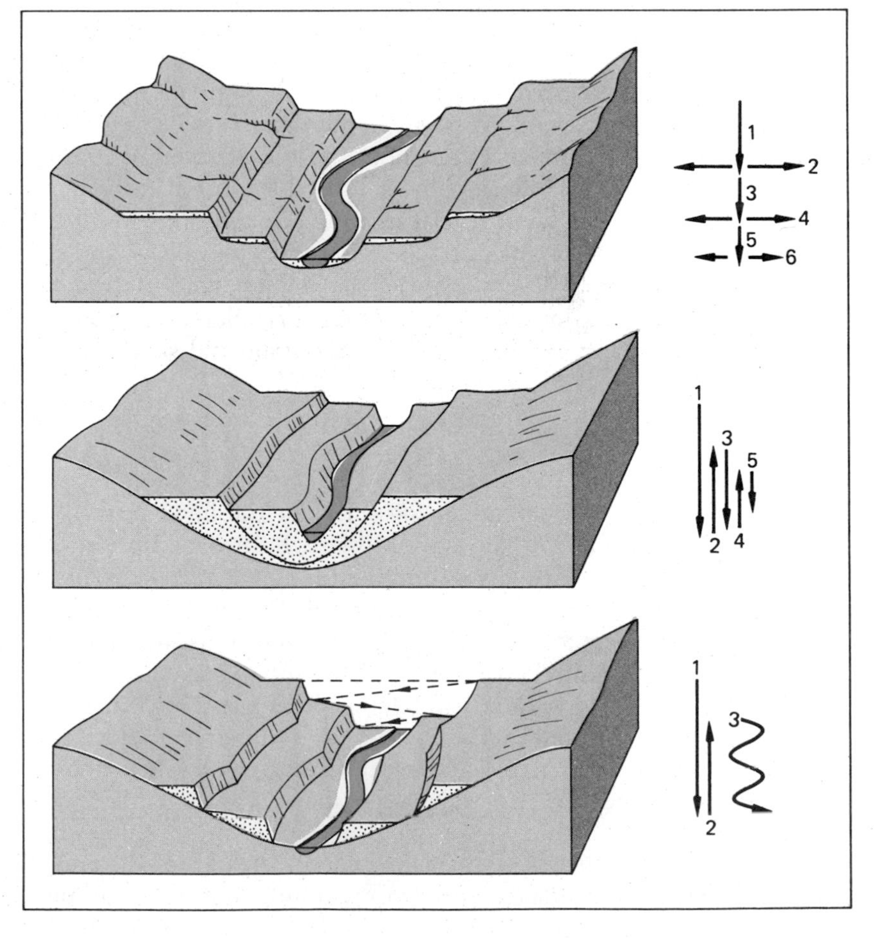

Figure 12.19 Stream terraces vary significantly in origin, as these diagrams illustrate. To the right of each diagram is an indication of the nature of the river movements necessary to create the associated terraces: downward arrows indicate stream incision; upward arrows, aggradation; horizontal arrows, lateral planation; sinuous arrows, simultaneous incision and lateral planation. The terraces in the top diagram have been produced by periods of downward stream incision (1, 3, 5) separated by intervals of valley widening by lateral planation and slope erosion (2, 4, 6). These terraces are erosional and are said to be *paired* since those on opposite sides of the stream match in elevation. The center diagram shows paired fill terraces produced by two separate phases of alluvial aggradation (2, 4) followed by renewed valley excavation (3, 5). In the bottom diagram, valley deepening (1) is followed by aggradation (2) producing an alluvial fill. Subsequently the stream cuts downward at the same time that it is planing laterally in the fill (3), producing *unpaired* terraces.

incision or to elevate it by deposition. Where a river cuts downward into a broad valley floor, the flood plain no longer functions as an overflow channel; instead, it becomes an abandoned, flat *terrace* overlooking the river, as illustrated in Figure 12.19.

A temporary phase of aggradation, indicating decreased stream energy or increased work in the form of sediment transport, can also produce a stream terrace. In such a case, the terrace is composed of alluvial fill deposited in the valley. To form a terrace, the fill must subsequently be trenched by the stream, indicating a return of its erosional vigor. Fill terraces are especially common where streams have been temporarily aggraded by masses of sediment delivered by Pleistocene glaciers, but they may also be produced by tectonic or climatic changes and by up-and-down fluctuations in the stream's base level.

River terraces are important indicators of environmental change; they signify that for some tectonic, climatic, or hydrological reason, the river has changed its tendency from lateral planation to incision, or from incision to aggradation to renewed incision, all in attempts to maintain equilibrium. Some rivers are bordered by great flights of terraces, suggesting intermittent downcutting separated by periods of lateral planation at successively lower levels, or repeated episodes of cutting and filling. The rivers of New Zealand are famed for their magnificent terraces, which reveal intense tectonic activity and a complex history of environmental disturbances.

Alluvial Fans

When a stream descending a steep mountain canyon suddenly issues onto an open plain, it usually loses a great deal of its kinetic energy. This may be due to either an abrupt decrease in its gradient or the spreading of waters previously confined in a narrow rock channel. In either case, the stream bed load tends to be deposited in the form of a low cone, issuing from the canyon mouth and spreading over the plain, as shown in Figure 12.20, p. 362. This deposit is known as an *alluvial fan*. Although the initial deposition at the apex of the fan may be due to a break in stream gradient, later deposition over the fan surface is largely due to the dispersal of water in small distributary channels and to seepage losses into the fan itself, thus diminishing the volume and therefore the velocity of the flow.

Alluvial fans are most conspicuous in arid regions where the vegetation cover is sparse and the infrequent heavy rainfalls remove large quantities of loose rock debris from unprotected slopes. The resulting stream floods, which may last only a few hours, are very highly charged with sediment, which is usually carried only as far as the canyon mouth, making the apex of the fan much thicker and higher than the margins. Often the flows are quite viscous due to their high content of solid matter, becoming mudflows or debris flows rather than stream floods. The density of debris flows is many times that of water flows, so that boulders weighing several tons are easily

Figure 12.20 This alluvial fan is located on the east side of Death Valley, California; its size can be judged from the road crossing it. Rains are infrequent in this region, but torrential storms occur in the mountains from time to time, carrying down large quantities of coarse sediment to be deposited over the fan.

transported by them and are often seen scattered over alluvial fan surfaces. In mountainous deserts, adjoining alluvial fans form continuous embankments, or *alluvial aprons* (also known as *bajadas*), between the mountains and the adjacent lowlands (see Chapter 14). In California's Death Valley, these alluvial aprons rise as much as 600 meters (2,000 ft) over a horizontal distance of 8 km (5 miles). There is often an abundance of groundwater present at the base of alluvial fans, which makes them favored spots for settlement and agricultural development in dry regions. A prime example is Salt Lake City, which sprawls over the slopes of a large alluvial fan adjacent to Utah's Wasatch Mountains. The Mormon settlers who founded Salt Lake City also established several other settlements in similar settings, including San Bernardino, California, and Las Vegas, Nevada.

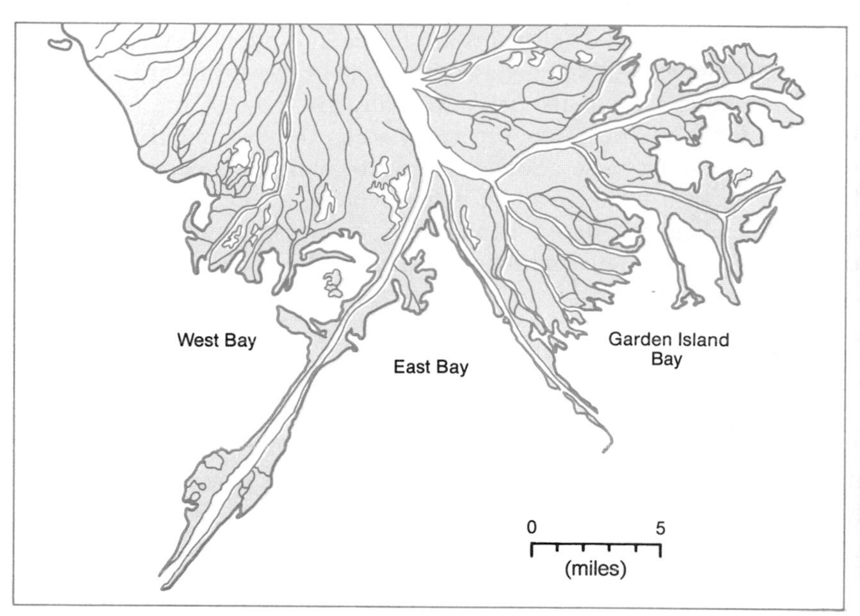

Figure 12.21 **(left)** The delta of the Mississippi River in Louisiana resembles a bird's foot because the delta is subsiding, leaving only the crests of its natural levees above sea level. The subsidence is due to compaction of sediments and also general tectonic sinking of the floor of the Gulf of Mexico.

(right) The fan-shaped delta of the Nile River in Egypt resembles the triangular shape of the Greek letter "delta," which accounts for the origin of the term. The delta of the Nile has a different shape from the delta of the Mississippi because of more vigorous erosion by currents of the Mediterranean Sea and the fact that the delta is not subsiding. The construction of the Aswan Dam has changed the sedimentation pattern of the Nile, and its delta is in danger of accelerated erosion.

Deltas

When rivers flow into larger bodies of water, such as lakes and ultimately seas, their velocity is checked and they lose their load-transporting ability. Their bed load and much of their suspended load is dropped at the river mouth in the form of a *delta*. The name was first coined some 2,500 years ago by the Greek historian Herodotus to describe the characteristic shape of this depositional feature, which often resembles the Greek letter Δ. Delta configuration, however, varies considerably, from the classic examples of the Nile and Niger rivers of Africa to the distinctive bird's-foot form of the Mississippi delta (see Figure 12.21). Each delta has its own characteristic shape, usually determined by the balance between the river's deposition of sediment and the removal of the sediment by wave erosion and offshore currents, which constantly transport material away from the delta fringes, distributing it along coasts or moving it into the ocean's depths. Where wave action or coastal currents are vigorous, delta formation may be prohibited entirely.

When a river enters its delta, its discharge becomes divided among several distributary channels. The branching pattern of the distributary channels is determined by prior underwater depositional patterns. The Mississippi delta actually is an ancient feature beginning about 1,000 km (625 miles) inland from the mouth of the stream, filling a former bay of the Gulf of Mexico that extended to Cairo, Illinois. The exterior bird's-foot delta has been built in the last 400 years, after a river avulsion near the present site of New Orleans. The bird's-foot form results from tectonic subsidence of the ocean floor in the Gulf of Mexico and the compaction of the delta sediments, which submerge all but the crests of the natural levees built along the delta distributary channels.

Like flood plains, delta plains are fertile and often very densely populated. The famed Nile delta is the home of two-thirds of Egypt's population and most of its cities and industry. Loss of life can be heavy in deltas during floods, especially when the distributary channels change their courses. Some of the worst disasters occur when hurricanes cause flooding along deltaic coasts, a fate that has frequently befallen the dense populations occupying the combined deltas of the Ganges and Brahmaputra rivers of India and Bangladesh, and also the inhabitants of our own Gulf Coast from Florida to Texas.

SUMMARY

Running water gains the ability to do work when its potential energy is converted to kinetic energy by the force of gravity. Much of the energy of flowing water is used to overcome internal friction and friction with the bed and banks of stream channels, and some is dissipated as heat in turbulent flow. The remaining energy can be used to transport solid material in suspension or as bed load. The greater the velocity of flow, the larger the particles the stream can transport.

The energy of a stream depends upon the interaction of many variables, including the discharge, sediment load, width, depth, and slope of the channel, and bed roughness, all of which affect the stream velocity. In general, stream processes tend to balance stream energy and the work to be done in the form of water and sediment transport. Both vertical and lateral adjustments of stream channels are means of equating stream energy to work. In general, stream flow tends to produce meandering channels, but where a coarse bed load must be transported, braided channels evolve. Streams create a variety of distinctive landforms by downward and lateral erosion and by the deposition of sediment. Valleys, flood plains, natural levees, terraces, alluvial fans, and deltas are the principal landforms directly associated with the fluvial processes. All of these forms are consequences of the tendency of streams to maintain equilibrium in the fluvial system.

Further Reading

Belt, C. B., Jr. "The 1973 Flood and Man's Constriction of the Mississippi River." *Science,* 189:4204 (August 1975): 681–684. This article analyzes the effects of human modification of the Mississippi River channel, which caused a flow of only moderately high magnitude to produce a flood of catastrophic proportions.

Chorley, Richard J., ed. *Water, Earth, and Man.* London: Methuen & Co., 1969. (588 pp.) There are several chapters on channelled flows, river regimes, and floods in this collection of articles on water in all its forms of occurrence, written by experts on the various topics.

Crickmay, C. H. *The Work of the River: A Critical Study of the Central Aspects of Geomorphogeny.* London: Macmillan Co., 1974. (271 pp.) This is an unusually interesting analysis of fluvial and slope processes, written in an engaging manner. The treatment is advanced but nonmathematical.

Dury, George H., ed. *Rivers and River Terraces.* New York: Praeger, 1970. (283 pp.) This collection of influential writings concerning fluvial processes and resulting landforms shows the progress of thought on these topics over the past 100 years.

Gregory, K. J., and **D. E. Walling.** *Drainage Basin Form and Process.* London: Arnold, 1973. (456 pp.) This is the most detailed current treatment of fluvial processes, reflecting English work in the field and updating Leopold, Wolman, and Miller. Most examples are from the British Isles.

Kemper, J. P. *Rebellious River.* Boston: Bruce Humphries, Inc., 1949. (279 pp.) The author recounts the story of man's early attempts to control the Mississippi River, spotlighting the errors made from a lack of understanding of a complex physical phenomenon.

Review Questions

1. What factors control drainage pattern? drainage density? the relationship between stream order and stream numbers?
2. To maintain constant stream velocity (and therefore transporting power), a decrease in stream depth must be offset by a change in one of the other factors affecting flow velocity. What sort of changes could compensate for a decrease in stream depth so that flow velocity remains constant?
3. Why is turbulent flow necessary for sediment transport?
4. How does the process of aggradation increase stream energy?
5. Describe the channel changes that normally accompany the passage of a major flood crest.
6. Why do stream gradients normally become less steep in a downstream direction?
7. Which requires a greater expenditure of energy, a gradient decrease produced by stream incision or a gradient decrease produced by increased stream sinuosity (increased lateral planation)?
8. Why are shallow, braided channels more efficient for the transport of coarse bed load than are deep, meandering channels?
9. Give a full definition of a "graded" stream.
10. Why does tectonic uplift cause stream incision?
11. The lower Mississippi River has followed several different courses to the Gulf of Mexico. Although the old channels have been silted up, the different locations of the ancient streams are clearly evident. What would be the principal evidence of the Mississippi River's former courses?
12. In what geographical locations would you expect to encounter underfit streams? fluvial terraces?
13. Why are the boulders found on some alluvial fans much larger than those of alluvial fill terraces?
14. Under what conditions would river deltas grow? shrink? be in a steady state?

Leopold, Luna B. *Water: A Primer.* San Francisco: W. H. Freeman & Co., 1974. (172 pp.) This is a simple but technically sound introduction to the general principles of hydrology and the action of water on slopes and in channels, by the former chief hydrologist of the U.S. Geological Survey.

———, **M. Gordon Wolman,** and **John P. Miller.** *Fluvial Processes in Geomorphology.* San Francisco: W. H. Freeman & Co., 1964. (522 pp.) A landmark when first published, the book presents the findings of several decades of research on fluvial processes by U.S. Geological Survey field workers. It is somewhat technical but highly readable and most informative.

Morisawa, Marie. *Streams, Their Dynamics and Morphology.* New York: McGraw-Hill Book Co., Earth and Planetary Science Series, 1968. (175 pp.) This brief paperback presents most of the fundamentals of fluvial processes in nontechnical terms in a concise format.

———, **ed.** *Fluvial Geomorphology.* Binghamton State University of New York: Publications in Geomorphology, 1974. (314 pp.) This is a collection of articles on topics of current interest assembled for a 1973 symposium on fluvial geomorphology. The subject matter is diverse, including an interesting contribution on the greatest flood known, which created Washington's Channelled Scablands.

Russell, Richard J. *River Plains and Sea Coasts.* Berkeley and Los Angeles: University of California Press, 1967. (173 pp.) Many interesting facets of stream behavior are explored, with most examples being drawn from observation of the Mississippi River—especially good on flood plains and deltas.

Schumm, Stanley A. *River Morphology.* Stroudsburg, Pa.: Dowden, Hutchinson and Ross, 1972. (429 pp.) This is another collection of classic articles on fluvial processes published between 1850 and 1971. Some are technical, others are in the explanatory-descriptive vein. Each constitutes a major contribution to fluvial studies.

Case Study:

Water Management on the Mississippi

From time to time, man has to deal with natural disasters. Earthquakes, volcanoes, floods, droughts, hurricanes, and tornadoes can strike unexpectedly, often wreaking havoc on lives and property. But often people increase the danger by building homes, planting crops, and erecting cities in problematic locations; for example, by building on a flood plain or delta.

Riverbanks have always been attractive areas for settlement, and river floods have brought bounty as well as tragedy: the yearly flooding of the Nile, for example, replenished the fertile soil along its banks. But most floods are catastrophic, and because precipitation and consequent flooding are generally unpredictable from year to year, attempts are made to control the flow and distribution of river waters with dams, levees, canals, and spillways. Even with such controls, nature occasionally surprises us.

Historically, the best way to avoid the floods of the Mississippi River that occurred every few years was to move to high ground. But as the valley became more settled, people were no longer content to adjust their affairs to the natural workings of the river. So the goal became to eliminate nearly all floods.

Until recently the principal defense against floods was the construction of artificial levees to augment the natural levees formed by the river during overbank flooding. However, confining the river within its normal channel, and out of its normal flood plain during peak discharge, raises its height considerably. A break in the levee under those conditions can produce a flood of disastrous proportions. In the 1930s, the opening of the Bonnet Carré spillway just north of New Orleans, Louisiana, inaugurated a new approach to flood control on the lower Mississippi. The spillway is a controllable crevasse 3 km wide and 7 km long leading from the Mississippi above New Orleans to the lower level of Lake Pontchartrain. During periods of great discharge, river water can be diverted to the lake through the spillway, which eases the pressure on the levees downriver at New Orleans and in the lower delta. The Bonnet Carré spillway has been used four times since the 1930s, successfully averting floods on each occasion.

By 1950 hydrologists and engineers reported that when the Mississippi was high, increasing volumes of water were flowing through the Old River channel into the Red and Atchafalaya rivers. The Atchafalaya River represents a much shorter and steeper route to the Gulf, and it threatened to divert much, if not all, of the Mississippi River into its own channel by about 1980. So the Old River Diversion Control was built in the late 1950s to allow only about one-third of the Mississippi flow into the Atchafalaya at all times. However, during periods of very high water, not all of the flow could be led past New Orleans safely, so the Morganza floodway was also built to divert floodwater from the Mississippi into the Atchafalaya basin.

The greatest flood in the Mississippi Valley in recent years occurred in 1973. Residents along the river knew high waters were coming months in advance. The question was: How high would the water rise?

The flood story of 1973 actually began in late 1972. Rainfall had been particularly high in the central Mississippi Valley during the summer and fall. Farther south, rainfall became heavy in the late fall and early winter. By winter, the lower Mississippi was unusually high and there was already talk of possible spillway openings. Spectacular deluges occurred during the spring, particularly in the south, and excessive rain continued to fall over the entire basin. The water came dangerously close to the tops of levees. In February the Bonnet Carré spillway was opened, and in mid-April plans were made to open the Morganza floodway for the first time, to release water into the Atchafalaya basin to reduce pressure on the weakened Old River Diversion Control upstream.

There is no perfect flood control system: relieving pressure in one area means adding pressure to another. Opening the Morganza floodway appeared to be the way to protect the greatest number of lives and property, but taking the pressure off Old River and New Orleans meant more water downstream in the Atchafalaya River. There was concern in the Atchafalaya basin that sediment and silt-laden waters from opening the Morganza floodway might fill in the shallow lakes and swamps. Several hundred people as well as livestock were evacuated from the Morgan City area in the Atchafalaya basin, and it was feared that the entire city would be flooded. Oil and gas wells were flooded and their production was stopped, at least temporarily. Farmland was under water, so crops of cotton and corn could not be planted that spring. However, varieties of late season soybeans were planted instead. Short-term adverse effects were also felt by the crawfish industry in the Atchafalaya area.

Opening the Bonnet Carré spillway affected the ecology of Lake Pontchartrain, for a time at least, by adding sediment and fresh water that changed the salinity and damaged the oyster population. However, there was some speculation that over time the ecology of the lake would be improved by the flushing action of the water, because it washed out debris and water-clogging vegetation and brought in a fresh supply of nutrients.

The opening of floodways such as the Morganza and the Bonnet Carré prevented the main river from flooding, so that flooding occurred only on the Mississippi's tributaries. Because water in the main stream was so high, the other rivers were unable to flow into it. Consequently, they overflowed their banks locally, forming "inland seas" of backwater flooding in the basins and valleys.

Backwater flooding occurred mainly in Mississippi and Louisiana. Farther north, the flooding was worse; the heaviest flooding occurred around St. Louis, Missouri. In one spot—Kaskaskia Island in Illinois—the main river overtopped the levees. In some places, levees broke or washed out, and the water poured through them; in other places, the levees held but were not high enough, and the water cascaded over them. Near St. Louis, people fled their homes, moved back to their sodden, muddy houses when the rivers dropped, and then had to leave again, and yet again, as new crests came down the river.

The rains and the flooding finally stopped in early summer. The flooding had lasted months, and four separate flood crests had flowed down the Mississippi. The long-term results of the 1973 flood were difficult to assess. Parts of nine states—Kansas, Iowa, Illinois, Missouri, Kentucky, Tennessee, Arkansas, Mississippi, and Louisiana—had been declared disaster areas. At least two dozen people were killed by the floods, and about 35,000 were evacuated from their homes. In Louisiana alone, about 3.7 million acres of land were under water.

Because the waters subsided so slowly, it was too late that season to plant a cotton crop in the fertile delta area of Mississippi. Soybean, corn, and sugarcane crops were also affected because planting was delayed. The heavy losses in food crops and livestock led to higher food prices throughout the country. Thousands of homes had to be rebuilt, roads and levees had to be repaired, and massive clean-up projects were undertaken.

But even with the great losses, many engineers believed that the flood control measures proved as effective as could be expected. The levees and flood walls prevented the Mississippi River itself from flooding except in one place, and water was kept out of the major cities. Most of the flood warnings came early enough that people could be evacuated, and loss of life attributed to the flood was relatively small. After the flood, study of the positive and negative effects of flood control systems continued.

During 1974, the river was again higher than normal, and even higher in 1975, when some flooding occurred. However, the next year, precipitation over much of the Mississippi River basin was far below normal. Beginning in August 1976 and continuing into the summer of 1977, the river fell to record or near-record lows from St. Louis to the Gulf. The low water severely hindered navigation. Towboat operators were forced to reduce the number of tows by about half, and each barge was loaded only to about two-thirds of capacity. The travel time and cost of shipping bulk goods such as grain on the river increased. In order to maintain navigation, channels were dredged and reservoir storage released to raise low-water levels. Because of climatic variability, water-resource planners have had to cope with both high-water and record low-water problems within two years.

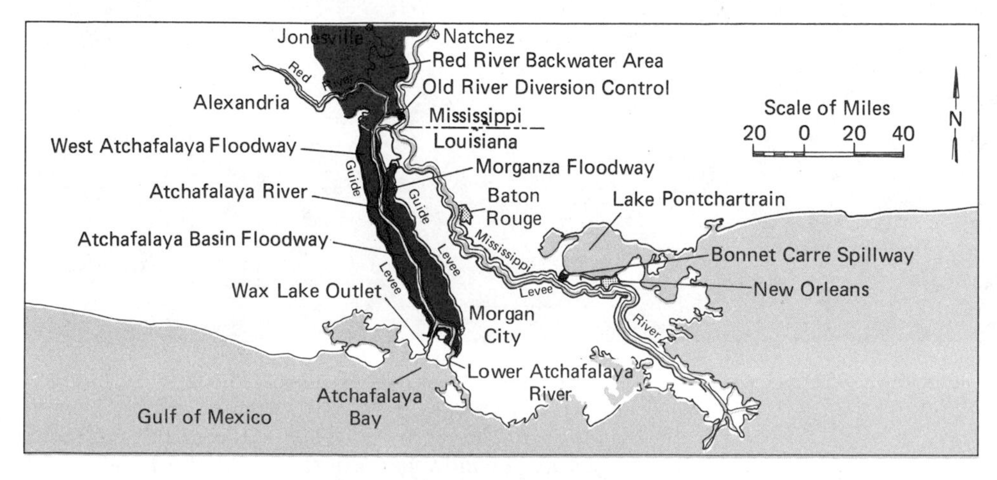

Lower Mississippi River Flood Control Plan.

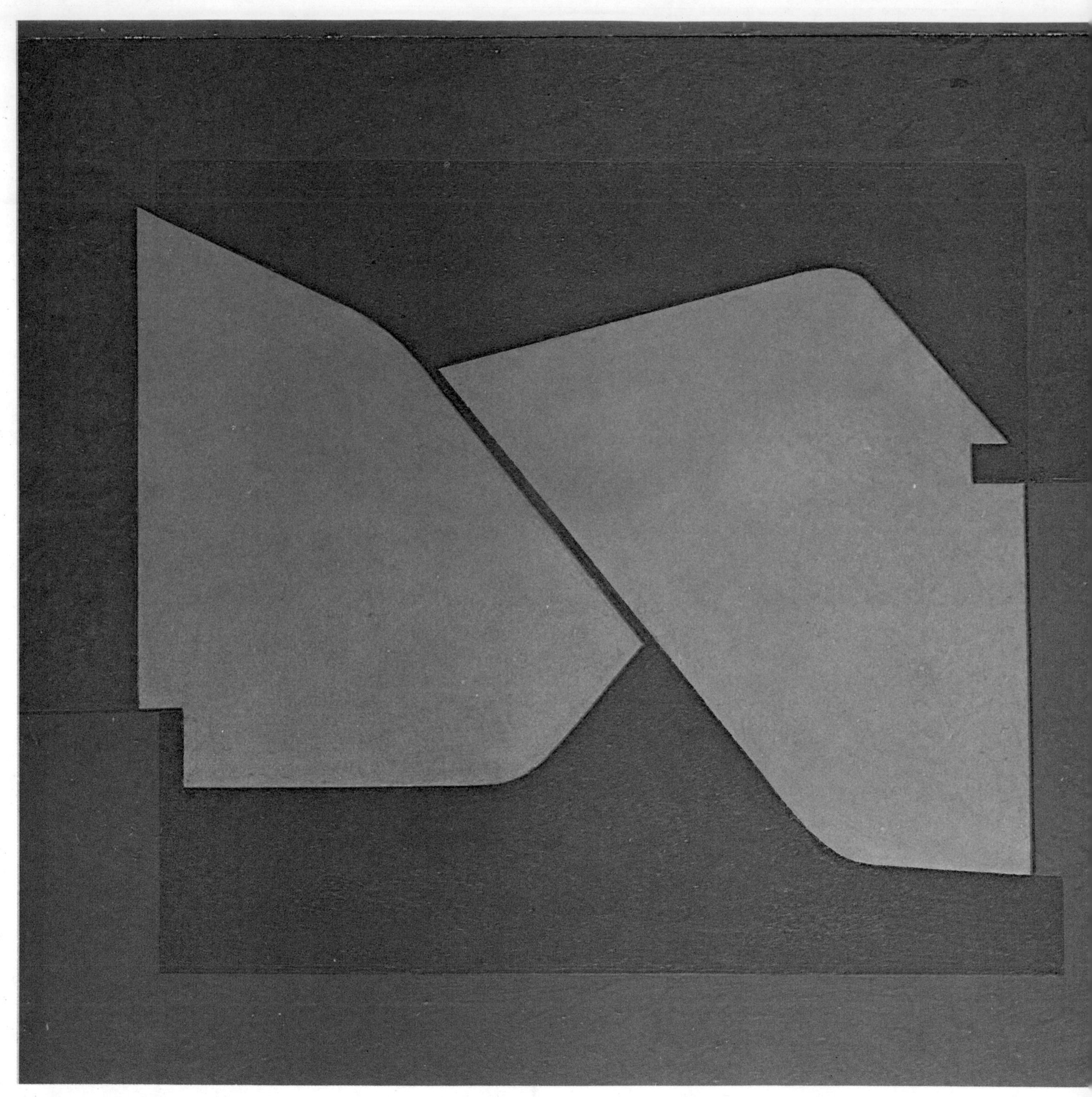

Kandahar by Victor Vasarely, 1951.

The land surface may be altered abruptly by a volcanic eruption or a sudden earthquake, or slowly by viselike compression. Such dramatic processes of change are powered by the earth's internal energy. Landforms created by movements of the earth's crust are subject to continued change by weathering and erosion driven by solar and gravitational energy.

THIRTEEN
Geologic Structure and Landforms

AMONG THE MANY WINERIES in California, one near the town of Hollister holds a special fascination for Pacific Coast geologists. They are interested in this establishment for reasons that have little to do with its bottled products. What attracts their attention is the fact that one of the winery buildings is slowly being torn apart by movements of the earth's crust. In addition to keeping an eye on this strange occurrence, the same scientists are carefully recording other phenomena in the vicinity: the gradual horizontal displacement of natural landscape features, vertical movements of the land surface, variations in the water level in deep wells, water pressure in the pores of rocks far below the earth's surface, and the electrical resistivity and magnetic properties of subsurface rocks. Abrupt changes in any of these may indicate the imminent danger of a violent rupture of the earth's crust—an earthquake.

In many other countries, researchers are watching similar phenomena. On the largest island in the Hawaiian chain, scientists continuously monitor an array of seismographs and measure minuscule changes in the slope angles of the volcanoes visible beyond the windows of the Hawaiian Volcano Observatory. A slight increase in slope angle indicates that a volcano is "puffing up," and trembling lines on the seismographs mean that subsurface molten rock is flowing preparatory to an eruption. In western Iran, geologists are observing an upward bulging of the land surface that has already bowed 1,700-year-old irrigation canals 6 meters (20 ft) above their former level. Along the Norwegian coast, fishermen as well as scientists note the growing navigational hazards as countless rocky islets expand in size year by year and new ones constantly appear, as the Norwegian continental shelf slowly rises out of the sea.

Very clearly the earth's internal forces remain as active as ever, pushing the crust this way and that, rupturing and wrinkling it, and occasionally exploding outward, bringing the molten material of the planetary interior into the human world. Most of this activity results from movements of the major lithospheric plates, described in Chapter 11, which produce areas of crustal compression, tension, and shearing, in addition to the volcanism caused by sea floor spreading and the subduction process. Even small-scale local crustal motions may be indirect expressions of global plate movements, although the exact connections are difficult to detect. All of this crustal activity is of more than academic interest, for over the past 100 years alone, nearly 1 million people have been killed—in many cases more than 100,000 at a time—as a consequence of sudden convulsions of the earth's crust. Volcanic eruptions and earthquakes have frequently taken tens of thousands of lives almost instantaneously. It is little wonder that scientists of many nations are coordinating their efforts to understand the processes at work in the earth's crust and to minimize their destructive effects upon the earth's inhabitants.

Catastrophic earthquakes and volcanic eruptions have clearly played a role in the formation of some of the earth's grandest scenery, but most landforms associated with crustal motions are the result of small, unspectacular movements repeated again and again over millions of years. The rocks that form the earth's loftiest point, the 10-km-high (6-mile) summit of Mount Everest, were formed on the floor of the sea. In fact, virtually all of the rocks found at high elevations originated on the sea floors or many kilometers below the earth's surface. Nevertheless, the crustal movements that create such heights are often so slow as to be imperceptible without the aid of scientific instruments capable of detecting extremely minute changes in elevation and position. Only occasionally are we directly aware of these slow displacements, as at the famous winery building near Hollister. In this case, a man-made feature is gradually being destroyed by slow horizontal slippage of rocks along the San Andreas fault, a 1,000-km-long (650-mile) fracture in the earth's crust. Movement along this fault also produces conspicuous horizontal offsets of fences, railroad tracks, curbstones, and lines of orchard trees, as well as mountain crests and stream valleys. At the winery, the rate of movement, which has been measured carefully since 1957, averages 1.3 cm per year.

Figure 13.1 (opposite) The layers of sedimentary rock visible in this photograph of the Canadian Rocky Mountains have been subjected to intense deformation by movements of the earth's crust. The once-horizontal sedimentary rock strata have been compressed and folded by stresses caused by motion of crustal plates.

The San Andreas fault system is one of the world's most famous *geologic structures.* The term "structure," in the geological sense, refers to the large-scale arrangement of rock masses (see Figure 13.1), as well as to both the gross and detailed physical and chemical

characteristics of individual rock types. It is an axiom of geomorphology that geologic structure in both of these senses—the nature of the rocks present and their spatial and geometrical arrangement—is a fundamental factor in the diversity of the earth's scenery. It is geologic structure that accounts for the landform contrasts seen in areas of uniform climate—the contrasts that are most important in any single area. Accordingly, this chapter focuses on the important relationships between landforms and the structural influences that account for much of the earth's scenic variety.

THE STRUCTURAL FOUNDATION

We inhabit the surface of the lithosphere, the outer shell of the earth composed of either solid or fragmented rock material. Rocks differ widely in mechanical strength, chemical stability, permeability, and the nature and mobility of their weathering products, all of which are factors in the creation of landforms. Furthermore, rock masses vary in their spatial and geometrical arrangement—ranging from orderly laminations on the regional scale to massive county-sized lumps and slabs, to violently disordered, wrinkled, splintered, and interlayered masses of diverse materials. Rocks are formed in many ways: by the crystallization of molten material, by sedimentation that produces layers of new rock from the weathered and eroded debris of older preexisting rock, and by processes of transformation related to tremendous heat and pressure deep within the earth itself. However, all rocks are composed of minerals, and this is where our discussion of structural influences on landscapes must start.

Rocks and Minerals

A rock held in one's hand may seem a rather unexciting object. But put that rock under a microscope, or better yet, take a very thin translucent slice from it and put *that* under a microscope, and it is easy to see why some people choose to become geologists. Figures 13.4, 13.5, and 13.9 provide examples of the beauty of rock *thin sections.* More importantly, these thin sections, enlarged about ten times, reveal that all rocks are composed of much smaller particles that either interlock or are cemented together. Each of these particles, which seems to glow brilliantly when viewed under polarized light, is one of about 2,000 naturally occurring inorganic substances that have distinctive, almost unvarying chemical and physical characteristics. Each particle is a *mineral.* Minerals are natural combinations of chemical elements in solid form. They are constructed of atoms having a characteristic three-dimensional arrangement that gives them an individualized crystal form, luster, color, hardness, and specific gravity (density relative to an equal volume of water). Some minerals are very familiar, such as *halite* (common salt, NaCl), ice (H_2O), *geothite* (iron rust, FeO(OH)), and *calcite* (whitish deposits in household water pipes and the icicle-like stalactites and stalagmites of caves, $CaCO_3$). The high-density metals—iron, aluminum, copper, lead, nickel, gold, silver, and so on—rarely form separate minerals, but are usually combined in mineral form with nonmetallic elements from which they must be separated by industrial "refining" processes. The minerals that yield these metals are known as *ores* of the metals.

All minerals are combinations of elements that are held together by electrical bonds. Thus the chemical basis of halite, or common salt, is the attraction of positively charged sodium ions for negatively charged chlorine ions. The most reactive ions are those whose outer electron shells are incomplete, tending either to lose electrons to ions of other substances or to gain electrons from them. The most reactive of all ions, as well as the most abundant (see Table 13.1), are those of silicon (Si) and oxygen (O); consequently, these dominate in mineral structure, producing two large families of minerals: the *silicates* and the *oxides*. Silicate minerals are by far the most common, constituting 92 percent of the material of the earth's crust. Most rock-forming minerals are silicates.

In the silicate minerals, combinations of the most common positively charged metallic ions (sodium, magnesium, aluminum, potassium, calcium, iron) are bound to a negatively charged silicate group (SiO_2, SiO_3, SiO_4). In fact, the fundamental building block of the silicate minerals is a tetrahedral atomic structure in which one silicon ion is surrounded by four oxygen ions, with positively charged metallic ions attracted to the faces of the tetrahedron (Table 13.2, pp. 372–373). In some minerals, the silica tetrahedrons remain separate from one another and are held together merely by the attractive force of the positive ions between them. However, in most silicate minerals, the tetrahedrons are linked at their corners to form chains, rings, sheets, or complex lattices. Linkage is by the sharing of oxygen ions at the apexes of the tetrahedrons. The susceptibility of the silicate minerals to chemical alteration is largely dependent upon the degree and strength of tetrahedral linkages. When the linkages are weak, the atomic structure of crystals can be expanded and weakened, and ions can move in and out of the structure, replacing one another and thereby altering the mineral. Consequently, minerals can be ranked according to their chemical stability. It follows that rocks that contain the least stable minerals are the most prone to chemical decomposition at the earth's surface, while rocks dominated by stable minerals are resistant to chemical decay.

Table 13.1. Major Elements of the Earth's Crust

Element	Weight (percent)	Volume* (percent)
Oxygen (O)	46.60	93.77
Silicon (Si)	27.72	0.86
Aluminum (Al)	8.13	0.47
Iron (Fe)	5.00	0.43
Calcium (Ca)	3.63	1.03
Sodium (Na)	2.83	1.32
Potassium (K)	2.59	1.83
Magnesium (Mg)	2.09	0.29
Totals	98.59	100.00

*Computed as 100 percent, hence approximate. After Brian Mason, *Principles of Geochemistry*, John Wiley & Sons, Inc., 3rd ed., 1966.

Igneous Rocks

Molten rock-forming material, or *magma*, at a temperature of 900° to 1,200°C (1,600° to 2,200°F), is present everywhere below the earth's crust at depths of about 70 km (40 miles). Sometimes this material forces its way up through a vent and "extrudes," or spills out, at the surface as red-hot lava (Figure 13.2, p. 373). A much greater volume of fluid magma "intrudes" itself into the upper part of the earth's crust and solidifies there, without reaching the surface. In either case, the product is *igneous rock* (from the Latin *ignis*, fire), meaning rock formed by the cooling and solidification of once-molten material. As the fluid magma cools, individual minerals "precipitate" out in crystal

Table 13.2 Major Rock-forming Minerals and Their Occurrence in Rock Types				
Mineral Group	Mineral	Generalized Chemical Composition		Atomic Structure
		Positive Ions	Negative Group	
Silicates	Olivine	Mg, Fe	(SiO_4)	Single Tetrahedron
	Garnets	Mg, Al, Ca, Fe		Single Tetrahedral Chains
	Pyroxenes	Na, Mg, Al, Ca, Fe	(SiO_3)	
	Amphiboles	Na, Mg, Al, Ca, Fe	(Si_4O_{11}), (OH)	Double Tetrahedral Chains
	Micas	Mg, Al, K, Fe	(Si_2O_5), (OH)	Tetrahedral Sheets
	Clay Minerals	Al, K		
	Plagioclase Feldspar	Na, Al, Ca	(SiO_2)	Three-dimensional Tetrahedral Frame-works
	Orthoclase Feldspar	Al, K		
	Quartz	Si	O	
Carbonates	Calcite	Ca	(CO_3)	Three-dimensional Calcite, Oxygen, and Magnesium Lattices of Varying Form
	Dolomite	Ca, Mg		

After A. Lee McAlester, *The Earth,* New York: Prentice Hall, 1973.

form at successively lower temperatures (Figure 13.3, p. 374). The minerals that crystallize at the highest temperatures are composed of simple unlinked tetrahedrons and are easily altered when subsequently exposed at the earth's surface. The minerals that form last, at the lowest temperatures, have the strongest tetrahedral linkages and are the most resistant to later alteration.

Magma that crystallizes far below the earth's surface does so very slowly, often requiring thousands of years to solidify. With this much time available for chemical reactions to occur, the resulting individual mineral crystals commonly grow to the size of human teeth. Igneous rock formed at a depth of many kilometers is especially coarse-grained and is known as *plutonic* rock (after *Pluto,* the Latin god of the lower world). Similarly, a large individual mass of magma that has intruded itself into the crust is a *pluton.* The most common plutonic rocks are those of the granite family (Figure 13.4, p. 376). Any igneous rock formed below the earth's surface is known as an *intrusive* igneous rock. Erosion frequently exposes these originally deep-seated rocks, and they are an important influence on landforms in many areas.

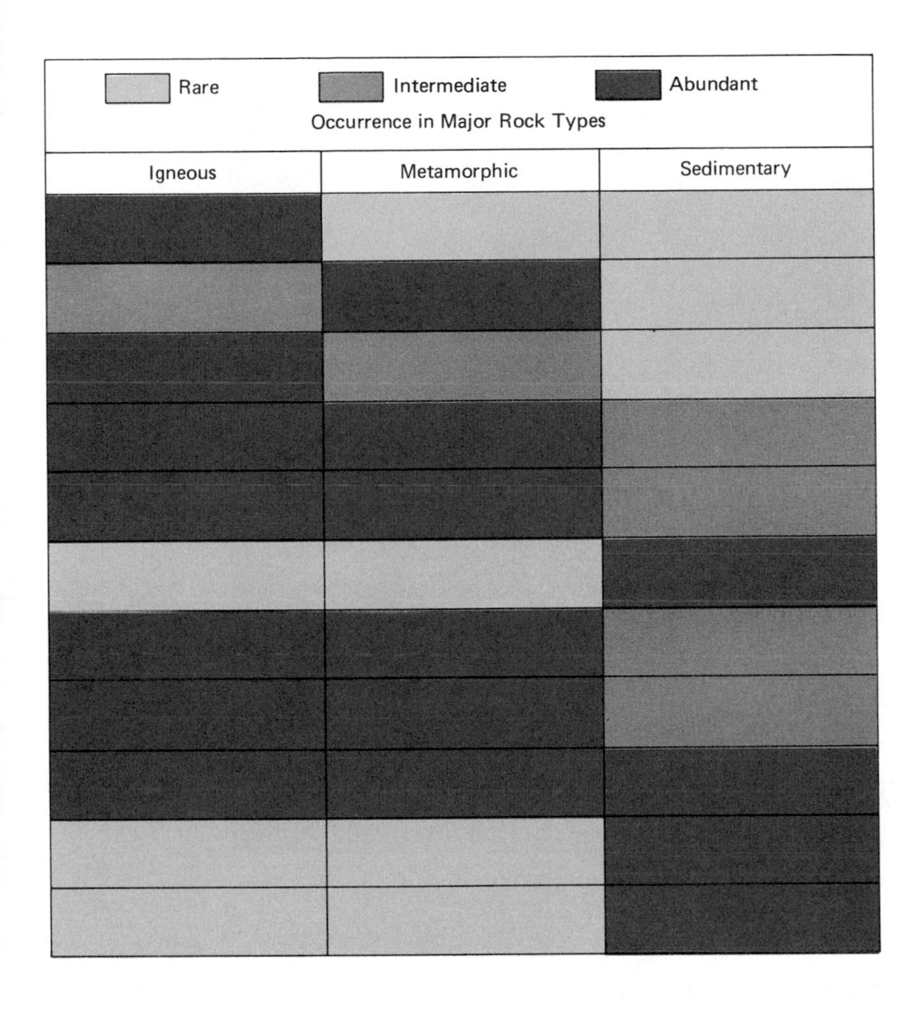

Figure 13.2 Red-hot lava erupted at the earth's surface at temperatures of 900° to 1,200°C (1,600° to 2,200°F) rapidly congeals into extrusive igneous rock (foreground). "Fire fountains" such as this one seen in Hawaii commonly accompany extrusions of basaltic lava and may erupt for periods of several minutes at a time, rising and falling, and often being intermittently active for several days and occasionally even months.

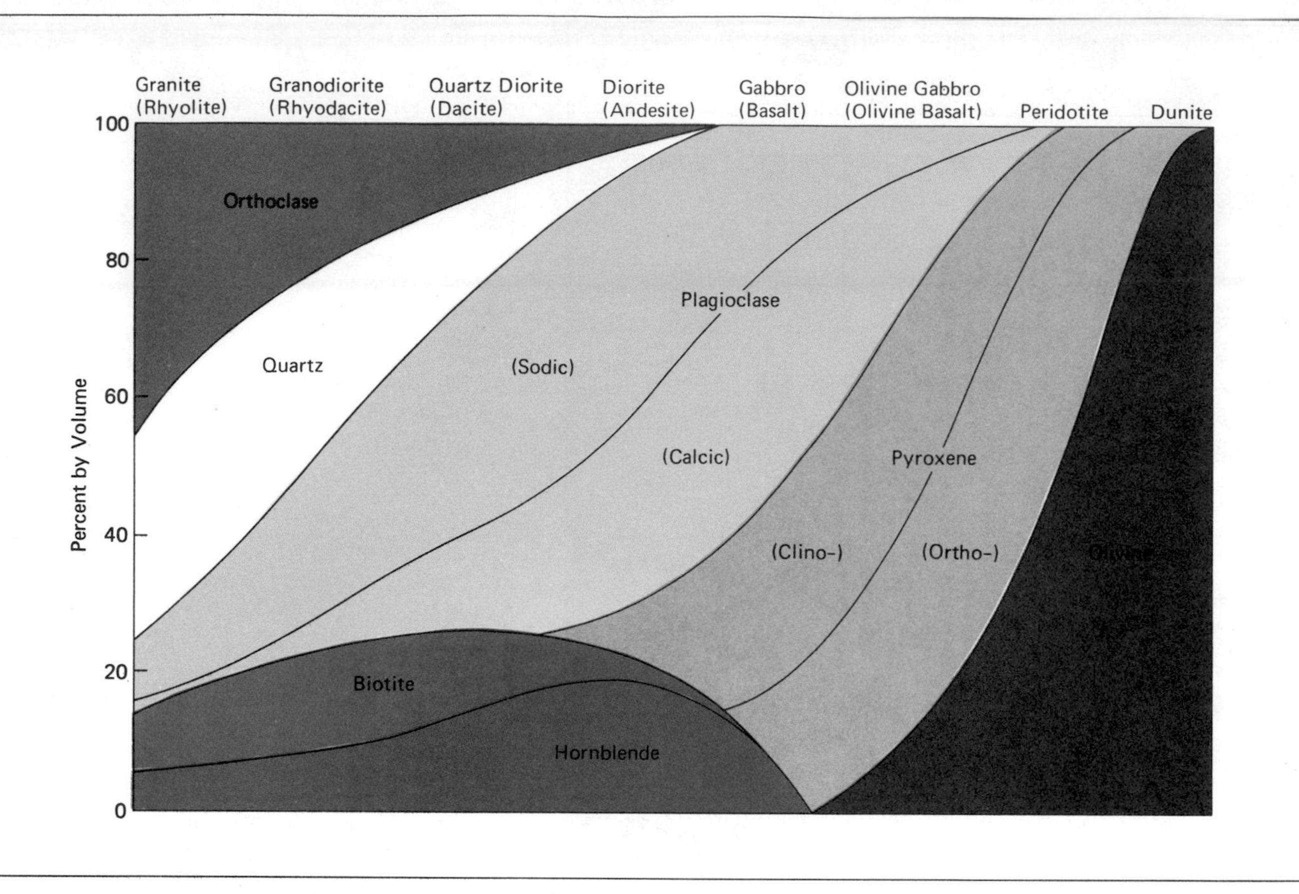

Figure 13.3 Igneous rocks are classified by the proportions of their constituent minerals. Here, the approximate mineralogical composition of the more common igneous rocks is shown in terms of the percentage of the total volume occupied by each mineral. Thus a granite should contain approximately 40 percent orthoclase, 30 percent quartz, 15 percent plagioclase, 10 percent biotite, and 5 percent hornblende. Intrusive rock types are labeled across the top, with their extrusive equivalents in parentheses. Peridotite and dunite are rare, probably being formed only in the earth's upper mantle.

When molten magma forces its way to the earth's surface, it produces a volcanic eruption. This may create two contrasting types of *extrusive* igneous rock: lava and tuff. *Lava* is fluid or semifluid magma that flows out, or is extruded, at the surface. Since the average temperature at the earth's surface is hundreds of degrees lower than that deep within the earth's crust, lava cools and crystallizes much more rapidly than subsurface magma. The rate of cooling of igneous rock is clearly visible in the size of the crystals composing the rock. Lava can always be distinguished from intrusive igneous rock by the smaller size of its crystals (Figure 13.5, p. 377). Lavas of the *obsidian* type have a glassy texture due to quick chilling that prevents any mineral segregation. Because of their sharp edges, obsidian flakes were among the earliest tools utilized by primitive people. However, most lavas consist of a scattering of large (1 to 2 mm long) mineral crystals *(phenocrysts)* imbedded in a fine-grained matrix *(groundmass)* of other crystals that are too small to distinguish with the naked eye.

Explosive volcanic eruptions hurl rock particles and bits of lava into the air, often in enormous volumes. These rain down to produce a surface deposit of *volcanic ash,* which may later become consolidated to form a type of rock known as *tuff.* Other widespread tuffs are the result of extremely violent volcanic eruptions that produce avalanches of fine incandescent particles that move through the air as a "glowing cloud" rather than as a surface lava flow. These settle to produce a distinctive type of tuff that is welded together by heat.

Every intrusive rock has an extrusive chemical equivalent, as shown in Table 13.3. This table also shows that a relatively small number of minerals dominate the composition of most igneous rocks. The most common igneous rocks are the extrusive *basalts* that compose the ocean floors and that also appear widely on the land, the *andesites* that form the bulk of the earth's large volcanic cones, and the intrusive *granites* that form the cores of the continents and are also widely exposed where the continental crust has experienced great upheaval in major mountain systems. Widespread tuffs, such as those of the Yellowstone region, are generally *rhyolitic* in composition.

Although igneous rocks are not the type most widely exposed on the surfaces of the continents, they form about 80 percent of the earth's crust. They underlie all the ocean floors, and over vast areas on the continents, they form a "basement" tens of kilometers in thickness, which is masked by a thin veneer of sedimentary rock types. As indicated in Chapter 11, there is an important distinction between the

Table 13.3. Mineral Compositions of Igneous Rocks

Intrusive Rock	Mineral Composition		Extrusive Rock
	Abundant	Less Abundant	
Granite	Quartz, potassium and sodium feldspars	Biotite, hornblende, and muscovite	Rhyolite
Syenite	Potassium feldspar	Sodium feldspar, biotite, hornblende, muscovite, and less than 5 percent quartz	Trachyte
Diorite	Sodium and calcium feldspars (plagioclase) and hornblende	Biotite and pyroxenes; quartz usually absent	Andesite
Gabbro	Calcium (plagioclase) feldspar, pyroxenes, and olivine	Hornblende	Basalt
Peridotite	Olivine and pyroxenes	Oxides of iron; feldspars usually absent	No extrusive equivalent

Figure 13.4 *Granite* is the most common intrusive igneous rock on the continents.

(top left) The thin section (enlarged about ten times) shows that granite is a coarse-grained rock consisting of many different minerals that have an interlocking structure. Silicate minerals are the main constituents of granite; when granite weathers, it first decomposes into sand. Further chemical alteration transforms some minerals to clay.

(top right) This dome of granite exhibits the characteristic way granite exposed at the earth's surface splits into sheets, or *exfoliates* due to release of confining pressure.

(bottom) The Sierra Nevada range in California, shown here in the vicinity of Mount Whitney, consists largely of granite. The boulders in the foreground are separated from massive parent rock by chemical weathering along the joints of the granite, while the vertical slabs in the background are produced by frost weathering.

Figure 13.5 *Basalt* is an extrusive volcanic rock (or lava) and is the predominant igneous material of the ocean basins.

(top left) The ten-times enlargement of a thin section shows basalt to consist largely of three different silicate minerals and volcanic glass.

(top right) This solidified volcanic lava flow on Kilauea crater, Hawaii, shows the rumpled surface of highly fluid basalt.

(bottom) The Devil's Postpile in the Sierra Nevada of California consists of prismatic columns of andesite, an extrusive rock intermediate in chemical composition between granite and basalt. The columns were caused by thermal contraction as the rock solidified.

igneous rocks of the continents and those forming the ocean floors. The main constituents of igneous rocks making up the continental crust are silica (silicon dioxide) and a smaller proportion of aluminum compounds. The general name for such silicic rock, which has an average density of about 2.8, is *sial* (silicon plus aluminum). This group is dominated by intrusive rock types, of which granite is representative. The average basaltic rock constituting the ocean floors has a density of about 3.0 and consists largely of silica and magnesium compounds in approximately equal proportions, together with aluminum, calcium, and other lesser constituents. Such chemically basic rock has been given the general name *sima* (silicon plus magnesium). As we saw in Chapter 11, the basaltic layer is quite young relative to the basement rocks of the continents and is of mainly extrusive origin, having been created by volcanism along the oceanic ridge systems. Beneath the basalt of the ocean floors, and extending under the continents as well, is a layer of even denser material (3.3) in the upper part of the earth's mantle. This is ultra-basic sima.

When a mixture of undifferentiated rock-forming material is heated to melting in the earth's interior, the melt eventually becomes differentiated into low-density sial and denser sima, with the sial tending to float to the top of the sima (see Figure 1.4, pp. 10–11.) In the same way, the low-density sialic continents appear to float approximately four-fifths submerged in a denser substrate of sima. The concept that the earth's crust is in hydrostatic equilibrium, with the continents "floating" in the denser material of the mantle, is known as *isostacy* (from Greek, *isos,* equal; *stacia,* standing). Isostacy is an important factor in landform development, for it implies that removal of weight, as by erosion, causes the surface to rise, or "float" higher, while addition of weight, as by sedimentation, causes the crust to sink to restore hydrostatic equilibrium (see Figure 13.6).

Figure 13.6 Isostacy is a factor in landform development due to its influence on surface crustal motions. Isostacy refers to the buoyant equilibrium of the earth's crust in which masses of low density sial "float" about four-fifths submerged in a denser substrate of sima. Each segment of a sialic area (density about 2.8) displaces an equal weight of sima (density of about 3.0). For equilibrium to prevail, any large scale, upward projection in the sialic areas must be matched by a downward projection of sialic material. Thus all mountain systems have a deep subcrustal "root" of sial, as shown in the figure. In actuality, the land surface and the base of the root are separated by a vertical distance of 25 to 65 km (15 to 40 miles), but they are shown separately here, with much vertical exaggeration in the scales of each.

The initial land surface (following uplift) and the base of the sialic root are shown here in solid lines. Surface erosion during and after uplift transfers material (and weight) from the high elevations to the low ones, producing the configuration shown with a dashed line. This transfer disturbs isostatic equilibrium. Removal of mass from the eroding area causes it to be too light for the depth of its root. As a consequence, the eroded area tends to rise. The areas receiving sediment increase in weight and sink. Lateral flow of sima in the upper mantle is presumed to be the mechanism permitting sinking in one area and vertical uplift in another, producing the surface contour shown with the dotted line. Due to this *isostatic compensation,* much more time is required and a far greater volume of material must be transferred in order to permanently remove a mountain range than would be the case without isostatic compensation. This is because isostatic uplift itself triggers erosion, which, in turn, produces more isostatic uplift. Once begun, erosion is somewhat of a self-stimulating process.

Sedimentary Rocks

Rock weathering constantly produces fragmented mineral material that is transported away by the energy of running water, wind, waves, glaciers, and the direct force of gravity. All of these erosional systems lose energy at some point, resulting in the deposition of their sediment loads. Streams carry rock debris to lakes, inland seas, and the oceans, where sediment accumulates in layers. Sediment is deposited on the land as well—at the bases of cliffs, on flood plains, in crustal depressions, at the peripheries of glaciers, and where sand-carrying winds decelerate. Sometimes the resulting sediments remain at or near the surface of the earth as unconsolidated deposits: the various types are summarized in Table 13.4, p. 380. Sediments deposited on the continents consist mainly of gravel, sand, and silt. Marine sediments formed close to shorelines consist of sand and silt; farther from shore we encounter clays, and beyond them the limy skeletal remains of microscopic marine plants and animals. If these continental or marine sediments are deeply buried by other deposits, they may become *lithified,* or converted into rock. This process involves compaction and the deposition of mineral cements (principally calcite, silica, and iron oxide) carried by water circulating between the sediment particles. Rocks formed from such loose deposits of rock fragments and organic debris are known as *sedimentary rocks.*

Sedimentary rocks form in layers, each layer, called a *bed* or *stratum,* representing a period of sediment deposition (Figure 13.7, pp. 380–381). The separations between strata, known as *bedding planes,* signify periods of nondeposition. Often the strata on either side of a bedding plane differ considerably in character, indicating a significant change in either the sediment supply or the energy of the depositional environment. Coarse sediments reflect a nearby sediment source and a vigorous transporting medium that keeps the fine material moving onward. Fine-grained deposits indicate a low-energy depositional environment in which even fine material cannot be transported.

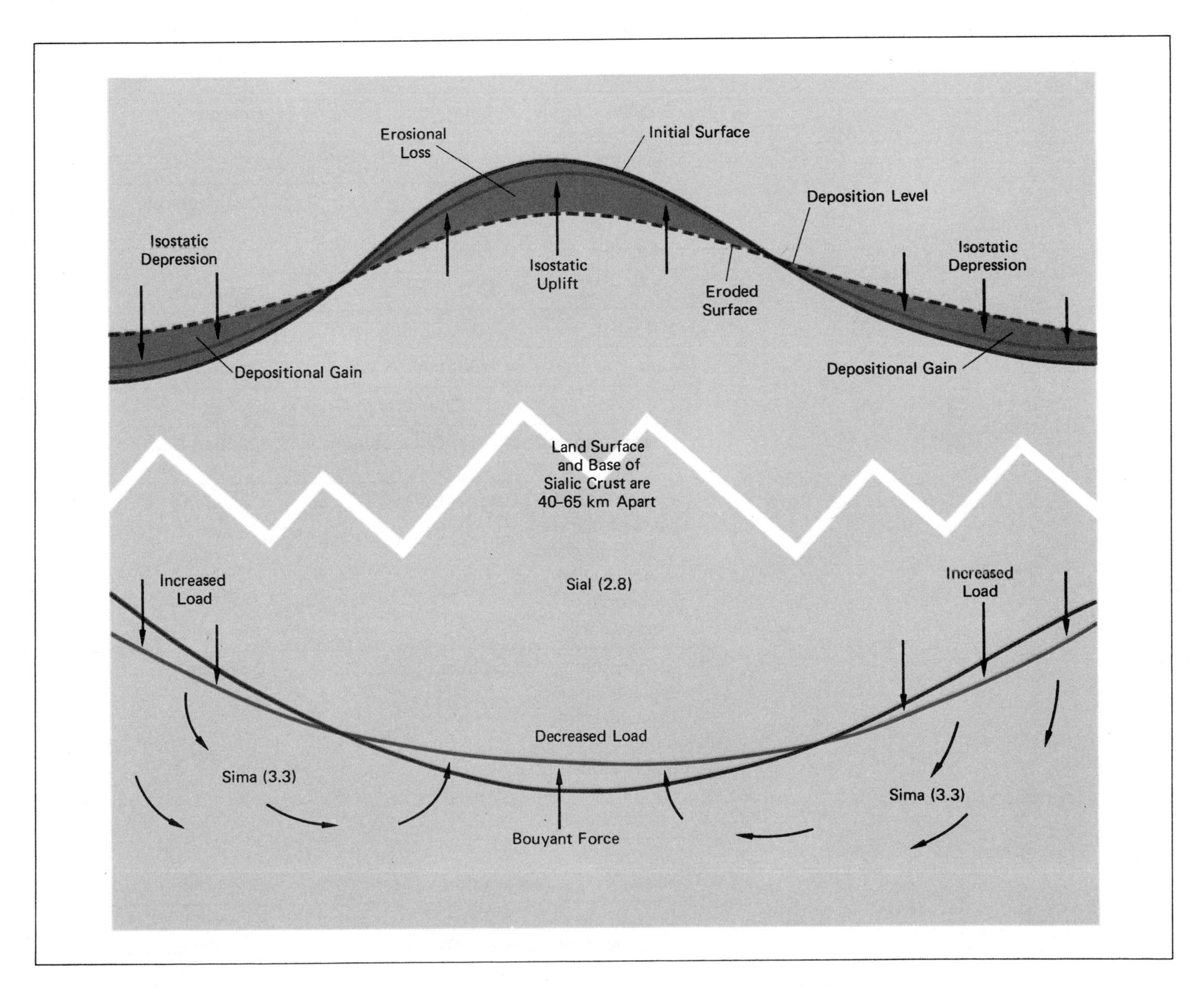

A single layer of sedimentary rock can vary considerably in character from point to point as a consequence of horizontal variations in the material forming the original deposit. In the case of a river flood plain, we commonly find the material composing the flood plain surface changing in a downstream direction, often grading from gravel to sand, to silt, and finally to clay. This reflects the attrition (breakup) of the stream's load as it is carried downstream to the sea. Later burial and lithification of these sediments would produce a stratum of rock that changes laterally from *conglomerate* (cemented gravel), to *sandstone* (cemented sand), to *siltstone* (cemented silt), to *mudstone*

Table 13.4. Common Sedimentary Rocks

Detrital in Origin		
Unconsolidated Sediment	**Grain Size**	**Lithified Rock**
Angular boulders, cobbles, pebbles	>2 mm	Breccia
Rounded boulders, cobbles, pebbles	>2 mm	Conglomerate
Sand	0.02–2.0 mm	Sandstone
Silt	0.002–0.02 mm	Siltstone (Mudstone)
Clay	<0.002 mm	Shale
Chemical in Origin		
Unconsolidated Sediment	**Mineral Composition**	**Lithified Rock**
Calcareous parts of marine organisms and direct calcium carbonate precipitates	Calcite ($CaCo_3$)	Limestone
Magnesium replacement of calcium and direct magnesium carbonate precipitates	Dolomite ($CaMg(CO_3)_2$)	Dolomite
Amorphous silica	Chalcedony, Quartz (SiO_2)	Chert (Flint)
Compacted plant remains	Carbon (C)	Bituminous Coal

(cemented mud), to *shale* (cemented clay). In the sea, well beyond the mouth of the river, the nutrients included in the solid and dissolved load of the stream would have nourished the microscopic floating marine plants and animals known as plankton. The skeletal remains of these organisms produce a calcareous ooze that later becomes lithified as calcium carbonate rock, or *limestone,* which is the source of cement. All the concrete of our highways, buildings, sidewalks, and patios thus consists of once-living matter. A small amount of limestone also seems to have been formed by direct precipitation from seawaters saturated with calcium carbonate. Chemical changes in calcareous deposits in lakes and on the sea floors sometimes transform them from calcium carbonate to magnesium carbonate, which forms the rock known as *dolomite.* Dolomite is a principal source of magnesium, which has many industrial uses.

In some places, the carbon content of marine organic deposits was abundant enough to produce accumulations of *hydrocarbons* (combinations of hydrogen and carbon) in the form of petroleum and natural gas. These low-density substances often migrate upward from the level at which they are formed, to fill the microscopic openings in more porous coarse-grained rocks, especially sandstone.

Economically, the most valuable of all rock types is *coal,* a sedimentary rock that originates as luxuriant vegetation growing in freshwater lagoons and swamps. To be preserved, this organic material must accumulate in an oxygen-poor stagnant-water environment and be acted upon by forms of bacteria that can thrive without oxygen. To be transformed into coal, the resulting organic complex must be compressed by deep burial. Coal is usually found interbedded with sandstone and shale. The first stages of coal formation are probably occurring today in such areas as the Dismal Swamp of Virginia and North Carolina.

Figure 13.7 Sedimentary rocks consist of fragments of rock debris or organic material compacted and cemented together by various chemical substances, most commonly silica, calcite, and iron oxide.

(left) Sandstone, one of the most common sedimentary rocks, is usually composed of grains of silicate minerals cemented together by other minerals, as shown in this enlarged thin section. In general, sandstone is any rock composed of fragments that are in the size range of sand grains. When sandstone weathers, the cementing material disintegrates, releasing the individual grains. Sandstone is formed by the consolidation of beds of sand deposited by wind or water both on the land and in the sea.

(right) This photograph illustrates the laminated nature of sedimentary rocks, which form *strata* that vary in thickness and physical and chemical characteristics. See also Figure 13.9, p. 385, which shows the remarkable exposure of sedimentary rock strata in the upper part of Arizona's Grand Canyon; the terraced slopes indicate marked variations in the various strata's resistances to weathering and erosion.

Figure 13.8 *Limestone* is formed primarily by organic processes. Most limestones are formed by the consolidation of shells of marine animals that have accumulated in shallow seas. Limestone consists largely of compacted and cemented calcium compounds; it is susceptible to erosion by solution, particularly by acidic water. In humid regions where the bedrock consists of layers of thick, bedded limestone, chemical erosion creates a distinctive landscape known as karst topography. Groundwater dissolves the limestone along its joints, which produces surface sinkholes and subsurface caverns.

(below) A portion of the karst region of southern Indiana. The landscape is dotted with sinkholes, which appear as depressions.

(opposite) This narrow passageway in New Mexico's Carlsbad Caverns shows the classic features produced by solution in thick beds of jointed limestone. The open passage is dissolved out along the plane of a vertical joint by groundwater draining laterally through the joint system at a time when the water table was much higher than at present. When the water table fell below the level of this passage, air entered and soil moisture percolating downward toward the lowered water table dripped from the cavern roof. This moisture itself carries calcium carbonate dissolved from the rocks above. Evaporation from water slowly dripping from the cavern ceiling caused precipitation of some calcite, which accumulated in downward-growing, icicle-like *stalactites.* Evaporation from water splashing on the cavern floor precipitated more calcite, producing upward-growing, columnar *stalagmites.* In time the stalactites and stalagmites shown may join to form columns that reach from the floor to the ceiling of the cavern.

The various sedimentary rocks differ notably in their physical and chemical character and in their effect on the landforms sculptured in them. Due to its extremely fine texture, shale is relatively impervious to water; therefore it produces a high proportion of surface runoff, resulting in a high drainage density (see Chapter 12). However, shale is mechanically quite weak, tending to fall apart in thin flakes. Since it is a weak material that produces above-average runoff, shale is the most erodible of all rock types. Offering little resistance to erosional stress, it generally produces gentle slopes.

Sandstone consists mainly of quartz and felspar grains cemented together; it is often rather porous, which permits water to soak into it so that sandstones generate below-average runoff. Thus many sandstones erode rather slowly. Thick sandstone beds usually form ledges or cliffs rising above shale slopes. Conglomerates, which consist of cemented gravel or cobbles, are generally even more resistant than sandstone due to greater mechanical strength, and they usually produce bold ledges where exposed.

Limestone is peculiar in that its basic mineral, calcite ($CaCO_3$), is soluble in slightly acid water. Since soil moisture commonly contains humic acids derived from the decay of vegetation, most groundwater has the ability to dissolve limestone. Such solution is concentrated wherever moisture enters most easily, as along rock joints and bedding planes and especially at joint intersections. Limestone solution is visible at the surface in the form of *sinkholes,* formed either by direct removal of matter at joint intersections or by collapse into underground voids produced by subsurface solution (see Figure 13.8). Some areas of Indiana have 1,000 sinkholes per sq mile. Portions of Kentucky and Florida are also conspicuously pitted with sinkholes, which often divert minor surface streams to underground routes.

The most spectacular features produced by limestone solution are the *underground caverns* formed wherever thick masses of limestone are present in an area of high topographic relief. Where high relief exists, water tables slope steeply toward streams. Resulting rapid groundwater flow accelerates the solution process. Of the 48 conterminous states, all but Vermont and New Hampshire have one or more limestone caverns, and some have hundreds. Certain North American caverns are among the world's most extensive, with more than 100 km (60 miles) of passageways, including rooms tens of meters high and hectares in area. Many caves are beautifully decorated with hanging icicle-like *stalactites* and upthrusting columnar *stalagmites* (see Figure 13.8). These and a variety of other types of cave deposits are composed of calcium carbonate deposited where carbonate-saturated water seeps into the cave and evaporates, leaving its chemical load behind. Landscapes whose form is dominated by solution effects in limestone are known as *karst* landscapes, the name being derived from the Karst region of Yugoslavia, where solution features are particularly well developed and have been described in detail.

Metamorphic Rocks

Tectonic movements frequently cause rock masses to be dragged deep down into the lowest portions of the earth's crust, where pressures and temperatures are far greater than at the earth's surface. This occurs most widely in the great subduction zones at the boundaries of converging crustal plates, where one plate is forced to descend beneath another. Under the enormous stresses experienced in these environments, rock material can deform and flow in a plastic manner, or it can be partially or wholly melted and recrystallized. It is this *metamorphosis*, or reconstitution, of rock material that transforms limestone to *marble*, shale to *slate*, sandstone to *quartzite*, granite to *gneiss*, and lavas to *schists* (Table 13.5, p. 384). What distinguishes a metamorphic rock is evidence of high-pressure flow of the rock material. Normally the minerals will be "smeared out," or oriented along visible planes of flow (Figure 13.9, p. 385); this is called *foliation*. Where the parent rock contained a variety of different minerals, as in the case of granite, these will become segregated into wavy bands of contrasting color that produce a distinctive gneissic foliation.

Complete melting of preexistent rocks generates new fluid magma, which may force its way upward into the crust as an igneous pluton. The process of metamorphism is transitional to complete melting. Many of the very ancient granites of exposed continental basements—the so-called crystalline shield areas—show evidence of having been formed by extreme metamorphism, but not complete melting, of former sedimentary rocks. Thus these particular granites are not truly igneous rocks in the sense of having had an origin in liquid magma.

Metamorphic rocks are denser and less porous than their parent rocks and are therefore more resistant to weathering and erosion. Their greater mechanical strength and resistance to moisture penetration and accompanying chemical and mechanical weathering

Table 13.5. Structure and Composition of Metamorphic Rocks

	Metamorphic Rock	Texture	Mineral Composition	Derived From
Foliated	Slate	Fine-grained; smooth, slaty cleavage; separate grains not visible	Clay minerals, chlorite, and minor micas	Shale
	Schist	Medium-grained; separate grains visible	Various platy minerals, such as micas, graphite, and talc, plus quartz and sodium plagioclase feldspar	Shale, basalt
	Gneiss	Medium- to coarse-grained; alternating bands of light and dark minerals	Quartz, feldspars, garnet, micas, amphiboles, occasionally pyroxenes	Granite
Nonfoliated	Quartzite	Medium-grained	Recrystallized quartz, feldspars, and occasionally minor muscovite	Sandstone
	Marble	Medium- to coarse-grained	Recrystallized quartz or dolomite plus minor calcium silicate minerals	Limestone or dolomite

outweigh any increase in surface runoff their new textures may produce. Metamorphic rocks vary in resistance according to their chemical makeup and their "massiveness" (absence of internal separations). The earth's oldest known rocks (about 3.8 billion years in age) are all metamorphic types, exposed where erosion has removed many kilometers of overlying younger rocks. These ancient rocks are thought to be the deep roots of former mountain ranges that were removed by erosion hundreds of millions (or even billions) of years ago.

Differential Weathering and Erosion

Each rock type has a particular response to specific weathering and erosion processes. Figure 13.10, p. 386, displays diagrammatically the effects of rock type and arrangement on the landforms in a section of the Hudson River valley. The great irregularity of relief and the diversity of landform patterns found within this region are quite striking. This variety is the consequence of differential weathering and erosion in rocks of varying chemical and mechanical resistance. The linear ridges have developed where rocks that are resistant to chemical and physical decay have been exposed at the surface; these ridges are associated with chemically resistant sandstones and mechanically resistant granite and basaltic lavas. The major lowlands have formed where mechanically weak shale and chemically weak limestone are easily removed.

In the region illustrated in Figure 13.10, differential weathering and erosion are possible because tectonic activity has deformed a series of rock layers, inclining them so that edges of varying material

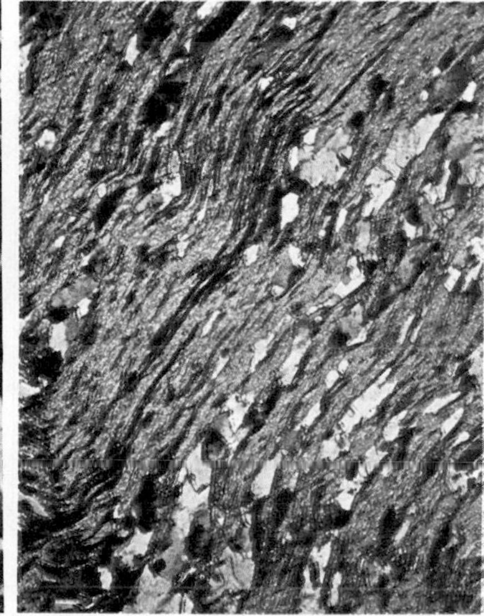

Figure 13.9 Metamorphic rocks form mainly by the transformation of preexisting rocks under conditions of heat and high pressure.

(top left) This outcrop of *gneiss* exhibits the swirled patterns often seen in metamorphic rock.

(top right) *Schist* is a crystalline rock that has an internal structure dominated by a layered arrangement of platy minerals as seen in this thin section enlarged ten times. Because weathering tends to be most effective between the sheets, schists tend to break into thin flakes.

(bottom) The upper portion of the Grand Canyon of the Colorado River in Arizona consists of sedimentary rock. The inner gorge in this section of the Grand Canyon is cut into ancient schist and granitic intrusions.

can be exposed at the earth's surface. Farther west (left) in the Catskill Mountains, where the deformation dies out, there is little structural influence on the landscape. Here the rocks are nearly flat-lying and no strong contrasts in erodibility are exposed. As a consequence, the landscape in the western area has a uniform character. Figure 13.11, p. 387, is an example of differential weathering and erosion on a smaller scale. In arid regions, even minor differences in rock resistance become very apparent due to the absence of vegetation and a soil cover.

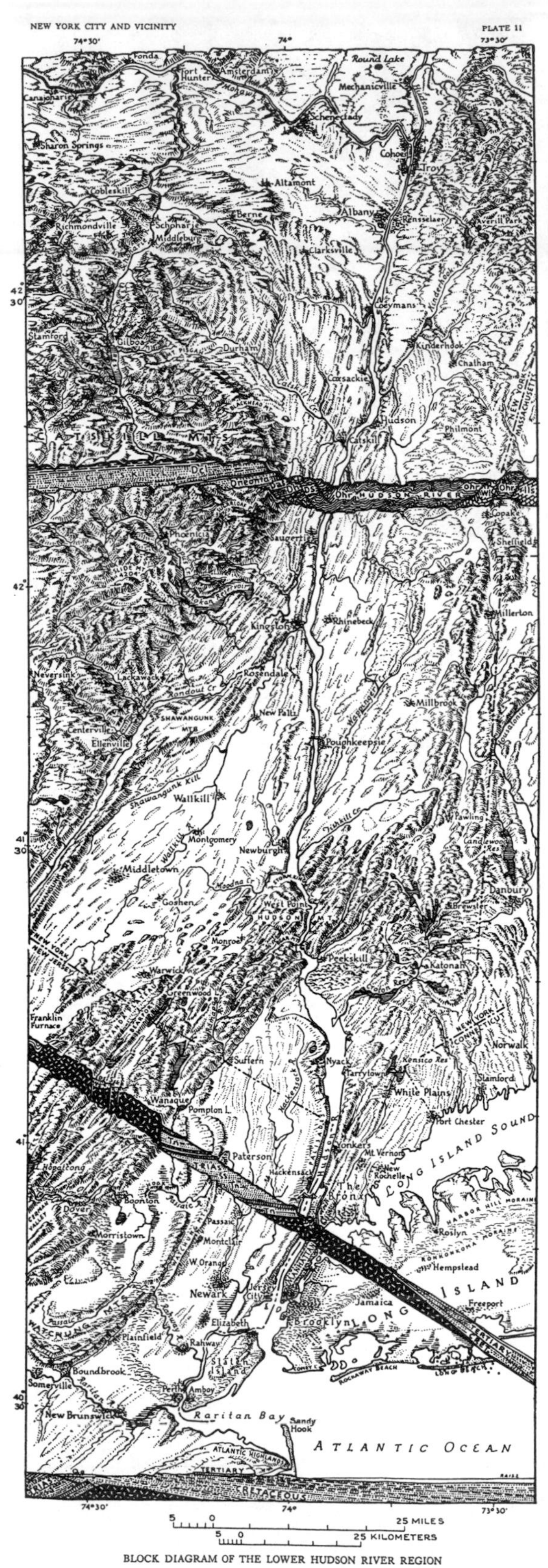

BLOCK DIAGRAM OF THE LOWER HUDSON RIVER REGION

In areas dominated by igneous and metamorphic rocks, which are usually massive, differential weathering and erosion are somewhat less conspicuous than in areas of bedded sedimentary rocks. Nevertheless, there are significant variations in resistance in igneous and metamorphic rocks. The rock itself varies from point to point chemically and texturally. The more chemically basic the rock (the more iron, calcium, and sodium it contains), the more prone it is to chemical decay; rocks with a high content of silica and potassium are generally more resistant. Likewise, dense, fine-textured rocks are less permeable and therefore less prone to interior chemical decay than are coarse-grained rocks.

The spacing of rock joints controls the access of water to the interior of a rock mass; accordingly, closely jointed rock decays more rapidly than sparsely jointed rock. Since there are no bedding planes in igneous and many metamorphic rocks, jointing becomes a dominant factor in surface form. As rock fragmentation and erosional removal proceed, depressions, and eventually valleys, form in the areas of high joint density, leaving the less jointed rock in high-standing blocks. Sometimes these effects leave great spires or naked domes of solid granitic or gneissic rock rising above jointed and decayed rock of valleys and hillslopes. Such a landscape is well displayed around Rio de Janeiro, Brazil (Figure 13.12, p. 389).

The layering of sedimentary rocks of different character provides a maximum opportunity for differential weathering and erosion. As discussed in Chapter 11, each rock type tends to develop a specific slope angle that balances erosional energy to the resistance of the material composing the slope. A view into the Grand Canyon of the Colorado River (Figure 13.9, p. 385) shows how this proceeds in an arid climate. Here chemical weathering is very slow, and the erosional stress must be extremely high to detach large masses of solid rock. As a result, mechanically resistant sandstones and limestones form nearly vertical walls; removal occurs by collapse when these cliffs are undercut by the more rapid erosion of weak shale beds at their base.

The Rock Cycle

The discovery of the subduction process has provided a crucial link in our understanding of the rock cycle (Figure 13.13, pp. 390–391). The majority of the rocks exposed at the earth's surface are sedimentary types produced by the everyday activities of erosional and depositional systems—the steady delivery of sediment from high to low places and the continual "snowfall" of marine microorganisms that maintains a blanket of organic ooze on the ocean floor. We have seen that despite the 4.6-billion-year age of the earth, all of the ocean basins are younger than 200 million years. We now know that the older ocean floor material disappears by subduction in oceanic trenches. Much of this material descends to depths of 70 to 250 km (40 to 150 miles) below the earth's surface, where it finally melts. The resulting low-density fluid magma subsequently rises toward the surface or is squeezed upward by tectonic pressures. The new magma either solidifies within the crust as an intrusive pluton or is extruded at the surface as lava. Erosion of the surface lava, and subsequent erosion down into the subsurface pluton, produce new rock fragments that are carried to new depositional sites to eventually create new sedimentary rock. This is the *rock cycle,* in which the same atoms that have been part of the earth since its formation are used over and over to create generation after generation of rock material.

VOLCANISM

The most spectacular part of the rock cycle is the eruption of molten rock material. This is part of the awesome process of *volcanism,* which produces new extrusive igneous rock at the earth's surface. Explosive volcanism actually begins with the melting of older rocks that have been forced downward into the earth's hot interior in the subduction process. But not all volcanism is violent. One can view the eruptions of the earth's most active volcano, Hawaii's Kilauea, from the crater rim in absolute safety. Some volcanic eruptions devastate

Figure 13.10 (opposite) This *physiographic diagram* by the master cartographer Erwin Raisz reveals how geologic structure and a diversity of rock types affect the scenery of the lower Hudson River region. In the upstream area, the cross section shows that gently dipping resistant sandstones form the Catskill Mountains, with metamorphic rocks underlying the Hudson River valley. In the cross section farther south, the highlands are shown to be produced by granitic rocks (darkest symbol), with basaltic lavas making ridges in the lowland west of the river. (From the Report of the International Geological Congress, Washington, 1933)

Figure 13.11 Differential weathering and erosion due to variations in rock type are especially clear in arid regions. Here in Monument Valley, Utah and Arizona, massive sandstone forms vertical walls. The lower layers of thinly bedded sandstones and shales produce much gentler slopes, although even here the more resistant layers make small cliffs.

enormous areas and take thousands of lives; others, which may produce a far greater volume of lava, are actually quite harmless. What accounts for the difference?

Sources of Volcanic Energy

The degree of violence of volcanic eruptions is largely a consequence of the chemistry of the magma feeding the eruption. Particularly important is the silica (SiO_2) content of the magma. Magma produced by the melting of older rocks contains more than 65 percent silica. Such silicic magma has a low density and a high content of gases that explode outward when their pressure exceeds the weight of rocks confining them. A different type of magma is present in the permanent layer of fluid simatic rock lying below the earth's crust, in the upper part of the earth's mantle. This material is chemically basic, having a silica content of 50 percent or less, and it is hundreds of degrees hotter than the silicic magma produced by the melting of preexisting rocks. Paradoxically, the hotter the magma, the less violent its eruptive force. This is because volcanic eruptions are powered by the explosion of pent-up gases rather than by liquid magma itself.

When chemically basic magma reaches the surface at a temperature of about 1,200°C (2,200°F), magmatic gases are still dissolved in the melt, making it extremely fluid. It pours out freely with little explosive activity, although impressive steam clouds may be present. But when siliceous magma arrives at the surface, at a temperature seldom exceeding 900°C (1,600°F), it always does so in explosive style. Due to the lower temperature of siliceous magma, its gaseous constituents begin separating from the molten material while it is still far below the surface. This gas, trapped in bubbles and pockets within the pasty magma and compressed between the magma and the rocks above it, is under enormous pressure. Nevertheless, much of the separation of gas is itself a consequence of the gradual reduction in pressure as the magma moves upward from the earth's interior. A surface eruption occurs when the pressure of these pent-up volcanic gases (including water vapor) overcomes the confining pressure of the overlying rocks. The first eruption of gases and steam at the surface reduces the pressure within the magma, which triggers a chain reaction—causing more gas to form, more explosive release of gas, which lifts liquid magma into the air, more pressure reduction, more gas separation, more explosive release, more magma lifting, and so on. Anyone who has quickly opened the pressure cap of a hot automobile radiator has seen scalding water shoot out with great force. Under pressure, water remains liquid well above its normal boiling point, but if the pressure is suddenly released, some of the water instantly vaporizes and expands with explosive force. It is the same with molten magma. Thus a major eruption of siliceous magma may produce one huge explosion after another for hours or even days at a time.

The force of such volcanic eruptions is incredible. When Krakatau, a volcanic island between Java and Sumatra, exploded in 1883, the noise was heard in Australia more than 3,000 km (1,900 miles) distant, and a seismic sea wave, or *tsunami,* was generated that drowned some 30,000 persons along the coasts of Southeast Asia. Krakatau itself was blown completely away, so that only open water remained where the volcano formerly rose. Similar devastating eruptions have occurred at various times in other parts of the world. In fact, any of the world's beautiful cone-shaped active volcanoes could conceivably produce such a cataclysm.

Eruptive Styles and Volcanic Landforms

Volcanic eruptions produce two products that variously combine to create the landscape features of volcanic regions. Fluid magma that issues from a vent constitutes lava, which may cover vast areas. However, all volcanic eruptions also hurl solid material into the air. These particles fall back to earth to form deposits of *pyroclastic* ("fire-broken") *ejecta.* The proportion of pyroclastic debris is vastly larger in the case of siliceous volcanism than where chemically basic magma is involved. Pyroclastic debris includes both solidified clots of lava and fragments of older rock torn loose by the force of a violent eruption.

Figure 13.12 The spectacular rock cones, or "sugarloaves," of Rio de Janeiro, Brazil, result from weathering and erosion in well-jointed gneissic rock. Unjointed masses resist subsurface chemical weathering and are left standing above surrounding decayed and eroded rock.

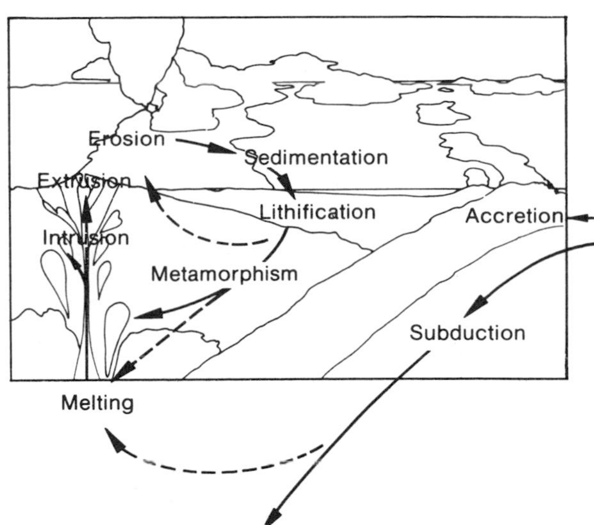

Figure 13.13 During the course of geologic time, rock materials pass from one form to another in the *rock cycle.* This schematic painting, and the inset diagram above, illustrate the principal phases of the rock cycle. At the lower left of the figure, molten rock, or magma, from the interior of the earth is shown welling up toward the surface. Solidified magma forms igneous rock. Sometimes the rock solidifies beneath the surface, forming intrusions, or plutons, as shown in the figure. Erosion of the surface cover may lead to the subsequent exposure of the intrusive rock masses. Some of the molten rock is also directly extruded at the surface as volcanic lava.

Rocks exposed at the surface are subject to erosion, principally by flowing water. The fragments and grains of rock are carried by water or wind to the sea or to inland basins, where they collect in layers of sediment. In the presence of heat or pressure, or by the chemical action of a cementing agent, sediment layers become consolidated, or lithified, into beds of sedimentary rock.

The motion of lithospheric plates plays an important role in the rock cycle. The downward subduction of a plate near a continental boundary, depicted in the lower right of the figure, forms a sea trough that slowly subsides, allowing great masses of sediment to accumulate. The downward movement of plates in subduction zones also returns rock material to the interior of the earth, where it can be melted to form new magma. Deformation near the boundary of colliding plates generates the pressure required to transform igneous and sedimentary rock into metamorphic rock.

The larger debris, or *scoria,* falls closest to the vent and produces a *scoria,* or *cinder, cone* composed of *volcanic bombs* (blobs of fluid lava), solid blocks, and finer *lapilli* (particles 2 to about 60 mm in diameter). Sometimes the magma is so gaseous that it rises to the surface as a froth, like the foam on beer. This froth solidifies as spongy *pumice,* which is light enough to float on water. The finest airborne debris, volcanic ash, may be carried hundreds of kilometers from the site of the eruption. Since the age of this material can be determined by radiometric dating techniques, it often forms a very valuable tool in dating soil and landform development over vast areas.

Although volcanism is a world-wide phenomenon, occurring on the land and in the seas, the distribution of volcanoes is quite systematic, as indicated in Figure 13.15, p. 394. Large-scale volcanism is associated with active lithospheric plate boundaries. Where crustal plates are being pulled apart, basic magma rises from the mantle to fill the voids and form ridges between the diverging plates. Thus the oceanic ridges are volcanic features, with the deep ocean floors composed of basalts that have been moved laterally from the ridges in the sea-floor-spreading process. Volcanic eruptions of the oceanic type are seldom highly explosive, although they may produce immense volumes of lava and great quantities of black volcanic ash. Many oceanic volcanoes are not situated on the continuous oceanic ridge system, but are scattered over the ocean floor. The majority of these have been inactive for many millions of years. Most of them probably originated on the great submarine ridge system and have gradually been carried away from it as part of the sea-floor-spreading process. Elsewhere, active volcanoes lie at the end of linear chains of submarine peaks, suggesting horizontal movement of crustal plates over "hot spots," or convective "plumes" in the upper part of the earth's mantle. The extinct volcanoes in such chains are thought to have moved past the

Figure 13.14 These photographs show a variety of constructional volcanic landforms.

(a) The simplest form is the *scoria cone,* produced by the fall of pyroclastic materials that have been hurled into the air by the force of escaping magmatic gases and steam during a volcanic eruption. The example here is Paricutín volcano, which rose from a Mexican cornfield in 1943 and erupted sporadically until 1952.

(b) Basaltic *shield volcanoes* such as Hawaii's Kilauea (foreground) and Mauna Loa (horizon) have very gentle slopes, making their true sizes hard to appreciate. Mauna Loa rises almost 3,000 meters (9,800 ft) above the summit of Kilauea. In the foreground is the summit crater of Kilauea volcano.

(c) *Strato-volcanoes* such as Mount Shasta in northern California are among the most impressive of all constructional landforms. Cones of this type are composed mainly of andesitic lava and pyroclastic material resulting from explosive eruptions. Note the younger satellite cone to the left of the main summit. Only the upper portion of the volcano is visible in this view. The base diameter of Mount Shasta is 32 km (20 miles) and the peak rises some 3,300 meters (11,000 ft) above its base.

(d) Volcanic calderas are large circular craters produced by the collapse of the summits of volcanoes during extremely violent eruptions. Crater Lake in Oregon, shown here, was produced by cataclysmic eruptions of a large volcano of the Shasta type some 6,600 years ago. The caldera is about 10 km (6 miles) wide.

(e) The flood basalts shown here comprise the upper 150 meters (500 ft) of the east wall of the erosional trench of Grand Coulee in the state of Washington. Such great piles of superimposed lava flows occur in areas of crustal tension and rifting.

hot spot; the active volcanoes mark its present location. The Hawaiian chain is the outstanding example (see Figure 11.6, p. 303), rising above sea level due to the enormous quantity of lava extruded. Currently, the island of Hawaii, situated at the southeast end of the Hawaiian chain, experiences periodic eruptive activity from three separate volcanoes, one of them born as recently as 1969.

The great Hawaiian volcanoes—dormant Mauna Kea (whose last eruption was more than 2,000 years ago) and active Mauna Loa, Kilauea, and the infant Mauna Ulu—are wonderful examples of oceanic volcanism. Hawaiian eruptions take place in vast craters or along linear fissures, with very hot basaltic lava often flowing many kilometers before solidifying. This creates landforms known as *shield volcanoes,* which rise very gently to broad rounded summits (see Figure 13.14). The shield volcanoes of Hawaii reach higher above the sea floor than Mt. Everest ascends above sea level. Although they rise some 9,000 meters (30,000 ft) above their oceanic base, the slopes of the great Hawaiian shield volcanoes rarely exceed an angle of 10 degrees. Shield volcanoes are found in most parts of the world in which basaltic lavas are erupted.

Two types of basaltic lava issue from the Hawaiian volcanoes—the rough, clinkery *aa* (ah-ah) type (Figure 13.16, p. 396) and the satiny smooth but often wrinkled *pahoehoe* (pah-*ho*-ay-*ho*-ay) type (Figure 13.17, pp. 398–399). The hottest lava produces flows of the pahoehoe type, in which gas remains dissolved in the lava, making it extremely fluid. Pahoehoe flows change to the aa type when the gas begins to escape, causing the lava to congeal and move as a creeping mass of red-hot slag. Clots of liquid lava and froth hurled into the air by the "fire fountains" present during Hawaiian eruptions produce basaltic lapilli and pumice that may form a deep blanket of shiny black particles around the eruptive vent.

d

e

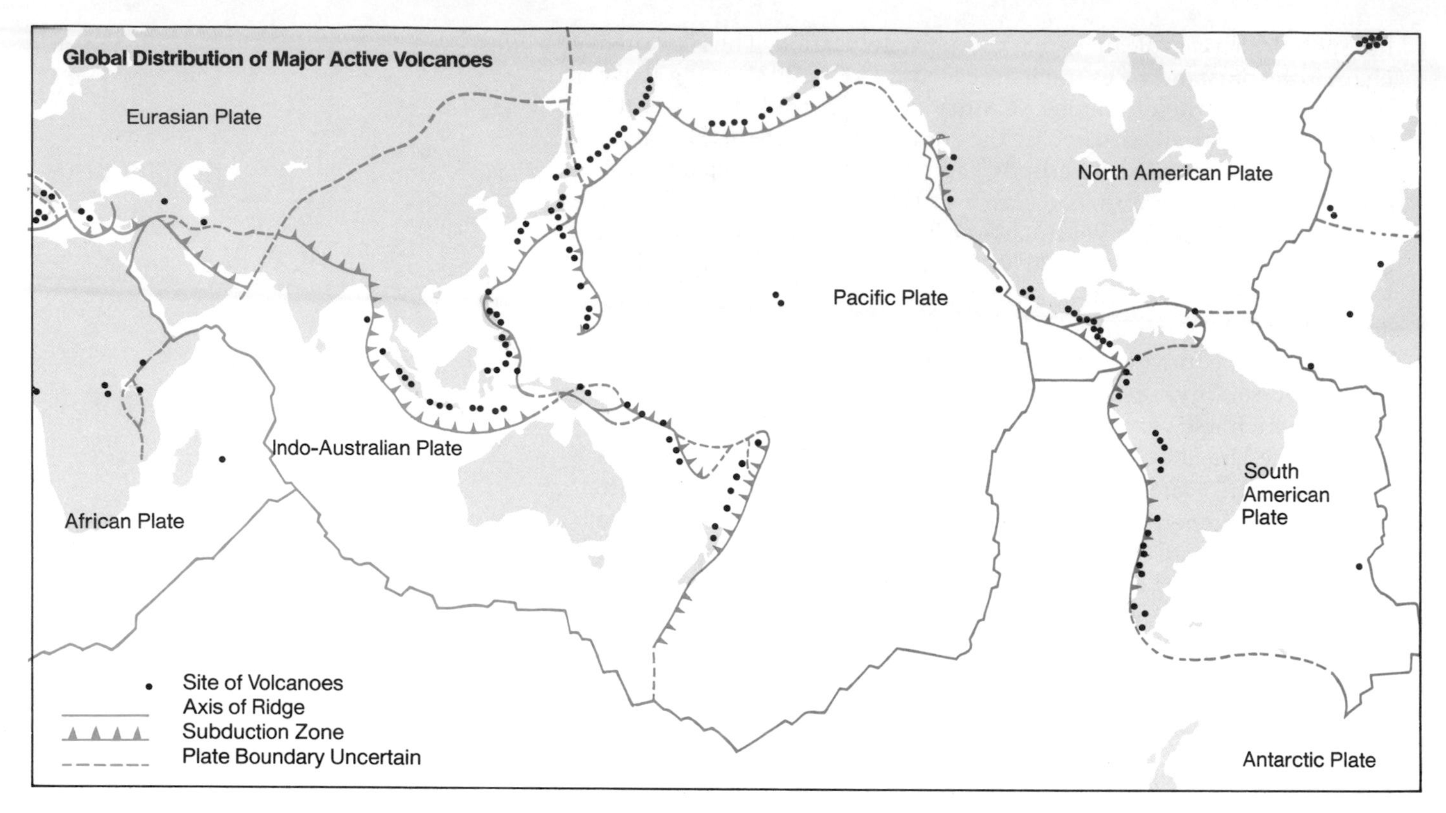

Figure 13.15 This map shows many of the known active volcanoes of the world; for clarity, some volcanoes have been omitted from regions where many are present. The map also shows the principal plate boundaries, oceanic ridges, and subduction zones associated with the earth's crustal plates. Regions of volcanic activity are frequently located near active plate boundaries, where molten rock from the earth's interior wells up through fissures. Note the large number of active volcanoes around the Pacific Ocean basin, which is ringed by oceanic trenches. The volcanoes in Iceland and some of the volcanoes in the Atlantic Ocean are associated with the spreading center along the Mid-Atlantic Ridge. Volcanoes in the Caribbean, in East Africa, and in the eastern Mediterranean appear to be associated with plate boundaries as well. Conversely, the east coasts of North and South America, where there are no plate boundaries, are devoid of volcanic activity. Isolated volcanic regions far from plate boundaries, such as the Hawaiian Islands, are due either to "hotspots" in the earth's mantle, or fractures in the lithospheric plates, which allow molten rock from the earth's interior to reach the surface. (After Bullard, 1962)

The volume of lava produced by the frequent Hawaiian eruptions seems enormous, but the greatest lava outpourings probably occur along the oceanic ridges, and occasionally on the continents. Again, basalt is the rock produced. In the Columbia Plateau region of Washington and Oregon and the Snake River Plain of Idaho, and in India, Brazil, Patagonia, South Africa, and Antarctica, vast areas are covered with *flood basalts.* In these areas, the land surface has been built up by repeated enormous floods of lava interbedded with volcanic ash, stream and lake bed deposits, and loess. Some individual basalt flows cover thousands of square kilometers, are tens of meters deep, and have volumes of tens to hundreds of cubic kilometers. In 1783 such a lava flood was witnessed in Iceland when a 16-km-long (10-mile) fissure vented some 12 cu km (3 cu miles) of basalt. The volume of basalt that poured out to form the Columbia Plateau is estimated at 400,000 cu km (100,000 cu miles). In some areas, considerable ash is interbedded with the flows; elsewhere, little ash is present, indicating that these enormous outpourings generally were not violent eruptions. Floods of basalt imply direct connections through the continental crust to the earth's mantle. Those of Brazil, India, Patagonia, and Antarctica may have been triggered by the initial ruptures of the larger ancient continent from which these fragments were detached during the birth of the Atlantic Ocean. Most continental areas of flood basalts have been at least partially dissected by stream erosion, so that the edges of the separate lava flows, one atop the other, are clearly displayed along the walls of canyons, such as those of the Columbia and Snake rivers.

Nonbasaltic explosive volcanism is found where lithospheric plates are converging, forcing rocks to descend and melt in subduction zones. Figure 13.13, pp. 390–391, indicates the relationship between subduction, the formation of plutons, and surface volcanism. The melting of older rock material generates magma that is generally dioritic in composition (Table 13.3, p. 375) and erupts at the surface as the lava known as *andesite.*

As we have seen, magma formed in the subduction zones is generally much more siliceous than magma derived from the mantle, and it arrives at the surface in a much more violent fashion. The first products of violent surface eruptions are scoria cones like Paricutín, born in a Mexican cornfield in 1943 and built to a height of 400 meters (1,300 ft) in 8 months (Figure 13.14, pp. 392–393). Lava flows emanating from the crater within such cones and gushing from fissures lower on the cone's flanks gradually veneer the loose scoria with solid rock. This andesitic lava is too depleted of gas to produce pahoehoe or even aa flows; it generally moves as a rubble of hardened blocks over a pasty liquid interior. The result is *block lava.* As lava is being extruded at the surface, magma forcing its way into cracks in the volcanic cone solidifies in networks of intersecting *dikes,* further reinforcing the cone's structure. If eruptions continue periodically over hundreds of thousands of years, a giant *strato-volcano* may result, composed of multitudes of andesitic lava flows, scoria beds, and interior dikes (see Figure 13.14). Such a cone may be quite complex, including a variety of lava types and many different eruptive centers in addition to the high central vent. All of the world's cone-shaped strato-volcanoes are geologically young features, showing little erosional modification. Most have been built within the past million years. Older volcanoes have been largely destroyed by erosion. Thus any intact cone-shaped volcano must be considered potentially dangerous.

Large volcanoes pose many hazards. Flows of lava move too slowly to be a real threat to human life but are destructive to property. Equally damaging are volcanic mudflows, known as *lahars.* These result from rainfall on fresh volcanic ash, from lava eruptions under glaciers or onto snowfields, and from slope failure on volcano summits weakened by the action of acidic fumes vented from subsurface magma. Still more dangerous are explosive outbursts of scoria and ash, which may drown the surrounding countryside. Most dangerous of all is the *nuée ardente* (glowing cloud) type of eruption, in which red-hot ash flows down the flanks of a cone at speeds exceeding 160 km (100 miles) per hour, incinerating everything in its path. Such an eruption is known as the *Peléean* type, after the volcano Mt. Pelée on the Caribbean island of Martinique. In 1902 this volcano produced a nuée ardente that in an instant reduced the city of St. Pierre to a broken shell and snuffed out the lives of all but two of its 30,000 inhabitants. Nuée ardente eruptions are associated especially with a peculiar type of volcano known as a *plug dome* (Figure 13.18, p. 400). Plug domes are formed when a mass of unusually stiff and highly siliceous magma, either rhyolite or dacite in composition, pushes to the surface. The pasty lava congeals as it rises, jamming the surface vent so

that enormous pressure builds beneath the volcano. This is released in periodic explosions of catastrophic force. In some instances, the amount of solid material ejected in such a blast is enormous, having a volume of many cubic kilometers. The only volcano to have erupted in historic times in the conterminous United States is a plug dome—Lassen Peak, in northern California, whose last period of activity was from 1914 to 1917.

Many giant strato-volcanoes include subsidiary plug domes, or have at times behaved like plug domes, producing such violent explosions that their entire summits have been destroyed. Crater Lake, in Oregon's Cascade Range, half fills a vast pit 10 km (6 miles) wide and 1,200 meters (4,000 ft) deep, produced by the collapse of a major volcano following a cataclysmic eruption about 6,600 years ago. Vast craters of this type are known as *calderas.* They result from subsidence of the upper part of the volcanic cone into its subsurface magma reservoir after the latter is partially emptied by an unusually violent eruption. Calderas, often containing lakes, are present in most areas of andesitic volcanism. Mt. Vesuvius, rising behind Naples, Italy, was decapitated by such an eruption in A.D. 79. The Roman town of Pompeii was buried by ash falls during this eruption, with many of its inhabitants apparently suffocated by volcanic fumes. Hollow molds of about 2,000 bodies have been found in the ash covering Pompeii. Since the famous Roman scholar Pliny the Elder was one of those killed in this eruption, which was described in detail by his nephew Pliny the Younger, violent eruptions that decapitate volcanoes or cause their collapse are known as *Plinian* eruptions.

It is difficult to say whether a volcano is active, dormant (potentially active), or extinct. Prior to its eruption in A.D. 79, Vesuvius had been inactive for many hundreds of years. It should be noted that at Krakatau, Crater Lake, and Vesuvius, new volcanic peaks have been built up inside the large calderas produced by the eruptions noted above. The new cone at Vesuvius, less than 2,000 years old, almost completely overwhelms the old caldera rim, rising high above it, while at Crater Lake the often-photographed Wizard Island is but the tip of a 600-meter (2,000-ft) volcano that just breaks the lake surface and remains far below the surrounding cliffs of the caldera. The new cone of Krakatau has been built in less than 100 years. Vesuvius is obviously active and can often be seen either erupting or venting steam and fumes. Crater Lake has shown no activity in historic time, but is it inactive or merely dormant? And what about Lassen Peak, or any of the peaks in the continuous line of volcanoes in the Cascade Range, from Lassen's giant neighbor, Mt. Shasta, to Washington's Mt. Baker, which has been venting fumes with increasing vigor since 1975?

The Cascade volcanoes are something of an enigma because, despite the presence of an oceanic spreading center offshore, there is no oceanic trench off the continental margin and no real evidence of an active subduction zone beneath the Cascade Range. Elsewhere around the rim of the Pacific, oceanic trenches are present and active subduction seems to be demonstrated by the abundance of earthquakes that originate many tens of kilometers below the surface. Volcanism is widespread along the so-called "Pacific Ring of Fire" or

Figure 13.16 (opposite top) Lava assumes different forms depending on its temperature, composition, and rate of cooling. This advancing *aa* lava flow, photographed on Hawaii, is characterized by chunky, angular blocks mixed with still-molten lava. Temperatures measured in these basaltic lavas are about 1,200°C (2,200°F). When an *aa* flow solidifies, an almost impassable field of jagged lava is formed.

(opposite bottom) When basaltic lavas pour out onto the sea floor, the rapid cooling develops large globular, pillowlike masses. The rapid cooling forms a glassy margin, and shrinkage during cooling forms radial cracks, or joints, in the pillow. These accumulations of submarine basalt are called *pillow basalts.*

Figure 13.17 (top) Due to the gases dissolved in it, *pahoehoe* lava, shown here at Kilauea crater in Hawaii, is much more fluid than *aa* lava even though the general chemical composition is similar. Note the thin layer of solidified crust on the molten rock. When the lava solidifies, the surface is left in the form of smooth billows or ropes.

(bottom) Shiprock, in northwest New Mexico, is a striking example of a *volcanic neck.* Note the extensive volcanic dikes radiating from Shiprock.

(opposite) Flat-topped *table mountains,* such as this one in southern California, form when the land surrounding a nearly horizontal lava flow is removed by erosion. This occurs when lava flows enter valleys cut in erodible rocks. The lava itself erodes very slowly since its great permeability causes rainwater to soak into it rather than running off over its surface.

"Andesite Line," which circles the Pacific from the Andes to New Zealand, with nonvolcanic interruptions from northern Mexico to central California and in British Columbia.

All volcanoes eventually become inactive, and their imposing structures are worn away by the forces of erosion. Figure 13.17 illustrates two volcanic landforms that may persist in the landscape long after the original surface structures have been removed. *Volcanic necks* and *dike ridges* are former subsurface magma conduits, and they stand forth as pinnacles and walls where they are more resistant to erosion than the rocks surrounding them. Volcanic necks are abundant in northern Arizona and the Four Corners area of Arizona, New Mexico, Utah, and Colorado. *Table mountains* are left where ancient stream valleys served as channels for thick lava flows that were less erodible than the original valley walls. Lavas generally erode slowly due to their extreme permeability, which permits water to soak into them rather than running off and exerting erosional stress upon their surface. Table mountains are particularly common in the western foothills of California's Sierra Nevada. There some of the lavas of

the table mountains covered and preserved gold-bearing stream gravels that would otherwise have been removed by later stream incision; thus table mountains played a role in the story of gold in California.

Volcanism and Man

From the human point of view, the most significant contribution of volcanism has been the provision and renewal of mineral-rich material that produces excellent agricultural soils. This is especially true in the wet tropics, where soils are generally infertile due to the leaching of bases (see Chapter 10). The richness of young volcanic soils in the Philippines, Indonesia, and South America's Andes Mountains have lured dense populations to the flanks of volcanoes, close to the hazards of explosive blasts, lava flows, ash falls, and lahars. The quality of soils derived from volcanic materials depends on the base content of the eruptive products. Soils derived from base-rich basaltic lava and ash are significantly more productive than those developed on nutrient-poor siliceous material.

Another significant aspect of volcanism is its role in the formation of the ores of metallic minerals, such as iron, copper, lead, zinc, tin, tungsten, nickel, and chromite, as well as gold and silver. Metallic ores are deposited at contacts between plutonic and other rock types, particularly where contact metamorphism occurs, and also where hot, richly mineralized solutions moving outward from deep-seated

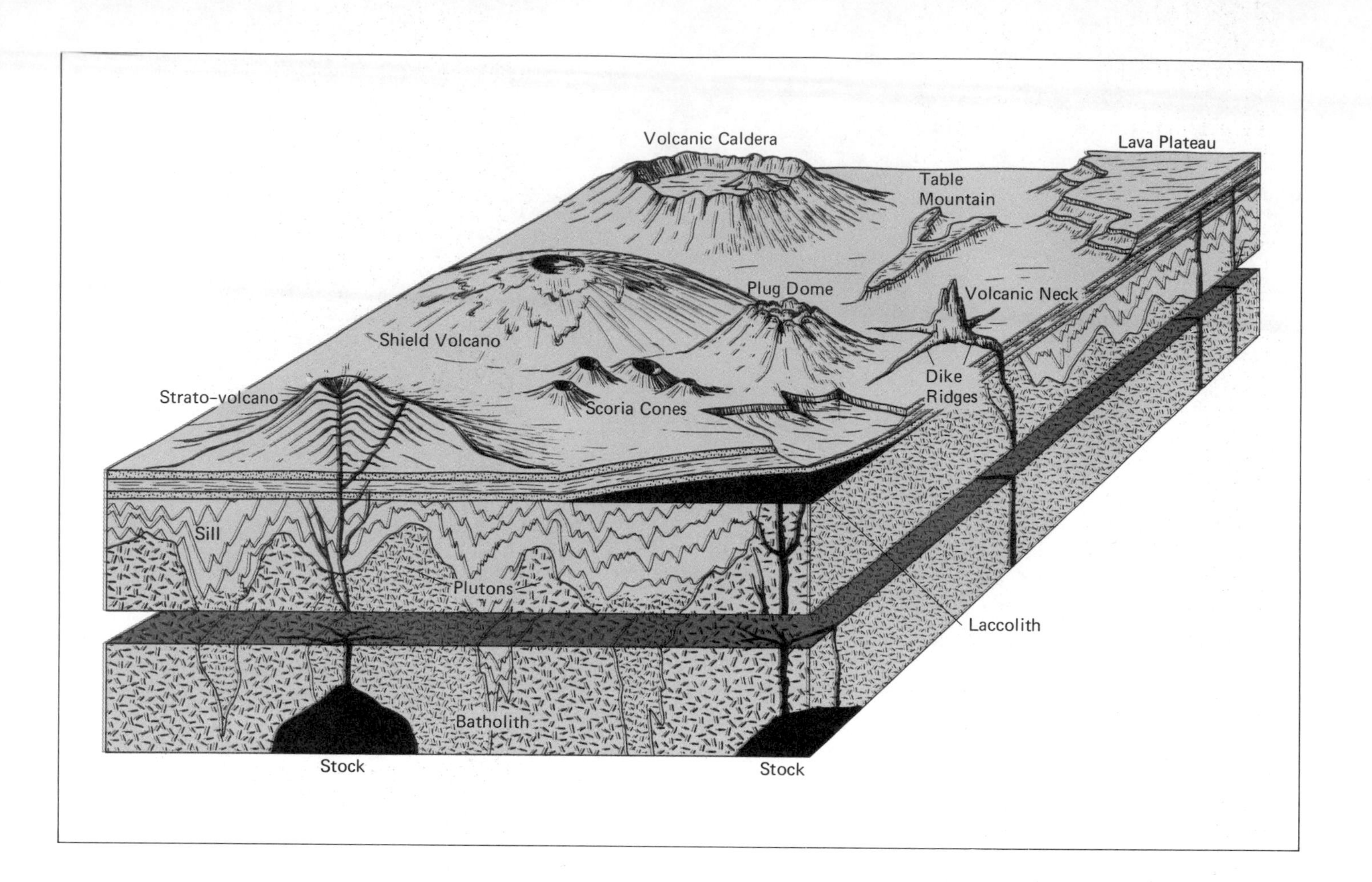

magma bodies or rocks undergoing metamorphosis work into fissures in rocks closer to the surface, reacting with them chemically. Ore deposition is especially likely when siliceous magma comes into contact with limestone or dolomite, in which case metallic ions in the magmatic solution tend to be exchanged for calcium or magnesium ions in the "host" rock. Ore formation does not occur automatically, however, and in fact is highly localized due to factors that are as yet far from completely understood.

Pockets of hot magma within the earth's crust pose the threat of disastrous surface volcanism, but they may also be made to serve mankind, for they are a potential source of *geothermal* energy. Except in a very few regions, the thickness of crustal material overlying magma reservoirs is such that only a negligible amount of heat energy is actually conducted upward to the earth's surface. But in a few areas, we can see surface manifestations of hot subsurface magma bodies. Such natural phenomena as geysers, in which boiling water periodically erupts at the surface, steam vents, and sulfur springs, with their smell of rotten eggs (hydrogen sulfide fumes), are all signs of a near-surface volcanic heat source (Figure 13.19, p. 402). All of these phenomena are common in areas of active volcanism, but they are also seen where surface volcanism has not occurred for thousands of years.

To tap substantial amounts of geothermal energy directly from the zone of fluid magma beneath the earth's crust would require wells many kilometers deep and potentially hazardous to drill. However, in regions of volcanic activity and areas where the crust is deeply fissured, hot springs and jets of steam bring heat energy to the surface naturally. Hot springs result when groundwater rises to the surface after coming into contact with hot rock or steam given off by magma. In some areas, steam generated by magma or by the heating of groundwater is channelled upward naturally through fissures; it may vent directly into the atmosphere, or it may remain at subsurface levels within the reach of ordinary wells, as is the case in portions of Iceland, Italy, New Zealand, and the western United States.

Inhabitants of Iceland have long used the water from hot springs to heat their homes and even to make their fields warmer for agriculture. Nearly half of the homes in Iceland and almost all of the buildings in the capital of Reykjavik are heated by hot water piped from nearby hot springs. The use of hot springs for heating is restricted to locations within a few kilometers of the springs because of heat losses along the delivery pipes.

Several countries use geothermal energy to generate electricity. Electric power can be distributed to a wide area to make the best use of energy from localized steam fields. The geothermal power plants in the Larderello area of Italy northwest of Rome make an important contribution to the Italian economy. In 1960 the Larderello stations generated 4 percent of Italy's total electric power output.

The first commercial geothermal station in the United States is located in a faulted region 150 km (90 miles) north of San Francisco. The plant produces electric energy at the rate of several hundred million watts, somewhat less than 1 percent of the total rate of electric power consumption in California. The Imperial Valley of southern California is believed to have the potential to supply a significant proportion of the state's future power needs from geothermal energy.

Figure 13.18 (opposite) This diagram summarizes the variety of surface and subsurface forms produced by volcanism. The cross section shows that a complex of intrusive igneous rock bodies underlies this area. The separate intrusions, called plutons, collectively form a *batholith* (such as the Idaho batholith or the Sierra Nevada batholith). Individual volcanoes are fed by small intrusions of magma that form subsurface *stocks* and *laccoliths.* A stock cuts up through the structure, whereas a laccolith is injected between older rock layers, doming up the surface above it. A *sill* is a sheet of magma injected between rock layers. A *dike* is a sheet of magma that cuts through older rock. Erosional lowering of the land surface may leave dikes standing up as walls of rock. Volcanoes often form where large dikes cross, concentrating the upward flow of magma.

Several types of volcanoes are shown. *Scoria cones* result from brief explosive eruptions that produce a ring of pyroclastic debris around the volcanic vent. *Plug domes* are produced where pasty siliceous magma is forced out at the surface. They are steep-sided, with a crest that may bristle with hardened masses of lava that have been pushed out as plugs. *Shield volcanoes* have gentle slopes and broad summits produced by huge floods of free-flowing basaltic lava. Often they have a large summit crater. *Strato-volcanoes* are high cones composed mainly of andesitic lava and pyroclastic material. Lavas and scoria are interbedded. *Volcanic calderas* are created by the collapse of strato-volcanoes during violent eruptions. Many have a summit lake and a new cone within the caldera, as shown. *Lava plateaus* result from lava floods fed by large eruptions from lengthy fissures rather than central vents. *Table mountains* are left when the land surrounding a lava flow is eroded away, producing a relief inversion in which an old lava-filled valley is transformed into a ridge. *Volcanic necks* are found where erosion has removed a volcano and lowered the land surface below the volcano's base. The volcanic neck is the lava-filled vent that fed the missing volcano. Volcano necks stand up as spires where the surrounding rocks are more erodible than the volcanic rock.

CRUSTAL DEFORMATION

Although volcanic activity has produced some of the earth's most majestic individual mountains, it is clear that all of the great mountain ranges—even those capped by volcanoes—have been formed by larger scale deformations of the earth's crust. Some mountain ranges, such as the Alps of Europe and the Himalayas of Asia, consist primarily of sedimentary rock; others are composed of intrusive igneous rocks that were formed far below the earth's surface. Prior to their deformation in the mountain-building process, the original sedimentary rock masses that constitute mountain ranges like the Alps and Himalayas were generally 10,000 meters (33,000 ft) or more thick, many times the thickness of sedimentary rock found in regions where mountain building has not been active.

The masses of sedimentary rock found in major mountain ranges originated from both shallow- and deep-water deposits. These deposits

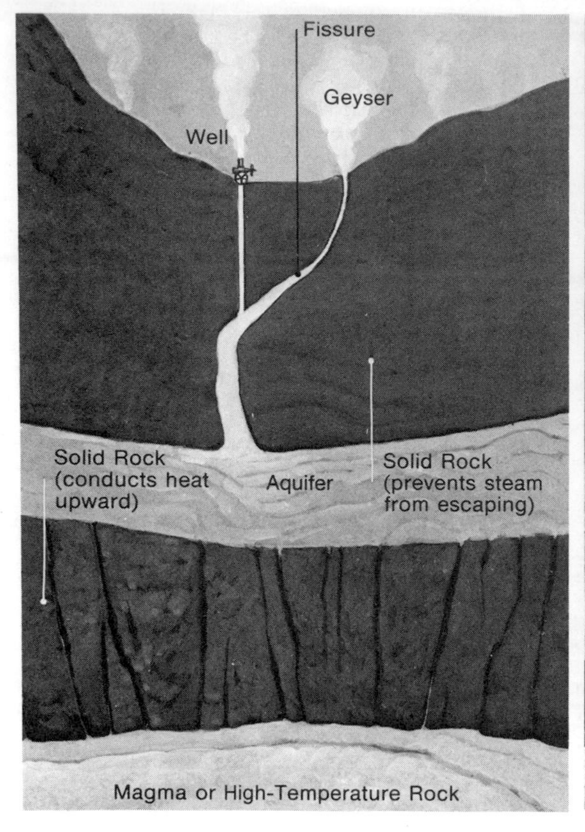

Figure 13.19 (left) This schematic diagram shows the type of geologic structure in steam fields that can be readily exploited for power generation. A confining layer of rock covers an aquifer permeated with water. Water is heated by contact with hot subsurface rock, and steam is generated. The steam comes to the surface through fissures or is led to the surface through wells sunk into fissures or into the aquifer. The high-pressure steam may then be used to operate electric generators. Geothermal energy is a comparatively clean and inexpensive source of power, but its use is confined to regions having subterranean heat sources.

(right) The hot water issuing from Castle geyser in Yellowstone National Park, Wyoming, is driven upward by steam generated from groundwater in fissured hot rock deep below the surface.

accumulate in slowly sinking troughs, or *geosynclines,* that develop along continental margins. Such sediment accumulations are shown in Figure 11.4, pp. 300–301. Parts of these geosynclines are situated on *continental shelves,* the submerged shallow-water regions fringing the emerged portions of the continents. The deep-water portions of the geosynclines are located farther from the land, in oceanic trenches and adjacent *continental rises,* which are vast submarine embankments of sediments lying against the edges of the sialic continental slabs (see Figure 11.4). The limestone portions of the Alps and Himalayas represent former shelf deposits, while the interbedded sandstones and shales appear to have accumulated on a continental rise or in an oceanic trench. The sands and muds from which these deep-water rocks have formed appear to be channelled to the deep ocean through the submarine canyons described in Chapter 16.

Most geosynclines seem to have been created by the descent of oceanic plates under lower density continental plates in the subduction process that produces oceanic trenches. Regions of recently active mountain building, such as the coastal ranges of western North and South America and the Himalayas of Asia, coincide with boundaries of actively converging plates. As the lower plate is subducted downward, the mass of sediments trapped in the oceanic trench is piled up and compressed against the front edge of the higher plate (see Figure 11.4). This piling up of material also exerts force against the sediments on the continental shelf, compressing and deforming these deposits, although somewhat less violently than the continental rise and trench deposits.

We have seen that melting of the descending plate generates magma that wells up to the surface, so that a chain of active volcanoes is frequently formed in the early stages of mountain building. Intrusive igneous rocks such as granite are therefore found among the deformed sedimentary rocks in mountain chains. However, these plutonic rocks are exposed only after deep erosion into the mound of deformed sediments, and before they can be exposed there must be strong uplift of the deformed mass. The necessary vertical movement may be as much as 15 to 20 km (9 to 12 miles). The uplift of rock deformed in the crushed geosyncline probably comes about when the rate of subduction and compression slackens, allowing the thick wedge of low-density sedimentary rock to float upward to reestablish isostatic equilibrium. The compression associated with subduction may also squeeze the sedimentary rock upward, creating mountainous surface relief. Most of the earth's high mountain systems are continuing to rise today, a matter of perhaps 1 cm per year, 5 to 10 meters (15 to 35 ft) per 1,000 years, countering the tendency of the erosional forces to wear them away. In fact, the rate of uplift in the most active mountain systems is three to four times the maximum known rates of surface lowering by erosion, which are in the neighborhood of 1 meter (3 ft) per 1,000 years.

The most recent period of mountain building on the earth's surface began about 20 million years ago with the initial compression and uplift of the Alps, Himalayas, and coastal ranges of the western United States. The Appalachian Mountains of the eastern United States were uplifted much earlier, approximately 250 million years ago, when the African continent pushed against the edge of North America, eliminating a former ocean that had existed between the two continents. This ancestral Atlantic Ocean began reopening some 50 million years later, initiating the current phase of sea floor spreading that has produced the ocean basins we know today (see Figure 11.3, p. 299).

Although the great mountain systems formed by large-scale crustal compression and later massive upheaval are among the most imposing of the earth's relief features, they are not the only products of crustal deformation. In other areas, the crust is rising and falling due to gradual vertical or tilting movements, such as the isostatic adjustments affecting areas recently released from the weight of continental ice sheets. Scandinavia and eastern Canada, whose ice sheets disappeared about 8,000 years ago, are still rising at rates of up to 20 meters (65 ft) per 1,000 years. Areas deglaciated within the past few hundred years, such as the coastal portions of Greenland, are rising much faster, as much as 12 cm (5 in.) per year. In the past 100 million or so years, the plateau region of southern Utah and northern Arizona has been arched upward more than 4 km (2.5 miles), the latest phases permitting the Grand Canyon of the Colorado River to be trenched into it. The Rocky Mountains of Colorado and Wyoming, unlike most major mountain systems, are not at lithospheric plate boundaries, but lie in the midst of the North American plate and seem to have been formed by massive vertical uplift without prior lateral compression.

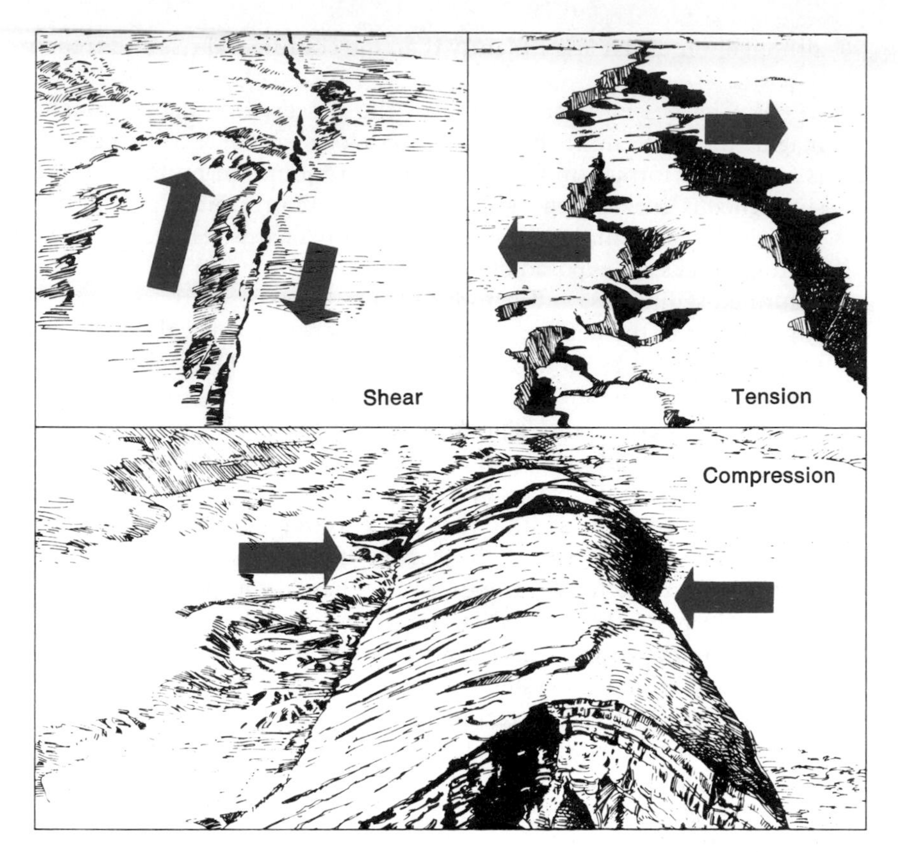

Figure 13.20 Shear, tensional, and compressional forces on the earth's crust. These three drawings show some of the possible landforms that can result from these forces or combinations of them.

(top left) Shear produces horizontal displacement, causing offset stream valleys, interruption of groundwater flow, and demarcations in vegetation.

(top right) Tension has produced a down-dropped block, forming a steep-sided valley.

(bottom) Compression has folded the earth into a linear ridge.

Elsewhere, the crust has been pulled apart by divergent motions that have subjected rock masses to tension. The best example is the oceanic ridge system, but the process is widespread on the continents as well. Rocks are strained by tension where an area has been stretched by being arched upward, or where two rigid blocks are moving laterally with respect to each other, which may either crumple or tear apart a "softer" area between them. Figure 13.20 illustrates some of the possible effects of shear, tensional, and compressional forces on the earth's crust.

Compression and tension are the principal stresses that produce rock deformation. They cause rock strain of two types: wrinkling, or *folding,* and rupture, or *faulting.* Compression may cause both folding, in which rock layers bend about several parallel axes without rupturing, and faulting, in which the rocks rupture and move as separate blocks, slivers, or slices. Tension can produce rock failure only in the form of faulting.

Where rocks are folded or faulted, landforms of unusual complexity often result. Some of these forms may be due to crustal motion, as in the fault-block landscape of Nevada, but more commonly the individual relief features are a consequence of differential erosion in rocks of varying types that become exposed as a consequence of the crustal movements.

Although cross sections revealing geologic structure are common on maps and in books, in real landscapes we can observe only the surface manifestations of the underlying structure. We can deduce the subsurface structure by studying the nature of the exposed rocks and by analyzing the effects of erosion on the structure. Correctly interpreting structure is often of great economic importance because geologic structure controls the location of mineral and fuel deposits useful to man. In the past few decades, improved instrumentation has greatly aided geologists in determining subsurface rock structures. Modern geologists use measurements of local magnetism, the local force of gravity, and the speed and intensity of seismic waves induced by explosions to clarify the nature of geologic structures and associated ore bodies.

Landforms Associated with Fault Structures

The Great Basin, centered in Nevada, is conspicuous on any physiographic map of the western United States (see Figure 13.21). It is dominated by numerous north-south ranges of mountains, "looking like an army of caterpillars crawling north," in the words of one observer of an early map of the region. Many of the mountain ranges in the Great Basin and those forming its borders are blocks of the earth's crust that have been tilted and elevated along faults. On its western margin, the Sierra Nevada is a massive tilted block, with a gentle slope to the west but a steep 3,000-meter-high (10,000-ft) wall facing east, overlooking the sunken block known as Owens Valley Four meters (13 ft) of vertical fault slippage here in 1872 produced the most violent earthquake in California's history. Some 80 km (50 miles) to the east, across another pair of fault-block ranges, is the deep

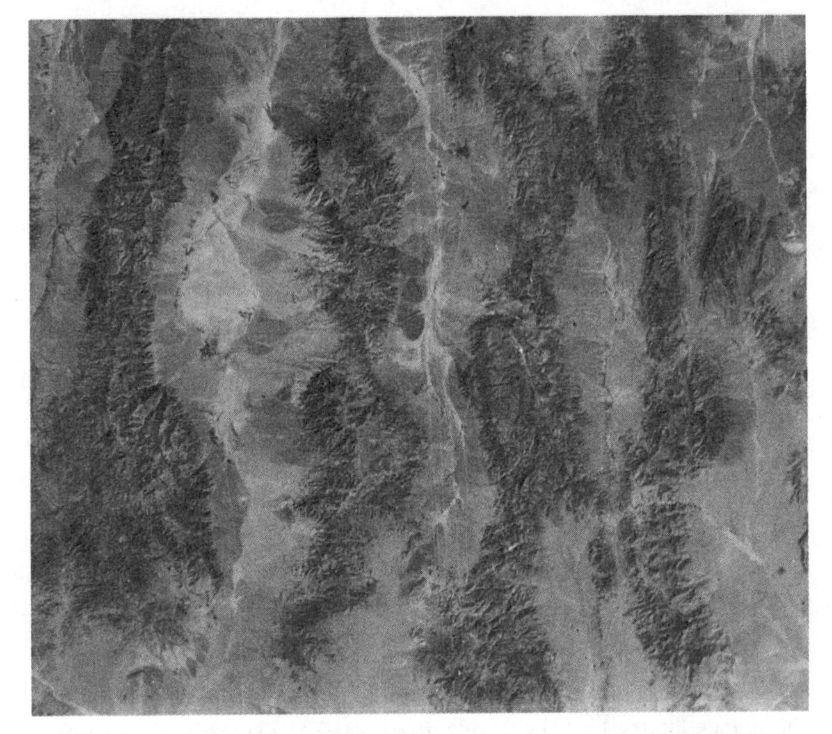

Figure 13.21 This Landsat image from space shows the characteristic topography of the Great Basin region in central Nevada. The width of the view is about 160 km (100 miles). The well-defined base lines of the Shoshone and Toiyabe ranges in the left (west) part of the image clearly indicate the fault origin of the ranges. The ranges in the central and right portions have less-well-defined base lines, suggesting less recent faulting or more erodible rock types. The light spots are dry lake beds.

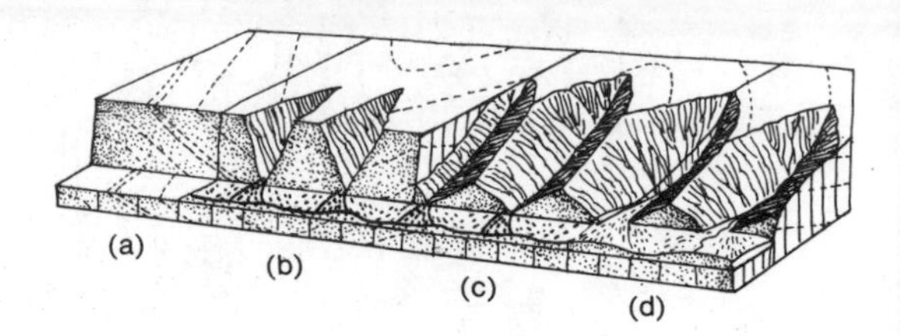

Figure 13.22 (left) In the absence of erosion, an uplifted fault block would form a continuous, straight *fault scarp,* as shown at (a). Processes of erosion result in the dissection of the fault scarp as shown at (b), (c), and (d). Note the straight base line and the *triangular facets* (truncated spur ends) in (c) and (d). **(right)** The east front of the Sierra Nevada is a fault scarp that has been dissected by erosion. Note the straight and regular appearance of the mountain range and its abrupt rise from the neighboring plain. Rising out of the lowland in front of the Sierran scarp is a second, much lower fault scarp.

trench of Death Valley, a block that has subsided well below sea level. At the eastern margin of the Great Basin, Utah's Wasatch Range is also a tilted block, with its western edge uplifted along a major fault that shows much evidence of geologically recent activity. The Great Basin as a whole appears to be a formerly high area that has been stretched in an east-west direction, resulting in its collapse along a host of faults, with most of the relief being created by the downward movement of sinking strips, called *grabens,* between blocks that remained high-standing. Thus the scenery of this region is a product of crustal tension.

Block faulted mountain ranges, such as the Sierra Nevada, display a number of distinguishing features. Particularly striking is the abrupt manner in which the steep mountain faces, or *fault scarps,* rise from the level of adjacent basins. Although erosion has deeply dissected the fault scarps, their remnants remain in the form of *triangular facets* (see Figure 13.22), and here and there areas of rock may bear polished flat surfaces, or *slickensides,* caused by the abrasion of one fault block slipping past another.

Mountains created by uplift along faults have remarkably straight base lines, with no projections, or *spurs.* Ranges with straight base lines or fault planes on both sides appear to have been vertically uplifted and are called *horsts.* Those having a straight base line or fault scarp on only one side are tilted blocks. Sometimes the base line of a bold mountain range is observed to cut across both resistant and easily eroded rock. In such a case, it is safe to assume that faulting rather than erosion created the existing relief.

Major faults penetrate deep into the earth's crust, and they act as channels to conduct heat and water upward from the interior of the earth. Hot and cold springs are therefore frequently found along fault lines; magma may also force its way up along such fractures, resulting in volcanic activity at the surface.

Faults that result from tension are called *normal faults.* The fault plane (see Figure 13.23) extends downward at a steep angle. In normal faulting the block resting on the inclined fault plane slips downward with respect to the adjacent block (Figure 13.24, p. 408). Movement down the plane, or *dip*, of the fault is called *dip slip*. Normal dip-slip faulting has produced the majority of the mountain ranges of the Great Basin region.

Where rocks fracture due to compression, which causes the crust to be shortened, fault planes vary in angle from oblique to nearly horizontal. An obliquely dipping fault resulting from compression is called a *thrust fault* or *reverse fault*, because in compression the block lying above the fault plane is thrust up and over the block below the fault plane—the motion being the reverse of normal dip-slip faulting (Figure 13.24). Large and small reverse faults are common in major mountain chains. The destructive earthquake experienced by California's San Fernando Valley in 1971 was produced by about 1 meter (3 to 4 ft) of sudden upward motion on a thrust fault at the base of the San Gabriel Mountains.

The most awesome of the faults resulting from compression are nearly horizontal *overthrusts*, in which slices of rock thousands of meters thick are sheared off and forced over the top of adjacent rock masses, often traveling tens of kilometers horizontally. In the Alps, three or four such slabs, each having the area of a state like Massachusetts or Connecticut, are often piled on top of one another, with older rocks sometimes thrust over much younger ones. The incredible force required to produce such deformation is striking evidence of the enormous energy within the earth. Overthrusts of great magnitude are not uncommon; they appear to have been a normal part of the development of major mountain chains since the earliest geologic times, and they are part of the structure of most major mountain systems. However, their influence on mountain scenery is primarily through the type of rock they carry into an area. Most overthrusts have been subsequently folded, so that their landforms are similar to those developed in folded rock structures, often touched up by glaciation where they occur in the high young mountains of the world.

Many of the earth's most important faults have been produced by *shear*, in which adjacent crustal segments move past one another horizontally rather than vertically. Faults of this type, which exhibit displacement parallel to the surface trace of the fault plane, or the fault *strike* (Figure 13.23), are called *strike-slip faults* (see Figure 13.24). No doubt the most thoroughly investigated of the world's strike-slip faults is the famous San Andreas fault, which extends north and south for 1,000 km (650 miles) in western California. The San Andreas fault system is not just a single fracture, but consists of numerous largely parallel faults located in a zone many kilometers wide.

A distinctive set of landforms is associated with strike-slip faults. They are referred to collectively as *rift topography*. The line of the San Andreas fault is marked by linear erosional valleys, closed depressions, and elongated hills that parallel the fault zone (Figure 13.25, p. 410). The straight valleys are caused primarily by rapid erosion of rock crushed along the fault plane due to the friction of rock masses being forced past one another. The jostling of the earth in a large fault zone sometimes causes a local area to sink below the surrounding land or become hemmed in on all sides by higher ground. The resulting

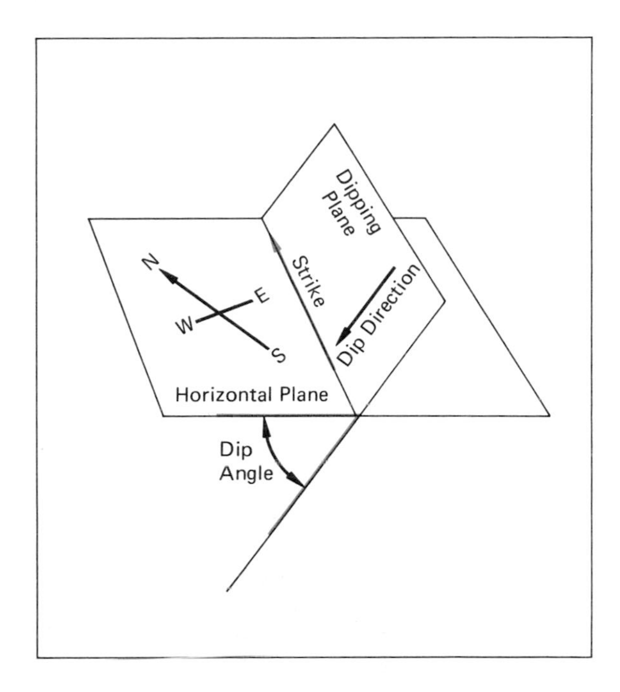

Figure 13.23 The attitude of any inclined plane, such as a fault or a tilted layer of rock, may be described precisely in terms of the *strike* and *dip* of the plane. The strike is the compass bearing of the line of intersection between the inclined plane and a horizontal plane. It is read in degrees east or west of a north-south direction. The dip is the angle the inclined plane makes with a horizontal plane; it is given in degrees together with a compass direction. In this diagram, the strike is about N 30° E and the dip is to the northwest at about 50°.

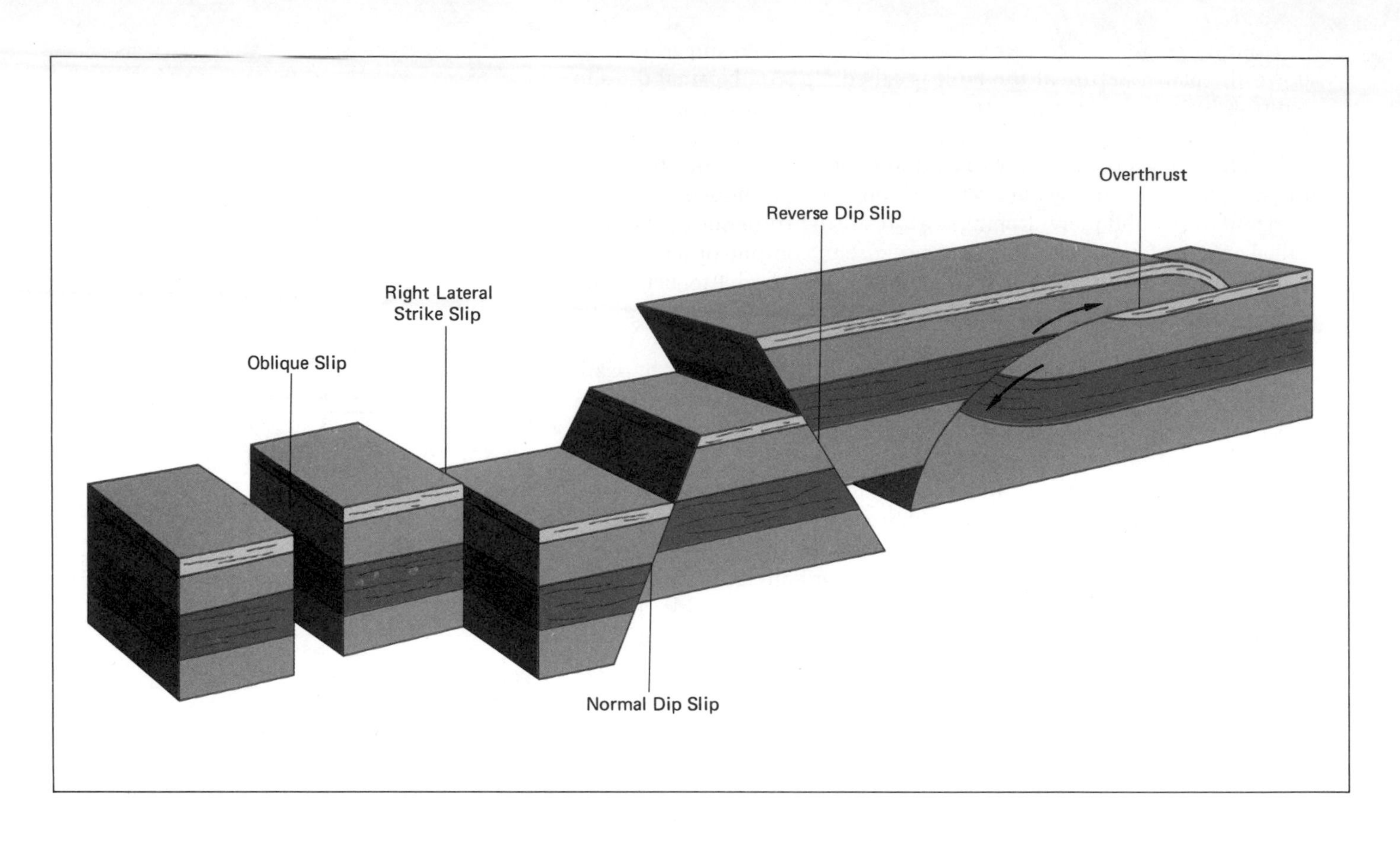

depression acts as a local drainage sump for the higher land around it, forming a *sag pond*. Sag ponds often appear in the mountainous areas along the San Andreas fault, and their origin is apparent when one sees their linear distribution on a map or from the air. In many areas, sag ponds are the clearest evidence of the fault location. The linear hills of rift topography are ridges forced up by pressure as well as slivers of rock dragged along in the fault zone. Springs are also common along the San Andreas fault; in fact, it was the presence of a spring that determined the location of the winery mentioned at the beginning of this chapter. Many of the springs are caused by displacement of the local water table, but some are hot and mineralized, indicating penetration deep into the earth's crust.

One of the interesting features of strike-slip faulting is the horizontal offset of features such as ridge lines and stream beds that cross the fault. Since displacement along the fault occurs in small increments, streams crossing the fault tend to remain in their old channels above and below the fault line. In time, the streams crossing an active fault may experience a horizontal offset of hundreds to thousands of meters.

With increased studies of the ocean floors and better understanding of the mechanism of sea floor spreading, the earth's large strike-slip faults are now appreciated as vital features in the motion of the major lithospheric plates. In many areas, adjacent plates are neither converging nor diverging, but appear to be sliding laterally past one another. This is because plate motion away from spreading centers cannot be uniform due to the spherical shape of the earth. The problem of crowding in areas where spreading centers converge is obvious. To accommodate sea floor spreading on a spherical surface, the individual oceanic plates have become segmented by strike-slip faults at right angles to the oceanic ridges, as can be seen in Figure 11.5, p. 302, which also shows that the spreading centers are themselves repeatedly offset along strike-slip faults. Offset of a sea-floor-spreading center is especially marked in the eastern Pacific Ocean, where the Northern American continent has pushed westward against the oceanic ridge system known as the East Pacific Rise. A series of fractures have developed that transform sea floor spreading to strike-slip faulting. Such faults have recently been termed *transform faults.* California's San Andreas fault system appears to be a transform fault—an unusual one in that it cuts through the edge of a continent rather than being restricted to the sea floor. Californians jolted by the next large earthquake on the fault will probably take little comfort from the knowledge that they are being affected directly by the seemingly remote but inexorable process of sea floor spreading.

Figure 13.24 (opposite) Faults are the result of the rupture of rocks due to crustal tension, compression, and shearing. Fault motion is described by the relative displacement or *slip* of the rock units in contact along the fault. Motion may be parallel to either the dip or the strike of the fault plane, or oblique to both.

In the center of the diagram is a *normal fault,* in which the type of motion is *dip slip,* with the block resting on the fault plane slipping down relative to the adjacent block. This type of motion results from tension and produces crustal extension. The faults on the right of the diagram result from compression and produce shortening of the crust. In both *overthrusts* and *reverse* faults, the rock mass above the fault plane moves up and over the rock mass below the fault plane. In all three faults the motion is parallel to the dip of the fault plane, but in the latter two the direction is the reverse of normal dip slip motion so that the type of motion is *reverse dip slip.*

Shearing stress involves side-by-side forces exerted in opposite directions. This results in horizontal motion parallel to the fault strike, or *strike slip.* The direction of the strike slip is designated in terms of the movement of the block on the far side of the fault. In this case, looking across the fault from either side, we see that the far block has moved to the right, so that the motion is *right lateral strike slip.*

The final fault on the left shows displacement oblique to both the strike and dip of the fault plane. Both strike slip and dip slip are involved in *oblique slip.* Whereas strike slip faults are always vertical and dip slip faults are always inclined, oblique slip can occur on both vertical and inclined faults.

Landforms Associated with Fold Structures

It may seem strange that solid rock strata thousands of meters in depth can be bent almost double, but fold structures are clearly visible in many rock exposures. Rock is brittle and tends to crack under sudden impact, such as a hammer blow, but prolonged stress, particularly at the high temperatures and pressures that prevail deep in the earth's crust, can slowly deform rock layers without breaking them.

Slow compressive stress can produce in rock every configuration you can make by pushing your hands over a loose tablecloth. The complexity of geometrical forms possible is overwhelming, but some basic elements are generally recognizable in all folded rock structures. Where layers of sedimentary rock have been forced into a series of wrinkles, or folds, the ridges are called *anticlines* and the troughs are termed *synclines* (see Figure 13.26, p. 411). Each anticline and each syncline has a longitudinal *axis* that marks the plane of maximum curvature of the rock layers forming the crests of the ridges and the deepest parts of the troughs. Between the anticlinal and synclinal axes are the *limbs* of the fold, in which the rock layers dip at various angles from the horizontal. These terms refer only to geologic structure, not to the visible surface landforms, which result from erosion of these structures.

The specific landform that develops at any point in an area of folded sedimentary rock depends on the relative resistance of the stratum involved and its thickness and geometry at that location. The scenery that develops regionally depends on the size and shape of the folds, the number, thickness, and relative resistances of the strata included, the rate of tectonic uplift, the history and nature of the drainage pattern, and the climatic controls of the forces of erosion.

Figure 13.25 (above) A portion of the San Andreas fault forms the dominant straight-line feature in the center of this photograph of the Carrizo Plain in California. The San Andreas fault is a plane of shearing at the boundary between the Pacific and North American lithospheric plates.

(right) This orange grove is located along the San Andreas fault. Note the offset of the orchard rows due to motion along the fault. The right lateral strike-slip fault has displaced the trees by about 2 meters (7 ft).

In real landscapes, anticlinal structures rarely produce simple ridges, and synclinal structures are seldom seen as deep troughs. Generally the anticlines are peeled down like an onion, exposing layer after layer of rock, some hard, some soft (Figure 13.27, p. 412). The

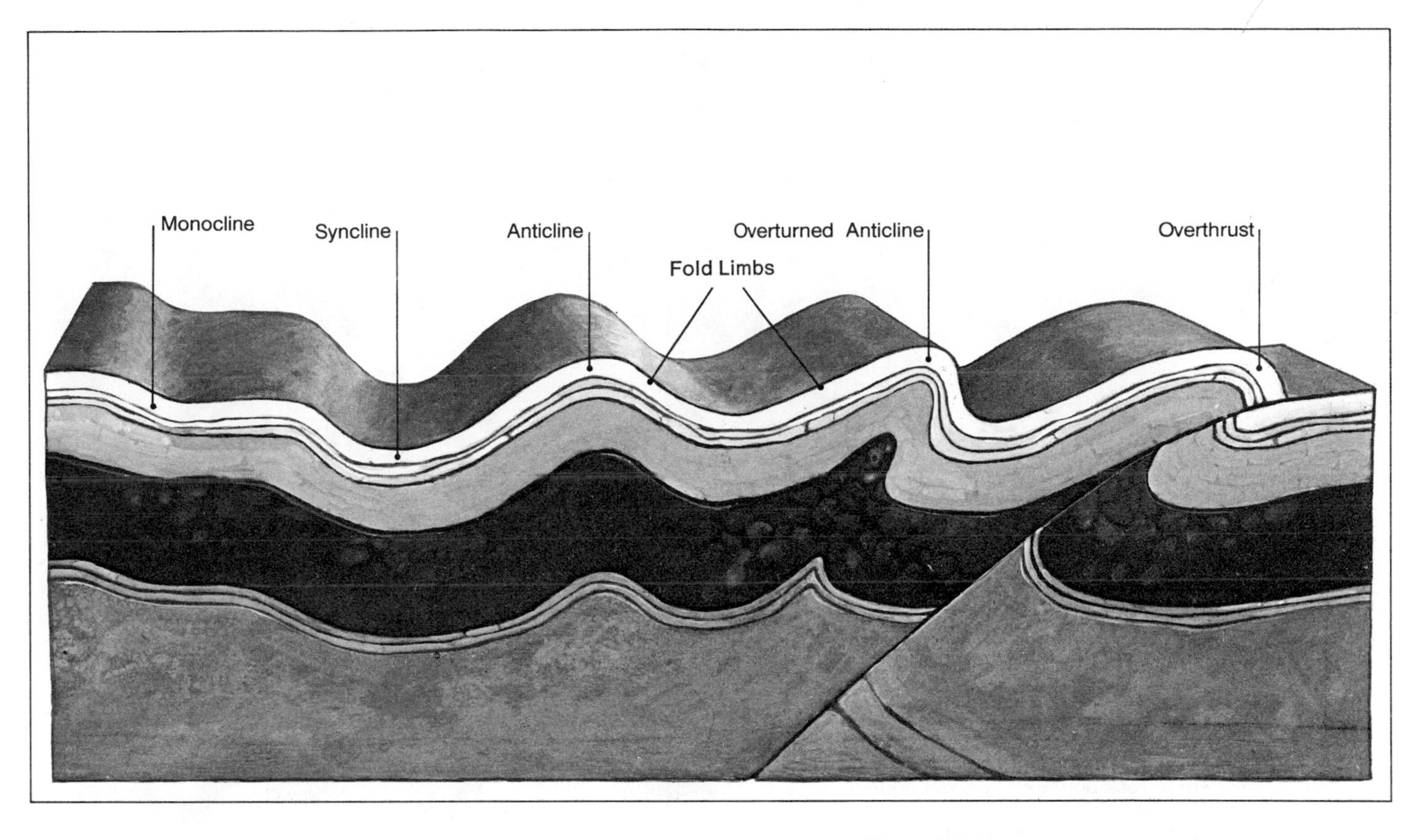

Figure 13.26 This diagram illustrates the geometry of folded rock strata and the terms used to describe it. A terracelike flexure in bedded rocks is a *monocline.* A downfold whose sides, or *limbs,* are rotated upward on either side is a *syncline,* and an arch whose limbs are rotated downward is an *anticline.* Where folding is extreme, the limbs of the fold may be asymmetrically inclined, and the strata on one limb of the fold may be oversteepened or even overturned. If the rock strata are subject to great pressure, an overturned fold may rupture along a zone of weakness, and one limb of the fold may be thrust over the other limb, forming an *overthrust.* Fold structures are rarely seen intact as in this drawing, but they are geomorphically significant because distinctive landforms develop by erosion in them, as seen in Figures 13.27 and 13.28, p. 412.

scenery is dominated by the edges of various inclined rock strata that are resistant to erosion. These edges form *homoclinal ridges* (ridges made by rocks of one-directional dip on the limbs of the folds). Figure 13.28, p. 412, shows how a layer of rock can make a *mesa,* a *cuesta,* or a *hogback ridge* as its inclination increases. Both cuestas and hogbacks are regarded as homoclinal ridges.

The simplest example of the erosion of a series of anticlinal and synclinal folds having level crests and troughs is shown in Figure 13.29, p. 413. The anticlines will be attacked by erosion as soon as they start to form, exposing the varying rock strata in their cores. If one or more of the folded layers is thick and resistant to erosion, the more rapid erosion of the less resistant layers will create a landscape of parallel ridges of resistant rock separated by valleys etched out in the erodible layers. As erosion continues, the edges of hard rock will be worn back in a down-dip direction. This is primarily due to the removal of the erodible material exposed under the edge of resistant rock, which leaves the resistant rock unsupported and prone to collapse. In the case of a dome, homoclinal ridges retreat outward from the dome's center. All anticlines are, in fact, elongated domes that die out eventually. This causes the homoclinal ridges to converge and meet along the fold axes, as in Figures 13.27 and 13.30, p. 413. The latter figure shows the variety of ridge and valley shapes generated in one area of folded strata as a consequence of the specific geometry of the folds and the varying resistance of the rock layers involved. Note that it is difficult to distinguish between the anticlines and synclines;

Figure 13.27 (left) Folded by compression of the crust, then left projecting above the surrounding terrain as less resistant materials weathered and eroded away, Sheep Mountain, Wyoming, now rises as a resistant core of rock bordered by lower hogbacks of resistant strata.

(right) Another view of Sheep Mountain, showing hogbacks in more detail.

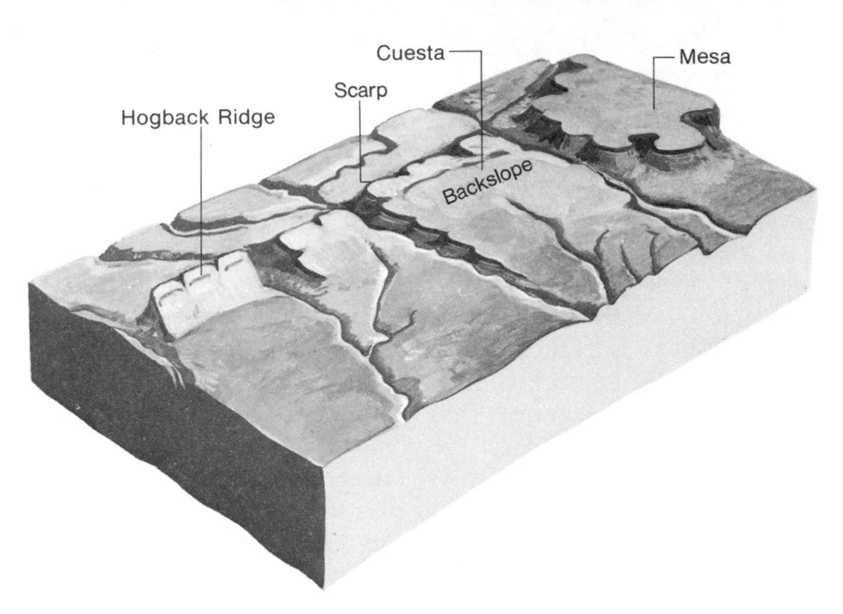

Figure 13.28 *Mesas, cuestas,* and *hogback ridges* are typical landforms that can be produced by differential erosion in regions where a resistant caprock overlies less resistant layers. A flat-topped mesa is formed where the rock strata are horizontal. A cuesta is formed where the strata dip at an angle of a few degrees, so that a cuesta possesses a steep face, or *scarp,* and a gently dipping *backslope.* A hogback ridge is formed where the strata dip steeply; thus a hogback ridge presents two steep slopes of different origin. One slope consists of the dipping caprock, and the other is the erosional scarp that cuts across the caprock and the underlying layers.

all has been converted to homoclinal ridges and erosional valleys. Although the nested hairpin ridges of central Pennsylvania are the classic example of differential erosion in folded rock strata, similar landscapes are found in many other parts of the world. Where soils are thin due to the absence of a protective vegetation cover, the scenery is much more severe, though similar in overall character. In such areas, steep homoclinal ridges become sharp hogbacks. These may be quite *serrate* (jagged), as in Figure 13.31, p. 414.

In central Pennsylvania, streams form a trellis-shaped drainage pattern (see Chapter 12). Tributary streams flow along parallel valleys in erodible rock between the long ridges of resistant rock before they join the master streams, which, surprisingly, seem to ignore geologic structure, often cutting across hard layers in deep gorges. Clearly the geologic structure strongly controls much of the drainage pattern in this region, but not its most important elements. Near Harrisburg, the Susquehanna River cuts through five successive high ridges of resistant sandstone (see Figure 13.30). This anomaly has never been fully explained, although the following hypothesis has been postulated. In the geologic past, a long period of erosion reduced central Pennsylvania to very low relief, with the Susquehanna becoming established on a flat plain, probably on a blanket of sediment that covered the underlying leveled rock edges (Figure 13.32, p. 415). Later, when the land was slowly uplifted, the river was able to maintain its preexisting channel by cutting down into the harder rocks beneath the sediment blanket. According to this theory, which cannot be proved, the general erosion of less resistant rocks by streams and their tributaries gradually etched out the pattern of ridges and valleys seen today. This hypothesis regards the Susquehanna as having been *superimposed* downward upon the structures it crosses.

Similar stream courses transverse to geologic structures abound in regions of fold mountains, often producing gorges far deeper and more spectacular than those of the Susquehanna. The Indus and Ganges rivers of India both cross the folded overthrusts of the high Himalaya range in valleys more than 5 km (3 miles) deep. In some cases, such transverse streams are thought to be older than the structures they cross, being called *antecedent* streams. Elsewhere, transverse streams are believed to have eroded headward through the structures by a series of drainage captures. Seldom is there any clear evidence of what really happened in such cases.

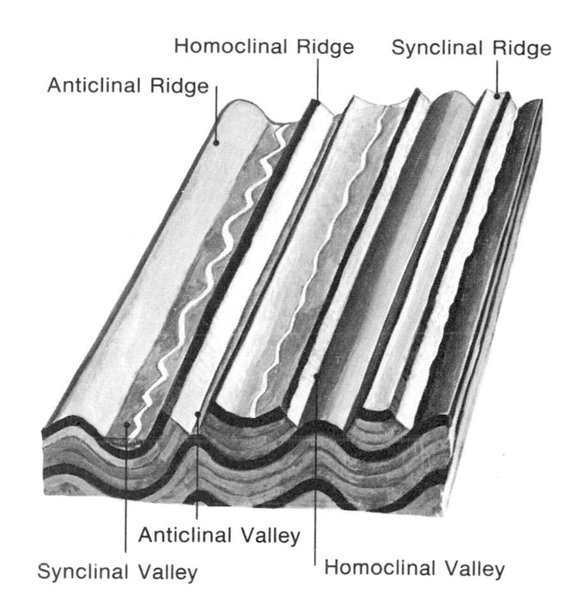

Figure 13.29 This schematic diagram illustrates the variety of landforms that can be produced by erosion of folded rock strata that crop out at the earth's surface. The dark-shaded rock layers are assumed to be more resistant than the other layers. Erosion of the weaker rock overlying hard rock on an anticline forms an *anticlinal ridge.* A *synclinal valley* of hard rock is formed similarly when weaker rock is eroded from a syncline. If a portion of the top layer of resistant rock over an anticline has been worn away, *homoclinal ridges* are formed by the exposed edges of rock strata, with an *anticlinal valley* between them. A *homoclinal valley* is formed in soft rocks between a homoclinal ridge and the steeply sloping limb of an exposed anticline or between homoclinal ridges that dip in the same direction. A *synclinal ridge* is formed where a remnant of resistant rock in a syncline acts as caprock to retard the erosion of the underlying layers. Ridges and valleys developed in structures of folded rock are common in many landscapes, such as the Appalachian Mountain region of the eastern United States. (See also Figure 13.30.)

Figure 13.30 Folded strata form a distinctive ridge and valley landscape in southeastern Pennsylvania. This aerial view near Harrisburg was made with side-looking airborne radar (see Appendix II); the image spans a distance of approximately 50 miles. The ridges are formed of resistant sandstone, and the valleys are cut into erodible shale or limestone. Note the nested hairpin ridges formed where anticlines and synclines rise and fall along the strike of the folding. The gaps in the ridges through which the Susquehanna River flows are clearly visible in the image.

Figure 13.31 Mount Rundle near Banff, Alberta, in the Canadian Rockies is an immense homoclinal ridge composed of limestone that dips to the west (right in this view).

SUMMARY

Geologic structure is a major influence on the development of landforms. Structure includes both the specific character of rock material and the large-scale arrangement of rock masses. Rocks are composed of minerals and vary according to the characteristics of their mineral components. Rock originates in several ways: igneous rock crystallizes from a molten condition; sedimentary rock is formed by the lithification of deposits of loose rock debris and organic material both on the land and in the sea; metamorphic rock is a product of the recrystallization of preexisting rocks under great pressure or heat. Chemical composition and physical structure determine the relative ease with which rocks can be weathered and removed by erosion. Differential erosion, particularly of layered or jointed rock, is important in the appearance of landforms. The rock cycle is an endless sequence of material transformations by which the same atoms are used over and over to form new rock material from the debris of older rocks and from the melting and reconstitution of rock material in subduction zones.

Volcanic activity is a major source of new rock and an important influence on landscapes. Volcanism occurs at active lithospheric plate boundaries. Nonviolent basaltic volcanism is present along the oceanic ridges and provides the rock material of the sea floors. Violent volcanic eruptions are concentrated above subduction zones, in which descending lithospheric plates generate chemically siliceous new magma, mainly of the andesitic variety. The violence of volcanic eruptions is a consequence of the gases trapped in subsurface magma.

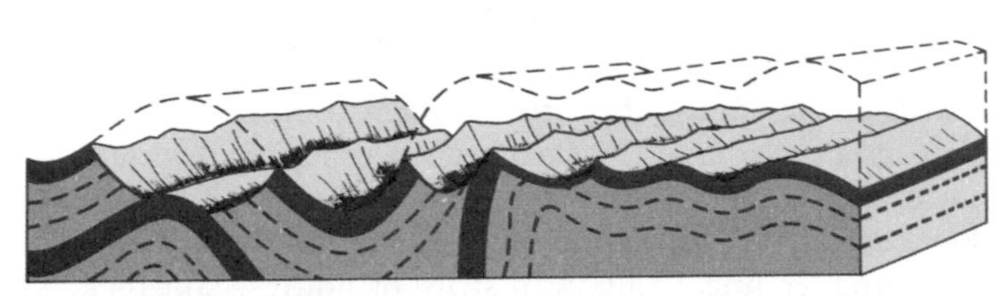

(a) Original eroding folds

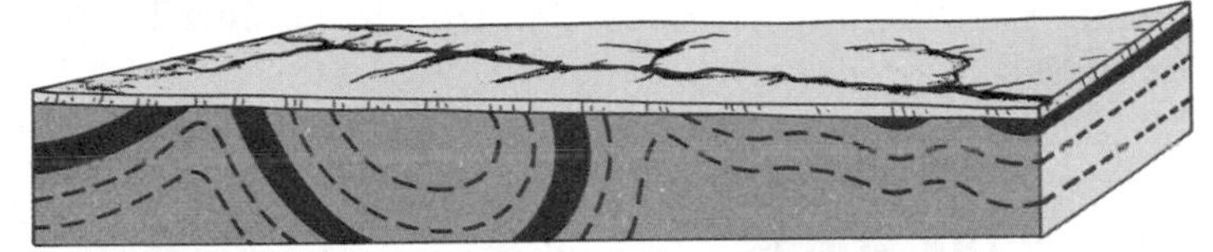

(b) Region worn to plain and covered with sediments

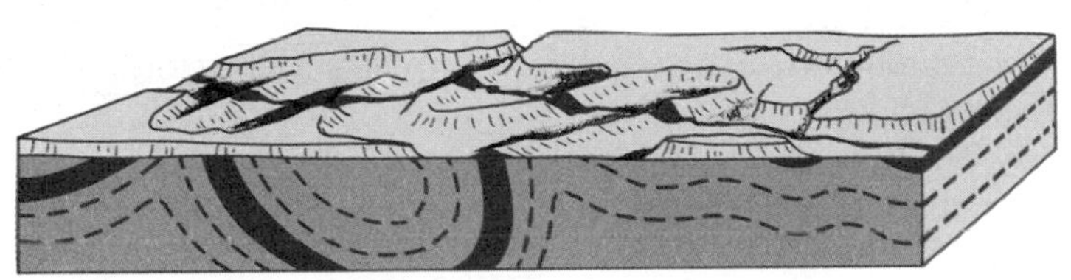

(c) Drainage incised in sediment cover to expose rock edges

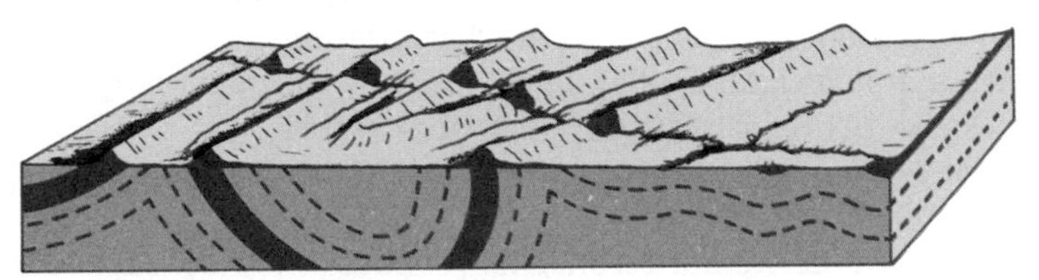

(d) Covermass removed and watergaps cut across resistant edges

Figure 13.32 These diagrams illustrate one hypothesis explaining how streams such as the Susquehanna River in the Appalachian Mountains of Pennsylvania have cut watergaps through bold ridges of resistant rock. A region underlain by folded rock strata (a) becomes worn to a plain of low relief (b). The plain becomes covered with sediments upon which streams develop independently of the rock structures beneath the newer sediments. Renewed uplift (c) causes these streams to incise through the sediment blanket, superimposing themselves across the edges of the tilted rock layers beneath. Erosion of the weaker rock (d) produces valleys between ridges of resistant rock. If the major streams are able to incise downward in pace with the removal of the weaker rock, they may maintain their original courses through notches they cut in the harder rock ridges. Each lowland composed of weak rock is occupied by a stream that flows parallel to the trend of the folding, eventually entering a major stream transverse to the fold trend.

Strato-volcanoes and plug domes are built by violent eruptions, whereas shield volcanoes are the product of quiet effusions of basaltic lava. Basaltic plateaus, table mountains, dike ridges, and volcanic necks are erosional remnants of past volcanic activity. Volcanic regions are often the locations of metallic minerals and also contain potentially valuable geothermal energy resources. The latter have been developed for heating and electric power generation in a few areas and will contribute increasing supplies of energy in the future.

Where the earth's crust is being compressed, rocks may fail either by bending, producing folds, or by rupturing, producing faults. Where the crust is being stretched, it collapses in normal dip-slip faults that may create tilted blocks or horst and graben structures. These may be expressed in the landscape as mountain ranges that rise very abruptly above adjacent basins. Movement on strike-slip faults is horizontal rather than vertical. Strike-slip faults have their own peculiar landforms, such as sag ponds, offset streams, and linear rift topography. Horizontal compression of the earth's crust produces reverse dip-slip faults, overthrusts, and folds. The structural ridges and valleys in folded rock strata are called anticlines and synclines. Erosion in these structures produce linear valleys in strips of nonresistant rock, with resistant layers standing out in relief as homoclinal ridges. In such areas, the topography reflects the varying resistance of different rock layers rather than the original ridges and troughs produced by compressive stress.

Review Questions

1. What is the principal factor causing silicate minerals to vary in chemical stability?
2. How can one distinguish between intrusive and extrusive igneous rocks?
3. How can it be that sedimentary rocks are the most widely exposed type on the continents, yet igneous rocks form 80 percent of the earth's crust?
4. How are sial, sima, and isostacy interrelated?
5. How can a single layer of sedimentary rock change horizontally from limestone to shale, then sandstone, then conglomerate?
6. Under what conditions, and in what locations, is rock metamorphism most likely to occur? How can metamorphic rocks be distinguished from other rock types?
7. What are the factors that create differential weathering and erosion?
8. Why is it that the hottest magma produces the least explosive surface volcanic activity?
9. Volcanism seems to occur both where crustal plates are pulling apart and where they are pushing together. Explain this apparent contradiction.
10. How many types of volcanoes can you distinguish? How do they differ from one another?
11. What three landforms result from differential weathering and erosion in volcanic regions?
12. In what ways has volcanic activity been beneficial to humans?
13. What are the geomorphic features that identify normal faults? strike-slip faults?
14. How do the individual mountains in regions of folded strata differ from those produced by faulting?
15. In what locations on the earth's crust would structures produced by folding, thrust faulting, and overthrusting be encountered?

Further Reading

Bullard, Fred M. *Volcanoes of the Earth.* Austin: Univ. of Texas Press, 1976. (579 pp.) This is the most recent of a large number of authoritative texts on volcanic processes and resulting landforms.

Eardley, A. J. *Structural Geology of North America.* 2nd ed. New York: Harper & Row, 1962. (743 pp.) This book is useful for its many cross sections illustrating the details of geologic structure in all parts of North America.

Hunt, Charles B. *Natural Regions of the United States and Canada.* San Francisco: W. H. Freeman & Co., 1974. (725 pp.) In Hunt's scheme natural regions are primarily differentiated by geologic structure, which in North America creates the main contrasts in land surface configuration. This book presents good general descriptions of the physiography of each of the natural regions of North America, including information on the climate, vegetation, and human use of each region.

Hurlbut, Cornelius S., Jr. *Minerals and Man.* New York: Random House, Inc., 1970. (304 pp.) Hurlbut's informative book on the occurrence, characteristics, and uses of minerals is renowned for its superb color photographs of mineral specimens.

Lobeck, A. K. *Geomorphology.* New York: McGraw-Hill Book Co., 1939. (731 pp.) Lobeck's textbook was constructed on the premise that a picture is worth a thousand words. This book includes about 200 pages on folding, faulting, and volcanism; these pages consist mainly of photographs and diagrams that have retained their value through the years. The material on folding is especially helpful.

Shelton, John S. *Geology Illustrated.* San Francisco: W. H. Freeman & Co., 1966. (434 pp.) Nowhere is there published a better collection of photographs illustrating structurally controlled landforms. The text is very readable and original in approach.

Shimer, John A. *Field Guide to Landforms in the United States.* New York: Macmillan, 1971. (272 pp.) The first portion of this handbook outlines the nature of the various structurally determined physiographic regions of the United States, with portions of Erwin Raisz's detailed physiographic diagram of the United States used as index maps of each region. Following this is a catalogue of individual landforms, each illustrated by an excellent line drawing.

Sparks, B. W. *Rocks and Relief.* New York: St. Martin's Press, 1971. (404 pp.) More than half of this book by a British geomorphologist deals with the geomorphic influence of geologic structure. It is the most thorough treatment to date of the effect of rock types on landforms. The treatment is advanced and detailed. The last half of the book describes the various geologic systems (time-stratigraphic units) in the British Isles.

Sweeting, Marjorie M. *Karst Landforms.* New York: Columbia University Press, 1972. (362 pp.) This text by an English geomorphologist deals exclusively with the distinctive landforms that develop where limestone is the bedrock.

Thornbury, William D. *Regional Geomorphology of the United States.* New York: John Wiley & Sons, Inc., 1965. (609 pp.) Thornbury's treatment of regional geomorphology is organized by structural provinces, and is a little more advanced than the book by Hunt, stressing the varied hypotheses in explanation of problematical geomorphic features. An outstanding contribution of the book is the very complete bibliography presented for each chapter.

Twidale, C. R. *Structural Landforms.* Cambridge, Mass.: MIT Press, 1971. (247 pp.) This well-illustrated text by an Australian geomorphologist deals with landforms associated with granitic rocks, joint systems, faults, and folded strata.

Way, Douglas S. *Terrain Analysis: A Guide to Site Selection Using Aerial Photographic Interpretation.* Stroudsburg, Pa.: Dowden, Hutchinson, and Ross, 1973. (392 pp.) Written as a guide to site selection for development purposes, the book is especially useful for its descriptions of the landforms developed on various geologic materials in different climatic environments. Well illustrated with diagrams and stereo aerial photographs.

Wilcoxson, Kent H. *Chains of Fire.* New York: Chilton Book Co., 1966. (235 pp.) This stimulating book on volcanism, written for the general public, includes many eye-witness accounts of volcanic eruptions of various types. Excellent reading.

Case Study:

Earthquakes

A little after 02:00 [2 A.M.] on December 16, the inhabitants of the region suddenly were awakened by the groaning, creaking, and cracking of the timbers of their houses and cabins, the sounds of furniture being thrown down, and the crashing of falling chimneys. In fear and trembling, they hurriedly groped their way from their houses to escape the falling debris. The repeated shocks during the night kept them from returning to their weakened and tottering dwellings until morning. Daylight brought little improvement to their situation, for early in the morning another shock, preceded by a low rumbling and fully as severe as the first, was experienced. The ground rose and fell as earth waves, like the long, low swell of the sea, passed across the surface, bending the trees until their branches interlocked and opening the soil in deep cracks. Landslides swept down the steeper bluffs and hillsides; considerable areas were uplifted; and still larger areas sank and became covered with water emerging from below through fissures or craterlets, or accumulating from the obstruction of the surface drainage. On the river, great waves were created which overwhelmed many boats and washed others high upon the shore, the returning current breaking off thousands of trees and carrying them into the river. High banks caved and were precipitated into the river; sandbars and points of islands gave way; and whole islands disappeared. *(Earthquake History of the United States,* 1973)

Obviously, this report must be from fault-shattered California or some other part of the tectonically-active mountainous west of North America. Or is it? In fact, the site of this description is New Madrid, Missouri—then a community of log cabins next to the Mississippi River—the date: December 1811. Further shocks occurred there in January and February of 1812. Reports of similar events are also available from Quebec (1638) and Charleston, South Carolina (1886), with only somewhat less catastrophic effects having occurred again in Missouri (1895), western Texas (1931), western Ohio (1937) and southern Illinois (1968), to name only a few normally placid geologic settings that have experienced sudden seismic shocks.

Unlike volcanic eruptions, which have restricted geographic distributions and localized destructive effects—or even floods, storm waves, landslides, snow avalanches, and sudden ground subsidence, each of which requires a particular physical setting—earthquakes can occur anywhere. Furthermore, unlike most other environmental hazards, they generally occur without warning. By their direct and indirect effects, including fires, flooding, landslides, and so-called "tidal waves," earthquakes have caused more destruction and taken more lives than any other single type of natural catastrophe.

What are earthquakes? And how can we account for their wide distribution? The slow distortion of rock masses by tectonic stress causes a build-up of potential energy in the rock much as in a compressed or extended spring. When the deformation, or strain, surpasses the elastic limit of the rock, it ruptures suddenly or slips along preexisting faults, releasing the stored energy in the form of seismic shock waves that speed outward from the point of rupture. Several types of shock waves are involved, including rapidly moving body waves of varying character that pass through the mass of the earth and slower surface waves that follow the earth's outer skin. The surface waves create the ground motion we call an earthquake. The point below the earth's surface at which the energy is released is the earthquake *focus,* or *hypocenter.* The point on the earth's surface directly above the focus is the earthquake *epicenter.* It is at the epicenter that shock waves are most strongly felt.

The type of ground motion produced and the damage occurring during an earthquake vary according to the depth of the focus, distance from the epicenter, and nature of the local geologic material. In a small tremor, the sensation produced ranges from a gentle horizontal shaking to one or more sharp jolts. Near the epicenters of large earthquakes, one has the feeling of being in a rough sea in a small boat, with the ground visibly rising and falling in moving crests and troughs similar to ocean waves. Normally, the shaking lasts no more than a few minutes. However, smaller aftershocks, sometimes numbering in the hundreds, may continue over a period of days to months, indicating continuing subsurface rock displacements.

The violence of an earthquake is measured on two different scales that specify the magnitude and surface effect of the energy released. The magnitude scale devised in 1935 by Charles F. Richter of the California Institute of Technology is a logarithmic one in which each successively higher unit represents about 32 times the energy release of the preceding unit. Thus the difference in energy released by earthquakes of magnitudes 4 and 8 on the Richter scale is not 2 times, but more than a million times. The largest earthquakes felt since the invention of *seismographs* (instruments that record earthquake intensity) have attained magnitudes of about 8.9 on the Richter scale. Any earthquake of magnitude 8 or above is considered catastrophic.

The earthquake intensity scale, which measures the surface effects of earthquakes, was developed by the Italian geophysicist G. Mercalli in 1905. The Modified Mercalli scale of 1931, which is now in general use, rates these effects from I to XII (see table); earthquakes of intensity XII are strong enough to throw objects into the air and to destroy most human constructions. There is no correspondence between the Richter and Mercalli scales because earthquake intensity at any moment can vary from point to point with no change in earthquake magnitude, due to the manner in which the local geologic material transmits seismic shock waves. For example, there is normally less ground motion on solid rock near the earthquake epicenter than on unconsolidated alluvium or artificial land fill some kilometers from the epicenter. Unconsolidated deposits magnify the ground shaking and often collapse, behaving almost as a fluid in severe earthquakes, as was the case at Anchorage during Alaska's 1964 Good Friday earthquake. This phenomenon was also conspicuous at Lisbon in 1755 and San Francisco in 1906, and in the New Madrid earthquakes of 1811–1812, in which there was catastrophic shaking and rupturing of the ground on the Mississippi River floodplain but little effect in the adjacent bedrock hills. Artificial land fills are a ubiquitous part of our expanding urban scene, especially in hilly and coastal regions. Fortunately, North America has experienced few severe earthquakes since the explosive expansion of cities and their suburbs in the present century. Unfortunately, every day brings us closer to the first great earthquake that will strike one of our modern urban centers, as has recently befallen Managua, Nicaragua (1972), Peking, Tangshan, and Tientsin, China (1976), and Guatemala City, Guatemala (1976).

We have already seen that earthquakes can be anticipated in areas overlying subduction zones and lithospheric plate boundaries, and also along the oceanic ridge system. In all of these locations, great masses of rock are being moved past one another as part of the recently discovered process of sea floor spreading. We may expect repeated news of earthquake disasters from Peru, Chile, Central America, California, Alaska, Japan, the Philippines, New Zealand, Indonesia, the Mediterranean, Turkey, Iran, the Himalayan foothills, Burma, and China. But some of history's most devastating earthquakes cannot be explained in terms of the gross motions of lithospheric plates—for example, those of the Mississippi Valley, Charleston, and Quebec, all apparently in the middle of the North American plate. And there is the greatest earthquake in European history, the Lisbon quake of 1755, which in 6 minutes took hundreds of thousands of lives and destroyed castles, fortresses, cathedrals, and mosques all over Portugal, southern Spain, Morocco, and Algeria, the ruins of which remain tourist attractions today. The epicenter of this earthquake was apparently offshore in the Atlantic, in a region where there is no major tectonic feature to explain such a geologic cataclysm. In time, geologists will uncover the reasons for earthquakes in such areas. For example, recent evidence suggests that the Mississippi Valley, a midcontinental area of low relief but one of frequent seismic activity, may be an ancient continental rift similar to those presently containing the Red Sea and the East African lake system. Several types of evidence suggest that this billion-year-old rift, which is now buried by later sediment accumulations, has been reactivated by the intrusion of mantle material at depth, followed by isostatic adjustments that may be triggering earthquakes such as those of 1811–1812 and many smaller subsequent tremors.

While ground motion during earthquakes causes enormous damage, and often virtually complete destruction of villages or even cities, subsidiary effects may be equally dangerous. The enormous fires that are triggered by earthquakes—as in San Francisco in 1906 and Tokyo in 1923—may be more destructive than the ground shaking. The New Madrid earthquakes caused changes in elevation of 2 to 6 meters (6 to 20 ft) over an area of 100,000 sq km (40,000 sq miles), creating temporary knickpoints in the Mississippi River and resulting in permanent submergence of 600 sq km (250 sq miles) of virgin forest by groundwater outflow and surface stream flooding. In coastal and undersea regions, earthquakes resulting from sudden movement along faults produce seismic sea waves, or tsunamis (see Chapter 16), popularly called tidal waves, which have frequently swept over populous coastal regions, causing enormous loss of life. Earthquake tremors may set up resonance effects in large and small water bodies, causing them to slosh back and forth rhythmically, often with destructive effects on docks and ships at anchor. During the Lisbon earthquake of 1755, these resonance effects, termed *seiches,* were observed as far away as Switzerland, the British Isles, and Scandinavia.

Almost all earthquakes trigger landslides, ranging from gigantic rock slides in mountain regions to stream bank cavings along flat floodplains. While the latter may create destructive waves in rivers, the former have often obliterated entire communities. The most tragic recent example occurred in Peru in May 1970, when an earthquake in the Peru-Chile oceanic trench produced an ice and rock avalanche from glacier-clad Mt. Huascaran in the Peruvian Andes that completely entombed one moderate-size town and a large portion of another, costing the lives of 18,000 people. The same hazard exists around Washington's ice-covered Mt. Rainier, as shown by the enormous prehistoric landslide deposits in many of the valleys radiating from the great peak.

Landslides, devastating in themselves, often create still further dangers. In 1959 an earthquake sent a great rockslide crashing into the Madison River Canyon in Montana, just outside Yellowstone National Park. While this slide snuffed out the lives of 26 campers, an even greater hazard was created, for, as in most such cases, the slide blocked the Madison River, causing it to back up in a rapidly rising lake. Downstream from the canyon is a lowland dotted with ranches and several small towns. Had the lake risen to the top of the landslide dam and spilled over, it would have washed away the dam, producing a flood of cataclysmic proportions in the valley beyond. Men and heavy earth-moving equipment were rushed to the scene to labor around the clock to excavate a controlled spillway for rising Earthquake Lake. Fortunately, the work was finished in the nick of time: 24 days after the slide occurred the lake overflowed, but under control, and the second disaster was averted. In the minds of the workers and the people downstream during those three weeks was the memory of a similar landslide dam formed on the Gros Ventre River, south of Yellowstone, only 34 years earlier. Two years after this slide blocked the river, the resulting lake spilled over, sending a wall of water, and the bodies of drowned persons and livestock, down to Kelly, Wyoming. The highest flood crest ever recorded resulted from the washout in 1895 of a landslide dam 274 meters (900 ft) high in the Himalayan foothills in India; the resulting flood flow was 73 meters (240 ft) deep. The danger there was so obvious that the downstream area was evacuated well before the overflow.

It would be no comfort to victims of earthquakes, volcanic eruptions, hurricanes, and tornadoes to be told that each of these phenomena is the result of processes that help to maintain equilibrium in the earth's intertwining physical systems, but such is indeed the case. Some disequilibrium is building even as you read this, and before many months have passed, it will be compensated by a seemingly violent event that will take lives, destroy property, and be recorded in newspaper headlines as another natural catastrophe.

Modified Mercalli (MM) intensity scale of 1931

I Not felt except by a very few under especially favorable circumstances.

II Felt only by a few persons at rest, especially on upper floors of buildings. Delicately suspended objects may swing.

III Felt quite noticeably indoors, especially on upper floors of buildings, but many people do not recognize it as an earthquake. Standing motor cars may rock slightly. Vibration like passing of truck. Duration estimated.

IV During the day felt indoors by many, outdoors by few. At night some awakened. Dishes, windows, doors disturbed; walls make cracking sound. Sensation like heavy truck striking building. Standing motor cars rocked noticeably.

V Felt by nearly everyone, many awakened. Some dishes, windows, etc., broken; a few instances of cracked plaster; unstable objects overturned. Disturbances of trees, poles, and other tall objects sometimes noticed. Pendulum clocks may stop.

VI Felt by all, many frightened and run outdoors. Some heavy furniture moved; a few instances of fallen plaster or damaged chimneys. Damage slight.

VII Everybody runs outdoors. Damage negligible in buildings of good design and construction; slight to moderate in well-built ordinary structures; considerable in

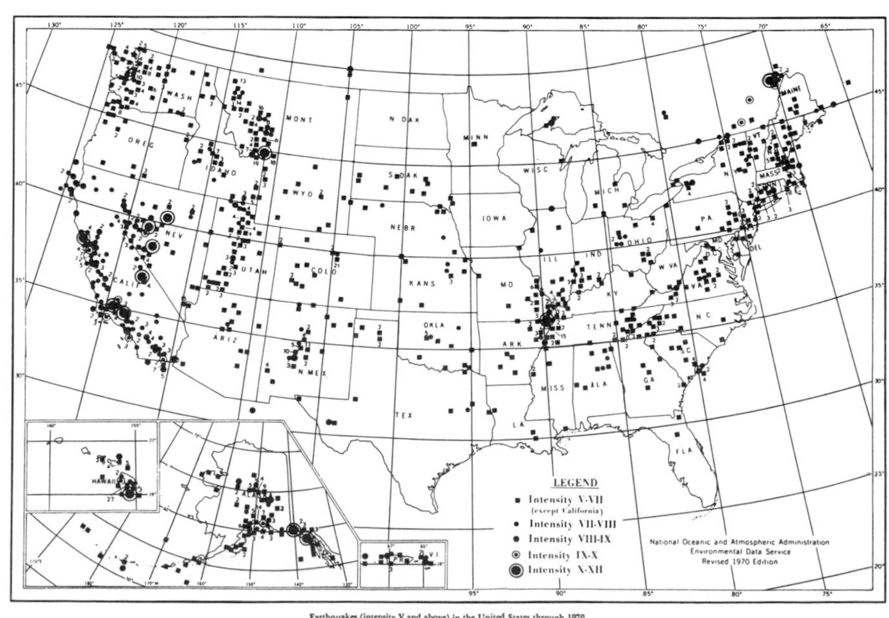

Earthquakes (intensity V and above) in the United States through 1970.

Historic earthquake epicenters in the United States are recorded on this map according to their intensities on the Modified Mercalli scale. It can be seen that no region of the United States is safe from earthquake damage. When looking at such a map, one must imagine a halo of destruction around each epicenter, often covering thousands of square kilometers. The Mississippi Valley earthquakes of 1811–1812 were felt over some 2,600,000 sq km (1,000,000 sq miles) to distances as remote as Washington D.C., with chimneys being toppled in Pennsylvania, Ohio, Georgia, and South Carolina, as well as in states nearer the epicenter. (From *Earthquake History of the United States,* 1973)

poorly built or badly designed structures; some chimneys broken. Noticed by persons driving motor cars.

VIII Damage slight in specially designed structures; considerable in ordinary substantial buildings, with partial collapse; great in poorly built structures. Panel walls thrown out of frame structures. Fall of chimneys, factory stacks, columns, monuments, walls. Heavy furniture overturned. Sand and mud ejected in small amounts. Changes in well water. Persons driving motor cars disturbed.

IX Damage considerable in specially designed structures; well-designed frame structures thrown out of plumb; great in substantial buildings, with partial collapse. Buildings shifted off foundations. Ground cracked conspicuously. Underground pipes broken.

X Some well-built wooden structures destroyed; most masonry and frame structures destroyed with foundations; ground badly cracked. Rails bent. Landslides considerable from river banks and steep slopes. Shifted sand and mud. Water splashed (slopped) over banks.

XI Few, if any, (masonry) structures remain standing. Bridges destroyed. Broad fissures in ground. Underground pipelines completely out of service. Earth slumps and land slips in soft ground. Rails bent greatly.

XII Damage total. Practically all works of construction are damaged greatly or destroyed. Waves seen on ground surface. Lines of sight and level are distorted. Objects are thrown upward into the air.

SOURCE: Bolt, Bruce, et al. 1975. *Geological Hazards.* New York: Springer-Verlag, p. 9.

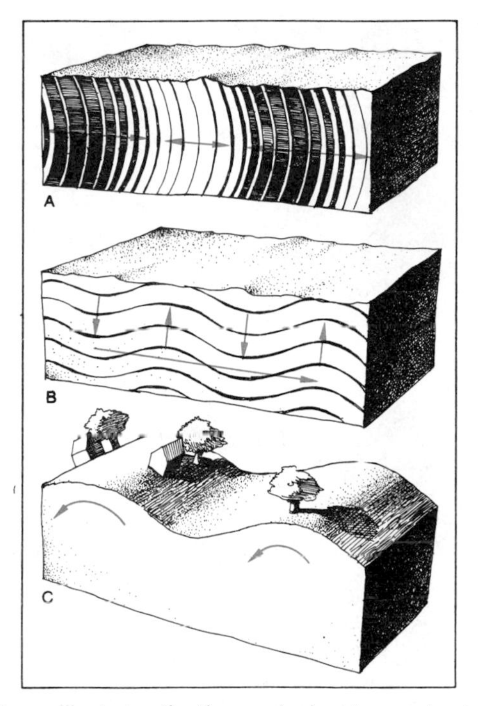

This figure illustrates the three principal types of seismic waves: (a) fast-moving *P,* or *primary,* waves that are similar to sound waves in that they are transmitted by alternately compressing and relaxing the material in the direction in which they are traveling; (b) *S,* or *secondary,* waves (also called *shear* waves) that vibrate material in a direction perpendicular to the wave path; and (c) slower *L,* or *surface* (long), waves that travel along the earth's outer skin with a low frequency but high amplitude, causing the surface to rise and fall as oscillating "rollers" pass through it, the motion being similar to that transmitting waves in the oceans.

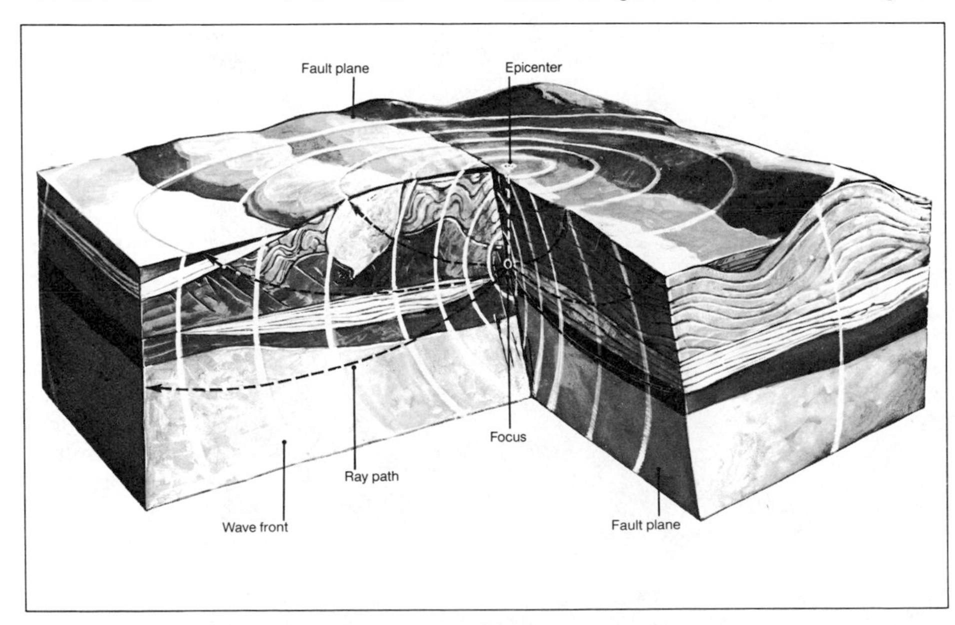

This cutaway diagram illustrates the relationship between the earthquake *focus,* or *hypocenter,* on a fault plane where seismic shock waves originate, and the earthquake *epicenter* on the earth's surface. Seismic shock waves radiate outward from the focus as expanding spheres.

The towering pillars of limestone in the karst region of China are an example of a unique landform produced by certain combinations of weathering processes with underlying rock structures. Climatic controls influence many of the processes that act on landforms, giving slopes their characteristic shapes.

Winter Landscape from the Ch'ing Dynasty. (Courtesy of the Smithsonian Institution, Freer Gallery of Art, Washington, D.C.)

FOURTEEN

Climate and Landforms

TO ONE ACCUSTOMED TO the smooth, forested hillslopes of the eastern United States, the first sight of the angular, rock-dominated country of the Southwest may be as astonishing as a view of another planet. In the East, bedrock is rarely visible, being hidden under a cover of soil and vegetation. In the Southwest, colorful landscapes dominated by naked rock are unusual enough to have attracted millions of tourists from other regions and even from foreign lands. However, there is nothing exceptional about the rock types and geologic structures of the Southwest. They may be found in many areas of much less spectacular scenery. Why, then, are the landscapes of the East and the Southwest so different?

The answer, of course, is climate. Contrary to the ideas of early writers on landforms, there are no "normal" landscapes associated with particular rock types, because the behavior of rock material varies as the stresses applied to it vary, and these stresses reflect the complexity of world climates that have defied easy classification, as seen in Chapter 7. Moreover, at any locality these stresses have

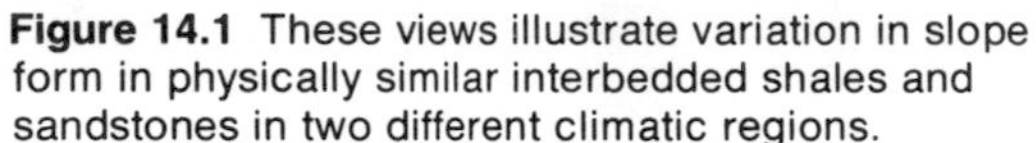

Figure 14.1 These views illustrate variation in slope form in physically similar interbedded shales and sandstones in two different climatic regions.

(left) In California's Coast Ranges, the grass cover developed in response to an annual rainfall of about 50 cm (20 in.) protects a thin soil, increases moisture infiltration into the ground, and reduces erosive surface runoff.

(right) In southern Utah, a rainfall of about 25 cm (10 in.) is insufficient to produce vegetative protection of slopes, resulting in intricate gully erosion and exposure of bare rock ledges that shed coarse waste onto the slopes. The nature of the rainfall regimes in the two areas is also a factor. In California, most of the precipitation falls in low-intensity drizzles associated with the passage of winter frontal systems. In southern Utah, summer thunderstorm activity produces brief episodes of high-intensity precipitation on bare ground.

changed through time due to temporal changes in regional climates—a significant factor in the earth's history over the past few tens of millions of years.

The effect of varying climates on landforms may be illustrated in terms of one common rock type, granite. Near Atlanta, Georgia, granite forms the smooth, bald dome of Stone Mountain (Figure 11.15, p. 316). But not far away in the Blue Ridge Mountains of Virginia and the Carolinas, granite produces forested narrow-crested ridges. The variation in this case is a matter of geologic structure. Stone Mountain is a mass of granite that is almost free of joints, whereas the granite of the Blue Ridge is more normally jointed and is covered with a weathered mantle. However, in the deserts of Australia, Arabia, North Africa, and the southwestern United States, normally jointed granite produces jumbles of rounded boulders. In the lofty heights of the Alps or Sierra Nevada, jointed granite often develops intricate angular or prismatic forms. In the White Mountains of New Hampshire, jointed granite produces bumpy, uneven slopes. And in a tropical rainforest, jointed granitic bedrock may never be exposed, for granite rapidly decays to clay in a warm and continuously moist climate. Figure 14.1 shows how slope forms can vary where similar rock types are exposed to different energy and moisture conditions.

The same rock type and geologic structure can produce a wide variety of landscapes because gross geologic structure and the physical and chemical properties of rock are only two of many factors that control the appearance of landforms. How relevant the physical or chemical qualities of rock material are depends upon the nature of the stresses being applied to the material. These stresses may themselves be either physical or chemical. It is climate that determines the nature of the atmospheric stresses affecting rock material, and it is the nature of these stresses in relation to the regional tectonic history and material resistance that determines the appearance of landscapes.

This chapter discusses the basic ways in which climate affects the geomorphic processes shaping landforms and concludes with a survey of the most distinctive landscapes associated with the dominance of particular climatic types.

CLIMATIC INFLUENCES ON GEOMORPHIC PROCESSES

Geomorphic processes differ strikingly in both type and relative intensity from one climatic region to another. Processes that dominate in some environments, such as the transport of sediment by wind in dry, vegetation-free areas, or frost-heaving of the ground in cold-winter regions, may be insignificant or altogether absent elsewhere. Such differences in geomorphic processes can lead to the development of distinctive landforms characteristic of a particular climatic type.

Climate and Weathering

One of the most fundamental ways that climate affects the development of landforms is through its control of weathering processes. Limestone blocks quarried to provide facing stone for the great pyramids in the desert landscape of Egypt have weathered so little in 4,000 years that tool marks are still visible on them. On the other hand, limestone grave markers in moist New England have corroded to the point of illegibility in only 300 years. The obvious disparity in rates of weathering in moist and dry regions is significant, since weathering is the starting point in the erosional development of landscapes.

Weathering processes, both mechanical and chemical, are controlled by temperature and moisture availability. High temperatures and abundant moisture favor chemical reactions that result in rock decomposition. Nearly all chemical reactions require moisture because in solution atoms become ionized and interact vigorously, combining and replacing one another. High temperature speeds up chemical reactions because it causes the ions to move more rapidly and thus to collide more frequently. The rate of many chemical reactions roughly doubles for every 10°C (18°F) increase in temperature. Low temperatures or water deficits are the factors producing mechanical weathering. Temperatures below the freezing point of water cause the growth of ice crystals in rock pores and more massive ice veins in rock joints, both of which exert expansive stresses leading to rock disintegration. In very dry climates, the evaporation of water laden with various salts causes the precipitation of saline crystals, whose growth, like that of ice crystals, can cause the disintegration of weakly cemented rock. In all cases, water is vital for chemical reactions to occur and for mechanical disintegration due to changes of state (crystallization) that apply pressure to confining materials.

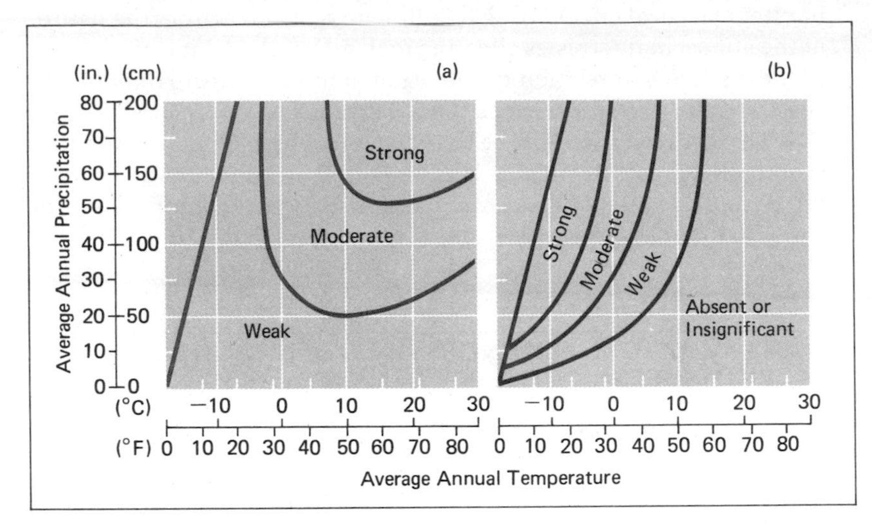

Figure 14.2 (a) This schematic diagram suggests the relative importance of chemical weathering in various regimes of temperature and precipitation. Low temperature and high precipitation cannot occur together; thus this impossible combination is isolated by the diagonal line on the graph. Chemical weathering is most active where temperatures are high and moisture is abundant. The reactive H+ and OH− ions of rainwater are primarily responsible for chemical weathering, which tends to be most intense near the surface of the ground and to decrease with depth. Although chemical weathering is weakest in dry regions, it may still be an important form of weathering. Chemical weathering dominates in many hot, dry regions where mechanical weathering processes may be comparatively inactive.

(b) This schematic diagram suggests the relative importance of mechanical frost weathering in various climatic regimes. Frost weathering occurs where free moisture is available to penetrate rock joints and pores, and where the temperature drops below freezing from time to time. It is unimportant in hot, dry regions or in regions that are perpetually warm. The diagram does not take into account relatively minor processes of mechanical weathering that may be locally important, such as salt crystallization. (Modified from Peltier, 1950)

Figure 14.2a illustrates the relative importance of chemical weathering under various combinations of temperature and rainfall. In hot, moist climates, the mantle of decomposed rock material is frequently tens of meters thick as a consequence of intense chemical weathering and because tropical regions have experienced the same climates for tens of millions of years. In the humid midlatitude regions, energy and moisture conditions have been subject to considerable fluctuation over the past few millions of years, with large-scale removal of weathered mantles occurring in some areas due to periods of accelerated slope erosion as well as to glacial action. Chemical weathering is of least significance in hot, dry climates, where little water is available to ionize the atoms, and in cold climates, where chemical reactions are slow and water is frozen much of the time.

Figure 14.2b suggests the relative importance of mechanical weathering in different climatic regions. Frost action, the principal process of mechanical weathering, is unimportant in warm regions and becomes increasingly significant in cooler regions that have a surplus of moisture.

Figure 14.3 summarizes the relative importance of the various weathering regimes under different temperature and moisture combinations. The diagram indicates that weathering processes are relatively inactive in dry climates, both hot and cold. A diagram that classifies weathering processes according to climatic factors should be interpreted only as a broad generalization and a framework for thinking about the landscapes of different regions. Bear in mind that some regions have different weathering regimes at different times of the year, and that most regions have experienced different weathering regimes sequentially as a consequence of climatic changes over the past few million years. Figure 14.3 omits the effect of seasonal variations in temperature and precipitation, which may have geomorphic significance, and does not consider the detailed processes involved in weathering. However, the general relationships depicted are valid.

Vegetation and Erosional Intensity

Climate also affects the development of landforms through the influence it exerts on the character of vegetation and degree of vegetative cover in a region. Plants have a direct influence on weathering through their nutrient cycles, acid secretions, and decay products, all of which affect the chemistry of the water in contact with soil and subsurface rock. However, the most important effect of vegetation on landform development is through its modification of the rates of erosion by wind and water.

A dense vegetative cover holds soil in place and helps maintain a permeable surface—by the chemical effect of organic humus, by the action of roots in keeping soil particles separated, and by preventing raindrop splatter from sealing the open spaces between surface soil particles. Because the presence of vegetation tends to increase the rate of water infiltration into the soil, a vegetative cover significantly reduces surface runoff. This, in turn, reduces erosion. As indicated in Chapter 11, surface flow of water rarely occurs on well-vegetated slopes; under such conditions, water moves downslope between soil particles as throughflow rather than on the surface as overland flow. In sparsely vegetated, arid regions, the amount of precipitation is small; however, the rate of precipitation often exceeds the infiltration capacity of the ground, so that surface runoff occurs. As a conse-

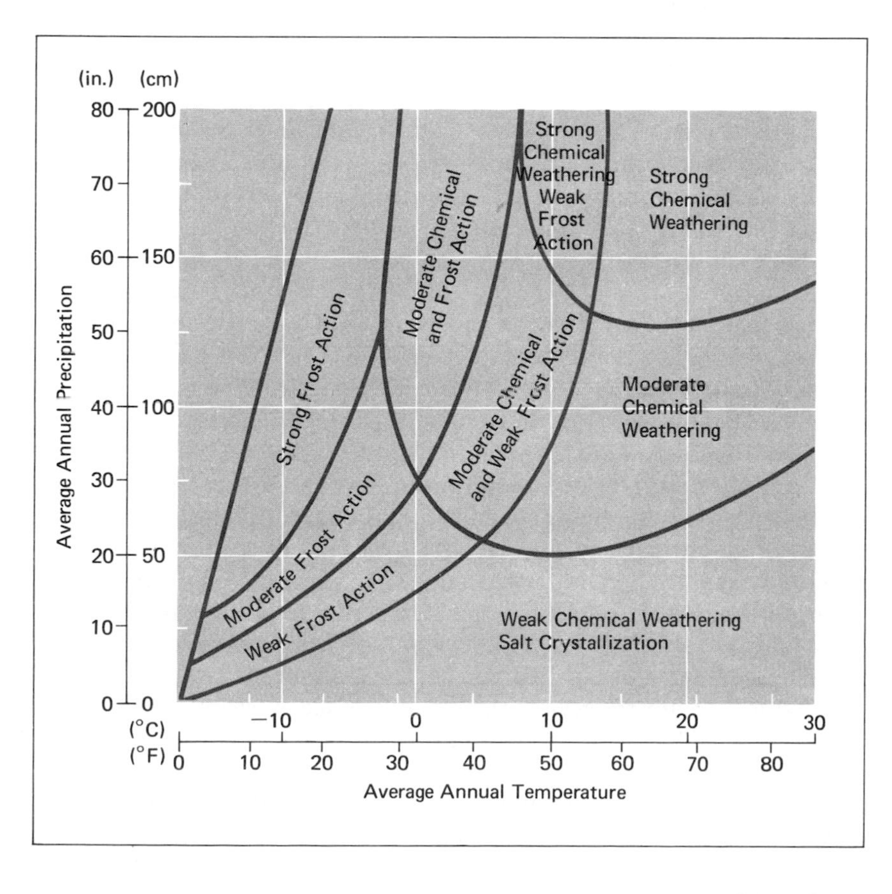

Figure 14.3 This schematic diagram suggests the dominant weathering process in various climatic regimes. Chemical weathering and frost weathering cannot occur unless moisture is available; thus weathering is retarded in dry regions, where precipitation and surface runoff are meager. However, even in regions where weathering is slight, the continuous action of relatively weak processes can eventually produce characteristic landforms. Many parts of the earth's surface have been subjected to different weathering regimes through time because of changes in climate. The landforms in such regions may have been formed partly by weathering processes that are no longer active there today. (Modified from Peltier, 1950)

Figure 14.4 This view illustrates landscape transformation in the humid climate of Tennessee due to removal of the original vegetation cover. The area formerly supported a mixed hardwood and pine forest. Deforestation originally occurred to provide fuel for a copper smelter built in 1854. Toxic fumes from this and subsequent smelters prevented vegetative regeneration, keeping the land in a nearly barren condition for more than 100 years despite the region's abundant rainfall. An artificially established grass cover offers some protection, but it has been largely removed by gully expansion, as this view indicates. Although the locality receives about 140 cm (55 in.) of rainfall annually, the landforms resemble those seen in deserts.

quence, evidence of erosion by running water is much more conspicuous in vegetation-free regions than in areas supporting a vegetative cover. Where hillslopes are not protected by a vegetative cover, loose material produced by weathering is flushed away almost as quickly as it forms, prohibiting soil development and producing an angular rock landscape in which even minute variations in rock type and structure are apparent (Figure 14.1, p. 420). Thus the presence or absence of vegetative protection is a principal factor in landscape character.

If hillslopes in a humid area are cleared of vegetation, erosion soon begins to carry off soil material, initiating a network of ever-deepening gullies, as in Figure 14.4. The sediment yield from bare soil is 50 to 100 times greater than the sediment yield from nearby grass-covered areas of the same size. Agricultural land usually remains free of plant cover for part of the year, and erosion accelerates during such periods. Studies of lake sediments near ancient Roman towns indicate that the annual sediment yield from adjacent land increased ten fold soon after the towns were founded and agricultural activity began. All around the Mediterranean basin, from Morocco to Turkey, phases of fluvial deposition that produced fills in erosional valleys tell a similar story. Recent estimates of human effects on rates of erosion indicate that the average erosion rate over all the earth's land areas has doubled or tripled because of such activities as clearing land for agriculture, logging operations, and sealing the land surface with concrete and asphalt pavement, all of which increase the amount of surface runoff.

Figure 14.5 illustrates how the rate of surface erosion, or *denudation,* is affected by precipitation and vegetation. The example considered is a moderate (10°) slope in a hot climate where the average monthly temperature is about 25°C (77°F). Potential evapotranspiration in such a climate is approximately 80 to 100 cm (30 to 40 in.);

thus where the annual precipitation is less than 50 cm (20 in.) vegetation tends to be sparse. As the diagram shows, denudation rates are highest in semiarid areas where the amount of moisture is insufficient to maintain a dense cover of vegetation but where rainfall occurs often enough to produce frequent periods of runoff. Denudation rates also increase rapidly when annual precipitation is greater than 150 cm (60 in.). Then precipitation greatly exceeds potential evapotranspiration, and surplus moisture becomes available to erode material mechanically or to carry it away in dissolved form.

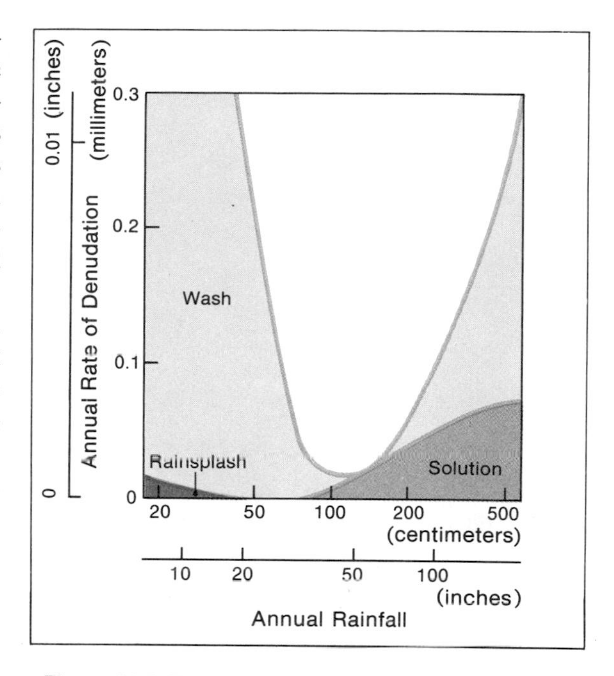

Figure 14.5 This diagram illustrates how the rate of lowering by erosion, or denudation, is affected by precipitation and vegetation. The example considered is a 10° slope in a warm region where the average annual temperature is 25°C (77°F). Soil wash by surface runoff is the most important process of denudation. The rate of lowering is least where the annual precipitation is about 100 cm (40 in.) because in such regions the increase in soil permeability due to organic activity allows little surface runoff to be generated by the moderate rainfall. In addition, vegetative protection of the soil surface reduces the erosive impact of rainfall. The rate of denudation is greater in drier regions because the vegetation is sparse and surface runoff readily erodes the exposed bare ground. Rainsplash on bare ground also plays a role. In very moist regions, more moisture is available than the soil can absorb, and large moisture surpluses are generated. The abundant runoff of such regions carries away material in solution as well as by soil wash.

The absence of a protective cover of vegetation exposes the land surface to assault by another force that is ineffective wherever a plant cover is well developed. This force is the wind. Where a continuous cover of vegetation exists, it screens the land surface against the force of the wind. But where the full force of the wind impinges directly on bare dry soil, the shear stress exerted upon the land surface may be sufficient to move particles the size of coarse sand. Dry sand with a diameter of 0.1 mm begins to move when the wind velocity within a meter of the ground surpasses 16 km (10 miles) per hour; coarse sand (2.0 mm diameter) is put in motion by a 50 km (30 miles) per hour wind. The presence of moisture between soil particles increases their cohesion and resistance to wind erosion. Thus, while it is safe to plow moist soils, there is always danger of wind erosion when dry soils are broken by the plow.

A bare surface over which the wind sweeps may be thought of as a vast stream bed subjected to shear stress by a moving fluid. Loose particles put into motion by the wind behave similarly to those entrained by streamflow. The finest particles are carried in suspension in the form of dust; heavier particles move by saltation and traction, just as in the streamflow shown in Figure 12.8, p. 344. However, while wind can attain far greater velocities than running water, the density and viscosity of air is a tiny fraction of that of water, giving it little buoyant force. Thus, despite the fact that its flow is turbulent, wind cannot move particles larger than sand, and it can rarely lift sand-size particles more than half a meter above the ground. Close observation of sand movement under wind stress shows that saltation and traction are closely linked. Particularly conspicuous is the slow forward migration of the whole sand surface in a process known as *creep* (not to be confused with soil creep) produced by the oblique rain of saltating particles that knock surface particles forward as part of the traction process. Where wind velocity decreases, particles settle out of the sand flow and deposition occurs.

Processes related to wind action are known as *aeolian* processes, after the Greek god of the winds, Aiolis (Latin, Aeolus). Aeolian erosion and deposition are natural occurrences wherever loose, unprotected sediments are present, such as along shorelines and in areas of recent alluvial or glacial deposits. However, the largest scale aeolian effects are encountered in the world's desert regions, where extreme moisture deficits produce vast expanses of bare dry soils.

Aeolian erosion and deposition are important gradational processes in deserts. However, the wearing away of solid rock by natural sandblast is rarely seen, even in deserts, for it requires soft rock, hard quartz sand, and exceptionally strong winds. Wind erosion consists principally of the removal, or *deflation*, of fine rock debris provided by other gradational processes, such as weathering or fluvial deposition. The deflation process leaves no spectacular landforms, except in a few places where easily eroded former lake beds have been deeply grooved, producing rare landforms known as *yardangs*—the sharp-crested ridges between the deflation furrows. Since deflation is normally areal rather than linear, there may be little by which to gauge how much a surface has been lowered by the process. Often the only evidence of deflation is a lonely bush perched on a column of soil, all the surrounding soil having been removed by wind erosion.

When strong winds rush over dry, unvegetated surfaces, as during the passage of cyclonic disturbances, the finest debris is lifted thousands of meters in great dust storms (see Figure 14.6). This dust, composed of silt-size particles, settles back to the surface far from its source, forming a blanket of fine material known as *loess*. Loess deposits on the plains east of the Rocky Mountains and in the midwestern United States are derived from deserts that once existed to the west (as in the great relict sand dune region of Nebraska) and from areas of Pleistocene glacial deposits to the north and east. South of the limit of the Pleistocene ice sheets of Eurasia, a continuous belt of loess extends from northern France to the Russian Ukraine. In both locations, the loess offers exceptional agricultural opportunities due to its high calcium carbonate content and the unleached young soils developed upon it. The most spectacular of all loess landscapes are those along the middle portions of the Yellow River (Hwang Ho) in China.

Figure 14.6 These two views illustrate erosion and deposition by the wind. On the left is a severe dust storm, photographed in Colorado in 1935. Here very strong winds associated with frontal disturbances are seen eroding bare agricultural land, stripping away the most fertile part of the soil—its A-horizon. This hazard exists wherever the natural vegetation has been removed from soil that is periodically dry. On the right is a 4- to 5-meter (12- to 15-ft) bank of *loess* in Tennessee. Loess is composed of dust-size mineral particles that settle out of the atmosphere downwind from an area of aeolian deflation. This loess is typical in that its fine texture, cohesiveness, and internal structure cause it to stand in near-vertical banks and to be gullied where vegetative protection is absent. The depth of loess in the Mississippi Valley region varies from centimeters to as much as 15 meters (50 ft). Most Mississippi Valley loess was derived from aeolian deflation of barren late Pleistocene glacial deposits.

There, deposits of loess a hundred or more meters in depth have been deeply trenched by gully erosion and burrowed into by thousands of generations of humans who have lived in shelters hewn out of the soft material. The loess of this region is derived from dust storms to the north and west in the Gobi Desert of Mongolia and northwestern China.

LOCAL AND REGIONAL CLIMATIC DIFFERENTIATION

Varying climatic influences upon gradational processes can be seen in geomorphic features of all scales, from individual boulders to hillside slopes and mountain walls. The following offers some examples.

Microrelief

The Goblet of Venus is an unusual sandstone formation located in semiarid country near Blanding, Utah (see Figure 14.7). What process could produce a landform like the Goblet? Such grotesque *hoodoo rocks* in arid regions are often mistakenly attributed to the erosive power of sand grains carried by the wind. In truth, however, aeolian abrasion is rarely capable of producing such a form. The Goblet is, in fact, a product of differential weathering controlled by microclimate.

Figure 14.7 This hoodoo rock, known locally as the Goblet of Venus, stood near Blanding, Utah, until it was toppled several years ago. Its narrow stem was created by the chemical action of trickling water and by salt crystallization, which is concentrated at the base of such rock projections.

Even in arid regions where the annual rainfall is only 10 to 20 cm (4 to 8 in.), water is the principal agent of weathering. When infrequent rains fall on an upward-projecting mass of rock, some of the rain water runs down the sides of the mass and concentrates at its base. The shaded lower parts of the rock dry more slowly than the exposed upper parts. This allows minute physical and chemical changes to occur on the damper surfaces, partially dissolving the cementing substances that bind the grains of the rock together. With enough repetition of the process, the grains will loosen and fall away. The base of the original block becomes indented and transformed into a narrow stem, because it stays wet the longest after rainfalls, making it the place where weathering can be most effective. The microrelief of the rock surface forming the stem of the Goblet is direct evidence that grain loosening by weathering is the major geomorphic process; the surface is rough and granular rather than smooth and polished, as would be the case had sandblasting been in effect.

Asymmetry of Slopes

Hoodoo rocks in arid regions illustrate the effect of the microclimate in shaping a single rock outcrop. Microclimate also influences the form of larger features, such as valley walls. Outside of tropical latitudes, the slope angles on the opposite sides of east-west running valleys often differ, although there is no corresponding difference in rock type. In the middle latitudes of the northern hemisphere, slopes that face north are about 5° steeper on the average than south-facing slopes. Where the underlying bedrock and geologic structure are not factors, the asymmetry of opposing valley slopes must be due to

Figure 14.8 The dissected slopes on Black Tail Butte, Wyoming, all show an asymmetry between the north-facing side and the south-facing side. The north-facing slopes are thickly covered with coniferous forest, whereas the drier south-facing slopes have only sparse clumps of shrubs and grasses. The nature of the vegetative cover influences the development of the slopes. Water runoff will tend to erode the exposed south-facing slope more rapidly than the north-facing slope, which is protected from rainsplash by its vegetative cover and which produces less surface runoff because of its deeper and more permeable soil.

microclimate differences. Such differences do not exist in the low latitudes, where both the north and south sides of ridges experience similar climates throughout the course of a year.

In the northern hemisphere, south-facing slopes receive more solar radiation than north-facing slopes, and thus south-facing slopes tend to be warmer, drier, less thickly vegetated, and covered by thinner soils than north-facing slopes. In Virginia, the south-facing slopes of some valleys are forested mainly by pine, while the steeper slopes on the opposite side of the valley support oak trees as well as pine. In central California, south-facing slopes are commonly covered by grass or chaparral, while steeper north-facing slopes are oak-clad.

In midlatitude regions, there is a tendency for the slope with the more abundant vegetation and the deeper and more permeable soil to be steeper than the opposing dry slope, as in Figure 14.8. The complicated interplay of available moisture, vegetation type, and soil permeability affects the amount of water runoff and the erosional regime on opposing slopes. Where there is a strong contrast in soil permeability, it seems that surface erosion reduces the angle of the drier, less porous slope, while the stream in the valley between the opposing slopes may migrate laterally toward the more permeable slope, causing it to be undercut and kept relatively steep. In cooler climates, the difference in snowmelt or the thaw of frozen ground on opposite sides of a valley can induce asymmetry. The most striking asymmetry of all is seen in high mountain regions where glaciers form in the cool valley heads on the north and east sides of the ranges, often causing mountain ridges to be deeply gouged by glacial erosion on one side while forms produced by fluvial erosion are dominant on the other side (see Chapter 15).

Climamorphogenetic Regions

On a still larger scale, entire landscapes may exhibit distinctive characteristics due to the geomorphic effects of certain climatic regimes. This is especially true where the climate is "aggressive"—characterized by strong action of one or another of the weathering or erosional processes. Figure 14.9a, p. 430, illustrates an attempt to use the factors of temperature and precipitation to define *climamorphogenetic regions* that could be shown on a world map—that is, areas in which certain combinations of climatically controlled geomorphic processes predominate and give a distinctive character to the landscape. The processes at work in the nine climamorphogenetic regions shown in the figure are listed in Figure 14.9b; five of the most strongly contrasting regions are discussed later in the chapter.

If climate were the dominant factor controlling the form of landscapes, it would be possible to associate a distinctive set of landforms with each of the climamorphogenetic regions. However, no landscape owes its character entirely to climate, or to rock type, geologic structure, or tectonic condition; the four factors cannot be completely disentangled. So it is not surprising that similar landforms may arise in different morphogenetic regions. Furthermore, because the climate in a given region is not necessarily constant, temperature and moisture conditions may change more rapidly than landforms can respond. Consequently, the landforms in a region may reflect climates of the past rather than the present. The best example is the unusual scenery in regions formerly glaciated but ice-free for the last 10,000 years. The concept of climamorphogenetic regions only suggests the climatically controlled processes that are presently active in a given region; it does not consider all the controls of landscape development and changes in climate through time.

In fact, many landforms are not uniquely associated with any particular climatic region, but are widely distributed. In general, landforms are not as sensitive to variations in climate as are soils or vegetation. Although geomorphic processes may differ in intensity from one climatic region to another, the landforms shaped by the different processes are sometimes similar. Because we live in a time closely following a period of rapid and major climatic fluctuations, it is difficult to speak with certainty about the precise relationship between climate and landform. The relationship is a complex one, and strong differences of opinion persist regarding the exact role of climate in landform development, present and past.

Nevertheless, at the climatic extremes, we find many landforms that are peculiar to but one climamorphogenetic region. Some of these are illustrated in Figures 14.17 to 14.28, pp. 442–453. It is impossible to imagine a thousand square kilometers of sand dunes developing anywhere but in a desert climate with active wind transport. Such distinctive forms have great importance because we sometimes encounter them in regions where they could not form at present, as in

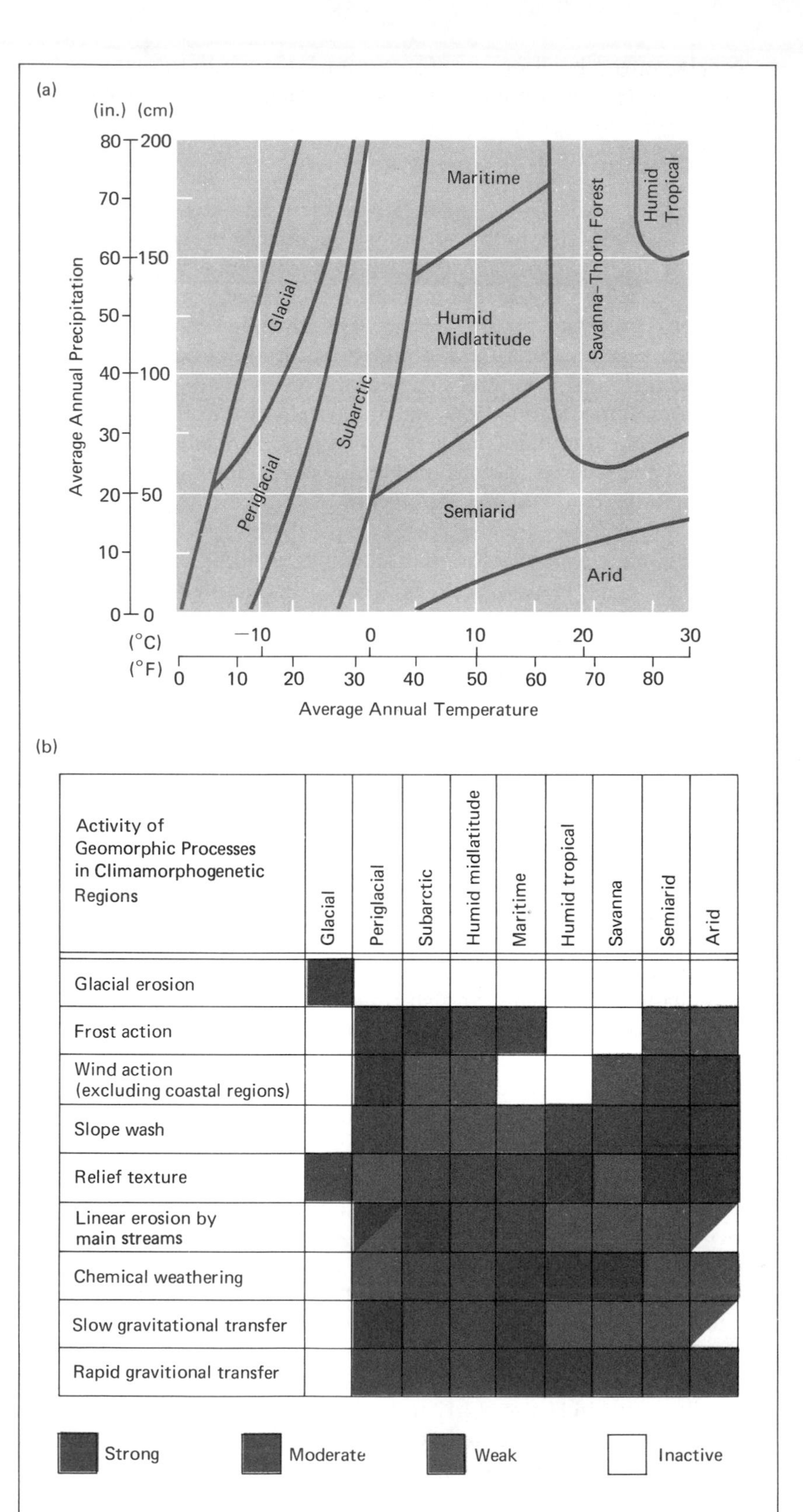

Figure 14.9 (a) As this schematic diagram suggests, each region on the earth can be characterized as a climamorphogenetic region possessing a certain combination of weathering and erosion processes that influence the development of landforms in the region. The boundaries in the diagram should be considered only as qualitative guides. The processes considered in classifying the morphogenetic regions include glacier and snow action, mass movement, frost weathering, and transport by water and wind (Figure 14.9b). Because of climatic change the morphogenetic character of a region may vary with time. The northeastern and north-central United States are now included in the humid midlatitude morphogenetic region, but 10,000 years ago they were glacial or periglacial.

(b) This chart summarizes the processes that produce distinctive climamorphogenetic regions. It must be understood that variations in geologic structure, rock type, and tectonic movements can produce major landform variations within each climamorphogenetic region. The gross relief contrasts between plains, hills, and mountains are often created by tectonic activity, and only the details of their erosional sculpture may show the influence of climate upon the landscape.

This chart shows the processes associated with existing climatic types. Many regions display evidence of processes associated with past climates unlike those of the present. Thus certain desert regions are gashed by deep canyons, but the canyons are older than the existing desert climate and were cut during past periods when more humid climates prevailed in the area.

In some climamorphogenetic regions, the effectiveness of particular geomorphic processes varies according to rock type and the existence of potential energy created by relatively recent tectonic uplift. In periglacial regions, linear erosion is generally very effective due to the abrasive tools carried by large streams in flood; however, where stream gradients are already flat, with the stream beds close to sea level, no further incision can occur, and streams wander through vast open lowlands. On the chart the presence of two different patterns for one process in a particular morphogenetic region indicates that the process varies significantly in effectiveness due to nonclimatic factors such as the above.

Relief texture is not a geomorphic process, but it is a measure of the tendency for linear stream erosion to dissect high-standing land surfaces. High relief texture indicates intricate dissection resulting from strong linear erosion by a dense drainage network. Low relief texture implies that drainage density is not high or that streams of low order do not possess the energy or abrasive tools (bed load) needed for incision to occur.

the case of the relict dune fields south of the Sahara near Timbuktu and Lake Chad, and also in the heart of Nebraska. This helps us to understand how climates have changed over time.

GEOMORPHIC PROCESSES IN DIFFERENT CLIMATES

At the end of this chapter, we present "A Gallery of Landscapes." The landscapes shown are products of some of the more extreme combinations of temperature and moisture conditions found on our planet. We have included more than one landscape in some of the morphogenetic regions to illustrate that although those landscapes differ because of geologic structure and rock type, they nevertheless show the dominance of certain climatically controlled processes and are unlike landscapes developed in similar structures and rock types under different climates.

Before illustrating these landscapes, it is necessary to make some general statements about the nature of the geomorphic processes and landforms in each of the five most distinctive climamorphogenetic regions.

Landforms of Periglacial Climates

Periglacial regions have near-glacial climates dominated by subfreezing temperatures throughout much of the year. While the term suggests areas peripheral to ice sheets, it also includes much larger cold regions that are too dry to support ice sheets. Such low-temperature regions are distinguished by a sparse vegetation cover, surface soils that freeze and thaw seasonally, permanently frozen subsoils, and active frost weathering capable of shattering rock both at and below the land surface. A particular feature of periglacial regions is the repeated disturbance of the soil by freezing and thawing. The resulting volume changes contribute to massive gravitational transfer on slopes. Added to this is the process of *solifluction*—the periodic movement of lobes of soil saturated with thawed water that is unable to percolate downward due to the permanently frozen subsoil, or *permafrost* (see Figure 11.19, p. 322). Landscape features commonly developed in areas of periglacial climates are illustrated diagrammatically in Figure 14.17, p. 442–443.

Hillslopes in periglacial regions often are crowned by an upper exposed rock cliff that is much shattered by frost action. Its upper surface may form a *periglacial block field,* or *felsenmeer* (German, *rock sea*). At the cliff base, a comparatively steep (30° to 40°) dry talus slope merges into a more gentle, wetter layer where solifluction commonly occurs. The solifluction slope is usually convex near its crest and concave toward its base, flattening to an angle of 10° or so at its lower end. Although the solifluction slope may be irregular in detail, the long-term effect of solifluction and associated mass wasting processes is to smooth the landscape, producing hillcrests that are broadly

Figure 14.10 This landscape in southern Ireland was smoothed by periglacial slope processes that were active in this region during Pleistocene time. Such landscapes rarely display linear erosion or rock outcrops, but their soils may be well supplied with angular rock fragments. Slopes of the type shown are common in both England and Ireland south of the limit of Pleistocene glaciation.

convex and slopes that are concave, as in Figure 14.10. Mass movement on this scale inhibits the development of rills and gullies, so that while much water moves over the surface during the thaw season, runoff channels may be poorly defined and the drainage density (channels per unit of area) is uncommonly low. First-order valleys are often steeply sloping flat-floored *dells* that are aggraded with colluvium brought from adjacent slopes but which have no stream channel at all. However, due to the coarse rock waste supplied by frost shattering and carried down into lowlands by mass wasting processes, the larger streams in hilly periglacial regions are abundantly supplied with tools and have great cutting power during their annual floods resulting from the spring thaw. As their discharge declines, such streams deposit their coarse load in bars of sand and gravel. Thus many of the larger streams are intricately braided.

During the long winter freezes, moisture-retentive silty soils in low-lying areas contract and crack, with veins of ice gradually filling the cracks. These ice veins grow into networks of *ice wedges,* each as much as a meter across and 3 or 4 meters deep. Sometimes areas of many tens of square kilometers are covered with systems of *ice-wedge polygons* that can be fully appreciated only in an aerial view. Masses of clear ice formed in boggy lowlands may grow into lens-shaped ice blisters known as *pingos,* which domc up areas of ice wedge polygons to form isolated circular mounds 10 or more meters high and as much as 100 meters (300 ft) across. Conversely, areas of thawed permafrost subside to form depressions that hold *thaw lakes,* which are present by the thousand in some regions.

In areas of higher or more rolling topography, frost cracking of the ground combines with a sorting action produced by repeated freeze and thaw to produce *sorted polygons,* in which ice wedges are replaced by concentrations of rock slabs that spill out at the surface to form *stone rings;* these are dragged downslope by mass wasting processes to form *stone garlands* and *stone stripes.*

During the Pleistocene glacial periods, periglacial conditions were much more extensive than at present, and periglacial landforms were developed in many areas that now have more moderate climates. These unusual landforms persist in mild climatic areas of western Europe and the eastern United States, and in many high-altitude regions, and are major evidence of the nature of climatic change over the past 10,000 years.

The periglacial climamorphogenetic region poses difficult problems for modern technological man. Formerly the haunt of seal hunters and reindeer herders, this region began to undergo increasing development in the twentieth century. For North America, change began with the building of military and communications facilities in the 1940s and 1950s, and the pace has been accelerated by the development of oil resources discovered in northern Alaska in the 1960s. Construction of roads, railroads, bridges, pipelines, and buildings in the periglacial region is made extremely difficult by the nature of ground that is permanently frozen at depth. The upper level of permafrost tends to thaw when structures are built on it. Heated buildings transfer warmth downward, and road and railroad fills trap outgoing longwave radiation, causing a rise in ground temperature. In either case, thawing produces differential ground movement, especially local subsidence that causes buildings to sink or twist off their foundations. Telephone poles, bridge abutments, and piers emplaced in the thawed layer during the warm season are frequently heaved out of the ground during hard winter freezes. Solifluction movement on slopes sometimes damages roads and rail lines. Consequently, all construction in this zone must be very carefully engineered to minimize disturbance of the permafrost and to resist destruction by subsidence and the flow of liquefied soil. This, of course, enormously increases the cost of development in this environment.

Landforms of Humid Midlatitude Climates

The familiar landscapes of the eastern United States and most of Europe are examples of landforms developed in humid midlatitude regions. In humid midlatitude climates, slope erosion by running water is relatively ineffective because the vegetation cover minimizes erosion by rainsplash and slopewash (overland flow). However, the number of stream channels and the intricacy of stream dissection, or *relief texture,* are much greater than in periglacial regions though less than in the humid tropics (Figure 14.11, p. 434). This reflects the fact that mass wasting, though present, is not as rapid as in regions of cold climate and fluvial *linear erosion* assumes more general importance. Slopes are soil-covered for the most part, and both chemical and mechanical weathering occur seasonally.

Figure 14.11 This view in Tennessee is representative of landscapes produced by fluvial dissection in a humid midlatitude climate. Erosion by running water is more apparent than in periglacial regions but the drainage density is lower than in vegetation-free areas and regions of extremely high rainfall.

In humid midlatitude regions, permeable soils commonly produce broadly rounded hillcrests; less permeable materials develop narrower ridge crests. The convexity of soil-covered hilltops and ridge crests is attributed to the process of soil creep. Even where resistant rock materials develop steep slopes and erodible materials gentler ones, the ubiquitous creep process nevertheless tends to blur distinctions between the different rock types. Sharp slope breaks due to changes in rock type are not a normal feature of this climamorphogenetic region.

The more or less intermediate condition of the humid midlatitude region is evident in the fact that it grades into wetter and drier and colder and warmer regions on its various perimeters. All of the processes found in the more extreme regions also occur in the moist midlatitude region, each of them developed weakly or moderately rather than being a dominant factor. Thus the humid midlatitude region is sometimes referred to as the *moderate* climamorphogenetic region. This region is the one to which William Morris Davis devoted his attention in his cycle of erosion. However, it should not be considered the "normal" region just because it is the most familiar to the majority of Americans and Europeans. The earth's land areas are vast, and the humid midlatitude region occupies but a fraction of them. In fact, the landscapes considered normal by Davis have experienced the influences of several dissimilar climamorphogenetic regimes and

Figure 14.12 This surface in Morocco is a *desert pavement* composed of closely packed gravel and larger rocks. It is a consequence of erosion of alluvial deposits by rare torrential flows of water that strip away all of the finer rock particles, peeling down the surface until it is armored by a continuous blanket of fragments too large to be flushed away. Thus the pavement is also called *desert armor.* Such surfaces are widespread in desert regions where there is no vegetative protection of the land surface.

contain many relict features that are difficult to relate to present geomorphic processes in the regions in which they occur. Evidence of both the periglacial and humid tropical geomorphic systems is widespread in the humid midlatitude landscapes of Europe and North America, and the occurrence of both climatic types in these areas during the period of landscape formation is well established.

Landforms of Arid Climates

In arid regions, such as the southwestern United States, the vegetation cover is limited—consisting of sparse grass or widely spaced shrubs—and bare rock exposures or accumulations of rock fragments dominate the scenery (see Figure 14.12). As there is little or no soil to smooth over variations in rock type, the slopes exhibit greater angularity than those of other regions. The sparse vegetation and compact soil make water infiltration into the soil slow, so runoff is heavy during occasional rainstorms. In dry regions, surface water erosion is more effective than creep, and the drainage density and rate of erosion by flowing water are high. However, the duration and distance of transportation of rock debris by ephemeral flows of surface water are small. As a consequence, in desert landscapes depositional landforms are unusually conspicuous: aggraded dry stream beds, alluvial fans, talus

Figure 14.13 Areas of granitic rocks in arid regions typically develop ramplike erosional *pediments* surrounding abruptly rising rock islands, or inselbergs. The sharpness of the slope break from pediment to inselberg is evident in this view of two inselbergs and associated pediments in the Mojave Desert of southern California.

accumulations, and sand dunes. Too often deserts are assumed to be great seas of sand. This is an erroneous view, for desert landscapes consist of diversified surface types, with the formation of sand dunes requiring special conditions, to be outlined later.

Although some hillcrests in arid regions are convex, in this case they are rounded by rainsplash erosion and soilwash rather than by creep. In the absence of a soil cover and the smoothing creep process, subtle variations in rock type result in conspicuous differences in slope angle and shape. Differential weathering and erosion reach their maximum development in arid regions.

Typical slopes that develop in sedimentary rock in dry regions are marked by a steep cliff at the summit, formed by the exposed edge of a resistant rock layer, followed by a comparatively straight talus slope with an angle of about 35°. The talus is only a thin layer of rock debris that veneers softer sedimentary rock below the resistant cap rock. Approximately equal quantities of material enter the talus slope segment from the cliff above and are flushed or blown away in finer form from the slope foot.

The lower reaches of slopes in dry regions often form an extensive gently inclined surface called a *pediment.* Pediments are erosion surfaces left where hillslopes and cliffs have been worn laterally back by slope erosion. Pediments differ from peneplains in that they seem to be produced primarily by slope erosion by running water rather than by a process of stream dissection and lowering of the dissected landscape as a consequence of mass wasting processes. Erosional pediments are the dominant landform in dry regions in terms of surface area. Although many pediments are cut only across easily eroded rock, others truncate the same resistant rock that creates bold hill-forms that rise sharply above the pediment surface (see Figures 14.13 and 14.24, p. 449). Pediments of the latter type are most commonly developed in granitic or gneissic rock.

It is difficult to account for certain characteristic landforms of desert regions in terms of the geomorphic processes presently active. Many of the landforms seen in deserts seem to be relict forms inherited from previous periods in which more humid climates prevailed. The development of vast pediments, which extend many kilometers outward from eroded hill masses, seems to require major climatic changes, with slope retreat and pedimentation having occurred in a semiarid soil-covered landscape that was subsequently eroded down to the solid rock at the former lower limit of subsurface chemical weathering. The erosional acceleration occurred as a result of climatic changes that greatly reduced vegetative protection of the land surface.

The vast areas of sand dunes, known as *ergs*, that are features of some deserts, may also be products of climatic changes. In the deserts

Figure 14.14 Vast accumulations of sand in great dune fields are known as *ergs*. The largest ergs are found in the deserts of Arabia and North Africa. However, in Nebraska we can see an ancient erg, here lightly dusted with snow, that shows this region to have been a desert as recently as 15,000 years ago. The Nebraska Sand Hills cover an area of some 62,000 sq km (24,000 sq miles). The individual crescent-shaped mountains of sand in this view are about 1.6 km (1 mile) long from horn to horn, and as much as 100 meters (330 ft) high. Each of these mounds is a *draa* that is crossed by many smaller dune ridges.

Figure 14.15 (opposite) Morphology of sand dunes, with arrows showing the direction of the effective winds. In (a), (b), (c), and (d), typical map patterns of the dune types are shown at the top, with oblique views below; (e) and (f) show the map pattern only.

(a) Individual dunes with elongated tips, or *horns,* pointing downwind are known as *crescentic* or *barchan* dunes. The steep slip face is concave. This type of dune is found in dry, vegetation-free environments, and is usually seen migrating across a nonsandy surface of gravel or clay. Barchan dunes may move several tens of meters each year. Symmetrical barchans indicate a nearly constant wind direction.

(b) Single dunes whose horns point upwind are known as *parabolic* dunes. Often they assume a hairpinlike form. Their slip face is convex. Such dunes form where sand blows into a moist area that has a vegetative cover. Vegetation or dampness in the lower part of the dune retards motion there, so that the dry crest pushes ahead of the base, causing the horns to "drag" behind. Parabolic dunes are found along coastlines and often push into forests, engulfing and killing the forest trees. Similar dunes also occur where older vegetation-covered sand accumulations become active again due to natural or artificial destruction of the vegetation. In such instances, the newly active parabolic dunes are called *blowouts.*

(c) Large erg areas sometimes include regularly spaced "sand mountains" as much as 200 meters (650 ft) high; these are called *star dunes* or *rhourds.* Rhourds seem to be fixed in position, and their formation, which is not well understood, seems to require winds from several directions. They are seen in Arabia and the Sahara.

(d) Ergs commonly contain dune ridges transverse to the wind direction that appear to be composed of linked barchans. These transverse dune systems are known as *aklé* dunes. They frequently incorporate parabolic segments between the linked barchans. Aklé dune systems reflect effective winds from a single direction.

(e) Where sand supplies are especially abundant, *longitudinal* dunes are often encountered. Some of these are hundreds of kilometers long and as much as 200 meters (650 ft) high. These dunes are thought to be the resultant of strong winds from different directions, or at least to express the effect of multidirectional winds. Their morphology varies from smooth *whalebacks* to knife-edged *seif* dunes. Many have a fairly complex topography, including variously oriented subsidiary crests and slip faces.

(f) In some ergs, transverse and longitudinal dune systems appear to cross, producing *grid* dunes. Either seasonal or long-term changes in wind direction may be responsible for such patterns.

(g) This diagram shows the migration of a dune as sand is transferred from the backslope to the slip face, causing the dune to be displaced in the direction of the slip face. The internal structure of dunes consists of steeply inclined laminations of sand produced by accretion on the slip faces. The result is *aeolian crossbedding,* which can be seen in many ancient nonmarine sandstones.

of North Africa and Arabia, individual sand seas, with their dunes resembling petrified waves, cover tens of thousands of square kilometers. An ancient erg, now grass-covered, occupies a large portion of western Nebraska (Figure 14.14, p. 437). To produce an erg, several things are required: an abundant source of sand, a delivery system that feeds sand into an accumulation area, and a weakening of the delivery system that causes massive deposition forming the erg. Climatic change may be a further requirement, for some desert sand accumulations are enormous, whereas provision of sand by desert weathering processes is extremely slow. Thus erg formation may require a period of humid climate weathering and stream transport of sand out of the source region (of decomposing sandstone or granitic rock), followed by a climatic change toward aridity, permitting the wind to become an effective agent of erosion and deposition. Any deceleration of sand-carrying wind will lead to deposition. Once an aeolian sand deposit is initiated, it is self-stimulating because sand saltates easily over clay, hard-packed soil, or rock, but not over dry sand itself. A sheet of dry sand traps incoming sand; when this occurs on a vast scale, with great quantities of sand brought into the area, an erg results. Major ergs have probably grown in several separate stages including climatic changes back and forth between sand generating and sand transporting and accumulating geomorphic systems.

Ergs contain dunes of many types, reflecting variations in the sand supply and the nature of the local sand-driving winds. The most general division is between transverse and longitudinal dunes. *Transverse* dunes form ridges perpendicular to the direction of the effective (sand-driving) winds, and they indicate that the sand-driving winds usually come from one direction. Where the effective winds come from different directions at different times, *longitudinal* dunes form. Their direction seems to be the resultant of the different wind vectors; thus their crests are oblique to most sand-driving winds. Figure 14.15 illustrates several of the dune types found in desert erg systems.

The most recent direction of sand flow in a dune field can be ascertained by observing the smallest accumulation forms—the closely spaced sand ripples that crawl over the dunes. The migration of sand ripples brings sand up to the dune crest, where the sand falls onto the steep *slip face* of the dune, which is always at the angle of repose of dry sand, about 34°. The ripples are themselves asymmetrical, their steep face (only millimeters high) on the downwind side. They move in various directions from day to day, often approaching the dune crest obliquely. The dune itself may be advancing due to transfer of sand from its backslope to its slip face, or it may be stationary with its crest tending to migrate back and forth slightly with changes in wind direction. Many stationary dunes are relict forms that are no longer subject to the same winds and sand inputs that established them initially.

In large ergs, sand dunes usually aggregate into larger sand mountains, known in the Sahara as *draa.* Draa ridges covered with smaller dunes are 100 or 200 meters high, with crests 2 km or more apart. Draa may themselves be transverse, longitudinal, or nondirectional in orientation.

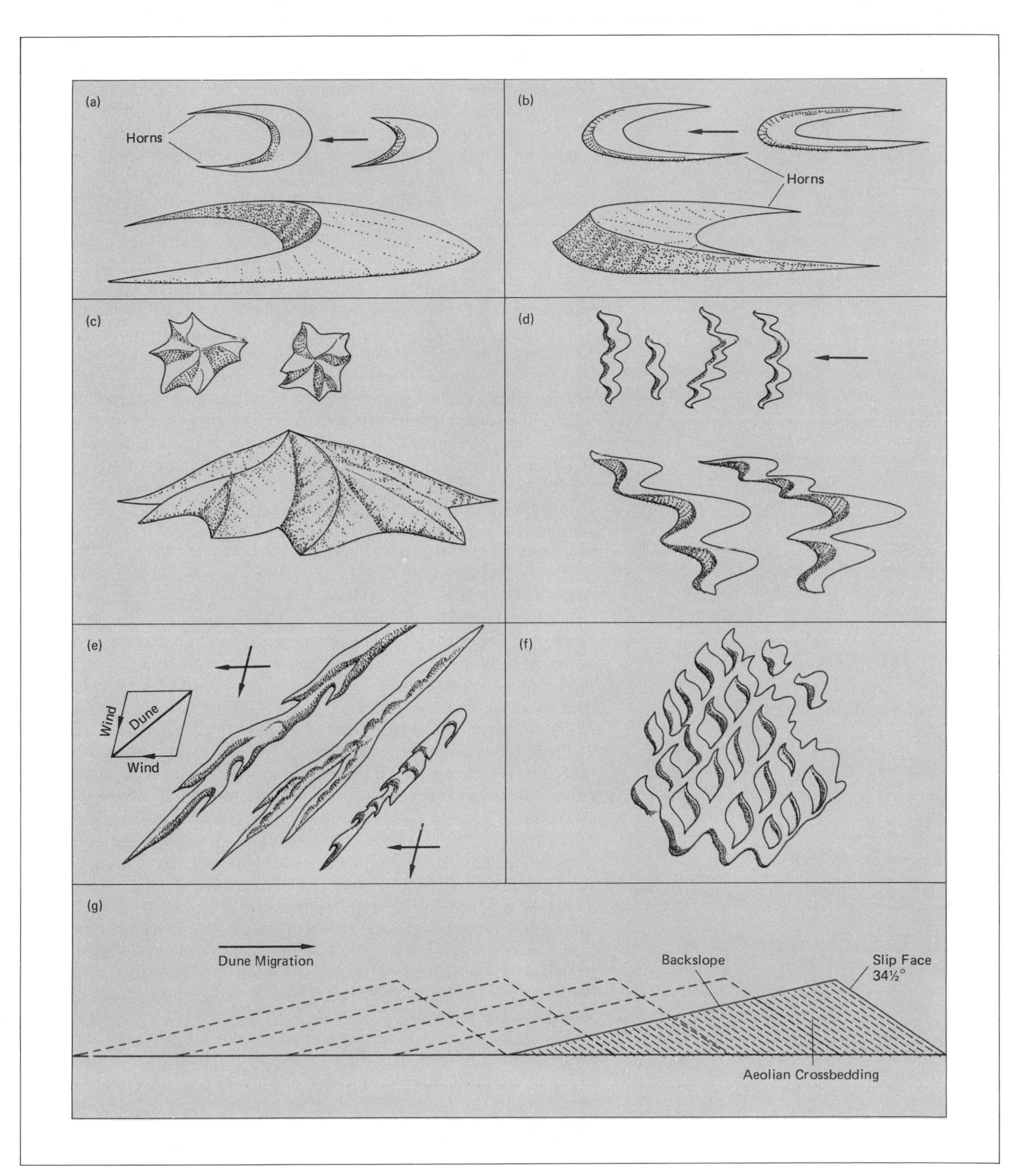
(a)
Horns
(b)
Horns
(c)
(d)
(e)
Dune
Wind
Wind
(f)
(g)
Dune Migration
Backslope
Slip Face
34½°
Aeolian Crossbedding

Landforms of Tropical Climates

Low-latitude regions display two characteristic climates and two characteristic landscapes. In the equatorial region, we find vast lowlands covered with lush tropical rainforests, which occasionally extend up the slopes of bordering mountains. Here conditions are warm and moist all year around, with monthly rainfall exceeding potential evapotranspiration. We shall refer to this as the humid tropical climamorphogenetic region. Peripheral to the equatorial lowlands are the tropical savannas and scrub forests consisting of vast tree-studded grasslands and dense tangles of thorny brush (see Chapter 8). In the savanna and scrub forest regions, precipitation is less than evapotranspiration for several months of the year, but rainfall in the high-sun season is heavy. The almost daily rainfalls of the tropical rainforests generate large amounts of runoff, producing the world's largest rivers. In areas of steep relief, the drainage density is extremely high and the landscape may become a maze of serrate ridges. Due to the heavy seasonal precipitation, the savannas and scrub forests likewise generate considerable seasonal runoff. But during the dry season, water tables in these regions sink far below the surface, causing all but the largest stream channels to dry up.

Although the abundant rainforest vegetation of the humid tropical region is concentrated in the equatorial lowlands, it sometimes extends upward onto steep mountain slopes, offering them some protection from severe gully erosion by surface water runoff. Dense vegetation on tropical riverbanks also seems to limit the ability of streams to cut laterally to the extent seen in the midlatitudes. Soil wash and removal of material by solution are nevertheless important erosional processes in both regions, and the sliding of water-saturated soil on steep slopes is conspicuous in areas where rainforests reach into highland zones.

On horizontal erosion surfaces in moist tropical climates, the weathered mantle is generally thick, sometimes attaining depths exceeding 100 meters (330 ft). Chemical weathering is very aggressive, quickly decomposing rock and permitting soluble bases to be leached away, if they are not taken up quickly by vegetation and kept in the nutrient cycle. The remaining concentration of less soluble oxides of iron or silica produces subsoils that tend to harden when exposed to air. Thus deforested humid tropical landscapes and savannas develop a crust of reddish *laterite* overlying a white or mottled substrate of thoroughly decomposed rock material. This is especially the case in savannas and scrub forests, where water table fluctuations seem to contribute to the process of laterite formation. The laterite crust may be several meters thick and often produces conspicuous ledges and mesas (see Figure 14.25, pp. 450–451).

In many savanna areas, the large-scale relief features are abrupt escarpments and steep-sided domes composed of poorly jointed granitic or gneissic rock that has resisted subsurface chemical decay (see Figures 14.25 and 14.26). Waterfalls are common, and deeply incised gorges are rare along large streams because thorough chemical decay of rock material prevents the streams from acquiring cutting tools in the

Figure 14.16 The Jari River in Brazil flows southward into the Amazon River, and is typical of streams in the humid tropics in being interrupted by several waterfalls and cascades. The valley of this large stream is also poorly defined and the channel is often divided by forested islands. All of these features suggest that due to the intense chemical weathering affecting the rocks of the region, the stream lacks a coarse bedload and the abrasive tools necessary for vertical incision and the production of a smooth longitudinal profile cut through diverse materials.

form of rock cobbles (see Figure 14.16). European geomorphologists often refer to the "paralysis" of linear erosion in the humid tropics, where stream loads consist primarily of quartz sand and dissolved material.

Other Climamorphogenetic Regions

With the exception of the glacial morphogenetic region, in which the geomorphic processes are unique and important enough to merit separate treatment (Chapter 15), the remaining climamorphogenetic regions are transitional between various of the five major regions described above. Thus the *subarctic* region is the zone of transition between the periglacial and humid midlatitude regions; the *maritime* region is a still wetter mild-winter version of the moist midlatitude region; and the *semiarid* region is transitional between the arid region and either the humid tropical or humid midlatitude region.

Figure 14.17 Cold climate landscape. Where temperatures remain below freezing the larger part of the year, landscapes take on a highly distinctive appearance. Below the top meter or so of the land surface, the water in all pore spaces in rock and soil remains frozen solid throughout the year. Permanently frozen rock and soil is known as *permafrost.* Permafrost reaches to depths exceeding 600 meters in northern Alaska and Siberia. Above the permafrost, the surficial blanket of soil thaws each summer, becoming saturated with water that cannot escape downward due to the frozen substrate. This water here and there lubricates the soil to the degree that the soil slips down as a mass, a few centimeters in a matter of a week or two. The water-saturated soil moves in lobes that are conspicuous on hillslopes, as the inset (bottom right) indicates. This type of movement is known as *solifluction.* Solifluction lobes are a few centimeters to a meter or so in height. Studies show that despite the appearance of great activity produced by these solifluction lobes, most are inactive, movement occurring only in a few places during each thaw. However, considered over a long period of time, the entire soil cover is draining downslope at a far more rapid pace than that produced in warmer regions by the creep process. Thus the portion of the soil that freezes and thaws annually, expanding, contracting, and moving downslope, is known as the *active layer.* The long-term effect of the solifluction process is to smooth the landscape, filling preexisting depressions and peeling down projections. Small valleys are infilled with *colluvium* (slope deposits), forming flat-floored *dells* with poorly developed watercourses. Water leaks through the colluvial valley fills.

Bare rock within the active layer or projecting above it is subject to intense frost weathering during the long winter season. The expansion of water in passing from the liquid to the crystal state exerts enormous pressures on confining walls, causing rocks to split and, occasionally, to crumble. Thus projecting solid rock masses are rapidly reduced to rubble, which becomes incorporated into the solifluction lobes. High areas are therefore lowered effectively. Broad summits are covered with angular rock rubble produced by the intensity of the freezing process in exposed sites. The result is a "sea of rocks," or *felsenmeer.* At the edges of rock exposures, large talus accumulations reflect the downslope movement of frost-riven blocks.

The annual freeze and thaw process produces a host of unusual minor landforms in addition to solifluction lobes. Repeated volume changes in the weathered mantle have the effect of sorting out coarse and fine material. Due to the efficiency of freezing in springing loose slabs of rock, the weathered mantle on hills contains an abundance of coarse debris. This becomes shunted away from the fines, which accumulate in masses surrounded by rings of slabs that are more or less "on end." The result is the stone rings shown in the inset (top center). On slopes these rings are drawn out downhill into "garlands" and "stripes" by movements of the active layer. In lowlands composed of silty allu-

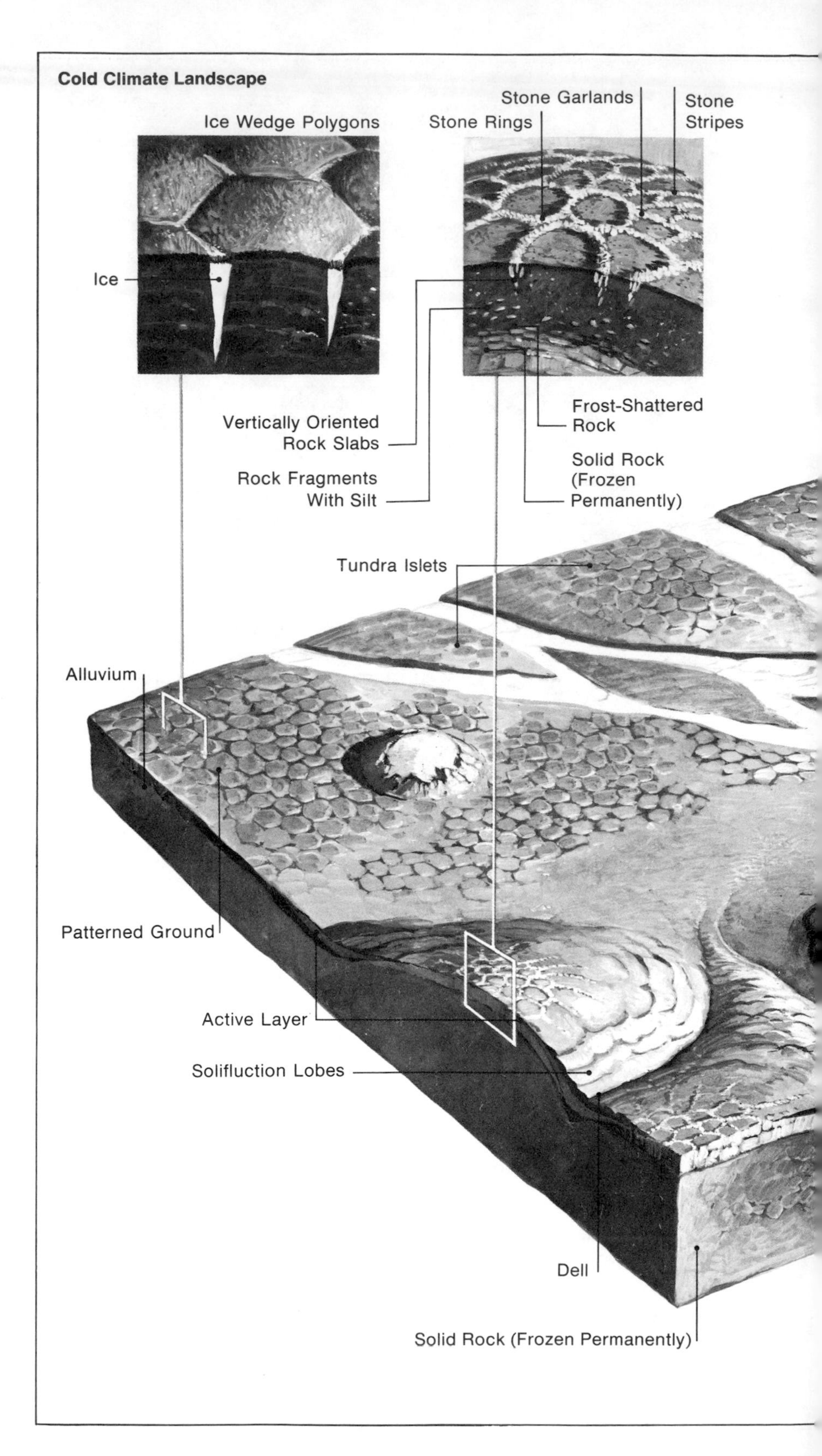

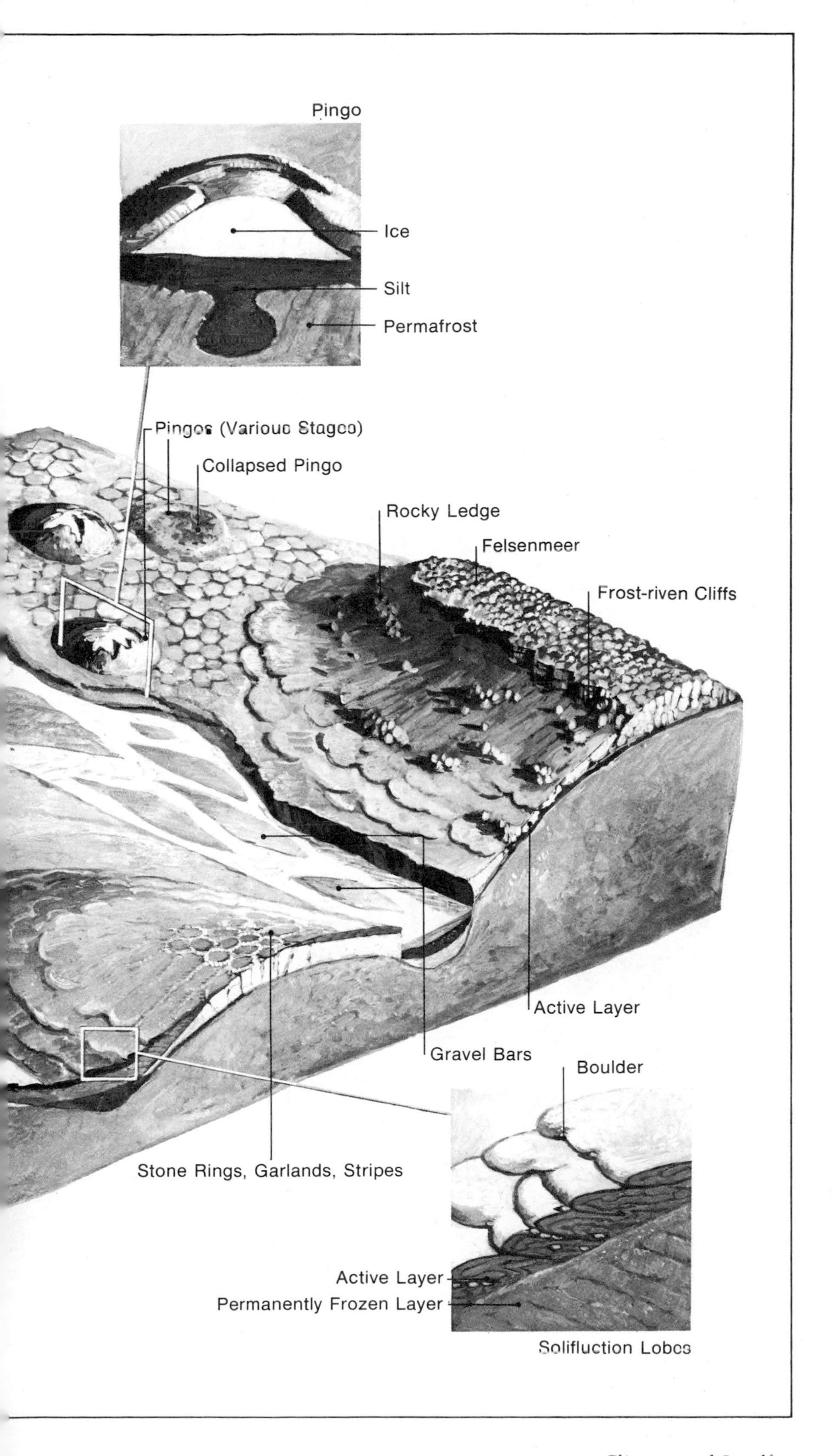

vial deposits, another type of *patterned ground* is developed. During the freezing process, water-soaked silt first expands, but at prolonged low temperatures, eventually begins to contract. Wet ground, frozen in the cold season, thus cracks, producing a network of polygonal fissures. Ice begins to form in these and eventually produces clear wedges, often a meter across and 5 meters deep (see inset at top left). These are *ice-wedge polygons.* Once formed they are permanent, continuing to grow very slowly. Clear ice also forms in the ground in lenslike masses that heave up the overlying sod (see inset at top right). This unusual, but not uncommon, phenomenon is known as a *pingo.* Pingos develop in old lake beds or marshes that are filling with sediment and vegetation. At a certain point, the encroachment of permafrost into the fill traps a lens of unfrozen water and fill. Freezing of the water in turn draws moisture from the fill to produce a growing mass of ice that eventually domes up the recently developed sod above it. If a pingo is destroyed by thawing, it leaves a depression ringed by an earth rampart. A large pingo may be 100 meters across and 30 meters high.

Other oddities of landscapes formed by *cryergic* (low-temperature) processes are overhanging (frozen) riverbanks and unusually smooth stream profiles due to the abundance of fresh rock waste, which provides streams with abrasive tools.

Figure 14.18 The small, ice-cored domed hill is a pingo near the delta of the Mackenzie River in the Northwest Territories, Canada. The surface of the ground exhibits the polygonal pattern often found in cold climate regions underlain by perennially frozen ground.

Figure 14.19 Desert landscape—relief by geologic movement. In this landscape, the difference in elevation of the highest mountain crests and the basin floors is a consequence of rupture of the land surface. The basin is an area that has dropped with respect to the high land adjacent. The line along which the rupture and displacement occur is a fault. The vertical displacement may attain thousands of meters. Such geologic structures may occur in any climate. What is distinctive in arid regions are the *playas, alluvial fans,* well-preserved evidences of *fault movement,* and *angularity* of form. In a humid area, such a basin would fill with water, overflowing eventually to produce an outlet channel to the sea. The lake would be eliminated by infilling with sediment to a point where a graded outlet channel would cross it, carrying off immediately all runoff from the surrounding mountains. In arid regions, water collects in the basin after rains, producing a temporary lake. This soon evaporates, leaving a dry lake bed, or playa. This may be a mud flat, or it may develop a crust of saline minerals precipitated during evaporation of the lake water or brought to the surface by capillary rise from a body of saline groundwater that is near the surface. Where a mud flat is present, the water table is very low and very large lakes have not been present recently (in the last several thousand years). The minerals of saline playas are zoned in order of solubility from carbonates at the edge, through sulphates, to chlorides (such as pure rock salt) in the center. These evaporite minerals are extremely valuable and contribute upwards of $50 million annually to the economy of California alone.

The alluvial fans reflect rapid torrential runoff from slopes lacking vegetative protection. Loose material is flushed off during cloudbursts, moves into flooding channels, and is carried out of the mountains. Where the bedrock channel ends, the floodwater spreads and percolates into the gravelly debris around the basin periphery. The resulting loss of discharge causes the coarse sediment load to settle out, adding to the volume of the fan. Two types of fan are present in this view. On the left, large fans drown the spurs of the down-tilting block, filling the canyons with alluvium. This is because filling of the basin creates a constantly rising base level. Across the basin the fans are small. They lie against the young fault scarp, and the valleys feeding them are actually "hanging" due to rapid uplift along the fault. The outer "toes" of these fans are drowned by alluvium due to subsidence along the right side of the basin block, and this partly accounts for their small size. Ordinarily, these fans are small, because their catchment areas, on the fault scarp, are smaller than catchments on the backslope of the block. The inset shows how fault uplift often creates a wineglass valley by rejuvenating erosion, causing slotlike gorges to be cut below older V-shaped valleys. The bowl of the wineglass is the open upper valley; the stem is the new erosional slot; the base is the alluvial fan below.

Fan surfaces in deserts are interesting due to their microtopography. Frequently, hardened streams of mud studded with sizable boulders (1 to 3 meters in diameter) are present, indicating that the fan is built largely by *debris flows*—mixtures of mud, gravel, and boulders that rumble out of the canyons during infrequent very heavy rains that fall on barren sur-

Desert Landscape in Areas of Block Faulting

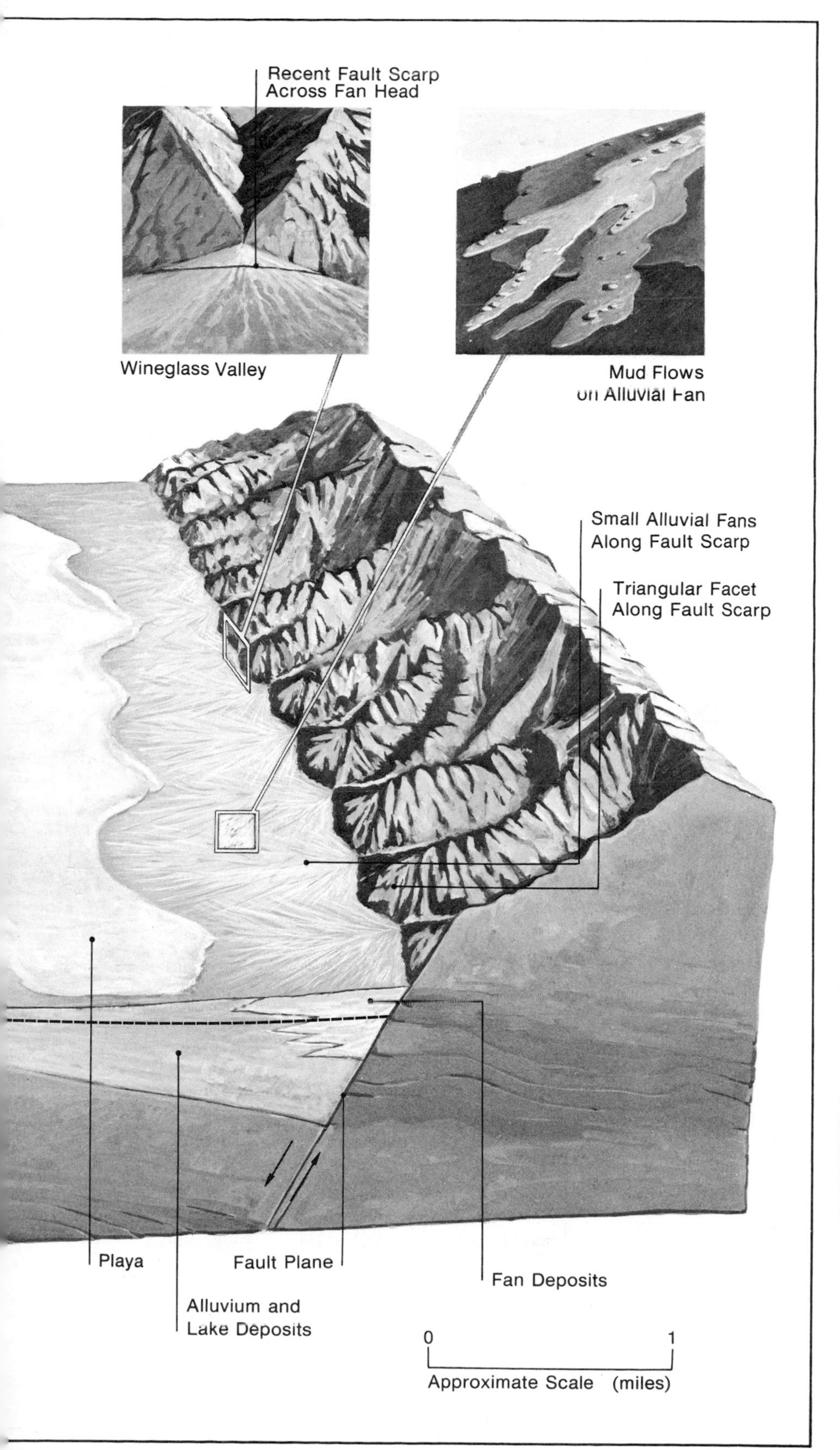

faces of loose material. The inset (top right) shows a recent debris flow covering a slightly older one. Surfaces of different ages are apparent in the *color* they display because the desert rocks tend to become coated with a patina of iron and manganese oxides with the passage of time—*desert varnish.* Thus new deposits are light in color, and successively older ones are increasingly darker. All sizable alluvial fans are a mosaic of tones as a consequence.

The prevailing angularity of form results from the absence of a smoothing cover of soil. Chemical weathering is almost eliminated due to lack of moisture, and the absence of a protective vegetative cover precludes soil accumulation, loose material being flushed away by the infrequent rains as fast as it is formed by the weathering processes.

Fault scarps composed of both bedrock and alluvium are clearer than in humid areas because the limitation of weathering and absence of soil creep allow unusually long preservation of newly formed surfaces, even where they are merely banks of gravel and mud. Scarps commonly cut across the upper parts of alluvial fans, indicating recent uplift of the mountain block above them.

Figure 14.20 Alluvial fans slope from bordering mountains toward the basin of Death Valley, California. The view shows the large fans on the west side of Death Valley.

Figure 14.21 Desert landscape—relief by erosion. This landscape is created by erosion by running water in an area of very gently dipping sedimentary rocks of varying resistance. The basic element is the *cuesta,* a sloping plain that terminates on its high side in a bold escarpment. The plain is the top of a layer of resistant rock, usually sandstone or limestone, and the escarpment marks the edge of the layer. In such a landscape, we find escarpments wearing backward through time, always in a down-dip direction. In the diagram, the dip is to the right, and the main cliffs are being pushed back to the right by erosion. Cliff retreat commences at structural highs, or domes, which are the first areas to be cut through by erosion, producing "holes," which subsequently enlarge. Cliffs are also produced where streams cut across the landscape, as in this view. The main streams may flow obliquely to the structural slope as a consequence of having been present here *before* tilting commenced. Such streams are *antecedent* to the geological movement. If they flow across a structural slope, along the strike, their canyons are asymmetric: surface water will run down into them on one side, producing tributary gullies, but not on the other, leaving a straight cliff there. The Grand Canyon of the Colorado in Arizona is a notable example. Commonly springs are present on the up-dip side of the canyon as groundwater leaks through the inclined permeable sandstone or limestone layers. Large springs create great cavernous alcoves, which have been used by humans for shelter— whole villages have been constructed within them, as by the Pueblo Indians of the southwestern United States (see inset at top right).

The cliffs actually retreat by being undermined by the more rapid erosion of less resistant rocks at their base. The shale in this view is rapidly removed, causing the more resistant *caprock* (sandstone) to collapse (see inset at top left). Collapse of rock faces occasionally produces an opening through a narrow wall of rock, forming a *natural arch.* These should not be confused with *natural bridges,* which are created where two meander loops along a canyon stream wear away the rock between them. Both arches and bridges are rare except where the sandstone is very thick and massive.

Mesas result where once-continuous layers of resistant rock have been fragmented by erosion into widely separated flat-topped remnants. *Buttes* are remnants too small to preserve a flat summit. Both mesas and buttes commonly form erosional outliers along the faces of retreating cuestas.

Where the less resistant rocks, usually shale or marl, have been stripped off over a wide area, exposing the top of a little-eroded resistant layer, a *stripped plain* results. Stripped plains commonly expose joint sets, and these may be exploited by weathering and erosion to form very rough rock badlands consisting of humps, fins, and other strange bedrock forms.

Sandstone that collapses due to undermining often shatters into loose sand. Streams in such areas also convey a large quantity of sand, which may blow out of the dry stream beds between the rare water flows. As a consequence, sand dunes are common in these landscapes. In the extreme deserts, such as the Sahara, the flow of sand driven by the prevailing winds creates vast areas of sand accumulation,

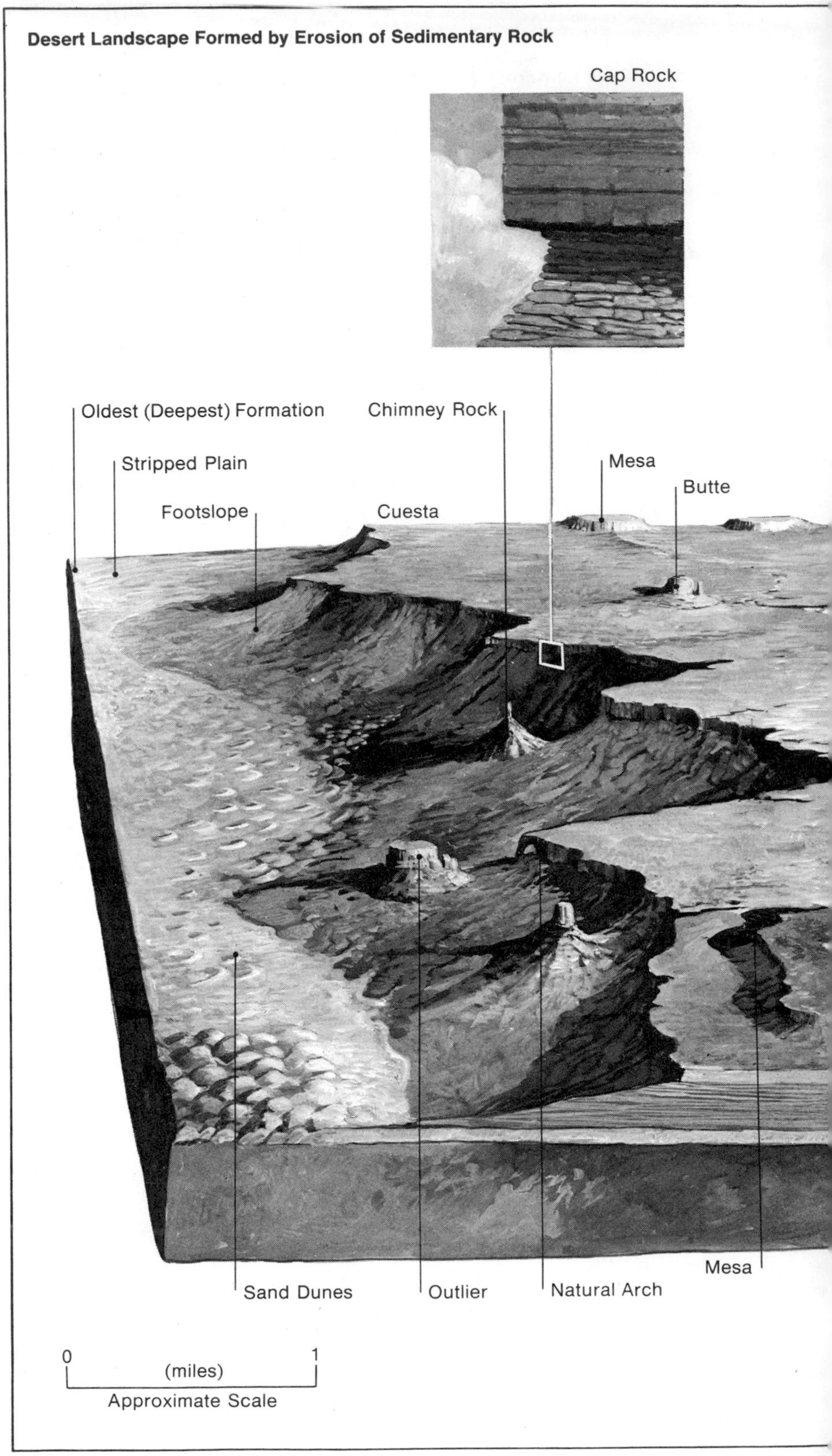

Desert Landscape Formed by Erosion of Sedimentary Rock

Cap Rock
Oldest (Deepest) Formation
Chimney Rock
Stripped Plain
Mesa
Butte
Footslope
Cuesta
Sand Dunes
Outlier
Natural Arch
Mesa
0 (miles) 1
Approximate Scale

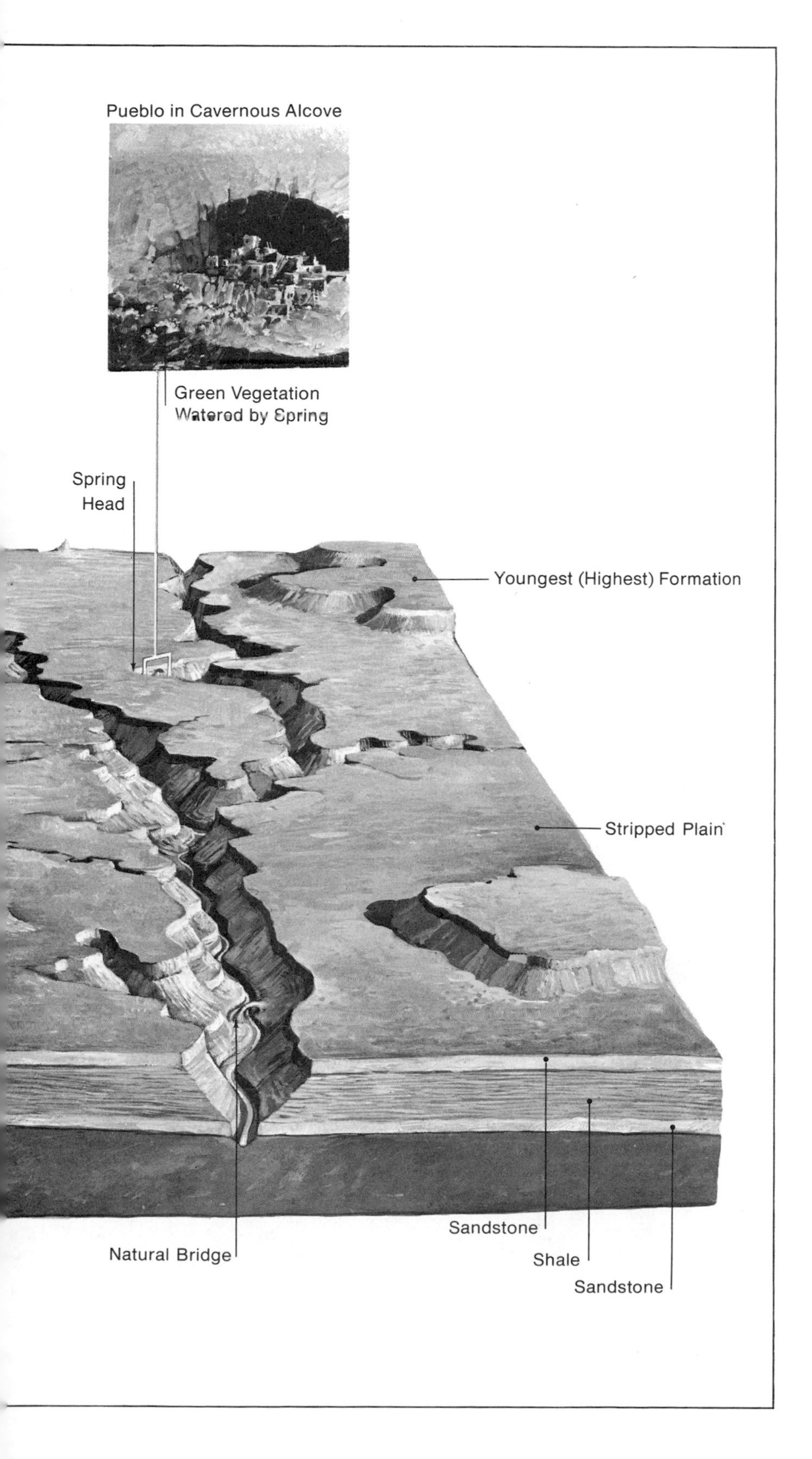

known as *ergs*. Dry sand moves when subjected to a wind exceeding about 17 km per hour. The smallest relief forms produced are sand *ripples*. These crawl over dunes, causing them to advance by dumping sand over their crests. In the world's great sand ergs, forms larger than individual dunes occur, having characteristic sizes and regular spacing. These forms are called *draa;* they are whale-backed mountains of sand, and active dunes crawl over them in the same way that ripples creep over dunes.

Layered structures like those illustrated also are common in humid areas. However, the humid landscape result is different. Cliffs are less distinct due to a blanket of soil that creeps downslope, protected by a vegetative cover. Chemical decay affects nearly all rock types so that lithological variations are less distinct. Natural bridges and arches are rare due to weakness of rock resulting from deep chemical weathering. Dunes are absent because vegetation colonizes debris accumulations, preventing detachment of particles by wind. The often highly colored bare rock that lends fascination to the desert scene is altogether obscured by a soil cover strongly anchored by vegetation.

Figure 14.22 Mesas and buttes dominate the landscape in Monument Valley in southeastern Utah. Vertical jointing in horizontally bedded rocks leads to the development of the vertical rock faces characteristic of landforms in arid regions with sparse vegetation. Landforms in regions of sand accumulation are depicted in Figure 14.15, p. 439.

Figure 14.23. Desert landscape—relief by erosion in crystalline rocks. Where granitic or gneissic rocks appear in deserts, the landscape is commonly one consisting of long, smooth ramps leading up to extremely rough bouldery slopes. The projecting relief forms rise very abruptly above the ramplike surfaces, resembling islands projecting above the sea; hence these summit relief forms are called *inselbergs* (island mountains). The lower portion of the ramp is an alluvial apron, but the upper part is an erosion surface, known as a *pediment,* that truncates solid granitic rock. The summit relief clearly is a remnant of a long period of erosion, but unlike a monadnock on a Davisian peneplain, it is very distinct from the erosion surface below it, which it meets at a sharp angle in most cases. These steep-sided residuals (inselbergs) are thought to have been formed by the retreat of slopes at a constant angle, which is unlike the mode of slope development thought to characterize most nondesert landscapes, in which slope angle is imagined as diminishing through time. Thus in humid regions the landscape is thought to "wear *down*" whereas in arid regions it is believed to "wear *back.*" The crucial factor in the two types of development is thought to be the presence or absence of a mantle of soil subject to the gravitational creep process. Soil covers slopes in humid areas, being protected by a vegetative cover; this soil creeps downslope as a consequence of gravitational force. This causes material to collect at the slope foot, anchoring it in place while removal continues on the higher part of the slope. Both on theoretical grounds and by actual measurement, creep causes slopes to decrease in angle as time passes, if the material delivered to the slope foot is not removed. The absence of a protective vegetative cover in arid regions prevents soil formation, as weathered material is washed away as fast as it is formed. Slopes retreat through the erosive effect of running water rather than by soil creep; there is no accumulation at the slope foot; and this leads to parallel slope retreat, both theoretically and by measurement in the field.

In the landscape shown, the bouldery slopes are a consequence of the tendency of granitic rocks to be well jointed, forming a rigid mass of plane-faced blocks, which weathering tends to loosen and round at edges and corners. The steep slopes have been regarded as the angle of repose of the loose blocks produced by the weathering process.

The lower blocks of the inselberg slopes are commonly perforated by cavities produced in the weathering process. The exteriors of the boulders may be very dark due to patination by desert varnish.

According to recent investigations of such landscapes, their development may have been misunderstood by geomorphologists. Evidence of the presence of a soil formerly covering them suggests that they are relict from a more humid landscape, having been altered to their present form by relatively recent erosion under desert conditions—the erosion being triggered by loss of the former vegetative cover. Nevertheless they evidence parallel slope retreat, even in the former soil-covered landscape, indicating efficient removal of material arriving at the slope foot.

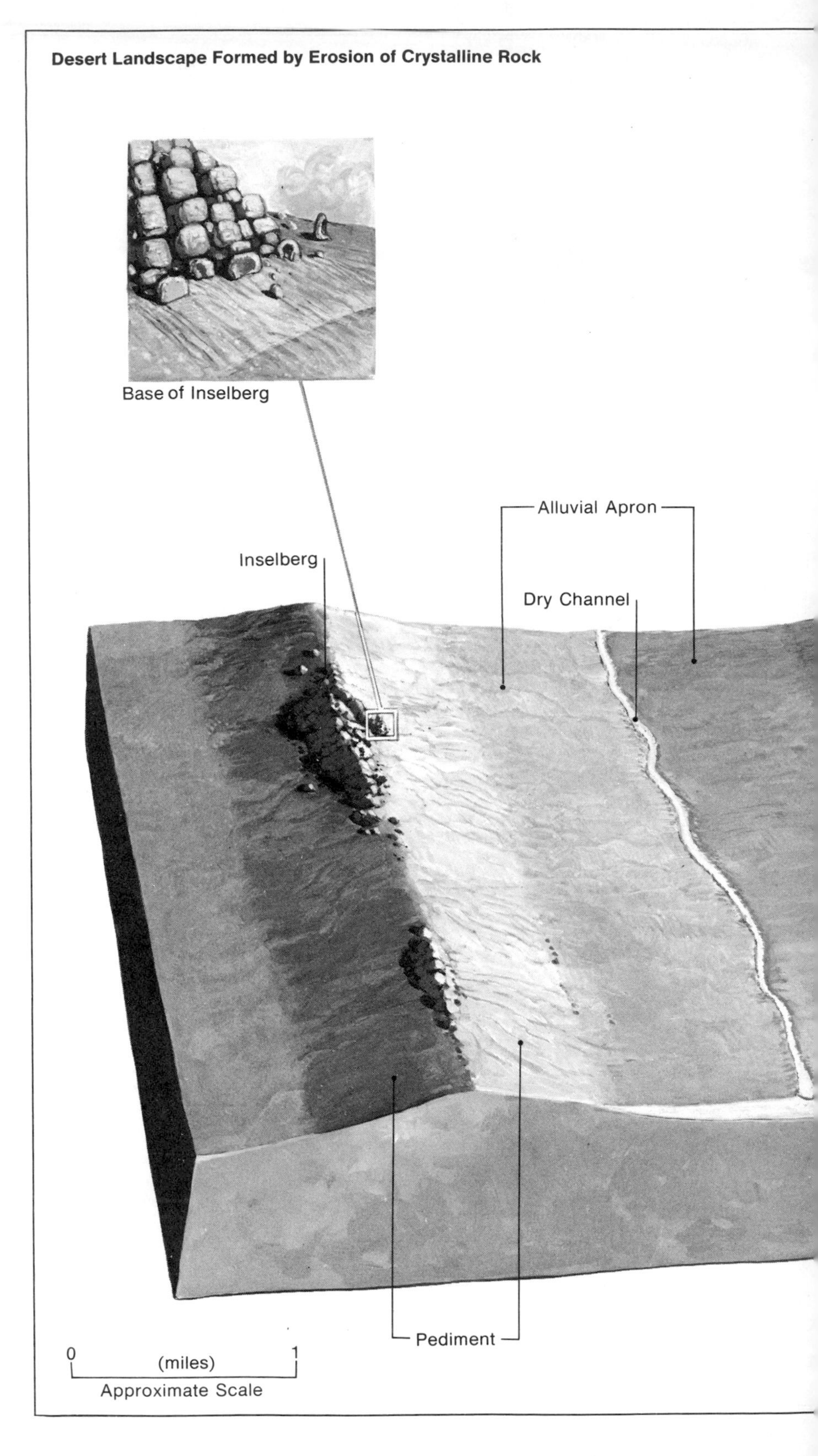

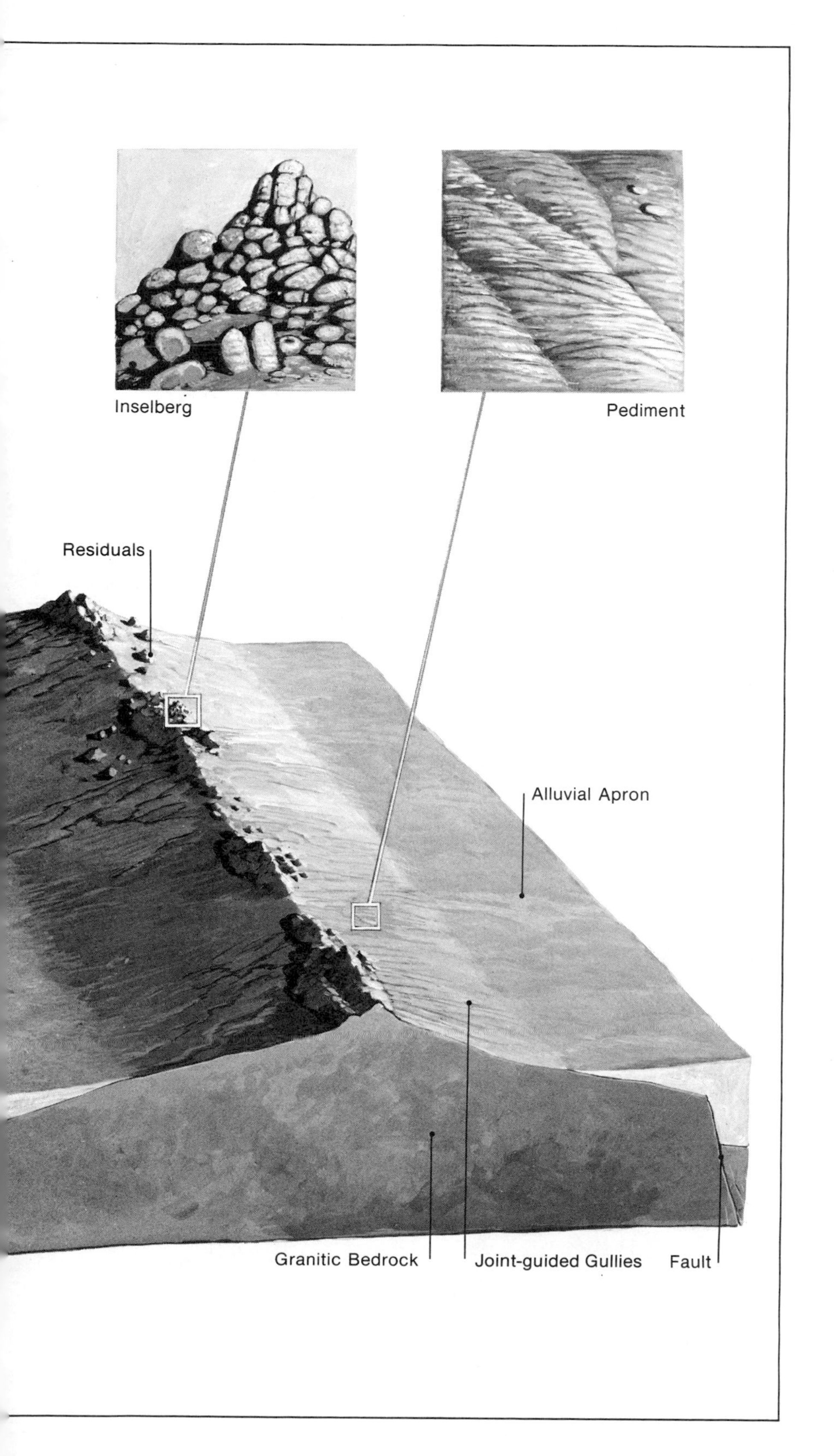

Figure 14.24 Most of the area of desert landscapes formed by erosion in crystalline rock consists of gently sloping pediments and rocky inselbergs, such as these examples located in the Mojave Desert east of Victorville, California.

Figure 14.25 Tropical savanna landscape. This landscape is widespread in the seasonally wet tropics on stable continental areas wherever crystalline rocks are dominant, as in Brazil, much of Africa south of the Sahara, India, and Australia. The term "savanna" refers to the vegetation association consisting of drought- and fire-resisting tree species rising from a sea of grass. The important geomorphic features are very *flat plains,* into which only shallow saucer-shaped valleys are incised by ephemeral streams; *escarpments,* over which streams cascade spectacularly; and enormous rock domes, termed *bornhardts* (after the German who brought those in Africa to the outside world's attention). This peculiar landscape is the consequence of an aggressive climate and the peculiarities of granitic rock. The mode of landscape development has been characterized as *double planation,* as there are two separate interfaces along which geomorphic activity takes place. One is the *land surface,* subject to rainwash, gravitational transfer, and stream erosion. Erosional removal lowers the land surface and wears back projections upon it. The second interface is that at which decayed rock meets fresh, unaltered rock—the *weathering front*—which lies below the land surface. The unusual aspect of this landscape is the aggressive chemical weathering under warm tropical conditions that decomposes rock in the subsurface as fast or faster than surface erosion can wear down the land. Boreholes in some locations in Africa indicate the weathering front to be 100 meters or more below the surface. They also show it to be highly irregular. Where the granite is closely jointed, the rock is deeply decayed, as water can penetrate it effectively. But where jointing is sparse, water is excluded and chemical decay cannot begin. Bornhardts, rising above the surface as much as 500 meters, are cores of monolithic granite that resisted chemical decay below the surface. They have been left behind as the land surface wears down and escarpments retreat. *Tors* are piles of boulders isolated by chemical decay below the land surface but exposed by erosion before they are completely decomposed.

The flat surrounding plains are armored with a *laterite* crust—a residue of oxides and hydroxides of iron and aluminum resulting from aggressive chemical alteration of the decomposed rock, such that all bases and even the silica are released and leached away. Mobilization of iron and aluminum and their fixation in the crust seem to require a water table that fluctuates in level seasonally, as it does in the savanna regions—the iron and aluminum being derived from the zone in which the fluctuation in level occurs. Because the water table varies in height most under hills and least under valleys, there is more chemical erosion under hills, so that their crowns literally sink as subsurface removal occurs. This is probably the cause of the conspicuous flatness of the laterized plains.

A last peculiarity of savannas is the presence of ungraded streams and the absence of gorges and canyons of any significant extent. Of course the two are associated. Canyons are cut by streams carrying abundant tools for abrasion, which permit irregularities in stream beds to be filed smooth. The absence of canyons and presence of ungraded profiles, including many waterfalls and cascades, are a consequence of a paucity of abrasional tools. Streams in

many tropical savanna regions are poor in cobbles and boulders because thorough chemical decay destroys coarse granitic rock waste before it reaches stream channels, which therefore transport sand and silt exclusively. Rocks other than granite and related types do not decay as massively in tropical climates, although they may decay equally rapidly. Whereas chemical weathering eventually causes a granitic block to collapse into sand by permeating its structure, it will attack a basaltic block on the surface only, constantly reducing its size but permitting a durable kernel to persist in its interior. Thus basaltic outcrops are an excellent source of tools for abrasion by streams whatever the climate (see Figure 14.27, p. 452).

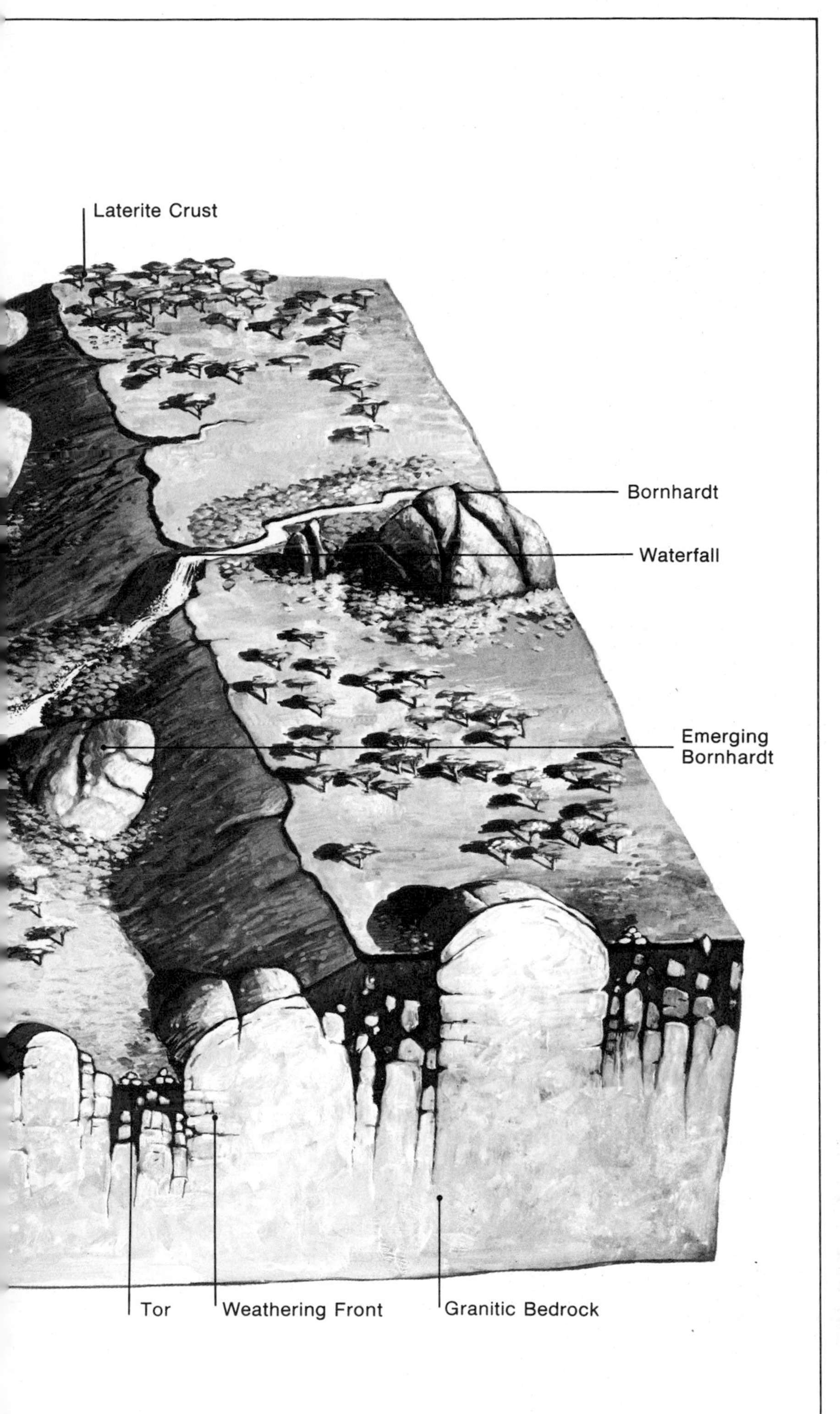

Figure 14.26. The massive granite dome of this bornhardt in Rhodesia, Africa, rises abruptly from the flat plain of the surrounding savanna.

Figure 14.27 Tropical high island landscape—volcanic dome. The distinctive aspect of humid tropical landscapes is the effect of water in very copious amounts, combined with high temperatures. Chemical weathering is very active, producing soils even on near-vertical slopes. The soil is often saturated with moisture, leading to landsliding, and water runoff attains enormous volumes that produce highly intricate sculpture of steep slopes. Areas of granitic, metamorphic, and sedimentary rocks are generally maturely dissected, having a very high drainage density and slopes steeper than those in higher latitudes.

Certain geological configurations produce truly unusual landscapes in the tropics, such as those illustrated, which occur widely. The figure shows the eroded flank of a volcanic dome, such as those composing the Hawaiian Islands. The landscape is peculiar in several ways relating to both the climate and the geological components present. The valley walls are as steep as 80°, yet covered with vegetation rooted in a shallow soil. Due to saturation with moisture, landslides are frequent and their scars are everywhere, exposing the bare basaltic lava composing the dome. The valley walls are fluted, each flute becoming a waterfall or cascade during the frequent rains. These flutes have been characterized as "vertical valleys." Where adjacent valleys have expanded so that their walls merge at the top, extremely jagged crestlines result. The main streams drop into the troughlike valleys over a series of waterfalls, each one terminatling in a small*plunge basin.* The ephemeral waterfalls along the valley sides also drop from one plunge basin to another. Above the abrupt valley heads there may be an upland swamp; that on the Hawaiian Island of Kauai may be the wettest place on earth (annual precipitation may approach 1,500 cm). Many of the valleys have flat floors resulting from subsidence of the islands and infilling by alluvium. It appears that the precipitous nature of the valleys dissecting these volcanic domes may be due to the low water table resulting from the permeability of the lava flows composing them—thus they are almost tantamount to gigantic spring alcoves, expanding due to sapping by spring action at their base.

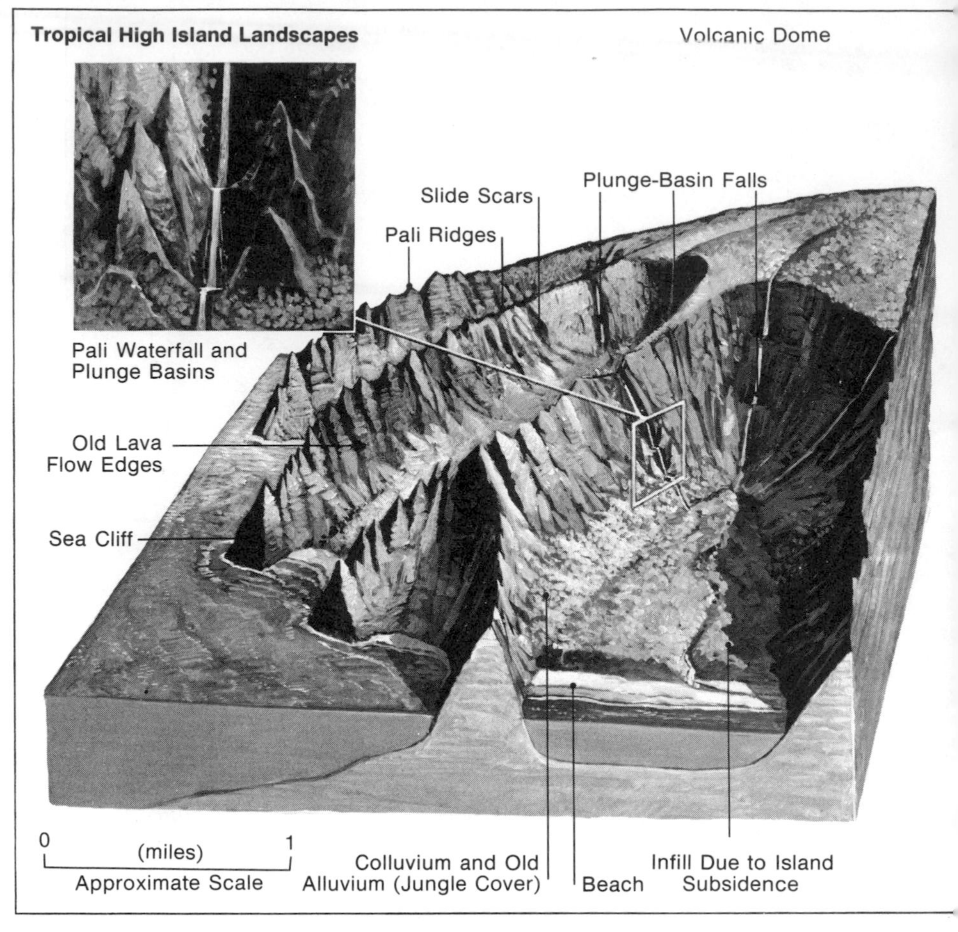

SUMMARY

Climate is a crucial factor in the processes of weathering and erosion that shape most landforms. Climatic effects on landforms are visible on a variety of scales, from rock microclimates to entire landscapes. Sometimes the combination of geomorphic processes characteristic of a particular climatic region develops highly distinctive landforms. This is especially true of the extreme climates that are dominated by cold, dryness, or high humidity. However, landforms of similar appearance occur in several different climatic regions, even though the various processes of slope formation may differ in intensity from one region to another. Many landforms cannot be classified according to climatic regions, but the collection of landforms that compose an entire landscape is usually a fairly clear expression of the region's existing climate or prior climatic history. Thus it is possible to recognize a number of climamorphogenetic regions in which the landscape reflects the domination of certain climatically controlled processes of weathering and erosion. The humid midlatitude morphogenetic region, which has been the source of many of our ideas concerning landform development, is but one such region. It is distinctive in being somewhat intermediate in character, experiencing a wide range of geomorphic processes but being dominated by no single one.

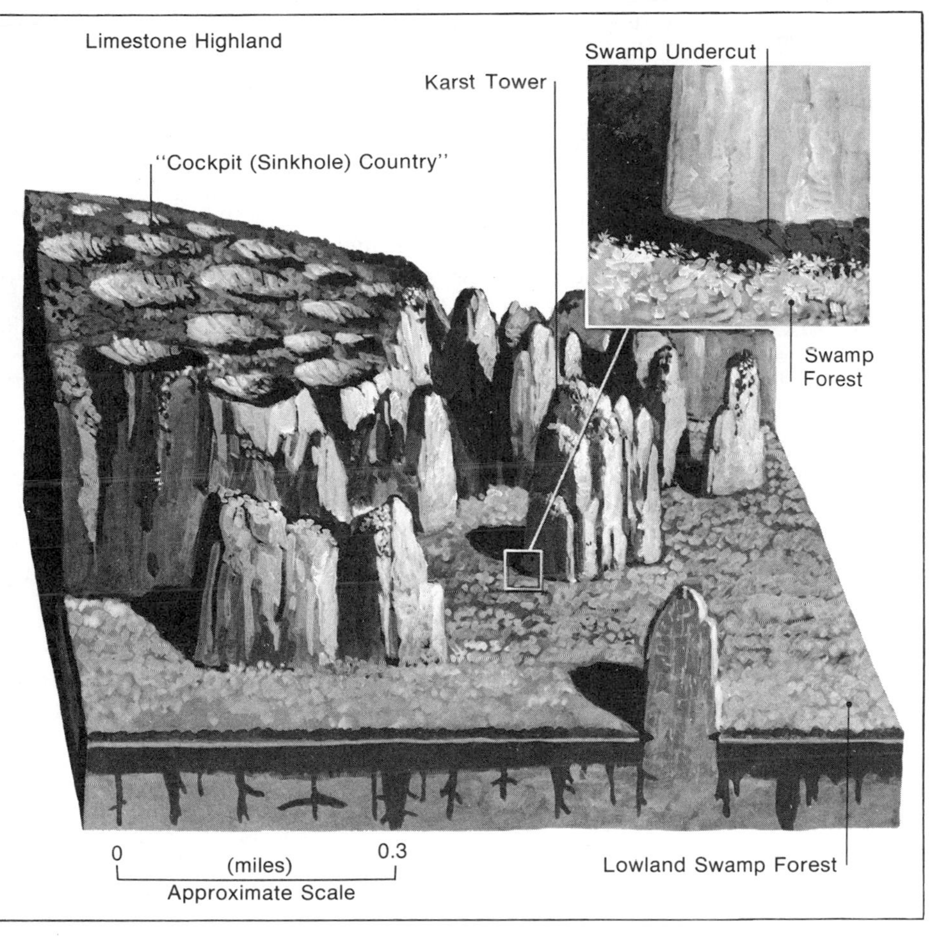

Figure 14.28 Tropical limestone highland. Another highly distinctive landscape is found where massive limestones appear in moist tropical areas of strong relief. Plateau tops are pockmarked with large sinkholes, produced by the solution of limestone along intersecting joints. Due to its content of organic acids, groundwater in the tropics is particularly aggressive toward limestone, which it corrodes rapidly. Between the sinkholes is a maze of often sharp-crested limestone ridges. This terrain has been termed *cockpit karst.* Solution working deep into the joints at the plateau edge detaches enormous slabs and pinnacles, which become isolated as vertical-sided mountains hundreds of meters in height. These have been made famous in Oriental art. They are common in southwestern China, North Vietnam, Malaysia, and Indonesia. Once isolated, these *karst towers* are kept steep-sided by undercutting and collapse. They rise from swampy lowlands, in which the water has been made acidic by contact with peat and other products of vegetative decay. This water corrodes a notch, sometimes 5 or 6 meters deep, around the karst towers. The overhangs collapse periodically, after which a new notch is formed, keeping the towers permanently "girdled." Tower karst develops only where limestone free of impurities is divided by well-defined but widely spaced joints. Where the limestone contains impurities of clay, marl, or chert, the towers are replaced by less imposing cone-shaped hills, which are common in Southeast Asia, Puerto Rico, and Cuba. Thus most tropical karst topography may be resolved into cockpit, tower, or cone karst, none of which is present in higher latitudes.

Review Questions

1. Climate affects geomorphic processes both directly and indirectly. What does this mean, specifically?
2. Why are "balancing rocks" and "toadstool rocks" more common in desert landscapes than elsewhere?
3. What would be the geomorphic effect of an increase in the vegetation cover of a formerly arid region?
4. What is the relationship between draa and loess?
5. What is a relict landform? What sort of relict landforms might you expect to find in a humid midlatitude climatic region? In a semiarid region? In a desert region?
6. What climamorphogenetic region should have the greatest diversity of relict landforms? Why is this so?
7. Although quite distinct in appearance from all other climamorphogenetic regions, desert regions resemble periglacial regions in certain geomorphic respects and humid tropical regions in other respects. What are the similarities in each case? Do they proceed from similar causes?
8. Does a climamorphogenetic regime that has one exceptionally aggressive process "pay" for this by excessive weakness in some

other process? In other words, is there some connection between the strengths of the different geomorphic processes in a climamorphogenetic region?

9. What impressive landscape feature seems to require a climatic change in order to develop?
10. One could imagine a bornhardt dome appearing as a consequence of any of three different processes. Explain.
11. The Potomac River carves deep gorges through the hard sandstones of the Appalachian Ridge and Valley Province, but only 24 km (15 miles) from the head of its estuary at Washington, D.C., it tumbles steeply down over a surface of ancient gneissic rock at the Great Falls of the Potomac. Knowing nothing else about the river or the region, how might this be explained. Is more than one explanation possible?

Further Reading

Birot, Pierre. *The Cycle of Erosion in Different Climates.* Berkeley and Los Angeles: University of California Press, 1968. (144 pp.) This is a highly condensed but information-packed outline of the nature of the erosional systems associated with different world-wide climatic regions—sparsely illustrated with diagrams.

Budel, J. "Climatogenetic Geomorphology," in Edward Derbyshire, ed. *Climatic Geomorphology.* New York: Barnes and Noble, 1973, pp. 203–227. This is a vigorous statement of the aims and value of climatogenetic geomorphology, by one of the pioneers in this field of research.

Butzer, Karl W. *Geomorphology from the Earth.* New York: Harper & Row, 1976. (463 pp.) This recently published textbook devotes about 100 pages to landform systems in different climatic regions. The treatment is up to date and more geographical in tone than that of most geomorphology texts, but the brevity of the author is at times frustrating.

Cooke, Ronald U., and **Andrew Warren.** *Geomorphology in Deserts.* Berkeley and Los Angeles: University of California Press, 1973. (384 pp.) This is a very thorough and most readable presentation of research findings concerning present geomorphological processes in desert environments. The last third of the book, devoted to the subject of sand dunes, is rather technical.

Ferrians, O. J., R. Kachadoorian, and **G. W. Greene.** "Permafrost and Related Engineering Problems in Alaska." *U.S. Geological Survey Professional Paper No. 678* (1969). (37 pp.) This treatise offers an interesting and well-illustrated account of the unusual problems encountered in the construction and maintenance of roads, railroads, bridges, and buildings in areas underlain by permafrost.

Garner, H. F. *The Origin of Landscapes.* New York: Oxford University Press, 1974. (734 pp.) Much of this large and well-illustrated volume is concerned with the geomorphic effects of varying climates, with considerable attention given to the specific role of climatic change in the formation of landscapes.

Holzner, L., and **G. D. Weaver.** "Geographic Evaluation of Climatic and Climato-genetic Geomorphology." *Annals, Assoc. of American Geog.,* Vol. 55 (1965): 592–602. This is a review article on the aims and accomplishments of climatic and climatogenetic geomorphology—good for an understanding of the historical development and interrelationships between the two fields.

Peltier, Louis C. "The Geographical Cycle in Periglacial Regions as It Is Related to Climatic Geomorphology." *Annals, Assoc. of American Geog.,* Vol. 40 (1950): 214–236. This article is important as the pioneer American statement proposing alternatives to the "normal" cycle of erosion. Peltier's preliminary morphogenetic regions and their climatic parameters, while far from satisfactory, continue to be repeated in modern texts.

Smith, H. T. U. "Dune Morphology and Chronology in Central and Western Nebraska." *Journal of Geology,* Vol. 73:4 (1965): 557–578. This is an interesting analysis of the complex history and morphology of Nebraska's Sand HIlls, the largest erg in North America, though no longer a desert region.

Stoddart, D. R. "Climatic Geomorphology: Review and Re-assessment." *Progress in Geography,* Vol. 1 (1969): 161–222. Stoddart's 60-page article reviews the evidence for and against climatic geomorphology as an alternative to other schools of thought in landform studies. The main controversy is whether unique climates produce unique landscapes, or whether climatically induced morphological differences are more apparent than real. A wealth of excellent information is assembled and subjected to a critical eye.

Thomas, Michael F. *Tropical Geomorphology.* New York: Halsted Press, 1974. (301 pp.) This work focuses upon the bornhardt landscapes of savanna regions and is really a book on the behavior of granitic rocks in tropical wet and dry climates. The author is a leading investigator in this field, and the book mainly concerns his own findings.

Tricart, Jean. *The Landforms of the Humid Tropics, Forests, and Savannas.* New York: St. Martin's Press, 1973. (352 pp.) Tricart presents, in a wholly qualitative manner, many interesting observations on the uniqueness of geomorphic processes in humid tropical and savanna regions.

______, and **A. Cailleax.** *Introduction to Climatic Geomorphology.* New York: St. Martin's Press, 1973. (295 pp.) Tricart and Cailleux present the European view in favor of studies of geomorphic processes as conditioned by climate and the biosphere. The work overtly challenges the Davisian system, which the authors believe still impedes the progress of landform analysis. The treatment is interesting, nonquantitative, and easily comprehended.

Washburn, A. L. *Periglacial Processes and Environments.* London: Edward Arnold, 1973. (320 pp.) This is the most thorough current analysis of geomorphic processes and problems in periglacial environments, by a leading contributor and field investigator. Photographic illustrations are outstanding.

Maligne Lake, Jasper Park by Lawren S. Harris. (The National Gallery of Canada, Ottawa)

Many of the earth's most spectacular landforms were created by the power of glaciers. Flowing ice is one of the most forceful agents of geomorphic change, and several regions of the earth still bear the chiseled marks made by ancient glaciers.

FIFTEEN

Glaciers and Glacial Landforms

Glaciation Present and Past
Pleistocene Glaciation
The Indirect Effects of Pleistocene Glaciation
Origin of the Ice Ages

Anatomy and Dynamics of Glaciers
The Making of Glacial Ice
The Glacier Budget
Movement of Glaciers

Glacial Modification of Landscapes
Glacial Erosion
Glacial Deposition

Landforms Resulting from Alpine Glaciation

Landforms Resulting from Continental Glaciation

OUTSIDE OF THE TROPICAL ZONE, our planet's temperatures are such that during the low-sun season, the precipitation that falls from clouds often consists of frozen water, or snow. In the higher midlatitude regions and beyond, snow may blanket the ground for days, weeks, or even months before it melts and flows off as liquid water. But it does melt every year, exposing bare rock or vegetation-covered soil in the warm season. However, this has not always been true. In the recent geological past, only a little more than 10,000 years ago, snow that fell in the higher midlatitudes failed to disappear in the warm season. Either summer temperatures were lower or much more precipitation occurred in the cold season than is presently the case. Whatever the reason, several times in the past 2 million years, snow has accumulated from year to year for thousands of years, increasing in depth until it became metamorphosed into glacial ice that spread outward and eventually engulfed half the surface of North America, with the same phenomenon occurring simultaneously in northern Europe.

We live in an unusual near-glacial age today. Only 15,000 years ago the sites of New York, Chicago, Detroit, Minneapolis, Seattle, Vancouver, Toronto, and Montreal were buried under as much as

2,000 meters (6,500 ft) of ice. This may occur again; indeed, some scientists predict it will occur within the next 5,000 to 10,000 years. A mere decrease of 5° to 7°C (9° to 13°F) in the earth's average temperature, along with an increase in winter precipitation, would be enough to cause ice sheets to reappear and begin moving into the densely settled portions of Europe and North America.

Throughout most of the earth's history, there have been no great accumulations of ice on the land. But today, even though we do not think of ourselves as living in a glacial age, three-fourths of the earth's freshwater supplies are stored in glacial ice. The world's existing glaciers contain an amount of water equivalent to 5,000 years flow of the Amazon River or 60 years of rain and snow over the entire planet. By comparison, the storage of water in lakes, rivers, swamps, and man-made reservoirs is paltry.

Most of the earth's freshwater reserve is impounded in the great Antarctic and Greenland ice sheets—70 to 75 percent in the Antarctic ice sheet alone. During much of the preceding 2 million years, two additional continent-size ice sheets were present, covering northern Europe, Canada, and 2.6 million sq km (1 million sq miles) of the United States, while smaller ice fields nearly submerged the high peaks in the world's great mountain ranges. Interestingly, one of the earth's coldest land areas, eastern Siberia, did not support a large ice sheet at this time; due to its aridity, this region merely became a cold desert.

While existing glaciers are important in diverse ways, glaciers of the past are of great significance to mankind because they have greatly altered the environments we now inhabit. Over vast areas in the higher latitudes, and also at high altitudes, slowly moving currents of ice have scraped away nearly all of the soil and weathered mantle that formerly covered the land; elsewhere, the decay of enormous masses of debris-laden ice has completely submerged the older topography under a blanket of glacial "drift," creating undulating plains of clay, sand, and gravel where once there were hills and valleys. In some areas, the effect of glaciers was detrimental to human use of the land; in others, it was highly beneficial. This chapter explores these effects and the processes that created them.

GLACIATION PRESENT AND PAST

Glaciers originate in regions where more snow falls each year than melts or evaporates. In most midlatitude areas that receive snow, the snowfalls begin in late autumn and melt away entirely by spring. But farther poleward, or at higher elevations, where summer temperatures are mild, snow persists for a longer portion of the year. In Alaska, Greenland, Iceland, the Canadian Arctic Islands, the Canadian Rockies, the European Alps, the Himalayas of Asia, the Andes of South America, the New Zealand Alps, and Antarctica, winter snows do not melt completely during the succeeding warm season. Under such conditions, snow accumulates from year to year and is gradually transformed into ice by processes to be described later. When a critical depth of ice accumulates—some 50 meters (150 ft) or so—the ice begins to deform as a consequence of its own weight and its low internal strength. A *glacier* is a mass of ice that is "flowing" due to gravitational stress on a solid material that becomes plastic under stress and that deforms rather easily.

Continuous ice sheets of continental dimensions presently exist in polar regions, and mountain glaciers can be found even in the tropics. Mount Kenya, located on the equator in eastern Africa, carries a number of small glaciers at altitudes above 4,500 meters (15,000 ft), as does Chimborazo, which rises to 6,271 meters (20,577 ft), 1½° south of the equator in the Andes Mountains of Ecuador.

Glaciers vary enormously in size. The ice sheet that covers Antarctica extends over an area of 12 million sq km (4.6 million sq miles), whereas thousands of glaciers in high mountains cover less than 1 sq km. More than 1,000 glaciers exist in the conterminous United States, most of them located in the Cascade Mountains of Washington. Nearly all of the glaciers in the United States outside of Alaska are small; their combined area is barely 500 sq km (200 sq miles). Nevertheless, they provide about 2.1 billion cu meters (1.7 million acre ft) of meltwater annually. Melting of glacial ice occurs mainly in the summer months, just when water demands for irrigation are highest and streams fed by precipitation and groundwater inflow are lowest. Alaska's glaciers cover more than 50,000 sq km (20,000 sq miles), and include one glacier system, the Malaspina, that is larger than the state of Rhode Island, having an area of over 4,000 sq km (1,500 sq miles).

Existing glaciers take a number of different forms. Dwarfing all other types are the great *continental ice sheets,* which are enormous unconfined blankets of glacial ice that completely submerge the land surface over areas of hundreds of thousands to millions of square kilometers. Two of these exist today, the Greenland and Antarctic ice sheets. Much smaller *highland icecaps* are ice sheets situated at high elevations in mountain regions, where they submerge hundreds to thousands of square kilometers of mountainous topography. They are conspicuous features in Iceland, the Canadian Arctic Islands, and the Canadian Rockies. More common in high mountains are *alpine glaciers,* which are streams of ice confined by valley walls as they drain from high mountain crest lines to lower elevations. Alpine glaciers are especially well developed in Alaska, the Canadian Rockies, the European Alps, the Andes of South America, and the Himalaya and Karakorum ranges of Asia.

Alpine glaciers always originate in high-altitude rock-walled amphitheaters called *cirques* (French, from Latin *circus,* or circle). Cirques themselves are distinctive erosional features created by frost action and the movement of glacial ice. In many high mountain regions, the only glaciers present are *cirque glaciers,* masses of ice that are restricted to these basins and do not enter the valleys at lower elevations. Where ice does spill outward from cirques or highland ice caps into the valleys below, it forms channelled flows called *valley glaciers.* Occasionally a valley glacier enters a larger ice-free valley or issues onto an open plain, where it spreads out in a pool of ice known as a *piedmont glacier.* Thus alpine glaciers can take three forms:

Figure 15.1 (opposite top left) Glaciers form where winter snow persists throughout the year. If the annual accumulation of snow exceeds annual losses due to melting and evaporation, a glacier eventually is created. Glaciers may be unconfined, as the *highland ice cap,* which submerges the older erosional topography in a mountain region, or confined, as *cirque* and *valley glaciers.* All glacial ice drains to lower elevations under gravitational stress. The combination cirque and valley glacier is known as an *alpine glacier.* The valley glacier descending from the unconfined ice cap is an *outlet glacier.* In some regions, the ice from one or more valley glaciers spreads out over a plain and forms a *piedmont glacier.*

(opposite top right) The surface of this valley glacier in the French Alps is broken by numerous crevasses.

(opposite bottom) The Malaspina Glacier in Alaska forms an extensive piedmont ice sheet. The ice flow is from right to left in this picture.

cirque glaciers, valley glaciers, and piedmont glaciers (see Figure 15.1). We will return to the landforms produced by glaciers of all types later in this chapter.

Pleistocene Glaciation

The present geologic epoch, called the *Holocene,* or *Recent,* is usually regarded as beginning about 10,000 years ago at the close of the last major period of glaciation, popularly known as the Ice Age. The preceding *Pleistocene epoch* began almost 2 million years ago and included at least four and perhaps seven or more major episodes of glaciation interspersed with warmer *interglacial* periods. Other glacial epochs have occurred during the earth's history, separated by warmer intervals of hundreds of millions of years.

Figure 15.2, p. 462, compares the extent of existing glaciation in the northern hemisphere with that during the late Pleistocene. At their maximum extent, Pleistocene glaciers covered some 26 million sq km (10 million sq miles), equivalent to one-third of the earth's present land area. A generally accepted explanation of the climatic changes causing this awesome refrigeration of our planet has not yet been produced (see Case Study following Chapter 7). But evidence of the glaciers is inescapable, consisting of vast ice-scoured landscapes and areas whose scenery is dominated by glacial deposits produced in a variety of ways. The agricultural heartland of North America is developed on soils whose richness is largely a consequence of their youth, being formed on late Pleistocene glacial deposits that have been subjected to leaching and eluviation for only 10,000 years. The Pleistocene ice sheets in North America terminated along a line roughly demarcated by the present Missouri and Ohio rivers, both of which channelled glacial meltwater along the ice margin.

During the Pleistocene epoch, glaciers also occupied high mountains around the world, even in such places as Hawaii. Pleistocene glacial erosion completely transformed the scenery of the world's high mountains, giving them their picturesque "alpine" appearance. The Pleistocene ice sheets and valley glaciers left a magnificent legacy of landscapes especially suited for human recreation and enjoyment. The astonishing pinnacle of the Matterhorn in Switzerland, the fjords of Norway, the ski slopes of Tuckerman Ravine in New Hampshire, the deep gouge of California's Yosemite Valley, the sandy arm of Cape Cod in Massachusetts, and the 10,000 (and more) lakes of Minnesota—all were fashioned by Pleistocene glaciers.

The Indirect Effects of Pleistocene Glaciation

The effects of glaciation are not limited to scouring and deposition in the area directly covered by ice. Either the glaciers themselves or the climatic refrigeration accompanying them had a variety of effects upon landscapes. Meltwater from the massive accumulation of ice during the Pleistocene generated runoff that enlarged some valleys, filled others with sediment, and resulted in conspicuous stream ter-

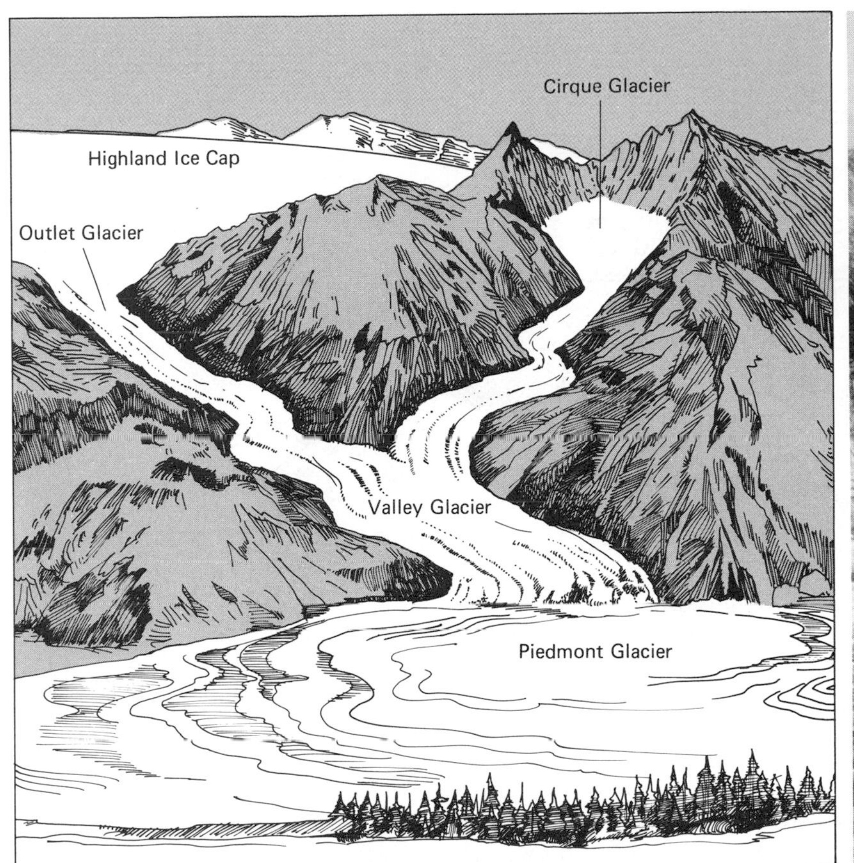
Cirque Glacier
Highland Ice Cap
Outlet Glacier
Valley Glacier
Piedmont Glacier

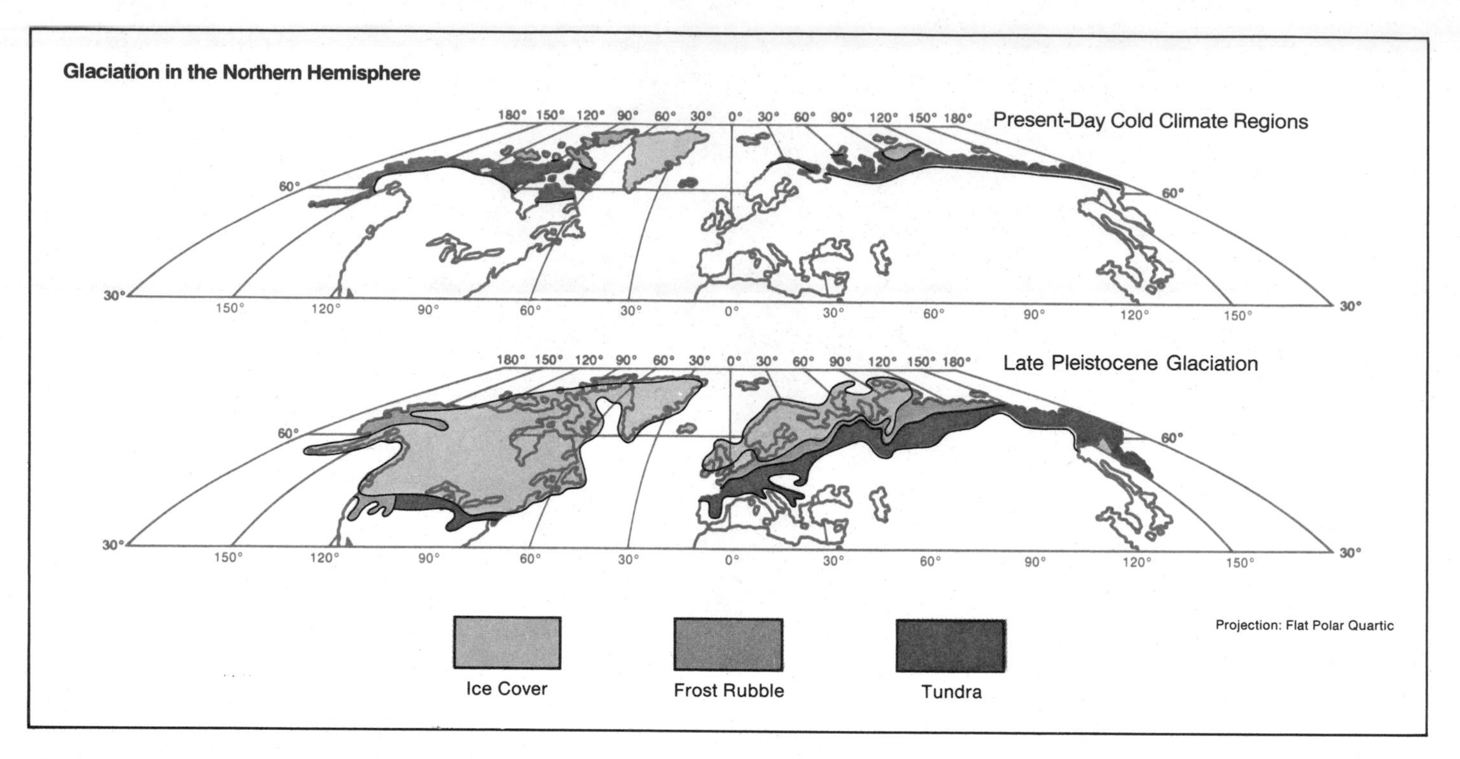

Figure 15.2 (top) At present, regions in the northern hemisphere that exhibit glaciation and other features characteristic of cold climates generally are restricted to the high latitudes. Isolated regions of glaciation also occur at high altitudes in mountain ranges such as the Alps, the Rocky Mountains, and the Alaskan mountains.

(bottom) During the last major period of glaciation, approximately 15,000 years ago during the late Pleistocene, large parts of northern North America and northern Europe were subjected to the direct or indirect effects of glaciation. Many sections of the northeastern and north-central United States exhibit landforms characteristic of glacier action, and periglacial regions not directly covered by ice sheets display the effects of cold climates in the form of smoothed slopes, solifluction deposits, patterned ground, rock streams, and dells, as well as accumulations of *glacial outwash* (alluvium deposited by streams of glacial meltwater). (After Davies, 1969)

races far beyond the glacier boundary. Drainage patterns were altered as the advance of the ice sheets blocked river systems and forced tributary streams to overflow into new paths. Portions of the Ohio and Missouri rivers were created as a series of spillways that carried water from one blocked drainage to another along the front of the ice.

Low temperatures peripheral to the ice sheets accentuated the breakup of rock by frost weathering. The ground adjacent to the ice became frozen solid, only the upper few inches thawing each summer. Consequently, slopes peripheral to the glaciated area were smoothed by the process of solifluction, producing the distinctive periglacial landscape (see Chapter 14). Periglacial processes were especially intense in the central European highlands, from France to Rumania, which were hemmed between the Scandinavian ice sheet to the north and the valley glaciers issuing from the Alps to the south.

The average thickness of the Pleistocene continental ice sheets is estimated to have been 1 to 2 km (0.6 to 1.2 miles), with a maximum thickness of about 4 km (2.5 miles). The weight of this ice—approximately 900,000 kg (1,000 tons) for each square meter of area—was enough to make the earth's semiplastic crust sink isostatically under the burden. The ice load depressed the crust by as much as 1,200 meters (4,000 ft) under the thickest parts of the continental ice sheets. Since the retreat of the ice sheets, the crust has been slowly returning to its original level, but full recovery is far from complete today. Parts of Scandinavia and North America are currently rising as much as 2 cm per year, with perhaps 100 to 200 meters (approximately 300 to 600 ft) of further uplift yet to be accomplished in some areas.

For ice sheets to grow on the land, water must be removed from the seas. Consequently, a world-wide lowering of sea level accompanies every period of glacial advance. During the Pleistocene glaciations, sea level was lowered 100 to 150 meters (330 to 500 ft), based on the estimated volume of the ice sheets, higher groundwater levels, and water stored in Pleistocene lakes. This drop is evidenced by the fact that river valleys extend into submerged coastal shelves down to a depth of about 100 meters (330 ft) below present sea level. Likewise, the average depth of the outer edge of the continental shelves is about 130 meters (430 ft), which coincides with estimates of Pleistocene sea lowering. Thus the continental shelves are regarded by some oceanographers as terraces produced by wave erosion during Pleistocene sea-level changes.

The lowering of the sea caused rivers to cut downward and considerably expand their valleys near their previous outlets. The return of the sea to its present level drowned these enlarged valleys, producing broad, deep estuaries that now form some of the world's best harbors, such as those at New York, Norfolk (Virginia), San Francisco, Seattle, Anchorage (Alaska), Leningrad, Lisbon, Manila, and Tokyo–Yokohama. If all the existing glacial ice were to melt away completely, sea level would rise another 70 meters (230 ft), which would submerge most of the world's largest cities and send arms of the sea far up the heavily populated valleys of the Mississippi, Ganges, Indus, Nile, Yangtze, and Yellow rivers, as well as many other interior lowlands.

The oceans' waters are shallow along most coastlines. If sea level were to drop by 130 meters (430 ft), as it did during the Pleistocene, a considerable amount of new land would be exposed, as shown in Figure 15.3, p. 464. The Florida peninsula would double in area, for example, and land bridges would join Ireland, England, and France, as well as most of the islands of Indonesia and Malaysia. The exposed land undoubtedly played a role in prehistoric times by providing new migration routes that permitted humans and animals to pass freely from one landmass to another, and low-lying refuge areas that enabled plants to survive the rigors of glacial climates. Peat deposits and mammoth teeth have been found in fishing grounds on the continental shelf off New England, indicating that these were subaerial rather than submarine plains during the Ice Age.

Origin of the Ice Ages

The mechanism that initiates and terminates periods of glaciation remains a puzzle. Ice sheets are in a delicate balance between growth and retreat, and relatively small changes in climate can cause the balance to shift either way. Some of the hypotheses advanced to account for climatic change include changes in the positions of the continents and in ocean circulation patterns due to sea floor spreading, the effect of increased altitude of the landmasses after a period of geologic upheaval, and variations in the amount of solar radiation the earth receives—variations caused by changes in the earth's relationship to the sun, the atmosphere's content of carbon dioxide and volcanic dust, and the sun's energy output.

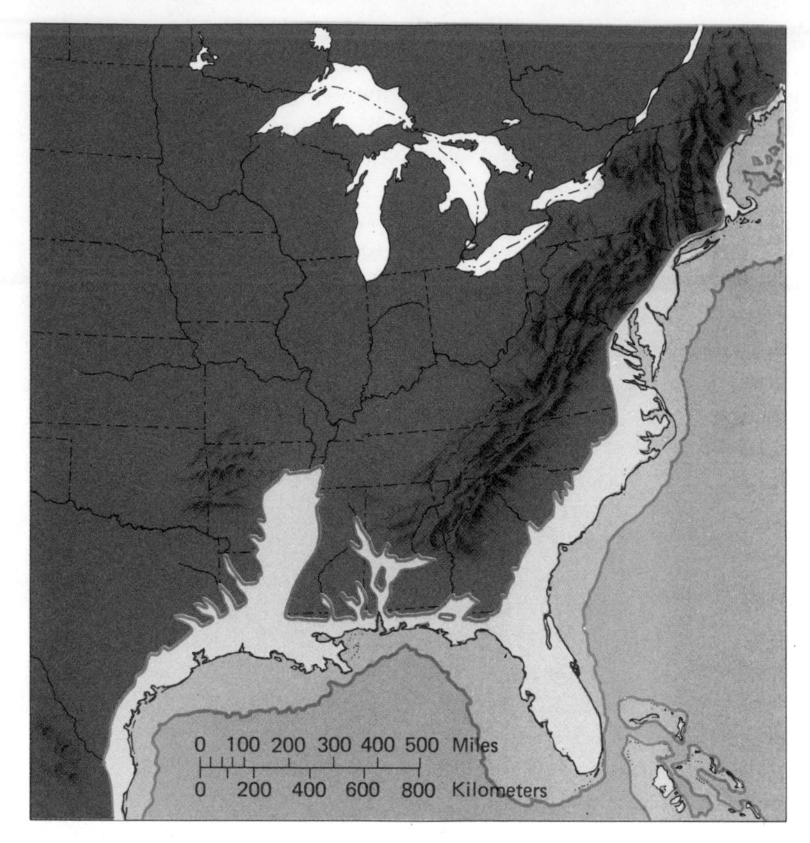

Figure 15.3 This map of the Atlantic and Gulf coasts of the United States shows the degree to which the storage of water as glacial ice affects the position of coastlines. The darkest blue tone represents ocean areas deeper than 130 meters (430 ft), which were unaffected by Pleistocene ice volumes. The middle tone shows presently submerged areas that were exposed as subaerial plains during low stands of the sea that accompanied maximum advances of Pleistocene ice sheets. The lightest tone indicates land areas that would be submerged if the present continental ice sheets in Greenland and Antarctica were to melt, raising sea level by about 65 meters (210 ft).

A successful theory must account for the growth and subsequent decay of ice sheets several times in less than 2 million years, as well as for glaciation that occurred hundreds of millions of years ago. It seems probable that several mechanisms periodically combine to produce unusual climatic cooling. The quest for a truly convincing explanation of the climatic changes causing the Ice Ages continues today.

ANATOMY AND DYNAMICS OF GLACIERS

Glaciers are formed above the annual *snowline,* the elevation at which some winter snow is able to persist throughout the year. Presently, the snowline reaches sea level only in polar areas; farther from the poles, the snowline is found at progressively higher elevations. In California's Sierra Nevada, the snowline is above 4,500 meters (15,000 ft) on slopes exposed to insolation and between 3,600 and 4,000 meters (12,000 and 13,000 ft) on shaded north- and east-facing slopes. Since this range rarely rises above 4,300 meters (14,000 ft), glaciers are present only in shaded spots on the cool northeastern faces of the highest crests. Northward, as the snowline descends, we find that even the south- and west-facing slopes of Mt. Rainier in Washington's Cascade Range rise well above the snowline, although the summit of Mt. Rainier at 4,392 meters (14,410 ft) is 26 meters (84 ft) lower than ice-free Mt. Whitney in the Sierras. Mt. Rainier's upper slopes are completely ice clad, with glaciers streaming to elevations as low as 1,200 meters (4,000 ft).

The Making of Glacial Ice

Excessive snowfall is what produces glaciers, but glaciers themselves are made of ice. What happens to snow that accumulates above the snowline of a mountain? Snow usually arrives at the surface in the form of lacy hexagonal ice crystals; but after a time, compaction, local melting and refreezing, and vaporization of the fine crystal lattice convert the initial delicate snowflakes to rounded granules of ice. The time required to produce these granules depends on the climate. In the extreme cold of Antarctica, snow remains fluffy for years. At higher temperatures, snow is wet and heavy, and the transformation of snow crystals to ice granules requires only a few days.

Newly fallen snow has a specific gravity (weight relative to water) between 0.05 and 0.15 because the uncompacted snowflakes have much air space between them. With time, and under the pressure of overlying snow, the ice granules formed from snowflakes gradually pack closer together. In midlatitude areas, after one summer the granular material has coalesced somewhat and attained a specific gravity of about 0.55. Such material is called *firn*. The appearance of firn signifies that the granules have begun to join with one another, but open interconnecting pores still exist among them. In tens or hundreds of years, depending on summer temperatures, the pores gradually disappear, until, at a specific gravity of about 0.85, the firn has become solid ice. Then, as air bubbles gradually disappear from the ice, the specific gravity increases to 0.9 or so, producing true glacial ice.

Since a layer of firn develops in one season in midlatitude regions, the upper parts of glaciers often show a succession of distinct annual layers. Dust, dirt, and pollen accumulate on the snow surface each summer and become locked in the ice to separate the annual layers. Old glaciers and ice sheets are repositories of climatic history. Although annual layers become indistinguishable with time, the ice can be cored and dated either by counting annual temperature-controlled variations in the content of the heavy isotope of oxygen (O^{18}) or by applying carbon-14 dating to the carbon dioxide dissolved in the ice. By studying the pollen and dust in each layer, scientists can gain knowledge of climatic variations through time. Periods of volcanic activity can be identified by deposits of volcanic ash in the ice. Even human pollution of the atmosphere is recorded: analyses of glacial ice in Greenland have substantiated the recent world-wide increase in atmospheric lead.

The Glacier Budget

Glaciers are dynamic systems, expanding, contracting, thickening in some sections, thinning in others, but constantly feeding ice forward to replace that lost by melting or evaporation below the snowline. Whether a glacier advances or recedes depends on its *mass budget*—the balance between the input of snow above the snowline and the loss, or *ablation*, of glacial ice by melting or evaporation below the snowline.

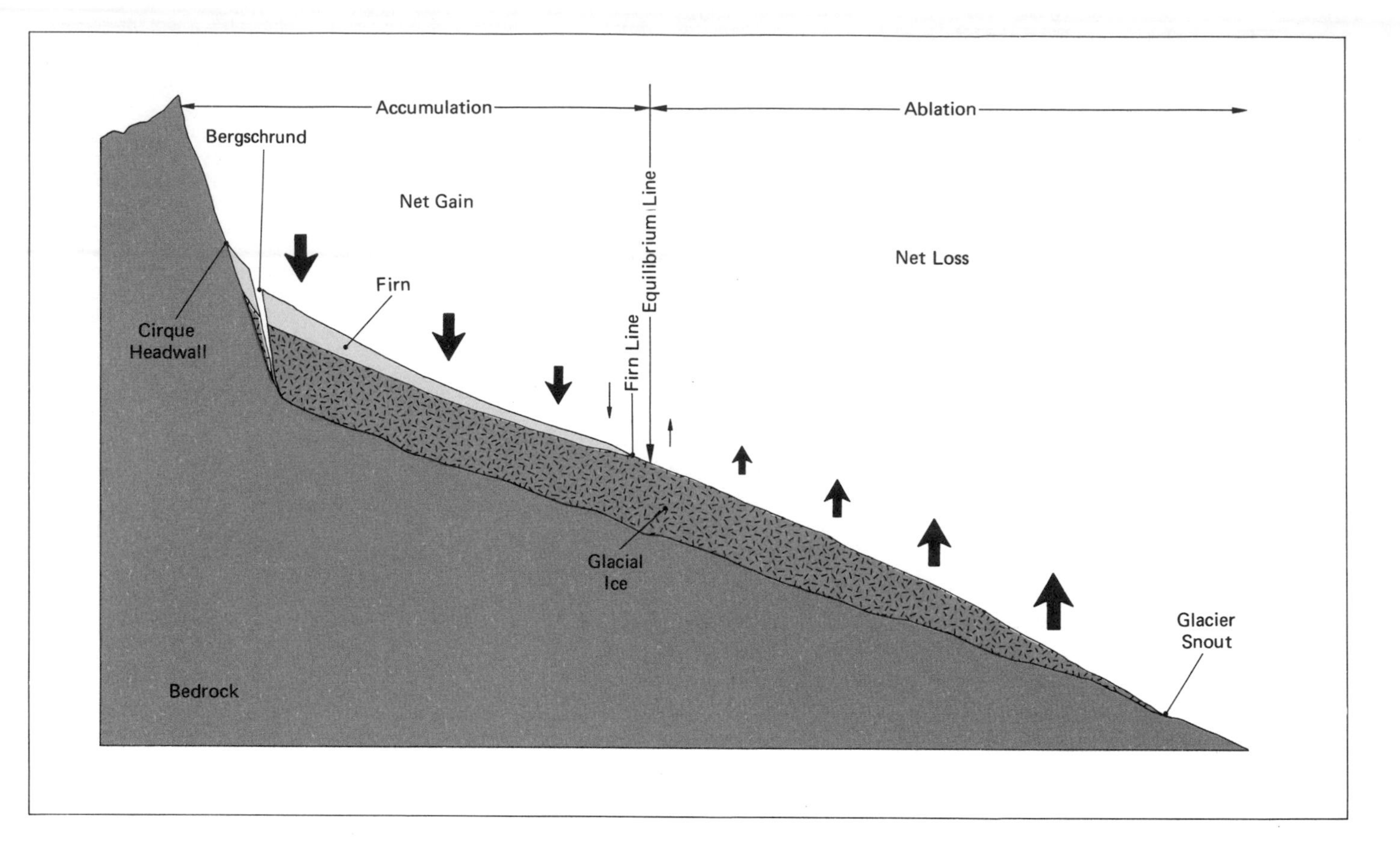

Figure 15.4 This schematic diagram illustrates the principal features of a cirque glacier. Arrows show relative gains and losses due to accumulation and ablation. The *firn line* marks the boundary between the annual accumulation of firn and glacial ice; the glacier's surface is white upslope from the firn line and bluish on the downslope side. The boundary between the region of accumulation and the region of ablation is slightly downslope from the firn line because melting and refreezing cause the lower edge of the snow to become superimposed ice. Alpine glaciers exhibit a deep crevasse, called a *bergschrund,* at their heads. It has been suggested that glacial erosion of the rock headwall occurs because water from melting snow and ice seeps into the crevasse and in some manner helps to pry rocks loose when it freezes.

All glaciers may be divided into two portions—an upper *accumulation zone,* which has a net annual gain of material, and a lower *ablation zone,* which has a net annual loss of material due to melting and evaporation (see Figure 15.4). The *equilibrium line* separates the two zones and is the line at which the input of snow exactly balances ablation losses. The equilibrium line more or less coincides with the *firn line,* which is the annual snowline on the glacier itself. Above the firn line, the surface of the glacier is snowy and white; below the firn line, the glacier has the blue-gray color of glacial ice. During the winter, snowfalls cause accumulation to exceed ablation on a glacier, but in summer, when snowfalls are less frequent or absent altogether, ablation is dominant. Thus glaciers are nourished in winter and decay in summer, as illustrated in Figure 15.5. Whether a glacier experiences net growth or decay depends on the balance between annual accumulation and annual ablation.

In an active glacier, ice is always moving forward to the downslope end of the glacier; this is true whether the glacier terminus (or *snout*) is advancing, retreating, or fixed in position. Where the position of the glacier terminus is stable from year to year, the amount of ice arriving at the margin by flow from the accumulation zone is exactly balanced by that removed by melting and evaporation. An increase in the rate of ice arrival or a decrease in the ablation rate will cause the glacier margin to advance. Such a glacier would have a

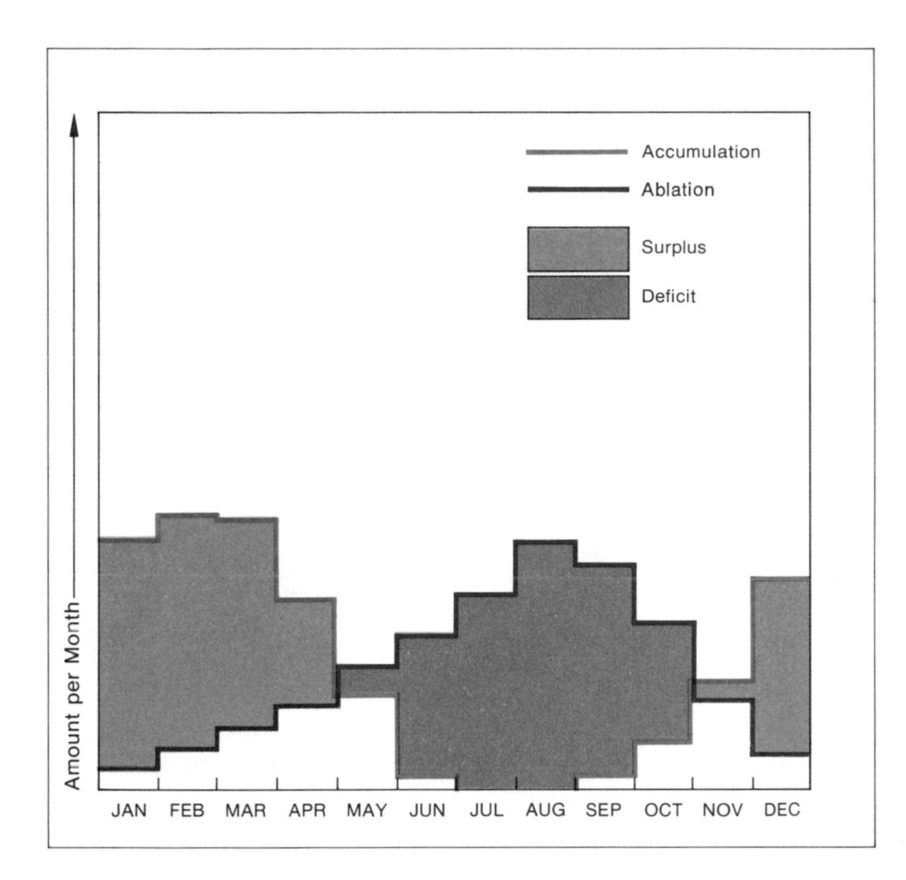

Figure 15.5 This diagram illustrates a typical glacier budget. During the winter months, the accumulation of new snow on the glacier exceeds the loss of ice and snow by melting and evaporation. During the warm summer months, ablation exceeds accumulation, and the glacier experiences a deficit. Note that both accumulation and ablation occur during most months; the glacier may be losing ice near its snout by melting while gaining new snow in its upper, colder reaches. The mass of the glacier increases during periods of surplus and decreases during periods of deficit. If the net annual surplus exceeds the net annual deficit on the average, the glacier grows and advances.

positive mass budget. If the rate of ablation exceeds ice replacement, the ice front retreats and a *negative mass budget* exists. Glaciers in the Canadian Rockies, Alaska, and Scandinavia lose a depth of about 12 meters (40 ft) of ice to ablation each year. To hold their position, this loss must be replaced by inflow from above the equilibrium line. Most of these glaciers have been retreating since about 1900, indicating that negative mass budgets have prevailed throughout the present century.

Movement of Glaciers

Glacial motion, like the flow of water, is a form of deformation of material under gravitational stress. However, due to the much greater viscosity of ice, its mode of flow is quite different from that of water. When a mass of ice attains a thickness of about 50 meters (150 ft), it begins to spread and flow outward as a consequence of its own weight. This internal deformation is very gradual, amounting to a few centimeters per day, and is accomplished by deformation both within and between the individual ice crystals composing the mass. As Figure 15.6, p. 468, suggests, the flow is laminar. The temperature of the ice composing most midlatitude glaciers is seldom much below the melting point, and the melting point varies with the confining pressure. Thus the movement of glacial ice tends to be abetted by repeated recrystallization as a consequence of partial melting under the stresses

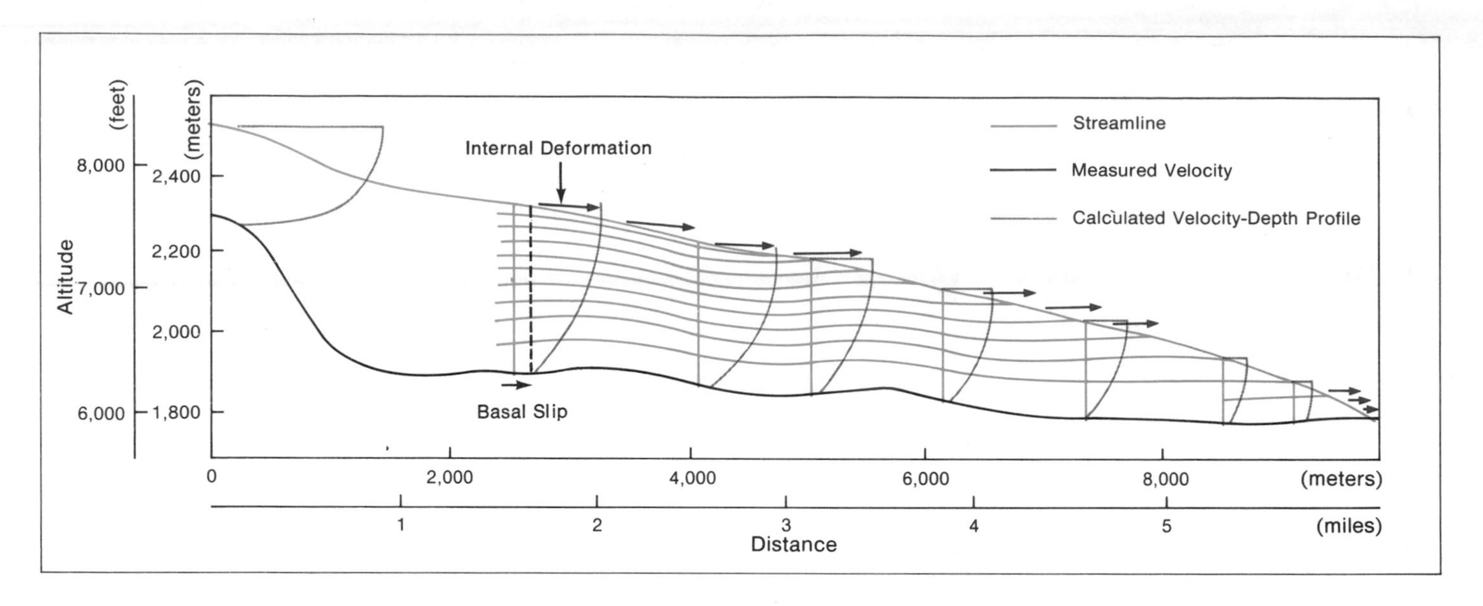

Figure 15.6 The movement of a temperate glacier is accomplished partly by basal slip over its bed and partly by internal deformation of the ice. As the diagram indicates, the process of basal slip causes the glacier to move as a whole, such that the distances traveled by the base and by the surface are the same. In the part of the motion due to internal deformation, however, the ice flows in such a way that the surface moves more rapidly than deeper portions of the glacier.

This diagram, in which the vertical scale has been exaggerated, shows the ice velocity at various points in the Saskatchewan Glacier. Only velocities at the surface of a glacier can be measured directly; the internal velocities indicated in the diagram are inferred. The measured velocity of this glacier's surface in its thicker upper portion is of the order of 400 meters (1,300 ft) per year. Near the snout, the velocity is approximately 100 meters (330 ft) per year. The streamlines, which show the trajectories of points within the ice, show that a point on the upstream portion of the glacier's surface tends to move slightly downward, whereas points on the downstream portion of the surface move slightly upward. Near the snout of the glacier, most of the movement is by water-lubricated basal slip. (After Meier, 1960)

within the ice mass. In addition to internal deformation, glaciers move by *basal slip,* in which the ice mass slides over its bed on a lubricating film of water produced by pressure melting at the base of the glacier. This permits more rapid flow than does internal deformation alone. Nevertheless, the cumulative rate of ice flow in a midlatitude alpine glacier is typically 1 meter (3 ft) or less a day.

The nature and rate of movement within a glacier are closely related to the temperature of the glacial ice, which may be at the pressure melting point or considerably below it. In *temperate glaciers,* meltwater produced by pressure melting or summer melt from the surface prevades the body of the ice. When this water refreezes, it liberates the latent heat of crystallization, which keeps the glacier "warm"—near the pressure melting point. Under such conditions, water-lubricated basal slip is important. In polar regions, however, summer melting is minimal, and ice temperatures remain far below the freezing point throughout the depth of the ice. The result is *cold,* or *polar, glaciers,* such as those in Antarctica, which are usually frozen to their beds, at least on their accessible margins. Glaciers of this type are relatively incapable of basal slip. The distinction between *thawed-bed* and *frozen-bed* glaciers is important in geomorphology because a glacier that is frozen to its bed tears out large masses of rock or frozen soil as it pushes forward, whereas a glacier that moves by basal slip over a thawed bed erodes mainly by abrasion—the filing action produced by the debris frozen into the base of the moving ice—and the crushing of rock projections under the pressure of the ice and its debris load. Complicating the situation is the fact that some midlatitude glaciers are of the polar type temperature-wise, and there is evidence that high-latitude ice sheets may be warm-based (thawed-bed) types behind their outer margins. This is important in the following discussion of glacial erosion.

Occasionally, an alpine glacier moving by basal slip exhibits behavior called a *surge,* in which the ice margin advances as much as 20 meters (65 ft) a day. There are several explanations for glacial surges. Some occur when the ice is literally floated away from its bed by an accumulation of meltwater under pressure, or when a lake dammed back by an ice stream seeps out under the glacier. Less spectacular advances occur following periods of unusually high snow accumulation. In such cases, the glacier thickens temporarily in the accumulation area, producing a "bulge" or wave of ice that subsequently moves through the glacier at a speed greatly exceeding the normal ice velocity. When the bulge reaches the glacier terminus, the ice front begins to push forward. This merely signifies that suddenly more ice is arriving than is melting away.

Internal deformation within a rapidly moving glacier produces strains that frequently cause the ice to split open in fissures known as *crevasses* (Figures 15.1, p. 461, and 15.7, p. 470). In confined valley glaciers, the fastest ice currents are at the center of the ice surface, with forward movement of the ice margins retarded by friction with the valley walls. The resulting shearing within the ice stream opens up crevasses, which are typical features along glacier margins. Where a valley glacier accelerates in speed due to steepening of the glacial bed, a series of crevasses will arc across the ice surface, being concave in the downstream direction. Sometimes the glacial bed steepens so abruptly that the ice surface becomes totally broken up in an *icefall* (Figure 15.7). Such areas are dangerous, since icefalls are composed of enormous unstable blocks that often tumble over with crushing force.

GLACIAL MODIFICATION OF LANDSCAPES

Like streams of water, flowing ice can modify landscapes by both erosion and deposition. However, glaciers seem to be far more powerful than running water as agents of geomorphic change. A thousand-meter-thick alpine glacier can remove rock material from both the floor and walls of a large valley or canyon, whereas a stream is confined to its valley floor. Furthermore, the special properties of glacial ice enable it to remove and transport debris using mechanisms not available to running water. Like streams, glaciers have their greatest erosional potential when their depth and velocity are highest, but unlike streams, they do not drop their load of debris wherever their velocity decreases. In fact, large-scale glacial deposition occurs only where the ice terminates or where great masses of ice stagnate with no forward motion at all.

Glacial Erosion

Glacial ice by itself has little abrasive capability. Ice cannot generate enough pressure to break away sound rock because slowly moving ice yields by melting as it is pressed against any solid obstacle. However, a glacier that is frozen onto its bed will tear out rock and soil material as it is thrust forward by the pressure of the ice behind it. Ice that freezes onto rocks and pulls them loose soon becomes armored with rock

Figure 15.7 These ice streams descending steeply from the summit of Mont Blanc in the French Alps are broken by crevasses. In the center of the photograph, the glaciers appear to be passing over convexities in their beds, producing small icefalls.

debris. Thus it is analogous to coarse sandpaper, with the ice acting as the glue that holds the abrasive particles in place. A rock-studded glacial bottom, or *sole,* is an effective file. The faster the glacier moves, the greater the number of abrasive tools scraped over a given surface, and therefore the greater the potential for erosion.

All glacial erosion can be attributed to the processes of *plucking* or *quarrying* (tearing out), *abrasion* (filing down), and *crushing* of rock projections in the glacier's path. Recent investigations suggest that plucking is best accomplished in the frozen-bed zone of a glacier, near the glacier margin where the ice is not thick enough to interfere with the constant flow of heat outward from the earth's interior. Thick ice traps this heat, causing the ground under the ice to thaw and the base of the ice to melt (see Figure 15.8). Abrasion is favored in the thawed-bed zone, where basal slip occurs due to the presence of water at the base of the ice. Crushing occurs in the thawed-bed zone when the subglacial surface is highly irregular, and it is concentrated on the downslope side of projections. In general, plucking roughens the surface under the ice and provides the glacier with tools for abrasion. Abrasion tends to smooth rock surfaces and erode grooves parallel to the direction of ice flow.

Glaciers probably remove far more material by plucking than by abrasion, particularly where the ice flow encounters unconsolidated (though frozen) soil and where well-developed jointing divides rock into blocks of movable size. Any process that assists in the breakup of rocks promotes erosion by glacial plucking. Mechanical weathering due to hard freezing is believed to occur in advance of expanding ice sheets, fragmenting the rock soon to be subjected to glacial erosion. Similarly, once a glacier erodes a critical thickness of soil and fragmented rock, the reduced confining pressure on the underlying rock may cause it to expand and crack from internal stresses. The jointing thus developed provides new fragmented rock to be quarried by the glacier.

The effects of glacial erosion can be seen on many scales, from an individual rock outcrop to an entire countryside. One of the first things a glacier accomplishes when it invades a new region is removal of the soil and weathered mantle. This demonstrates that glacial erosion is more potent than the combination of fluvial erosion and mass wasting, which removes all weathered material only where there is no vegetative protection of the surface. Glacial removal is accomplished by the quarrying process in the frozen-bed zone. The resulting exposure of the underlying bedrock enables the glacier to begin removal of solid material that would remain relatively untouched by other erosion processes. Abrasion produces scratching, grooving, and smoothing of rock faces (Figure 15.9, p. 472). Crescent-shaped chips and gouges, called *friction cracks,* produced by the pressure of rock upon rock are characteristic small-scale effects of glacial abrasion on individual rock outcrops. Sometimes rock surfaces are sculptured into smoothly contoured furrows and concavities; either flowing meltwater under the glacier or a slush of mud mixed with ice crystals could be the erosional agent responsible for such forms. The difference between abrasion and plucking is quite evident on single rock outcrops, which often display *stoss and lee* topography. As Figure 15.10, p. 473, indicates, abrasion tends to smooth and streamline the upslope (stoss) face of rock knobs, while plucking and crushing steepen and roughen the downslope (lee) side. Quarrying, crushing, and abrasion may occur sequentially at the same locality.

Severe ice erosion is evident where the landscape exhibits numerous rock basins that fill with water (see Figure 15.11, p. 473). Some rock troughs have been deepened more than 600 meters (2,000 ft) by glacial erosion, as in the case of Norway's Sogne and Hardanger fjords and the deep gouge of California's Yosemite Valley, each of which contains completely enclosed rock basins of at least this depth. Glacial erosion excavated some 400 meters (1,350 ft) of bedrock to

Figure 15.8 This is a theoretical model of the process of glacial erosion by a continental ice sheet that is frozen to its bed along its periphery (frozen-bed zone) but moves by basal slip where the ice is thicker (thawed-bed zone). According to this hypothesis, the outward flow of heat from the earth's interior (about 4 calories/sq cm/year at the surface) is impeded by a thick insulating ice cover, causing the base of the glacier to warm and the ground under the ice to thaw. Where the ice is thinner, the earth's internal heat passes outward freely, the ground is not warmed, and the ice freezes onto the surface. In the frozen-bed zone, glacial erosion is by plucking or quarrying of frozen material. In the thawed-bed zone, glacial erosion is by crushing and abrasion. In the frozen-bed zone, debris can be lifted well up into the ice by thrust faulting within the ice. In the thawed-bed zone, debris remains concentrated in the lowest portion of the glacier. (After Clayton and Moran, 1974)

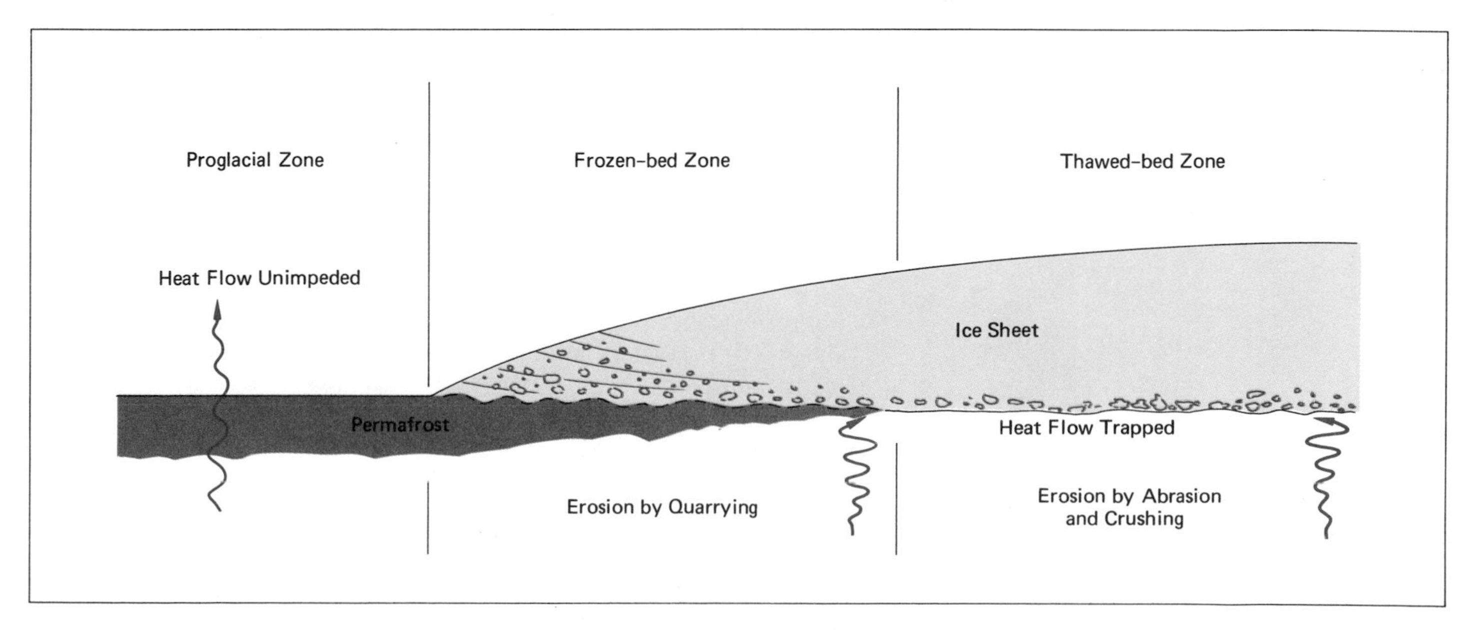

Figure 15.9 The ice of a glacier is armed with rock fragments of various sizes, which can abrade the bedrock under the flowing glacier. Glacial abrasion has polished and striated the rock shown in the photograph. The rock also exhibits a prominent gouge caused by the crushing and plucking action of the glacier.

form the basin now occupied by Lake Superior, and ice-scoured areas of Canada, New England, and Minnesota contain hundreds of thousands of large and small lakes produced by glacial quarrying.

The present rate of glacial erosion is not easy to measure because the area of active erosion is covered by ice. One method is to measure the rate at which sediment is carried away from the glacier by meltwater streams. Available data indicate that glacial erosion is 20 to 50 times more rapid than erosion of nearby unglaciated areas. Glacial abrasion of marks purposely chiseled into bedrock or of objects artificially placed in glaciers indicates removal rate of 1 to 40 mm (0.04 to 1.5 in.) per year, the variation being related to differences in rock resistance to abrasion.

Glacial Deposition

In the absence of glacial deposition, there would be no Denmark, no Long Island (New York), and no incredibly productive agricultural land in Saskatchewan, Manitoba, the Dakotas, Iowa, Illinois, Indiana, and Ohio. Both Denmark and Long Island are ridges of glacial debris that rise above sea level; the North American agricultural heartland is blanketed by glacial deposits that cover an older hilly landscape like that seen farther south in Missouri and Kentucky.

Glaciers have been called "dirt machines," being likened to conveyor belts that constantly feed rock debris forward to the glacial margin. Most of the debris carried by glaciers is concentrated in the lowest few meters of ice and, in the case of alpine glaciers, along the ice margins. Since flow in glaciers is laminar rather than turbulent, there is normally little upward or lateral transport of material into the body of a glacier. The exception is near the glacier margin, where the ice is thin and may be frozen to the glacial bed. Here the pressure of the thicker upslope ice may cause the glacier to develop thrust faults that push basal ice up over slower-moving marginal ice, thus thickening the debris zone (Figure 15.8, p. 471). At the glacier margin, the debris is released and either overridden by advancing ice or, if the ice front is stationary, piled into a ridgelike glacial dump. Even when the ice front is retreating, the "conveyor belt" remains in action, continuing to transport debris forward to the glacier margin. Forward transport of debris stops only when the ice thins to the point where flow cannot be maintained over obstacles in the glacial bed. When this occurs, the ice mass disintegrates by melting in place.

The general term for all types of material deposited directly or indirectly from glaciers is *glacial drift.* (The term dates from the time, about 150 years ago, when it was supposed that the debris covering so many landscapes in the northern hemisphere was left by the great flood of the time of Noah.) Glacial drift is quite variable in character, for it can be produced in many different ways. When ice advances, it eventually becomes so laden with debris that it begins to release some of its load, which is plastered over the subglacial surface in the form of *glacial till.* Material lodged directly by advancing ice is a distinctive

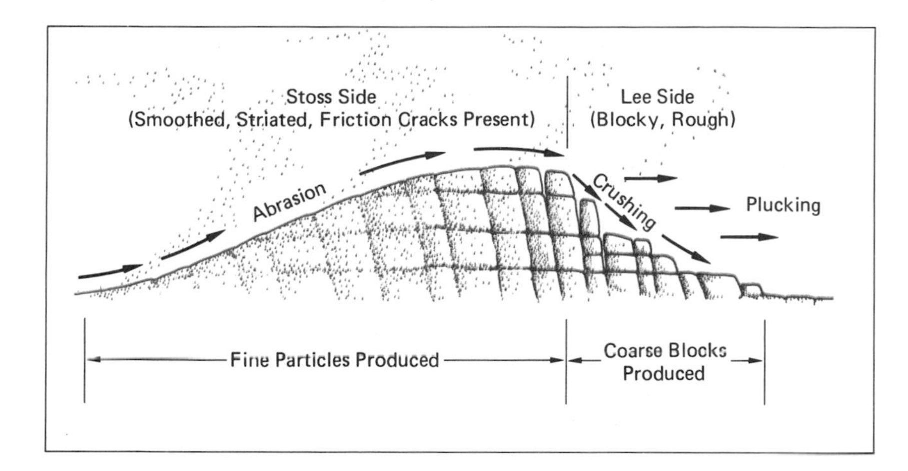

Figure 15.10 (left) In this diagram illustrating the characteristic action of glaciers on outcrops of jointed rock, the direction of glacier flow is from left to right. The upslope *(stoss)* side of the rock is smoothed and polished by abrasion, and the down-slope *(lee)* side is roughened as the glacier crushes projections and plucks large rocks from the outcrop.

Figure 15.11 (below) Ice scouring in crystalline rocks tends to produce multitudes of basins that later fill with water, as seen in this aerial photograph covering an area of about 26 sq km (10 sq miles) in northern Canada. White areas are frozen lake surfaces.

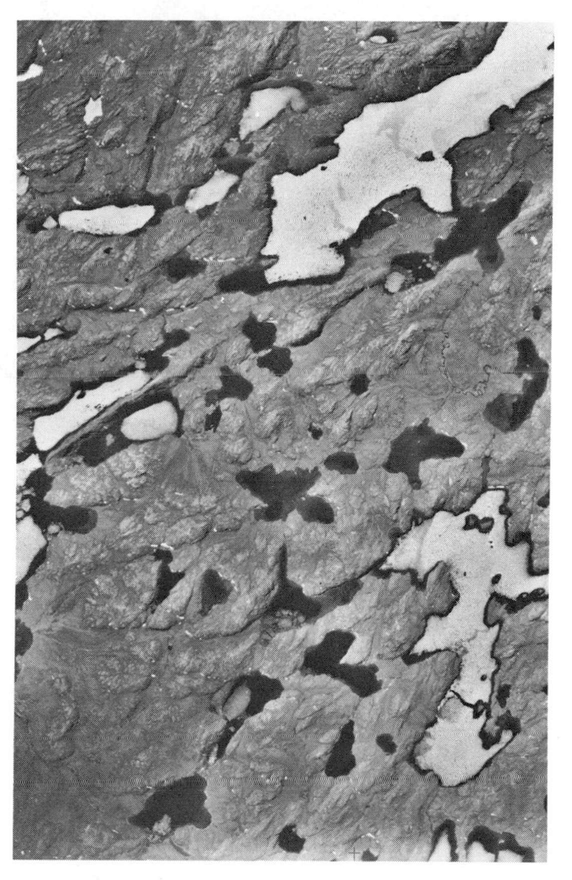

mixture of coarse and fine material, including fragments of all sizes from boulders to clay—indeed, it was once called "boulder-clay." Till is most distinctive by the absence of any form of sorting or bedding (see Figure 15.12a, p. 474). Usually some boulders within the till are faceted or striated by abrasion during transport.

When glacial ice has pushed to its limit and subsequently begins to retreat, its marginal portions sometimes stagnate, melting in place, with great volumes of debris collecting amid the chaotic topography of the decaying ice. Much of this debris is washed from the ice or deposited by meltwater streaming from the ice. The material is clearly sorted and crudely stratified into lenses and layers of silt, sand, gravel, and cobbles (see Figure 15.12b). As the supporting ice surfaces melt away, this material slumps and slides so that its layers become deformed. Such sorted, bedded, and deformed deposits are called *ice-contact stratified drift.*

When the ice margin retreats, enormous volumes of debris are carried away by braided meltwater streams. The coarser particles are deposited in vast alluvial aprons extending out from the periphery of the ice. The material deposited is called *glacial outwash.* Till, ice-contact stratified drift, and outwash are all regarded as glacial drift.

Occasional components of glacial drift are enormous surface boulders known as *glacial erratics.* These range up to the size of houses and are rock types not otherwise found in the local area. Some have ridden hundreds of kilometers with the ice, and their known sources help establish the flow patterns of continental ice sheets.

Specific landforms are associated with the various types of glacial drift. The general term for a deposit of drift lodged by glacial ice is *moraine.* Moraine includes both till and ice-contact stratified drift, but not outwash deposits. An *end moraine* is a ridge of drift that marks the position where the glacier terminus remained stationary for a period of time or where it pushed up a ridge of drift before retreating. Where a morainic landscape lacks conspicuous linear features, it is called *ground moraine,* a *till plain,* or a *drift plain.* Where outwash

a

b

Figure 15.12 (a) At this location in California's Sierra Nevada, *glacial till* lies atop sheeted granitic bedrock. The till contains fragments of all sizes, from boulders to clay, and is distinctive by its lack of bedding or sorting.

(b) In contrast to till is the deposit of *ice-contact stratified drift*, in which discrete lenses of sand, gravel, and cobbles are visible. This stratified drift is exposed in an end moraine in central New York State.

collects around the margin of a continental ice sheet, the result is an *outwash plain.* Outwash plains that are confined by valley walls in areas of both alpine and continental glaciations are termed *valley trains.*

The depositional topography produced by glaciers includes a host of small-scale landforms made by advancing, retreating, and stagnating ice. As we shall see in the following pages, those produced by alpine glaciers differ noticeably from those created by continental ice sheets.

LANDFORMS RESULTING FROM ALPINE GLACIATION

Alpine glaciation has created some of our planet's most magnificent scenery—from the inhospitable towering masses of Mount Everest and its satellite peaks to the beautiful ice-clad pinnacles and knife-edged summits of Switzerland. Alpine glaciation either partially or totally remodels the upland topography of mountain regions and also transforms the valleys leading away from the high summits.

The basic element in alpine glacial sculpture is the cirque, the amphitheaterlike rock basin that is the first product of mountain glaciation (see Figure 15.13). The precise mechanics of cirque formation remain unclear. What is evident is that alpine glaciers are born where climatic changes lower the snowline below mountain summits. Occupied by firn that becomes transformed into glacial ice, hillside depressions and the heads of stream valleys are deepened and expanded into steep-walled bowls with ice-scoured floors. In some mountain regions, such as the Uinta Mountains of Utah and the Bighorn Range of Wyoming, there are areas in which glacial erosion never proceeded beyond this point. Such areas are pockmarked with steep-walled upland amphitheaters that are separated by broad mountain ridges. Elsewhere, cirques expanding toward one another from the opposite sides of ridges have reduced the ridges between them to serrate knife-edges of rock, known as *arêtes* (see Figures 15.14, p. 476, and 15.15, p. 477). Where three cirques intersect, three sharp-crested arêtes result, which come together in a steep rock pinnacle, or *glacial horn.* The spectacular soaring peaks of alpine topography are horns formed by cirque growth from three or four sides. Each face of such a peak is a remnant of a cirque headwall.

In most previously glaciated mountains, ice streams slowly cascaded from the cirques in which they were born and invaded the valleys below. The results are nearly as impressive as glacial modification of uplands. Valley glaciers both deepened and oversteepened the prior stream valleys, converting them to *glacial troughs* (see Figures 15.15 and 15.16, p. 478). The interlocking spurs that reflect the sinuous paths of deeply incised mountain streams were trimmed back. Due to the rigidity of the ice creeping through the valleys, these spurs bore the brunt of glacial scour. Thus valleys were expanded and simplified into open grooves as they were being deepened. The result is a valley cross section resembling a wide V with the sharp point replaced by a smooth curve, the cross-sectional shape being known as a *cate-*

Figure 15.13 The steep-walled basins in this photograph of the Wind River Mountains, Wyoming, are *cirques* formed by the quarrying action of glacial ice. The steep walls of a cirque are believed to be produced by collapse when frost action at the bottom of the bergschrund undercuts the cliffs. Small lakes, or *tarns,* are present in the cirques and appear in the glacially scoured area in the foreground. This part of the Wind River Range preserves large areas of the rolling preglacial topography between the separate cirque basins.

nary curve (such as that made by a flexible cord held at the two ends and hanging freely between).

Due to the greater volume of ice channelled, glacial deepening of major valleys is more rapid than deepening of tributary valleys by smaller ice streams. Retreat of the ice leaves the tributary valleys hanging along the margins of the major glacial troughs, and streams occupying these valleys must descend to the floor of the trough by picturesque cascades, or *hanging valley waterfalls,* as shown in Figure 15.15.

Because glaciers do not evenly excavate the underlying weathered or jointed rock, valley glaciers commonly produce irregular longitudinal valley profiles. The result may be a *glacial stairway,* with high steps and waterfalls in the center of the trough and lakes, known as *tarns,* in deeply eroded basins. Strings of *paternoster lakes,* such as those visible in Figure 15.13, are quite common in alpine topography, being named for their resemblance to the beads on a rosary.

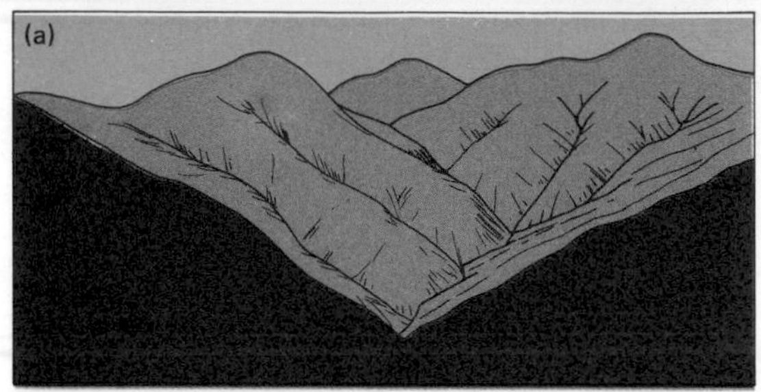

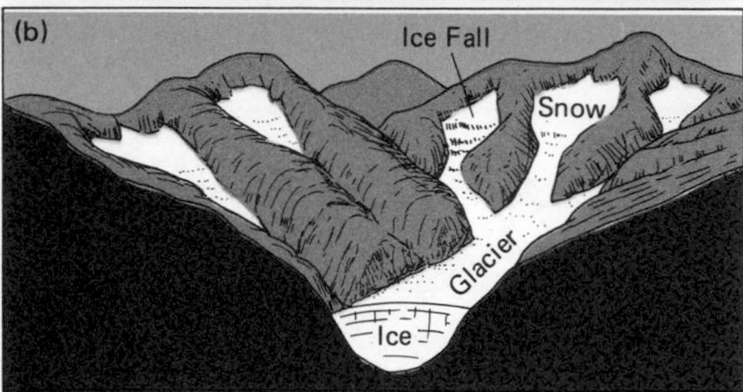

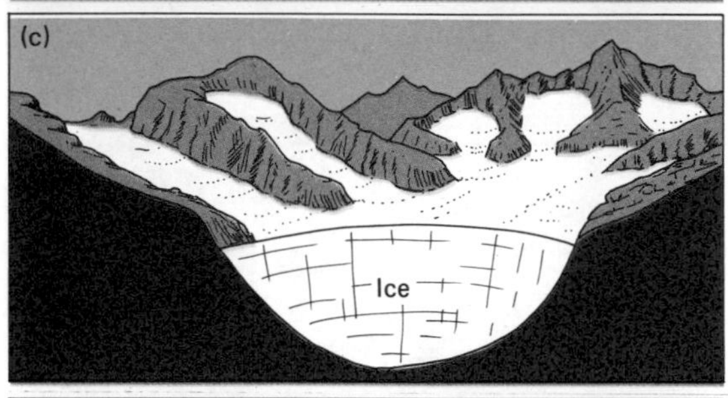

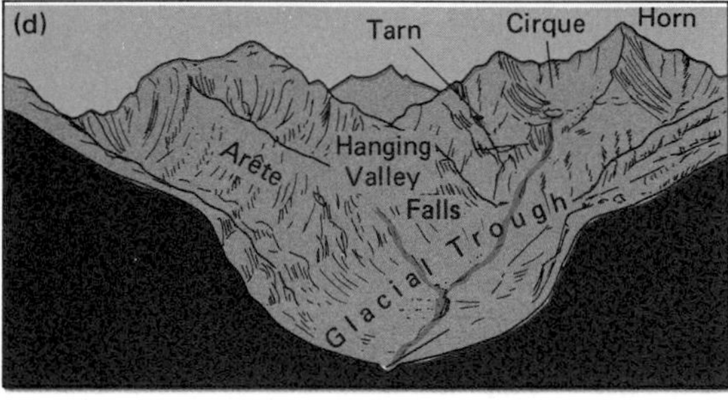

Figure 15.14 This series of diagrams shows the transformation of a fluvially dissected mountain region (a) into a glacially eroded landscape (d). At the highest levels, snowfields develop into *firn* basins that gradually produce glacial ice (b). Cirque growth by glacial scouring and frost action creates sharp crested *arêtes* that come together in horns (c). At lower elevations, ice streams convert sinuous fluvial valleys into open *glacial troughs.* Valleys are deepened and expanded, and their walls are made much steeper by glacial erosion (c). The greater modification of trunk valleys by large ice streams leaves smaller tributary valleys hanging along their margins, producing *hanging valley* waterfalls and cascades (d). Uneven glacial scouring of jointed bedrock excavates basins that later fill with *tarn lakes.* The result is alpine scenery (d). (After James, 1966)

Downward erosion by valley glaciers is best displayed in the fjord landscapes of Greenland, Norway, eastern and western Canada, Alaska, Chile, and New Zealand (Figure 15.16, p. 478). *Fjords* are glacial troughs that have been cut far below present sea level, resulting in deeply penetrating arms of the sea surrounded by rock walls 1,000 meters (3,300 ft) or more high. Proof of the glacial origin of fjords lies in the fact that their longitudinal profiles include many deep basins that could not have been excavated by fluvial processes. Norway's famed Sogne Fjord reaches 180 km (112 miles) inland. Many fjords, such as the Sogne, have depths of more than 1,000 meters below sea level. This is not surprising, because a glacier with a density of 0.9 entering the sea (with a density of about 1.0) can continue to erode downward until it is nine-tenths submerged, after which it will float. Thus a thousand-meter-thick ice mass will still have the potential to erode its bed in 800 meters (2,700 ft) of water. When thinking of the depth of fjord excavation, one must remember that although sea level was some 130 meters (430 ft) lower during the maximum glacial advances, the land surface was depressed even more than this under the weight of the ice. Present-day glaciers are continuing to fashion fjords in Greenland, Antarctica, Alaska, and the Canadian Arctic Islands.

The depositional landforms produced by alpine glaciation are no match in scenic grandeur for alpine erosional forms, but they are nevertheless quite conspicuous. The largest depositional features produced by valley glaciers are *lateral moraines* built along the sides of the ice streams. These are narrow-crested ridges of till banked against valley walls or issuing from mountain canyons, where they frequently rise more than 300 meters (1,000 ft) above surrounding lowlands. Due to the waxing and waning of glaciers during the Pleistocene, smaller end moraines built by successions of valley glaciers commonly make several concentric loops, which may close across valleys, impounding lakes between the successive morainic crests (Figure 15.17, p. 479). The larger end moraines mark the terminus of a major ice advance and are called *terminal moraines.* The smaller morainic loops behind them mark pauses in the retreat of the ice—periods when the ice held its ground for a few years—and are termed *recessional moraines.* The climatic cooling that has occurred since the postglacial temperature maximum about 5,000 years ago has produced glacial readvances in all of the world's high mountains, creating *neoglacial* (post-Pleistocene) *moraines,* seen mainly in cirques. These are ridges usually composed of coarse rock rubble, lacking the matrix of fine debris seen in most Pleistocene till. Present glaciers have retreated well behind their various neoglacial moraines as a consequence of the warming trend of the twentieth century.

Alpine glaciers also produce large amounts of glacial outwash, which aggrades stream valleys beyond the glacial margin (Figure 15.18, p. 479). The deposits consist of sand and gravel, changing to silt at greater distances from the sediment source. In periods between glaciations, such as the present, the sediment supply is cut off, stream energy increases due to the decrease in load, and streams are able to incise the outwash deposits, converting them to terraces. It was obser-

Figure 15.15 (top left) Severe frost weathering of the rock walls above cirque glaciers may erode the walls between adjacent cirques to the narrow, ragged rock ridges, or *arêtes,* seen here in the French Alps.

(bottom left) The deepening of the principal stream valleys in mountain regions by glacial erosion is more rapid than deepening of tributary valleys that channel smaller ice streams. After the glaciers melt away, the tributary valleys may be left as *hanging valleys,* which join the deep glacial trough high above the valley floor. The waterfall in the photograph is a stream falling from a hanging valley in Yosemite National Park, California.

(right) The Matterhorn in Switzerland is the classic example of a *horn* formed by erosion where the headwalls of three or more cirques intersect.

vation of the degree of weathering and soil development on a series of outwash terraces along rivers flowing north from the Alps that in 1909 allowed Albrecht Penck, the father of Walther Penck (see Chapter 11), to establish the history of multiple glaciations in the Alps.

LANDFORMS RESULTING FROM CONTINENTAL GLACIATION

The dual effects of continental glaciation are epitomized by the Canadian Shield area on one hand, and the landscape of the midwestern United States north of the Ohio and Missouri rivers on the other. The two landscapes present a very conspicuous contrast.

The Canadian Shield is a vast area of ancient crystalline rocks, granites and gneisses especially, surrounding Hudson Bay and extending outward to the great line of lakes—Great Bear, Great Slave, Athabaska, Winnipeg, Superior, Huron—and the St. Lawrence River. These lakes, which occupy depressions created by glacial excavation, separate the area of major ice scouring in North America from the area dominated by glacial deposition. The region between Hudson Bay and this vast circle of large lakes has been stripped of its soils and weathered mantle, and has been subjected to hundreds of thousands of years of glacial quarrying and abrasion. The resulting landscape is one of ice-scoured rock knobs and quarried-out depressions strewn with rock-bound lakes of all sizes and shapes. Any good atlas map of Canada makes clear the pitting of this landscape and the resulting count-

Figure 15.16 (left) Valley glaciers follow preexisting fluvial valleys as they flow toward lower elevations. The glacier scours the valley floor and walls, removing irregularities and forming a smooth *glacial trough.* The glacial trough in the photograph is the Lötschental in Switzerland.

(right) If a valley glacier reaches the sea, it can continue to scour out its channel below sea level until the ice is nine-tenths submerged, when it will finally float. When the glacier melts or retreats, it leaves a deep rock-walled ocean inlet, or *fjord,* such as Milford Sound, New Zealand.

less large and small lakes. Stream patterns are rambling and unsystematic, the old patterns having been deranged by ice erosion. In general, drainage lines are somewhat rectangular, following joints in the exposed crystalline bedrock.

There are deposits of till within this glacially stripped region, but they are thin and discontinuous. The glacial deposits are sandy or gravelly and consist of ice-contact stratified drift resulting from ice stagnation, or *fluted moraine* that has been striated by active ice (see Figure 15.19). Denuded of soil and with a short growing season, the region has little productive potential other than the growth of undemanding coniferous forest. However, the forests, the many lakes, and the picturesque scenery of this wilderness give it great recreational potential. Similar landscapes occur south of the St. Lawrence River in New England and northern New York, where crystalline and metamorphic rocks also prevail. Here the relief is greater than over much of the Canadian Shield due to a more mountainous preglacial landscape.

A somewhat different erosional topography appears in central New York State south of Lake Ontario, where the continental ice sheet moved into a fluvially dissected plateau composed of sedimentary rocks. In this region, drainage divides were breached and stream valleys aligned with the major ice currents were greatly deepened, producing fjordlike configurations in New York's picturesque Finger Lakes. Even though 300 km (180 miles) inland, the floors of some of these glacial troughs were deepened well below sea level. Other valleys oriented across the line of ice movement were clogged with till derived from the soft rocks in the Great Lakes lowlands. Glacial retreat from this landscape left recessional moraines and flooded the valleys and glacial troughs with outwash, producing much flat agricultural land that was not a part of the preglacial landscape.

Over much of the Midwest, by contrast, glacial deposition proceeded on an immense scale. Lobes of the continental ice sheets scraped through the soft rocks of the lowlands now containing lakes

Figure 15.17 (left) Convict Lake, on the east face of the Sierra Nevada, California, lies in a glaciated valley dammed by deposits of glacially transported rock debris, known as *moraines.* Although these landforms are about 20,000 years old, they still bear the unmistakable marks of glacier action.

Figure 15.18 (below left) Outlet glaciers from a highland ice cap on Baffin Island in the Canadian Arctic have built the outwash fan in the foreground. Note the discoloration of the sea produced by the arrival of silt-laden glacial meltwater.

Figure 15.19 (below right) Vast areas of northern Canada have a surface of *fluted moraine* produced by active ice flow that leaves a thin layer of striated drift. Note the many lakes in this landscape. Some of these result from erosion; some from irregular deposition of glacial drift.

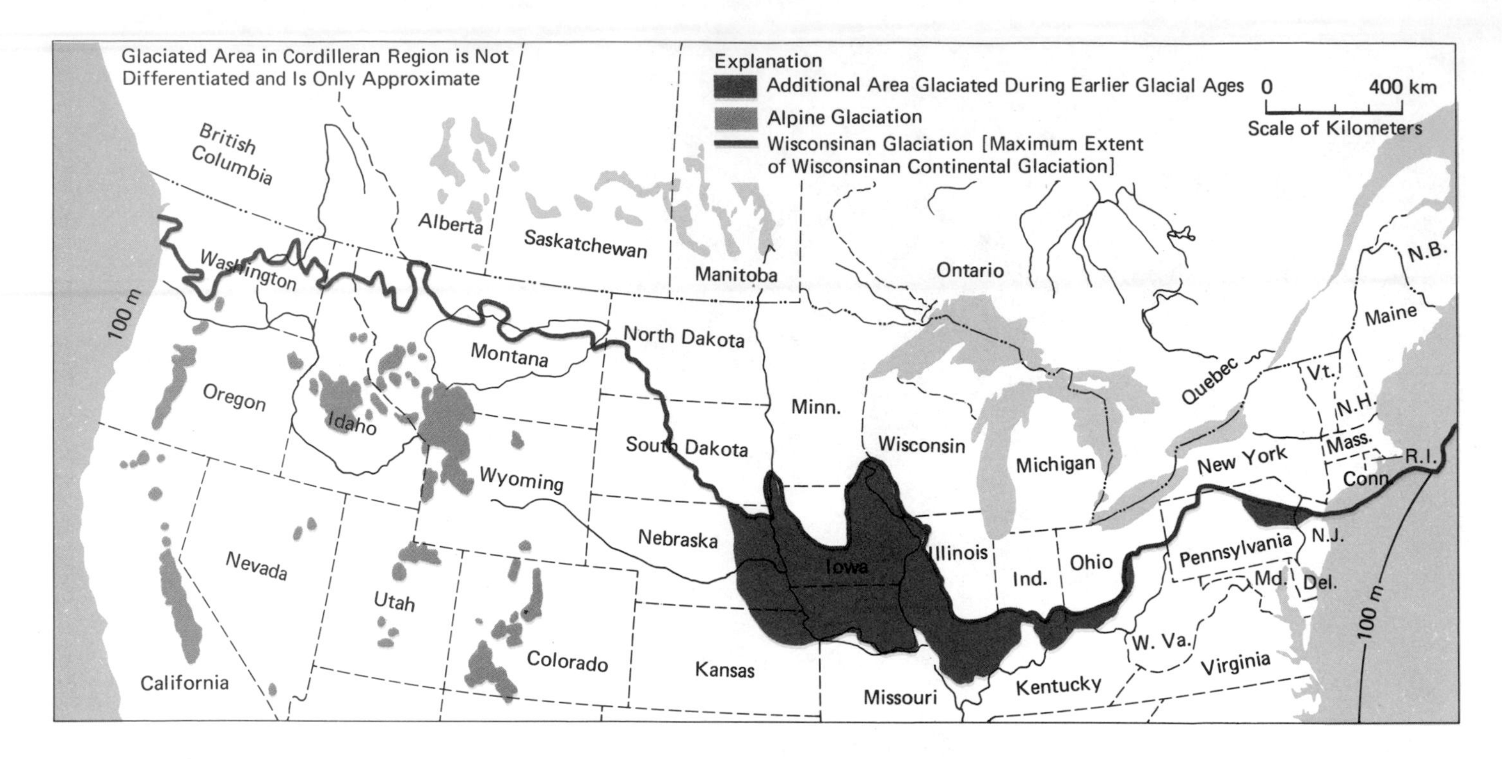

Figure 15.20 This map shows the southern limit of Pleistocene continental glaciation in North America, as well as the major areas of alpine glaciation. The limit of the latest Wisconsinan glaciation, which terminated about 10,000 years ago, is differentiated from the limit of earlier Pleistocene glaciations, the most recent of which ended about 120,000 years ago. (After Flint, 1971)

Erie, Huron, Michigan, and Superior and spread the resulting fine rock debris over the North American heartland from Alberta and Saskatchewan to eastern Ohio (see Figure 15.20). As noted previously, drift deposits in this area are deep enough to obscure completely the preexisting topography. An example of the preglacial landscape is preserved in the so-called Driftless Area of Minnesota, Iowa, and Wisconsin—a triangle with Dubuque (Iowa), Winona (Minnesota), and Wassau (Wisconsin) at its apexes. The Driftless Area is a region of low hills including some delicate rock formations (see Figure 15.21) that could not have survived ice scouring. Although probably touched by the earliest Pleistocene glaciation, this area remained unglaciated throughout the majority of the Pleistocene due to its slightly higher elevation, which caused the Lake Superior and Lake Michigan ice lobes to diverge around it.

The depth of glacial deposits in some areas of Illinois can be measured in tens of meters, to a maximum of 100 meters (over 300 ft) where preglacial stream valleys have been completely submerged in glacial drift. The old stream gravels in the larger of the buried valleys are of considerable importance as aquifers, supplying well water for towns and cities in the region. From Alberta to Ohio, vast areas are gently rolling till plains, while others are nearly flat outwash plains (see Figure 15.22). The drift surfaces are diversified by a host of small-scale landforms composed of different types of glacial deposits. The largest relief forms are end moraines, which mark positions where the ice front lingered as debris was conveyed forward to it. Both terminal and recessional end moraines are present. They are arranged in southward-bulging arcs, indicating that the ice front was lobate in

form (see Figure 15.23, p. 482). The end moraines usually consist of irregular rolling (rather than hilly) topography, including many depressions that produce marshes and lakes. Most end moraines rise less than 100 meters (300 ft) and are 10 km or more (over 6 miles) wide. Thus end moraines produced by continental glaciation include huge volumes of drift but are much less conspicuous than the sharply defined lateral moraines built by alpine glaciers. The largest end moraines formed by continental ice sheets are *interlobate moraines,* created where two adjacent ice lobes brought debris into a central dumping ground between them.

Drift plains show a variety of landforms smaller than the great end moraines. Particularly distinctive landforms produced by advancing ice are *drumlins* (see Figure 15.24 , p. 482), which are elongated streamlined hills composed of every type of drift. Drumlins occur in great swarms, with the long axis of each one nicely aligned with the direction of ice flow. These streamlined forms clearly result from glacial readvance into an area of prior glacial deposition. They are a type of longitudinal shear feature. The greatest drumlin fields in the United States are located in central New York and eastern Wisconsin, but the best known individual drumlins are Bunker Hill and Breed's Hill in Boston, Massachusetts, where British and Colonial troops clashed in 1775 in the first real battle of the American Revolutionary War.

Advancing ice also creates "rippled" till surfaces in many areas, known in North America as *washboard moraines.* These are transverse shear features. The ridges, which are transverse to the ice flow, are only a few meters high and tens to hundreds of meters apart. Their

Figure 15.21 Castle Rock, in the Driftless Area of southwestern Wisconsin, is formed by erosion of sandstone. It would have been destroyed if the ice sheets of the late Pleistocene had invaded this region.

Figure 15.22 This aerial view of a Wisconsinan drift plain in northern Illinois shows the overall flatness of the topography produced by glacial deposition. Undrained depressions are occupied by lakes. The small-scale surface irregularities of this drift surface are barely visible.

Figure 15.23 (top right) End moraines in the midcontinent region of North America. Each moraine, shown in the dark tone, marks a position at which the front of the continental ice sheet remained stationary long enough to build up an exceptional accumulation of glacial drift. Some of these end moraines are *terminal,* marking the farthest forward advance of the ice at a certain period; many are recessional, indicating long pauses in the retreat of the ice front. The lobate form of the ice front is conspicuous in the pattern of end moraines. (From the Geological Society of America)

Figure 15.24 (upper left) This hill, shaped like the reverse side of a spoon, is a *drumlin,* photographed in western Ireland. The blunt end of the drumlin faces toward the ice currents that modeled it. Drumlins are composed of drift deposits that are subsequently eroded by a readvance of glacial ice.

(bottom left) The long sinuous ridge in the center of the photograph is an *esker* in northern Canada near Great Bear Lake. This deposit marks the course of a subglacial stream channel and is composed of ice-contact stratified drift, principally sand and gravel. (From the Canada Department of Mines and Resources)

Figure 15.25 (bottom right) The left half of this view, in the state of Washington, shows the irregular topography produced by the stagnation of the ice border in this region. The low hills are composed of material collected between masses of "dead ice," which subsequently melted, forming the depressions between the hills.

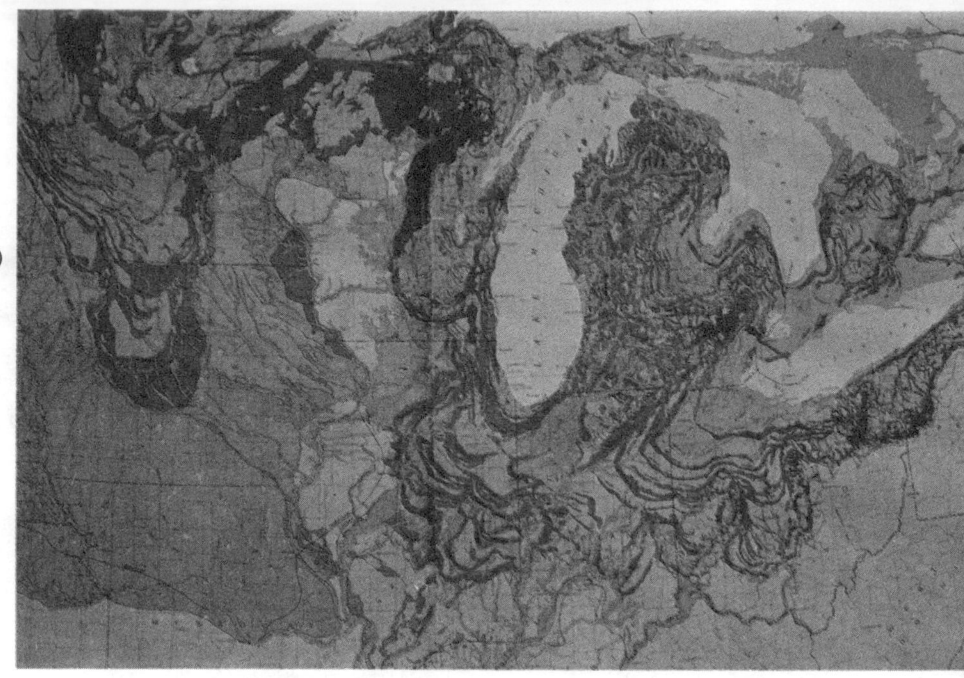

origin remains a matter of speculation, although it is widely held that they form at the base of thrust planes where active ice overrides stagnating ice.

Ice-contact stratified drift is seen in landforms that result from ice stagnation, where the peripheral zone of a glacier melts in place with no forward ice flow occurring. Stratified drift is deposited unsystematically around such "dead ice," collecting in crevasses and depressions on the ice surface (see Figure 15.25). Final disappearance of the ice leaves irregular mounds of drift, called *kames.* Closed depressions in drift surfaces mark places where ice masses have melted after being wholly or partially buried by drift. These depressions, forming ponds and marshes, are known as *kettles.* Many end moraines are *kame moraines* composed of *kame and kettle topography.* Where ice masses have been floated out onto outwash plains and partially buried, they eventually melt, producing *pitted outwash plains.* Stratified drift also may accumulate along the margins of stagnating ice lobes that are hemmed in by valley walls. Disappearance of the ice then leaves *kame terraces,* consisting of slumped and deformed drift with kettles scattered over the terrace surfaces.

Figure 15.26 This diagram represents the typical landforms encountered on drift plains in the midwestern United States and Canada. The length of this block would be on the order of 100 km (60 miles). The sizes of features are exaggerated for the sake of legibility (drumlins are only 1 to 3 km long). The drift covers an irregular bedrock topography. The largest features are the *end moraines* and *interlobate moraines* constructed of drift delivered to the ice margin during a long period in which the ice front is stationary. The topography of the end moraines is often of the *kame and kettle* type, composed of ice-contact stratified drift. In front of the end moraine is commonly encountered a flat *outwash plain* with a scattering of *kettle lakes,* indicating positions of melted ice masses that were partly or fully engulfed by sand and gravel outwash carried away from the ice front by meltwater streams. Behind the end moraine may be an area of ice stagnation topography, including *eskers* built of sand and gravel deposited by streams flowing under the melting glacier. The otherwise regular surfaces of *till plains* may be diversified by *washboard moraines,* consisting of inconspicuous low ridges transverse to the ice flow, and streamlined *drumlins* aligned with the ice flow and composed of glacial drift that has been scoured by advancing ice. In front of this block (to the left), the entire sequence might be repeated, indicating an older position of the ice margin.

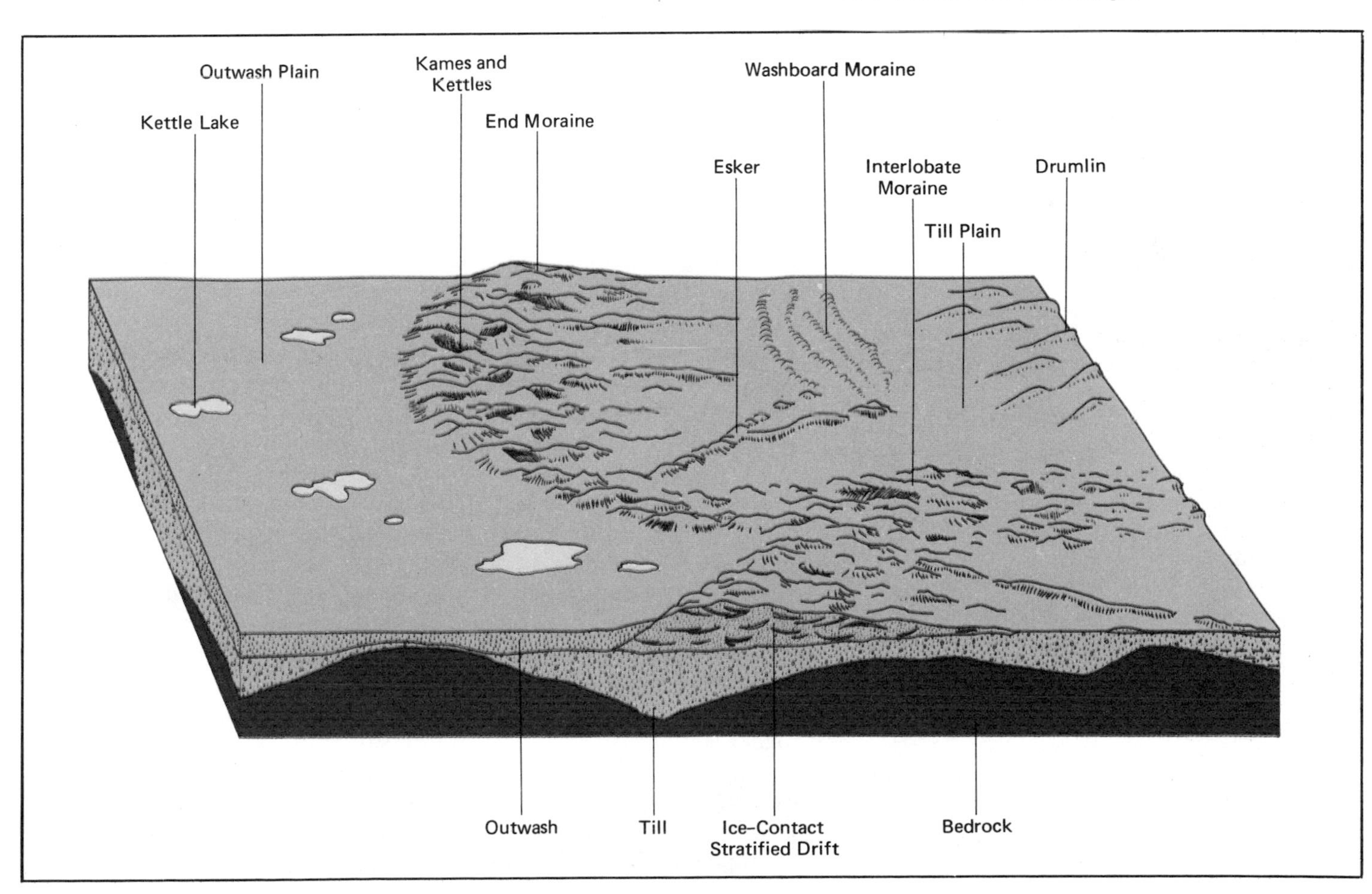

Further Reading

Andrews, John T. *Glacial Systems: An Approach to Glaciers and Their Environments.* North Scituate, Mass.: Duxbury Press, 1975. (191 pp.) This treatment of glaciers, glacial landforms, and glacio-eustatic and glacio-isostatic effects is concise and advanced; it is for students with an understanding of mathematics.

Coates, Donald R., ed. *Glacial Geomorphology.* Binghampton: State University of New York, 1974. (398 pp.) This symposium volume includes several outstanding papers on glacial models and glacial erosional and depositional landscapes. Many of the concepts presented are new and original.

Embleton, Clifford, ed. *Glaciers and Glacial Erosion.* New York: Crane, Russak, and Co., 1972. (287 pp.) This collection of previously published articles contains many highly readable "classics" in the literature of glacial geomorphology.

Embleton, Clifford, and **C. A. M. King.** *Glacial Geomorphology.* New York: Wiley, 1975. (573 pp.) This is a very complete treatment of all aspects of glacier structure and dynamics, processes of glacial erosion and deposition, and the landforms resulting from each.

Flint, R. F. *Glacial and Quaternary Geology.* New York: Wiley, 1971. (892 pp.) Flint's book is the standard American source on all aspects of glaciation and the Pleistocene. The treatment is broader but less detailed than that of Embleton and King.

Paterson, W. S. B. *The Physics of Glaciers.* Elmsford, N.Y.: Pergamon Press, Inc., 1969. (250 pp.) Paterson's concise text concentrates on the structure and dynamics of glaciers. The treatment is advanced.

Post, Austin, and **E. R. LaChapelle.** *Glacier Ice.* Seattle: The Mountaineers, 1971. (110 pp.) The highlight of Post and LaChapelle's book is its excellent large-format photographs of glaciers. The text, dealing with the various types and structures of glaciers, is also excellent.

Perhaps the most unusual forms related to ice stagnation are *eskers,* irregular sinuous ridges of ice-contact stratified drift (see Figure 15.24b, p. 482). Eskers are stream deposits accumulated in tunnels under or within stagnant glacial ice. Eskers many tens of kilometers in length and 10 to 30 meters (30 to 100 ft) high are especially common in New England and eastern Canada, and many can also be seen in Michigan, Minnesota, and Wisconsin. Some eskers in Finland are hundreds of kilometers in length, indicating the stagnation of enormous expanses of inactive ice. Figure 15.26, p. 483, illustrates typical landforms found on drift plains in North America.

The latest deposits (originating 10,000 to 20,000 years ago), which are well developed in Wisconsin and therefore termed *Wisconsinan* in age, are the only ones to preserve the original glacial depositional topography, including multitudes of small depressions that hold lakes or marshes. All older till sheets, known in order of increasing age as *Illinoian* (termination approximately 120,000 years ago), *Kansan* (approximately 600,000 years ago), and *Nebraskan* (perhaps 1.5 million years ago), are undulating, well-drained, stream-dissected plains. The differentiation of the various Pleistocene deposits is based on their degree of weathering and soil development, and the actual numbers and dates of the various glacial stages within the Pleistocene is not a matter of total agreement at present. Analysis of the deposits left by the Wisconsinan glaciations suggests that each of the major glacial stages included several ice advances and retreats over periods of many tens of thousands of years.

SUMMARY

Alpine glaciers and continental ice sheets form at high altitudes and high latitudes where cool summers permit some winter snow to persist through the following warm season. Snow accumulates in the cooler source regions of glaciers and is transformed into firn, which in turn gradually develops into dense glacial ice. Beyond the accumulation area, ice ablates from the glacier by melting and evaporation. Relative rates of accumulation and ablation govern glacial expansion and contraction. Glaciers move under gravitational stress, both by internal deformation and by slippage over their beds. Glaciers are efficient agents of erosion, and they transport large amounts of rock debris, which they obtain by picking up loose material from weathered surfaces and by abrading and plucking or quarrying bedrock. Abrasion is accomplished by rock fragments held in the ice, and quarrying proceeds where the glacier is frozen to its bed. Glacially eroded rock debris is deposited in moraines composed of till and ice-contact stratified drift. Meltwater from glaciers carries glacially eroded rock debris far beyond the glaciated area.

At present, glaciers are found only in high mountains or in the higher latitudes, but during cold intervals within the Pleistocene, glacial ice covered vast areas of North America and Europe and also expanded in high mountains on all of the continents. Pleistocene glaciers left a distinctive imprint on the land; many landforms, such as cirques, drumlins, and eskers, are produced only by glacial action.

Glaciers are responsible for such features as the basins occupied by North America's Great Lakes, the fjords of Norway, and the till plains of the North American heartland. While glacial erosion stripped vast areas of their soil, the subsequent deposition of glacial till and outwash created extremely favorable conditions for agriculture in warmer areas that are climatically better suited for intense human utilization.

Review Questions

1. What is the evidence that present climates are considerably cooler than those that have prevailed throughout most of the earth's history?
2. What different types of glaciers are found in high mountain regions?
3. Pleistocene climates produced several unusual effects on the earth's landscapes in addition to glacial erosion and deposition. What were these effects?
4. How is it possible for a glacier margin to retreat when there is a constant forward flow of glacial ice?
5. What is the principal difference between temperate and cold glaciers? What is the geomorphic importance of this distinction?
6. Why does ice-scoured bedrock commonly display asymmetric forms that are steeper on one side than on the other?
7. It is usually possible to determine the limit of the latest Pleistocene glacial advance by merely looking at a map of moderate scale. Why?
8. How do glacial till, glacial drift, and moraine differ?
9. Sketch three successive stages in the development of a horn peak.
10. How do glacial troughs differ from the youthful stream valleys usually seen in mountain regions?
11. How do lateral moraines produced by valley glaciers differ from end moraines constructed by continental ice sheets?
12. What landforms are encountered in areas of deposition by continental ice sheets? How do those landforms produced by active ice differ from those created in areas of ice stagnation?

Schultz, Gwen M. *Glaciers and the Ice Age; Earth and Its Inhabitants During the Pleistocene.* New York: Holt, Rinehart and Winston, 1963. (128 pp.) This small book outlines the nature of the Ice Ages, glaciers and glacial landforms, and traces the evolution of the theory of the Ice Ages. A portion of the book is devoted to a discussion of the emergence of man during the Pleistocene.

Sharp, Robert P. *Glaciers.* Condon Lectures, University of Oregon, Eugene, Oregon, 1960. (78 pp.) This small paperback is a good nontechnical introduction to the principles of glacial flow.

Sugden, David E., and **Brian S. John.** *Glaciers and Landscape: A Geomorphological Approach.* London: E. Arnold, 1976. (376 pp.) Although the subjects covered in this book are generally similar to those treated by both Flint and Embleton and King, Sugden and John's more selective treatment is highly original and thought-provoking.

Early Morning After a Storm at Sea by Winslow Homer. (The Cleveland Museum of Art; gift from J. H. Wade)

Atmospheric energy generates water waves that transport energy over vast distances, expending that energy against coastlines, driving them back in some places and building them out elsewhere, molding them to accord with prevailing wave climates. The shoreline is a place of constant activity and change—one of the best locations for observing the dynamic relationship between landforms and the processes that create them.

SIXTEEN

Marine Processes and Coastal Landforms

SINCE THE INVENTION OF ships with sails and oars, which probably occurred about 3,000 B.C., cities based on trade have been situated on coastlines. However, archaeologists find the apparent distribution of ancient seafaring settlements puzzling. Many coastal areas with fertile agricultural hinterlands have no townsite associated with them. The ancient Mycenaeans of southern Greece, for example, were great traders and founders of maritime cities, yet several coastal regions in their territories lack any trace of Mycenaean settlements. Did they never exist, or have they vanished?

A recent geomorphic study seems to illuminate this question. Analysis of the coastal landforms of southern Greece indicates that waves have battered this shoreline back some 800 meters (0.5 mile) in the past 6,000 years. Thus it is quite possible that Mycenaean settlements were indeed present, but were long ago chewed away by the sea. At one lonely point on the Greek coast, some 200 km (125 miles) southwest of Athens, there is dramatic evidence of the erosive power of the sea. Here a fragment of mosaic pavement extends to the edge of

Figure 16.1 Wave erosion of weak chalk formations along the coast of England has caused sea cliffs to retreat into a rolling, fluvially dissected landscape. In such settings, the rate of cliff retreat may be as much as 2 meters (6 ft) a year.

a 15-meter-high (50-ft) cliff overlooking the sea. This is the last remnant of the floor of a church built during the sixth century, almost 2,000 years *after* the Mycenaean period. Clearly, wave erosion has driven the cliff back to (and through) the church, which no doubt was originally constructed at a safe distance from the cliff.

The coastal zone is a particularly dynamic region on our planet, one of continual energy conversion where wind-initiated water waves and currents ceaselessly transform the edges of the land, molding shorelines to accord with motions of the sea surface (see Figure 16.1). Furthermore, two-thirds of the world's population resides close to the sea and is constantly affected by its behavior and its many influences. In this chapter, we shall examine the nature of the sea and its interaction with the land.

MOTIONS OF THE SEA SURFACE

Much of the sea's fascination, as well as its power to shape coastlines, proceeds from the fact that it is never still. Being a liquid with little internal cohesion or resistance to deformation, its surface is easily set in motion by the natural stresses constantly applied to it. The friction created by wind produces *waves* that give relief to the sea surface and *currents* that transport ocean water in slowly moving streams (see Chapter 5). The gravitational attractions of the sun and moon, along with the centrifugal force created by rotation of the earth-moon system, produce the rise and fall of the water surface that we call *tides.* Waves, currents, and tides are daily phenomena along the earth's coasts; however, there have been much larger and longer-term changes in the level of the sea, significantly altering the sizes and shapes of the continents themselves.

Long-Term Changes in Sea Level

If a map of the world could have been made 20,000 years ago, its land and water areas would hardly have resembled those of today, as Figure 15.3, p. 464, indicated. There was then no separation between Asia and North America, which were joined by a bridge of land some 2,000 km (1,200 miles) broad. The Atlantic coasts of the Americas lay about 100 km (60 miles) farther east, and one could have strolled from France to Ireland through a tundra landscape, having to cross only a few large streams. Humans saw these conditions, for our species was already distributed around the world at the height of the Pleistocene Ice Age, when the level of the seas was some 130 meters (430 ft) lower than at present as a consequence of water storage on the land in the form of the great continental ice sheets.

Sea level appears to have reached its present position approximately 6,000 years ago (see Figure 16.2), following the complete disappearance of the North American and Scandinavian continental ice sheets. Melting of the existing Antarctic and Greenland ice sheets would raise the level of the oceans another 70 meters (230 ft) or so, drowning the coastal plains and river valleys regarded as home by the majority of the earth's population. Elevation of the seas to the height

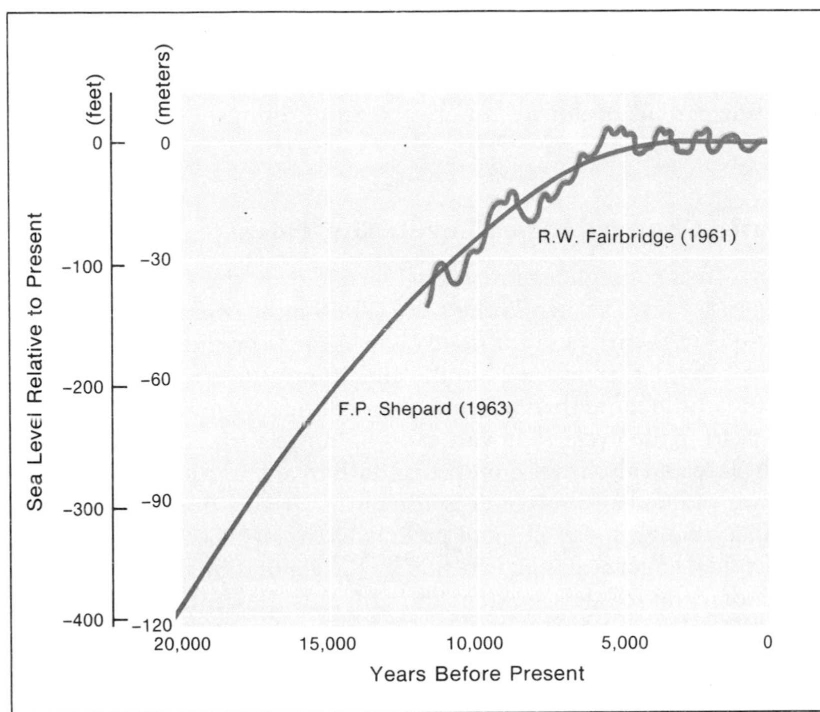

Figure 16.2 As this graph indicates, the level of the sea has risen by at least 120 meters (400 ft) since the late Pleistocene glacial maximum. Melting of the ice sheets returned large quantities of water to the oceans. The estimates of two different scientists are shown, one (Shepard) assuming a fairly steady rise of sea level as the ice sheets melted, and the other (Fairbridge) suggesting that sea level fluctuated considerably during its rising trend. Numerous coastal landforms that were exposed 20,000 years ago are now deep under water; many of the world's major harbors are located in wide river valleys that were submerged by the rising seas. (After Bird, 1969)

of Mount Ararat, where Noah's ark was said to have come to rest, would require more than twice the water existing in the earth's hydrosphere and is therefore a physical impossibility.

How stable is the level of the sea today? This is a matter of some disagreement among scientists. Three different views are held: that the sea is still rising very slowly, that its level has oscillated some 1 to 4 meters (3 to 13 ft) during the past 6,000 years, and that sea level has been relatively steady over this time. Since there have been noticeable fluctuations in climate, causing changes in the size of alpine glaciers, even over the past 500 years, it seems that slight oscillations of sea level are to be expected, as suggested in Figure 16.3. What makes the current trend difficult to establish firmly is the fact that many coastal regions are slowly rising, sinking, or tilting tectonically, so that changes in sea level relative to the land over historic time vary considerably from place to place. Thus there are two phenomena that can affect the level of the sea relative to the land: *eustatic* changes, meaning simultaneous world-wide changes in sea level due to changes in the volume of seawater or the capacity of the ocean basins, and *tectonic* changes, meaning local changes in the elevation of the land due to crustal motion. The latter includes isostatic adjustments due to the weight of seawater lying on the continental shelves and the compaction of sediments deposited on the sea floor.

We have already discussed, in Chapter 15, the indirect benefits of Pleistocene eustatic changes in the creation of estuarine harbors and temporary migration routes for plants and animals. In the same vein, alluviation at the heads of estuaries created by eustatic or tectonic events builds deltas into them, producing flat plains that are ideal for

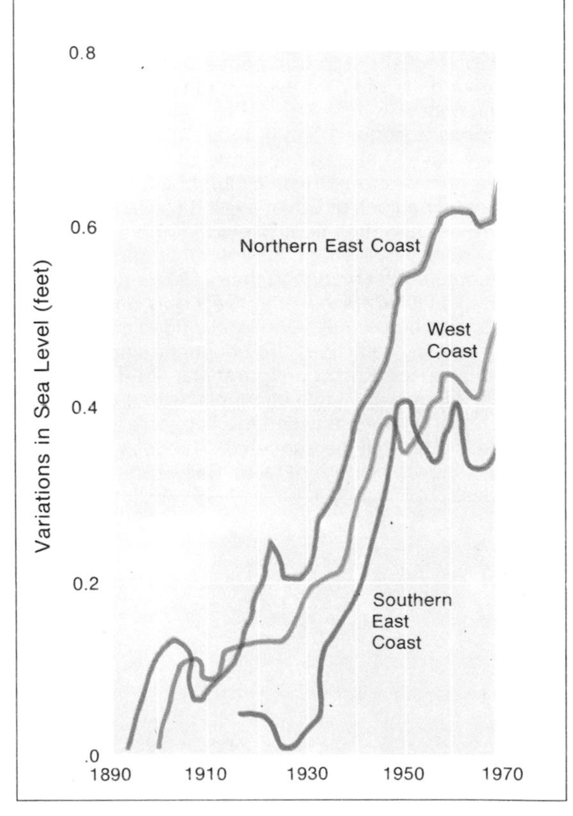

Figure 16.3 The height of sea level relative to the coastlines of the United States has been increasing during this century. Part of the increased height is attributed to local lowering of the land. (National Oceanic and Atmospheric Administration)

Figure 16.4 (opposite top left) (a) Tidal bulges of water, here greatly exaggerated in height, are produced on the earth because of the gravitational effect of the moon and sun and the centrifugal force produced by rotation of the earth-moon system. As the earth rotates, two tidal bulges sweep across it, producing twice-daily high tides in general. Detailed analysis shows that tides on the earth also possess a once-daily component produced because of the positions of the moon and sun relative to the earth's rotation.

(b) The highest high tides and the lowest low tides, or spring-tides, occur during new moon and full moon, when the moon and sun act together.

(c) The lowest high tides, or neap-tides, occur during the moon's first and last quarters, when the moon and sun are arranged as shown in the diagram.

(opposite top right) This diagram illustrates how the gravitational attraction of the moon creates one tidal bulge on the side of the earth facing the moon, while the centrifugal force produced by rotation of the earth-moon system creates a tidal bulge on the side of the earth farthest from the moon. Three successive positions of the earth and moon are shown, numbered 1, 2, 3, with arrows indicating the path followed by the center of each. Note that the earth and moon rotate together about the center of gravity of the earth-moon system at *C,* which lies just below the earth's surface on the side closest to the moon. It is the centrifugal force *(CF)* created by the earth's rotation about this off-center axis that forces water toward the point on the earth that is farthest from the moon.

(opposite bottom) The graphs show the tide heights relative to average sea level at Dover, England, and Victoria, British Columbia, for a period of 15 days. The tides at Dover exhibit the usual pattern of twice-daily high tides. Note that the heights of the daily tides vary through the month because the earth, moon, and sun change their relative positions. The tides at Victoria show effectively only one high and one low tide per day. Once-daily tides can occur where the twice-daily tides are unusually small because of special land configurations. If tides arrive at a location by way of two channels, with a relative time delay of 6 hours between the separate flows, the twice-daily tides may be canceled. The tides at many locations show a mixture of twice-daily and once-daily tides.

subsequent agricultural development. Much of the lower Mississippi River valley consists of a complex of alluvial infills of slightly differing elevations, resulting from Pleistocene eustatic reversals that produced alternating periods of downward erosion and infilling.

Daily Changes in Sea Level: The Tides

Of more direct consequence to coastal dwellers are the smaller daily changes in sea level known as the *tides.* Tides most commonly cause two periods of rising water, or *flood tides,* and two periods of falling water, or *ebb tides,* in every 24 hours and 50 minutes—the sea rising slowly for a little more than 6 hours, then falling for another 6 hours, then repeating the cycle. The vertical tidal range varies greatly from place to place and also from month to month in any one location, and the timing of flood and ebb tides is also quite variable from one place to another, though consistent and predictable in any specific locality.

The tidal effect is actually created by the gravitational attractions of the moon and to a lesser extent the sun, plus the centrifugal force generated as the earth and moon rotate about a common point at the center of gravity of the earth-moon system (see Figure 16.4).

The point on the earth that is closest to the moon is most strongly affected by the moon's gravitational force; water on that side of the earth is pulled toward the moon, creating a tidal bulge. Moving away from this point toward the earth's center, the moon's gravitational pull diminishes; and at the earth's center, lunar attraction becomes balanced by an opposing force that pulls in the opposite direction. This critical opposing force is the centrifugal effect of rotation of the earth-moon system. It is the balance between these opposing forces that keeps the earth and the moon from either crashing together or flying apart. While the moon revolves around the earth and the earth rotates on its own axis, the earth and moon also revolve together about a common center of gravity for the earth-moon system as a whole. This common center of gravity falls just inside the mass of the earth on the side facing the moon, far from the earth's center. Thus, in addition to rotating on its axis, the earth also revolves like an eccentric gear, as Figure 16.4 illustrates. The enormous excess mass on the far side of the uncentered pole of rotation sets up centrifugal force directed away from the moon. This forces ocean water outward in the second tidal bulge, 180° from the tidal bulge created by lunar gravity. Of course, the two tidal bulges draw water from the areas between them, so that low tides occur in regions at 90° from the moon (or when the moon is near the horizon). The earth rotates through each of these watery bulges every 24 hours and 50 minutes.

The most obvious clue to the cause of tides is the fact that the interval between successive high tides is 12 hours and 25 minutes, causing high and low water to recur 50 minutes later each day. The moon likewise lags behind on its course by 50 minutes each day. This relationship was first noted by Arab geographers about 1,000 years ago; 700 years later Isaac Newton explained ocean tides as being caused by the gravitational attraction of the moon. The moon's mass also pro-

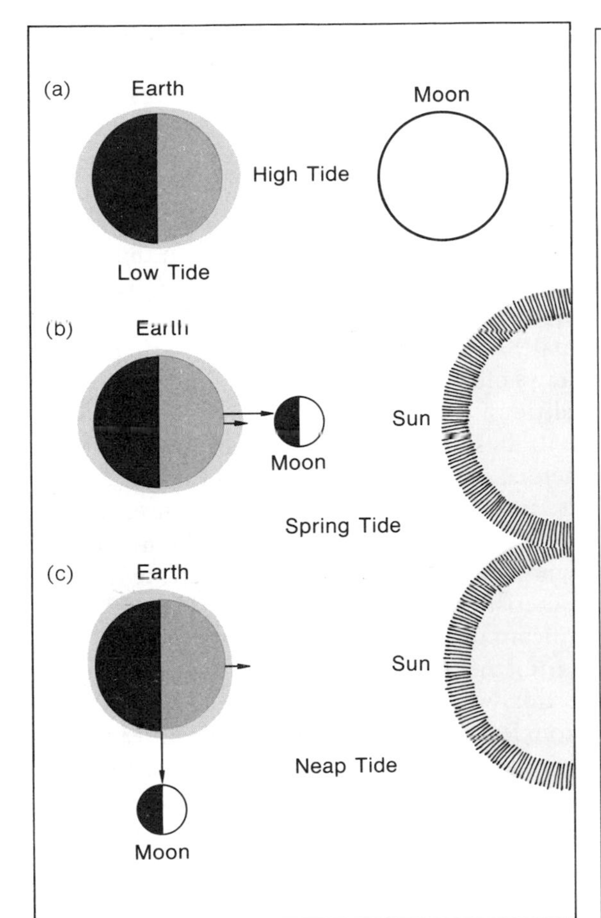
(a)
Earth
Moon
High Tide
Low Tide
(b)
Earth
Sun
Moon
Spring Tide
(c)
Earth
Sun
Neap Tide
Moon

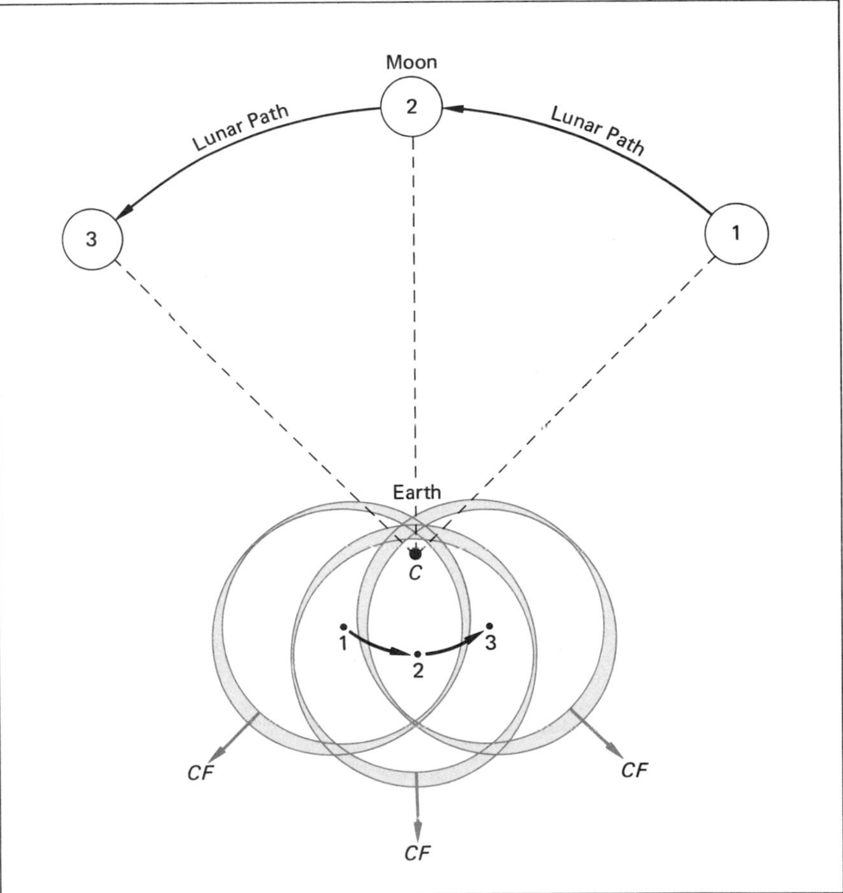
Moon
2
Lunar Path
Lunar Path
3
1
Earth
C
1
2
3
CF
CF
CF

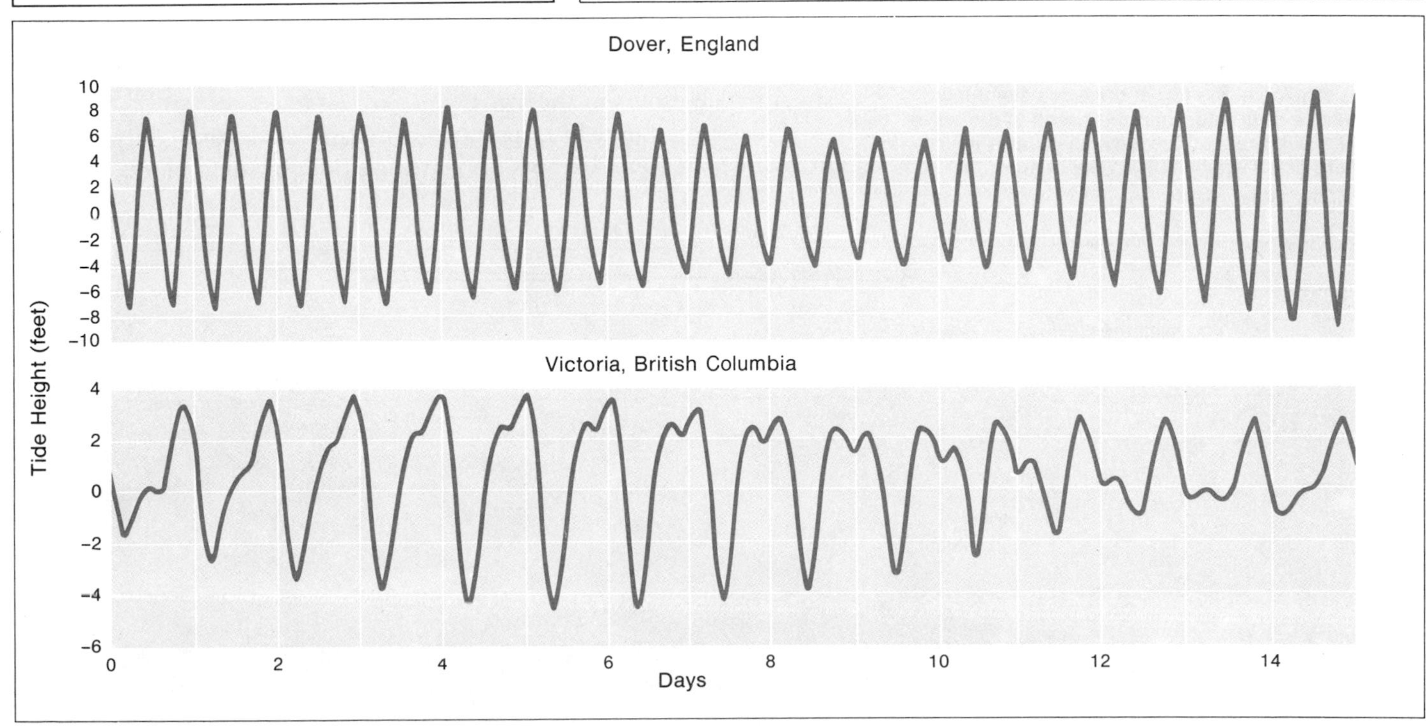
Dover, England
Victoria, British Columbia
Tide Height (feet)
Days
10
8
6
4
2
0
-2
-4
-6
-8
-10
4
2
0
-2
-4
-6
0
2
4
6
8
10
12
14

duces tides in the solid earth, but the rigidity of the planet is such that these are barely detectable.

The sun also has an effect on the tides; however, the sun's great distance from the earth reduces its gravitational attraction to about half that of the much smaller but closer moon. The greatest tidal range in any region, known as *spring tide*, occurs when the sun aligns with the earth and moon to accentuate either the outward gravitational pull on the earth's water (time of new moon—moon between the sun and earth) or the centrifugal effect (full moon—earth between the sun and moon). Conversely, the tidal range is at a minimum *(neap tide)* when the sun and moon make a 90° angle with the earth, so that the solar gravitational effect draws off some of the crest of both the lunar and centrifugally produced tides.

The oceanic response to these tide-producing forces is complex, being affected by many factors, including the patterns of ocean currents, variations in water density and atmospheric pressure, and interference by landmasses of varying sizes and shapes. The moving water masses also set up complex resonance effects that create local peculiarities in tidal characteristics. This can be seen in Figure 16.4 (bottom). Tides are insignificant in the Mediterranean Sea due to the small size of its opening into the Atlantic Ocean at the Straits of Gibralter. However, where tidal waters crowd into the English Channel, their range increases from less than 0.25 meter in the open sea to 7 meters (23 ft) at the narrowest part of the Channel. Tidal ranges reach their maximum where the flood tide is channelled into narrowing estuaries; as the mass of water is increasingly confined, it surges strongly upward. The Bay of Fundy in Nova Scotia, Canada, is famed for the world's largest tidal range, the average spring tide range being 15.4 meters (50.5 ft). This is a consequence of special circumstances: steep, rocky walls that confine the water, a channel of uniform width that abruptly splits into two narrow arms, and rapidly shallowing (or "shoaling") bottoms in the arms.

Tides are extremely important in marine navigation. Ships can enter or depart some harbors only at high tide. In some places, tidal

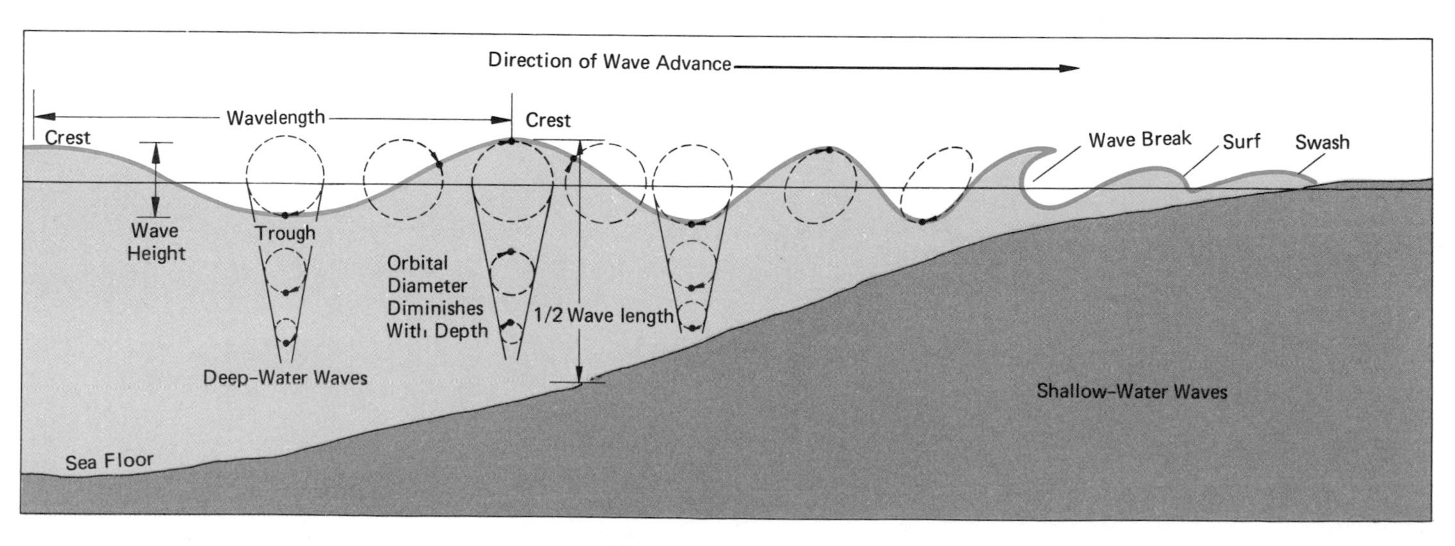

Figure 16.5 The nearly circular motion executed by individual water particles is what transmits forward wave motion. When a wave crest passes, the surface water particles at that point are at the top of their orbits. When a wave trough passes, the same surface water particles are at the bottom of their orbits. Orbital motion of water particles is negligibly small at depths greater than half a wavelength. Waves in water deeper than half a wavelength, which are called *deep-water waves,* are not influenced by the presence of the ocean bottom. As the wave enters shallow water near a shore, the ocean bottom begins to interact with the wave motion. The wave becomes steeper and the wave length becomes shorter. The water particles at the top of the wave eventually reach a speed greater than the speed of the wave, and the wave breaks. The wave dissipates its remaining energy as it washes up on the beach.

currents are very powerful and constitute a hazard. In ancient times, fleets and the armies they carried were sometimes lost when strong tidal currents tore loose and devastated ships at anchor or washed away vessels beached on the land. And during World War II, weather and tide conditions determined the exact timing of the Normandy landings, the Allied invasion of occupied France.

Momentary Changes in Sea Level: Waves in the Open Sea

The waves that make the sea surface so fascinating to watch are a marvelous example of the energy transformations that are a feature of all of the earth's physical systems. The significance of all waves—whether they are sound waves, light waves, or water waves—is that they carry energy from one place to another. Thus the energy that initiates wave motion is transmitted by the waves to some other point where it is ultimately transformed into heat and work of some kind.

The vast majority of water waves are initiated by atmospheric energy in the form of tangential wind stress on the water surface. As a water wave passes, it initiates an orbital motion of water particles that produces an up-and-down oscillation of the water surface, as diagrammed in Figure 16.5. Wave disturbance is passed from particle to particle in such a way that the passage of a water wave involves a rise in the water surface, producing a *wave crest,* followed by a sinking of the surface to produce a *wave trough.* There is little forward displacement of the water itself, for what is transmitted is energy, not mass. This is obvious when a wave passes under a bit of scum or a floating object, which merely rises and falls as the wave moves by it. The vertical distance between the crest and the trough is called the *wave height.* The distance between wave crests is known as *wave length,* and the time required for successive crests to pass the same point is the *wave period.* The speed of a wave in deep water is a function of wave length: the greater the wave length, the more rapid the wave transmission.

Water waves are of two types: *forced waves* (often simply called *sea*), which have sharp peaks and broad troughs and are raised directly by wind stress on the water surface; and *swell,* which consists of round-crested linear waves traveling outward from the source of disturbance. The size of forced waves increases with wind force and duration and the extent of water over which the wind can build the wave, which is known as the *fetch* (see Figure 16.6). During storms in the open sea, the height of forced waves is commonly about 6 meters (20 ft), but waves as high as 30 meters (100 ft) have been observed under exceptional conditions.

The vast majority of waves that roll in against our shores are swell generated by strong winds in some far-off place. Typical swell with wave lengths of 100 meters (330 ft) moves at about 15 meters per second (30 miles per hour). Swell moving from different directions often intersects, producing wave patterns of varying complexity. Although wind can generate waves with a wide range of wave lengths, only those with long wave lengths (100 to 200 meters, or 330 to 660 ft)

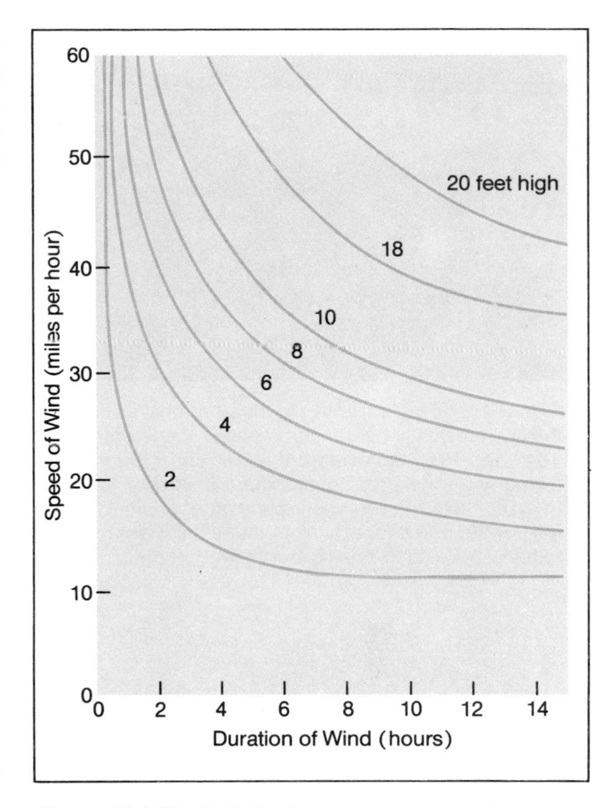

Figure 16.6 The height of waves on open water depends on the fetch, the speed of the wind, and the length of time the wind has been blowing. These calculations are based on a distance from the windward shore that is great enough to preclude the influence of fetch on wave height. (After Sverdrup and Munk, 1947)

Figure 16.7 This attractive park along the waterfront in Hilo, Hawaii, was developed after the tsunamis of 1946 and 1960 devastated the portion of the town formerly on the site. Development of shoreline parks in urban areas is a reasonable way of minimizing loss of life and property from tsunamis that are certain to strike such coasts from time to time.

are able to travel great distances to become swell. The reason is that as water moves up and down in wave motion, a small part of the wave energy is dissipated as heat resulting from internal friction between water particles. Because waves of short length have less energy to begin with and oscillate the sea surface more frequently than long waves, they dissipate their energy more rapidly. After a group of waves having different lengths has traveled a long distance, the waves of shorter length have largely died out, leaving only the longest waves. Swell of moderate size is capable of traveling half way around the world before all of its energy is absorbed. Thus nearly all moderate swell lives long enough to impact against a coastline. When we watch waves rolling against our coasts, we should keep in mind that they very likely were born in storms raging in the seas off Japan, Iceland, or even Antarctica.

Seismic Sea Waves, or Tsunamis

Not all waves are initiated by wind stress on the water. The most hazardous of all waves are known popularly (but improperly) as "tidal waves," although they are quite unrelated to true tidal action. They are also commonly called *tsunamis,* the Japanese word for exceptionally large waves that strike the land. These greatest of all waves are actually shock waves in water, triggered by earthquakes, landslides, and volcanic eruptions on the ocean floor, and the scientific term for them is *seismic sea waves.* Oceanic earthquakes are frequent, for the greatest concentration of large earthquakes is in the subduction zones, which create the oceanic trench system. The Pacific Ocean is ringed with earthquake-producing oceanic trenches, and thus the shores of the Pacific experience frequent seismic sea waves. In contrast, there are no trenches in the Atlantic basin, except for those of the Caribbean area, and consequently seismic sea waves are infrequent on Atlantic coasts, most of those recorded having been triggered by earthquakes or volcanic disturbances along the oceanic ridge system.

Seismic sea waves are an awesome phenomenon. In the open sea, they are barely detectable, having enormous wave lengths—exceeding 100 km (60 miles)—and heights of less than a meter. However, commensurate with their wave lengths they move at incredible speeds—as much as 800 km (500 miles) per hour. All water waves slow and bunch together approaching the land, when they begin to be retarded by friction with the sea floor. The principle of energy conservation causes wave height to increase as wave speed slows. In the case of seismic sea waves, this causes a series of rapidly moving mountains of water to lift suddenly out of an ordinary sea, often rising to heights of 10 to 30 meters (30 to 100 ft) as they impact against the land. In some cases, a great wave trough appears before the first crest, causing the sea surface to sink and recede far from the shore. This happened at Hilo, Hawaii, in 1923. Many people ran out to pick up fish stranded on the newly exposed land and were caught by the monstrous wave crest that suddenly materialized to engulf them.

Historically, seismic sea waves have caused as much destruction and loss of life as have major earthquakes and volcanic eruptions on the land. The 30,000 people killed as a consequence of the eruption of

Krakatau (Indonesia) in 1883 died in the seismic sea waves the volcanic eruption generated. Hawaii lies in the path of tsunamis emanating from three sources: the Peru-Chile trench, the Aleutian trench, and the western Pacific trenches. Fortunately, tsunamis can be predicted, and loss of life averted, but waterfront property damage can be reduced only by restricting building in the hazard area (see Figure 16.7). An international tsunami warning system is in effect over much of the Pacific Basin. Hawaii has about 10 hours to prepare for tsunamis radiating from the Peru-Chile trench; California has 4 hours to ready itself for tsunamis from the Aleutian trench. (Everywhere, of course, there are a few foolish people who go down to the beach to see the predicted "tidal wave" and are themselves never seen again!) At Crescent City, California, a tsunami generated by the 1964 Alaskan earthquake ran inland a distance of 500 meters (⅓ mile), submerging 30 city blocks.

Scientists have calculated the expectable magnitudes and frequencies of tsunamis at various locations under average conditions over long periods of time. At Hilo, a dangerous 10-meter-high (30-ft) tsunami can be expected once in every 100-year period on the average, with some tsunami damage anticipated every 10 years or so. In parts of Japan, a 10-meter-high tsunami can be expected once in every decade. As with all probability calculations, such figures cannot be a basis for the prediction of events in any specific year; Hawaii suffered two devastating 10-meter-high tsunamis only 14 years apart, in 1946 and 1960.

WAVE ACTION AGAINST THE LAND

This chapter began with an illustration of the significance of coastal erosion over time. How can waves, which roll harmlessly past boats, swimmers, and children in inflated inner tubes, be so destructive to the land itself? In deep water, waves striking a flat cliff or sea wall are deflected back out to the sea without any commotion at all. Waves are, in fact, not mechanically destructive until they "break."

Wave Break

In the open sea, waves glide onward unaffected by the ocean floor, for the orbital motion of the water transmitting waves reaches downward only about half the wave length. But when waves move into shallow water and wave motion begins to be affected by friction against the sea floor, wave characteristics are modified. Waves then change from deep-water types with periods dependent upon wave length to shallow-water types with velocities dependent upon water depth. When the forward progress of waves is retarded by friction with the sea floor, the waves begin to slow, to bunch together, and, as in the case of tsunamis, to increase in height. Actually, as shown in Figure 16.5, friction causes the orbit of the surface water particles to be deformed from a circle to a forward-leaning ellipse. Two things can happen to this moving ellipse. It can collapse in a mass of foam, producing a

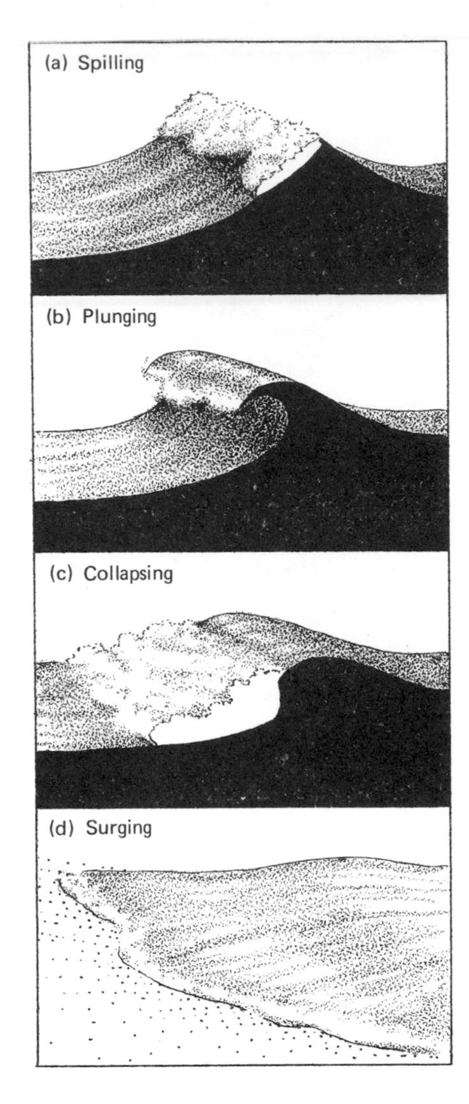

Figure 16.8 The manner in which waves break affects beach erosion as well as the stability of man-made structures such as breakwaters and seawalls. Breaker type is a response to wave height and steepness and beach slope.
(a) *Spilling breakers* result from the downward slumping of the crests of steep waves. In (b) *plunging breakers* steep wave crests curl over the front face of the wave and fall vertically. In (c) *collapsing breakers* the lower part of the wave front becomes vertical and the wave collapses in a mass of foam. When wave height is small, waves slide up the beach without collapse of the wave form, producing (d) *surging breakers.*

surging, collapsing, or *spilling breaker,* or it can topple over forward, with its crest arching down to produce a *plunging breaker* (see Figure 16.8). Surfers attempt to ride the smooth moving hill of water just ahead of the toppling crest, or "curl," of a plunging breaker. The way a wave breaks depends in part on the obliqueness of wave approach to the shore and the steepness of the sea bottom. An oblique approach and gentle slopes favor spilling breakers, whereas a direct approach and steep slopes cause the sudden wave retardation that produces plunging breakers.

Beach Drifting and Longshore Drifting

If the wave breaks on a beach, the mass of water foams forward as *swash* or *uprush,* entraining fine sediment and pushing it up the slope of the beach under turbulent transport conditions. This forward movement is opposed by friction and gravity. As the speed of the swash diminishes due to these opposing forces, a point comes at which forward momentum and gravitational pull down the beach slope are in balance. The uprush stops, much of the water sinks into the beach, and the remainder turns in its path and slides back down the beach slope as *backwash,* dragging fine sediment with it. Each particle of water and every grain of sediment that is driven obliquely onshore by uprush and dragged down by backwash are also shunted laterally along the shoreline in zigzag fashion—a process called *beach drifting* (see Figure 16.9). Beach drifting is very important because it helps to provide the constant flow of sand necessary to maintain beaches and the associated depositional features to be discussed.

In the offshore zone of wave break, the water is kept in an agitated condition and is often turbid with fine suspended sediment being swirled this way and that. The sediment may be in suspension only momentarily after each wave impact, but while in suspension it moves with the water carrying it. Since waves generally meet coastlines at a slight angle, they push water against the shore obliquely. This continual input of water cannot heap up vertically, and so much of it drains off laterally, parallel to the coast in the general direction of wave advance. The result is a *longshore current.* The oblique onshore piling of water by the arrival of one breaker after another causes the sediment-laden water in the zone of breaking waves to migrate slowly along the shore in the direction of the longshore current, producing *longshore drifting* of sediment. Sediment transport by longshore drifting is particularly active when large waves are arriving at the shore either as swell or as forced waves (storm surges) driven onshore by local winds. The combined processes of beach and longshore drifting produce a constant flow of sediment along smooth coastlines—a flow that waxes and wanes from day to day, depending upon the state of the sea, but which is responsible for the maintenance of all depositional landforms seen in coastal regions.

Wave Refraction

As waves enter shallow water, the line of the waves is seldom exactly parallel to the shore; one portion of each wave strikes the shore before

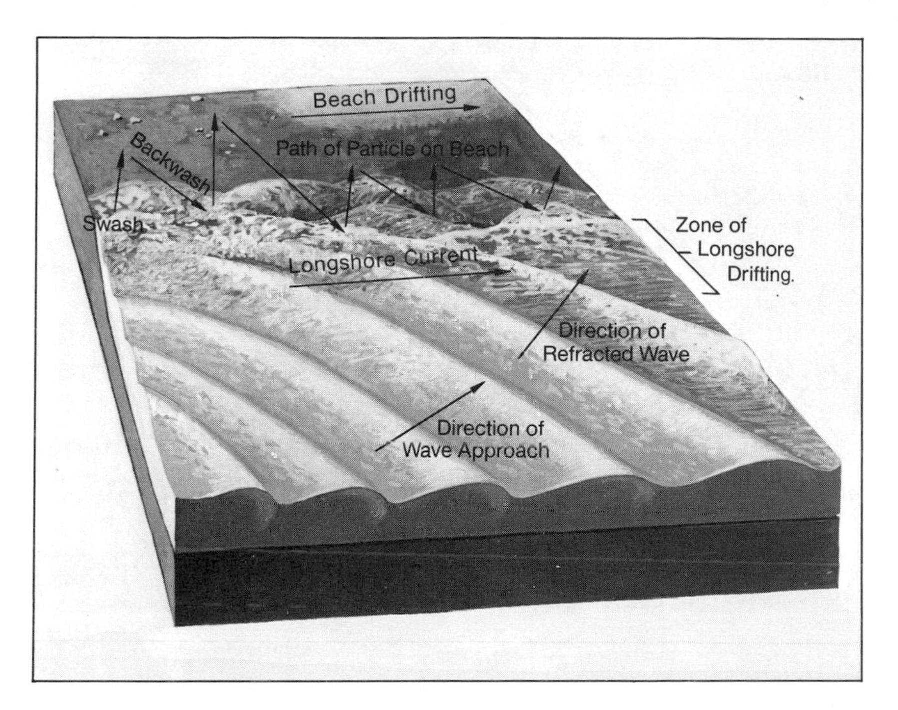

Figure 16.9 Waves that strike a beach obliquely cause transport of material parallel to the beach. Some of the material is shunted along by *swash* and *backwash*, as wave break causes water to wash up the beach obliquely and then return downslope. This produces the movement of particles known as *beach drifting*. The longshore current established by the oblique waves also transports material parallel to the beach in the process of *longshore drifting*. Most of this material moved in the surf zone is transported in suspension.

the rest of the wave. But even before this portion of the wave meets the shore, it is affected by friction with the sea floor so that its forward velocity decreases while the part of the wave that has not yet "felt bottom" moves on ahead. This change in velocity causes the wave to pivot slightly at the point where it is first affected by friction, bending the line of the wave, as illustrated in Figure 16.10, p. 498. This phenomenon is known as *wave refraction*. Wave refraction is important in the development of coastal landforms because it affects the direction of energy transmission by incoming waves.

If a coast is irregular, waves are first slowed by friction as they approach projecting headlands. As Figure 16.10 shows, the resulting wave refraction causes wave energy to become focused on the headland. In bays between coastal promontories, refraction causes the lines of wave approach to diverge, spreading each unit of wave energy over a larger area, which creates a low-energy wave environment. Given sufficient time, the above-average wave energy at headlands and below-average wave energy in bays would smooth the coast by wearing back the headlands and allowing sediment to accumulate in the bays. However, the straightening process is only a general tendency that may be interrupted by sea-level changes and is retarded by the varying resistances of diverse rock types.

Coastal Erosion

Along a large proportion of the earth's coastlines, waves do not break upon sand or gravel beaches but impact directly against steep slopes that are progressively worn back by wave attack. Most of the removal is accomplished when storm-driven waves crash against the land, reaching higher and striking with much greater force than average

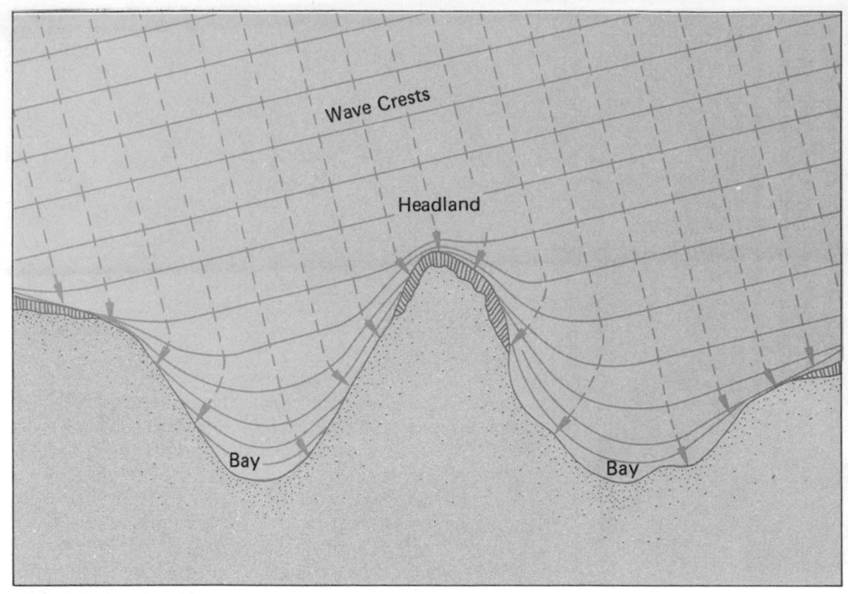

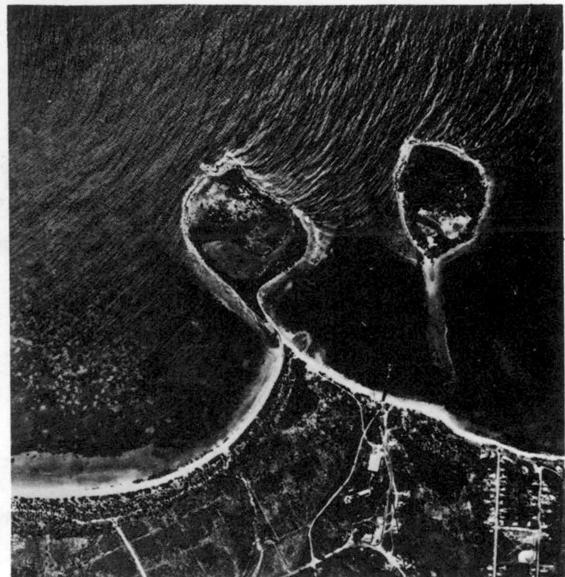

Figure 16.10 (left) Wave motion is altered in direction, or *refracted,* when it interacts with shorelines. The diagram shows waves refracted at a headland or coastal promontory. Solid lines represent wave crests; dashed lines show direction of wave energy transmission. Compartments bounded by wave crests and energy vectors contain equal amounts of wave energy. Refraction causes waves to concentrate their energy on headlands, accelerating erosion and causing the formation of such features as sea stacks and sea caves excavated into the sides of the headland. In bays, wave energy is dissipated by divergence of wave trajectories. Therefore bays become areas of sediment deposition.

(right) This photograph shows wave refraction along a portion of an irregular coast.

waves. Wave erosion is accomplished by a distinctive combination of mechanisms. Some removal is by chemical corrosion of rock that is constantly wetted, and some is a result of the crystallization of salts in rock pores repeatedly dampened by salt spray. Part of the work is done by abrasion, as rock debris is hurled against cliffs by wave impact. Lighthouse windows 30 meters (100 ft) above sea level commonly have to be defended against wave-tossed rock shrapnel by wire screens. A significant part of the breakup of rock along coasts results from the enormous impact of tons of water crashing into the jointed rock during storms. However, one of the most effective mechanisms of wave erosion is the compression of air within pores, fissures, and joint partings in the rock when a wall of water smashes against it—followed by the explosive expansion of the air as the wave falls away. This pneumatic effect seems to be a major factor in the loosening of rock subject to storm wave attack.

A coast's vulnerability to wave erosion depends on the local geological materials, the fetch that controls the size of forced waves, the pattern and magnitude of swell from distant storms, and the possibilities for wave refraction at the coast. The rate of coastal retreat under wave attack may be as much as 2 meters (6.5 ft) a year where the rocks are composed of poorly cemented sand or gravels, or less than 1 meter in a thousand years where resistant rocks are present. During unusually intense storms, sandy cliffs have been cut back by more than 10 meters (33 ft) in a few hours.

Since coastal erosion is concentrated in the zone of wave impact, ranging from low tide to somewhat above high tide, the effect of wave erosion is to undercut sea cliffs by notching them at their base, as illustrated in Figure 16.11. The upper portions of the cliffs retreat by collapse due to lack of support. The backward retreat of sea cliffs leaves behind a submarine rock platform called a *wave-cut platform* or *abrasion platform,* whose seaward-sloping surface may be littered with debris removed from the land. The angle between the sea cliff and abrasion platform is known as the *shoreline angle.* As shown in Figure 16.12, the abrasion platform may be skirted by a deposit of sand that extends still farther seaward as a *wave-built terrace.* It was once supposed that areas the size of the British Isles could be, and had been, planed off at sea level by marine erosional processes, prior to later tectonic uplift, with traces of several partial planations also being well preserved. However, as the width of a wave-cut platform expands, an increasing amount of wave energy is lost by friction with the surface of the platform. Accordingly, the strength of wave attack against the land becomes progressively less as the platform grows, until there is no further effective wave erosion at the shoreline angle. In fact, it is now believed that the broad abrasion platforms seen in some areas must have been produced during periods of rising sea level, with the zone of frictional energy loss always being a narrow one close to the shoreline angle.

Where coasts rise steeply and are irregular, with promontories or headlands projecting beyond embayments, wave refraction at the headlands concentrates wave attack upon the projections from two sides. Selective wave erosion along lines of weakness in these projections may separate them into clusters of isolated rock stubs known as *sea stacks* (see Figure 16.12). Undercutting along the sides of promontories often develops *sea caves,* which sometimes link together from the two sides to convert the promontory into a *sea arch* (see Figure 16.13, p. 500). Collapse of the arch leaves an isolated rock pinnacle, producing a single sea stack.

It is not uncommon to find unmistakable wave-cut platforms together with former sea stacks on dry land well above sea level, forming a terrace along the coast. Such *marine terraces* could indicate

Figure 16.11 Wave impact and abrasion by storm-wave-tossed rock shrapnel carves a *wave-cut notch* at the base of actively retreating sea cliffs. The wave-cut notch in the foreground extends all along the base of the cliffs seen here, which are cut into basaltic rock on the west coast of the Hawaiian island of Kauai.

Figure 16.12 This diagram illustrates the principal features that may be seen along cliffed coasts. The major elements are *sea cliffs* that are kept steep by being undercut at the base along a *wave-cut notch,* and a *wave-cut platform* that extends seaward from the base of the sea cliff. The cliff and platform meet at the *shoreline angle,* which is usually at the high-tide level. Features occasionally seen are sea caves and erosion remnants such as *sea arches* and *sea stacks.* Depositional *wave-built terraces* may also be present. The existence and character of these subsidiary forms depend on the nature of the rock present. (Modified from Butzer, 1976)

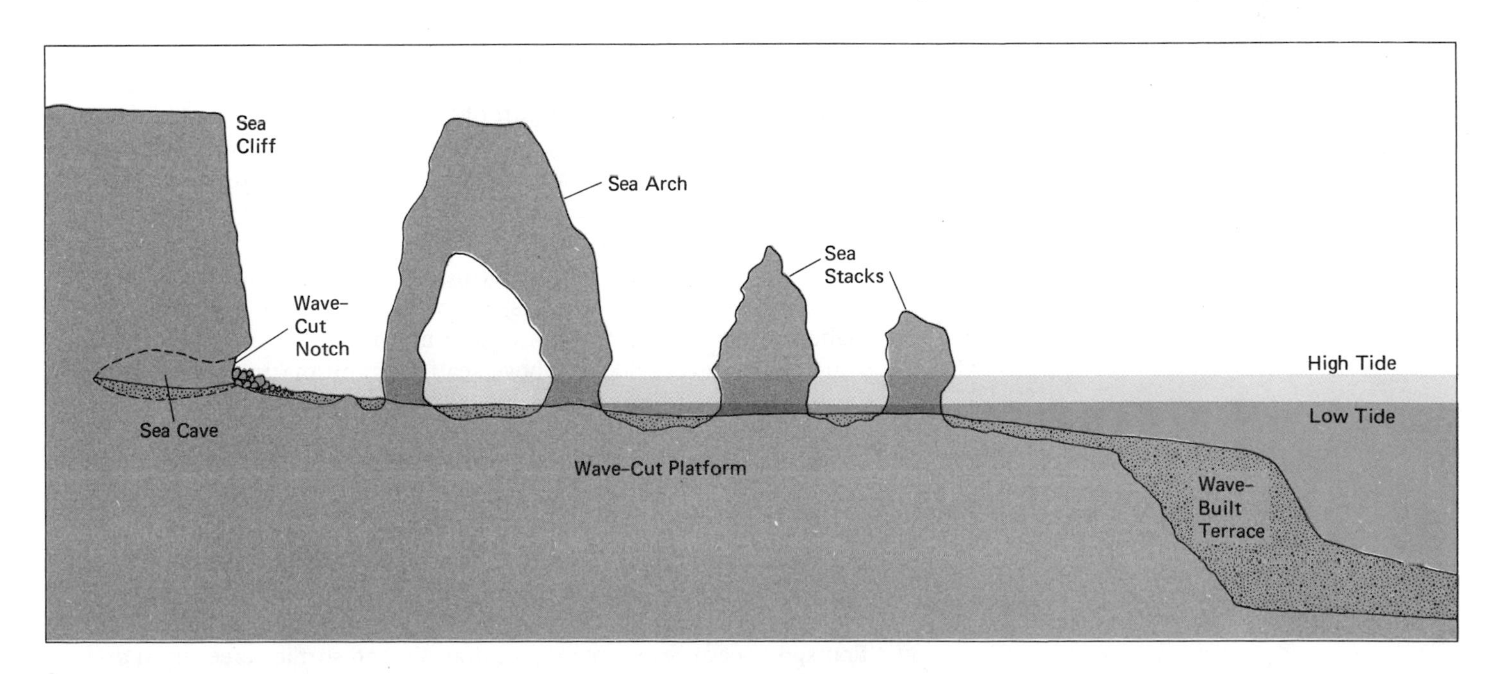

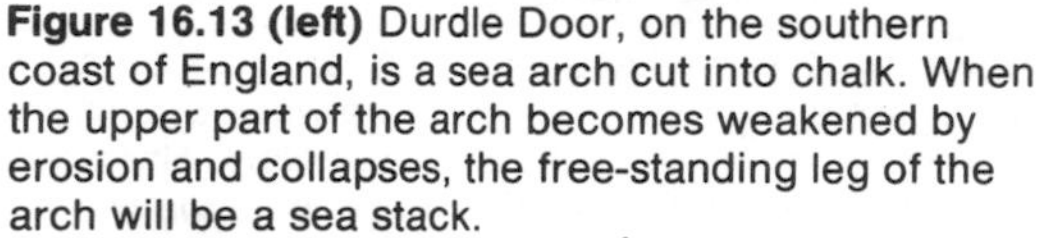

Figure 16.13 (left) Durdle Door, on the southern coast of England, is a sea arch cut into chalk. When the upper part of the arch becomes weakened by erosion and collapses, the free-standing leg of the arch will be a sea stack.

(right) The chalk cliffs at Étretat on the coast of Normandy, France, were cut by waves to form sea arches and sea stacks. These cliffs are cut in the same geologic formation as Durdle Door, but are on the opposite side of the English Channel. Note the vertical face of the cliffs; wave action at their base undermines the rock and causes the collapse of the upper portions. The cliffs thus maintain a vertical face as they retreat.

Figure 16.14 This photograph of the Oregon coast shows the exposed surface of a wave-cut platform at low tide, and above it a marine terrace. This terrace was an older wave-cut platform produced at sea level and subsequently uplifted tectonically, initiating platform development at a lower level. As this low headland is worn back by wave erosion, sediment is transferred into the bay in the background, where it accumulates in a broad beach. Thus this view illustrates the general process of coastal straightening by wave action.

either a drop in sea level or uplift of the land relative to the sea. Since the phenomenon is generally a local one, the first possibility is usually excluded. Marine terraces are especially common along tectonically active coasts, such as the Pacific Coast of the United States, where many areas have clearly been lifted out of the sea (see Figure 16.14). Some localities along the California coast exhibit as many as six clearly distinguishable terraces, arrayed like a giant staircase. The highest terraces that are definitely of marine origin in California lie some 600 meters (2,000 ft) above sea level, and marine shells on the lowest terraces are more than 100,000 years old.

Coastal Deposition

Sediment transport under marine conditions follows the same principles as sediment transport by running water, wind, and (to a degree, at least) glacial ice. When the shear stress of water moving in a wave or current exceeds the inertia of a particle, the particle is set in motion. Transport occurs by suspension, saltation, and surface creep, as it does

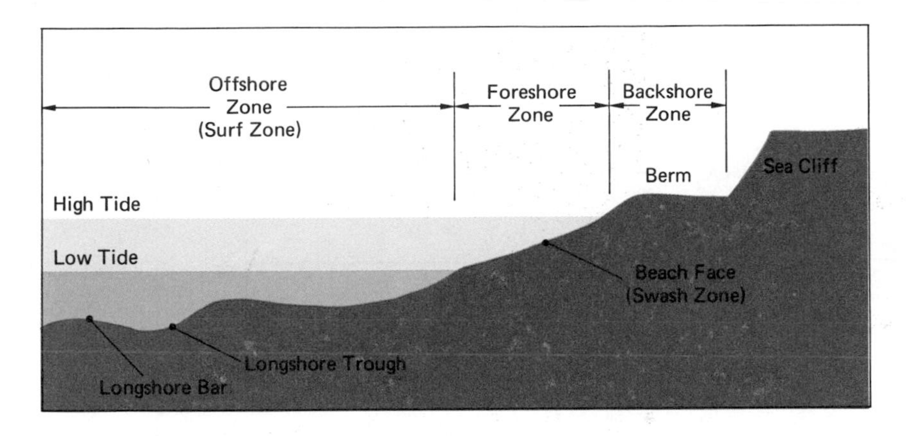

Figure 16.15 A *beach* is the zone of transition between the land and sea. It extends from the low tide level to the upper limit reached by the highest storm waves, which is the area subject to alternate erosion and deposition of sand. The actual profile of a beach is constantly changing. The diagram shows the principal geomorphic divisions of a beach. The *berm* is the nearly flat portion at the top of a beach; it is covered with material deposited by waves and constitutes the *backshore*. The *foreshore* extends from the edge of the berm to the low-tide line. Within this zone is the *beach face,* which is the area subject to swash and backwash. The *offshore,* which is permanently under water, contains *bars* and *troughs.* This is the zone of wave break and surf action. (After Shepard, 1967)

with running water or wind. It is easy to see sand and gravel moving in all of these ways in the swash zone, where the collisions produced, especially by fine gravel, make a clearly audible sound. Deposition occurs where wave energy decreases, permitting sediment to come to rest.

Every marine depositional form has a characteristic sand (or gravel) budget in which gains must balance losses if the form is to persist. Many marine depositional forms periodically grow and shrink due to shifts in the balance between sediment input and outgo. Although waves strike coasts from varying directions from day to day, every coast has an average "wave climate" and an average direction of sediment flow. Thus, except during severe storms, there is normally sediment arriving at one end of a depositional form and leaving at the other end. Any interference with the flow of sediment at either end of the system results in a change in the form of the deposit.

Beaches

Beaches are exposed areas of sand or gravel (here called "shingle") deposited by wave action between the level of low tide and the highest levels reached by storm waves. On the landward side of the beach is material of a different type or a surface of different character, such as sand dunes, permanent vegetation, or a sea cliff. Beaches form along both irregular and straight coastlines. On irregular coasts, wave refraction causes sand or shingle to collect in the embayments between projecting headlands, creating *bayhead beaches.* Beaches also develop in continuous fringes along straight coasts where unresistant geological materials are present. Figure 16.15 indicates the main features of beach morphology: the sloping beach *face,* which is exposed to wave uprush; the *berm,* or higher, nearly horizontal area of dry, loose sand; the *foreshore zone* between high and low tides; the *backshore zone* above high tide; and the *offshore zone* that is submerged even at low tide.

Beaches are among the most sensitive indicators of changes in geomorphic systems. We usually enjoy them in summer, when they are well filled with sand due to the absence of destructive storm waves (Figure 16.16, p. 502). However, even in summer, beaches must be

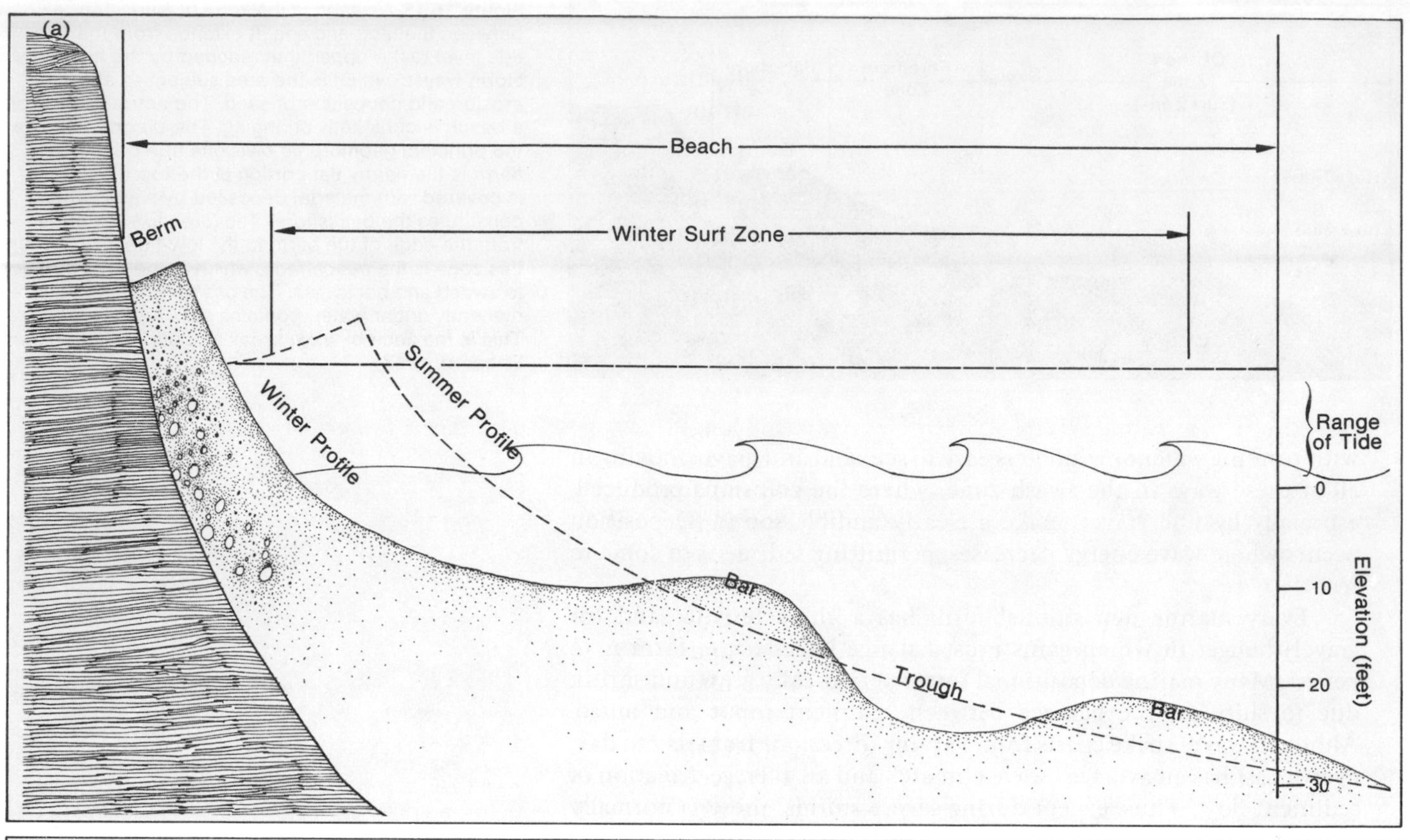

(a)
Beach
Berm
Winter Surf Zone
Summer Profile
Winter Profile
Range of Tide
Bar
Trough
Bar
Elevation (feet)
0
10
20
30

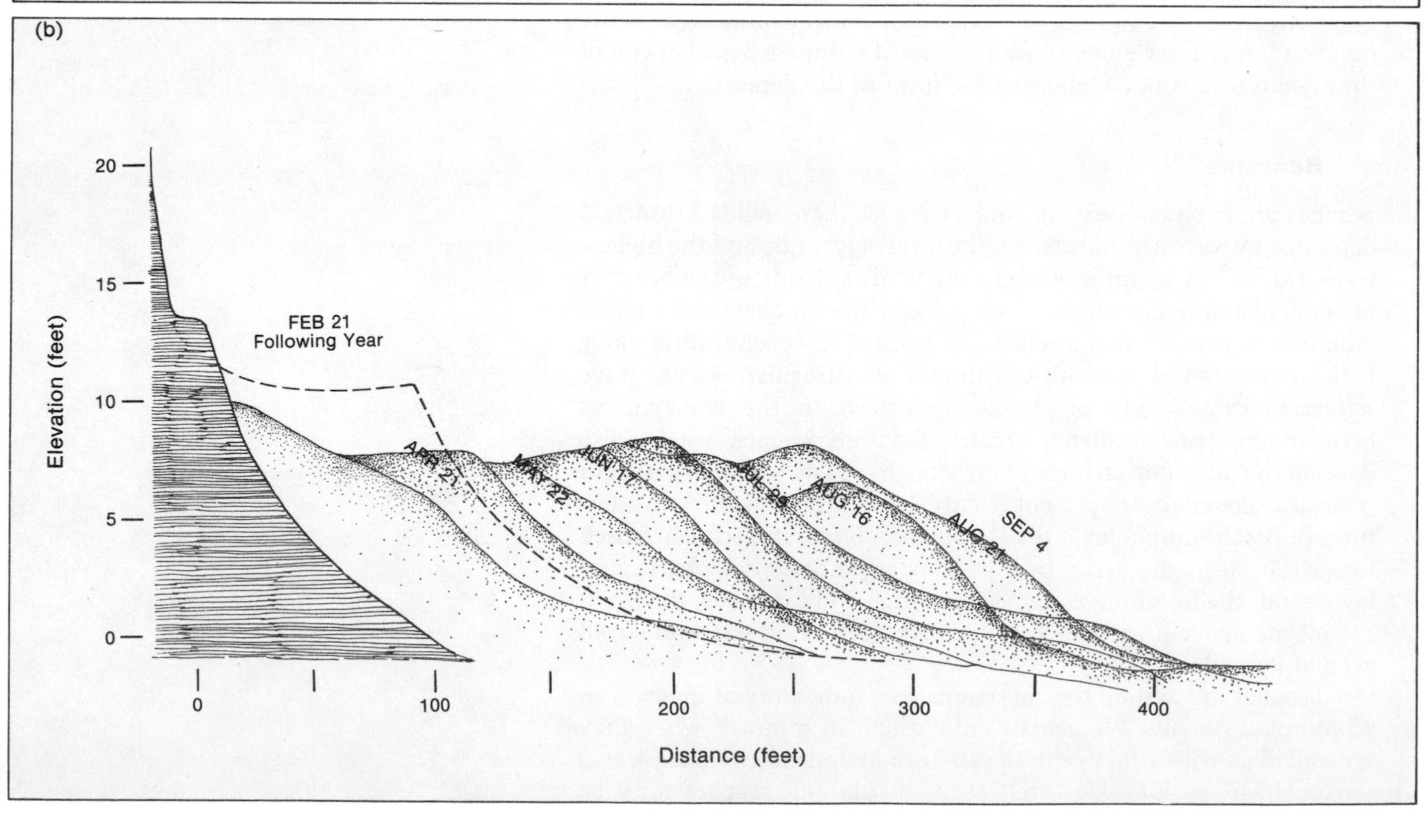

(b)
FEB 21
Following Year
APR 21
MAY 22
JUN 17
JUL 29
AUG 16
AUG 21
SEP 4
Elevation (feet)
20
15
10
5
0
0
100
200
300
400
Distance (feet)

steadily supplied with sediment. The sediment may be generated by wave erosion of adjacent headlands, by wave attack on the sea bed itself, or by rivers carrying products of erosion from the land. Along a coast that has a prevailing longshore drift, sand eroded from one beach is often passed along to the next beach. On the southern coast of Long Island, New York, sand produced by wave erosion of glacial drift moves from east to west along the beaches. Loss seems to be outrunning accumulation, however, and some beaches are retreating 1 or 2 meters (3 to 6 ft) per year. Certain beaches elsewhere have been shrinking in recent years because the construction of upstream dams has seriously reduced the amount of sand supplied to the coast, or because housing developments on backshore dunes have cut off part of the sand supply for the beach system.

One solution to beach erosion is to interrupt the longshore drifting with a *groin* that extends into the water at right angles to the beach. The longshore current loses speed at the groin and drops part of its sediment—just as a stream drops part of its sediment when its velocity decreases. Inevitably, the greater accumulation promoted by the groin in one place robs beaches farther down the coast of incoming sediment, and those beaches erode all the more rapidly, to the anguish of resort owners. Miami Beach must import sand to balance erosional losses, despite efforts of individual hotels to maintain their beaches by constructing groins.

A change in the size of breakers, or a change from long waves of low height to short waves of greater height, greatly accelerates the loosening and entrainment of the particles composing the beach, which are carried in suspension seaward into the zone of breaking waves. Large, steep waves can peel a beach downward 2 meters (6.5 ft) or more in a day, with the beach material being deposited as sub-

Figure 16.16 (opposite) (a) The shape of a beach changes throughout the year. During winter, destructive waves remove material from the beach, narrowing the berm. The material is deposited in offshore bars. During summer, the process is reversed, and the berm grows at the expense of the bars. The vertical scale in the diagram is exaggerated 25 times for clarity.

(b) The growth of the berm on the beach at Carmel, California, is shown in this series of measured profiles. By February of the following year, most of the berm had been once more cut back. The vertical scale is exaggerated 10 times. (After "Beaches" by Willard Bascom. Copyright © 1960 by Scientific American, Inc. All rights reserved)

Figure 16.17 Beaches are not fixed landforms, but continually change as material is deposited or swept away. During summer **(left),** this beach at La Jolla, California, is covered with sand deposited by constructive waves, but during winter **(right),** destructive waves sweep the sand away, leaving a beach of rocks too large for the waves to remove. (After *Geology Illustrated* by John S. Shelton. W. H. Freeman and Company. Copyright © 1966)

merged *offshore bars* in the zone of wave break (see Figures 16.15, p. 501, and 16.16, p. 502). In areas affected by seasonal storminess, beaches tend to be stripped downward and steepened during the storm season—usually winter in the middle latitudes. Some beaches periodically lose all of their sand (see Figure 16.17, p. 503). They retain only large cobbles that are too heavy to be entrained even by the largest storm breakers. During the season of calms, the sand stored offshore slowly works back toward the beaches, refilling them to their former level.

Some especially attractive beaches have been produced by unusual circumstances and have very short life expectancies. In Hawaii, basaltic lava flows that run into the sea often explode violently when they come into contact with the seawater, which is instantly vaporized. This produces an accumulation of black cinders and mineral crystals, which wave action subsequently forms into black sand beaches. Since there is no continuous supply of the black sand, the beaches shrink a little every year. The famous Kalapana Black Sand Beach, a major tourist attraction on the island of Hawaii, was becoming distressingly narrow until it suddenly disappeared by subsiding below sea level during an earthquake in 1975.

Spits and Barrier Islands

Beach drifting and longshore drifting frequently produce sediment accumulations that extend outward from initial shorelines. Such features fall into two broad categories: those that are attached to the land and those that are detached, forming islands. Linear sediment accumulations that are attached to the land at one or both ends are termed

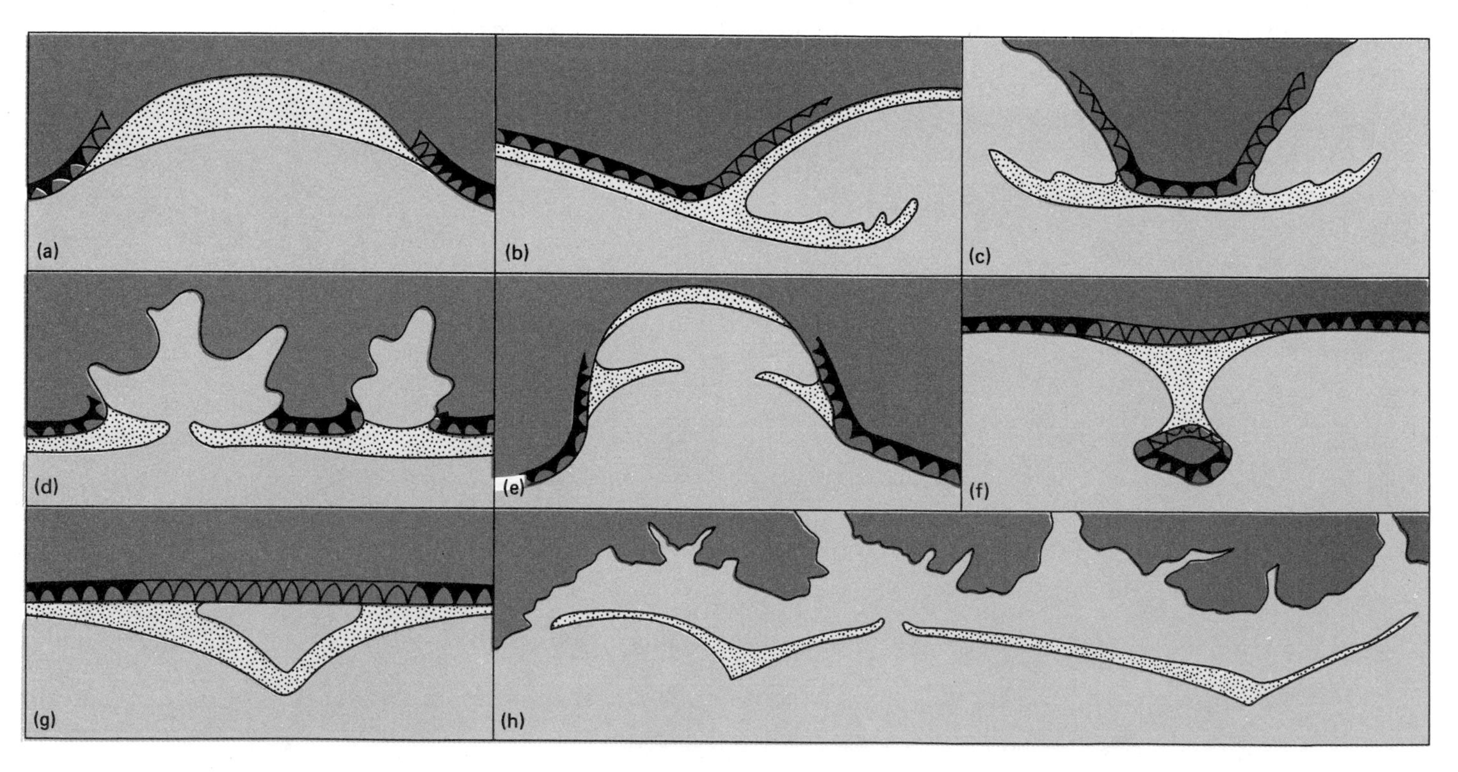

Figure 16.18 These diagrams illustrate the most common depositional forms produced by wave action. The dark sawtooth symbol indicates an active sea cliff; the light sawtooth symbol shows an inactive sea cliff. Land areas are dark. Depositional forms are yellow.

(a) *Bayhead beach* with sediment deposition resulting from a low-energy wave environment.

(b) *Recurved spit* formed by sediment transport to the right, prolonging the previous line of the coast.

(c) *Winged headland* with sediment moved both ways from the eroding cliff as a consequence of changing directions of wave approach.

(d) *Baymouth* or *barrier spits* straightening an initially irregular coast of submergence. Changes in wave approach produce spits extending from both sides of the bays, eventually closing them off and converting them into lagoons.

(e) *Mid-bay spits* with sediments accumulating before reaching the head of the bay.

(f) *Tombolo* formed by the deposition of sediment to the lee of an island and eventually linking the island to the mainland.

(g) *Cuspate spit* developed by reversals in the direction of longshore drifting along a straight coast.

(h) *Barrier island* developed along a coastline that is low-lying but irregular, suggesting recent submergence. The points on such barrier islands are called *cuspate forelands.* Cape Hatteras, North Carolina, is an outstanding example of such a form.

Figure 16.19 Deposition by longshore currents has caused the gradual extension of this curved spit near the mouth of the Colorado River in Mexico. The bay behind the spit may eventually be closed off, forming a *lagoon*.

spits. Sediment embankments completely separated from the land generally take the form of *barrier islands* (also known as *barrier beaches*).

Some spits form where sediment is moving along a linear coast that changes direction sharply, causing wave refraction and creating a low-energy environment in which sediment can accumulate. Imagine the processes of longshore and beach drifting along a straight coast. If the coastline abruptly turns inland, as in a bay, the wave fronts begin to pivot and diverge at this point, decreasing wave energy and sediment transporting ability. The sediment moved by beach and longshore drifting comes to rest where the incoming waves begin to pivot around the corner. The resulting deposit becomes the new shoreline, which causes the point of wave pivot to be shifted laterally. This, in turn, produces more deposition. In time, a linear spit forms, usually with a curved end, as at Cape Cod, Massachusetts, where the spit has been extended northward from the end of a wave-eroded glacial moraine.

Spits also form in the low-energy wave environments in the lee of coastal islets or large sea stacks. Here the convergence of waves behind the obstacle sweeps sediment together from two sides, often producing a type of spit known as a *tombolo,* which ties the island to the land (see Figure 16.18). The construction of spits from one or both sides of small bays often closes off the bays, transforming them into *lagoons* (see Figure 16.19). The lagoons eventually fill with fine sediment and are colonized by vegetation. This is one of the most important processes in the straightening of irregular coasts by marine action. Spits likewise tend to form at river mouths, being built of fluvial sand; however, river currents and floods tend to maintain openings through such spits, preventing lagoon formation.

Spits are especially subject to damage during hurricanes, which produce the largest waves experienced by most coasts in the low and middle latitudes. Many spits preserve a complex microtopography revealing that they are compound features, having been partially eroded by storm waves and rebuilt many times during the course of their growth.

Many of the world's coastlines are paralleled by linear sandy islands known as barrier islands, some of them hundreds of kilometers

Figure 16.20 This view from the Apollo spacecraft shows the barrier island system off the irregular submerged coast of North Carolina. Cape Hatteras is near the center and Cape Lookout is at the bottom of the picture. Outgoing tides are visibly washing fine sediment seaward through tidal inlets, which are a normal feature of barrier island systems. However, the effect of storm waves is to erode the seaward side of the islands and to move sediment across to the landward side, causing the islands to migrate toward the mainland (see the Case Study following this chapter).

long. These are often more regular than the coastlines behind them, as is the case around the Gulf of Mexico from Texas to Florida, and along the Atlantic Coast from Florida to New York. Until fairly recently it was presumed that these offshore sand accumulations were produced by wave erosion, which was thought to throw up a ridge of debris where wave motion first "feels" bottom on the continental shelves. However, there is increasing evidence that barrier islands have a complex history, possibly originating as ancient dunes formed behind beaches during Pleistocene low stands of the sea. Apparently, these sand banks were moved landward by wave action during the postglacial rise in sea level, and have greatly increased in size since current sea level was attained. Occasionally, differently oriented barrier islands, reflecting different angles of wave attack, link to form a prominent point, or *cusp,* as at Cape Hatteras, North Carolina.

Offshore barrier islands have great value as summer recreation areas, and they also produce a sheltered waterway between them and the mainland. In the United States, their presence has permitted the development of an intracoastal waterway system that is navigable all the way from Massachusetts to the Mexican border, with only short interruptions in New Jersey and Florida.

CLASSIFICATION OF COASTS

Due to the many variations in coastal scenery, a number of schemes have been developed for classifying coastal landforms. These classifications are a good indication of the complexity of coastal landforms and the processes that create them.

Tectonic Classification of Coasts

The most general morphological contrast in coastal type is between regular coasts that present little vertical relief and are fronted with continuous straight beaches, and perhaps barrier islands, and those that are steep and irregular, with deep marine embayments and many rocky promontories in which waves have cut high sea cliffs. The Atlantic and Gulf coasts of the United States are typical of the first type (except in the New England region), while the Pacific Coast of North America clearly exemplifies the second type. A further contrast between the two coastal types appears in their submarine extensions. The Atlantic coast is bordered by a broad continental shelf, whereas the Pacific coast has only a narrow shelf. Farther south, the Pacific coast of South America is bordered by an oceanic trench.

The gross contrast between the Atlantic and Pacific shores of the United States is a consequence of the movements of lithospheric plates. The shores of the Pacific Ocean in general are areas of strong tectonic activity due to the convergence of crustal plates, with the lighter continental plates pushing over the denser plates composing the Pacific sea floor. We have previously spoken of the compression, volcanism, and broad uplift triggered at this active interface. By contrast, the Atlantic shores are an interface between land and water only. The continents and the floor of the Atlantic Ocean are welded together and are moving as a unit—the only separation being at the mid-ocean ridge system and in the Caribbean region, where subduction and volcanism are occurring. Thus the Atlantic coasts often remain low and smooth. The exceptions, particularly in the north, are in areas where ancient mountain systems and plateaus developed long ago at former plate boundaries.

In terms of their relationships to the motions of lithospheric plates, coasts have been defined as collision (or subduction) coasts, trailing-edge coasts, and marginal sea coasts. *Collision,* or *subduction, coasts* are the Pacific type in which plate convergence and tectonic activity create strong surface relief and a steep coast. *Trailing-edge coasts* are low-lying due to the lack of tectonic activity, as our Atlantic shores. *Marginal sea coasts* are separated from the deep ocean basins

Figure 16.21 In this area a landscape previously subjected to erosion by a continental ice sheet has been submerged by the postglacial rise in sea level. The long, parallel projections of land and elongated islands that jut into the ocean off the coast of Maine were continuous ridges before being submerged. Submergence of the coast filled the valleys with water and exposed the ridges to wave erosion.

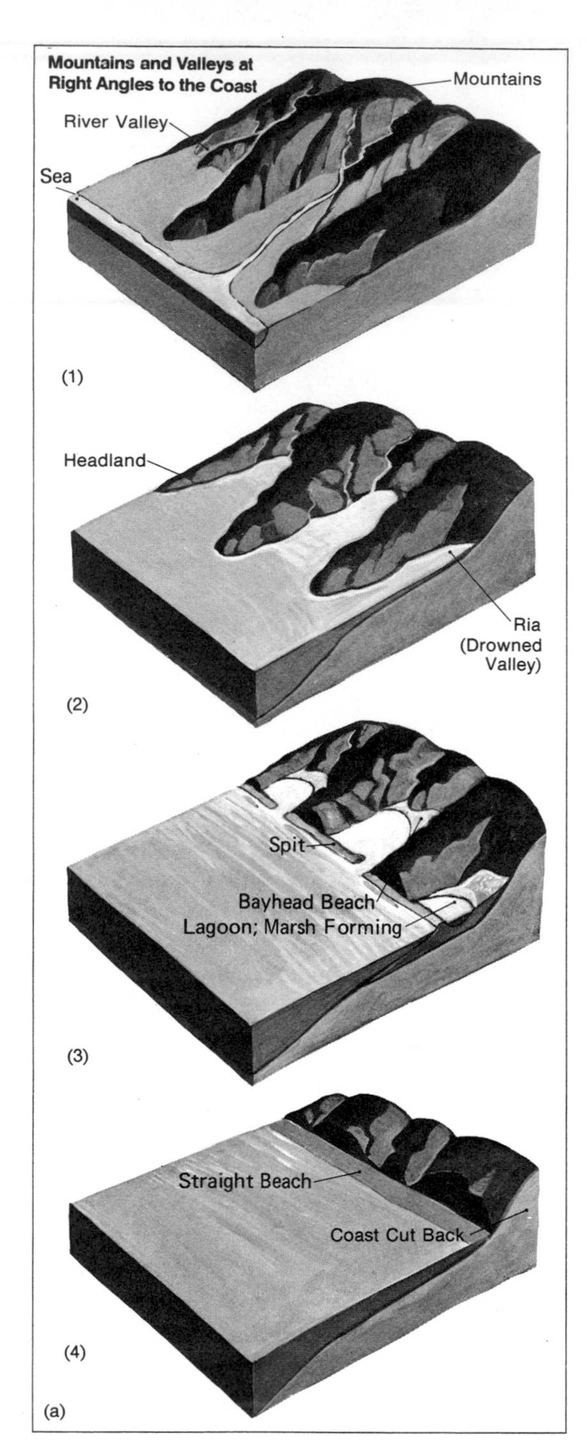

by island arcs such as the Caribbean Antilles or the Japanese or Indonesian archipelagoes (island chains). Generally, marginal sea coasts, such as those of the Gulf of Mexico, have broad continental shelves and low subaerial relief similar to trailing-edge coasts.

Genetic-Descriptive Classification of Coasts

The most straightforward coastal classifications are purely descriptive, involving such categories as high-cliffed, low-cliffed, lagoonal, deltaic, estuarine, fjorded, barrier island, fringing reef, barrier reef, and so on. This type of classification is possible on any scale, with the terms used as examples being general enough to permit differentiation of the world's coastlines on a single map.

Several popular classification schemes combine genetic and descriptive approaches. In such systems, the broadest distinctions are genetic, with minor subdivisions made in descriptive terms. In addition to the broad distinction between the marine and terrestrial origins of coastal forms, genetic classifications have been concerned with whether the coast appears to have *prograded* (advanced seaward) by deposition of deltas, sand spits, and lagoonal fills; *retrograded* (retreated) by erosion; or developed its characteristics as a consequence of *submergence* or *emergence* of the land relative to the sea. Many morphological subdivisions are possible within such genetic schemes; for instance, a submergent coast could involve drowned glacial or fluvial relief, either of which could have been dominated by either erosional or depositional landforms (see Figure 16.21, p. 507). In this regard, it must be reiterated that most coasts have been submerged in the past 15,000 years due to the rise in sea level resulting from melting of the Pleistocene continental ice sheets. The only exceptions are a limited number of areas in which the rate of tectonic uplift has equaled or exceeded the rate of eustatic change.

Coastal Submergence and Emergence

Although morphological classification of coasts as submergent or emergent has lost its popularity due to recognition of the general world-wide submergence of the past 15,000 years, the original concept retains some validity. Emergence clearly is required for the creation of marine terraces, which are common on tectonically active coasts. On tectonically quiescent coasts, emerged areas of sea floor should be low-lying, producing a broad, flat coastal plain. In the case of the Atlantic and Gulf coasts of the United States, there has indeed been long-term slow emergence, with Pleistocene eustatic changes superimposed on this general trend. Thus this region, whose estuaries indicate recent submergence, does present a landscape reflecting a generally emergent character. As marine terraces are found in tectonically active areas, features of emergence are found on all of the coastal types in the tectonic classification scheme.

The landforms resulting from coastal submergence are deeply penetrating estuaries, or *rias,* where the sea has invaded river valleys, along with cliffed headlands, and perhaps sea arches and stacks. As the headlands are battered back, their debris produces spits that close off bays, forming lagoons in the process of coastal smoothing (see Figure 16.22). The configuration of submerged coasts varies according to the nature of the prior subaerial landforms and the orientation of relief features with respect to the coastline. The rocky islets off the coasts of New England, Labrador, British Columbia, Norway, and Chile result from submergence of glacially scoured landscapes. Other highly irregular coasts, such as those of Greece and western Turkey, which are also fringed by rugged islands, result from submergence of mountainous topography, with the line of the coast cutting across the trend of mountain ridges. Where the shoreline of a submergent coast parallels mountain ridges, the coastline is less irregular, though steep, with the possibility of linear islands parallel to the mainland shoreline. The Yugoslavian coast of the Adriatic Sea is the outstanding example.

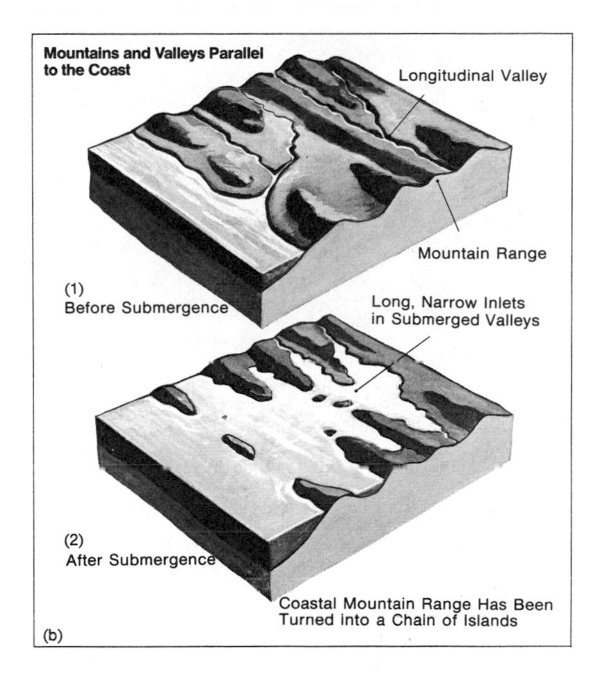

Figure 16.22 (opposite) (a) The sequence of schematic diagrams shows the typical evolution of a coastline formed where valleys and ridges transverse to the coast are submerged by a rise of sea level relative to the land.

(1) The initial coastline consists of valleys and ridges behind a narrow coastal plain.

(2) The relative rise of sea level forms drowned river valleys, or *rias,* and headlands jutting into the ocean.

(3) Wave erosion cuts back the headlands and forms vertical cliff faces. Deposition of sediment by currents builds spits and beach areas; spit growth eventually closes off the bays, forming lagoons.

(4) Continued erosion wears the coast back to a straight line bordered by vertical cliffs and a narrow beach. Stream erosion will reduce the elevation of the ridges and highlands of the land area.

(above) (b) (1) The diagram shows a coastal region where ridges and valleys are parallel to the coast.

(2) Submergence of the coastline by a relative elevation of sea level with respect to the land leaves numerous islands oriented parallel to the coast, such as those off the Dalmatian Coast of Yugoslavia. (After Bunnett, 1968)

Coral Reef Coasts

Aside from the fjord landscapes noted in Chapter 15, the only other coastal types that have a climatically controlled distribution are those composed of living organisms. The principal living organism that can produce solid masses of rocklike material on a large scale is *coral.* Corals are marine animals that build a stony exterior skeletal structure composed of calcium carbonate. Corals live in colonies composed of many species, each with its own unusual configuration, ranging from huge mushroom-like masses to delicate twiglike structures (see Figure 16.23, p. 510). Associated with corals are various forms of algae that produce limy encrustations. As these colonies and the marine plants associated with them die, new layers form, building up *coral reefs.* However, none of the reef-building coral species can survive in water temperatures lower than 18°C (65°F), and corals truly flourish only where water temperatures are between 25° and 30°C (77° to 86°F). Thus coral reefs are found only in the warm seas of tropical regions, seldom occurring poleward of the 30th parallel in the northern hemisphere. An additional requirement is clear water that is free of suspended sediment. Thus coral cannot survive near the mouths of large sediment-carrying streams, which accounts for their preference for island rather than mainland locations. Coral reefs are present in the West Indies, the Florida Keys, and as far north as Bermuda, which lies in the warm Gulf Stream just north of the 32nd parallel. The greatest development of coral reefs is in the tropical Pacific Ocean, from the Hawaiian Islands to the Great Barrier Reef of Australia.

Coral must anchor itself to a solid substrate, usually rock but sometimes sand. It rarely grows at a depth exceeding 45 meters (150 ft) due to its need for light. There are three types of coral reefs: *fringing reefs,* which are built out laterally from the shore; *barrier reefs,* which are separated from an island or landmass by a lagoon; and *atolls,* which enclose a lagoon, with no other land present (Figure 16.24, p. 510). Atolls are often thought of as circular, but they are normally very irregular in plan (Figure 16.25, p. 511). In all coral reefs, the outer

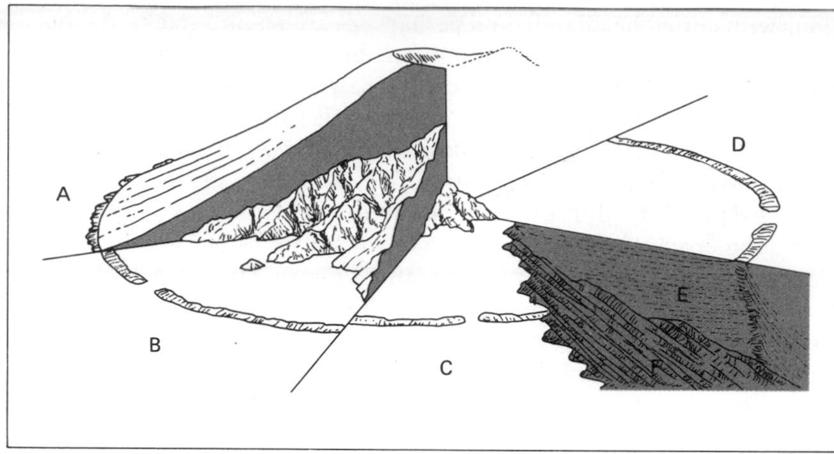

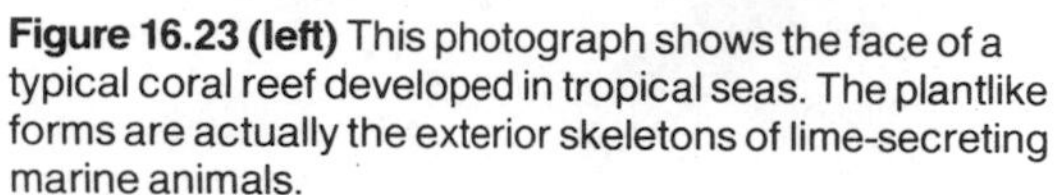
Figure 16.23 (left) This photograph shows the face of a typical coral reef developed in tropical seas. The plantlike forms are actually the exterior skeletons of lime-secreting marine animals.

Figure 16.24 (right) Deduced stages of an upgrowing barrier reef (B, C) around a subsiding volcanic island (A), producing an atoll (D) in which the thick reef limestone (E) rests unconformably on the volcanic rocks (F) (Adapted from Davis, 1915)

(seaward) slope is very steep, even overhanging, while the inner slope is quite gentle, forming a horizontal platform in the case of fringing reefs. Barrier reefs and atolls are always pierced by several openings through which boats may pass from the interior lagoons to the open sea. During storms, these passages may be very dangerous.

The formation of fringing reefs is not difficult to account for; in countless locations, coral growth can be seen extending out from the shore in reef platforms. However, barrier reefs and atolls require somewhat more explanation. Although there have been several hypotheses to account for the latter, modern investigations seem to substantiate Charles Darwin's theory that barrier reefs and atolls indicate subsidence of islands initially bordered by fringing reefs. According to Darwin, gradual subsidence of the central island caused the most vigorously expanding part of the fringing reef—its outer portion—to grow upward at a pace equal to island subsidence, thus remaining in the photic (sunlit) zone of the sea. In the case of coral atolls, subsidence has lowered the island completely below sea level, but the coral has been able to grow upward, sometimes hundreds of meters, as fast as the island sank. Some of the upward accumulation in both barrier reefs and atolls seems to be due to hurricanes and tsunamis, in which large waves tear loose subsurface coral and toss it onto the reef or into the lagoon behind.

Darwin's island subsidence theory has been supported by vertical borings made into several atolls. At Eniwetok Atoll in the western Pacific, borings penetrated 1,405 meters (4,608 ft) of shallow-water deposits that represent some 50 million years of marine-lagoonal deposition (from Eocene time to the present) and no deep-sea organisms were included at any level. There remain many unanswered questions regarding the complex structure and microtopography of barrier reefs and atolls, but their overall origin appears to have been resolved.

RELIEF FEATURES OF THE OCEAN FLOOR

While the constant change observable at shorelines makes the coastal zone one of the earth's most fascinating areas of study, the ocean

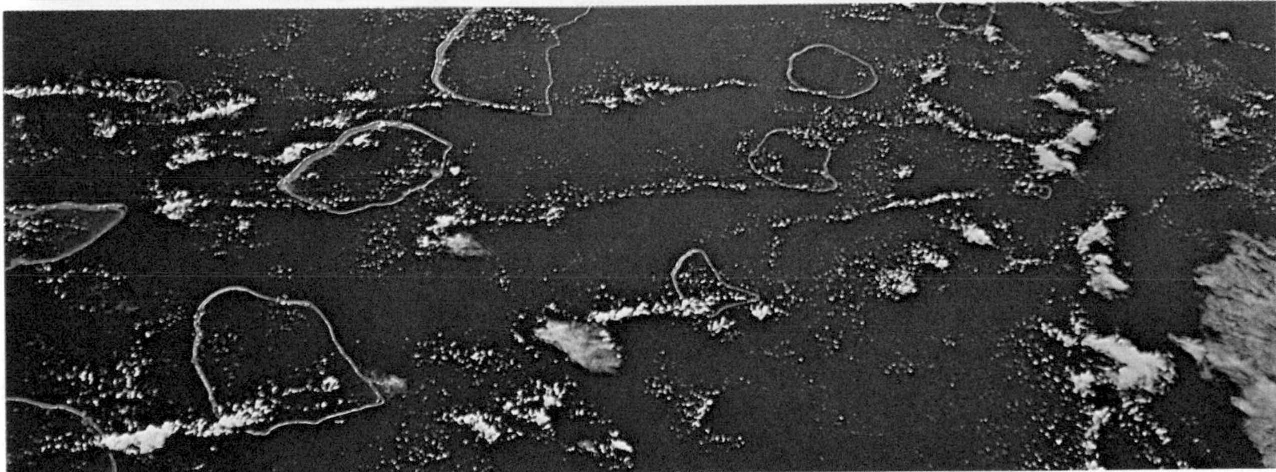

Figure 16.25 Coral builds outward from tropical coasts and upward when coasts begin to subside below the photic (sunlit) zone of the sea.

(top) A portion of the Great Barrier Reef off the northeast coast of Australia. This reef is growing upward as well as outward to keep pace with subsidence of the east coast of Australia.

(bottom) These atolls photographed from the Apollo 7 spacecraft are in the Tuamotu Archipelago in the South Pacific. The rings of coral have been able to grow upward fast enough to remain in the photic zone even though the volcanic islands forming their base have sunk completely below sea level. Note that the atolls are irregular rather than circular in plan.

floors themselves, once a mystery, are proving to have quite diversified relief, including many intriguing features that have yet to be explained.

Much of the ocean bottom lies at depths of 5,000 meters (16,000 ft) or more—well beyond the access of deep-sea craft, which have until recently been restricted to depths of 2,000 meters (6,500 ft) or less (see Figure 16.26, p. 512). However, in 1975 a manned deep submersible carried out investigations in the Caribbean at a depth of 3,660 meters (12,000 ft). The immense pressure at this depth—386 kg/sq cm (5,500 lbs/sq in.)—will make it difficult to go much deeper in the near future. Because the vast majority of the landforms of the deep ocean floor have not yet been looked at or touched, information about them must be gained by indirect means. The most widely used method for charting the relief of the ocean floor is to generate sound waves at the surface and measure the time it takes for the echo to return from the bottom, the technique having been devised to locate enemy subma-

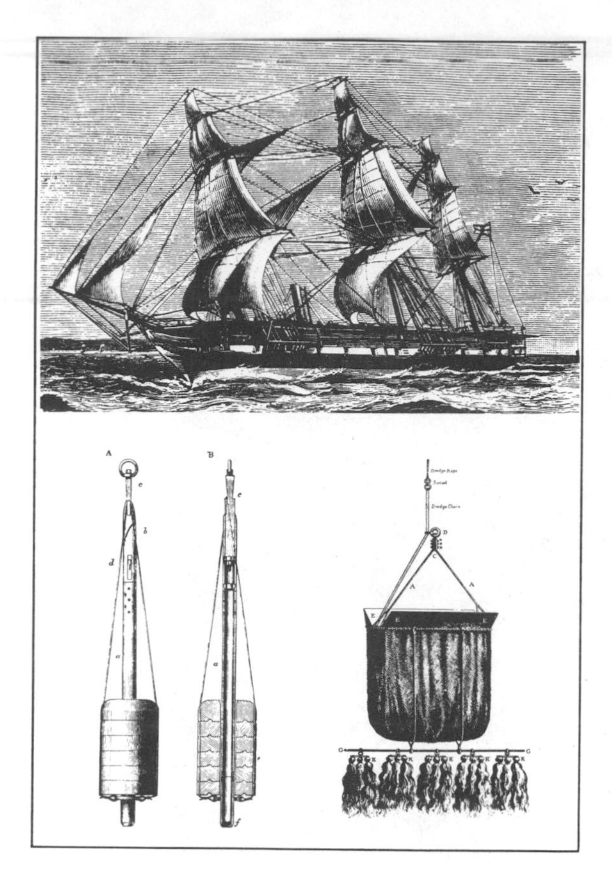

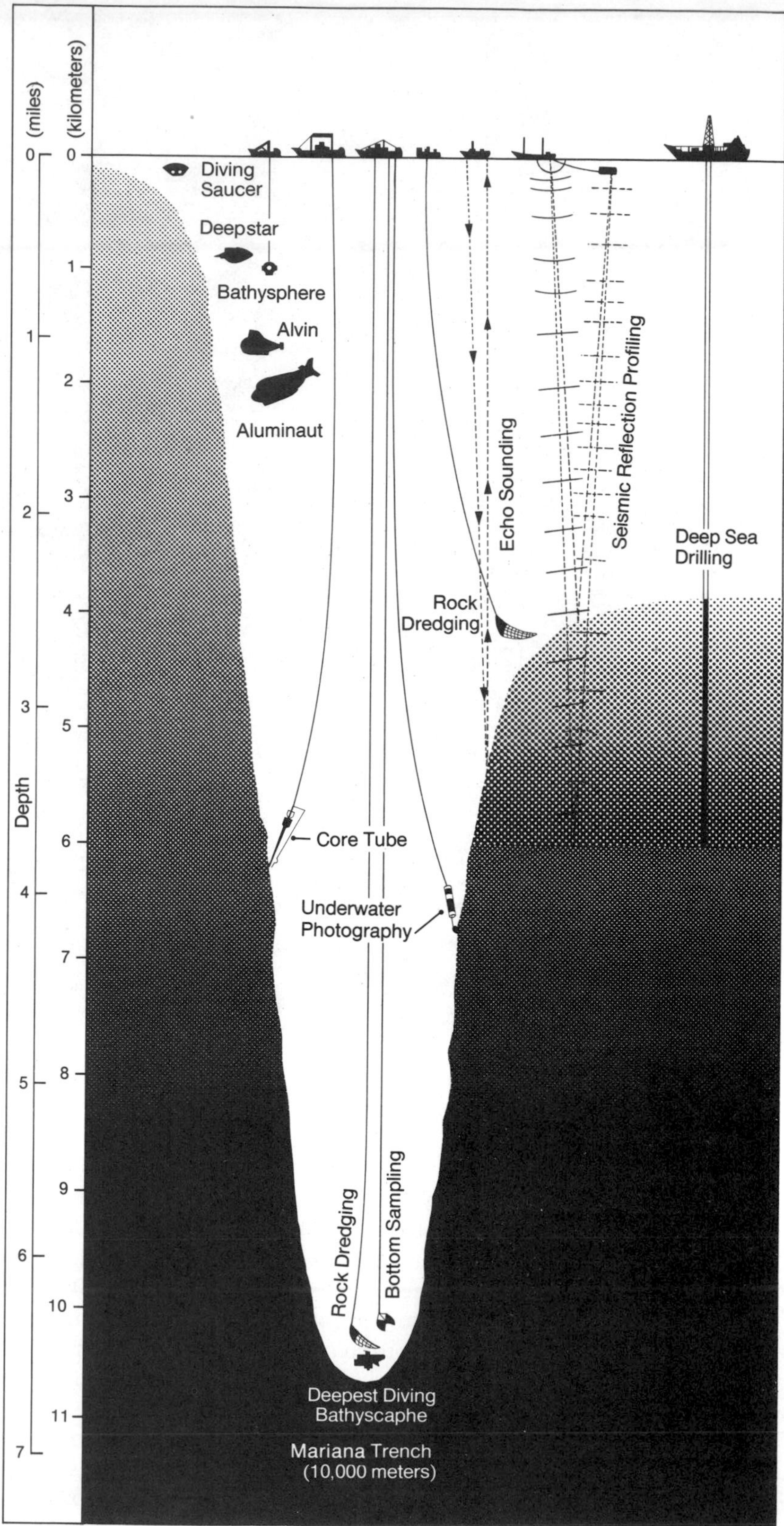

Figure 16.26 Detailed study of the ocean floor began 100 years ago with the voyage of the British research ship *HMS Challenger* **(top left)** in 1872. The *Challenger* made the first world-wide measurements, or *soundings,* of ocean depths. Some of its equipment **(above left)** included weights on lines used to measure ocean depths and dredges to bring up samples from the ocean floor. Only about 500 deep-sea soundings were made during the 3-year voyage because of the primitive and laborious methods available.

(right) Modern methods of oceanographic study include echo sounding and seismic profiling for the determination of ocean floor topography; deep-sea drilling for the retrieval of sediment samples from the ocean bottom; and the use of a variety of deep-sea and surface craft, underwater cameras, and dredging tools. (Adapted from B. C. Heezen and C. D. Hollister, *The Face of the Deep.* © 1971 by Oxford University Press, Inc.)

rines during World War II. Sound waves reflected from undersea mountaintops return to the surface sooner than sound waves that descend into trenches, so that a profile of the ocean floor can be traced automatically from echo data (see Figure 16.27). Corings and dredging devices lowered from ships especially equipped for oceanographic research also bring back samples of ocean floor material. The ongoing Deep Sea Drilling Project, utilizing the oceanographic research vessel *Glomar Challenger* mentioned in Chapter 11, retrieved cores from more than 400 boreholes in the deep ocean floor between 1968 and mid-1976. Some of these were drilled in water depths exceeding 6,000 meters (20,000 ft), a very difficult procedure.

In previous chapters, we have noted the major relief features of the ocean floors: continental shelves, continental slopes, continental rises, oceanic trenches, and oceanic ridge systems. All of these features can be explained to the satisfaction of most scientists.

Continental Shelves, Slopes, and Rises

The continental shelves have an average overall slope of less than $\frac{1}{10}$ of 1°, but they include considerable local relief in many areas (see Figure 16.28, p. 514). They are the drowned rims of the continents and appear to have been planed off by wave erosion during slow rises in sea level at the end of each Pleistocene glacial period. Thus they are at least in part abrasion platforms around the edge of the land. This surmise is supported by the depth of the outer edge of the shelves, which falls close to 130 meters (430 ft)—the approximate depth of sea level lowering during the Ice Ages. However, the elevation of the shelf edge varies considerably, and in some tectonically active areas, there

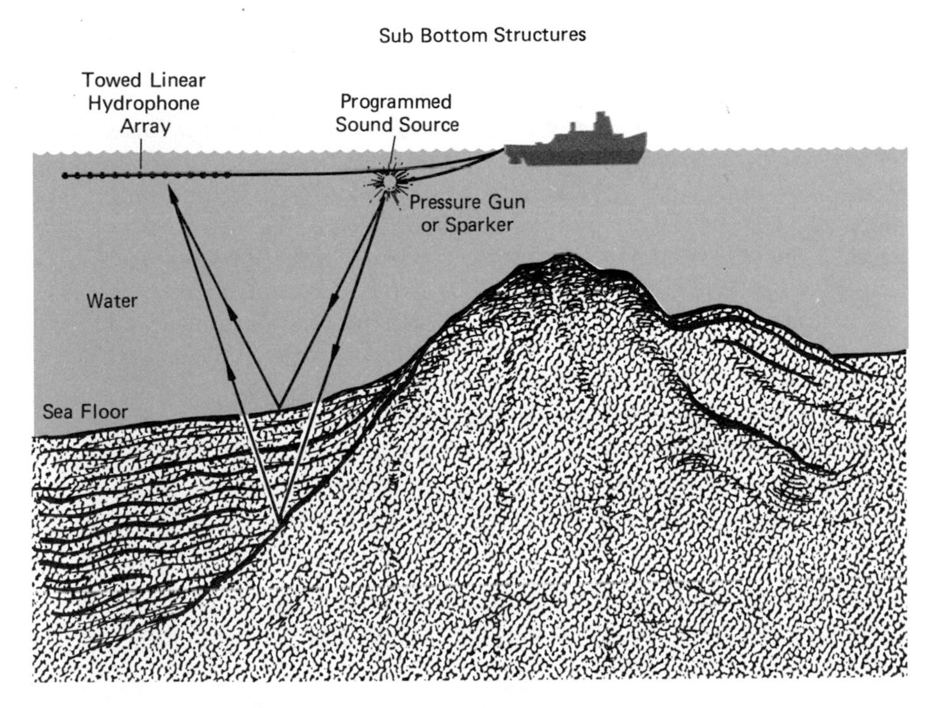

Figure 16.27 This figure illustrates the method by which information on the character of the sea floor is obtained. The oceanographic vessel tows a sound source that creates a loud noise by an electric impulse or an explosion of gas. This acoustic signal travels outward in all directions and is reflected back from the sea floor and, if strong enough, from separate layers of sediment on the sea floor as well as the rock beneath the sediments. The reflected signals are sensed by a towed hydrophone array. Usually the signals are sent out every second. The reflected signals drive a recording device that makes a continuous trace of the type shown here.

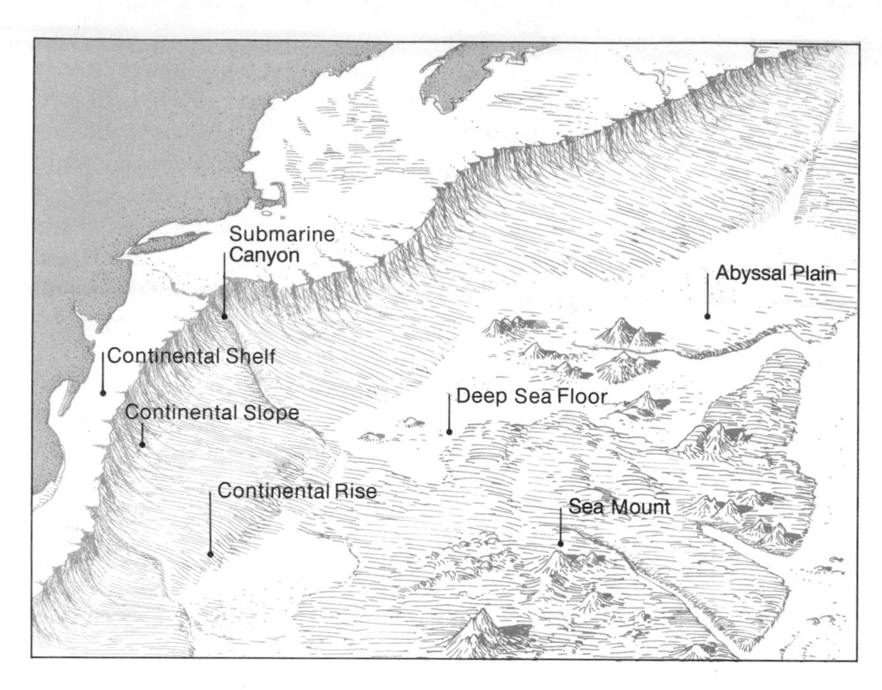

Figure 16.28 The topography of the ocean bottom off the coast of the northeastern United States shown in this figure exhibits several major divisions. The vertical scale is exaggerated approximately 20 times for clarity. Near the margin of the continent, the water is shallow and the ocean bottom forms a gentle slope called the *continental shelf.* At the edge of the shelf, the bottom slopes steeply down the *continental slope* to the deep sea floor. The *continental rise,* where the continental slope meets the sea floor, is a region where sediments accumulate. Note also the smaller features of relief such as the deeply trenched *submarine canyon* associated with the Hudson River and the relief forms on the sea floor. (After Heezen and Tharp)

is no true continental shelf at all. The average width of the shelves is 75 km (46 miles), but again there is considerable variation.

The continental shelves have long had great economic importance. Due to the shallow depths of water over the shelves, sunlight penetrates to the bottom over the inner portions, and nutrients are constantly supplied by erosion on the land. These two effects combine to support an abundant marine food chain. Thus the great fisheries of the world are associated with the continental shelves. More recently, large deposits of oil have been found in the sediments blanketing the shelves in the Gulf of Mexico, the North Sea, and other places. This has greatly intensified long-standing controversies among maritime nations about the ownership of the continental shelves.

The slightly steeper and more irregular *continental slopes* (average slope down to 1,800 meters [6,000 ft] is between 4° and 5°) descend to some 3,000 to 3,600 meters (10,000 to 12,000 ft), marking the actual edge of the continental plates, while the *continental rises* seem to be embankments of sediment washed from the land and resting against the continental slopes. It is the sediment blankets of the continental shelves and continental rises that are crumpled and upheaved by compression during the collisions of crustal plates that generate the earth's great mountain chains. Stretching out from the base of the continental rises are the deep ocean floors. Some portions of these, the so-called *abyssal plains,* are flat and featureless blankets of marine sediments that cover the older volcanic rocks of the sea floors. At one time, it was assumed that the sea floors were composed almost entirely of abyssal plains. However, the results of echo soundings show that this is far from true, and that there is a highly diversified relief beneath the waters of the seas. The greatest submarine relief features are the ocean trenches and ridge systems, which have been discussed previously. However, many interesting features of smaller scale are also present, including plateaus, great fault scarps, clusters of undersea mountains, and enormous canyons.

Submarine Canyons

The major enigmas of oceanic topography are the *submarine canyons* that gash the continental shelves and slopes throughout the world. Their existence has been known for nearly a century. According to soundings and inspections using scuba diving gear and submersible vehicles, submarine canyons are morphologically similar to ordinary fluvial valleys and canyons. They vary from broad troughs to vast steep-sided canyons having dimensions equal to those of the largest fluvial canyons on the continents (see Figure 16.29). Submarine canyons appear off the mouths of New York's Hudson River, the Columbia River (Washington), and the Zaire (Congo) River, but many others are not in the vicinity of any large terrestrial streams, and the highest-walled one of all extends between two low islands on the Bahama Bank between Florida and Cuba. Submarine canyons are extremely puzzling because many extend across the entire continental slope down to the deep sea floor at a depth of nearly 5,000 meters (16,000 ft)—far too deep to be drowned river valleys. Some portions are excavated in soft sediments, but others are cut through hard rock in steep-walled gorges. All of these canyons appear to channel great quantities of sediment from the continental shelves into the deep ocean basins. They may be the principal conduits for the debris that accumulates in the deep (oceanic trench) portions of geosynclines (see Chapter 13). Submarine canyons that are not located near the mouths of major rivers, such as those off the coast of California,

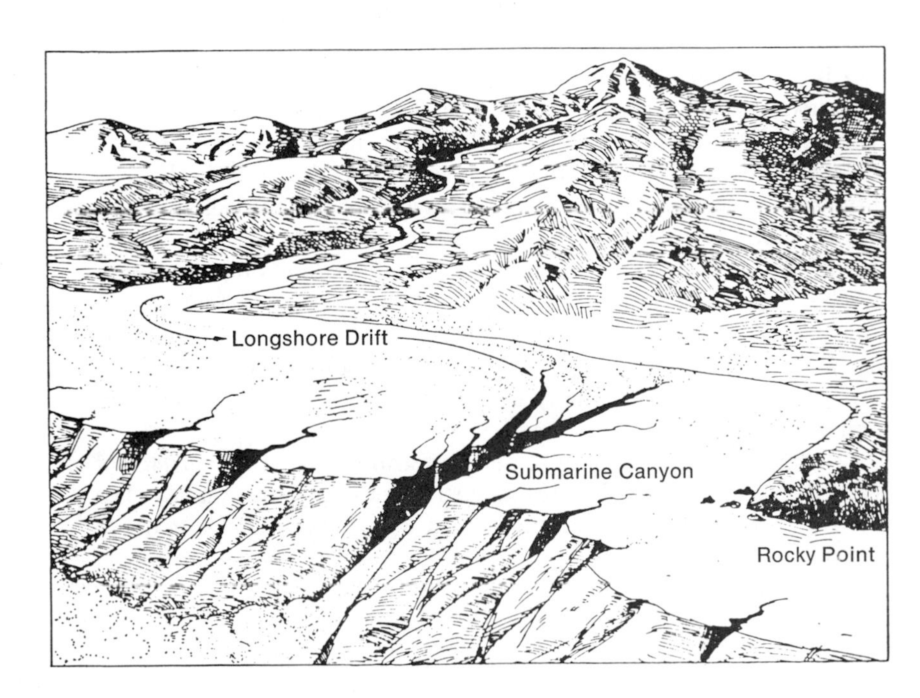

Figure 16.29 This diagram shows a submarine canyon cutting across a narrow continental shelf. These canyons are cut into bedrock and may attain the dimensions of the largest fluvial canyons on the land. They descend much more rapidly than do fluvial valleys, however. They appear to be a principal means of channelling marine sediment down to the continental rises. The longshore drift process feeds sediment into the heads of the canyons, which lie quite close to the land. The canyons then funnel this sediment into the ocean depths as submarine *turbidity currents.*

appear to trap sediment moved by longshore drift, subsequently funneling it down to the deep sea floor in great submarine landslides.

The most plausible explanation for the origin of submarine canyons is that they are products of erosion in the deep sea caused by turbidity currents. A *turbidity current* is a suspension of sand and silt that is cohesive enough to remain intact and sufficiently dense to flow down a continental slope to the sea floor. Clean washed sand and the remains of shallow-water organisms have been found in fan-like deposits at the foot of submarine canyons, providing evidence of long-distance sediment transport. Although turbidity currents have been witnessed, their ability to carry on erosion deep under water has not been fully demonstrated. However, the occasional rupture of submarine cables on continental slopes by turbidity currents seems to indicate a considerable destructive potential that over time could account for rock-cut submarine canyons. Nevertheless, much is yet to be learned about both the currents and the canyons.

Guyots

Another puzzling feature of the sea floor is the *guyot,* a high but conspicuously flat-topped submarine mountain. The form was discovered during World War II and has been named after a famous Swiss-American geographer of the last century. Hundreds of submerged volcanoes, or *seamounts,* are present in the world's oceans, particularly in the Pacific (see Figure 11.7, pp. 304–305). A large proportion of these are guyots. Although guyots have the appearance of having been planed off by erosion at the surface of the ocean, their summits are often 1,200 meters (4,000 ft) or more below the sea surface, far too deep for wave action, even during times of lowered sea level. Shallow-water fossils, including those of terrestrial plants, have also been found in deposits on the tops of some guyots.

It is generally agreed that guyots are the remnants of massive volcanoes that once projected above sea level and were subsequently truncated by erosion. However, subsidence must have followed to lower the truncated surface to such great depths. It has recently been proposed that volcanic islands were formed at oceanic ridges and then carried laterally down the flanks of the ridges into the ocean depths as part of the sea-floor-spreading process. This hypothesis of subsidence by lateral plate motion seems to be substantiated by age determinations on the volcanic rocks of the guyots, which indicate increasing age with increasing depth and distance from the oceanic ridge systems. It is probable that many coral atolls have been built up from summits of sinking guyots.

At present we have a more detailed knowledge of the surface of Mars than of the landforms of the earth's sea floors—our planet's last frontier. Future years are certain to turn up an ever-increasing list of mysteries of the deep and more problems to engage the attention of physical geographers and their colleagues in the earth sciences.

SUMMARY

The sea surface is in constant motion due to the action of waves, currents, and tides. Most waves and currents are produced by wind stress on the water surface, whereas tides are created by the gravitational attractions of the moon and, to a lesser degree, the sun, and by the rotation of the earth-moon system. Tides are not important agents of geomorphic change, but they allow wave action to reach higher areas on coastlines. Waves vary in size as a function of wind strength, duration, and fetch. The largest waves are produced by intense storms, such as hurricanes, and by submarine earthquakes and volcanic eruptions that produce rapidly moving seismic sea waves known as tsunamis. Most of the waves breaking against the shores of the land are swell traveling outward from distant storms.

Waves transmit energy from regions of energy input to far-off coastlines, where the energy is expended in the work of erosion and sediment transport. Wave motion is transmitted by oscillations of the sea surface produced by orbital motions of water particles. Where waves approach the land, they are slowed and refracted by friction with the sea floor. Along irregular coasts, wave refraction concentrates erosional energy on headlands and causes sediment to be shunted into bays, producing a tendency for the coast to become more regular. Wave break, swash, and backwash result in lateral movement of sediment in the processes of beach and longshore drifting. Beaches, spits, and barrier islands are dynamic landforms that are constantly changing in form due to shifts in the balance of sediment arrivals and losses.

Coastal landforms are related to movements of lithospheric plates, with high-relief collision coasts associated with converging oceanic and continental plates. Low-relief trailing-edge and marginal coasts occur where the continent is welded to the adjacent sea floor. Most coastal classifications are both genetic and descriptive in character. Although morphological classification of coasts as submergent or emergent is inconsistent with the general submergent trend of the past 15,000 years, features resulting from both submergence and emergence do exist, often in the same locality. Coasts that show mainly emergent characteristics are low-lying, with broad coastal plains. Coasts that display mainly submergent characteristics vary according to the nature of the submerged relief, but are usually irregular in plan and steeply rising from the shoreline.

Aside from the oceanic trenches and ridge systems that are vital in the motion of the crustal plates, relief features of the ocean floor are quite diverse, including continental shelves, slopes, and rises, submarine canyons, abyssal plains, and a host of smaller forms consisting of faulted and volcanic features such as seamounts and flat-topped guyots. The origins of some submarine landforms, especially submarine canyons and guyots, continue to be matters of active speculation.

Further Reading

Bird, E. C. F. *Coasts.* Cambridge, Mass.: M.I.T. Press, 1969. (246 pp.) Bird's small book offers a concise and well-illustrated introduction to coastal processes and landforms, with many examples being drawn from Australia.

Inman, Douglas L., and **Birchard M. Brush.** "The Coastal Challenge." *Science,* 181 (July 6, 1973): 20–31. This article details modern findings concerning the physical processes in coastal systems, and shows how human activities have changed coastal environments, often diminishing both their utility and aesthetic qualities.

King, Cuchlaine A. M. *Beaches and Coasts.* 2nd ed. New York: St. Martin's Press, 1972. (570 pp.) King combines theory, model experiments, and field observation in detailed analyses of coastal processes and landforms. This comprehensive work updates and quantifies the Zenkovich book listed below.

Russell, Richard J. *River Plains and Sea Coasts.* Berkeley: University of California Press, 1967. (173 pp.) About half of this engagingly written book is devoted to Russell's personal observations regarding shoreline processes. The author concentrates upon low-latitude coasts.

(continued on p. 518)

Case Study:

Problems at the Edge of the Land

Seacoasts are the summer playground of the world. In any coastal region accessible to large numbers of people with leisure time and money to spend on recreation, each warm weekend triggers a mass migration to the shore—for the purposes of boating, surfing, playing in the salt water, and lying in the sun. Local communities and national governments have catered to the desires of pleasure-seekers by developing beach areas, making them accessible by road, and creating protected marinas for boating enthusiasts. Is there anything wrong with this?

The problem is that once an investment is made in shoreline development, it must be protected from damage from any source, and the shoreline is as much a hazard zone as is a river floodplain. Sooner or later, a natural event occurs that becomes a catastrophe when lives and property are involved. Plans for coastline development run into a hard fact—today's shoreline is not a permanent feature. Our planet has just emerged from the trauma of the Pleistocene, with its climatic changes and sea level fluctuations. Our coasts are not yet adjusted to the prevailing sea level, which was established but a moment ago on the geologic scale of time. Everywhere the sea has been slowly rising for thousands of years. Along steep coasts this has flooded the inner rim of the continental shelf, and waves have steepened the coast even more, slowly driving back sea cliffs. But along coasts of low relief, the rise of the sea surface has been accompanied by a drastic horizontal retreat of the entire coastal zone and all the landforms that compose it. This phenomenon is still in progress.

As noted in the preceding chapter, there is a marked contrast between the low-lying Atlantic and Gulf coasts of the United States and the steeply rising Pacific coast. Both environments are a lure to people in search of beauty and entertainment, and both coasts have been developed for recreation and residential construction. On both sides of the continent, the natural system is colliding with human desires.

The Atlantic and Gulf coasts are guarded by a chain of coastal spits and barrier islands, reaching from Cape Cod in Massachusetts to Carolina's Outer Banks, then turning down the southeastern coast through Georgia's Sea Islands, Florida's Cape Canaveral, Palm Beach, Miami Beach, the Ten Thousand Islands, the Gulf Islands National Seashore, then, after the interruption of the Mississippi delta, Matagorda and Padre islands in Texas and their prolongation in Mexico. All these were once strips of sand with beaches hundreds of meters wide facing the sea. The beaches were backed by low grassy dunes, and behind the dunes was a zone of grass and shrubland, succeeded by salt marsh on the side of the island facing the mainland. The barrier islands have protected the mainland, bearing the brunt of wave attack and storm surges (water piled against the shore by hurricane winds). The barriers were frequently washed over by high seas and sometimes broken through during hurricanes, when great quantities of sediment were moved across them and deposited in the salt marshes on their landward side. Clearly, their tendency is to be moved landward, by erosion on one side and deposition on the other. In some locations, their measured shift over the past century has averaged close to 10 meters (33 ft) per year. Their migration is a normal accompaniment to a rising sea level on a low-lying coast. Along the mainland in Florida, Georgia, and the Carolinas, we can see a series of ancient barrier beaches that were driven back to the mainland during earlier interglacial periods.

But how can roads be maintained along one of these shifting barrier beaches—and parking lots, boat landings, motels, restaurants, and beach cottages? The only way is to stop the shifting. This can be done, at least for a while, by preventing overwash during storms. The technique is to heighten the dunes artificially, using sand-trapping fencing or dredged sand, and to plant them with stabilizing shrubs and dense grasses, creating a sort of sea wall behind which roads and other facilities can be located.

The effect of this modification should have been foreseen. In the man-made landscape, the energy of storm waves is concentrated on the beach in front of the artificially elevated dunes instead of being exhausted in overwash across the full width of the barrier island. Thus beaches have become exposed to wave energies they have never before experienced. First, they lose their finer particles; the beach profile becomes steeper; and then intensified backwash drags away coarser particles. The process feeds upon itself, and the beaches become progressively narrower and steeper. Today the once broad beaches of many artificially "defended" barrier islands are strips only a few tens of meters wide in front of wave-steepened dunes. Migration of the islands has been halted for the moment, but their principal attraction, the beaches, are disappearing. If present processes continue, the dunes themselves will be undercut and washed away, and the islands will retreat at a faster rate than ever before. Already sandbagging has been necessary to protect some dune areas, but this is only a temporary palliative, for sandbags and sea walls themselves produce greater beach steepening and are soon undercut and removed. Many solutions to this problem have been proposed, including elevated highways and other structures that would permit overwash, preserving the beaches, but this will not stop the inexorable migration of the island shorelines. The only real answer appears to be a more passive use of the barrier islands, maintaining them in their natural state as undeveloped beaches accessible by ferry from the mainland. This is the plan adopted for the Cape Lookout National Seashore Area, south of Cape Hatteras.

Along the Pacific Coast, the physical circumstances differ significantly. Here the problem is not the retreat of barrier islands, for such features do not exist along steep coasts, but the retreat of the mainland itself. Beach preservation is important here, for the surest way to minimize wave erosion of the land is to keep waves breaking on broad beaches rather than impacting directly against a sea cliff. Three types of human activities are causing increased erosion along the Pacific Coast: artificial destruction of coastal dunes that are reservoirs of sand that is recycled to the sea during storm surges; interference with the longshore flow of sand required to maintain beaches in a quasi-equilibrium condition; and interference with the delivery of sand from rivers flowing down to the coast.

Destruction of dunes has occurred in connection with sand mining and residential development. Here, as on the east coast, dunes have sometimes been leveled merely to provide cottage owners with unobstructed views of the sea. The classic examples of interference with longshore sand flow are the cases of the Santa Barbara and Santa Monica (California) breakwaters, both built in the 1920s to create protected harbors. The Santa Barbara jetty projects from the land in the form of an "L", while that at Santa Monica was constructed parallel to the shore to permit sand to pass by it. In both locations, the jetties reduced wave energy inside the harbor, causing sand to accumulate there rather than moving on. As would be anticipated, the Santa Barbara jetty also trapped sand moved against it by longshore drifting. In both instances, interruption of the sand flow caused alarming shrinkage of beaches farther down the coast. This exposed the land behind the beaches to increased wave attack, producing accelerated shoreline recession. The situation became so critical in both cases that permanent dredging operations had to be established in the harbors, with sand being pumped through pipelines and returned to the shoreline downcoast from the harbors. Unfortunately, the rocks exposed along much of the Pacific coast are poorly consolidated and highly susceptible to wave erosion, with normal rates of cliff retreat being in the neighborhood of 15 to 30 cm (6 to 12 in.) per year. Added to this is the tendency of California homeowners, beset by near-total summer drought, to water their lawns heavily, including those at the very brink of the sea cliffs. It has been demonstrated that lawn watering has triggered vast landslides in seaside residential areas, some individual slides having destroyed dozens of palatial homes.

The beaches of Coronado, near San Diego (California), are the ones most noticeably affected by interference with a riverine sand supply. The sand nourishing the beaches formerly drifted northward from the Tijuana River. The mouth of the Tijuana is in California, but the major portion of the stream lies across the border in Mexico, where it has been dammed, trapping its abundant sediment in a reservoir whose life will be brief—though not brief enough for the citizens of Coronado. Plans to dam large rivers in northern California raise the specter of massive beach deterioration in the future.

And so the struggle between man, determined to bend nature to his will, and the seemingly irresistible natural forces, continues. But, as we have attempted to make clear, the real struggle is to understand nature's ways, and to prevent deterioration in complex natural systems that have been disturbed by past human actions based upon a lack of knowledge of the mechanisms at work in the world we inhabit.

Shepard, Francis P. *The Earth Beneath the Sea.* 2nd ed. Baltimore: Johns Hopkins Press, 1967. (242 pp.) This very readable nontechnical treatment of the features of the sea floor is by a leading investigator of submarine canyons. As well as presenting facts and interpretations, Shepard indicates how data are collected beneath the sea.

Shepard, Francis P., and **Harold R. Wanless.** *Our Changing Coastlines.* New York: McGraw-Hill Book Co., 1971. (571 pp.) This large book is a complete inventory of the coastal morphology of the United States, including Alaska and Hawaii. The text is superbly illustrated with aerial photographs.

Steers, J. A., ed. *Applied Coastal Geomorphology.* Cambridge, Mass.: M.I.T. Press, 1971. (227 pp.) Steers has assembled articles by various authors to illustrate the nature of coastal changes over varying time intervals in different localities around the world.

Strahler, Arthur H. *A Geologist's View of Cape Cod.* Garden City, N.Y.: Natural History Press, 1966. (115 pp.) This small nicely written book details how glacial deposition and wind and wave action have fashioned the landforms of a popular tourist area. Very well illustrated and nontechnical.

Zenkovich, V. P., translated by D. G. Fry and edited by J. A. Steers and C. A. M. King. *Processes of Coastal Development.* New York: Interscience-Wiley, 1967. (738 pp.) This is a translation of a classic work by an outstanding Russian geomorphologist. The detailed treatment of shoreline processes and resulting landforms is technical but nonmathematical.

Review Questions

1. Explain the presence of simultaneous tidal bulges on opposite sides of the earth.
2. What factors produce the highest and lowest tides at any locality? What are such tides called?
3. What two types of waves are seen in the open sea? What controls the size of ocean waves?
4. Are tsunamis a greater hazard on the Atlantic, Pacific, or Gulf coast of North America? Explain.
5. In what two ways is marine sediment transported parallel to coastlines?
6. How does wave refraction affect wave energy and coastal erosion and deposition?
7. What erosional processes combine to cause the retreat of sea cliffs?
8. In what way is a beach similar to a glacier?
9. What is required to produce annual cycles of beach expansion and shrinkage?
10. In what locations do coastal spits tend to develop, both generally and specifically?
11. What are the principal features permitting recognition of submergent coasts?
12. Outline the historical development of a mountainous coastline that is fringed by marine terraces.
13. What is the relationship among fringing reefs, atolls, and guyots?
14. How might sediment from the land bypass the continental shelves to reach the continental rises and ocean trench systems?

APPENDIX I: Scientific Measurements

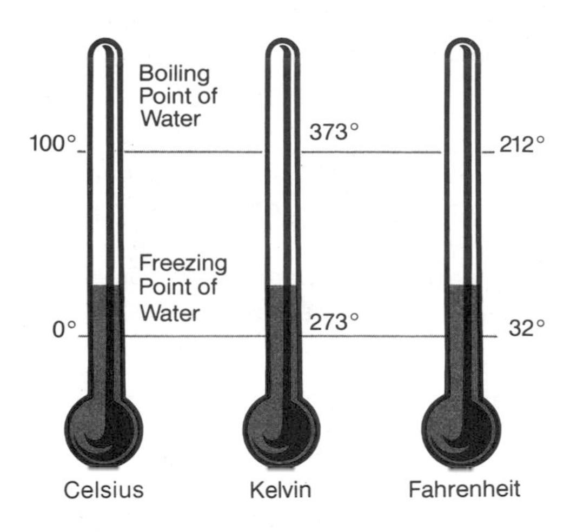

Figure I.1 The Celsius and Kelvin scales are used to report temperatures in scientific work. The freezing point of water corresponds to 0° on the Celsius scale, approximately 273° on the Kelvin scale, and 32° on the Fahrenheit scale. The temperature range between the freezing point of water and its boiling point is divided into 100 equal degrees on the Celsius and Kelvin scales, so that the boiling point of water can be expressed as 100°C or 373°K. The temperature of boiling water on the Fahrenheit scale is 212°F.

Since there are 100 degrees of temperature difference between the freezing and boiling points of water on the Celsius and Kelvin scales and 180 degrees of difference on the Fahrenheit scale, a temperature change of 1°C corresponds to a change of 1.8°F (9/5°F).

Units of Measure

The length, volume, mass, or temperature of an object must be expressed according to a definite system of units if the measurement is to be meaningful to others. In the *metric system,* the system of units used in science, length is commonly expressed in meters or centimeters (cm) (1 meter = 100 cm). One meter is about 3 in. longer than a yard, and 1 cm is approximately 3/8 in. One kilometer (km) (1,000 meters) is approximately 0.6 mile. Mass in the metric system is expressed in kilograms (kg) or grams (1 kg = 1,000 grams). One kg is equivalent to 2.2 pounds, and 1 gram is equal to about 0.035 ounce. A table of conversion factors is given in the next section.

Two scales are commonly used to measure temperature. On the *Celsius* (C) scale (formerly called the centigrade scale), the temperature at which pure water crystallizes, or "freezes," is taken as 0°C, and the temperature at which water vaporizes, or "boils," is fixed at 100°C. The interval from freezing to boiling is divided into 100 equal degrees. Normal room temperature of 72° Fahrenheit (F) is equivalent to approximately 22°C, and normal body temperature is equivalent to 37°C.

The second temperature scale, called the *absolute,* or *Kelvin* (K), has a different zero point from the Celsius scale. The temperature of a gas is a measure of the kinetic energy of its molecules; as a gas is cooled, the molecules move more slowly and have less energy. Theoretically, there is a temperature at which all motion would cease, and this point is taken as zero on the Kelvin scale. Zero on the Kelvin scale is equivalent to approximately –273°C (–459°F), so that the freezing point of water on the Kelvin scale is 273°K (see Figure I.1). Many scientific laws are simpler when expressed in absolute temperature because of its fundamental significance on a molecular level.

Metric to English Conversions

Length

1 kilometer = 1,000 meters = 0.6214 mile = 3,281 feet

1 meter = 100 centimeters = 1.0936 yards = 3.281 feet = 39.37 inches

1 centimeter = 10 millimeters = 0.3937 inch

1 micron = 10^{-6} meter = 10^{-4} centimeter = 3.937×10^{-5} inch

Area

1 square kilometer = 10^6 square meters = 0.3861 square mile = 247.1 acres

1 square meter = 10^4 square centimeters = 1.196 square yards = 10.764 square feet = 1,550.0 square inches

Volume

1 cubic kilometer = 10^9 cubic meters = 0.2399 cubic mile

1 cubic meter = 10^6 cubic centimeters = 1.308 cubic yards = 35.31 cubic feet = 61,024 cubic inches

Mass

1 metric ton = 1,000 kilograms = 2,204.6 pounds

1 kilogram = 1,000 grams = 2.2046 pounds

Time
1 day = 86,400 seconds
1 year = 3.156 × 10^7 seconds

Speed
1 meter per second = 3.281 feet per second
1 meter per second = 3.6 kilometers per hour = 2.237 miles per hour
1 kilometer per hour = 0.62 mile per hour
1 knot = 1 nautical mile per hour = 1.151 miles per hour

Pressure
1 atmosphere = 1,013.2 millibars = 760 millimeters of mercury = 29.92 inches of mercury

Temperature
°C = 5/9 (°F − 32°)
°F = 9/5 °C + 32°
°K = °C − 273.15°

Energy
1 calorie = 4.186 joules = 3.968 × 10^{-3} British Thermal Unit
1 langley = 1 calorie per square centimeter

Power
1 calorie per second = 4.186 joules per second = 4.186 watts
1 calorie per minute = 251 watts

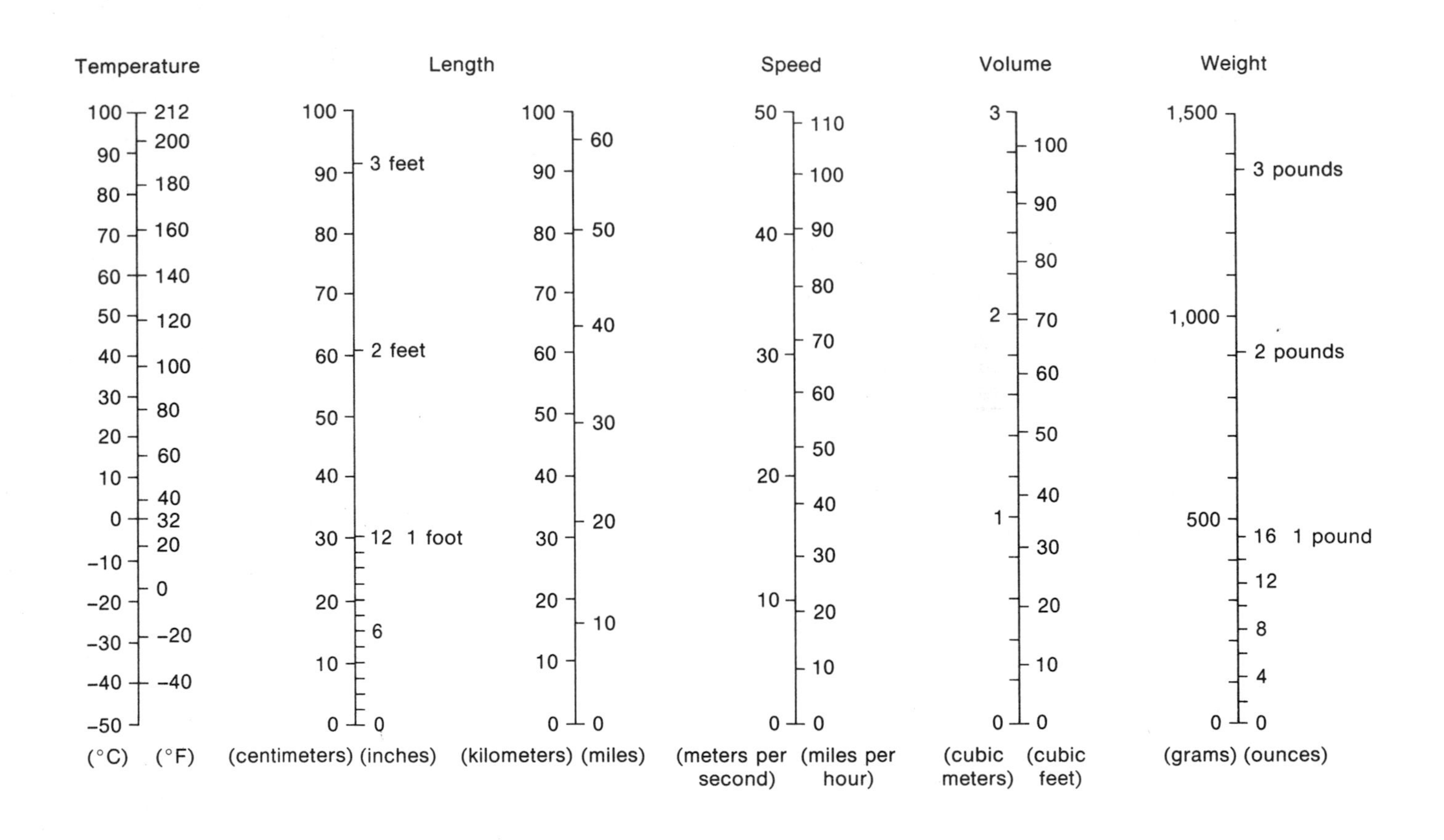

APPENDIX II: **Tools of the Physical Geographer**

Maps: A Representation of the Earth's Surface

Physical geographers are concerned with the surface of the earth—with the circulation of the atmosphere and the oceans, the distribution of climates, soils, and vegetation, and the shape and location of landforms. They use a number of special tools and techniques in their study of the earth; among these are maps for recording, interpretation, and analysis.

Modern maps exist in great variety, from simple street maps to complex navigational charts for jet aircraft. A map can convey a large amount of information in a way that is easily assimilated: a well-made map of climatic regions, for example, can make important climatic relationships much clearer than can lengthy written descriptions or tables of data. Maps have many other uses in geography: geographers may use maps on a local scale to study the particular features of landforms or small drainage basins, for example, and on field trips they draw their own maps to record significant observations. This section describes how maps display information, and it discusses the principles of their construction in order to show how to make the best use of maps.

A *map* can be defined formally as a two-dimensional graphic representation of the spatial distribution of selected phenomena. (A three-dimensional representation is more properly called a *model.*) A map is *planimetric;* that is, it shows *horizontal* spatial relationships on the earth. In addition to specifying position on the earth's surface, maps represent distance and direction from one location to another and indicate the sizes and shapes of various regions. With appropriate symbols, maps can also represent the vertical relief of landscapes, the steepness of slopes, and the location of rivers or of buildings, roads, and other man-made objects.

Scale and Distance. A map's *scale* gives the relationship between length measured on the map sheet and the corresponding distance on the earth's surface.

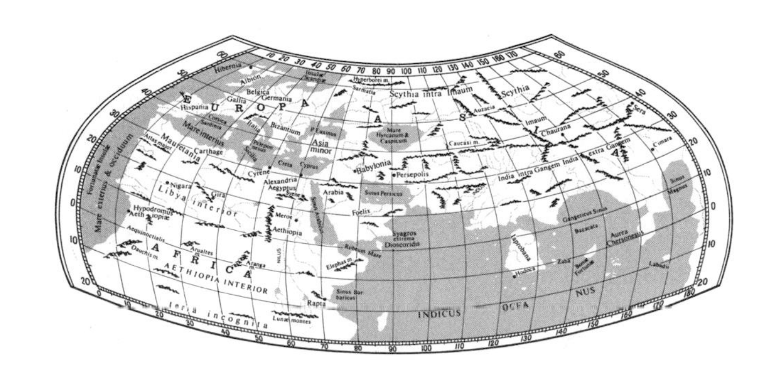

Figure II.1 (top) This Babylonian map from 500 B.C. is one of the earliest attempts to portray the world.

(bottom) Claudius Ptolemy's map of the world, reconstructed from his descriptions written in the second century, is a comparatively accurate representation that takes into account the spherical shape of the earth.

There are several ways to express the scale of a map. It may be given in a simple *verbal statement,* such as "1 in. equals 3 miles." Other lengths will bear the same proportion to distances on the earth, so that 2 in. on the map will correspond to 6 miles, and so on. The scale of a map also may be indicated by a *graphic scale* marked off in units of distance on the earth, as shown in Figure II.2a. One advantage of a graphic scale is that the scale remains correct if the map is copied in a larger or smaller size, whereas a verbal statement of scale becomes incorrect if map size is altered.

Scale is often expressed as a fraction, called the *representative fraction.* A representative fraction of 1/5,000 (commonly written 1:5,000) means that a length of 1 unit on the map represents 5,000 units of distance on the earth. A representative fraction makes no reference to any particular system of units, because it represents simply the ratio between length on the map and a corresponding distance on the earth. If the representative fraction is 1:5,000, for instance, 1 in. on the map represents 5,000 in. on the earth. Alternatively, 1 cm on the map represents 5,000 cm on the earth.

A verbal statement of scale can be restated as a representative fraction by converting both members of the statement to the same units; thus a scale of 1 in. to the mile is equivalent to 1:63,360 because there are 63,360 in. in 1 mile. Conversely, a representative fraction can be expressed as a verbal statement of scale by assigning units and applying conversion factors as required. A scale of 1:100,000 can be stated as "1 cm equals 1 km," for example. Table II.1 lists equivalents for selected scales. Note the ease of conversion between representative fractions and verbal statements of scale when the metric system is used.

When a portion of a globe is transferred to a flat map—which is known as *projecting* a spherical surface onto a plane—distortions inevitably occur. Consequently the scale of a map cannot be constant for every portion of the map. However, scale does not vary greatly on a flat map of a small region, and even the conterminous United States can be mapped in such a way that the scale does not vary by more than a few percent. Significant variations of scale may occur on global maps, however. For this reason, the map user must keep in mind that the stated scale for a given map is correct only for a limited portion of the map. The scale on a Mercator map of the world, for instance, is several times larger at higher latitudes than at the equator (Figure II.2b).

Table II.1. Map Scale Equivalents

Scale	Inches to 1 Mile	1 Inch Represents (feet)	1 Inch Represents (miles)	1 Inch Represents (kilometers)	1 Centimeter Represents (kilometers)
1:2,500	25.344	208	0.039	0.064	0.025
1:10,000	6.336	833	0.158	0.254	0.10
1:20,000	3.168	1,667	0.316	0.508	0.20
1:24,000	2.640	2,000	0.379	0.610	0.24
1:31,680	2.000	2,640	0.500	0.805	0.3168
1:50,000	1.267	4,167	0.789	1.270	0.50
1:62,500	1.014	5,208	0.986	1.588	0.625
1:63,360	1.000	5,280	1.000	1.609	0.6336
1:100,000	0.634	8,333	1.578	2.540	1.00
1:125,000	0.507	10,417	1.973	3.175	1.25
1:250,000	0.253	20,833	3.946	6.350	2.50
1:1,000,000	63×10^{-3}	83,333	15.783	25.400	10
1:5,000,000	12.7×10^{-3}	0.417×10^{6}	79	127	50
1:10,000,000	6.3×10^{-3}	0.833×10^{6}	158	254	100
1:50,000,000	1.3×10^{-3}	4.17×10^{6}	790	1,270	500
1:100,000,000	630×10^{-6}	8.33×10^{6}	1,580	2,540	1,000
1:250,000,000	250×10^{-6}	20.83×10^{6}	3,950	6,350	2,500
1:500,000,000	130×10^{-6}	41.7×10^{6}	7,900	12,700	5,000
1:1,000,000,000	63×10^{-6}	83.3×10^{6}	15,800	25,400	10,000

Figure II.2

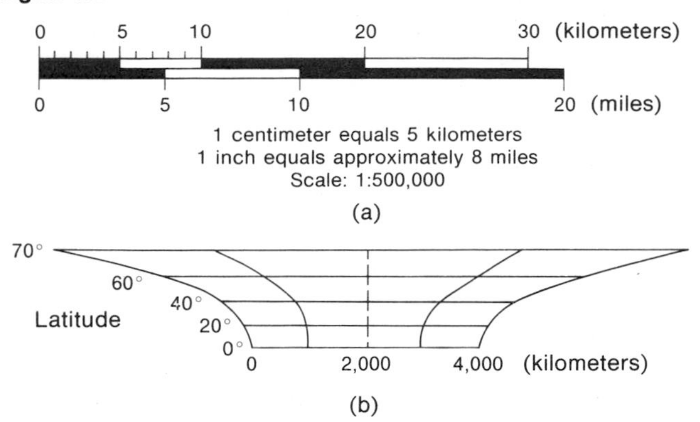

When the representative fraction is a small number, less than 1/1,000,000 (1:1,000,000), a map is called a *small-scale* map. Small-scale maps are used when a large portion of the earth's surface, such as a continent or an ocean, must be represented on a map of limited size. If a map has a representative fraction larger than 1/250,000 (1:250,000), it usually is called a *large-scale* map. Large-scale maps are capable of showing greater detail than small-scale maps. Maps are made at different scales to suit different purposes (see Figure II.3, p. 524). A geographer interested in studying a local landform or planning a day's field trip would need a map on a scale such as 1 in. to the mile (1:63,360) or larger. But to represent world climate distributions, a small-scale map with a representative fraction of 1:100,000,000 would be more suitable.

Location. The principal method for specifying location on the earth's surface is by the system of latitude and longitude. *Latitude,* the position of a place north or south of the equator, is expressed in angular measure relative to the center of the earth. The angle of latitude varies from 0° at the equator to 90° at the poles. *Longitude,* the position of a place east or west of a selected prime meridian, is expressed in angular measure that varies from 0° to 180° east or west of the prime meridian. The most commonly accepted prime meridian is the one on which the Greenwich observatory in England is located, but other prime meridians are sometimes used. The framework of lines representing parallels of latitude and meridians of longitude on a map is called the *graticule* of the map. Depending on the method chosen to construct a map, the lines of the graticule may be straight or curved, and they may or may not intersect at right angles to one another, although they do intersect at right angles on a globe.

The division of 1 degree of angular measure into 60 minutes and the subdivision of 1 minute of angular measure into 60 seconds make the system of latitude and longitude cumbersome for accurately specifying the position of objects. Therefore, alternative systems have been devised for giving location, based on the use of rectangular *grid* systems analogous to the grid system of x and y coordinates used for plotting data on graphs.

The first step in constructing a grid system is to choose a standard form of map that meets the needs of the user. (The advantages and drawbacks of different types of maps are discussed later in this appendix.) Once the map is chosen, a square grid is overlaid on the map and numerical coordinates are assigned to the reference lines of the grid. The coordinates are usually expressed in units of distance from a selected origin. The grid coordinates of any location can then be read from the map as illustrated in Figure II.4, p. 525. By convention, the coordinate to the east, or the *easting,* is specified first. Then the coordinate to the north, or the *northing,* is specified. The rule is to read toward the *right* and *up,* following the same order used for giving the x and y coordinates of a point on a graph. The grid coordinates of a location are often given as one number consisting of an even number of digits; the first half of the number gives the easting, and the second half, the northing.

One of the grid systems most frequently encountered consists of the Universal Transverse Mercator (UTM) and the Universal Polar Stereographic (UPS) grids developed for military and civilian use. The UTM grids extend between latitudes 80°N and 80°S, and the UPS grids cover the regions poleward from latitude 80°.

For the purpose of setting up the UTM grids, the globe between latitudes 80°N and 80°S is divided into 60 sections, each 6° of longitude wide. A square grid with 100,000-meter spacing is then superimposed on a transverse Mercator map of each section. The central meridian of each section is assigned an easting of 500,000 meters; the equator is assigned a northing of 0 meters for locations in the northern hemisphere and a northing of 10,000,000 meters for locations in the southern hemisphere. Each 6° section is divided into twenty 6° by 8° quadrilaterals, and for purposes of zone identication, each quadrilateral is assigned an index consisting of a number from 1 through 60 and a letter from *C* through *X* (with *I* and *O* omitted). Each of the 100,000-meter squares is further assigned a two-letter identification index (see Figure II.5, p. 525). The UPS grid zones are designated in a similar way.

The United States National Ocean Survey (formerly the United States Coast and Geodetic Survey) has designed a grid system for each of the states, called the *State Plane Coordinate System.* The basic grid square of

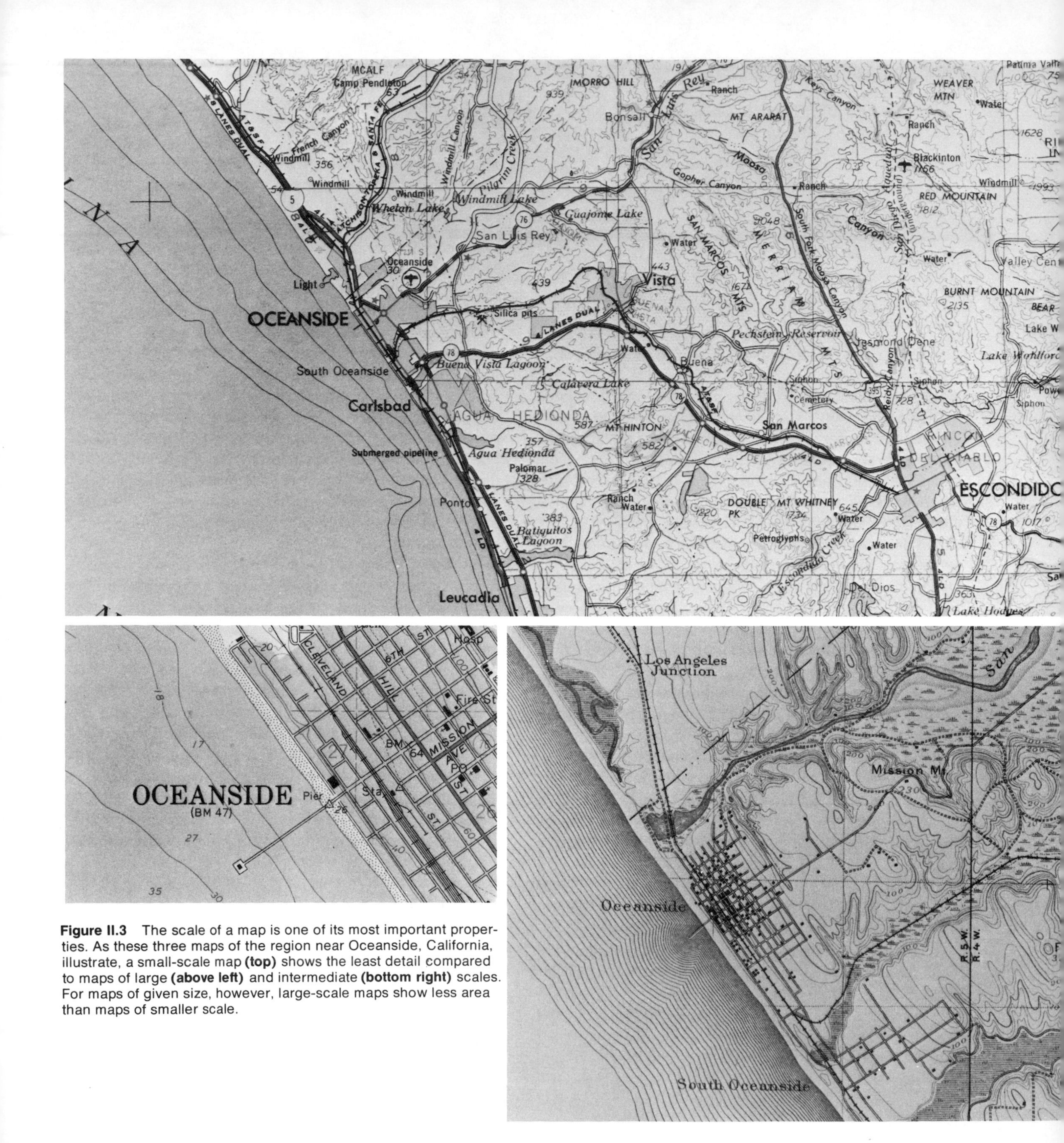

Figure II.3 The scale of a map is one of its most important properties. As these three maps of the region near Oceanside, California, illustrate, a small-scale map **(top)** shows the least detail compared to maps of large **(above left)** and intermediate **(bottom right)** scales. For maps of given size, however, large-scale maps show less area than maps of smaller scale.

Figure II.4

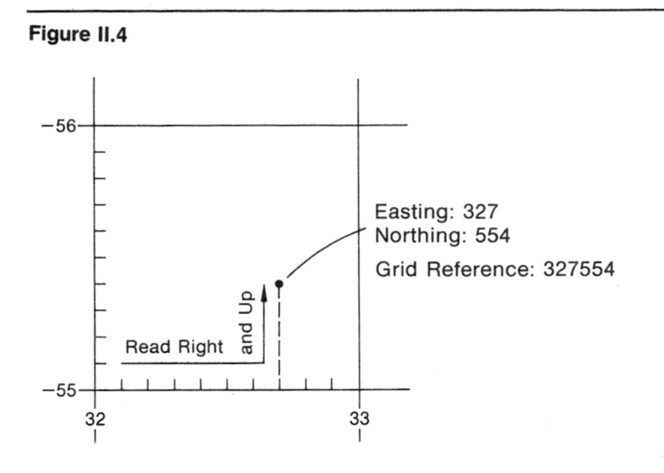

the state coordinates is 10,000 ft on a side; eastings and northings for the grid are listed in units of feet.

Some maps contain references to one or more grid coordinate systems as well as to the conventional system of latitude and longitude. The topographic maps prepared by the United States Geological Survey, for example, bear tick marks around the margins that indicate UTM coordinates, state plane coordinates, and latitude or longitude. On the topographic maps published by the United States Army Topographic Command (TOPOCOM), formerly known as the Army Map Service, the UTM grid is drawn on the map, with latitude and longitude given as tick marks on the margins.

Aeronautical charts employ a system of coordinates that is equivalent to latitude and longitude but uses a different method of notation, known as the World Geographic Reference System (GEOREF). According to the GEOREF system, longitude is divided into twenty-four 15° zones and latitude is divided into twelve 15° zones. A two-letter index specifies each 15° by 15° quadrangle, and each quadrangle is divided into 1° by 1° sections (see Figure II.6, p. 526). Four letters are necessary to specify a given 1° by 1° section: the first two provide the designation of the 15° quadrangle, and the last two, the designation of the 1° quadrangle. For accurate specification of location, each 1° interval is divided into 60 minutes of angular measure and subdivided into decimal fractions of a minute. Most aeronautical charts employ the GEOREF system rather than the UTM grid.

A modified grid system has been in use for many years in connection with the survey of public lands conducted by the Bureau of Land Management. The basic land unit of the survey, which was begun in the eighteenth century, is the *township,* a square plot 6 miles on a side. Townships are laid out with two sides along meridians and the other two sides along parallels of latitude. Because meridians converge toward the north, the north-south sides of the townships usually take a jog eastward or westward

Figure II.5 The UTM grid is subdivided successively into zones, quadrangles, and 10,000-meter grid squares.

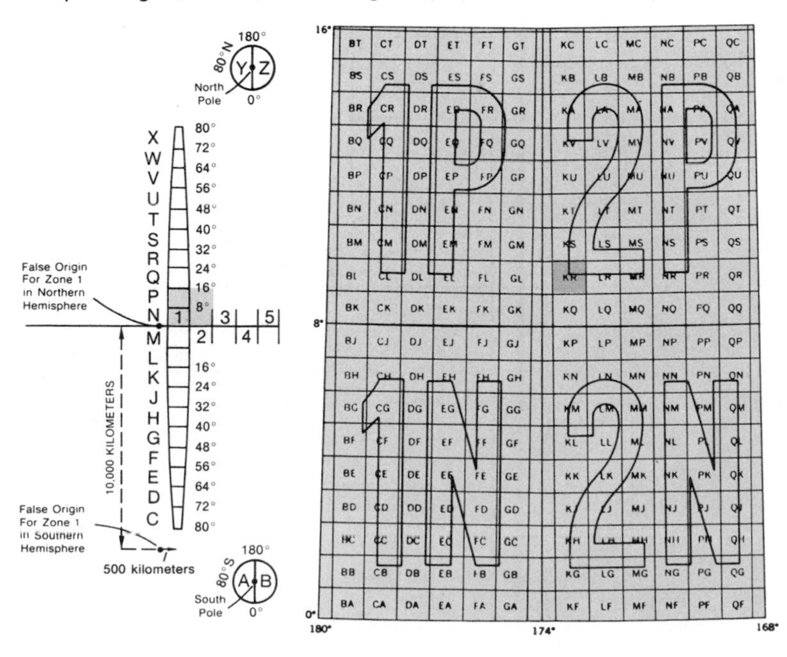

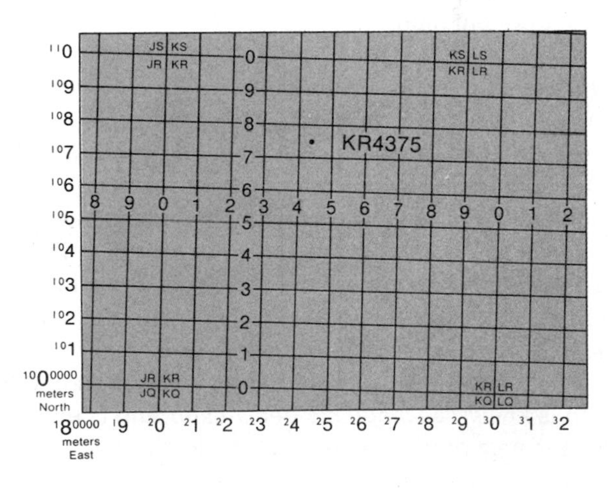

every 24 miles to maintain the size of the 6-mile square.

Townships are laid out with respect to a north-south *principal meridian* and an east-west *base line.* Different land surveys established thirty-one sets of principal meridians and base lines for the conterminous United States and five sets for Alaska. The location of each township in a survey region is given with respect to the point at which the principal meridian and the base line intersect. The coordinates that specify a particular township are read off as the number of townships north or south of the base line; the number of townships east or west of the principal meridian is called the *range.* The system for locating townships is an exception to the "right and up" rule of reading because northings are read before eastings. Townships are further subdivided into 36 squares, 1 mile on each side, which are called *sections;* sections are numbered 1 through 36 in a serpentine fashion, beginning in the upper right corner of the township (see Figure II.7).

Direction. By definition, meridians of longitude lie along a true north-south direction and parallels of latitude lie along a true east-west direction. Because of the distortions inherent in mapping the surface of a sphere onto a flat sheet of paper, meridians or parallels often vary in direction across a map sheet. There is sometimes no uniform direction relative to the map sheet that represents north, although the north-south line at any given location is always coincident with the local meridians. But for large-scale maps that cover only a local region, the distortions are small and the map sheet can be aligned with respect to a single standard direction to establish a sense of orientation.

Many large-scale maps indicate the direction of *true north* by means of a star-tipped arrow or the symbol *TN.* However, this direction is usually not the same as *magnetic north,* the direction in which a magnetic compass needle points. Large-scale maps usually indicate the direction of magnetic north by means of a half-headed arrow and the symbol *MN.* The earth's magnetic field is not uniform, and the magnetic poles do not coincide spatially with the geographic poles, so the relation between magnetic north and true north must be specified separately for each local region. The difference between magnetic north and true north is known as *magnetic declination* and is expressed in degrees east or west of the true meridian of a given location. Across the conterminous United States, for example, the magnetic declination varies from 0° to as much as 25° east or west, and in polar regions 90° or more is possible. Furthermore, the direction of magnetic north at a given location varies with time, often by as much as 1° in 20 years. For precision map work, therefore,

Figure II.6 The 15° square of the GEOREF system is subdivided into 1° squares.

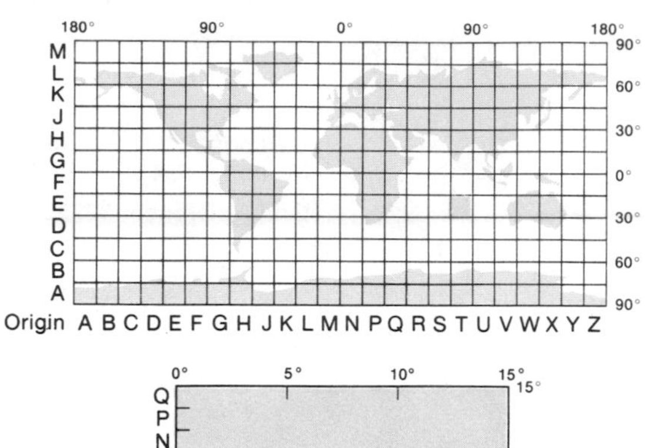

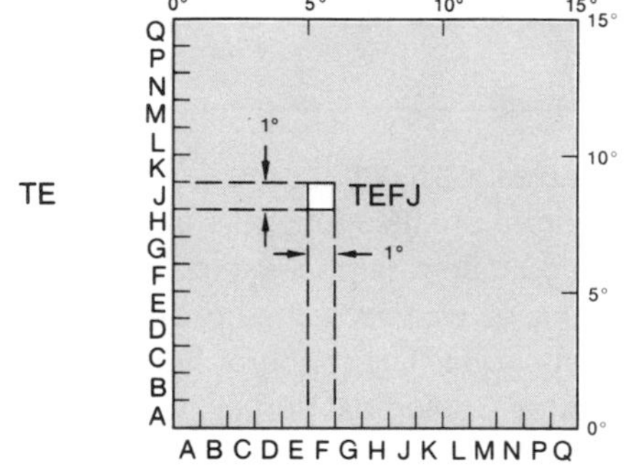

Figure II.7

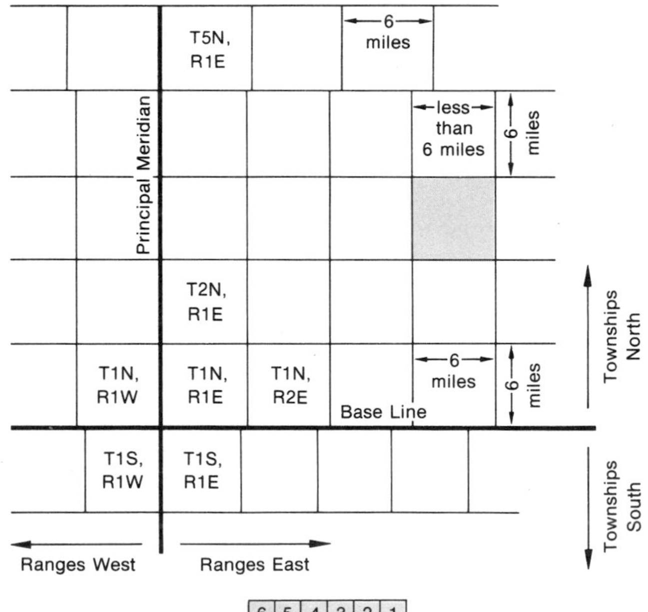

6	5	4	3	2	1
7	8	9	10	11	12
18	17	16	15	14	13
19	20	21	22	23	24
30	29	28	27	26	25
31	32	33	34	35	36

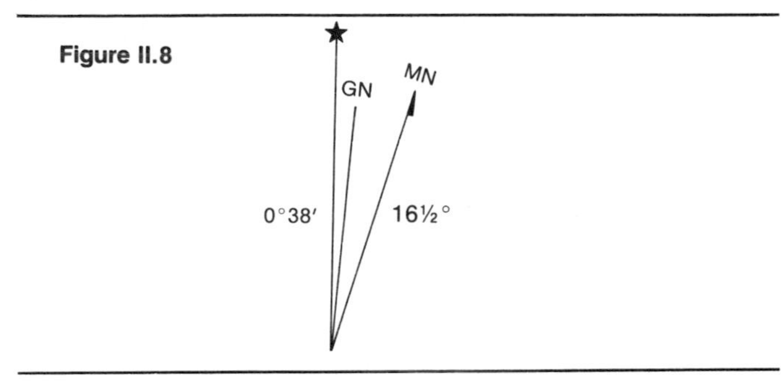

Figure II.8

reference should be made to recent compilations of magnetic declinations, such as those prepared by the National Ocean Survey.

Direction on a map may also be specified as *grid north,* the northerly direction arbitrarily determined by a particular grid system, and symbolized on the map by *GN.* Grid north generally does not coincide with either magnetic north or true north. The grid north directions specified by two different grid coordinate systems usually differ from one another as well.

Directions other than north can be expressed in terms of *azimuth,* which is the angle of the desired direction measured clockwise from a chosen reference direction and expressed in degrees between 0° and 360°. According to the choice of true north, magnetic north, or grid north as a reference, the corresponding azimuths are termed *true azimuth, magnetic azimuth,* or *grid azimuth.*

The Shape of the Earth. An exact determination of the earth's shape is required for the highest precision in map making and in position finding on the earth's surface. The earth bulges slightly at the equator, so that the diameter of the earth measured in a plane through the equator is 43 km (27 miles) longer than the polar diameter.

In the United States, Canada, and Mexico, the reference surface for map making is the *Clarke spheroid of 1866,* a smooth mathematical surface that closely approximates the shape of the earth. The Clarke spheroid of 1866 assumes an equatorial diameter for the earth of approximately 12,756 km (7,926 miles) and a polar diameter of approximately 12,713 km (7,899 miles). These values are in agreement with a 1968 determination of the earth's shape made by an orbiting satellite. The difference between the equatorial and polar diameters is so small that the earth is considered a sphere in most map work.

Table II.2 lists the lengths corresponding to 1° of longitude and 1° of latitude according to the Clarke spheroid of 1866. The length of 1° on a meridian is nearly constant from the equator to the poles, but the length of 1° on a parallel tends toward 0 at the poles. At latitude 60°, the length of 1° on a parallel is approximately half its value at the equator.

Map Projections

A model globe is the only way to represent large portions of the earth's surface with accuracy, because only a globe correctly takes into account the spherical shape of the

Table II.2a. Length of 1° on a Meridian (Clarke Spheroid of 1866)

Latitude	Kilometers	Statute Miles	Nautical Miles*
0°-1°	110.57	68.70	59.70
10°-11°	110.60	68.73	59.72
20°-21°	110.71	68.79	59.78
30°-31°	110.86	68.88	59.86
40°-41°	111.04	69.00	59.96
50°-51°	111.24	69.12	60.06
60°-61°	111.42	69.24	60.16
70°-71°	111.57	69.33	60.24
80°-81°	111.67	69.39	60.30
89°-90°	111.70	69.41	60.31

Table II.2b. Length of 1° on a Parallel (Clarke Spheroid of 1866)

Latitude	Kilometers	Statute Miles	Nautical Miles*
0°	111.32	69.17	60.11
10°	109.64	68.13	59.20
20°	104.65	65.03	56.51
30°	96.49	59.96	52.10
40°	85.40	53.06	46.11
50°	71.70	44.55	38.71
60°	55.80	34.67	30.13
70°	38.19	23.73	20.62
80°	19.39	12.05	10.47
89°	1.95	1.21	1.05

* 1 nautical mile is approximately equal to 6,076.1 feet, or 1.15 statute miles.
Source: Adapted from List, Robert (ed.). 1971. *Smithsonian Meteorological Tables.* 6th rev. ed. Washington, D. C.: Smithsonian Institution Press.

earth. A flat piece of paper cannot be fitted closely to a sphere without wrinkling or tearing, so small-scale maps that represent regions hundreds or thousands of miles in extent inevitably introduce noticeable distortions. Maps printed on flat sheets are far more portable and convenient to use than globes, however, and it is important for the geographer to be able to employ flat maps with an understanding of their properties.

The fundamental problem of map making is to find a method of transferring a spherical surface onto a flat sheet in a way that minimizes undesirable distortions. Any method of relating position on a globe to position on a flat map is called a *projection,* or *transformation.* A projection is a correspondence between a globe and a flat sheet such that every point on the globe or selected portion of the globe can be assigned a corresponding point on the sheet. Numerous projections have been devised for mapping, each with its characteristic advantages and distortions. Because no projection is free of distortion, the choice of a projection should be made with regard to its proposed application. Several different projections are used for the maps in this book: the flat polar quartic projection for global maps, the stereographic projection for maps of polar regions, and the Albers' conic equal area projection for maps of the United States. Each projection has advantages for its particular application, and no single projection is best for all uses.

The principles of projection are illustrated in Figure II.9 using the example of the gnomonic projection. This projection can be constructed by tracing the rays of light from a light source at the center of a transparent globe onto a plane that touches the globe at one point, called the *point of tangency.* Each point on the portion of the surface of the globe that is projected onto the plane is assigned a corresponding point on the plane, which constitutes a map. However, only a few projections, such as the gnomonic, can be visualized geometrically. Many projections can be expressed only as a mathematical rule that relates points on a globe to points on a flat sheet.

Projections and Distortions. The principles of projection ensure that every properly drawn map shows correct location on the earth's surface, regardless of the projection used. However, maps are often relied upon to show other properties, such as direction and distance from one location to another and the shape and size of areas. A single flat map cannot depict all of these properties without distortion, so every projection represents a compromise. The map user should realize where and to what extent inaccuracies are present in the projection he is using. The gnomonic projection illustrated in Figure II.9, for example, exaggerates distances far from the point of tangency

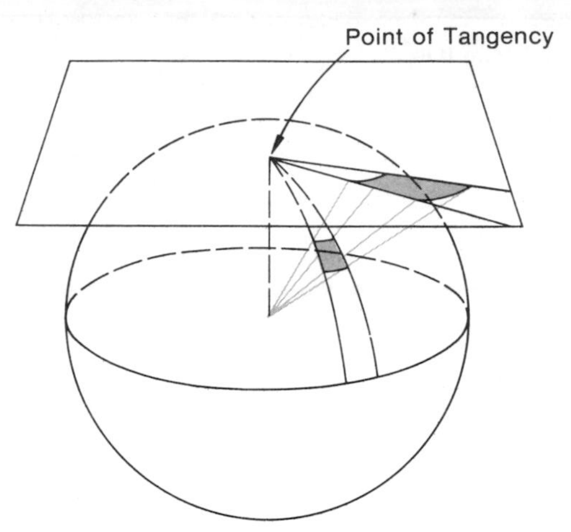

Figure II.9 Construction of the Gnomonic Projection

compared to distances near the point of tangency. Distortion of distance (scale distortion) is therefore greatest near the outer borders of a gnomonic projection and least near the point of tangency.

Scale distortion may lead to distortion of direction, shape, and size, depending on the projection. On some projections, the scale in a small region is not the same in all directions, which necessarily leads to distortion of direction. An azimuth of 45° with respect to a meridian on a globe, for example, will not be mapped as a 45° angle if the scale on the map is different in the east-west and north-south directions. Furthermore, distortion of direction implies that the outline of a small region will not be mapped accurately, and so distortion of shape will be present on such a map.

A number of projections, called *conformal* projections, have been devised so that scale is the same from every direction in a small region of the map. Shapes of small regions are therefore portrayed accurately also. However, the scale on a conformal map necessarily changes from one region to another. Regions that span a large portion of the globe tend to exhibit overall distortion when mapped on a conformal projection; each small section is accurately depicted, but the relative sizes of separate sections may be incorrect. The well-known standard Mercator projection is conformal, but it represents areas in the higher latitudes several times larger than areas of the same size near the equator.

Another important class of projections is the *equal area,* or *equivalent,* projections. On such projections, the scale in a small region is different in different directions, and shapes are distorted. However, the scale is designed to vary over the map in such a manner that the relative sizes of any two areas are correct. Figure II.10 illustrates how shapes can be distorted without altering their areas. Equal area projections are especially useful for displaying distributions in physical geography because areas are represented in their correct proportions and misleading relative comparisons are avoided. Many projections are neither conformal nor equivalent area but represent compromises to obtain adequate representation of shape without badly distorting size.

Figure II.10

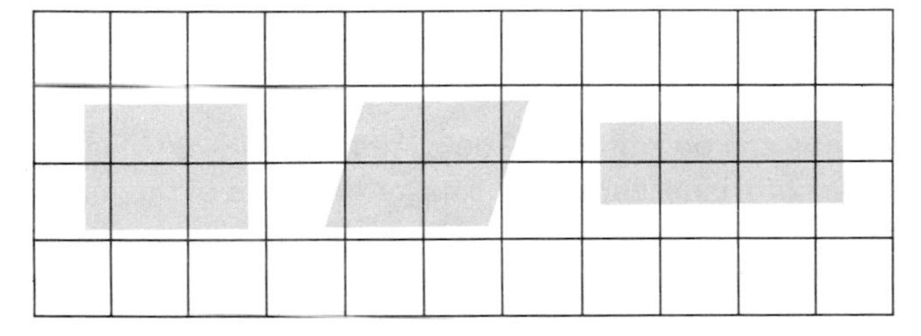

An impression of a projection's major properties can sometimes be obtained by seeing how the lines of latitude and longitude are portrayed. On a globe, parallels and meridians intersect at right angles. If a certain map shows them crossing at right angles, the projection may or may not be conformal, but if they are shown crossing at a different angle, distortion of direction is certainly present and the projection is not conformal. Distortions of shape or size can be seen by comparing the small quadrangles bounded by parallels and meridians on a projection to those on a globe.

Map projections can be made from a globe onto a plane, cylinder, or cone, which are the only surfaces that can be spread on a flat sheet without distortion. Projections are conventionally classified into families: *azimuthal,* or *zenithal,* projections of the globe onto a plane, *cylindric* projections onto a cylinder, and *conic* projections onto a cone. A fourth family, sometimes called *pseudocylindric* projections, is reserved for certain projections usually used to portray the entire globe. The projections in a given family tend to have similar properties and similar distortion characteristics.

Azimuthal Projections. Azimuthal projections are projections of a globe onto a plane tangent to the globe at some point. The point of tangency is often taken to be the north or south pole or a location on the equator, but in principle any point on the globe may be used. Most azimuthal projections can depict only one hemisphere of the earth or less at one time.

In azimuthal projections, azimuths measured around the point of tangency are mapped with no error. The distortion of any azimuthal projection is least nearest the point of tangency and increases with increased distance away from the point of tangency. Figure II.11, p. 531, exhibits some of the characteristics of selected azimuthal projections. Distortion patterns are depicted by degrees of yellow shading, with the deeper tints corresponding to regions of greater distortion.

Cylindric Projections. A cylinder closely fitted to a globe makes contact with the globe along a *great circle,* called the *circle of tangency.* Cylindric projections are usually designed so that the circle of tangency is the equator or a meridian. If the equator is chosen as the circle of tangency, the whole surface of the earth, except for the polar regions, can be shown on one map.

Most cylindric projections with the equator as the circle of tangency show parallels of latitude and meridians of longitude as sets of straight parallel lines intersecting at right angles. The best known such projection, the standard Mercator, is conformal; others are not. The standard Mercator projection is frequently used for navigation charts because a straight line on this projection represents a line of constant azimuth, which simplifies navigation.

Distortions on a cylindric projection are least nearest the circle of tangency and increase with increased distance from the circle of tangency. On a standard Mercator projection, for example, areas in the higher latitudes are grossly exaggerated. Conversely, on a cylindric equal area projection based on the equator, shapes at higher latitudes are badly distorted.

The base maps for the UTM grid system are prepared using a Mercator projection fitted to two arcs of tangency in each of the 60 separate sections between 80°N and 80°S latitude. No point on one of the 60 sectional maps is more than 150 miles or so from a meridian of tangency, and distortion is correspondingly small. Figure II.12, p. 532, shows the properties of several cylindric projections.

Conic Projections. A cone placed upon a globe contacts the globe along a circle of tangency. If the apex of the cone is above a pole, the circle of tangency coincides with a parallel of latitude, known as the *standard parallel* of the conic projection. Parallels of latitude are shown as curved arcs in such conic projections, and meridians of longitude as converging straight lines. The distortions of a conic projection are least nearest the standard parallel and increase away from the standard parallel. Conic projections are therefore useful in mapping countries, such as the United States, that extend primarily along the east-west direction. Conic projections are usually restricted to portions of the earth's surface that cover a limited range of latitude. Such a map usually does not extend more than 100° in longitude to avoid excessively curved parallels.

Conic projections of greater accuracy are produced by allowing the cone to cut through the globe, so that contact is made at two standard parallels. If the standard parallels are not too far apart, the whole region bounded by the parallels can be mapped to excellent accuracy. Albers' equal area conic projection, which was chosen for the maps in *The National Atlas of the United States,* can show the conterminous United States with a linear scale distortion that does not exceed 2 percent. For the conterminous United States, the standard parallels of the Albers' conic projection are chosen to be 29½°N and 45½°N. For Alaska, the standard parallels are 55°N and 65°N, and for Hawaii, 8°N and 18°N. The properties of several conic projections are shown in Figure II.13, p. 533.

Pseudocylindric Global Projections. A projection that shows the entire earth is useful for displaying global distributions of climate, soils, vegetation, and other quantities important in physical geography. Such a map should be an equal area projection so that relative areas can be compared without error. If the projection displays parallels of latitude as parallel straight lines, regions with the same latitude are aligned on the map, a useful property because many quantities of interest in physical geography depend on latitude.

Although significant distortion always accompanies any attempt to portray the entire earth on a single flat map, several projections have been developed that give creditable representations. On most of the commonly used global projections, the equator and the prime meridian are shown as straight lines that intersect at right angles at the center of the projection. In general, the regions of greatest distortion on such projections lie near the outer margins. Regions of least distortion are at the center of the projection, often at the intersection of the equator and the prime meridian or near the prime meridian in the midlatitudes. Figure II.14, p. 534, shows the properties of several pseudocylindric projections.

The *flat polar quartic* projection, an equal area projection, is the basis for many of the global distribution maps in this text (see page 79). In this projection, the poles are represented by lines one-third the length of the equator, and the boundary of the map is a complex curve. Distortion on the flat polar quartic projection tends to be smallest near the intersection of the equator and the prime meridian and greatest at high latitudes near the margins.

Interruption and Condensation of Projections. On a map showing global distributions, the landmasses of the earth are often of greater interest than the oceans. In such a case, the projection can be *interrupted* in the Atlantic, Pacific, and Indian oceans. The projection may then be reprojected to standard meridians constructed through each major landmass so that no land area is far from a standard meridian, which improves the overall accuracy of the map. If ocean areas are of no interest in a particular application, portions of the Atlantic and Pacific oceans can be omitted entirely, producing a *condensed interrupted map.* Condensation does not reduce the distortion of the map, but condensed maps make good use of available space in displaying landmasses.

Generalization in Maps

No map can be sufficiently detailed to show every minor aspect of a landscape. The smaller the scale of the map, the less the detail that can be presented on a map of given size. The map maker must therefore simplify the information that his map is to present.

The simplification process, which is known as *generalization,* entails omitting detailed information that would fill a map with unreadable clutter, but not at the expense of obscuring the general character of the represented landscape. A large-scale map is not merely an enlarged portion of a small-scale map; in general, a large-scale map shows a greater amount of detail than does a small-scale map of the same region.

Generalization is a process that depends on the skill of the map maker and his ability to retain essential features on a map while eliminating detail. A map maker cannot, for example, show the individual peaks and slopes of the Andes Mountains on a global map. However, by artful choice of symbols he can communicate impressions of their height and ruggedness.

Symbolization: The Representation of Relief

A variety of graphic symbols represents natural and man-made objects on maps. Some symbols have a pictorial quality: railroad tracks, for example, are often represented by a line with regularly spaced cross ticks. Other symbols are purely conventional choices, and because there is no standard system of symbolization for maps, the map legend or symbol table should always be consulted.

The graphic methods used to denote objects such as rivers, roads, and buildings on a map must often be symbolic generalizations because of the difficulty of showing the object in true scale. A stream 80 ft wide would have to be rendered as a nearly invisible line 1/1,000 in. wide if scale were followed exactly on a map with a scale

Figure II.11

Table of Azimuthal Projections

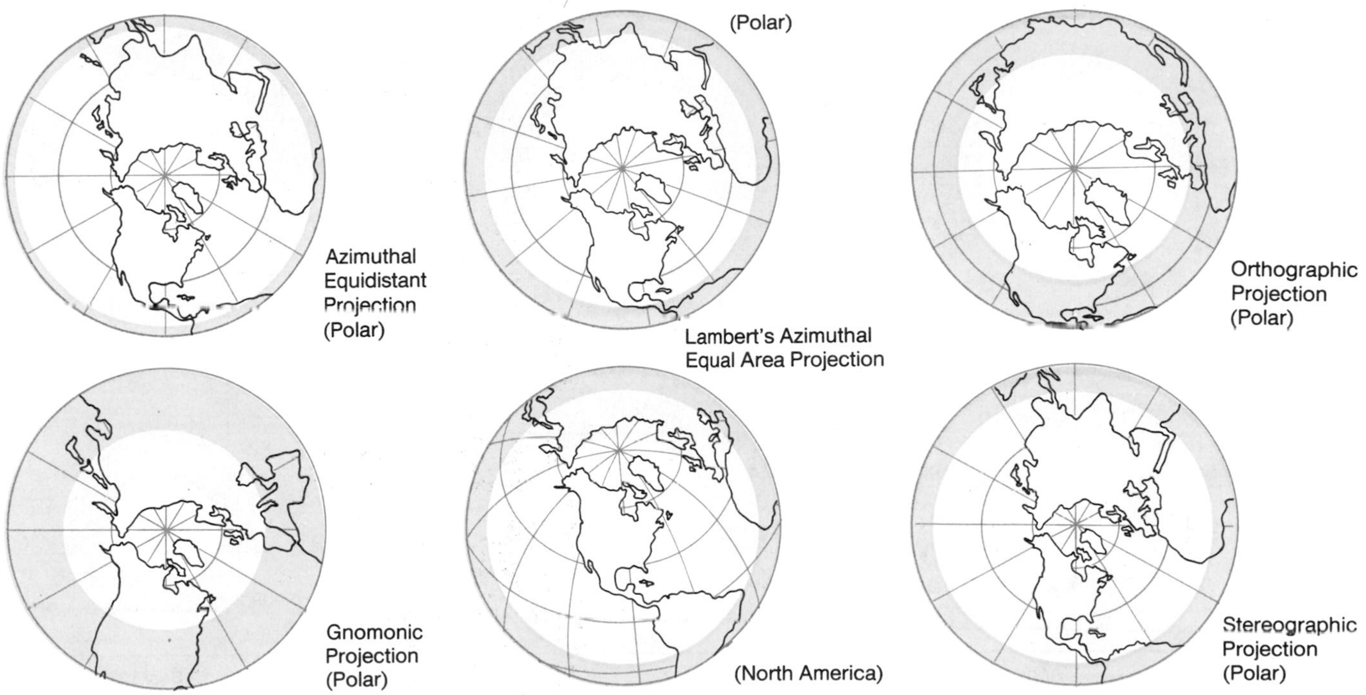

Azimuthal Equidistant Projection (Polar)

Lambert's Azimuthal Equal Area Projection (Polar)

Orthographic Projection (Polar)

Gnomonic Projection (Polar)

(North America)

Stereographic Projection (Polar)

Projection	Appearance	Properties	Distortion Pattern	Best Uses
Azimuthal Equidistant (Polar)	Meridians are straight lines outward from the pole; parallels are equally spaced circles concentric about the pole.	Scale is constant and correct along meridians. Directions from central point are correct.	Distortion increases slowly with increased distance from the center. Shapes are represented comparatively well, but areas are distorted.	Directions and distance to the center are undistorted; useful for charts of radio propagation to a given location or for showing relative distance from a given location.
Gnomonic (Polar)	Meridians are straight lines outward from the pole; parallels are circles concentric about the pole with rapidly increasing spacing outward.	Great circles anywhere on the map are represented by straight lines.	Distortion increases rapidly with increased distance from the center. Shapes become badly distorted.	Polar navigation charts and great circle navigation. (The shortest route between two points on the earth's surface lies on a great circle.)
Lambert's Equal Area (Polar)	Meridians are straight lines outward from the pole; parallels are circles concentric about the pole with slowly decreasing spacing outward.	The only azimuthal equal area projection. Directions from central point are correct.	Distortion increases moderately with increased distance from the center. Shapes are represented well.	Polar maps and maps of one hemisphere, especially where distributions are to be represented.
Orthographic (Polar)	Meridians are straight lines outward from the pole; parallels are circles concentric about the pole with rapidly decreasing spacing outward.	Directions from central point are correct. Gives appearance of earth as seen from deep space.	Distortion increases moderately with increased distance from the center.	Used primarily for illustrations, shows how the earth looks from outer space.
Stereographic (Polar)	Meridians are straight lines outward from the pole; parallels are circles concentric about the pole with moderately increasing spacing outward.	The only azimuthal conformal projection. Directions from central point are correct.	Area distortion increases rapidly with increased distance from the center. Shapes are represented well.	Basis map for the UPS grid system, poleward of latitude 80°.

Figure II.12

Table of Cylindric Projections

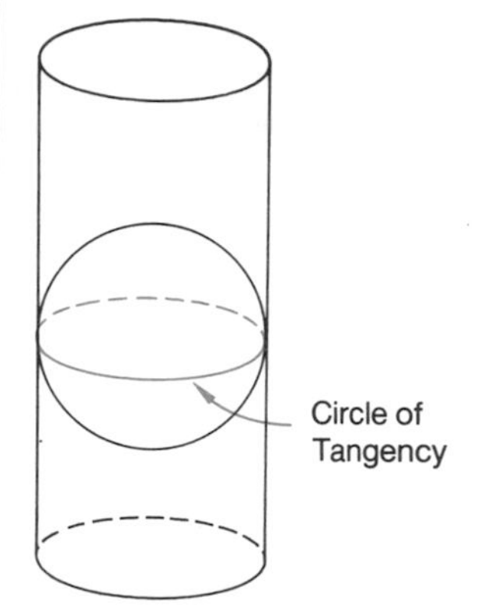

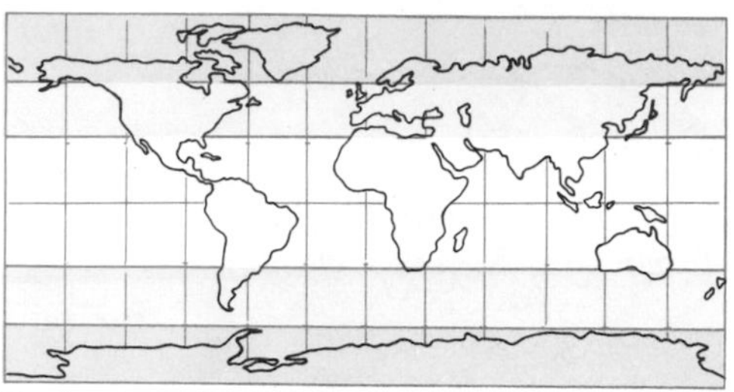

Equirectangular Projection

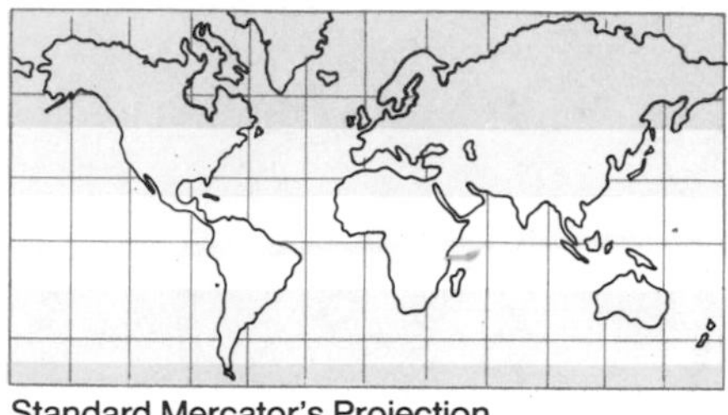

Standard Mercator's Projection

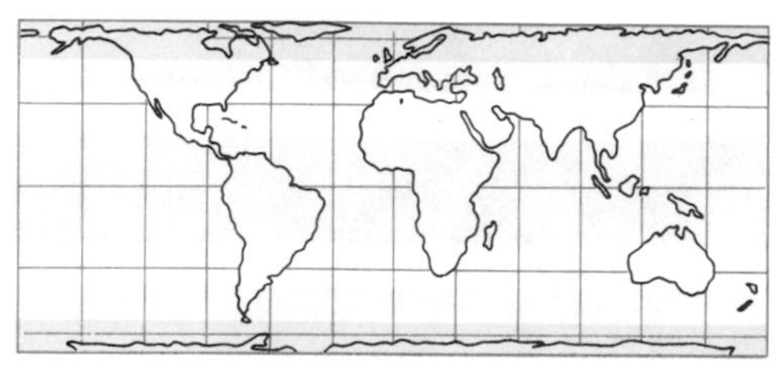

Lambert's Cylindrical Equal Area Projection

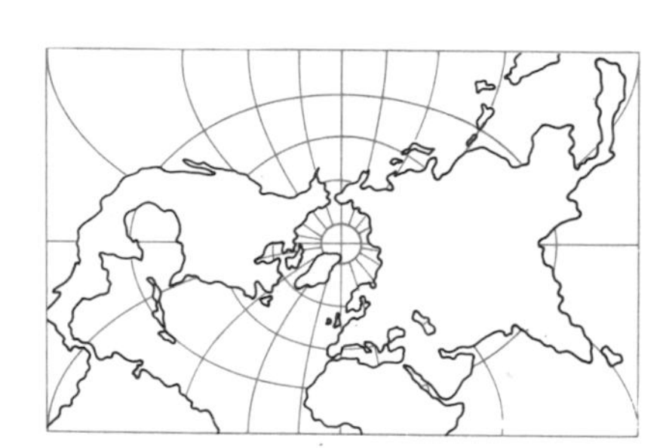

Transverse Mercator's Projection

Projection	Appearance	Properties	Distortion Pattern	Best Uses
Equirectangular	Parallels and meridians are equally spaced and form a square grid. Often a parallel is chosen to be the circle of tangency.	No major properties.	Scale is correct along the standard parallel and along meridians, but shape and area distortion increase with increased distance from the standard parallel. Shapes and areas distant from the standard parallel are badly distorted.	Used only for large- or moderate-scale maps of limited areas.
Lambert's Cylindrical Equal Area	Parallels and meridians form a rectangular grid. Employs two parallels equidistant from the Equator as circles of tangency.	An equal area projection. Can depict the whole earth except the polar regions.	Shape distortion increases with increased distance from the standard parallels. Shapes at high latitudes are seriously compressed in the north-south direction.	Not widely used because of severe shape distortion at high latitudes, but it would be satisfactory for presenting distributions in the lower latitudes.
Standard Mercator's	Parallels and meridians form a rectangular grid, with the Equator as the circle of tangency.	A conformal projection. A straight line on the map is a line of constant azimuth on the earth. Can depict the whole earth except the polar regions.	Area distortion increases rapidly with increased distance from the Equator. Shapes of small regions are represented well, but there are gross distortions of area at high latitudes.	Navigation charts.
Transverse Mercator's	The circle of tangency is a meridian or portion of a meridian. Most meridians and parallels are curved.	A conformal projection. Most straight lines on the map are not lines of constant azimuth on the earth.	Area distortion increases with increased distance from the central meridian. Scale distortion is constant along lines parallel to the central meridian. Shapes of small regions are represented well.	Basis map for UTM grid system, for some State Plane Coordinate grids, and for the British Ordinance Survey maps.

Figure II.13

Table of Conic Projections

Albers' Conic Equal Area Projection

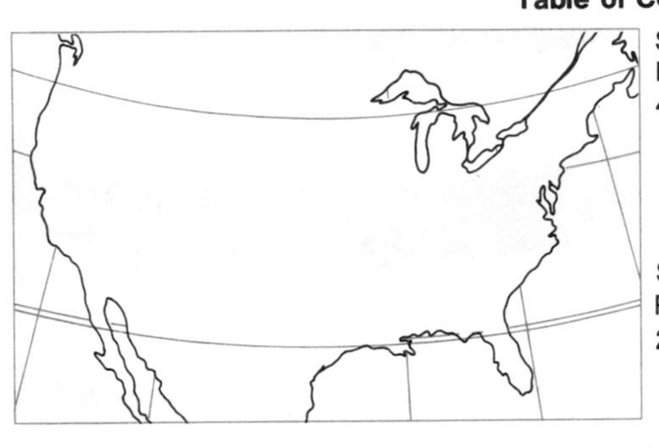

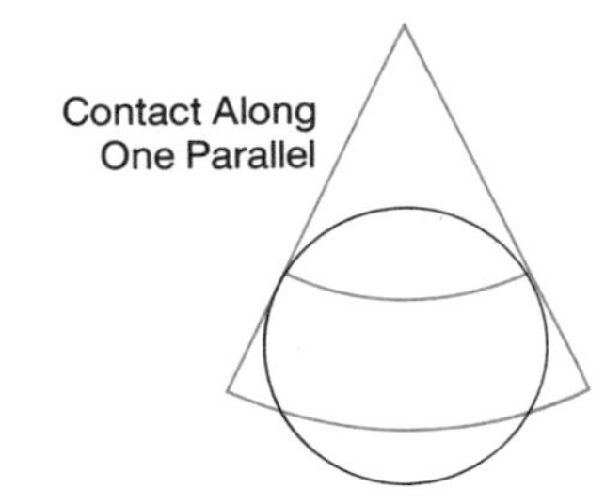

Lambert's Conic Conformal Projection

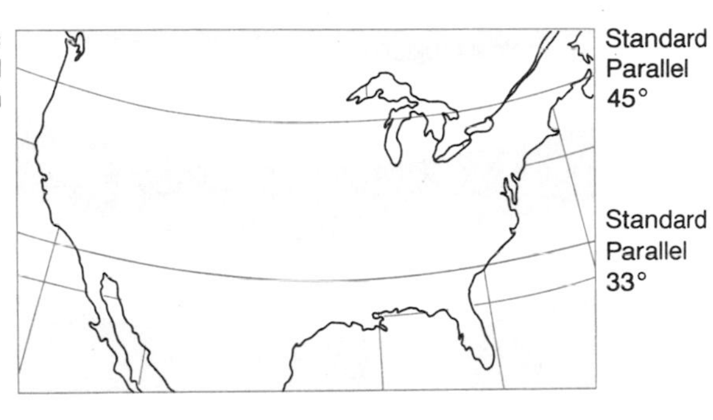

Polyconic Projection

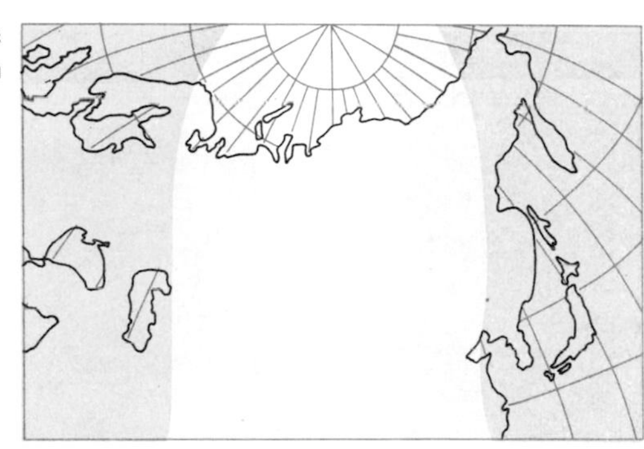

Projection	Appearance	Properties	Distortion Pattern	Best Uses
Albers' Conic Equal Area	Meridians are converging straight lines, and parallels are arcs of concentric circles. Two standard paralles are used; for maps of the conterminous United States, they are 29½°N and 45½°N.	An equal area projection.	Distortion increases slowly with increased distance from the standard parallels. Scale along meridians is slightly too large between the standard parallels and somewhat too small outside them. Distortion is small in normal applications.	An excellent projection for countries such as the United States, which extends primarily along the east-west direction. Usually applied over a restricted range of longitude. Distributions are shown without bias because it is an equal area projection. Chosen for the maps in the National Atlas of the United States.
Lambert's Conic Conformal	Meridians are converging straight lines, and parallels are arcs of concentric circles. Two standard parallels are used; for maps of the conterminous United States, they are often 33°N and 45°N.	A conformal projection.	Distortion increases slowly with increased distance from the standard parallels. Scale along meridians is slightly too small between the standard parallels and somewhat too large outside them. Distortion is small in normal applications, but slightly greater than for Albers' projection.	Used for countries that extend primarily in the east-west direction and for some air navigation charts because straight lines on the map represent great circles.
Polyconic	Meridians are curves converging toward the poles, except for one straight central meridian. Parallels are arcs of circles, except perhaps for one parallel, but each circle has its own center. Every parallel is a standard parallel for the projection.	An excellent compromise between a conformal projection and an equal area projection.	Distortion tends to increase with increased distance east and west of the central meridian. Distortion is small when small regions are mapped.	Basis for the topographic maps of the United States Geological Survey. Satisfactory for maps of small regions, particularly regions extending primarily north and south.

Figure II.14

Table of Pseudocylindric Projections

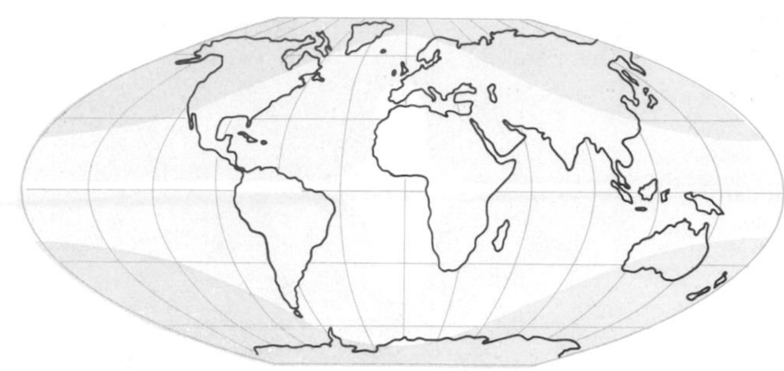

Flat Polar Quartic Equal Area Projection

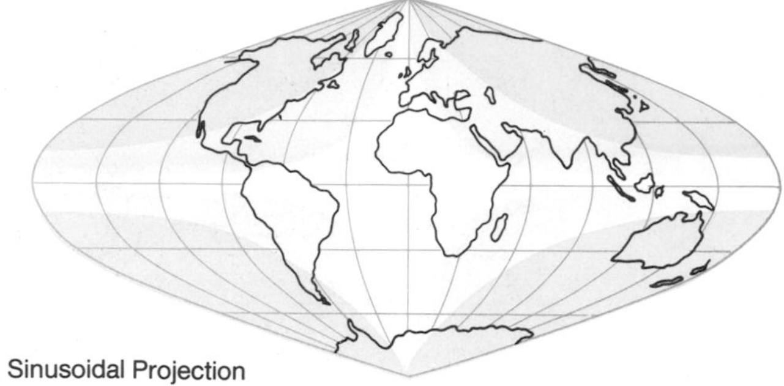

Sinusoidal Projection

Mollweide's Projection

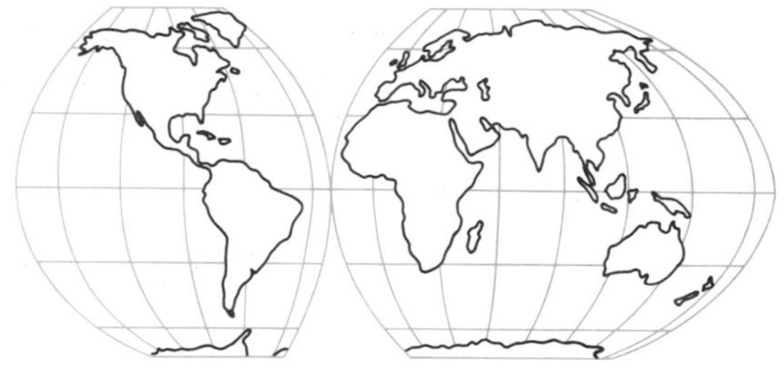

Interrupted Flat Polar Quartic Projection

Projection	Appearance	Properties	Distortion Pattern	Best Uses
Flat Polar Quartic Equal Area	Parallels are straight parallel lines. Meridians are curves in general but the central meridian is straight. The poles are represented by straight lines one-third the length of the Equator. Meridians converge toward the poles and are equally spaced along each parallel. The spacing between parallels decreases slightly with increased latitude. The boundary of the map is a complex curve.	An equal area projection.	Distortion is least nearest the Equator and central meridian and greatest at high latitudes near the boundaries.	Generally useful as a world map and for depicting global distributions.
Mollweide's	Parallels are straight parallel lines. Meridians are elliptical in general, but the central meridian is a straight line half the length of the Equator. The meridians converge to a point at each pole and are equally spaced along each parallel. The spacing between parallels decreases slightly with increased latitude. The boundary of the map is an ellipse.	An equal area projection.	Distortion is least in midlatitude regions near the central meridian and greatest at high latitudes near the boundaries.	Generally useful as a world map and for depicting global distributions.
Sinusoidal (Sanson-Flamsteed)	Parallels are straight parallel, equally spaced lines. Meridians and the boundary are sinusoidal curves. The central meridian is straight and half the length of the Equator. Meridians converge to points at the poles. The length of each parallel is equal to its length on a globe of corresponding size.	An equal area projection.	Distortion is least nearest the Equator and the central meridian and greatest at high latitudes near the boundaries.	Somewhat inferior to other projections as a global map, but useful for maps of individual continents.

of 1:1,000,000. Instead, the map maker must employ a narrow yet easily visible line to portray the river. Similarly, the individual turns and bends of a river are omitted on a small-scale map, leaving only the larger curves to denote the stream's course.

The physical geographer interested in landforms is especially concerned with the symbols that depict vertical relief on maps. *Contour lines,* special kinds of *shading,* and *color tints* are some of the methods for indicating relief on maps. For quantitative work, particularly with large-scale maps, relief is represented most accurately by contour lines.

A contour line on a map represents a line of constant altitude above a chosen reference level, called a *datum plane,* such as mean sea level, the average height of the surface of the sea. Consider the hilly island in Figure II.15. The figure shows horizontal planes at a regular vertical spacing, or *contour interval,* of 200 ft. Each

Figure II.15

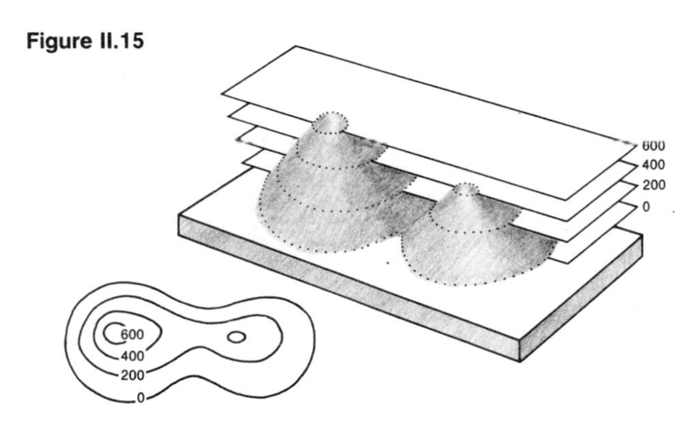

Figure II.16

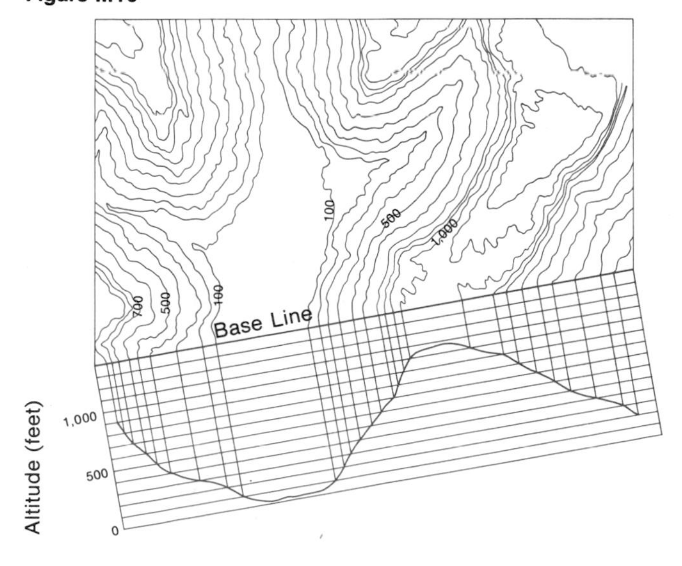

Figure II.17 The relation between contour lines and topography is illustrated in this landscape model. The model was cut from a block of plastic by a cutter set to trace contour lines from a map; the depth of cut was adjusted according to the elevation of each contour line. Although individual layers are distinguishable on the model, the closely spaced contours give a good impression of the topographic relief.

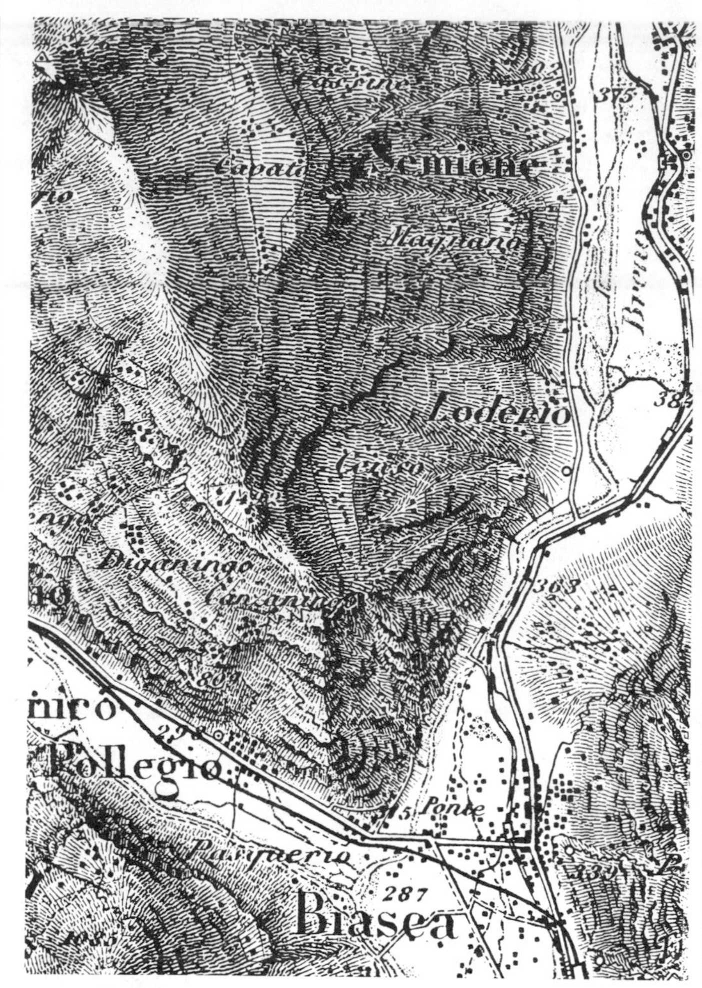

Figure II.18 Relief can be depicted on maps by hachuring **(left)** or by shaded relief **(right)**.

horizontal plane cuts the surface of the ground in a curve, which is the contour line for that altitude; every point on a given contour line is the same altitude above the datum plane.

As shown in Figure II.16, p. 535, contour lines can be transferred to a flat map to represent vertical relief. For accuracy without unnecessary clutter, the contour interval for a map should be chosen to be commensurate with the nature of the landscape being depicted; the contour interval will normally be larger for a mountainous region than for a gently sloping plain. The choice of contour interval also depends on the scale of the map. On a large-scale map, for example, the contour interval may be only a few feet, and the contour lines will show comparatively small features of slopes. A small-scale map of the same region will employ a larger contour interval, and in addition the small kinks and bends of the contours will be smoothed and averaged, so that less detail will be represented. To make contour lines easier to read, every fifth contour line is thickened and labeled with its altitude. The user of a contour map should bear in mind the value of the contour interval between adjacent contour lines.

As Figure II.16 indicates, the horizontal spacing of the contour lines on a contour map can be interpreted in terms of the local slope of the landform. Contour lines that are relatively close together represent a steeper slope than do contour lines that are relatively far apart. The convexity or concavity of a slope can also be inferred by inspecting the horizontal spacing of the contour lines. For quantitative purposes, a *topographic profile* of the landscape along a given direction can be prepared from a contour map according to the method illustrated in Figure II.16. The vertical scale of a topographic profile is usually exaggerated with respect to the horizontal scale in order to portray relief features more clearly. The altitudes of hilltops and the bottoms of depressions are sometimes stated explicitly as *spot heights* on a map.

In addition to contour lines and spot heights, a variety of qualitative or partly quantitative artistic techniques can be used on a map to indicate relief. In *hachuring,* a method popular among cartographers in the nineteenth century, a slope is depicted by straight line segments drawn along the direction of maximum steepness of slope, so that hachure marks are at right angles to contours. The marks are drawn so that their density, either in width or spacing, increases with increased steepness; steep slopes appear darker than gentle slopes in hachuring (Figure II.18 *left*). Hachured maps present a pleasing appearance when well drawn, but they are seldom used now because of the difficulty of reading slope angle from them and because of the labor involved in their construction.

A modern method of symbolizing relief on a map is to add shading to give an impression of height and depth. Such *shaded relief* is commonly drawn as though the area were illuminated from the upper left, or "northwest," corner of the map (Figure II.18 *right*).

Altitude tinting, the use of color tints for successive ranges of altitudes, is often employed, particularly on small-scale maps where contour lines cannot be read easily. Green is usually used for altitudes near sea level, and in a typical system, colors for higher altitudes progress through yellow, orange, red, and brown. The green tint used for low altitudes should not be taken as indicative of vegetation cover. The methods of hachure, shaded relief, and altitude tinting are sometimes used in combination on a contour map to depict the general character of a landscape while retaining the quantitative accuracy of contours.

Three-dimensional models with exaggerated vertical relief are mass-produced by molding thin sheets of plastic; on large-scale models, the quality of the molding is sufficiently good to represent the landscape accurately. Global relief maps are sometimes prepared from photographs of three-dimensional models, which are illuminated from the upper left to emphasize relief by light and shadows.

A *block diagram,* such as the one in Figure II.19, is a means of representing landforms and their underlying geologic structure in one drawing. A number of block diagrams are used in this text. Strictly speaking, a block diagram is not a map because it does not show position directly, but the landforms depicted on block diagrams are usually drawn with reference to contour maps of the region. The vertical scale on a block diagram is usually designed to be the same as the horizontal scale in order to

Figure II.19

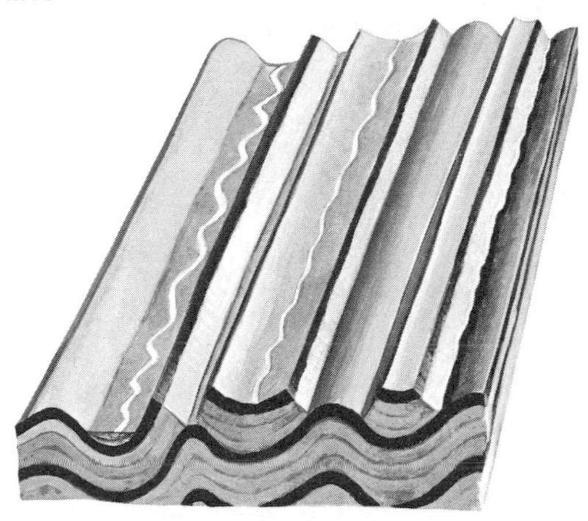

portray the geologic structures in proper relation to the landforms. The landform drawings in a block diagram often omit features such as vegetation to give the structures of interest greater emphasis.

Mapping in the United States

Regional and city planning and the development of a country's resources are closely tied to the availability of specialized maps. The study of geographic features such as soils, landforms, and water usage is also largely dependent on information compiled in the form of maps. In addition to maps of topography and location required for navigation and planning, government agencies prepare maps of water, soil, geologic phenomena, timber, and other features. Maps are also employed to report statistical, economic, and demographic information; a map devoted to a single topic, such as the distribution of population or rainfall, is known as a *thematic* map. Figure II.20, p. 538, gives examples of several kinds of maps.

The responsibility for mapping the United States is divided among many civil and military agencies, depending on the purpose of the map. Only a few examples are given here to indicate the extent of the mapping services carried on by the government.

The United States Geological Survey (USGS) publishes several series of topographic maps in different scales that cover the United States, Puerto Rico, and other territories. The USGS Topographic Maps are discussed in detail in the following section of this appendix because of their particular importance to physical geographers.

In addition to topographic maps, the Geological Survey publishes maps of national forests and grasslands, geologic maps, and hydrologic maps for water-resource planning. The hydrologic atlases prepared by the

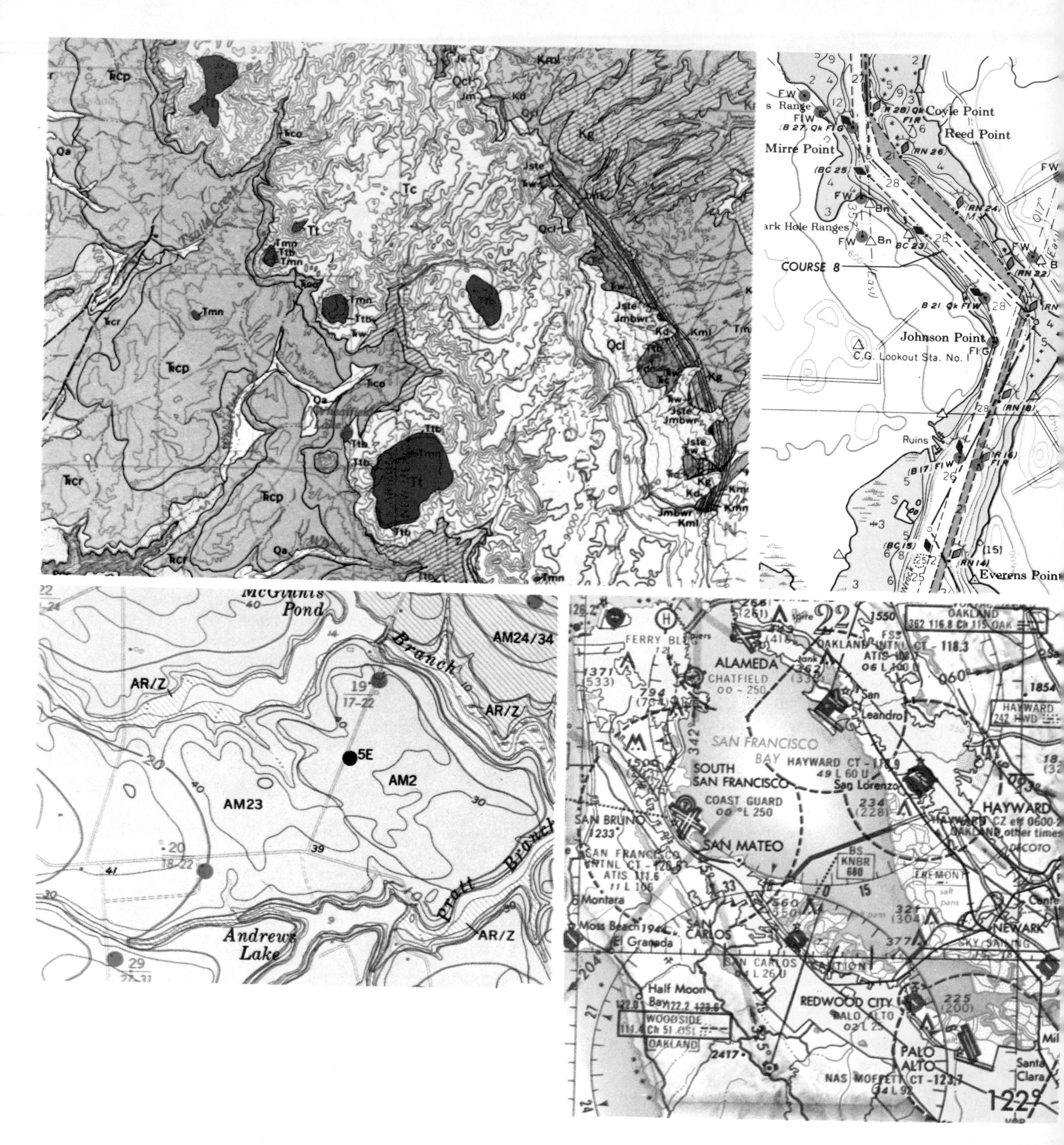

Tc
Tt
Ttb
Tmn
Trcp
Trcr
Trco
Tw
Qa
Qcl
Kml
Kd
Kg
Jste
Jmbwr
Coyle Point
Reed Point
Mirre Point
COURSE 8
Johnson Point
C.G. Lookout Sta. No. 1
Ruins
Everens Point
Pond
Branch
AM24/34
AR/Z
5E
AM2
AM23
Pratt Branch
Andrews Lake
OAKLAND
ALAMEDA
CHATFIELD
FERRY BLDG
SAN FRANCISCO BAY
SOUTH SAN FRANCISCO
COAST GUARD
San Leandro
San Lorenzo
HAYWARD
HAYWARD CT - 118.9
SAN BRUNO
SAN MATEO
KNBR 680
FREMONT
NEWARK
Montara
Moss Beach
El Granada
SAN CARLOS
Half Moon Bay
WOODSIDE
REDWOOD CITY
PALO ALTO
NAS MOFFETT CT - 123.7
122°

Figure II.20 (top left) Geologic maps are used to portray the distribution of rocks and deposits of varying ages and types.

(top right) Charts for river navigation show channels and buoy markers.

(bottom left) Maps showing the depth of the water table and conditions of surface drainage are useful in water-resource planning.

(bottom right) Aeronautical charts are used for aircraft navigation and show visual checkpoints and radio beacons.

Geological Survey detail the expected flow rate from wells sunk into groundwater supplies. Other maps in hydrologic atlases show the extent of floods for help in land-use planning. The Map Information Office of the Geological Survey is a useful source of information concerning the availability of maps and aerial photographs.

Maps related to navigation purposes are known as *charts.* The preparation of nautical and aeronautical charts is carried out by several agencies. Nautical charts of foreign waters are prepared by the Naval Oceanographic Office (formerly the Hydrographic Office of the Navy). This office also issues small-scale bathymetric charts that portray the topography of much of the ocean floor. The National Ocean Survey prepares nautical navigation charts of United States coastal and offshore waters, and some inland waterways, using the Mercator projection and a range of scales. The harbor charts prepared by the Ocean Survey may have scales as large as 1:18,000 to show accurate locations for important features such as main channels, marker buoys, and underwater cables. The Army Corps of Engineers issues nautical charts of the Great Lakes and navigation charts for numerous rivers and inland waterways.

Aeronautical charts are produced by several agencies, including the National Ocean Survey and the United States Air Force Aeronautical Chart and Information Center. The scales employed vary from 1:250,000 on charts for aircraft of moderate speed to 1:2,000,000 and smaller for the needs of global jet transport. Special features of aeronautical charts include identification of airports and air navigation radio beacons, and a simplified representation that emphasizes major relief features and landmarks easily visible from the air. Aeronautical charts are revised frequently from aerial photographs and can be helpful to geographers.

USGS Topographic Maps

A *topographic map* is a graphic representation of a portion of the earth's surface at a scale large enough to show human works such as individual buildings. Additionally, topographic maps show the configuration of the land surface. The United States has maintained a topographic map program since 1879 under the direction of the Geological Survey. The Geological Survey publishes the National Topographic Map Series, an invaluable source of information for physical geographers in the United States who are concerned with landforms or who must plan and execute field studies. The maps are compiled primarily from aerial photographs, but field surveys are sometimes used to verify details of photo interpretation.

A *map series* is a family of maps produced according to the same general specifications or with some features, such as scale, in common. The Topographic Series is itself made up of several series of maps. Each of the maps is drawn so that its sides coincide with parallels of latitude or meridians of longitude, and a similar symbolism is used in all series. Table II.3, pp. 540–541, lists the principal series comprising the Topographic Series, together with some of their properties. In addition to the series listed, there is a set covering national parks, monuments, and historic sites, a set for certain metropolitan areas, a set for rivers and flood plains, and sets for Puerto Rico, Antarctica, and other areas.

Using USGS Topographic Maps. The $7\frac{1}{2}$-minute quadrangle series and 15-minute quadrangle series are large-scale maps suited to field survey work and landform studies on a local scale. Such maps contain a large amount of information, often in abbreviated or symbolic form.

Topographic maps employ an extensive set of symbols to depict natural and man-made features. The USGS topographic map symbol table, reproduced in Figure II.21, p. 541, is not printed on the map sheets but is available separately. The colors on a topographic map are an integral part of the symbolization. As Table II.4, p. 544, shows, each color is restricted to a particular class of symbols: blue, for example, indicates water features, and a red tint indicates urban areas where only landmark buildings are mapped individually.

Field Work and Mapping

The physical geographer engaged in the study of streams, slopes, land use, or soils may find that the information he requires for the analysis of an area is not available from published sources. The smallest slope details may not be indicated on maps of the area of interest, for example. Or the geographer may be concerned with phenomena occurring during a short span of time, such as the erosion of stream banks during spring flood conditions, the study

Table II.3. Principal Map Series in the National Topographic Map Series

Series and Scale	1 Inch Represents	Projection	Angular Size (latitude-longitude)	Area Covered* (square miles)	Remarks
7½-Minute Quadrangle 1:24,000	2,000 feet	Polyconic	7½ minutes by 7½ minutes (⅛° by ⅛°)	49 to 70	Available for most of Hawaii, much of 48 states, and only a small part of Alaska. To meet national standards of map accuracy, at least 90 percent of mapped objects must be shown within 40 feet of true location. Also available without contour lines and woodland color tint.
15-Minute Quadrangle 1:62,500	0.986 mile	Polyconic	15 minutes by 15 minutes (¼° by ¼°)	197 to 282	Available for most of the United States except Alaska. Compiled from same data as 7½-minute series, but show less detail because of smaller scale. To meet standards of accuracy, at least 90 percent of mapped objects must be shown within 100 feet of true location. Also available without contour lines and woodland color tint.
Alaska 1:63,360	1 mile	Polyconic	15 minutes by 20 to 30 minutes	207 to 281	Covers most of Alaska. Comparable in scale and content to 15-minute series. More detailed 1:24,000 data are not available. Also available without contour lines and woodland color tint.
United States 1:250,000	3.95 miles	Transverse Mercator	Usually 1° by 2°	4,580 to 8,669	The conterminous United States and Hawaii are completely mapped in 473 separate maps. Alaska is completely mapped in 153 separate maps.
State Maps 1:500,000	7.89	Lambert conic conformal		Individual states of the conterminous United States	Also available for most states as base maps without relief, as topographic maps, and as maps with shaded relief. Base maps for the conterminous United States are also available with a scale of 1:1,000,000.

Series and Scale	1 Inch Represents	Projection	Angular Size (latitude-longitude)	Area Covered* (square miles)	Remarks
United States 1:1,000,000	15.78 miles	Lambert conic conformal	4° by 6° (4° by 12° for Alaska)	73,734 to 122,066	Maps in this series conform to specifications for the International Map of the World. Prior to 1962 the chosen projection was a modified polyconic, but the Lambert conic conformal has since been adopted. Available for Alaska and most of the conterminous United States.

*Area covered depends on latitude.
Source: United States Department of the Interior Geological Survey. 1970. *The National Atlas of the United States of America.* Washington, D.C.: United States Department of the Interior.

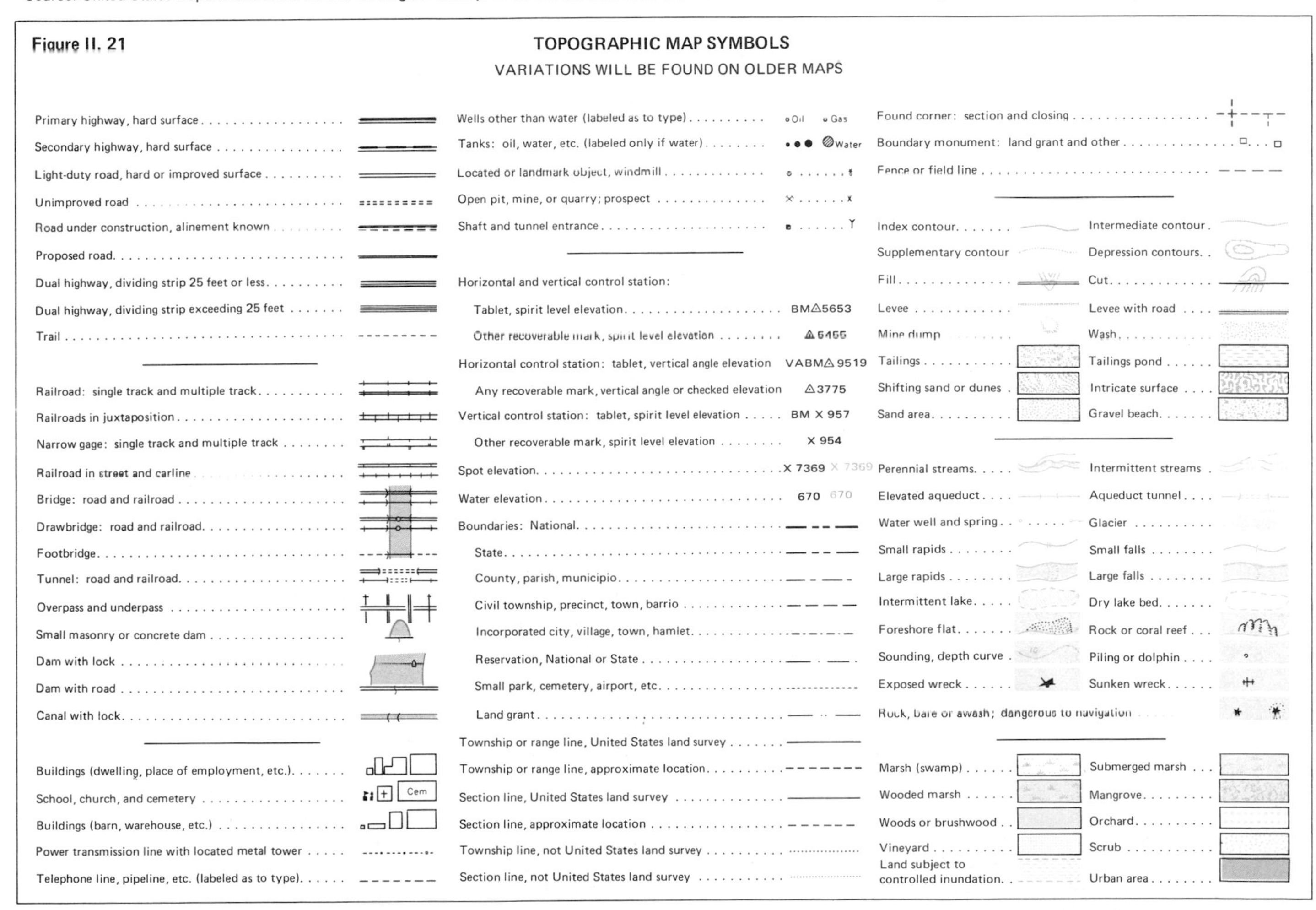

of which may require daily or weekly observation. In such cases, the geographer must make his own survey of the area. Because maps are often the simplest and best way to record and display the data of interest, geographers need to be familiar with the field study and mapping techniques suitable for their purposes. This section discusses some of the observational methods employed to study regions of small size using limited manpower and instrumentation.

An Approach to Field Work. After the geographer has decided that field study in a small region is necessary for the solution of a particular research problem, some preliminary planning must be done. Circumstances in the field may dictate later changes, but it is essential to begin a field study with a definite plan in order to avoid wasteful, undirected effort. The first step in planning field work is to choose a region for study suited to the problem at hand.

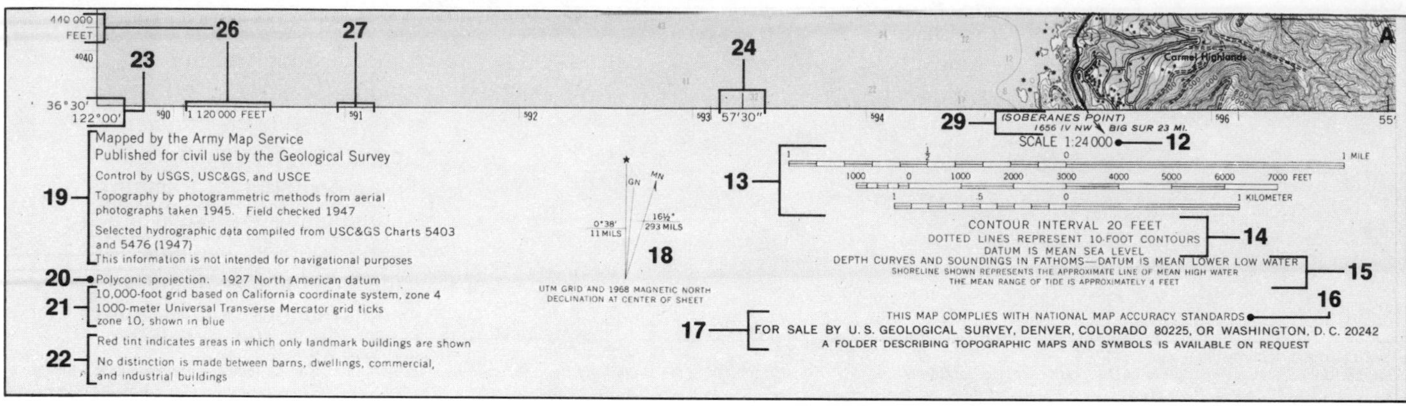

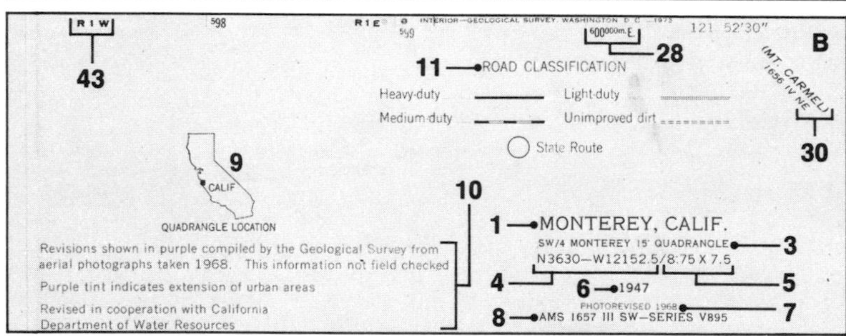

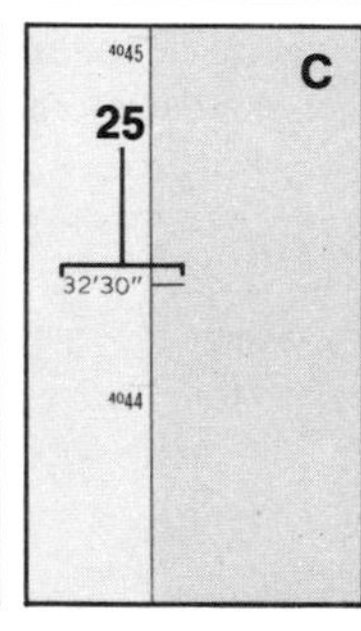

Figure II.22

The margins of topographic maps contain latitude and longitude markings, references to grid coordinate systems, scales, and much other helpful data. Figure II.22 takes the example of the Monterey, California, 7½-minute quadrangle topographic map and shows the location and meaning of several dozen pieces of information that are typically found on maps in the 7½-minute and 15-minute series. Items numbered on the map are explained in the following list:

1. Name of map sheet
2. Name of map series
3. This map is the southwest portion of the Monterey 15-minute quadrangle
4. Latitude and longitude of southeast corner of map
5. Extension of map in minutes of latitude and longitude
6. Date of edition
7. Date of revision
8. Army Map Service sheet number and series number
9. Location of quadrangle in the state
10. Note on revision
11. Road symbol table
12. Scale of map as a representative fraction
13. Graphic scales in several units of length
14. Note on contour interval and contours
15. Note on depth curves, soundings, and shorelines (1 fathom equals 6 ft)
16. Statement of map accuracy
17. Central offices from which map is available
18. Declination diagram showing true north, UTM grid north, and 1968 magnetic north (1 mil equals 3.375 minutes)
19. Credits, sources, and dates of preparation
20. Projection
21. Grids referred to on this map
22. Note on symbolization
23. Latitude and longitude reference for southwest corner
24. Longitude tick mark based on reference at left margin
25. Latitude tick mark based on reference at bottom margin
26. California state coordinate grid reference points in feet
27. UTM grid reference (blue tick marks)
28. UTM grid reference 600,000 meters east of origin
29. Name and Army Map Service sheet number of neighboring map sheet in same series to the south
30. Name and Army Map Service sheet number of neighboring map sheet in same series to the southeast
31. Green tint indicates woodland
32. Red tint indicates urban areas; only major buildings or landmarks such as churches and schools are shown
33. Isolated buildings are indicated in sparsely settled areas
34. Purple tint indicates areas that have been revised since the previous edition
35. Blue is used for water, such as intermittent streams and the ocean
36. Contour lines are brown; every fifth contour line is a bolder index contour marked with its altitude
37. Unchecked spot height
38. Checked spot height
39. Bench mark, a point at which the elevation has been measured precisely; part of the elevation control network that is used in surveying
40. Boundary of state reservation
41. Principal meridian for laying out townships for United States land survey
42. Spanish land-grant survey boundary (meets and bounds system)
43. Bureau of Land Management survey system: Range 1 West
44. Section number
45. Heavy-duty road
46. Unimproved dirt road
47. Quarry
48. Sand
49. Shoreline rocks
50. Bare rocks indicated by star symbol
51. Depth sounding in fathoms

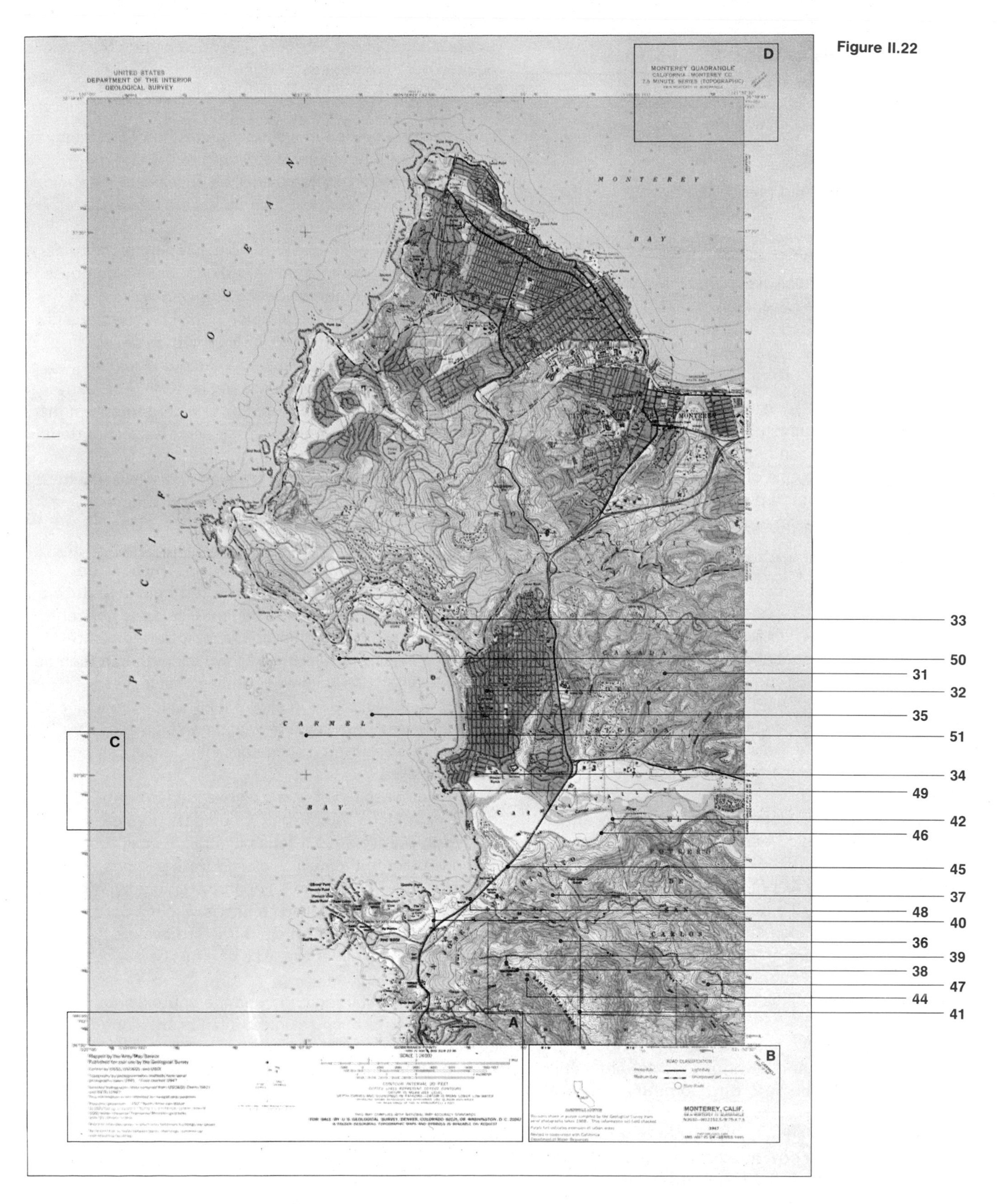

UNITED STATES
DEPARTMENT OF THE INTERIOR
GEOLOGICAL SURVEY
MONTEREY QUADRANGLE
CALIFORNIA - MONTEREY CO.
7.5 MINUTE SERIES (TOPOGRAPHIC)
D
M O N T E R E Y
B A Y
P A C I F I C
O C E A N
C A R M E L
B A Y
C
A
B
MONTEREY, CALIF.
33
50
31
32
35
51
34
49
42
46
45
37
48
40
36
39
38
47
44
41

Figure II.22

Table II.4. Color Code for USGS Topographic Maps

Color	Uses
Black	Man-made objects, names, notes, most labels
Blue	Water features, including streams, lakes, canals, and marshes
Brown	Contour lines and relief features
Green	Vegetation of various types: woodlands, orchards, vineyards, brush, and thick ground cover
Purple	On interim maps, areas that have been revised since the previous edition
Red	Urban areas, land boundaries, and some roads

Source: Adapted from USGS Symbol Table.

Maps and aerial photographs can be helpful in locating a suitable region, but first-hand inspection of possible sites may also be needed. The size of the site chosen depends on the nature of the research problem and on the available resources. A study of stream erosion processes, for example, might involve an intensive study of only a few hundred feet of stream bank, whereas a study of land use in farming country might entail a broad survey covering several square miles.

Usually the geographer is interested in studying the land use of an area, the spatial distribution of a quantity (such as a tree species), or some aspect of the physical landscape. Data from such field studies are often best compiled in the form of a map, and proper execution of the map is therefore central to the success of the study. The scale of the map is determined by the needs of the problem and the degree of detail that must be represented. A study of erosion processes on a slope, for example, may require the mapping and description of every few square feet of slope, whereas the basic unit of area in a local soil survey might be several acres in extent. If the area to be studied is a few square miles or so, the geographer may be able to record data directly on a map such as a USGS topographic map or on an aerial photograph of the region. Many of the map series in the USGS Topographic Map Series are available without contour lines and woodland tint, making them suitable as base maps for some studies. However, the scale of USGS topographic maps is not large enough to represent in detail areas of a few thousand square meters or less, and the geographer may have to prepare his own maps.

Before going into the field, the geographer must give some consideration as to how the data are to be selected and classified. Every map is the result of selection and generalization, and the geographer must ensure that the required data are obtained without making serious omissions and without collecting quantities of irrelevant information. If an extensive field study is planned, the data-gathering procedures should be tested beforehand on a small part of the region. The data from the test run should be analyzed as completely as possible to determine whether the proposed field techniques are adequate for gathering the required data.

Mapping Techniques. Simple mapping techniques suitable for field use include field sketching, the compass traverse, and use of the plane table. A *field sketch* is a panoramic sketch of a local region; it can be used to show landforms and the relation of landforms to their surroundings and can be labeled to map out spatial distributions of vegetation, soils, and so forth. For many purposes, a field sketch is superior to a photograph of the region because the geographer can emphasize important features in a sketch.

A field sketch need not be executed artistically to be of value in a field study; often a simple, uncluttered drawing is all that is required. The rule in sketching is to begin with the main outlines of a landscape, such as the skyline, and work toward smaller details. To show the correct relative locations of features, it is helpful to draw a horizontal line on the sketch pad to indicate the horizontal reference line, and a vertical line to indicate the central axis of the sketch. A ruler held at arm's length can then be used to estimate the relative distances of selected landmarks, as illustrated in Figure II.23. Once the major topographic features and a few landmarks have been sketched, the relative positions of other details can be located by eye.

A *compass traverse* is a simple map-making technique, which requires only a compass, to measure directions. A *traverse* is a series of straight lines running from one point to the next. A *closed* traverse ends at the original starting point, and an *open* traverse does not. A closed traverse is used when features within a particular area are to be mapped. An open traverse is more suitable when features on either side of a central line are to be recorded and mapped.

Once the starting point and the approximate route of the traverse have been selected, a landmark along the first leg of the route is sighted from the starting point. The azimuth of the landmark is then measured using a compass, taking precautions to keep the compass away from magnetic materials that might influence its reading. When the azimuth has been measured and recorded in the data book, the investigator then walks toward the landmark while pacing off the distance. If the investigator makes

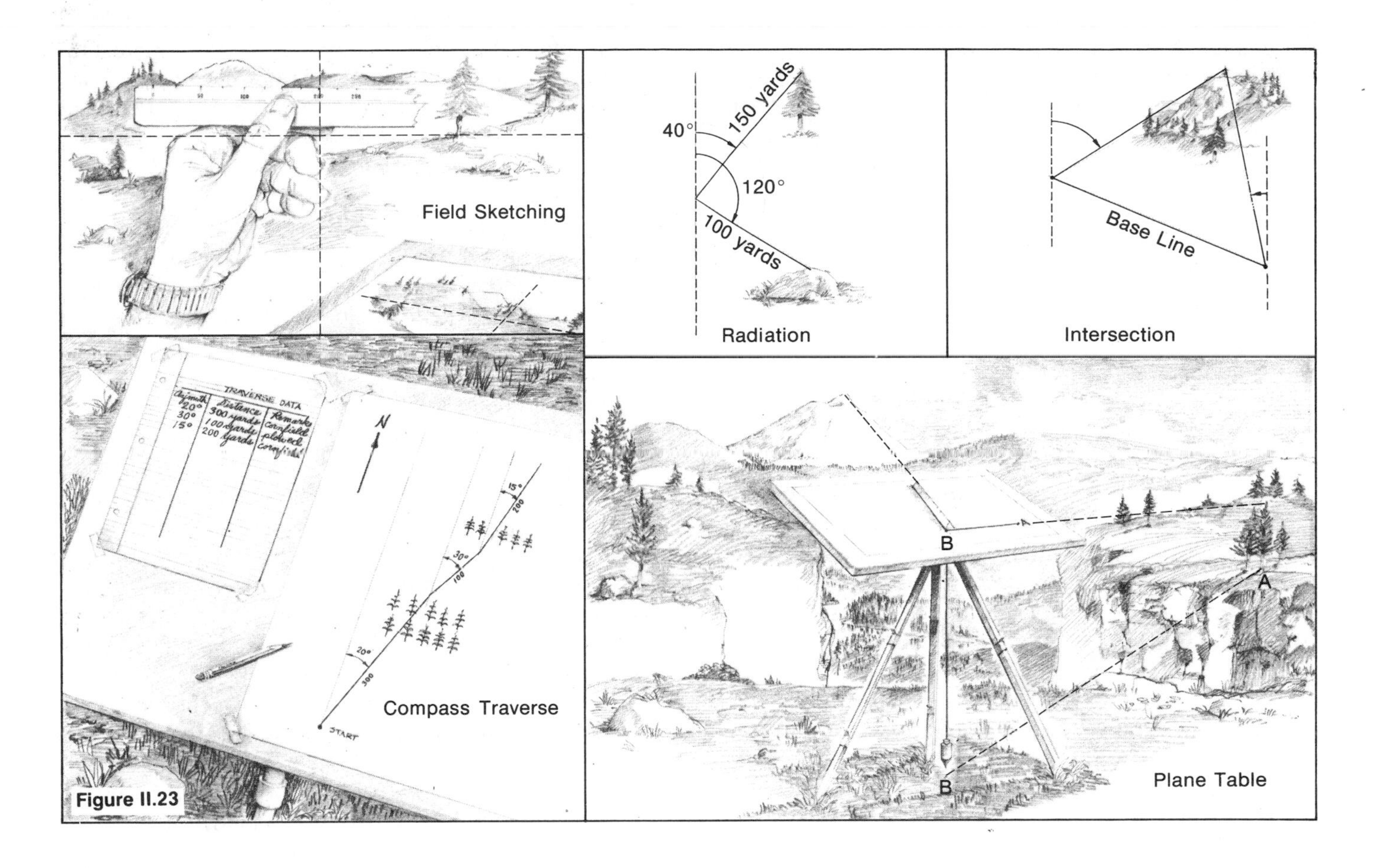

Figure II.23

observations along the traverse, the data and the location should be recorded. After reaching the end of the first leg, a sighting can be taken to a new landmark and the traverse can be extended.

The field record of a compass traverse consists of a list of azimuths, distances, and observations. To convert these data to a map, the legs of the traverse can be laid out on drawing paper to their correct scale and with their correct azimuths (see Figure II.23). Observational data can then be added.

Sometimes the position of an object away from the line of the traverse must be measured. One method of finding its location, known as *radiation,* is to take a sighting on the object from a known point on the traverse, measure the azimuth by compass, and pace off the distance to the object. Steps can be saved by employing the technique of *intersection* instead. As Figure II.23 illustrates, the position of the object relative to the traverse is completely specified by measuring the azimuth of the object from two known points along the traverse. For best accuracy, the two sighting points should be spaced so that the lines of intersection meet at the object with an angle of between 60° and 120°.

The methods of radiation and intersection are the basis for map making with a *plane table,* which is a horizontal drawing board supported by a stand or tripod. A map can be drawn while in the field using a plane table, allowing the work to be checked until satisfactory accuracy is obtained.

When using a plane table, the first step is to establish a *base line* of known distance and azimuth between two sighting points that command good views of the area to be mapped. After the base line has been drawn to an appropriate scale on drawing paper fastened to the plane table, the table is placed over one of the sighting points and oriented so that the base line on the map sheet is exactly parallel to the base line on the ground. Then a sighting device such as a triangular rule can be laid on the map sheet and aligned from the sighting point toward the object to be mapped (Figure II.23). In this way the line to the object can be drawn on the map at the correct azimuth without the need for compass measurements. The location can be fixed by employing the method of radiation, pacing off the distance from the sighting point to the object. Alternatively, the determination of azimuths from both sighting points fixes the location of the object by intersection. Normally the plane table must be moved and set up at several successive stations to complete the map. The position of each station must be precisely established on the map before the plane table is moved to it, and the orientation of the plane table at each station is checked both by compass and by sighting back toward the preceding station.

Remote Sensing

The study and mapping of the earth's surface through direct field work is unsatisfactory for many purposes because of the limited coverage possible, particularly in relatively inaccessible areas, and because of the great expense involved. New capabilities, generally known as *remote sensing,* have been developed in recent years for studying the surface of the earth. Remote sensing methods employ the interaction of electromagnetic energy with the surface of the earth. Special sensing devices can acquire data at a distance from the objects or phenomena being studied. In some methods, the electromagnetic radiation of sunlight, laser light, or microwave radar that is reflected from objects on the earth is sensed and recorded; in other methods of remote sensing, the infrared thermal radiation directly emitted by objects is detected.

Remote sensing equipment is usually mounted in aircraft or space satellites. The advantages of remote sensing include the ability to cover large regions in a very short time, the ability to provide data from very large to very small scale, the possibility of repetitive measurements to follow the progress of selected events, and the ability to penetrate areas inaccessible to ground survey. The techniques of remote sensing have been greatly extended recently in an effort to obtain the information needed to manage the earth's resources more efficiently, and much of the information being gathered is of direct interest to physical geographers. The capabilities of remote sensing technology include the detection of vegetation types and their seasonal changes, the measurement of surface water distribution and soil moisture, and the depiction of landforms and surface geologic structures.

Aerial photography, the first remote sensing method developed, is now used for most original mapping. Aerial mapping surveys generally employ the visible light reflected from the earth and detected with cameras using black-and-white film. Optical techniques allow the distortions inherent in a photograph taken at low or moderate altitudes to be rectified so that distant areas in the photograph have the same scale as areas directly below the camera. Two exposures can be used to record a stereoscopic (three-dimensional) image from which contour lines can be drawn. A single photograph may cover an area of a few square miles to hundreds of square miles, depending on the cameras and the altitude of the aircraft. Ground features as small as 1 cm in size can be detected on photographs made from aircraft flying at an altitude of a few kilometers, and satellite imagery can provide resolution of objects only a few meters in size from an altitude of several hundred kilometers. These advanced capabilities are gradually becoming available for civilian uses, such as detection of local sources of air pollution.

Aircraft fitted with detectors of longwave thermal infrared radiation can provide images of objects on the earth's surface with intensities proportional to their temperatures. Infrared imagery methods can be employed to detect ocean currents or the mixing of warm water from industrial effluents with cooler river water.

Application of remote sensing methods to the study of the earth has been revolutionized by using remote sensors in conjunction with space satellites. Surveillance of the earth's cloud cover from weather satellites has greatly improved the accuracy of weather predictions and the

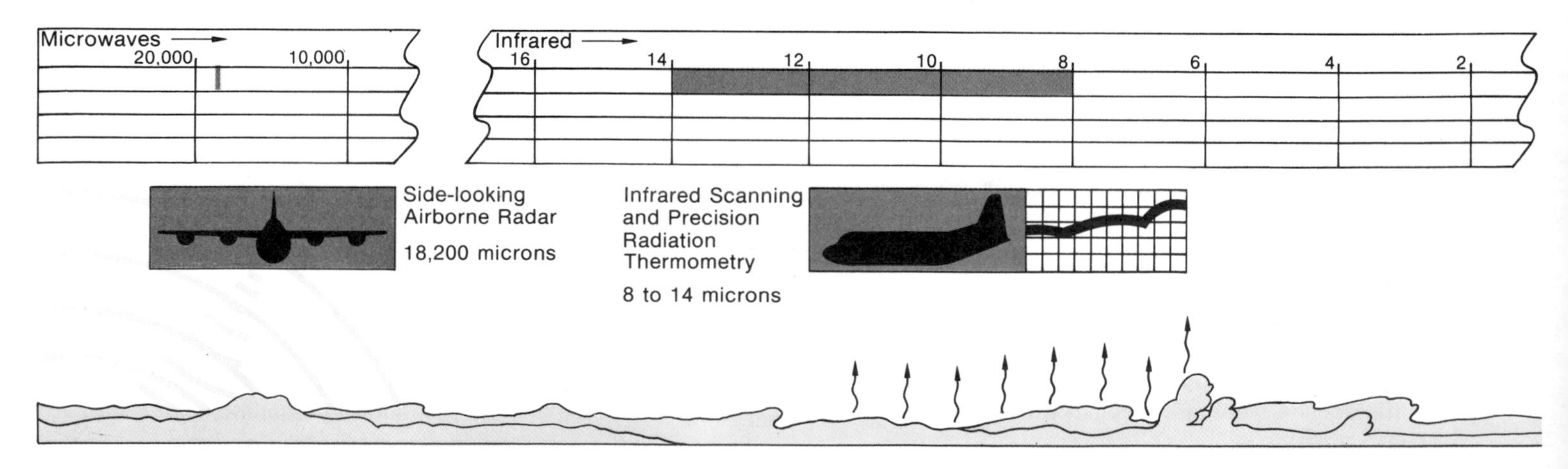

ability to track potentially damaging hurricanes and typhoons. Weather satellites usually orbit at altitudes of several hundred kilometers.

Sunlight falling upon the earth consists of electromagnetic radiation in the visible region and in the near-visible infrared region of the electromagnetic spectrum. Different objects may reflect a different relative proportion of each wavelength, so that the light reflected from an object, when analyzed in terms of wavelength, forms a means of identification known as the *spectral signature* of the object. Vegetation, for example, usually reflects proportionately more short wavelength infrared radiation than does bare soil. Furthermore, different types of vegetation can usually be distinguished by their signatures. Diseased plants or the lack of soil moisture can also be detected.

Several remote sensing techniques have been developed for using characteristic signatures to distinguish and identify objects on the earth's surface (see Figure II.24). *Color infrared film* is similar to ordinary color film, except that it depicts the previously invisible infrared radiation as red, the red colors as green, and the green colors as blue (see Figure II.25, p. 548). Vegetation usually appears red on a color infrared photograph of a region; the intensity of the color is directly proportional to the amount of chlorophyll in the vegetation. Color infrared photographs of test fields have proven that different crops and their stages of growth can be distinguished by such means.

Another way to make use of spectral signatures is to employ several cameras, each fitted with a filter that allows only wavelengths in a certain band to enter. Landsat, formerly Earth Resources Technology Satellite, ERTS-1, which views the earth from an altitude of nearly 1,000 km, employs four multispectral scanners and three television cameras, each sensing the same scene on a different band of wavelengths. These images can be artificially "combined" to produce a "color picture" of the area covered (see Figure II.26, p. 549). From its orbit between the poles, Landsat can scan a given portion of the earth once every 18 days to allow the progress of crops to be followed through the growing season; to monitor streamflow, snow depth, and soil moisture; and to detect other short-term changes in the environment.

Many remote areas in the tropics are almost continuously covered by clouds, hiding the ground from view. *Side-looking airborne radar,* which utilizes radio waves of short wavelength, can penetrate clouds and yield a high resolution picture of the surface. The radar apparatus is usually fitted to the side of the aircraft to give an unrestricted lateral view; radar waves emitted from the aircraft are partly reflected from the ground and detected by a receiving apparatus on the aircraft. Measurement of the time required for the radar waves to make the round trip is electronically translated into a measure of distance, producing a map of the region. Airborne radar has been used to map the Amazon Basin, Panama, and portions of Africa and New Guinea. Although radar methods do not differentiate between types of vegetation as clearly as multispectral methods, forest, range, and agricultural land can be separately identified.

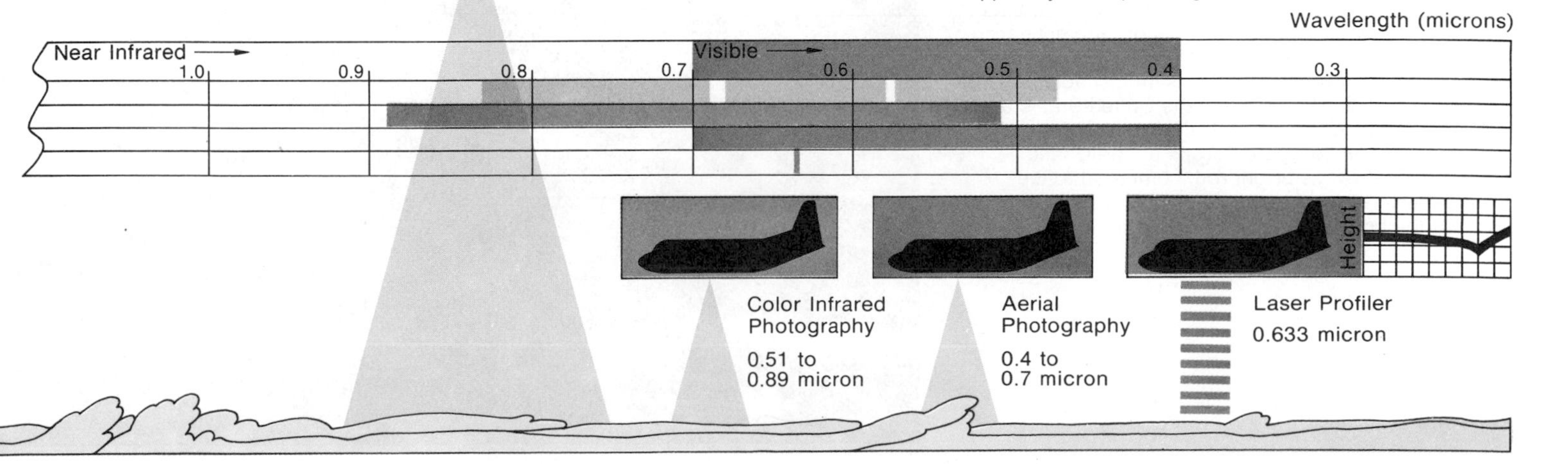

Figure II.24 Remote sensing techniques utilize many portions of the electromagnetic spectrum, as shown in this chart. The visible and near-infrared regions are used for photography and television, and the infrared is used for scanning surface temperatures. Relief features are mapped by laser profiling and radar.

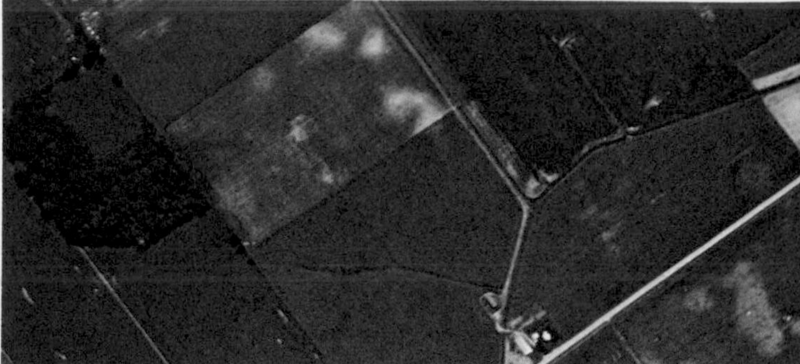

Figure II.25 (top) The aerial view on the right was photographed with ordinary color film. The same fields photographed with color infrared film are shown on the left; healthy vegetation appears bright red.

(above left) Stereo pairs are used to make topographic relief apparent in aerial photographs. When viewed properly, relief features in this view in eastern Utah stand out as a three-dimensional model. To see the stereo effect, the left-hand photograph should be viewed with the left eye, and the right-hand photograph with the right eye. A card held between the photographs helps to separate the images.

(bottom right) Vegetation on irrigated land is clearly differentiated from dry desert in this color infrared photograph of the Imperial Valley of California taken from earth orbit. The Salton Sea appears as a large dark region at the upper left. The pattern of agricultural fields is clearly distinguishable; note in particular the border between the United States and Mexico that is evident toward the bottom of the photograph. The well-irrigated crops on the United States side show as bright red, compared to the bluish hues representing vegetation on the Mexican side of the border.

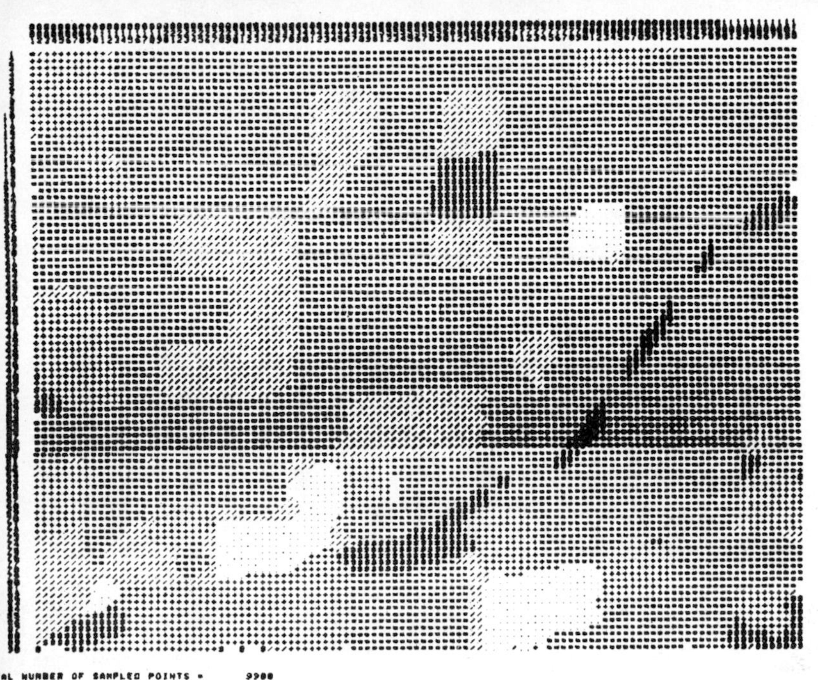

Figure II.26 (top left) Monterey Bay, California, appears at the left of this photograph taken by ERTS-1 from an altitude of 900 km (560 miles). The color print is made from a composite of three black-and-white photographs, each exposed to different spectral wavelengths; the image is similar to that produced by color infrared film. Vegetation areas stand out in red, and linear features associated with the San Andreas fault zone run from the upper left to the lower right. A patch of fog hides part of the coast south of the Monterey peninsula. A topographic map of the Monterey peninsula is shown in Figure II.22.

(top right) Relief features, including linear features in the San Andreas fault zone, show crisply in this side-looking radar image of part of the San Francisco peninsula. The thin linear feature at the lower right is the 2-mile-long linear accelerator used in atomic research at Stanford University.

(bottom left) The areas of different shading on this computer print-out correspond to water bodies and different vegetation types. The computer map was prepared directly by computer analysis of multiband images from ERTS-1. Accuracy of identification exceeded 95 percent.

APPENDIX III: Classification Systems

Köppen System of Climate Classification

The Köppen system of climate classification recognizes five principal climate regions, which are symbolized by *A, B, C, D,* and *E.* The *B* regions, which are the dry realms, are further divided into two major subregions, *BS* (steppe) and *BW* (desert), and the *E* region is divided into the regions *ET* (tundra) and *EF* (perpetual frost).

Figure III.1 shows Köppen's criteria for establishing the boundaries between the regions *A, C,* and *D;* the region *BS;* and the region *BW.* The solid line on each graph marks the boundary between the humid and dry realms. Each of the three sets of criteria is based on average annual temperature and precipitation. Which criterion is used depends on whether precipitation occurs evenly throughout the year or primarily during the summer or the winter. Köppen considered a region to have a dry winter if at least 70 percent of the precipitation occurs during the 6 summer months, and to have a dry summer if at least 70 percent of the precipitation occurs during the 6 winter months. Regions not fitting either category are considered to have an even distribution of precipitation. Summer is interpreted as the season when the sun is high over the horizon at midday in the particular hemisphere being considered, and winter is interpreted as the low-sun season.

The classification of a place into *A, C,* or *D* on the one hand, or *BS* or *BW* on the other, can be determined graphically by plotting the average annual temperature and precipitation on the appropriate diagram. The equation for each of the boundary lines is also given in the diagrams. In each case, the upper equation is in degrees Celsius and centimeters, and the lower equation in degrees Fahrenheit and inches. For example, when precipitation occurs primarily during the summer, the metric equation for the boundary between the humid and dry realms is $P = 2T + 28$, and the equation in English units is $P = 0.44T - 3$.

For a given average annual temperature, the regions *A, C,* and *D* receive more precipitation annually than do *BS* regions, which in turn receive more precipitation than *BW* regions. The regions *A, C, D, ET,* and *EF* are distinguished

Figure III.1

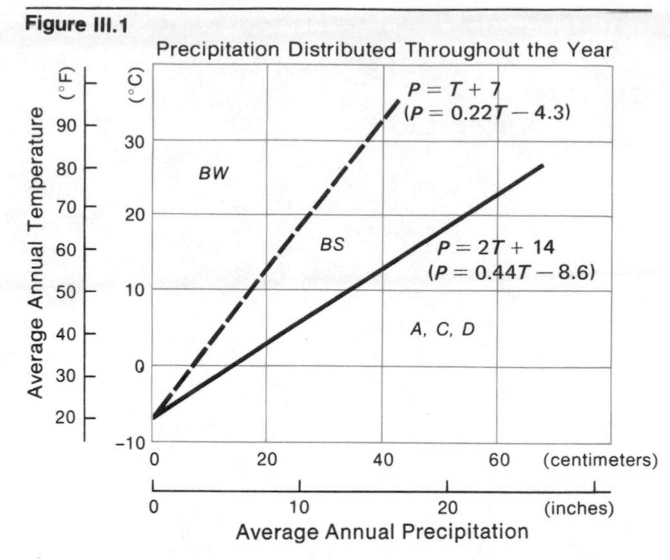

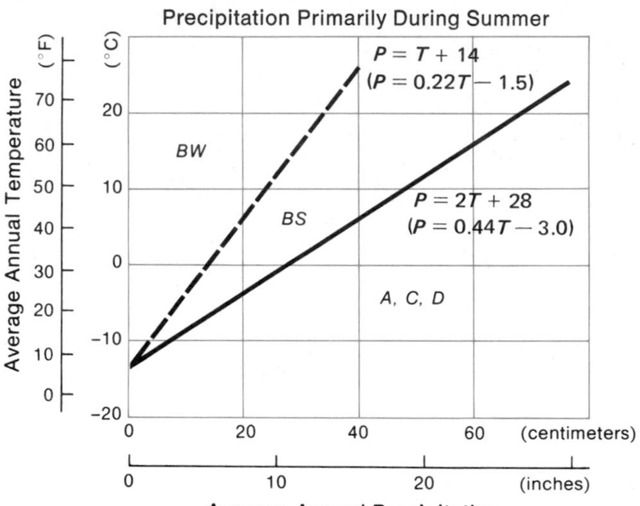

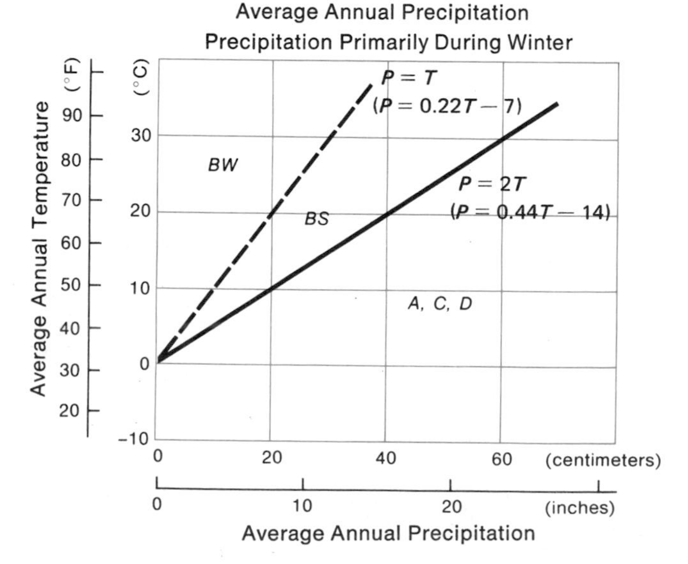

from one another according to various criteria based on temperature, with the warmest being *A* and the coldest being *EF*. The criteria are as follows:

A:	Average temperature of the coldest month exceeds 18°C (64.4°F).
C:	Average temperature of the warmest month exceeds 10°C (50°F). Average temperature of the coldest month lies between 18°C (64.4°F) and –3°C (26.6°F).
D:	Average temperature of the warmest month exceeds 10°C (50°F). Average temperature of the coldest month is below –3°C (26.6°F).
ET:	Average temperature of the warmest month lies between 10°C (50°F) and 0°C (32°F).
EF:	Average temperature of the warmest month is below 0°C (32°F).
H:	Unclassified highland climates.

Principal Subdivisions of A Climates. The *A* climate regions are subdivided into *Af, Am,* and *Aw* regions on the basis of the amount and seasonality of precipitation. If precipitation in the driest month exceeds 6 cm (2.4 in.), the region is classified as *Af,* as indicated in Figure III.2. *Af* regions, such as equatorial lowland rainforests, receive abundant moisture for plant growth throughout the year. If a region has a winter dry period during which precipitation for the driest month is less than 6 cm, the region is classified as *Aw* or as *Am.* The distinction between *Aw* and *Am* regions depends upon the relation between the amount of precipitation during the driest month and the annual precipitation, as Figure III.2 shows. Tropical wet and dry regions are classified as *Aw.* The *Am,* or monsoon, climate regions have enough annual precipitation so that the moderate winter dry season does not exhaust supplies of soil moisture, and plant growth is not seriously affected.

A fourth subdivision, the *As* region, would in principle be characterized by a summer dry season. However, *As* regions only occur locally near the equator, and the *As* classification is not significant on the global scale.

Figure III.2

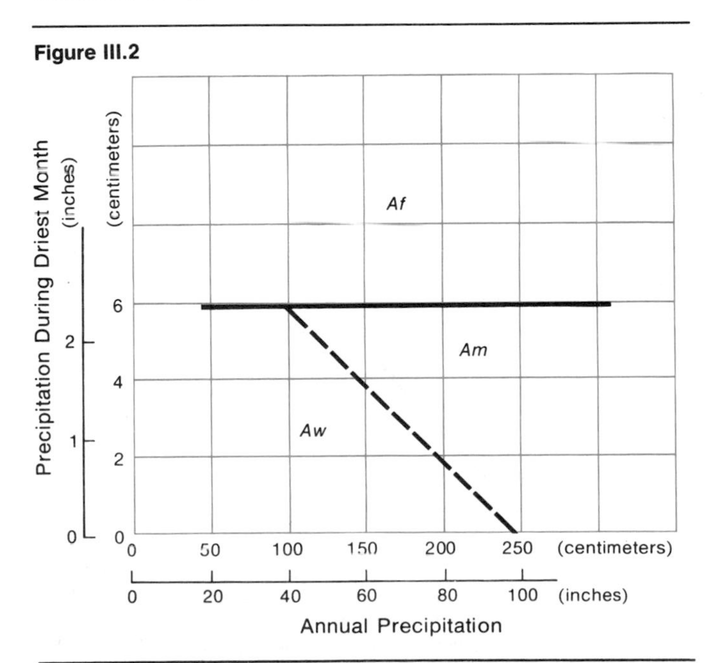

Principal Subdivisions of BS and BW Climates. The *BS* and *BW* regions are divided on a thermal basis, as specified by the additional notation *h, k,* or *k′*.

h (hot):	Average annual temperature exceeds 18°C (64.4°F).
k (cold winter):	Average annual temperature is less than 18°C (64.4°F) and average temperature of the warmest month exceeds 18°C.
k′ (cold):	Average annual temperature is less than 18°C (64.4°F) and average temperature of the warmest month is less than 18°C.

The deserts of North Africa and central Australia are examples of *BWh* regions. *BWk* regions occur in the high plateaus of central Asia.

Principal Subdivisions of C and D Climates. The *C* and *D* climate regions are further differentiated according to the seasonality of precipitation. A second letter of notation, *s,* denotes regions with a dry summer (*Ds* seldom occurs); *w* denotes regions with dry winter; and *f* denotes regions with no marked dry season.

Cs, Ds:	The driest month occurs during the warmest 6 months, and the amount of precipitation received during the driest month is less than one-third the amount received during the wettest month of the coldest 6 months. Also, precipitation during the driest month must be less than 4 cm (1.6 in.).
Cw, Dw:	The driest month occurs during the coldest 6 months, and the amount of precipitation received during the driest month is less than one-tenth the amount received during the wettest month of the warmest 6 months.
Cf, Df:	No marked dry season occurs, so that the *s* and *w* categories are not applicable.

The subdivisions *Cs, Ds, Cw, Dw, Cf,* and *Df* may be further differentiated according to the seasonality of temperature by adding a third letter of notation—*a, b, c,* or *d.*

- *a:* Average temperature of the warmest month exceeds 22°C (71.6°F).
- *b:* Average temperature of the warmest month is less than 22°C (71.6°F), but at least 4 months have an average temperature greater than 10°C (50°F).
- *c:* Average temperature of the warmest month is less than 22°C (71.6°F), fewer than 4 months have an average temperature greater than 10°C (50°F), and the average temperature of the coldest month is greater than −38°C (−36.4°F).
- *d:* Same as *c,* except that the temperature of the coldest month is less than −38°C (−36.4°F).

The Köppen system of climate classification possesses additional symbols to denote special features such as frequent fog or seasonal high humidity, but they are seldom used on the global scale. A global map of the Köppen climate classification is shown in Figure 7.5 on pp. 192–193.

Thornthwaite's Formula for Potential Evapotranspiration

The tables and graphs presented in this section make it possible to calculate potential evapotranspiration for a given month at a given location from Thornthwaite's formula, if the average monthly temperatures are known. Thornthwaite's formula is designed to be used with temperatures expressed in degrees Celsius, so temperature data on the Fahrenheit scale should be first converted to Celsius before beginning the calculation of potential evapotranspiration.

The first step in applying Thornthwaite's method is to calculate a monthly heat index, *i,* for each of the 12 months. The monthly heat index is defined according to the formula

$$i = \left(\frac{T}{5}\right)^{1.514}$$

where *T* is the long-term average temperature of the month in °C. Approximate values of the monthly heat index can be read from the graph in Figure III.3.

The sum of the twelve monthly heat indexes is the annual heat index, *I.* The annual heat index is representative of climatic factors at a given location because it is based on long-term averages. Thornthwaite found an empirical formula that gives potential evapotranspiration, *PE,* for a given month of a particular year in terms of *I.* His formula for *PE,* unadjusted for duration of sunlight, is

$$\text{Unadjusted } PE \text{ (centimeters)} = 1.6\left(\frac{10\,T}{I}\right)^{m}$$

where *T* is the average temperature in °C for the specific month being considered, and *m* is a number that depends on *I.* To a good approximation, *m* is given by the formula

$$m = (6.75 \times 10^{-7})I^3 - (7.71 \times 10^{-5})I^2 + (1.79 \times 10^{-2})I + 0.492$$

With these formulas, unadjusted potential evapotranspiration can be calculated using tables of logarithms or a computer. Alternatively, approximate values of unadjusted potential evapotranspiration can be read from the nomogram in Figure III.4 with enough accuracy

Figure III.3

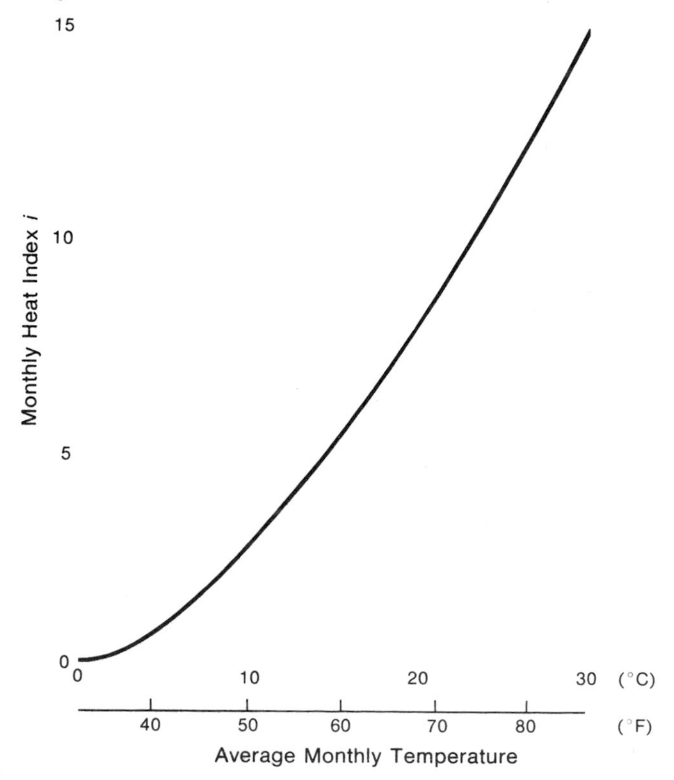

Figure III.4

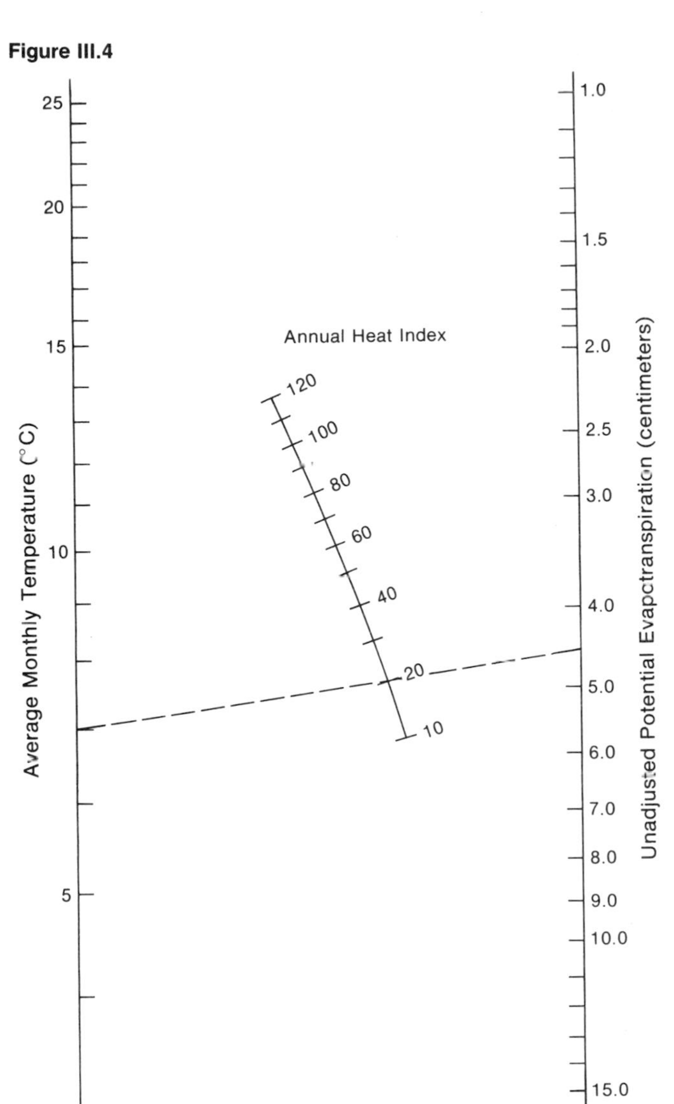

for most purposes if the values for the average monthly temperature and the annual heat index are known.

If the average temperature of the month is below 0°C, potential evapotranspiration is taken to be 0. If the average monthly temperature exceeds 26.5°C, unadjusted *PE* is given directly in terms of temperature according to the graph in Figure III.5.

Values of unadjusted *PE* must be corrected for the duration of daylight in order to obtain the desired final values. Unadjusted *PE* for a given month at a given location should be multiplied by the correction factor listed in Table III.1. The correction factors for latitude 50°N are used for all latitudes farther to the north, and the factors for latitude 50°S are used for all latitudes farther to the south.

Figure III.5

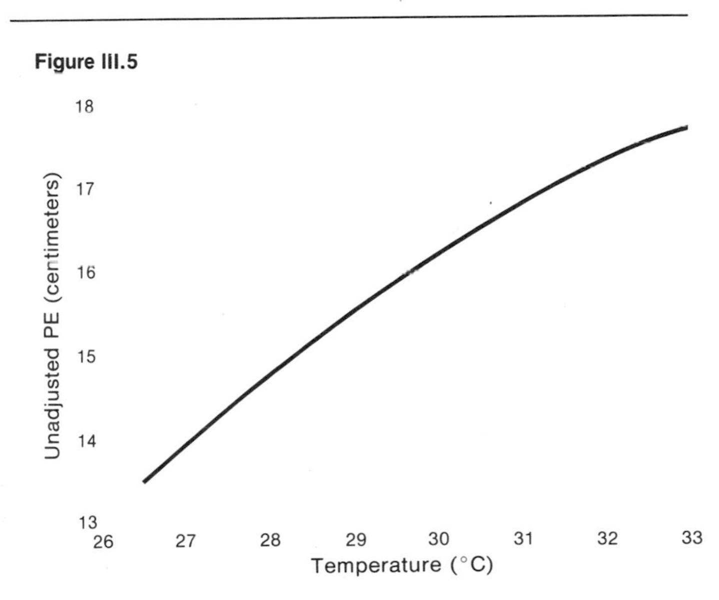

Table III.1 Daylength Correction Factors for Potential Evapotranspiration

Latitude	JAN	FEB	MAR	APR	MAY	JUN	JUL	AUG	SEP	OCT	NOV	DEC
0°	1.04	0.94	1.04	1.01	1.04	1.01	1.04	1.04	1.01	1.04	1.01	1.04
10°N	1.00	0.91	1.03	1.03	1.08	1.06	1.08	1.07	1.02	1.02	0.98	0.99
20°N	0.95	0.90	1.03	1.05	1.13	1.11	1.14	1.11	1.02	1.00	0.93	0.94
30°N	0.90	0.87	1.03	1.08	1.18	1.17	1.20	1.14	1.03	0.98	0.89	0.88
40°N	0.84	0.83	1.03	1.11	1.24	1.25	1.27	1.18	1.04	0.96	0.83	0.81
50°N	0.74	0.78	1.02	1.15	1.33	1.36	1.37	1.25	1.06	0.92	0.76	0.70
10°S	1.08	0.97	1.05	0.99	1.01	0.96	1.00	1.01	1.00	1.06	1.05	1.10
20°S	1.14	1.00	1.05	0.97	0.96	0.91	0.95	0.99	1.00	1.08	1.09	1.15
30°S	1.20	1.03	1.06	0.95	0.92	0.85	0.90	0.96	1.00	1.12	1.14	1.21
40°S	1.27	1.06	1.07	0.93	0.86	0.78	0.84	0.92	1.00	1.15	1.20	1.29
50°S	1.37	1.12	1.08	0.89	0.77	0.67	0.74	0.88	0.99	1.19	1.29	1.41

Source: Thornthwaite, C. Warren. 1948. "An Approach Toward a Rational Classification of Climate," *Geographical Review*, 38: 55-94.

Soil Classification Systems

The 7th Approximation soil classification system of 1960 has superseded the earlier 1938 United States Department of Agriculture classification system for use in the United States. However, many of the terms from the 1938 classification remain in common use.

The 1938 Soil Classification System. The 1938 classification divides soils into three orders: *zonal, intrazonal,* and *azonal.* The orders are divided into suborders, which are in turn divided into the great soil groups. The listing given here is a modification of the 1938 classification prepared in 1949 by James Thorp and Guy Smith, soil correlators for the Division of Soil Survey of the United States Department of Agriculture. The great soil groups are listed under their appropriate suborder.

Zonal Soils

1. Soils of the cold zone

    Tundra soils

2. Light-colored soils of arid regions

    Desert soils
    Red desert soils
    Sierozem
    Brown soils
    Reddish brown soils

3. Dark-colored soils of semiarid, subhumid, and humid grasslands

    Chestnut soils
    Reddish chestnut soils
    Chernozem soils
    Prairie soils
    Reddish prairie soils

4. Soils of the forest-grassland transition

    Degraded chernozem
    Noncalcic brown soils

5. Light-colored podzolic soils of timbered regions

    Podzols
    Gray podzolic soils
    Brown podzolic soils
    Gray-brown podzolic soils
    Red-yellow podzolic soils

6. Lateritic soils of forested warm midlatitude and tropical regions

    Reddish brown latosols
    Yellowish brown latosols
    Latosols

Intrazonal Soils

1. Halomorphic (saline and alkali) soils of imperfectly drained arid regions

    Saline soils (solonchak)
    Solonetz soils
    Soloth soils

2. Hydromorphic soils of marshes, swamps, seep areas, and flats

    Humic-gley soils
    Alpine meadow soils
    Bog soils
    Half-bog soils
    Low-humic gley soils
    Planosols
    Groundwater podzolic soils
    Groundwater lateritic soils

3. Calcimorphic soils

    Brown forest soils
    Rendzina soils

Azonal Soils (no suborders)

Lithosols
Regosols (includes dry sands)
Alluvial soils

The 7th Approximation. The following list gives the orders and suborders of the 7th Approximation. The suborders are further divided into a large number of great groups, as tabulated in 1960 and 1967 publications of the United States Department of Agriculture Soil Conservation Service. The descriptions presented here are adapted from *The National Atlas of the United States of America.* Some indication of soil use is also given. Table III.2 gives the prefixes used to differentiate the great groups.

Table III.2. Roots of Soil Terms in the 7th Approximation

Root	Derivation	Mnemonicon	Connotation
acr	Gr. *akros,* at the end.	acrolith	Extreme weathering
agr	L. *ager,* field.	agriculture	An agric horizon
alb	L. *albus,* white.	albino	An albic horizon
and	From *Ando.*	Ando	Ando-like

Root	Derivation	Mnemonicon	Connotation
anthr	Gr. *anthropos,* man.	anthropology	An anthropic epipedon
aqu	L. *aqua,* water.	aquarium	Characteristic associated with wetness
arg	L. *argilla,* white clay.	argillite	An argillic horizon
calc	L. *calcis,* lime.	calcium	A calcic horizon
camb	L. *cambiare,* to exchange.	change	A cambic horizon
chrom	Gr. *chroma,* color.	chroma	High chroma
cry	Gr. *kryos,* coldness.	crystal	Cold
dur	L. *durus,* hard.	durable	A duripan
dystr	Gr. *dys,* ill; *dystrophic,* infertile.	dystrophic	Low base saturation
eutr	Gr. *eu,* good; *eutrophic,* fertile.	eutrophic	high base saturation
ferr	L. *ferrum,* iron.	ferric	Presence of iron
frag	L. *fragilis,* brittle.	fragile	Presence of fragipan
fragloss	Compound of *fra(g)* and *gloss.*		See the formative elements *frag* and *gloss*
gibbs	From *gibbsite.*	gibbsite	Presence of gibbsite
gloss	Gr. *glossa,* tongue.	glossary	Tongued
hal	Gr. *hals,* salt.	halophyte	Salty
hapl	Gr. *haplous,* simple.	haploid	Minimum horizon
hum	L. *humus,* earth.	humus	Presence of humus
hydr	Gr. *hydro,* water.	hydrophobia	Presence of water
hyp	Gr. *hypnon,* moss.	hypnum	Presence of hypnum moss
luo, lu	Gr. *louo,* to wash.	ablution	Illuvial
moll	L. *mollis,* soft.	mollify	Presence of mollic epipedon
nadur	Compound of *na(tr)* and *dur.*		
natr	Ar. *natrium,* sodium.		Presence of natric horizon
ochr	Gr. *ochros,* pale.	ocher	Presence of ochric epipedon (a light-colored surface)
pale	Gr. *paleos,* old.	paleosol	Old development
pell	Gr. *pellos,* dusky.		Low chroma
plac	Gr. *plax,* flat stone.		Presence of a thin pan
plag	Ger. *plaggen,* sod.		Presence of plaggen horizon
plinth	Gr. *plinthos,* brick.		Presence of plinthite
quartz	Ger. *quarz,* quartz.	quartz	High quartz content
rend	From *Rendzina.*	Rendzina	Rendzina-like
rhod	Gr. *rhodon,* rose.	rhododendron	Dark-red colors
sal	L. *sal,* salt.	saline	Presence of salic horizon
sider	Gr. *sideros,* iron.	siderite	Presence of free iron oxides
sombr	Fr. *sombre,* dark.	somber	A dark horizon
sphagno	Gr. *sphagnos,* bog.	sphagnum moss	Presence of sphagnum moss
torr	L. *torridus,* hot and dry.	torrid	Usually dry
trop	Gr. *tropikos,* of the solstice.	tropical	Continually warm
ud	L. *udus,* humid.	udometer	Of humid climates
umbr	L. *umbra,* shade.	umbrella	Presence of umbric epipedon
ust	L. *ustus,* burnt.	combustion	Dry climate, usually hot in summer
verm	L. *vermes,* worm.	vermiform	Wormy, or mixed by animals
vitr	L. *vitrum,* glass.	vitreous	Presence of glass
xer	Gr. *xeros,* dry.	xerophyte	Annual dry season

Source: Modified from Steila, D. *The Geography of Soils.* Englewood Cliffs, N.J.: Prentice Hall, 1976.

1. **Alfisols:** Moderate to high alkalinity, gray surface horizon, subsurface clay; usually moist, perhaps dry for part of warm season.

   Aqualfs: Seasonally wet; for general crops (drained) or woodland (undrained).
   Boralfs: In cool to cold regions; woodland and pasture.
   Udalfs: In midlatitude to tropical regions; usually moist, dry for short periods; row crops, small grain, and pasture.
   Ustalfs: In midlatitude to tropical regions; reddish brown; dry for long periods; range, small grain, and irrigated crops.
   Xeralfs: In regions with rainy winters and dry summers; dry for a long period; range, small grain, and irrigated crops.

2. **Aridisols:** Semidesert and desert soils with definite horizons; low in organic matter; never moist for more than 3 consecutive months.

   Argids: Accumulation of clay in a horizon; range and some irrigated crops.
   Orthids: Accumulation of such salts as calcium carbonate or gypsum; range and some irrigated crops.

3. **Entisols:** No definite horizons (generally azonal).

   Aquents: Permanently or seasonally wet; limited pasturage.
   Fluvents (Alluvial Soils): Organic content decreases irregularly with depth, formed in loam or clay alluvial deposits; range, irrigated crops (dry regions), and general farming (humid regions).
   Orthents: Loam or clay; organic content decreases regularly with depth; range, irrigated crops (dry regions), and general farming (humid regions).
   Psamments: Texture of loamy fine sand or coarser; range and wild hay (Alaska), woodland, small grains (warm, moist regions), pasture and citrus (Florida), and range and irrigated crops (warm, dry regions).

4. **Histosols:** Wet organic peat and muck soils, formed in swamps and marshes; truck crops where drained.

5. **Inceptisols:** Weakly developed horizons; materials have been altered or removed but have not accumulated; usually moist except during part of warm season.

   Andepts: Formed in volcanic ash; woodland, range, and pasture.
   Aquepts: Seasonally wet, organic surface horizon; pasture and hay (Alaska) or woodland and row crops if drained (southeastern United States).
   Ochrepts: In crystalline clay minerals, light-colored surface horizons; woodland and range (Alaska, Northwest), pasture, wheat, and hay (Oklahoma and Kansas), and pasture, corn, and small grain (Northeast).
   Tropepts (Latosols): In tropical regions; pineapple and irrigated sugarcane (Hawaii).
   Umbrepts: In crystalline clay minerals, thick dark surface horizon; low in alkalis; woodland and range.

6. **Mollisols:** In subhumid and semiarid warm to cold climates; dark surface horizon rich in organic matter, highly alkaline.

   Albolls: Seasonal perched water table, bleached subsurface horizon; small grain, peas, hay, pasture, and range.
   Aquolls: Seasonally wet; gray subsurface horizon; where drained, small grains, corn, and potatoes (north central United States), and rice and sugarcane (Texas).
   Borolls: In cool and cold regions; small grain, hay, and pasture (north central United States), and range and woodland (western United States).
   Rendolls (Rendzinas): Subsurface horizon with large amount of calcium carbonate; cotton, corn, small grains, and pasture.
   Udolls: In midlatitude climates; usually moist, no accumulation of calcium carbonate or gypsum; corn, small grains, and soybeans.
   Ustolls: Mostly in semiarid regions; intermittently dry for a long period; wheat and some irrigated crops.
   Xerolls: In regions with rainy winters and dry summers; dry for a long period; wheat, range, and irrigated crops.

7. **Oxisols:** In tropical or subtropical lowland regions; mixtures of kaolin (aluminum clay) and silicon dioxide, low in weatherable minerals.

   Humox: Usually moist; high in organic content, low in alkalinity; sugarcane, pineapple, and pasture (Hawaii).
   Orthox (Latosols): Usually moist; moderate to low in organic content, relatively low in alkalinity; sugarcane, pineapple, and pasture (Hawaii).
   Ustox (Latosols): Partly dry for long periods; pineapple, irrigated sugarcane, and pasture (Hawaii).

8. **Spodosols (Podzols):** In humid, mostly cool, midlatitude regions; acid, low supply of alkalis, with compounds of aluminum or iron.

   Aquods: In seasonally wet regions; pasture, range, or woodland, and citrus and truck crops (Florida).

Orthods: Accumulation of organic matter and compounds of aluminum and iron in a horizon; woodland, hay, pasture, and fruit.

9. **Ultisols:** Usually moist but dry part of the year; low supply of alkalis, subsurface horizon of clay accumulation.

   Aquults: Seasonally wet; woodland, limited pasture; hay, cotton, corn, and truck crops where drained.
   Humults: In warm humid regions; high in organic content; small grain, truck, and seed crops (Oregon, Washington), and pineapple and irrigated sugarcane (Hawaii).
   Udults: In continuously moist regions; relatively low in organic content; general farming, woodland, and pasture, and cotton and tobacco in some regions.
   Xerults: In regions with rainy winters and dry summers; dry for long periods; low in organic content; range and woodland.

10. **Vertisols:** Clay soils; wide, deep cracks when dry.

    Pelluderts (Grumusols): Black or dark gray surface horizon.
    Torrerts (Grumusols): Usually dry; wide, deep cracks open most of the time; range and some irrigated crops.
    Uderts: Usually moist; cracks not open for more than 2 or 3 months at a time; cotton, corn, small grains, pasture, and some rice.
    Usterts: Wide, deep cracks open and close more than once each year, usually open for 3 months or more; general crops and range, and irrigated crops (Rio Grande valley).
    Xererts: Wide, deep cracks open and close once each year, remain open for more than 2 months at a time; irrigated small grains, hay, and pasture.

Index

b

c

d

e

f

h

j

k

l

m

n

O

p

q

r

t

u

x

y

z

Credits

Chapter 1
5—Tom Lewis; 6–7—Photo from Big Bear Solar Observatory, Sweden; 8—Doug Armstrong; 10–11—Robert Kinyon/Millsap & Kinyon after R. Phinney; 11—(right) John Dawson; 13—Butch Higgins; 14—(left) Warren Hamilton, (right) G. R. Roberts, Nelson, New Zealand; 16—Werner Wetzel; 18—From the experiments of E. H. B. D. Kettlewell, University of Oxford; 20—Bill Ralph; 22–23—John Dawson after *Adventures in Earth History,* edited by Preston Cloud, 1970, W. H. Freeman Company.

Chapter 2
32–33—John Dawson; 37—Tom Lewis; 39—Doug Armstrong adapted from *Handbook of Geophysics and Space Environment,* edited by Shea L. Valley, Air Force Cambridge Research Laboratories, U.S. Air Force, 1965; 40, 41, 44—Tom Lewis; 46—(top) Vantage Art Inc. after P. V. Hobbs et al., "Atmospheric Effects of Pollutants," *Science, 183,* p. 910, March 8, 1974; 46–47—John Dawson after Gilbert and Plass, "Carbon Dioxide and Climate," *Scientific American,* 1959; following page 51, CS1—(left) Doug Armstrong, (right) Doug Armstrong after A. Armstrong and John Thornes, *The Geographical Magazine,* © IPC Magazines, Ltd., 1972, by permission of New Science Publications; CS2—Wide World Photos.

Chapter 3
54—Doug Armstrong; 55—Doug Armstrong after G. M. B. Dobson, *Exploring the Atmosphere,* 1963, by permission of the Clarendon Press, Oxford; 58–59—John Dawson; 60, 61— Doug Armstrong after *Smithsonian Meteorological Tables,* edited by Robert J. List, 6th ed., 1971, by permission of The Smithsonian Institution; 63—Doug Armstrong after U.S. Air Force, 1965; 65—Tom Lewis after Herbert Riehl, *Introduction to the Atmosphere* © 1972 by McGraw-Hill Book Company; 66—*The National Atlas of the United States of America, 1970;* 67—Andy Lucas and Laurie Curran after Löf, Duffie, and Smith, *World Distribution of Solar Radiation,* Report no. 21, University of Wisconsin, 1966; 69—Vantage Art, Inc. after William D. Sellers, *Physical Climatology,* © 1965, The University of Chicago Press; 71—Doug Armstrong after F. K. Hare, *The Restless Atmosphere,* 1966, Hutchinson Publishing Group, Ltd.; 72—Doug Armstrong after Herbert Riehl, *Introduction to the Atmosphere,* © 1972 by McGraw-Hill Book Company; 73, 74—Doug Armstrong after William D. Sellers, *Physical Climatology,* © 1965, The University of Chicago Press; 75—Vantage Art, Inc. after J. R. Mather, *Climatology: Fundamentals and Applications,* © 1974 by McGraw-Hill Book Company; 76, 77—(top) Doug Armstrong; 77—(bottom) Doug Armstrong after Arnold Court, *Journal of Meteorology, 8,* 1951, American Meteorological Society; 79—Andy Lucas and Laurie Curran after Glenn Trewartha, *Introduction to Climate,* © 1968, McGraw-Hill Book Company and *Goode's World Atlas* © 1970, Rand McNally & Company; 80–81—Andy Lucas after *The National Atlas of the United States of America,* 1970; 83—Howard Sochurek.

Chapter 4
86—Brian Brake/Rapho/Photo Researchers, Inc.; 87—The Bettmann Archive; 88—Doug Armstrong after *Smithsonian Meteorological Tables,* edited by Robert J. List, 6th ed., 1971, by permission of The Smithsonian Institution; 90—John Dawson; 91—Doug Armstrong after *Smithsonian Meteorological Tables,* edited by Robert J. List, 6th ed., 1971, by permission of The Smithsonian Institution; 92—(top) Vantage Art, Inc. after Morris Neiburger et al., *Understanding Our Atmospheric Environment,* 1973, W. H. Freeman Company, (bottom) Doug Armstrong adapted from *Hydrology Handbook,* 1949, published by the American Society of Civil Engineers; 93—Doug Armstrong after R. Geiger, *The Climate Near the Ground,* © 1965, Harvard University Press; 94, 96—John Dawson; 98—Doug Armstrong after Hermann Flohn, *Climate and Weather,* © 1969 by H. Flohn, used with permission of McGraw-Hill Book Company; 99—David Cavagnaro; 100—(top) David Cavagnaro, (bottom) Andy Lucas and Laurie Curran adapted from Arnold Court and Richard Gerston, *Geographical Review, 56,* © 1966, American Geographical Society of New York; 102—Vantage Art, Inc. after Herbert Riehl, *Introduction to the Atmosphere,* © 1972 by McGraw-Hill Book Company; 103, 104—Doug Armstrong; 106–107—(a, c, d, e, f, g) R. A. Muller, (b) National Center for Atmospheric Research (NCAR), Boulder, Colorado; 108—John Dawson; 109—Vantage Art, Inc. after Horace R. Byers, *General Meteorology,* 3rd ed., 1959, McGraw-Hill Book Company; 110—(top) Mickey Osterreicher, (bottom) Nik Wheeler/Black Star; 112—(left) General Electric Company, (right) The Bettmann Archive, Inc.; 113—Andy Lucas after *Yearbook of Agriculture,* U.S. Department of Agriculture, 1941.

Chapter 5
119—Vantage Art, Inc.; 120—(top) Vantage Art, Inc. after Herbert Riehl, *Introduction to the Atmosphere,* © 1972 by McGraw-Hill Book Company, (bottom) Doug Armstrong; 121—Vantage Art, Inc. after Arthur N. Strahler, *Introduction to Physical Geography,* 1st ed., 1951, John Wiley & Sons; 123—Vantage Art, Inc.; 124—Andy Lucas after Herbert Riehl, *Introduction to the Atmosphere,* © 1972, McGraw-Hill Book Company; 125—Doug Armstrong after Jerome Namias, "The Jet Stream," *Scientific American,* © 1952, Scientific American, Inc.; 128–129—Andy Lucas and Laurie Curran after U.S. Weather Bureau, (bottom right, p. 128) after Hermann Flohn, *Climate and Weather,* © 1969 by H. Flohn, used with permission of McGraw-Hill Book Company; 129—Andy Lucas and Laurie Curran after W. Schwerdtfeger, "The Climate of the Antarctic," *World Survey of Climatology,* vol. 14, edited by S. Orvig, © 1970 by Elsevier Publishing Company; 131—Calvin Woo after Dieter H. Brunnschweiler, *Geographica Helvetica,* vol. 12, 1957; 132—Calvin Woo adapted from Hermann Flohn, *Climate and Weather,* © 1969 by H. Flohn, used with permission of McGraw-Hill Book Company; 134—Calvin Woo; 135—Vantage Art, Inc.; 136—Doug Armstrong; 138—Calvin Woo; 139—Andy Lucas after Hermann Flohn, *Climate and Weather,* © 1969 by H. Flohn, used with permission of McGraw-Hill Book Company; 141—(top) Tom O'Mary after *The Atlas of the Earth,* p. 30, © 1971, Mitchell-Beazley, Ltd., (bottom) Andy Lucas after *The National Atlas of the United States of America,* 1970; 142—National Oceanic and Atmospheric Administration; 144—Doug Armstrong after P. Weyl, *Oceanography,* 1970, John Wiley & Sons; 145—Andy Lucas after L. Don Leet and Sheldon Judson, *Physical*

Geology, 3rd ed., © 1965, by permission of Prentice-Hall, Inc.; 148–149—Andy Lucas and Laurie Curran after *Goode's World Atlas,* © 1970, Rand McNally and Company; following page 151, CS2—(top left) ESSA Weather Bureau, (bottom left) R. A. Muller, (right) Joe R. Eagleman, Vincent U. Muirhead, and Nicholas Willems, *Thunderstorms, Tornadoes, and Building Damage,* 1975, Lexington Books.

Chapter 6
155—Tom Lewis after R. L. Nace, *Water, Earth, and Man,* edited by R. J. Chorley, 1969, Methuen & Co., Ltd., Publishers; 156–157—John Dawson; 159—David Cavagnaro; 160—Doug Armstrong after *Yearbook of Agriculture,* U.S. Department of Agriculture, 1955, and E. E. Foster, *Rainfall and Runoff,* © 1949, Macmillan Co.; 161—Vantage Art, Inc after G. D. Smith and R. V. Ruhe, "How Water Shaped the Soil and the Land," *Yearbook of Agriculture,* U.S. Department of Agriculture, 1955; 162—John Dawson after Raphael G. Kazmann, *Modern Hydrology,* 2nd ed., © 1972 by R. G. Kazmann, used by permission of Harper & Row; 164—John Dawson; 165—Doug Armstrong after R. C. Ward, *Principles of Hydrology,* © 1967, McGraw-Hill Book Company (UK) Limited, used with permission; 166—Vantage Art, Inc.; 168—Andy Lucas and Laurie Curran adapted from *Geographical Review,* vol. 38, © 1948, American Geographical Society of New York; 169—Tom Lewis; 170—Doug Armstrong; 172, 174—Vantage Art, Inc.; 175—Doug Armstrong after C. W. Thornthwaite and J. W. Mather, *Publications in Climatology,* vol. 8, 1955; following page 177, CS1—(left) Steve Harrison and Louis Neiheisel; CS1–CS2—California State Department of Water Resources.

Chapter 7
182—Ron Wiseman; 184—Doug Armstrong; 188–189—Andy Lucas and Laurie Curran after Vernon C. Finch and Glenn Trewartha, *Physical Elements of Geography,* 1949, McGraw-Hill Book Company; 190—Vantage Art, Inc.; 192–193—Andy Lucas and Laurie Curran after "Köppen-Geiger-Pohl map (1953), Justes Perthes, and Köppen-Geiger in *Erdkunde,* vol. 8, and Glenn T. Trewartha, *An Introduction to Climate,* 4th ed., 1968, McGraw-Hill Book Company; 194—(a) G. R. Roberts, Nelson, New Zealand, (b) T. M. Oberlander, (c) Erich Hartmann/Magnum, (d) Marc & Evelyne Bernheim/Woodfin Camp & Associates; 195—(e) Dan Budnik/Woodfin Camp & Associates, (f) T. M. Oberlander, (g) Earl Dibble/Photo Researchers, Inc.; 196—(top) Vantage Art, Inc., (bottom) Doug Armstrong after D. Carter and J. Mather, *Publications in Climatology,* vol. 19, no. 4, 1966; 197—Andy Lucas and Laurie Curran after C. W. Thornthwaite and J. Mather, "The Water Balance," *Publications in Climatology,* vol. 8, 1955; 198—Vantage Art, Inc. after Douglas B. Carter, Theodore H. Schmudde, and David M. Sharpe, "The Interface: As A Working Environment: A Purpose for Physical Geography," *Tech. Paper no. 7,* Comm. on College Geog., Association of American Geographers, 1972; 199—Vantage Art, Inc. after Robert A. Muller and Philip Larimore, Jr., "Atlas of Seasonal Water Budget Components of Louisiana," *Publications in Climatology, 28,* no. 1, 1975; 200—Doug Armstrong after R. Muller, "Frequency Analyses of the Ratio of Actual to Potential Evapotranspiration for the Study of Climate and Vegetation Relationships," Proceedings of the Association of American Geographers, vol. 3, 1971; following page 201, CS3—Vantage Art, Inc. from *Understanding Climatic Change: A Program for Action,* National Academy of Sciences, 1975, Washington, D.C.; CS4—Vantage Art, Inc. from R. A. Muller and J. E. Willis, "Climate Variability in the Lower Mississippi River Valley," *Geoscience and Man,* 1978.

Chapter 8
205—John Dawson after E. P. Odum, *Fundamentals of Ecology,* 3rd ed., © 1971, W. B. Saunders Company; 206—Doug Armstrong adapted from Blumenstock and Thornthwaite, "Climate and the World Pattern," *Yearbook of Agriculture,* 1941, Government Printing Office; 207—Vantage Art, Inc. after John R. Mather and Gary A. Yoshioka, "The Role of Climate in the Distribution of Vegetation," *Annals of the Association of American Geographers,* vol. 58, 1968; 209—John Odam and Andy Lucas after Vernon C. Finch and Glenn T. Trewartha, *Physical Elements of Geography,* © 1949, McGraw-Hill Book Company; 210—Sergio Larrain/Magnum Photos; 212—Doug Armstrong after H. Nelson, *Climatic Data for Representative Stations of the World,* © 1968, by permission of University of Nebraska Press; 213—(top) Doug Armstrong after C. Troll, *Oriental Geographer,* vol. 2, 1958, by permission, (bottom) Doug Armstrong after D. Carter and J. Mather, *Publications in Climatology,* vol. 19, 1966; 214—James S. Packer; 215—(right) G. R. Roberts, Nelson, New Zealand, (left) John Lewis Stage/Photo Researchers, Inc.; 216—Doug Armstrong after H. Nelson, *Climatic Data for Representative Stations of the World,* © 1968, by permission of University of Nebraska Press; 217—Doug Armstrong after D. Carter and J. Mather, *Publications in Climatology,* vol. 19, 1966; 218—T. M. Oberlander; 219—(left) T. M. Oberlander, (right) Doug Armstrong after H. Nelson, *Climatic Data for Representative Stations of the World,* © 1968, by permission of University of Nebraska Press; 220—Steve Harrison and Louis Neiheisel after M. Hendl, *Einführung in Die Physikalische Klimatologie,* Band II, 1963; 221—Doug Armstrong after D. Carter and J. Mather, *Publications in Climatology,* vol. 19, 1966; 222—(left) Doug Armstrong after H. Nelson, *Climatic Data for Representative Stations of the World,* © 1968, by permission of University of Nebraska Press; 220—Steve Harrison and Louis Neiheisel after M. Hendl, *Einführung in Die Physikalische Klimatologie,* Band II, 1963; 221—Doug Armstrong after D. Carter and J. Mather, *Publications in Climatology,* vol. 19, 1966; 222—(left) Doug Armstrong after H. Nelson, *Climatic Data for Representative Stations of the World,* © 1968, by permission of University of Nebraska Press, (right) R. A. Muller; 223—(top) Doug Armstrong after C. Troll, *World Maps of Climatology,* © 1965, Springer-Verlag Publishing, (bottom) Doug Armstrong; 224—G. R. Roberts, Nelson, New Zealand; 225—Doug Armstrong after D. Carter and J. Mather, *Publications in Climatology,* vol. 19, 1966; 226—Vantage Art, Inc.; 227—(left) T. M. Oberlander, (right) Vantage Art, Inc. after Glenn Trewartha, *Introduction to Climate,* 4th ed., 1968, McGraw-Hill Book Company; 228—(left) Brian Hawkes/Carl Ostman Agency, (right) Doug Armstrong after C. Troll, *World Maps of Climatology,* © 1965, Springer-Verlag Publishing; 229—Barbara Hoopes; following page 231, CS2—James S. Packer.

Chapter 9
234—Tom O'Mary; 235—Tom Lewis; 236—Doug Armstrong after Jen-Hu Chang, *Climate and Agriculture,* © 1968 by Aldine Publishing Company, reprinted by permission of the author and Aldine; 237—Doug Armstrong

after D. M. Gates, *Advances in Ecological Research,* edited by J. Cragg, 1968, Academic Press, London; 239—John Dawson after E. P. Odum, *Fundamentals of Ecology,* 3rd ed., © 1971, W. B. Saunders Company; 240—Vantage Art, Inc. after Jen-Hu Chang, *Climate and Agriculture,* © 1968 by Aldine Publishing Company, reprinted by permission of the author and Aldine; 242—Andy Lucas and Laurie Curran after Jen-Hu Chang, reproduced by permission from the *Annals of the Association of American Geographers,* vol. 60, 1970; 244—John Dawson after Robert L. Smith, *Ecology and Field Biology,* 1966, Harper & Row, and R. F. Johnston, *Wilson Bulletin,* 1956, published by the Wilson Ornithological Society; 245—Doug Armstrong after L. B. Slobodkin, *Ecology,* vol. 40, 1959, by permission of Duke University Press; 246—Tom O'Mary after R. Whittaker, *Communities and Ecosystems,* 1970, Macmillan Company; 247—Tom O'Mary after H. Odum, *Ecological Monographs,* 1957, and E. P. Odum, *Fundamentals of Ecology,* 3rd ed., © 1971, W. B. Saunders Company; 248—Tom O'Mary after data from G. Woodwell and R. Whittaker, *American Zoologist,* vol. 8, 1968; 249—Doug Armstrong after U.S. Department of Agriculture; 251—Vantage Art, Inc., modified from *World Agricultural Situation,* U.S. Department of Agriculture, 1976.

Chapter 10

257—(top) Ward's Natural Science Establishment, (bottom) T. M. Oberlander; 258—John Dawson after Arthur N. Strahler, *Physical Geography,* 3rd ed., © 1960 by John Wiley & Sons, by permission; 260—John Dawson; 261—Doug Armstrong after E. M. Bridges, *World Soils,* © 1970, Cambridge University Press; 262—Roy W. Simonson, Courtesy USDA; 263—John Dawson; 264—(top) Doug Armstrong after Lyon and Buckman, *The Nature and Properties of Soils,* 4th ed., © 1943 by Macmillan Company, (bottom) Brian Bunting, *Geography of Soil,* 1965, Hutchinson University Library; 265—Vantage Art, Inc. modified from Robert Ruhe, *Geomorphology,* 1975, Houghton Mifflin, and Soil Survey Staff, 1951, 1962; 266—Doug Armstrong after E. Bridges, *World Soils,* 1970, Cambridge University Press; 267—Vantage Art, Inc. after U.S. Department of Agriculture; 269—Doug Armstrong adapted from Blumenstock and Thornthwaite, "Climate and the World Pattern," *Yearbook for Agriculture,* 1941, USDA, Government Printing Office; 270—Vantage Art, Inc.; 271—Dr. John S. Shelton; 272–273—Vantage Art, Inc.; 277—Andy Lucas after C. F. Marbut, "Soils of the United States," *U.S. Dept. of Agriculture Atlas of American Agriculture,* part 3, 1935; 280–281—Vantage Art, Inc. adapted from USDA, U.S. Conservation Service, 1972; 282—Vantage Art, Inc. modified from N. M. Strakhov, *Principles of Lithogenesis,* vol. 1, 1967, Plenum Publishing Corp., and P. W. Birkeland, *Pedology, Weathering, and Geomorphological Research,* 1974, Oxford University Press; 284–285—John Dawson, profile information adapted from S. R. Eyre, *Vegetation and Soils,* Edward Arnold Publishers; 286—Courtesy, U.S. Soil Conservation Service; 287–289—Reproduced from Soil Science Society of America, C. F. Marbut Memorial Slide Collection; 291—U.S. Soil Conservation Service.

Chapter 11

296—Grant Heilman Photography; 298—NASA; 299—*The Atlas of the Earth*, p. 36, © 1971, Mitchell-Beazley, Ltd.; 300–301—Calvin Woo from Robert S. Deitz, "Geosynclines, Mountains, and Continent Building," *Scientific American,* © 1972 by Scientific American, Inc.; 302—(top) Calvin Woo from John F. Dewey, "Plate Tectonics," *Scientific American,* © 1972 by Scientific American, Inc.; 303—Andy Lucas after data by Ian McDougall, *Nature Physical Science,* vol. 231, 1971; 304–305—Courtesy of the National Geographic Society; 306—T. M. Oberlander, 307—Aerofilms Ltd., London; 308—Vantage Art, Inc.; 310—(top and bottom left) G. R. Roberts, Nelson, New Zealand, (center left) Florence Fujimoto, (bottom right) David Miller; 311—(top) David Miller, (center left) G. R. Roberts, Nelson, New Zealand, (bottom) Florence Fujimoto; 313—Vantage Art, Inc.; 314—T. M. Oberlander; 315—Grant Heilman Photography; 316—(top) Warren Hamilton/U.S.G.S., (bottom) T. M. Oberlander; 317—(left) Butch Higgins, (right) G. R. Roberts, Nelson, New Zealand; 318—(top) Dr. John S. Shelton, (bottom) Warren Hamilton/U.S.G.S.; 319—T. M. Oberlander; 320—(left) John Dawson, (right) David Cavagnaro; 321—T. M. Oberlander; 322—P. S. Smith/U.S.G.S.; 323—U.S. Navy Office of Information; 324—T. M. Oberlander; 325—John Dawson after Longwell, Knopf, and Flint, *Outlines of Physical Geography,* 2nd ed., © 1941, John Wiley & Sons; 326—Vantage Art, Inc.; 327—John Dawson after C. D. Holmes, *American Journal of Science,* vol. 253, 1955; 328—Vantage Art, Inc.; 331—T. M. Oberlander.

Chapter 12

336—Doug Armstrong; 337—Vantage Art, Inc.; 339—Doug Armstrong after Luna B. Leopold, M. G. Wolman, and John P. Miller, *Fluvial Processes in Geomorphology,* © 1964, W. H. Freeman Company; 340—Doug Armstrong; 341—(top) Doug Armstrong after Luna B. Leopold, M. G. Wolman, and John P. Miller, *Fluvial Processes in Geomorphology,* © 1964, W. H. Freeman Company, (bottom) Vantage Art, Inc.; 344—Vantage Art, Inc.; 346—Doug Armstrong adapted from Marie Morisawa, *Streams: Their Dynamics and Morphology,* edited by P. Hurley, © 1968, McGraw-Hill Book Company, from F. Hjulstrom, *Studies of the Morphological Activity of Rivers as Illustrated by the River Fyris,* vol. 25, 1935, University of Upsala Geological Institite; 347, 349—Doug Armstrong after Luna B. Leopold, M. G. Wolman, and John P. Miller, *Fluvial Processes in Geomorphology,* © 1964, W. H. Freeman Company; 350—NASA; 352—(top) Dr. John S. Shelton, (bottom) G. R. Roberts, Nelson, New Zealand; 353, 354—Doug Armstrong after G. H. Dury from *Water, Earth, and Man,* edited by R. J. Chorley, © 1969, Methuen and Company; 355—(top) John Dawson after Marie Morisawa, *Streams: Their Dynamics and Morphology,* edited by P. Hurley, © 1968 McGraw-Hill Book Company, (bottom) from G. K. Gilbert, from O. D. Von Engeln, *Geomorphology,* 1942, Macmillan Company; 357—Department of the Interior, U.S. Geological Survey; 358—John Dawson after R. B. Bunnett, *Physical Geography in Diagrams,* 1968, © Longmans Group, Ltd.; 360—Vantage Art, Inc.; 362—Dr. John S. Shelton; 363—(left) Doug Armstrong after S. M. Gagliano et al., "Hydrologic and Geologic Studies of Coastal Louisiana," Department of the Army, 1970, (right) NASA; following page 365, CS2—Vantage Art, Inc. from U.S. Army Corps of Engineers, Department of the Army, Lower Mississippi Valley Division.

Chapter 13

369—Dr. John S. Shelton; 373—Warren Hamilton/U.S.G.S.; 374—Doug Armstrong; 376—(top left) M. E. Bickford, University of Kansas, (top right) G. K. Gilbert/U.S.G.S., (bottom)

U.S. Forest Service; 377—(top left) M. E. Bickford, University of Kansas, (top right) Warren Hamilton/U.S.G.S., (bottom) U.S. Forest Service; 379—Vantage Art, Inc.; 381—(left) M. E. Bickford, University of Kansas, (right) Dr. Warren B. Hamilton/U.S.G.S.; 382—(top) Dr. John S. Shelton, (bottom) T. M. Oberlander; 385—(top left) M. E. Bickford, University of Kansas, (top right) Warren Hamilton/U.S.G.S., (bottom) T. M. Oberlander; 386—Report of the International Geological Congress, 1933, Washington, D.C.; 387—T. M. Oberlander; 389—© Louis Villota/The Image Bank; 390–391—John Dawson; 392—(left) R. Segerstom/U.S. Geological Survey, (right) T. M. Oberlander; 393—Dr. John S. Shelton; 394—Andy Lucas after F. M. Bullard, *Volcanoes,* 1962, University of Texas Press; 396—(top) Charles A. Wood, (bottom) Warren Hamilton/U.S.G.S.; 398—(top) Ward's Natural Science Establishment, (bottom) Dr. John S. Shelton; 399—T. M. Oberlander; 400—Vantage Art, Inc.; 402—(left) John Dawson, (right) David Miller; 404—John Dawson; 405—T. M. Oberlander; 406—(left) John Dawson after C. A. Cotton, *Geomorphology,* 7th ed., © 1960, Whitcombe & Tombs, Ltd., (right) T. M. Oberlander; 407, 408—Vantage Art, Inc.; 410—Dr. John S. Shelton; 411—John Dawson; 412—(top left and top right) Dr. John S. Shelton, (bottom) John Dawson; 413—(photo) SLAR Imagery, Grumman Ecosystems Corporation, (diagram) John Dawson after W. D. Thornbury, *Principles of Geomorphology,* 1966, John Wiley & Sons; 414—T. M. Oberlander; 415—Vantage Art, Inc.; following page 417, CS3—Vantage Art, Inc. from *Earthquake History of the United States,* 1973, U.S. Dept. of Commerce, National Oceanic and Atmospheric Administration; CS4—John Dawson.

Chapter 14

420—T. M. Oberlander; 422, 423—Vantage Art, Inc. after Louis C. Peltier, "The Geographical Cycle in Periglacial Regions as It Is Related to Climatic Geomorphology," *Annals of the Association of American Geographers, 40,* 1950; 424—T. M. Oberlander; 425—Doug Armstrong; 426—USDA Soil Conservation Service; 427—Tom O'Mary; 428—Grant Heilman Photography; 429—Vantage Art, Inc., 432—T. M. Oberlander; 434—Warren Hamilton/U.S. Geological Survey; 435, 436, 437—T. M. Oberlander; 439—Vantage Art, Inc.; 441—Hilgard O'Reilly Sternberg; 442–443—John Dawson after Theodore M. Oberlander; 443—(right) R. Belanger/Bedford Institute of Oceanography, Environment Canada; 444–445—John Dawson after Theodore M. Oberlander; 445—(right) Dr. John S. Shelton; 446–447—John Dawson after Theodore M. Oberlander; 447—(right) Dr. John S. Shelton; 448-449—John Dawson after Theodore M. Oberlander; 449—(right) T. M. Oberlander; 450-451—John Dawson after Theodore M. Oberlander; 451—(right) Ministry of Information, Salisbury, Rhodesia; 452–453—John Dawson after Theodore M. Oberlander.

Chapter 15

461—(top left) John Dawson, (top right) Anitra Kolenkow, (bottom) Austin Post/U.S.G.S.; 462—Andy Lucas and Laurie Curran after J. L. Davies, *Landforms of Cold Climates, An Introduction to Systematic Geomorphology,* vol. 3, 1969, A.N.U. Press, Canberra; 464—Theodore M. Oberlander; 466—Vantage Art, Inc.; 467—Doug Armstrong; 468—Doug Armstrong after M. F. Meier, 1960; 470—Theodore M. Oberlander; 471—Vantage Art, Inc. after L. Clayton and S. R. Moran, "A Glacial Process-Form Model," in D. R. Coates, ed., *Glacial Geomorphology,* 1974, State University of New York, Binghamton, N.Y.; 472—G. K. Gilbert/U.S.G.S.; 473—(top) Vantage Art, Inc., (bottom) Air Photo Division, Energy, Mines & Resources © Canadian Government; 474—T. M. Oberlander; 475—Austin Post/U.S.G.S.; 476—Vantage Art, Inc. after P. E. James, *Geography of Man,* 3rd ed., 1966, Ginn and Company; 477—(top, bottom left) T. M. Oberlander, (right) William Burkhardt; 478—(left) William Burkhardt, (right) Alvin Lynch; 479—(top) Dr. John S. Shelton, (bottom) Air Photo Division, Energy, Mines & Resources © Canadian Government; 480—Vantage Art, Inc. after R. F. Flint, *Glacial and Quarternary Geology,* 1971, John Wiley & Sons; 481—(top) Robert J. Kolenkow, (bottom) T. M. Oberlander; 482—(top left) T. M. Oberlander, (top right) Geological Society of America, (bottom left) Canada Dept. Mines and Resources, (bottom right) John Shelton; 483—Vantage Art.

Chapter 16

488—Eric Kay/Östman Agency; 489—(top) Doug Armstrong after E. C. F. Bird, *Coasts, An Introduction to Systematic Geomorphology,* vol. 4, 1969, by permission of the M.I.T. Press, (bottom) Doug Armstrong after NOAA; 491—(top left) Tom O'Mary, (top right) Vantage Art, Inc., (bottom) Doug Armstrong; 492—Doug Armstrong; 493—Doug Armstrong after Sverdrup and Munk, "Wind, Sea, and Swell: Theory of Relations for Forecastings," *Technical Report of the U.S. Hydrographic Office,* no. 1, Publication 601, 1947; 494, 495—Theodore M. Oberlander; 497—Tom O'Mary; 498—(left) John Dawson after M. J. Selby, *The Surface of the Earth,* © 1967, Cassell and Co., London, (right) Department of the Navy; 499—(top) T. M. Oberlander, (bottom) Vantage Art, Inc. modified from Karl W. Butzer, *Geomorphology from the Earth,* 1976, Harper & Row; 500—(top left) C. M. Dixon/Östman Agency, (top right) Eric Kay/Östman Agency, (bottom) T. M. Oberlander; 501—Vantage Art, Inc. after Francis P. Shepard, *The Earth Beneath the Sea,* 2nd ed., 1967, Johns Hopkins Press; 502—John Dawson after Willard Bascom, "Beaches," *Scientific American,* © 1960 by Scientific American, Inc.; 503—Dr. John Shelton from *Geology Illustrated,* © 1966, W. H. Freeman Company; 504—Vantage Art, Inc.; 505—Dr. John S. Shelton; 506—NASA, Pilot Rock, Inc., ©1976; 507—Dr. John S. Shelton; 508–509—Vantage Art, Inc. after R. B. Bunnett, *Physical Geography in Diagrams,* © 1968, Longmans Group, London; 510—(left) John C. Hutchins/The Image Bank, (right) John Dawson after R. J. Chorley et al., *The History of the Study of Landforms,* vol. 2, *The Life and Work of William Morris Davis,* 1973, Methuen; 511—(top) John Lewis Stage/The Image Bank, (bottom) NASA; 512—Tom Lewis adapted from Bruce C. Heezen and Charles D. Hollister, *The Face of the Deep,* © 1971 by Oxford University Press; 513—Vantage Art, Inc.; 514—John Dawson after Bruce C. Heezen and Marie Tharp, Lamont Geological Observatory; 515—John Dawson.

Appendix

519—Tom Lewis; 521—(top) The British Museum, (bottom) *Atlas of the Universe,* p. 13, © 1971, Mitchell-Beazley, Ltd.; 523—Andrea Lindberg; 524—U.S. Geological Survey; 525—(top) Andrea Lindberg, (bottom) Andrea Lindberg after Erwin Raisz, *Principles of Cartography,* © 1962, McGraw-Hill Book Company; 526—Andrea Lindberg after Erwin Raisz, *Principles of Cartography,* © 1962, McGraw-Hill Book Company, and *Army Field Manual,* FM 21–26, Oct. 1960; 527, 528, 529—Andrea Lindberg; 531, 532, 533, 534, and 535 (top left) Andrea Lindberg after Arthur Robinson and Randall Sale,

Elements of Cartography, 3rd ed., © 1953, 1960, 1969 by John Wiley & Sons, reprinted by permission, (bottom left) Andrea Lindberg adapted from Whitwell Quadrangle, Tennessee, U.S. Geological Survey, (right) Gerald Ratto Photography; 536—(left) Reprinted from Arthur Robinson and Randall Sale, *Elements of Cartography,* © 1953, 1960, 1969 by John Wiley & Sons, reprinted by permission, (right) U.S. Geological Survey; 537—John Dawson; 538—*The National Atlas of the United States of America,* 1970; 541—U.S. Department of the Interior; 542—U.S. Geological Survey; 543—Steve Harrison and Louis Neiheisel, U.S. Geological Survey; 545—Tom O'Mary; 546–547—Calvin Woo; 548—(top) Courtesy of the Laboratory for Applications of Remote Sensing (LARS), (middle left) U.S. Geological Survey, (bottom right) NASA; 549—(top left) NASA, (top right) R. J. P. Lyon, Stanford University, (bottom) Courtesy of Forestry Remote Sensing Laboratory, University of California, Berkeley; 550–553—Andrea Lindberg.